Visual neuroscience

VISUAL
NEUROSCIENCE

Edited by

J. D. PETTIGREW

Department of Physiology and Pharmacology
University of Queensland

K. J. SANDERSON

School of Biological Sciences
Flinders University of South Australia
and

W. R. LEVICK

John Curtin School of Medical Research
Australian National University

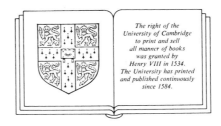

The right of the
University of Cambridge
to print and sell
all manner of books
was granted by
Henry VIII in 1534.
The University has printed
and published continuously
since 1584.

CAMBRIDGE UNIVERSITY PRESS
Cambridge
London   New York   New Rochelle
Melbourne   Sydney

Published by the Press Syndicate of the University of Cambridge
The Pitt Building, Trumpington Street, Cambridge CB2 1RP
32 East 57th Street, New York, NY 10022, USA
10 Stamford Road, Oakleigh, Melbourne 3166, Australia

First published 1986

Printed in Great Britain by the University Press, Cambridge

Library of Congress cataloguing in publication data

Main entry under title:

Visual neuroscience.

Bibliography.
1. Vision–Congresses.   2. Visual cortex–Congresses.
3. Visual pathways–Congresses.   4. Visual perception–
Congresses.   I. Pettigrew, J. D. (John Douglas),
1943–   .   II. Sanderson, Ken, 1946–
III. Levick, W. R. (William Russell), 1931–
QP474.V53 1986   599'.01823   85–19504

British Library cataloguing in publication data

Visual neuroscience.
1. Vision   2. Neurophysiology
I. Pettigrew, J. D.   II. Sanderson, Ken
III. Levick, W. R.
599.01'823   QP475

ISBN 0 521 25829 4

CONTENTS

Contents

CONTRIBUTORS

Lindsay M. Aitkin, *Department of Physiology, Monash University, Clayton, Vic. 3168, Australia.*

G. Baumgartner, *Department of Neurology, University Hospital Zürich, Rämistrasse 100, CH–8091, Zürich, Switzerland.*

H. B. Barlow, *Department of Physiology, Cambridge CB2 3EG, UK.*

Jean Bullier, *INSERM Unité 94, 16 Avenue du Doyen Lépine, 69500 Bron, France.*

W. Burke, *Department of Physiology, University of Sydney, Sydney, NSW 2006, Australia.*

B. G. Cleland, *Department of Physiology, University of Sydney, Sydney, NSW 2006, Australia.*

D. P. Crewther, *School of Optometry, University of New South Wales, Kensington, NSW 2033, Australia.*

S. G. Crewther, *School of Optometry, University of New South Wales, Kensington, NSW 2033, Australia.*

R. H. Day, *Department of Psychology, Monash University, Clayton, Vic. 3168, Australia.*

Bogdan Dreher, *Department of Anatomy, University of Sydney, Sydney, NSW 2006, Australia.*

Geoffrey H. Henry, *Department of Physiology, The John Curtin School of Medical Research, Australian National University, Canberra, Australia.*

Austin Hughes, *Director, National Vision Research Institute of Australia, 380–386 Cardigan Street, Carlton, Vic. 3053, Australia.*

Seisho Ito, *Department of Physiology, Akita University School of Medicine, 1-1-1 Hondo, Akita 010, Japan.*

John Irwin Johnson, *Anatomy Department and Neuroscience Program, Michigan State University, East Lansing, MI 48824-1316, USA.*

Jon H. Kaas, *Department of Psychology, Vanderbilt University, Nashville, TN 37240, USA.*

Hiroshi Kato, *Department of Physiology, Akita University School of Medicine, 1-1-1 Hondo, Akita 010, Japan.*

J. Kulikowski, *Visual Sciences Laboratory, Ophthalmic Optics Department, UMIST, P.O. Box 88, Manchester M60 1QD, UK.*

James W. Lance, *Department of Neurology, The Prince Henry Hospital, Sydney, Australia and School of Medicine, University of New South Wales, Kensington, NSW 2033, Australia*

W. R. Levick, *Department of Physiology, John Curtin School of Medical Research, Australian National University, Canberra, Australia*

D. M. MacKay, *Department of Communication and Neuroscience, University of Keele, Keele, Staffs. ST5 5BG, UK.*

D. M. McLean, *Department of Ophthalmology, University of Washington, Seattle, WA 98195, USA.*

Jeremiah I. Nelson, *Department of Physics, Philipps University, Renthof 7, D-3550 Marburg, Federal Republic of Germany.*

Tetsuro Ogawa, *Department of Physiology, Akita University School of Medicine, 1-1-1 Hondo, 010, Japan.*

Guy A. Orban, *Laboratorium voor Neuro- en Psychofysiologie, Katholieke Universiteit te Leuven, Medical School, Campus Gasthuisberg, B-3000 Leuven, Belgium.*

E. Peterhans, *Department of Neurology, University Hospital Zürich, Rämistrasse 100, CH-8091 Zürich, Switzerland.*

John D. Pettigrew, *Department of Physiology and Pharmacology, University of Queensland, St Lucia, Qld 4067, Australia.*

David H. Rapaport, *School of Anatomy, University of New South Wales, Kensington, NSW 2033, Australia.*

R. W. Rodieck, *Department of Ophthalmology, University of Washington, Seattle, WA 98195, USA.*

K. J. Sanderson, *School of Biological Sciences, Flinders University of South Australia, Bedford Park, SA 5042, Australia.*

Ann Jervie Sefton, *Department of Physiology, University of Sydney, Sydney, NSW 2006, Australia.*

Jonathan Stone, *School of Anatomy, University of New South Wales, Kensington, NSW 2033, Australia.*

R. von der Heydt, *Department of Neurology, University Hospital Zürich, Rämistrasse 100, CH-8091 Zürich, Switzerland.*

David I. Vaney, *The Physiological Laboratory, University of Cambridge, Downing Street, Cambridge CB2 3EG, UK.*

Heinz Wässle, *Max-Planck-Institut für Hirnforschung, Deutschordenstrasse 46, 6000 Frankfurt/M. 71, Federal Republic of Germany.*

PREFACE

The Lord Howe Celebration

This book's origins lay in a conversation between the editors at the National Vision Research Institute of Australia in 1980. We discussed possibilities for suitable locations to celebrate Peter Bishop's career in visual neuroscience and decided that the following criteria should be satisfied: the place should be conducive to an informal, family-style celebration akin to the tolerant way Peter had always run his research 'family'; there should be ample opportunity for extramural activities so that the scientific presentations could be balanced with outdoor activities in which attending family members could take part; there should be sufficient space to strike the right balance between pleasant confinement and lots of interesting outside activities, so that the celebrants could interact at all times without running the risk that participants would be diluted or swallowed up between presentations; and there should be a special atmosphere, tangible on arrival, to make the event all the more memorable. Not many places satisfied all these criteria but one stood out, Lord Howe Island, a few hundred miles off the Eastern Australian sea-board.

Accordingly, in the first week of September 1983, sixty friends, family, disciples and colleagues of Peter Bishop arrived there in numerous small-plane loads to celebrate a distinguished career in visual neuroscience. The formal scientific presentations were held in a small church hall within view of the coral lagoon and the towers of Mts Lidgbird and Gower. But there were numerous other scientific interactions, of the spontaneous kind, on the tracks and bike paths and beaches of this magnificent piece of the world heritage.

This book is a record of the presentations made at the Lord Howe I. celebration, except in the case of MacKay who could not attend but still felt that he would like to contribute to the book. Manuscripts arrived over a long span from the time of the celebration until late in 1984. Authors were instructed to cover the field as well as their own contributions to it, although the reader should be prepared for some idiosyncratic points of view from a group which has had such an enormous collective impact on the directions taken by research in visual neuroscience. In an attempt to link different chapters, these have been grouped into sections as they were presented at the celebration, each section with some background information in an attempt at integration.

The book also has some personal details of each author's career and the impact on it of Peter Bishop. Perhaps the most noticeable aspect of Peter Bishop's contribution to science has been his nurturing effect; so many young investigators and exciting new areas of research bloomed in his laboratories. Indeed, I know a number of scientists in areas unrelated to vision who have expressed their debt of gratitude for the encouragement and guidance they received from Peter concerning their research. This positive effect on the growth of individual scientists, irrespective of whether this growth might be exploitable in terms of Peter's own scientific interests or not, is such a recurring theme that we have asked each contributor to provide a biographical note with a specific reference to Peter's influence on his/her own career. These can be read at the beginning of the appropriate chapter. These notes, the impact of the many new ideas which originated amongst his disciples while under his direction, and the success of so many of these disciples (25 full chairs around the globe is remarkable by any standards), all support the presence of an unusual kind of nurturing influence, so I would like to give my own thoughts on its possible bases.

I believe that there were two important ingredients in Peter's nurturing formula: (1) a perfectly equipped laboratory environment sheltered from non-scientific interference; and (2) a confidence in the personal autonomy of the individuals working in that environment. Peter always worked tirelessly to provide the best working conditions and the best that technology could offer in the laboratory but was also happy with whatever direction the research took, playing more the role of the enchanted listener at the concert rather than the role of conductor which would have been justified by the time and effort he had put into gathering the orchestra's instruments and players. Not that he was incapable of grabbing the baton, or even one of the instruments themselves, to show how a piece might be played, like that night in 1968 when Peter showed Ken Sanderson and Ian Darian-Smith how to reveal the binocular influences he knew must be found in the physiology of lateral geniculate neurons (see Sanderson's note and Section IV for more details).

These two aspects of Peter's approach certainly influenced me tremendously, since I performed poorly when constrained by the strong expectations of others and became highly productive when given a genuinely exciting problem of my own and a superbly equipped laboratory with which to solve it. Peter did not foresee the discovery of disparity-detecting binocular neurons, but he recognized that simultaneous stimulation of both of a binocular neuron's receptive fields with the same stimulus was an important question requiring quantitative methods and a Risley biprism. By providing me with these tools as well as a good measure of personal autonomy in pursuing the question, he supplied the key factors in the eventual discovery.

There are many similar stories in this book, from visual neuroscientists who have had enormous influence on the development of their field, with ideas like feature extraction, parallel visual processing along the X, Y and W streams, binocular inhibition, and end-stopped receptive fields, to name just a few.

J. Pettigrew
St Lucia, 1985

SECTION

I

The retina

There can be little doubt that the biggest, and most
fruitful, change in visual physiology occurred in the
1950s when the delivery of stimuli became more
'natural'. During this decade there was a move away
from the construction of multibeam ophthalmoscopes
capable of shining different lights in various configurations 'down' onto the retina, as investigators realized
that much more complicated and subtle light patterns
could be provided by the outside world, so long as the
eye's dioptrics were in sufficiently good shape to focus
that world onto the retina. Visual neurons were soon
found which responded selectively to different features
within those patterns.

   The credit for this radical shift in thinking, and
the long string of experimental findings which followed,
is usually and quite justifiably given to 'What the frog's
eye tells the frog's brain', that colourfully titled and
provocative paper by Lettvin, Maturana, McCulloch &
Pitts in 1959. But like many ideas whose time had come,
the use of natural stimuli and doubts about the giant
multibeam ophthalmoscopes had already arrived in a
number of laboratories before Lettvin's experiments
with coloured photographs of the frog's world became
widely known. Indeed, the philosophy behind the
technique of looking at the behaviour of visual neurons
when presented with complex natural stimuli had
already been clearly expressed, even if it is modestly
hidden in the discussion rather than emblazoned in the
title, by Barlow in his 1953 contribution to the *Journal
of Physiology*.

   It is therefore appropriate that this section, and

1

the book, begin with a contribution by Barlow. Apart from this pioneering work on what might be called the 'psychology of the retina', Barlow's name is closely associated with a wide range of topics in visual physiology, such as the vital role played by inhibition in the elaboration of the receptive field, diffraction in eyes, dark light, dark adaptation, quantal sensitivity at the absolute threshold, and the efficiency of visual coding. Of all these contributions, arguably the single most influential contribution was the notion of feature extraction by individual neurons: different neurons are wired so they respond selectively to different, abstract features of the visual environment. First tentatively expressed in work on the frog, this idea became compelling with the publication of studies in collaboration with Levick and others, on single ganglion cells in the rabbit retina. Cells which had been previously thought to be responding unreliably could now be made to respond in a dramatic way if the appropriate visual feature was present (such as a particular direction of motion in part of the pattern), sweeping away misguided notions about the unreliability of nervous transmission which were common at that time.

Indeed, Barlow has championed the cause that sources of unreliability within the nervous system are usually small compared to those originating with the stimulus itself, whose physical limitations must be understood. In Chapter 1, Barlow takes up the issue of physical limitations within the stimulus and extends this to some of the possible limitations in the central processing machinery. Feature extraction is then taken up for more detailed treatment in subsequent chapters where you will find the beginning to answers to questions such as:

1. Why bother to carry out feature extraction in the eye when a more flexible examination of the permutations and combinations of all photo-receptors could be performed by the abundant neuronal resources found more centrally in the visual pathway? (Levick in Chapter 3 explains the problem posed by the bottleneck at the optic nerve and Pettigrew in Chapter 15 considers the needs for coordinating any elaborate feature extraction with the process of binocular interaction.)

2. What are the most important features to use when parsing up the visual world and how should one distribute the machinery for detecting those features across visual space (Wässle: Chapter 2)?

3. How does one design a retinal circuit so that one ganglion cell responds to one spatiotemporal pattern of photoreceptor activation (i.e. one feature) while a neighbouring ganglion cell responds to a different pattern in the same set of photoreceptors? (Vaney: Chapter 4).

Peter Bishop has been called the 'Gullstrand of the cat eye' and there is little doubt that his schematic eye for the cat facilitated the increasingly precise characterization of visual processing by this species. A number of findings depend critically on his schematic eye, such as the relation between ganglion cell densities and visual acuity (Chapters 2, 5 & 8) and measurements of binocular disparity and tilt of the horopter (Chapter 15). More recently it has become apparent that there is much to be gained by an examination of the way in which both the optical resolution and the neural resolution of the eye vary with eccentricity, an exercise requiring visual optics in the form of a very sophisticated schematic eye in addition to visual neuroscience. Austin Hughes takes up this topic in Chapter 5 having been encouraged in his pursuit of comparative visual optics by Peter's own interest and work in this area.

1

Why can't the eye see better?

H. B. BARLOW

Some people say that we hardly understand how the eye works at all, so it is perverse and premature to ask why it does not work better. I am a great admirer of this direct approach, but it is incomplete and liable to give unsatisfactory answers. Everybody must at some stage have dismantled a clock to find out 'how it works', but if this was your only question your answer was probably misleading. Thus, if it was an old-fashioned lever-escapement clock you probably found the mainspring and the chain of gears and said to yourself, 'Aha! the spring drives the hands through these gears'. That is quite right but what are the other gears and the escapement for? 'Oh they just slow it down', you probably said to yourself, perhaps with an uneasy feeling that there was more to it than that. There is, of course, more to it than that, for the whole evolution of timepieces has centred round the escapement, persuading it to divide up time into more exactly equal segments regardless of perturbing influences. You reach this heart of the matter, not just by asking, 'How does it work?', but by asking, 'Why doesn't it keep better time?'. The visual system is also complex; it performs a difficult task and we need to identify the difficulties before we can understand much about the system.

In this chapter I shall follow the same general plan as in an article I wrote 20 years ago on the physical limits of vision (Barlow, 1964b), but I shall deal only briefly with the parts that have stood the test of time and further experiment, shall correct some errors, and shall add material on new limiting factors, not

3

necessarily physical ones, that have become evident since that time.

What is needed for seeing?

Seeing is getting information about the environment from the light that enters our eyes. It is convenient to divide up this operation into three steps. The first step is to split up the entering light so that the amounts coming from different parts of the field, or from different parts of the spectrum, are absorbed preferentially in different structures. The next step is to generate signals dependent upon the amounts of light in each of these structures, and to transmit these signals to wherever the information is needed. This has to be done continuously, so that the signals refer to the amount of light absorbed in the recent past, for speed of response is an important requirement and photoreceptors cannot integrate indefinitely over time as a photographic plate can; this amounts to the additional requirement that the light be split up temporally, as well as spatially and spectrally. The final step is to collect and compare the signals from different structures, sometimes using signals that occurred at earlier times, and to make decisions about what there

is in the visual environment on the basis of these assembled signals.

In my 1964 article (Barlow, 1964*b*), the primary emphasis was on the second of these three steps, generating signals dependent on the amounts of light after it has been subdivided. It was argued that the quantity of light in each of the subdivisions was small because the eye, unlike most physical instruments, splits up the light in space, colour and time simultaneously, and cannot increase the amount available by integration along any of these dimensions. Because the quantities were small, noise from quantum fluctuations was likely to be important, and might indeed limit the extent to which splitting up was worthwhile. There are three problems with this argument. The first is that something other than quantum fluctuations becomes of dominant importance at levels of illumination only slightly above threshold; however, this does not completely invalidate the argument because, whatever its origin, this extra noise is likely to be more serious in small subdivisions than in large ones. The second objection is that the splitting up and the signalling do not necessarily occur serially; for instance the signals from a rod appear to be partially combined with those of its neighbours before its synaptic signal is generated (Fain, 1975; Fain, Gold & Dowling, 1976). This is certainly an interesting fact, but it does not effect the conceptual validity of the argument.

The third objection is more important: I overlooked the possibility that systems might integrate

Fig. 1.1. Contrast sensitivity for isoluminant red–green sinusoidal gratings compared with ordinary green gratings. For each spatial frequency the relative amounts of red and green were adjusted to ensure isoluminance (from Mullen, 1985).

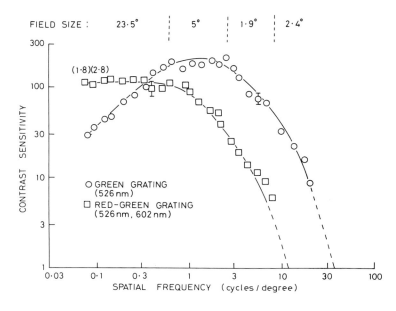

along one dimension while maintaining good resolution along another, and that parallel pathways might do this in different ways. This is certainly important; for instance, Mullen (1985) has shown that the colour system has poor spatial resolution and can therefore increase the amounts of light available by integrating in space (see Fig. 1.1), but this does not affect resolution based on luminance changes in the image, because this is mediated by a system operating in parallel. Similar considerations apply to the motion system (Burr & Ross, 1982). With these qualifications, the previous article seems to have been correct about the limits involved in generating these graded signals, but the evidence is now enormously improved and is worth summarizing in the next sections.

Generating graded signals

The psychophysical evidence 20 years ago suggested that rods responded to single quanta (Hecht, Shlaer & Pirenne, 1942), that equivalent signals occurred in the dark (Barlow, 1956), possibly from thermal isomerization of rhodopsin, and that the rate of such noise events was increased following lights that bleached small quantities of the photopigment, thereby elevating the threshold (Barlow, 1964a). Elegant experiments on photoreceptors have confirmed these conclusions. Fig. 1.2 shows directly that rods give easily detected responses to single quanta (Baylor, Lamb &

Yau, 1979). Furthermore, similar responses occur (Baylor, Matthews & Yau, 1980) at a very slow rate in total darkness and there is an increased rate of occurrence of these spurious signals (Lamb, 1980) early in dark adaptation (Fig. 1.3); however, this is unlikely to be the only cause of the elevation of threshold (Barlow, 1972). It was also thought that the lower sensitivity of cones might be attributed to the higher rate of thermal isomerization of their red-shifted pigments (Barlow, 1957). Quantal fluctuations cannot be detected in cones because another source of noise usually masks them; it is not known whether this is caused by thermal isomerization, but it is at least a possibility (Simon & Lamb, 1977). The situation in blue-sensitive cones is not clear; being shifted to the blue compared with rods, one might expect their pigments to be more stable thermally.

Little progress has been made in finding why measurements of quantum detection efficiency show that the visual system performs further and further from the quantal fluctuation limit as the luminance level increases (Barlow, 1962). Psychophysical measurements of quantum efficiency based on the desensitizing effect of adding noise to the visual stimulus lead to much higher values than those based on detection (Pelli, 1983), but low values of detective efficiency have been reported at the level of retinal ganglion cells in the cat (Levick *et al.*, 1983). The quantum efficiency of

Fig. 1.2. Responses of a toad rod to the absorption of single photons. Forty flashes of an intensity that caused an average of 0.53 isomerizations of rhodopsin per flash were delivered just before the times indicated by dots. In some trials no isomerizations occurred, in others one, and in others two or more; the frequencies of these fitted the Poisson expectations. (From Baylor *et al.* 1979.)

Fig. 1.3. The top two traces show events, recorded as in Fig. 1.2, occurring at a slow rate in total darkness: Baylor *et al.* (1980) have shown that these are probably due to thermal isomerization of rhodopsin. The lowest trace was taken 3 min after a bleach calculated to isomerize 9×10^6 rhodopsin molecules in the rod (about 0.4%); the increased fluctuations are thought to be caused by an increase in the rate of the spontaneous events (from Lamb, 1980).

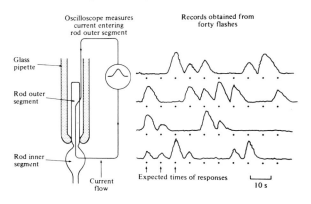

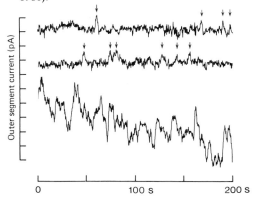

receptors when signalling suprathreshold light levels has not yet been measured.

Limitations to splitting up the light

In vertebrate eyes the light is split up according to (1) the position in the visual field from which it comes (i.e. its direction of entry), (2) its spectral composition, and (3) its time of arrival. There have been interesting developments in understanding the first and second of these, though the limits on the third are not fully understood.

Direction of entry

The ability to separate light according to its direction of entry determines the spatial resolution or acuity of an animal's vision. In vertebrate eyes, as opposed to the compound eyes of arthropods which will not be considered further in this chapter, splitting according to direction of entry is achieved by forming an image on an array of receptor cells. The quality of

this image is one obvious limiting factor, but it has quite recently become clear that there is another less obvious one, namely the ability of a receptor to restrict its sensitivity to the light that falls directly upon it. Let us first consider the image itself.

Image quality can be expressed in the form of the line spread function, which simply shows the distribution of light in the image of a very thin line. Alternatively, nearly the same information can be expressed in terms of the modulation transfer function (MTF), which shows the extent to which spatial sinusoids of different frequencies are demodulated, or lose contrast, in the image. Direct estimates of human image quality have been made by Krauskopf (1962), Westheimer & Campbell (1962) and Campbell & Gubisch (1966), using the technique of analysing the light reflected from the fundus and emerging from the eye. Examples of these for the human eye are shown in Figs 1.4 and 1.5, which show the best line spreads and MTFs achievable at three different pupil diameters.

Fig. 1.4. Line spread functions of intact human eyes at four pupil diameters obtained from the light reflected from the fundus through the pupil and corrected for double passage through the optics. The fine inner curves show the expected performance with diffraction-limited optics, while the points and the smoothed denser curves show the actual performance. Errors would broaden the curves, but there is very little room for improvement in the results for the 2.0-mm pupil; with larger pupils the eye performs substantially below the diffraction limit. (From Campbell & Gubisch, 1966.)

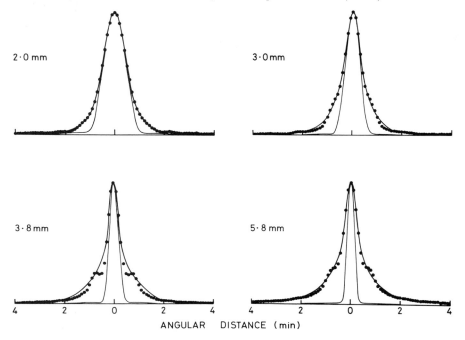

ANGULAR　DISTANCE　(min)

A factor which causes an inevitable loss of high spatial frequencies, and thus sets a physical limit to acuity, is diffraction at the pupil aperture. The cut-off, or highest spatial frequency, that can pass the pupil and be represented in the image, is simply d/λ cycles/radian, where d and λ are, respectively, the pupil diameter and the wavelength of the light. How closely do actual eyes approach this limit? A pupil diameter of 2.5 mm has a cut-off frequency at 78 cycles/degree for light of wavelength 550 nm; this is approached closely by human grating resolution in bright lights and also by Australian birds of prey (Reymond, 1984). One is led to entertain the hypothesis that image quality is the sole determinant of human resolution, and that it is close to the diffraction limit.

Figs 1.4 and 1.5 show the best estimates of the actual line spread functions and MTFs, compared with their theoretical limits. The actual spreads are not of course as narrow as the optima, nor do the MTFs lie on the theoretical curves, but they are quite close to them provided that the pupil diameter is small; at pupil diameters larger than 2 mm, the diffraction-limited line spread functions narrow and the cut-off spatial frequencies increase, but the actual line spread functions broaden and spatial resolution gets worse.

What about the first part of the hypothesis, that image quality is the main determinant of resolution? The comparison made in Fig. 1.6 shows that this hypothesis is wrong, for if psychophysical contrast sensitivity is limited by image quality at 10 cycles/degree, it falls below what image quality would allow at both higher and lower frequencies. There must be other limiting factors to account for the failure of the retina and postretinal mechanisms to make use of all the information available in the retinal image. At high frequencies, the probable factors are the difficulty of confining light within a single receptor, and the losses engendered by having too few sampling stations with large receptive fields that attenuate high frequencies. The latter factors must be responsible for the greatly reduced sensitivity to high spatial frequencies in the parafovea and periphery of the retina, but at the foveal centre the difficulty of confining light to a single receptor is the most likely candidate and is considered next. The reduced sensitivity at low frequencies probably results from lateral inhibition, which will be considered later.

Fig. 1.5. Modulation transfer functions of intact human optics obtained by the same method as in Fig. 1.4. For each pupil diameter, the spatial frequencies have been normalized by dividing by the diffraction limit for that diameter. (From Campbell & Gubisch, 1966.)

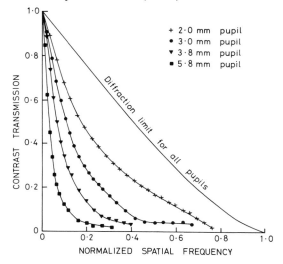

NORMALIZED SPATIAL FREQUENCY

Fig. 1.6. The continuous inner curve shows the contrast sensitivity of the human eye at high luminance levels plotted on double logarithmic coordinates. The dotted curve touching this at 10 cycles/degree is the MTF of the human eye. Clearly, the contrast sensitivity at higher and lower frequencies must be degraded by factors other than the quality of the image. (From Barlow & Mollon, 1982.)

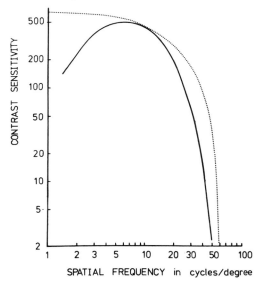

SPATIAL FREQUENCY in cycles/degree

Receptor resolution

In 1964 I thought that chromatic aberration was also important in setting a limit, since it prevented acuity being improved by enlarging lens aperture. There may still be something in this, but Enoch (1961) has pointed out that light must be confined within one receptor to prevent blurring within the retina, and Snyder (1972) has shown how receptors must cease to act as effective waveguides if they are below about 2 μm in diameter. Perhaps the most convincing indication that receptor size is important came from the examination of the eyes of a range of hawks done by Snyder & Miller (1977). The overall size of these eyes varied greatly, and in general every structure within them was scaled precisely in accordance with the overall size; but the foveal receptors were close to 2 μm in diameter and did not vary at all; the presumption is that they were already at the smallest effective size, even in the largest eye. If chromatic aberration had been the limiting factor, receptor diameter should have decreased in proportion to the square root of the eye's focal length, which was certainly not the case. The fact that receptor diameter, not chromatic aberration, sets the limit to acuity has important implications for the expected effect of eye size on acuity (see Table 1.1 below).

Why is it so important for the eye to split up light as finely as possible according to direction of entry? One can see a great deal, even with a very blurred image, and in some cases blurring actually aids visibility (Harmon & Julesz, 1973). But everything that can be seen in a blurred image could in principle be detected by performing the appropriate spatial

Fig. 1.7. The Shannon information per 1 cycle bandwidth is plotted against spatial frequency to show the potential importance of different frequency bands. The top two curves are for diffraction-limited images at two luminances. The third curve shows the substantial loss of high-frequency information in the actual retinal image; because the image is two-dimensional, the loss at high frequencies, shown in Figs 1.4 and 1.5, appears more important in this plot. The information available centrally can also be calculated from the contrast sensitivity function by assuming that Campbell and Robson's thresholds correspond to a D' value of 2.5; this again shows that the discrepancy at high frequencies, illustrated in Fig. 1.6, corresponds to a large loss of information. In contrast, the loss at low frequencies is negligible. Spectral and polarization information have not been taken into account. (Calculated from data of Campbell & Gubisch (1966) and Campbell & Robson (1968).)

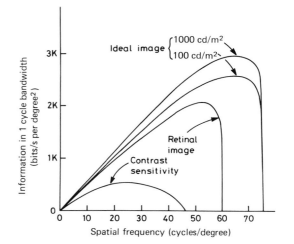

Table 1.1. *The effect of eye size*

Type of performance	Limiting factor	Measure used	Expected value	Expected value / Human value
Acuity	Diffraction at pupil	Grating cycles per radian	$\dfrac{d}{\lambda}$	$\dfrac{d}{2}$
Absolute sensitivity	Retinal noise	$\dfrac{1}{\text{min. no. of quanta}}$	$\propto \dfrac{D^2}{F}$	$\dfrac{1.5\,D^2}{F}$
Acuity and speed at low luminance	Quantal fluctuations	Cycles per radian or flicker fusion frequency	$\propto D$	$\dfrac{D}{5}$

D = maximum pupil diameter; d = minimum pupil diameter; λ = wavelength; F = posterior nodal distance.

summating operation on a sharp image, while the reverse is not the case. Splitting up as finely as possible therefore gives vision its versatility.

One way to make this clear is to calculate the amount of information per unit waveband, using Shannon's measure, at different spatial frequencies. Such calculations are plotted in Fig. 1.7, and these curves show immediately which spatial frequencies carry most information, while the areas under the curves show the total amounts of information available. The top curve is for a diffraction-limited image at 1000 cd/m^2; information about polarization is not used, and spectral information has been neglected. The next curve is for 100 cd/m^2 luminance; the curve is lower because the smaller number of quanta reduces the highest attainable contrast sensitivity. The third curve down represents an estimate made for the retinal image with a 2.4 mm pupil, as described by Campbell & Gubisch (1966), and the lowest curve is an estimate made for the contrast sensitivity function in bright light (from Campbell & Robson, 1968). The points to notice are as follows: (1) reducing light intensity by a factor of 10 has only a moderate effect, for it only reduces the dynamic range by $\sqrt{10}$, and information is proportional to the log of the dynamic range; (2) the eye's optics lose a lot of high-frequency information but preserve quite a large proportion of the low-frequency information that passes the 2.4 mm pupil; (3) the low-frequency cut shown in Fig. 1.6 has a negligible effect on the total information, because the amount available per unit waveband decreases so rapidly with spatial frequency; and (4) in contrast, the loss of high frequencies has a large effect in reducing the total amount that is available centrally. It is presumably the fact that the light entering the eye contains so much information at higher spatial frequencies that has incited the evolution of acuities almost up to the diffraction limit.

Colour

Within a central circle 1 degree in diameter, the entering light is split into almost 10 000 spatial subdivisions, whereas the trichromacy of colour vision implies that it is split into only three spectral subdivisions. The eye is obviously much less interested in colour than in spatial position and it is worth asking, 'Why not split the light into four or more spectral subdivisions?'.

Fig. 1.8d shows the Fourier transforms of the spectral sensitivities of the red, green and blue components of the human colour system, using the derivations of Smith & Pokorny (1975). The physical significance of these curves can be visualized as follows (Barlow, 1982). Imagine a suitably dispersed spectrum formed, for instance, by a prism. Now place in this spectrum a comb with teeth having a sinusoidal profile, and recombine the spectrum with a lens. If the sinusoidal comb had a low frequency, one tooth might block the blue and green while the yellow and red passed through the neighbouring gap and the recombined light would obviously look orange; shifting the position of the comb would reverse what was blocked and passed, and the recombined light would change to greenish blue. If, on the other hand, the comb was of high frequency or short wavelength, the teeth would obstruct some light of all colours, and the hue of the recombined light would be almost unaffected, whatever the position of the comb. We now see the relevance of Fig. 1.8d, for it shows how individual colour mechanisms would respond to comb-filtered spectra of various comb-frequencies. The ordinate shows the predicted percentage modulation of each system's output as the comb (assumed to pass 100% at the peaks in the gaps and to pass 0% at the troughs caused by the teeth) was moved through the spectrum. The scale of the abscissa is unfamiliar; recall that human colour vision is good only over the range 435 to 650 nm, which corresponds to 689 to 461 THz (Terahertz; $1 \text{ THz} = 10^{12} \text{ Hz}$) or a bandwidth of 228 THz. The abscissa gives cycles/1000 THz bandwidth, so to obtain the number of cycles of the comb within the range of good human colour discrimination one should divide the abscissa value by 4.4. It now becomes clear that the spectral sensitivities of the human colour mechanisms are so broad that they attenuate comb-filtered spectra by 40% even when there is only one complete cycle of the comb within the range of good colour discrimination; for higher comb-frequencies even more of the modulation is lost.

It is a well-known proposition that only $(2N + 1)$ sample points are needed to obtain all the information from a segment of waveform containing N periods of the highest frequency present. We have seen that frequencies higher than 4.4 cycles/1000 THz bandwidth, corresponding to one cycle in the range of good colour discrimination, are demodulated more and more strongly, so the three channels of our trichromatic system look like a reasonable choice; more channels, corresponding to more sample points, would pick up more information, but the higher comb-frequencies they might inform us about would be strongly demodulated by the broad spectral sensitivity curves.

That seems a plausible justification for tri-chromacy, but the argument has a corollary. If the spectral sensitivity curves could be narrowed, then high comb-frequencies would be less demodulated and tetrachromacy would be a better proposition. The dotted curve in Fig. 1.8d shows the Fourier transform of the spectral sensitivity for a chicken cone containing a coloured oil droplet, calculated by Bowmaker & Knowles (1977). The principal effect of the droplet is to narrow the sensitivity curve, thus enabling it to respond to higher comb-frequencies. Fig. 1.8c shows how a narrowed red-sensitivity curve might be fitted in with the other three human mechanisms. Liebman (1972) has suggested that turtles, which also have oil droplets in their receptors, have five spectral channels,

Fig. 1.8. Top left (a) shows a comb-filtered spectral energy distribution of frequency 5 cycles/1000 THz bandwidth and of 100% modulation. Bottom right (d) shows the Fourier transform of the spectral sensitivities of the human colour mechanisms (continuous curve) and of the much narrower spectral sensitivity of a chicken cone with a coloured oil droplet (dashed); the former would demodulate the spectral energy distribution of (a) by about 50%, but the latter would hardly affect it. Oil droplets might therefore make tetrachromacy worthwhile, and (c) shows how such a narrowed curve might fit in with the three known human mechanisms. The roach has the four pigments shown in (b); these curves are no narrower than those of the human pigments, but they span a bigger spectral range; this is another way of making tetrachromacy worthwhile, though there seems to be no behavioural evidence of tetrachromacy in any animal. (From Bowmaker, 1983.)

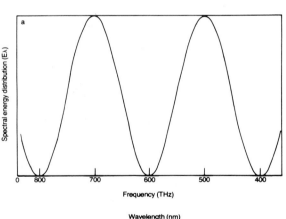

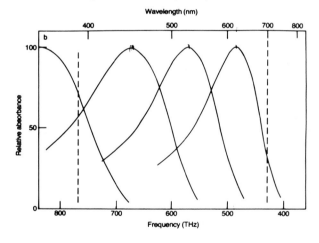

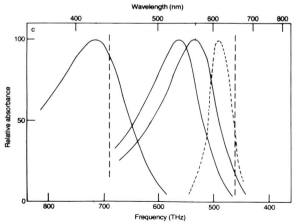

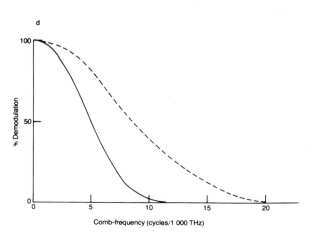

though I know of no behavioural demonstration of more than three channels in any animal.

Another case where tetrachromacy might be worthwhile is in animals which have a greater spectral range of useful vision than human beings. Bowmaker (1983) has evidence that this may occur, for he has found that the roach has the four different cone pigments shown in Fig. 1.8b; each of these has a reception band as narrow as that of the human pigment, but the positions of their peaks range from 355 to 620 nm, compared with the human extremes of 430 and 565 nm. Four different cone pigments in one retina do not prove tetrachromacy, but they certainly prompt one to look for behavioural evidence.

Another possible way of making polychromacy worthwhile would be to narrow the absorption spectra of the receptor photopigments. Little seems to be known about possible limiting factors here, but an interesting fact has recently come to light. If the absorption spectra of pigments with peaks at different positions are plotted on a scale of $(\lambda)^{1/4}$, then they become very nearly superimposable by a simple lateral shift (Barlow, 1982). This implies that the breadth of an absorption spectrum of this class of pigment varies with the 3/4-power of the wavelength of its peak absorption. The Vitamin A2-based photopigments appear to follow the same rule, but their absorption spectra are a little broader than those of Vitamin A1 pigments. It would be interesting to know the physical basis of these rules.

Can the contrast sensitivity function for comb-filtered spectra be measured psychophysically? Preliminary results (Barlow *et al.*, 1983) suggest that it does not follow the form predicted by the Fourier transforms of Fig. 1.8. Low comb-frequencies are attenuated and there is a peak at intermediate frequencies; this is presumably a direct manifestation of colour opponency, just as the attenuation of low spatial frequencies is thought to be the result of spatial opponency. This will be discussed further in the section on collecting and comparing.

Temporal factors

Fig. 1.9 shows the responses of a monkey rod to brief flashes of light of increasing intensity (Nunn & Baylor, 1982). Notice that they are extremely slow, especially at low intensities. Cones are about three times faster, but it is still surprising to find psychophysical evidence that changes in the timing of events can

be detected down to less than 200 μs (Burr & Ross, 1979). Faster receptors would seem to have obvious advantages for animals as well as sportsmen, so the question 'Why aren't they faster?' needs an answer.

The first point to make is that the prolonged responses to brief flashes shown in Fig. 1.9 provide a mechanism for temporal summation, which will improve the sensitivity of the eye to long-lasting stimuli. Anybody who has examined single frames of cinema film or video tape will appreciate how effective temporal averaging is in improving signal/noise ratios. It is not the only possible mechanism for doing this, for if the receptors generated rapid brief pulses, there would be nothing to stop these being integrated temporally at some central site. This would have the advantage that it would be optional; a parallel system that required speed rather than high signal/noise ratios could take advantage of the receptors' rapidity. One should notice that the responses become faster for strong flashes, and especially at high adaptation levels, which fits in with the psychophysical evidence, but one is left with the feeling that faster receptors would be advantageous in much prey-catching and evasive behaviour. Lamb (1984) found that the toad rod

Fig. 1.9. Responses of monkey rods, at 34 °C, to brief flashes of light. The intensity was increased by a factor of about 2 between each superimposed response. These are very much quicker than the responses of toad rods in Fig. 1.2; they would be quicker still at higher adaptation levels, and cones are about three times faster, but these still seem slow in comparison with the proved ability of the human system to detect time delays of the order of 200 ns; furthermore, faster reactions would obviously be advantageous where quick reactions are required. (From Nunn & Baylor, 1982.)

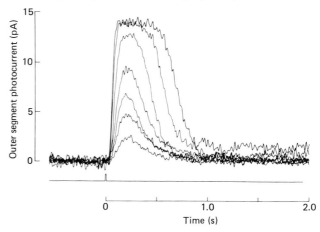

kinetics had a temperature coefficient (Q_{10}) of 2.2, and R.A. Cone has suggested that the speed of response is related to receptor size. Perhaps fundamental limits to photoreceptor speed will emerge when the steps between the absorption of light and changes in membrane currents are better understood.

The effect of eye size

The best that an eye can achieve in splitting up the light is strongly dependent on its size, for a larger pupil catches more quanta and allows a higher diffraction-limited acuity. Increasing pupil diameter by increasing numerical aperture alone will improve sensitivity but will not improve acuity for two reasons: (1) chromatic aberration will get worse; and (2) if there was an improved retinal image, this would require smaller receptors but, as we have seen, receptors have a lower limit for their useful diameter at about 2 μm (Snyder, 1972). The Abbé formula for microscope resolution (minimum resolvable distance = 0.61 λ/$n.\sin\theta$) suggests that to resolve 2 μm requires a numerical aperture (n) of 0.17, which implies that the pupil should subtend about 15 degrees at the receptors. It seems that there can be no selective pressure to increase numerical apertures much above this value for reasons of acuity, though it would be advantageous to do so to improve sensitivity.

Taking these limiting factors into account, one can draw some conclusions about the likely visual performance of an animal simply by observing the size of its pupil, as shown in Table 1.1. The diameter of the pupil in bright light (d) is the value relevant for the highest acuity attainable, and the fourth column gives the highest attainable spatial frequency in cycles/radian. One often wishes to compare an animal's performance with that of mans, so this ratio is given in the fifth column: thus a hawk with a pupil diameter in bright light of 4 mm would be expected to have a grating acuity about twice that of mankind, whereas for a lizard with a pupil diameter of 0.5 mm the expected acuity would be only one-quarter that of man. Note that the expected acuity is almost certainly not reached in the many species of animal that are not specialized for high resolution: dogs and lions, for instance, are probably not as good as people, though they have large pupils and the table suggests that they might be better; the same must be true for the large herbivores like cows, which have enormous pupils.

For sensitivity, the diameter at low luminances

(D) is relevant. A proportional sign is given because many specializations, such as reflecting tapeta, would influence sensitivity, and the body temperature would also be important where thermal decomposition is the main determinant of retinal noise. For simplicity these other factors have been neglected in the final column and it has been assumed that the proportionality factor is substantially the same as for man. Note that for the cat this column suggests a sensitivity some 10 times that of man, which is probably about right, but again the expected performance would not be reached in animals not well adapted to low luminances. The lowest row of the table is more tentative, for there are few conditions where vision is truly limited by quantal fluctuations. Nevertheless, this row indicates that animals with large pupils have scope for considerable improvement over people.

Collecting and deciding

Splitting up the light and producing graded signals for each of the subdivisions is a necessary prerequisite for making use of the light that reaches the eye, but it is only the first step towards finding out what is in the visual environment. The rest of the job is probably more difficult and is certainly less well understood. Furthermore, the limits of the next steps, as far as one can discern them, are not set by hard physical laws such as those governing quantal fluctuations, diffraction at the pupil, or the problem of holding light within the waveguide-like outer segment of a photoreceptor. Instead, they arise from the logic of the process of recognition and decision, and from the limited capabilities of neurons for performing the kinds of task that we can see are necessary. Let us consider the former first.

Logic of recognition

Our eyes identify familiar objects in the world around us and tell us where they are. Of course they also warn us of unfamiliar objects, guide our eye movements, tell us when to blink and allow us to admire sunsets, but object recognition is surely the basic skill and probably the most difficult one. Consideration of this topic usually starts with the invariance problem: the object has to be recognized in different three-dimensional orientations, over a range of positions and sizes, under varying conditions of illumination, against different backgrounds, and so forth. To meet this requirement it is thought that representations are formed and stored using descriptors and coordinate systems such that the various confounding factors do not change the representation, or only change a small number of its parameters which can be ignored if required. This sounds plausible but difficult, and I'm not sure how much success has been achieved in finding and implementing such representations. But there is another aspect that also needs emphasizing.

If such a representation is to work well, then it must be *complete* and *exclusive* in the following sense. The quantities stored should be determined by all the important parts of the object represented, not just by some of them, and they should be determined only by the object, and not by the rest of the visual scene. Considering the range of forms an object can assume in its image, this is clearly a difficult requirement to meet, but its desirability is quite easy to see: if parts of the object do not influence the representation, then those parts could be removed or changed without affecting the representation, and this obviously will not do if the parts concerned are important; and if the rest of the visual scene can influence the representation, then it will be unnecessarily susceptible to outside factors.

The requirements for invariance, completeness and exclusiveness mean that object recognition is a horrendously complicated matter. To do the job right the system has got to implement just the right logical function of a very large number of input variables, and the number of possible logical functions from which it must pick the right one is unimaginably large. It is inconceivable that a brain can always find the right logical function under all circumstances, but we do not know at all what restrictions apply; all we can say is that the logical complexity of the object recognition problem is so great that it surely limits what the brain can do, and we should be aware of this fact.

It is more useful to look at the requirements for the lower-level descriptors out of which high-level representations are made, for here we know better how they must be implemented. I think we can already discern two serious shortcomings in the neural equipment available to do this, namely the limited dynamic range of nerve fibres as communication channels and the limited range of connections any one cell can establish with others. In both cases a sceptic might say, 'If this property is so important it must be under intense evolutionary pressure, so why has it not been improved?'. Perhaps they have been evolved as

far as is allowed by some biophysical factors we do not know about, but a discussion of such possibilities would lead us too far from the point, and they will simply be taken as limits that cannot be improved upon.

To demonstrate their importance let us suppose that the system needs to know the numbers of quanta absorbed in blobs and patches of a great variety of sizes and shapes in a vast number of different positions and orientations in the visual field. Experience of image-processing tells one that obtaining this type of information is extraordinarily time-consuming, even on a modern computer. In such a machine the quantities would normally be represented in a word of at least 16 bits, whereas it is hard to believe (see below) that a nerve fibre can transmit more than about 6 bits, and it would require 100 ms to do even that. This is the reason for believing that the dynamic range of nerve fibres is an important limitation.

A computer suitable for image-processing would also have direct access to a wide range of addresses, perhaps using 20 bits or more for this purpose. The range of connectivity of an individual nerve cell limits how many other cells it receives information from and can transmit to, and it is hard to believe that this even approaches the 1 million implied by the 20-bit address range of the computer, hence the importance of connectivity as a limitation of neural computing. There are obvious problems with the analogies I have drawn, especially because of the parallel nature of neural computing, but the test of the case is to go ahead and look at the physiology of the visual pathway with these limitations in mind, and to see if they help to make sense of the arrangements we find. The limited dynamic range of nerve fibres seems to throw more light on retinal coding, whereas limited connectivity seems important for the cortex, so they will be considered separately, in that order.

Retinal coding and the dynamic range of nerves

We do not know if dynamic range is a problem in receptors, bipolars, horizontal and amacrine cells, but it must be serious for the ganglion cells. They characteristically pick up information from many receptors, and would require a greater dynamic range on that account, but they also have to signal by means of discrete all-or-nothing impulses rather than by graded potentials. It is hard to make a firm estimate of the dynamic range of nerve messages, mainly because we are ignorant about the possible importance of the precise timing of individual impulses, but the generally accepted view is that the important variable is the number of impulses in some short interval of time. If that interval was as long as 100 ms, the maximum number of impulses would be about 64, so the dynamic range would be equivalent to 6 bits. If the interval was only 10 ms, the dynamic range of each message would be equivalent to 3 bits or less, but there would of course be 10 times as many messages. In any case the dynamic range seems extremely small, compared with the range of quantal absorption rates in the receptors that have to be signalled, for this range must regularly exceed 10^{12} or 40 bits. The problem posed by this discrepancy between the wide range of the input and the narrow range of the output makes sense of many features of the retina.

Adaptation

Craik (1938) looked upon adaptation as a mechanism for adjusting the response range of the retina in accordance with the prevailing light level, and it has subsequently been shown how this is done by

Fig. 1.10. The responses of an ON-centre and an OFF-centre ganglion cell are shown after adapting to eight different levels separated by factors of 10. After adapting for a few minutes the luminance on the receptive field centre was shifted to a new value whose log is plotted as abscissa, and the response in the next 1.28 s is plotted as ordinate. Note that the ON and OFF systems each respond sensitively to only a narrow range of stimulus intensities, that the response ranges shift with adaptation level, and that having both ON and OFF systems doubles the range of stimulus intensities for which sensitive responses are given. (From Barlow, 1969.)

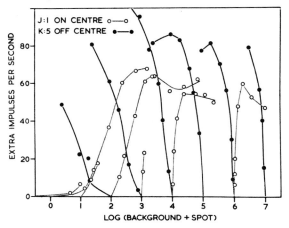

those who have recorded the response of individual nerve cells (Werblin, 1971). It was the wide range of the input, combined with the high incremental sensitivity of the eye, that led Craik to this view, but the poor dynamic range of nerve fibres clearly makes adjustments of sensitivity even more important. Fig. 1.10 shows very clearly how the range of a ganglion cell's response shifts with adaptation level.

Spatial opponency and the on–off system

To use a limited output range effectively it is obviously advantageous to remove any constant component. Lateral inhibition can be regarded as a means of finding the local mean value of luminance and subtracting it from the signal to be transmitted, thus conserving the output range for the useful signal (Barlow & Levick, 1976). Nerve signals cannot assume negative values, so subtracting from the mean entails signalling the upward and downward deviations from the mean separately; however, this brings the additional advantage of doubling the available dynamic range.

Colour opponency

The extensive overlap of the spectral sensitivities of the different primary colour mechanisms means that signals in them are necessarily strongly correlated. Buchsbaum & Gottschalk (1983) have worked out the implications of this, and shown that the colour opponent arrangements that are found to approximate quite closely to those that would be recommended for optimal coding of the information. Their argument gains strength from the limited range of the output, for this tells one why optimal arrangements are so important.

Signalling probabilities

It was suggested earlier that, ideally, nerve signals should inform their destinations of the exact numbers of quanta absorbed in particular blobs and patches of the image, but this is not strictly true. The aim of most central information-processing must be to build up the probability of making a correct deduction about the environment, and for this purpose what is needed are probability statements about the constituent parts, such as 'This part of the image is unusually bright (or unusually dim, or unusually red or blue in spectral content)'. In the retina the unusualness could be assessed from what has occurred locally in the image in the fairly recent past, and Fig. 1.10 shows the responses of an ideal device, signalling the log of the improbability of the Null hypothesis that there was no increase, or no decrease, in the quantal flux. It seems to me that the messages provided by retinal ganglion cells, as shown in Fig. 1.10, fit rather well the idea that they are signalling improbabilities (see Fig. 1.11) (Barlow, 1969).

There are two further comments to make about the limited dynamic range of the optic nerve fibres. First, it must be admitted that there are many aspects of retinal coding which it does not help us to understand, such as the distinction between X and Y units (Enroth-Cugell & Robson, 1966), or the reasons for coding more complex features such as direction of movement or oriented edges (Levick, 1967). Secondly, there is the interesting question of how this explanation of the selective advantage of adaptation, lateral inhibition, and colour opponency fits in with other explanations. It is sometimes suggested that these mechanisms exist because they are useful for the performance of some higher-level task, such as detecting edges or aiding image segregation, but this is a tautology: higher mechanisms have no alternative but to use the information provided by the retina. What one needs is an explanation in terms of image properties; the redundancy reduction hypothesis (Barlow, 1961, 1985) and the recent ideas of Srinivasan, Laughlin & Dubbs (1982) on predictive coding may provide this.

Fig. 1.11. Idealized ON and OFF units that signal $-\log p$, where p is the probability of the number of quanta shown on the abscissa being absorbed when the mean rate of absorption is the adaptation level: it is supposed that, as in Fig. 1.10, the retina is adapted to three different levels, then stepped to another level. These curves have points in common with the experimental results shown in Fig. 1.10. (From Barlow, 1969.)

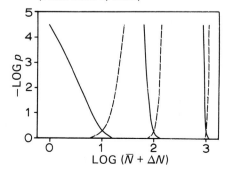

Attaching importance to the limited dynamic range of optic nerve fibres certainly does not conflict with these ideas, for it gives an added reason for the importance of predictive coding or redundancy reduction. These other ideas may, however, add a broader perspective to the problem.

Cortex and connectivity

The role of the cortex must be to derive knowledge of objects in the environment from the image it receives in the optic radiation from the lateral geniculate nucleus. Cortical neurons receive connections on their spines, dendrites, and cell bodies in one quite restricted region of the cortex, and transmit impulses down their axons to other parts of the brain, often to other quite restricted parts of the cortex. Because of their diversity it is a bit absurd to talk of a typical cortical neuron, but the above description is so general that no-one could quarrel with it seriously, and the important point to note is that both the neuron's sources of information and the destinations of its axon are restricted, when compared with the address range available in an image-processing computer. A cortical neuron may be unusual in being able to interconnect a very large number of other cells, but this large number is only a very small fraction of all the cells in the brain which it might be advantageous for it to connect with. As an exercise I shall assume that this limited capacity to interconnect is an overriding factor in the design of the cortex, and make some predictions about how it would have to set about doing its job if that were the case.

In discussing object recognition it has already been emphasized that the information does not arrive in neatly delineated orderly packages, but is liable to be disordered by changes of perspective, size, illumination, background, and so on. The amazing thing about perception is that, in spite of these disorganizing factors, it gives us an organized and orderly view of the objects around us; hence what the nerve cells in the visual system must do above all else is to find order and preserve it. The first requirement must surely be to preserve whatever order is already to be found in the image and prevent things that belong together being further separated; hence the requirement for an accurate topographical map in V1, for although information about objects is not well segregated in the image, there is certainly a higher prior probability of neighbouring points belonging to the same object than would be the case if the image had been fragmented by a disordered projection.

Since the cortex has to deal with objects, and since its neurons pick up information from regions of only a few square millimetres, it must find ways of bringing together all the information relevant to an object within small cortical areas. We have seen that neighbouring points in the image are likely to belong to the same object, but they are not certain to, because objects have edges and holes or may be partially obscured by other objects. In the same way, regions of the image of similar colour, texture, disparity, or direction of motion do not necessarily belong to the same object, but they are more likely to than are regions of the image that differ in these characteristics. Hence, in its efforts to organize the image into a representation of objects, the cortex needs first to detect these characteristics, because they are the local features which can link together the parts of the image that belong to the same object (Barlow, 1981).

This has been presented as a prediction from the limited ability of neurons to interconnect with other neurons, but I think it fits what we know of the physiology of V1, as well as much previous work on the psychological factors that cause segregation of the image into figure and ground. I will not repeat the evidence that neurons in V1 are selective for colour, motion and disparity, but the role of the simple cells as orientation or texture detectors needs amplifying. The orientation selectivity of cortical neurons was the aspect to which Hubel & Wiesel (1962) originally gave most emphasis, and it certainly seems to play a very important part in the organization of the primary visual cortex, yet at first sight it may not be obvious that orientation selectivity plays a key role in image segregation. This was, however, the first characteristic to be recognized as important in artificial intelligence. Guzman (1968), working on the problems of deriving the three-dimensional form and position of a number of blocks from a line drawing of them, came across the problem of identifying the two parts of a block whose contour had been obscured by another one. The colinearity of the two separated parts of the edges was what he called the 'linking feature' that enabled them to be reunited in the final representation; it was therefore the orientation and position of the edge-segments that enabled this to be done. Notice that in a computer program the importance of a linking feature is that it enables parts to be reunited after a

limited, instead of an exhaustive, search; if the address range of the computer was small, this would obviously be even more important, and that is the situation in the neural computer, whose elements can only interconnect a tiny proportion of the other cells in the brain.

A more up-to-date view of cortical neurons regards them as providing a representation of a local part of the image in terms of spatial frequencies (Glezer, Ivanoff & Tscherbach, 1973; Robson, 1975). It has long been recognized that cortical neurons vary in the size as well as the orientation of the stimulus they prefer, and this idea of local Fourier-type analysis incorporates this fact into a systematic model. How does this fit in? Because of its limited connectivity, a given neuron can only respond to properties of a small patch of the image, yet the role of a group of such neurons must be to provide a description of the local pattern of light and shade in that small patch that will be helpful, together with the descriptions of other patches, in recognizing objects. There are innumerable possible ways of describing such patches, but one in terms of spatial frequency has merit for several reasons. First, it is potentially a complete description of the patch (Sakitt & Barlow, 1982). Secondly, each element, which we suppose signals the coefficient of one of the frequency and orientation components, is derived from the whole of the patch so that all such messages have nearly equal signal/noise ratios. Thirdly, the components have generality so that the similarity or dissimilarity of descriptions from different parts can be compared. Finally, the search for similarities can be much simplified by proper arrangement of the destinations of the information about each component. For example, Marr & Hildreth's (1980) program for edge detection uses knowledge of edges of different degrees of blur but similar orientation, and it would be possible to bring this information together by the appropriate targeting of the destinations of the simple cells in the primary cortex. It should be added that information from continuations of the edge in either direction would also be useful, and this too could be achieved by the appropriate pattern of connections from the primary cortex. We obviously do not know whether or how this is done, but I think recognizing the connectivity problem makes us look at the operation of the visual cortex in a new light.

The next prediction is that V1 will not be the only visual area (Barlow, 1981): multiple visual areas will be necessary, and have of course been found

(Allman & Kaas, 1976; Zeki, 1978). In these secondary and tertiary areas, the magnification factor should be less if there is a well-defined topographical map, so that one cell can look at a larger region of the visual field; furthermore, one would expect to find evidence of information being brought together on non-topographical principles, for example according to orientation or colour or direction of motion. It is too early to say how far this prediction is correct, but some of the evidence is promising (Zeki, 1980).

It is apparent that what has so far been described will not take the cortex very far towards object recognition. The main hindrance is that each of the cues for segregating the information about the object from irrelevant information is uncertain and unreliable. An object cannot be reliably defined by its position in the image, by its texture, by its colour, by its direction of motion, or by its disparity. To have any chance of working, the system must use all these cues in combination, but the combination required will not be fixed: sometimes colour will be reliable by itself, sometimes it will not help, and so on for the other possible linking characteristics. It is crucially important to take account of the reliability of each message when combining them, if the messages have differing degrees of reliability, so it is highly desirable for each message to convey its own reliability. This is the reason why the signalling of improbabilities that was suggested for the retina seems an even better arrangement in the cortex. If the cortex uses such a probabilistic language, one can perhaps see why it is so adept at combining weak cues and recognizing which are strong cues.

Summary and conclusions

In the early stages of vision, physical factors set firm limits to what can be achieved, and the eye seems to approach them quite closely. The evolution of the eye has clearly been moulded to a great extent by these factors, the most important of which appear to be: diffraction as a limit to acuity; quantum fluctuations and the thermal stability of photopigments as limits to sensitivity; propagation down the waveguide-like outer segment of photoreceptors as a limit to their size and separation; and the narrowness of spectral sensitivity curves as a limit to the number of worthwhile colour channels. There are probably other important limiting factors lying behind the loss of quantum efficiency at high background levels, and the slowness of the receptor's response.

It is much harder to spot the limiting factors for the higher functions of vision, but it is certainly important to realize the tremendous difficulty of the task of object recognition set by the requirements for invariance in the face of perturbing factors, and for completeness and exclusiveness in the representation of the objects to be recognized. There are also two characteristics of neurons as computing elements that must create difficulties, namely the limited dynamic range of neural messages, and the limited range of connectivity of any given nerve cell.

Asking 'Why doesn't the eye see better?' will not answer all problems of vision, but it does direct our attention to interesting aspects, and it may help us to understand why the visual system is organized in the way it is.

References

Allman, J. M. & Kaas, J. H. (1976) Representation of the visual field on the medial wall of the occipito-parietal cortex in owl monkey. *Science*, **191**, 572–5

Barlow, H. B. (1956) Retinal noise and absolute threshold. *J. Opt. Soc. America*, **46**, 634–9

Barlow, H. B. (1957) Purkinje shift and retinal noise. *Nature*, **179**, 255–6

Barlow, H. B. (1961) Possible principles underlying the transformations of sensory messages. In *Sensory Communication*, ed. W. Rosenblith, pp. 217–34. Boston: MIT Press

Barlow, H. B. (1962) Measurements of the quantum efficiency of discrimination in human scotopic vision. *J. Physiol. (Lond.)*, **160**, 169–88

Barlow, H. B. (1964a) Dark adaptation: a new hypothesis. *Vision Res.*, **4**, 47–58

Barlow, H. B. (1964b) The physical limits of visual discrimination. In *Photophysiology*, vol. 2, ed. A. C. Giese, pp. 163–202. New York: Academic Press

Barlow, H. B. (1969) Pattern recognition and the responses of sensory neurons. *Ann. N. Y. Acad. Sci.*, **156**, 872–81

Barlow, H. B. (1972) Dark and light adaptation: psychophysics. In *Handbook of Sensory Physiology*, vol. 7, ed. D. Jameson & L. M. Hurvich, pp. 1–28. Berlin: Springer-Verlag

Barlow, H. B. (1981) Critical limiting factors in the design of the eye and visual cortex. The Ferrier Lecture. *Proc. R. Soc. Lond., Ser. B*, **212**, 1–34

Barlow, H. B. (1982) What causes trichromacy? A theoretical analysis using comb-filtered spectra. *Vision Res.*, **22**, 635–43

Barlow, H. B. (1985) Perception: what quantitative laws govern the acquisition of knowledge from the senses? In *Functions of the Brain*, Wolfson College Lectures, 1982, ed. C. Coen, pp. 11–43. Oxford: Oxford University Press

Barlow, H. B. & Levick, W. R. (1976) Threshold setting by the surround of cat retinal ganglion cells. *J. Physiol. (Lond.)*, **259**, 737–57

Barlow, H. B. & Mollon, J. D. (1982) (ed.) *The Senses.* Cambridge: Cambridge University Press

Barlow, H. B., Gemperlein, R., Paul, R. & Steiner, A. (1983) Human contrast sensitivity for comb-filtered spectra. *J. Physiol. (Lond.)*, **340**, 50P

Baylor, D. A., Lamb, T. D. & Yau, K-W. (1979) Responses of retinal rods to single photons. *J. Physiol. (Lond.)*, **288**, 613–34

Baylor, D. A., Matthews, G. & Yau, K-W. (1980) Two components of electrical dark noise in retinal rod outer segments. *J. Physiol. (Lond.)*, **309**, 591–621

Bowmaker, J. K. (1983) Trichromatic colour vision: why only three receptor channels? *Trends Neurosci.*, **6**, 41–3

Bowmaker, J. K. & Knowles, A. (1977) The visual pigments and oil droplets of the chicken retina. *Vision Res.*, **17**, 755–64

Buchsbaum, G. & Gottschalk, A. (1983) Trichromacy, opponent colour coding and optimum information transmission in the retina. *Proc. R. Soc. Lond. Ser. B*, **220**, 89–113

Burr, D. C. & Ross J. (1979) How does binocular delay give information about depth? *Vision Res.*, **19**, 523–32

Burr, D. C. & Ross, J. (1982) Contrast sensitivity at high velocities. *Vision Res.*, **22**, 479–84

Campbell, F. W. & Gubisch, R. W. (1966) Optical quality of the human eye. *J. Physiol. (Lond.)*, **186**, 558–78

Campbell, F. W. & Robson, J. G. (1968) Application of Fourier analysis to the visibility of gratings. *J. Physiol. (Lond.)*, **197**, 551–66

Craik, K. J. W. (1938) The effect of adaptation on differential brightness discrimination. *J. Physiol. (Lond.)*, **92**, 406–21

Enoch, J. M. (1961) Optical interaction effects in models of parts of the visual receptors. *Arch. Ophthalmol.*, **63**, 548–58

Enroth-Cugell, C. & Robson, J. G. (1966) The contrast sensitivity of retinal ganglion cells of the cat. *J. Physiol. (Lond.)*, **187**, 517–52

Fain, G. (1975) Quantum sensitivity of rods in the toad retina. *Science*, **187**, 838–41

Fain, G. L., Gold, G. H. & Dowling, J. E. (1976) Receptor coupling in the toad retina. Cold Spring Harbor Symposium. *Quant. Biol.*, **40**, 547–61

Glezer, V. D., Ivanoff, V. A. & Tscherbach, T. A. (1973) Investigation of complex and hypercomplex receptive fields of visual cortex of the cat as spatial frequency filters. *Vision Res.*, **13**, 1875–904

Guzman, A. (1968) Decomposition of a visual scene into three-dimensional bodies. *Proc. Fall Joint Computer Conference, San Francisco*, 1968, 33(I), 291–304

Harmon, L. D. & Julesz, B. (1973) Masking in visual recognition: effects of two-dimensional filtered noise. *Science*, **180**, 1194–7

Hecht, S., Shlaer, S. & Pirenne, M. (1942) Energy, quanta and vision. *J. Gen. Physiol.*, **25**, 819–40

Hubel, D. H. & Wiesel, T. N. (1962) Receptive fields, binocular interaction and functional architecture in the cat's visual cortex. *J. Physiol. (Lond.)*, **160**, 106–54

Krauskopf, J. (1962) Light distribution in human retinal images. *J. Opt. Soc. America*, **52**, 1046–50

Lamb, T. D. (1980) Spontaneous quantal events induced in toad rods by pigment bleaching. *Nature*, **287**, 349–51

Lamb, T. D. (1984) Effects of temperature changes on toad rod photocurrents. *J. Physiol. (Lond.)*, **346**, 557–78

Levick, W. R. (1967) Receptive fields and trigger features of ganglion cells in the visual streak of the rabbit's retina. *J. Physiol. (Lond.)*, **188**, 285–307

Levick, W. R., Thibos, L. N., Cohn, T. E., Catanzaro, D. & Barlow, H. B. (1983) Performance of cat retinal ganglion cells at low light levels. *J. Gen. Physiol.*, **82**, 405–26

Liebman, P. A. (1972) Microspectrophotometry of photoreceptors. In *Handbook of Sensory Physiology*, Vol. VII/I, *Photochemistry of Vision*, ed. H. J. A. Dartnall, pp. 481–528. Berlin: Springer-Verlag

Marr, D. & Hildreth, E. (1980) Theory of edge detection. *Proc. R. Soc. Lond., Ser., B*, **207**, 187–217

Mullen, K. T. (1985) The contrast sensitivity of human colour vision to red-green and blue-yellow chromatic gratings. *J. Physiol.*, **359** (in press)

Nunn, B. J. & Baylor, D. A. (1982) Visual transduction in retinal rods of the monkey *Macca fascicularis*. *Nature*, **299**, 726–8

Pelli, D. (1983) The spatio-temporal spectrum of the equivalent noise of human vision. *Invest. Ophthalmol. Vis. Sci.*, **24** Suppl. 46

Reymond, E. (1984) *Spatial visual acuity of Australian birds of prey*. Ph.D. thesis, Australian National University, Canberra

Robson, J. G. (1975) Receptive fields: neural representation of the spatial and intensive attributes of the visual image. In *Handbook of Perception*, *Vol. 5, Seeing*, ed. E. C. Carterette & M. P. Friedman, pp. 81–116, New York: Academic Press

Sakitt, B. & Barlow, H. B. (1982) A model for the economical encoding of the visual image in the cerebral cortex. *Biol. Cybernet.*, **43**, 97–108

Simon, E. J. & Lamb, T. D. (1977) Electrical noise in turtle cones. In *Vertebrate Photoreceptors*, ed. H. B. Barlow & P. Fatt, pp. 291–304. London: Academic Press

Smith, V. C. & Pokorny, J. (1975) Spectral sensitivity of the foveal cone photopigments between 400 and 600 nm. *Vision Res.*, **15**, 161–71

Snyder, A. (1972) Coupled-mode theory for optical fibers. *J. Opt. Soc. America*, **62**, 1267–77

Snyder, A. W. & Miller, W. H. (1977) Photoreceptor diameter and spacing for highest resolving power. *J. Opt. Soc. America*, **67**, 696–8

Srinivasan, M. V., Laughlin, S. B. & Dubbs, A. (1982) Predictive coding: a fresh view of inhibition in the retina. *Proc. R. Soc. Lond. Ser. B*, **216**, 427–59

Werblin, F. S. (1971) Adaptation in a vertebrate retina: Intracellular recording in *Necturus*. *J. Neurophysiol.*, **34**, 228–41

Westheimer, G. & Campbell, F. W. (1962) Light distribution in the image formed by the living human eye. *J. Opt. Soc. America*, **52**, 1040–5

Zeki, S. M. (1978) Functional specialisation in the visual cortex of the rhesus monkey. *Nature*, **274**, 423–8

Zeki, S. M. (1980) The representation of colours in the cerebral cortex. *Nature*, **284**, 412–8

2

Sampling of visual space by retinal ganglion cells

HEINZ WÄSSLE

Introduction

An image falling onto the retina is first degraded by the optics of the eye: scatter in the lens and in the vitreous humor, geometrical optical aberrations, and diffraction at the pupil all limit the optical resolution of the eye. The cat eye in that respect is slightly worse than the human eye and can resolve 3–6 min of arc (Wässle, 1971; Bonds, 1974; Robson & Enroth-Cugell, 1978) compared to 1 min of arc in man (Campbell & Green, 1965).

The image on the retina is sampled by the photoreceptor array and spatial resolution of this process is limited. Each individual photoreceptor does not respond to light incident at a point on the retina, but light is collected over a small area roughly equal to the cross-section of its inner segment. The retinal image is only sampled at spatially discrete locations by the cone array. This implies an upper limit to the spatial frequencies about which information is unambiguously available within the visual system (Shannon, 1949; Barlow, 1964; Snyder & Miller, 1977). Assuming this limit is defined by the cones, which, in the cat, according to Steinberg, Reid & Lacy (1973) have a peak density of 26.000 mm², the calculated value for spatial resolution is 3–4 min of arc. In the fovea of the macaque monkey, the peak density of cones is about 250.000 mm², which corresponds to a calculated spatial resolution of 1 min of arc.

In the cat, bipolar cell dendrites are connected to four to nine cones (Boycott & Kolb, 1973) and there is no evidence for a bipolar cell which contacts only one cone, such as the midget bipolar found in simians (Kolb, 1970). Because bipolar cells also occur in lower density than the receptors, this integration of cone information in the cat causes a loss of spatial resolution. The same holds for ganglion cells. As shown specifically in this chapter for brisk-transient (Y) units, ganglion cells collect light over the whole extent of their dendritic field. The dendritic tree defines the receptive field centre and, as 'sampling aperture', sets the limits of spatial resolution. For the brisk-sustained (X) class, the mosaic of ON- and OFF-centre cells will be analysed in detail. Their array defines, as 'sampling distance', which spatial frequency is unambiguously available within the visual system.

Spatial analysis of ON- and OFF-centre brisk-transient (Y) cells

The physiological properties of brisk-transient (Y) cells are summarized in Chapter 3. They were unequivocally identified morphologically (Cleland, Levick & Wässle, 1975; Saito 1983) and correspond to the alpha cell class (Boycott & Wässle, 1974). Alpha cells have larger cell bodies than all other ganglion cells (Fig. 2.1A) which permitted an enumeration of their retinal distribution from Nissl-stained material (Wässle, Levick & Cleland, 1975). They form an almost constant proportion of 2–5% of all ganglion cells.

An ON–OFF dichotomy is present for brisk-transient (Y) cells as for all other concentrically organized ganglion cell classes. Nelson, Famiglietti & Kolb (1978) showed from intracellular recordings and dye injection that ON- and OFF-centre cells have their dendritic branches at different heights in the inner plexiform layer (IPL). Peichl & Wässle (1981) confirmed this stratification hypothesis for brisk-

transient (Y) cells. In a small patch of retina, they recorded extracellularly from as many brisk-transient (Y) cells as possible. After the experiment, the whole-mounted retina was stained by a reduced silver method which reveals the cell body and all dendritic branches of alpha cells (Fig. 2.1B). The recording area was identified and it could be directly observed that all OFF-brisk-transient (Y) cells had their dendrites close to the inner nuclear layer, while all ON-cell dendrites were stratified at a level about 10 μm further towards the ganglion cell layer.

This combination of physiological and anatomical techniques can be used to identify a particular brisk-transient (Y) cell after the recording (Peichl & Wässle, 1983). Thus, the longstanding question (Kuffler, 1953; Gallego, 1954; Brown & Major, 1966; Dowling & Boycott, 1966; Creutzfeldt et al., 1970) regarding the extent of the congruence between the receptive field centre (RFC) and the dendritic tree of an individual ganglion cell can be answered. Because reduced silver stains all alpha cells of the retina completely (Wässle, Peichl & Boycott, 1981b), a topographical distribution of ON- and OFF-cells can now be given. Hence, parameters defining the spatial sampling of brisk-transient (Y) cells and their RFC and intercell spacing can be obtained.

Morphological identification of a brisk-transient (Y) cell

Fig. 2.2A indicates the general procedure used in the experiments. A brisk-transient (Y) cell was recorded and its RFC was carefully plotted on the tangent screen. The microelectrode tip and blood vessels in the vicinity were backprojected onto the screen with a fundus camera. After the experiment, the whole mount of the retina was stained with reduced

Fig. 2.1. (A) Low-power micrograph from a Nissl-stained whole mount of the cat retina; alpha cells can be recognized by their larger cell bodies. (B) Micrograph from a reduced silver-stained whole mount of the cat retina at the raphe; the focus is taken at the level of the cell bodies viewed from the vitreal side. The cell bodies, axons and dendrites of the alpha cells are fully stained; only the cell bodies of other ganglion cell classes are stained. By counterstaining a reduced silver-stained whole mount with Nissl staining, it was demonstrated that reduced silver really stains all alpha cells of the cat retina. (C) Micrograph from a Golgi–Cox-stained whole mount preparation of the cat retina. The retina is viewed from the vitreal side and the focus is taken at the inner plexiform layer, some 15 μm from the ganglion cell perikarya. The focal plane shows the dendritic trees of two ON-alpha cells overlapping the whole field. In the upper part of the field, the dendritic tree of an ON-beta cell is in sharp focus, while in the lower part the dendritic tree of an OFF-beta cell is out of focus. (D) The identical field to (C) but with the focal plane 10 μm further into the inner plexiform layer. Now the dendritic branches of the OFF-beta cell are in focus, while the dendritic trees of the ON-alpha and ON-beta cells are out of focus. (All scale bars represent 100 μm.) (From Wässle et al., 1983.)

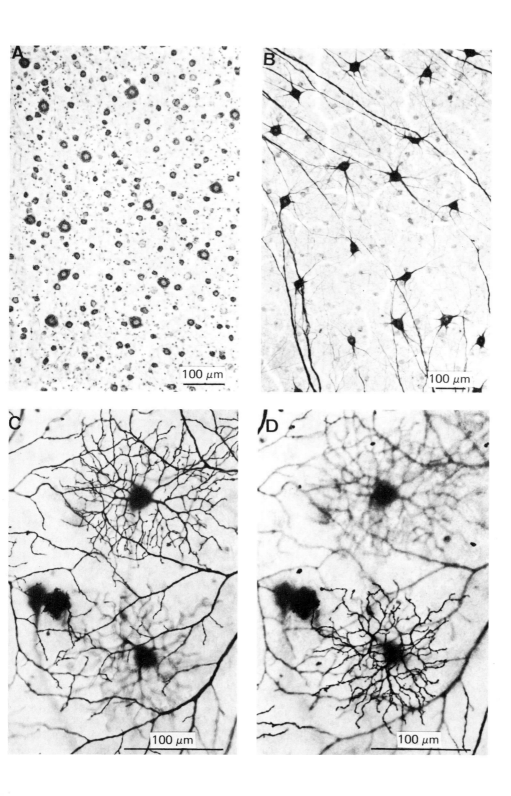

Fig. 2.2. Identification of a single recorded brisk-transient (Y) cell from the population of stained alpha cells. (A) Schematic representation of the method. In the middle, the eye penetrated by the recording microelectrode is shown. The receptive field centre (RFC) was plotted on the tangent screen indicated by the stippling. With the aid of a fundus camera, the recording electrode and blood vessels in the vicinity were backprojected onto the screen. After the experiment, the whole mount of the retina was stained with reduced silver and from the blood vessel pattern, both the recording site and the RFC could be inserted. (B) Screen protocol showing the RFC (hatched) of an OFF-centre brisk-transient(Y) cell, the blood vessels (dotted), blood vessel branching points (flower symbols) and the backprojected electrode tip position (star symbol). (C) Drawing of the OFF-alpha cell population from the reduced silver-stained whole mount of the retina. The area was identified by the blood vessels (stippled) and the two branching points (flower symbols). The RFC (hatched) and the electrode tip (star symbol) were inserted with respect to these landmarks. (D) Response profile of the recorded cell, showing a horizontal and vertical scan of a light spot (diameter 40 min, brightness 15 cd/m², background 7 cd/m²) through the RFC.

A

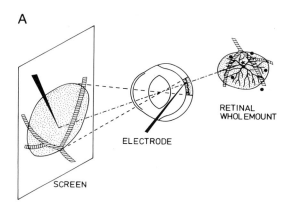

B

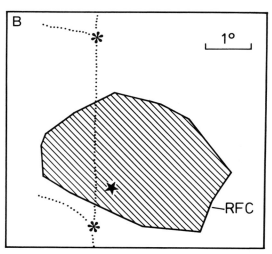

C

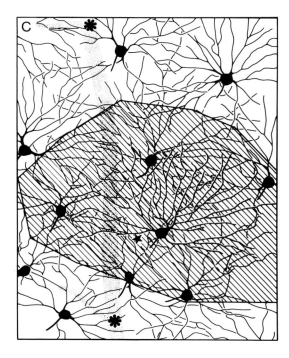

D

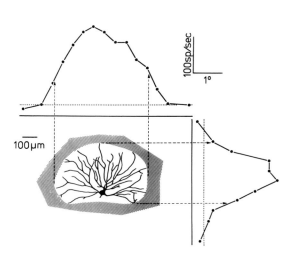

silver (Peichl & Wässle, 1981) and from the blood vessel pattern, both the recording site and the RFC could be superimposed. Fig. 2.2B shows the screen protocol containing the RFC of an OFF-brisk-transient (Y) cell, the electrode position during cell body recording from the unit, and the blood vessel in the neighbourhood. Fig. 2.2C shows the same area from the whole-mounted retina after staining with reduced silver and has all OFF-alpha-cells drawn in. Using the blood vessel pattern for reference positions, the recording position and the RFC could be juxtaposed. There is a slight ambiguity about the recorded cells when comparing soma position and elctrode position. This is resolved by comparing the RFC with the dendritic field positions. Since these only match for the cell to the right of the recording point, we conclude that this is the recorded OFF-cell.

The outline of a RFC is first determined by flashing a small spot of light and carefully observing the response on an audiomonitor. The border of the RFC is considered to be the locus at which the centre response could be heard on only 50% of the trials. The outline of the hatched area in Fig. 2.2 B–D shows this border. Then a horizontal and vertical scan of the response profile is taken. From poststimulus time histograms, the peak responses are measured and this gives a horizontal and vertical sensitivity profile as in Fig. 2.2 D. The dendritic field is inserted and it is obvious that the subjective plot of the RFC border has the general shape of the dentritic field, although it is somewhat larger. In the sensitivity profile we assume the border of the RFC, where the centre response significantly exceeds the maintained activity (indicated by the dotted lines in Fig. 2.2 D). The horizontal and vertical diameters measured in this way are larger than the corresponding dendritic field dimensions by a factor of 1.35, resulting in an RFC area 1.8 times larger than the dendritic field area. Altogether, 10 units were identified in this way (Peichl & Wässle, 1983) and for all, the size of RFC was substantially larger than the dendritic field size (mean ratio 1.95, S.D. 0.35 in area; or 1.4, S.D. 0.13 in diameter).

In summary, the results show that the dendritic tree of an alpha cell determines the position, shape and size of its RFC; however, the RFC is nearly twice as large in area.

Dendritic overlap and spacing of alpha cells

Fig. 2.3A shows all alpha cells stained in a 1 mm² field of retina. The dendritic fields of neighbouring alpha cells overlap and a dense plexus of dendritic branches is formed. If one selects the OFF-population (B) and the ON-population (C), a homogeneous overlay of the field with dendritic branches becomes apparent. The cell bodies are distributed non-randomly: this was shown by analysing their nearest-neighbour distribution, which proved to be Gaussian (Wässle *et al.* 1981*b*; Wässle & Riemann, 1978).

Dendritic fields were defined by connecting the outermost dendritic tips of every ON-alpha cell by a smooth outline (Fig. 2.3D). The dendritic fields of the 10 ON-alpha cells are similar in size (mean area 0.127 mm², S.D. 0.023). For the cell in the centre, the RFC (stippled area) was measured as described in the previous section. It exceeds in size the dendritic field area by a factor of 1.53 but has the same general shape. From Fig. 2.3, the parameters defining the spatial sampling of brisk-transient (Y) cells can be calculated. The density of ON-centre cells is 9 cells/mm², that of OFF-centre cells 10 cells/mm², and hence, they occur in approximately equal numbers. The number of RFCs of either type overlapping every retinal point (coverage) is the product of RFC area and density. From Fig. 2.3C, this number is 1.75 for the ON-centre brisk-transient cells, and from Fig. 2.3B, a coverage of 2.75 was obtained for OFF-centre cells. Previous, less direct estimates of the coverage factor for both types of brisk-transient (Y) cells together were 3–5.9 (Peichl & Wässle, 1979) and 3–6 (Cleland *et al.*, 1975).

In the central area, alpha cells have a peak density of about 180 cells/mm² (Wässle *et al.*, 1981*b*), which would give 90 mm² ON-centre and 90 mm² OFF-centre brisk-transient (Y) cells. In the central area, they have RFC sizes of 0.6–0.8 degrees (Cleland, Harding & Tulunay-Keesey, 1979; Peichl & Wässle, 1979), which could resolve 0.3–0.4 degrees. From this density, a minimum sampling distance of 0.5 degrees can be calculated providing there is unambiguous resolution of a 1 degree spatial period. The visual acuity of the cat measured behaviorly is 6–10 min and hence is much better than that of the brisk-transient (Y) cell resolution. Hence, these cells must play a role in the visual function, other than providing high spatial resolution.

Spatial analysis of ON- and OFF-brisk-sustained (X) cells

On the basis of dendritic morphology, cell body size and axonal size, Boycott & Wässle (1974) classified cat retinal ganglion cells into three morphological classes: alpha, beta and gamma cells. Gamma cells were found to be a heterogeneous category and one subclass of them has been defined (delta cells). In addition, Boycott & Wässle (1974) showed that the cell body sizes of beta and gamma cells have overlapping distributions.

Recently, more evidence is accumulating for additional morphologies of cat retinal ganglion cells. From Golgi-stained material, Stone & Clarke (1980) described a further group of gamma cells. Retrograde labelling of ganglion cells with horseradish peroxidase (HRP) (Kawamura, Fukushima & Hattori, 1979; Leventhal, Rodieck & Dreher, 1981), provides evidence for yet another cell type with a medium-sized cell body but loose dendritic branching (epsilon cell). This might correspond to a gamma cell with a medium-sized cell body (Stone & Clarke, 1980). Finally, 23 different types of ganglion cells were described in the cat retina from Golgi-stained material

Fig. 2.3. Alpha cells stained in a small field of the temporal retina using a recorded silver method. (A) All alpha cells within the 1×1 mm frame (dotted line) are drawn. (B) Only the OFF population is selected. (C) Drawing of all ON-alpha cells. (D) Dendritic fields are drawn surrounding every ON-alpha cell body. The stippled area in the centre illustrates the RFC of the underlying alpha cell recorded. (Scale bar represents 150 μm.)

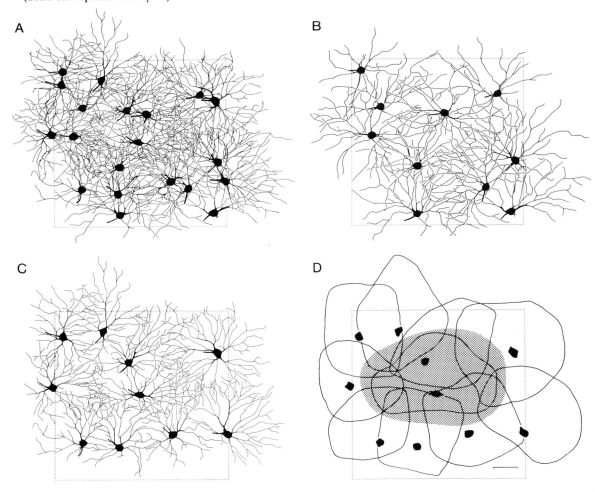

Fig. 2.4. (A) Micrograph of the ganglion cell layer after retrograde labelling of the ganglion cells with HRP from the lateral geniculate nucleus. The focal plane is taken at the ON-lamina in the inner plexiform layer, so the cell bodies are out of focus. Altogether 15 cells are labelled: one OFF-alpha cell; five ON-beta cells; four OFF-beta cells; and five gamma cells. (B) Drawing of the five ON-beta cells identified in (A). Although their dendritic fields are not completely stained, the beta morphology is apparent. (C) Micrograph of the same field as in (A) but with the focus 10 μm further towards the inner nucleus in the OFF-lamina. (D) Drawing of the four OFF-beta cells identified in (C). (From Wässle *et al.*, 1983.)

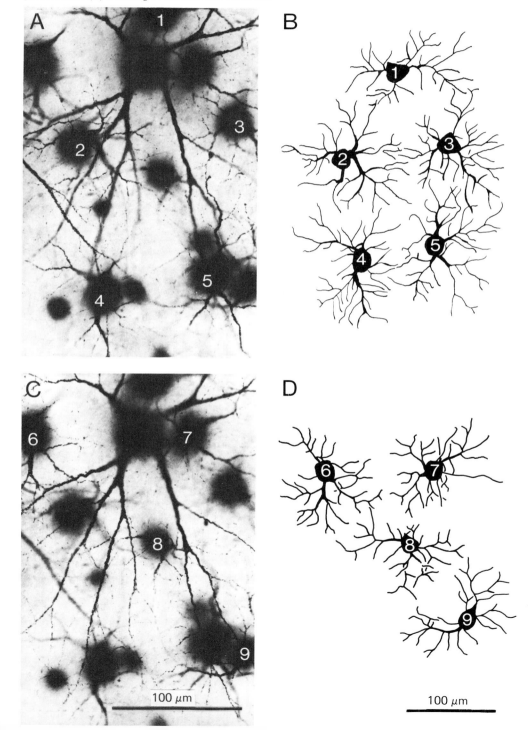

(Kolb, Nelson & Mariani, 1981). All authors agree that alpha and beta cells, which are the major topic of this chapter, are clearly distinctive and well defined.

There is overwhelming evidence that brisk-sustained (X) cells are the physiological correlate of beta-cells (Boycott & Wässle, 1974; Levick, 1975) and recently their identity has been corroborated from intracellular recordings in the optic tract (Sur, Humphrey & Sherman, 1982) and in the retina (Saito, 1983). In Golgi-stained whole-mount preparations, beta cells are characterized by a small bushy dendritic tree and have a medium-sized, round cell body (Boycott & Wässle, 1974). When several overlapping dendritic trees of beta cells are stained, there is a clear indication of differential stratification (Fig. 2.1C, D). Based on the results of Nelson *et al.* (1978) and Saito (1983), it will be assumed that beta cells with predominant dendritic branching in the *outer* part of the inner plexiform layer are OFF-centre, those with dendritic branching *closer* to the ganglion cell bodies are ON-centre.

Multiple injections of HRP solution into one lateral geniculate nucleus gave homogeneous staining of some 75% of all ganglion cells in the corresponding retinal hemifields. The filling of dendrites in HRP-labelled cells was good enough to distinguish beta cells from other ganglion cells with medium-sized somata (Fig. 2.4). All alpha, all beta and about half of the ganglion cells with small cell bodies were labelled. In this chapter, only the beta cell population is analysed in detail (Fig. 2.4B, D). Comparison of the dendritic morphology of HRP- (Fig. 2.4) and Golgi-stained (Fig. 2.1) beta cells shows that HRP stains only the central part of the dendritic tree of the cells. Fortunately this was sufficient to recognize beta cells and to reveal their differing branching levels, and therefore to separate ON- and OFF-beta cells (Fig. 2.5).

The mosaic formed by ON- and OFF-beta cells was analysed by measuring the distance to the nearest neighbour of each cell (Wässle & Riemann, 1978). As a first step, in Fig. 2.5A, the distance of every beta cell from the next beta cell was measured, regardless of physiological type. The histogram in Fig. 2.5D shows the obtained distances for nearest neighbours. For the great majority of cells (129 out of 136), the nearest neighbour is of the opposite functional type, which confirms the qualitative impression (Fig. 2.5A; Fig. 2.3A) that ON- and OFF-cells tend to be grouped into pairs. The overall histogram can be approximated by a Gaussian, although it is skewed to lower distances. There is considerable scatter of nearest-neighbour distances. Consequently, the mean/SD ratio of 2.7 is low, indicating a rather imprecise mosaic.

If the analysis is restricted to the ON- (Fig. 2.5B) or to the OFF-populations (C), then independent mosaics of distinctive regularity become apparent (E, F). The nearest-neighbour histograms are Gaussian and are closely similar for both ON- and OFF-beta-cells. For the ON-beta cells, the mean/SD ratio is 5.2 and that for the OFF-cells is 5.3. This shows that ON- and OFF-beta cells each form a regular mosaic and that these mosaics are independent, as previously shown for ON- and OFF-alpha cells (Wässle *et al.*, 1981*b*). A similar independence was shown for the mosaics of the two horizontal cell types (Wässle, Peichl & Boycott, 1978).

Superposition of two regularities creates not a random overall distribution, but a rather sloppy regularity. The idea that ON- and OFF-cells are independently arrayed was tested in the following way. The ON-mosaic and a mirror image of the OFF-mosaic were superimposed and the resulting nearest-neighbour histogram of this uncorrelated array was compared with the histogram in Fig. 2.5D: both the mean and the standard deviation closely agreed, indicating that the mosaics have a similar form. No statistically significant difference between the histograms was observed (sign reversal test, $P > 0.17$).

Spatial sampling and visual processing

Visual acuity

The cone mosaic has to be considered as the first limiting factor of spatial resolution. Unlike the case of many primates or birds, however, in the central area of the cat retina there are six to eight cones converging onto one beta-cell, the peak density of cones being 20.000–30.00 mm^2 (Steinberg *et al.*, 1973), and the peak density of beta cells being 4000–6000 cells mm^2. There are no midget bipolar cells to mediate the contact of a single cone with a single ganglion cell in the cat (Boycott & Kolb, 1973). The beta cells are the only ganglion cell candidates for a high-acuity system, because at each retinal location they have the highest density and the smallest dendritic fields. This assumption is further supported by the fact that all beta cells project to the thalamus and hence to the visual cortex (Illing & Wässle, 1981).

Fig. 2.5. Analysis of the beta–cell mosaic.
(A) Drawing of beta cells labelled in the upper temporal retina after HRP injection into the ipsilateral geniculate. (B) The mosaic formed by the ON-beta cells. (C) OFF-beta cells. The stippled circles in (B) and (C) indicate the dendritic field of a beta cell from Golgi-stained material at the same eccentricity. Thus neighbouring dendritic fields would overlap and the cells are understained by HRP. (D) Histogram showing the distribution of the distances from each cell to its nearest neighbour, taken from the heterogeneous ON–OFF mosaic in (A). The hatched bins indicate the few instances where the nearest neighbour was of identical type. (E) Nearest-neighbour histogram for the ON-beta cells from (B). (F) Nearest-neighbour distribution for the OFF-beta cells from (C). From Wässle *et al.*, 1981*a*.)

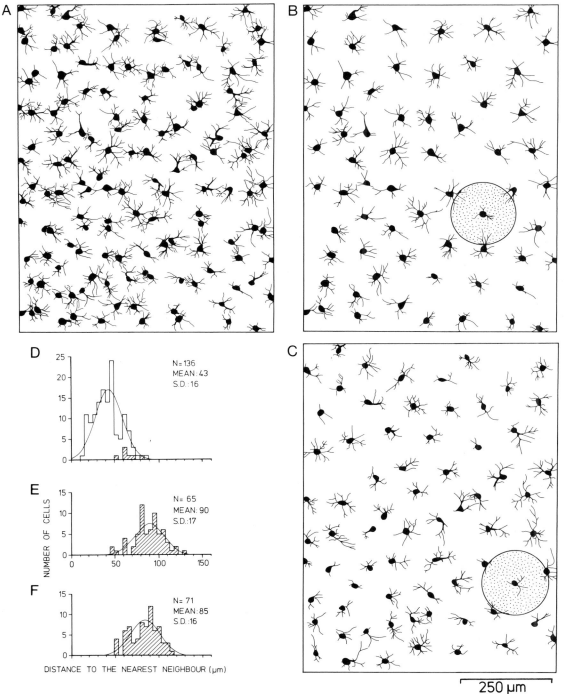

250 μm

One of the basic requirements for a system ensuring high spatial resolution is regular spacing of the sampling points. Irregularly spaced sampling points introduce positional noise which would reduce the visual acuity (French, Snyder & Stavenga, 1977). If the visual system could use the heterogeneous mosaic of ON- and OFF-beta cells in some sophisticated cooperation, it could increase its visual resolution by a factor of $\sqrt{2}$ in comparison to an independent operation of the ON- and OFF-channel, because the density of sampling points would be doubled (Hughes, 1980). We believe that this is not the case and assume that the independent mosaics of ON- and OFF-beta cells define the limits of visual discrimination. The heterogeneous mosaic of ON- and OFF-beta cells (Fig. 2.5A) does not show regular intercell spacing and the nearest-neighbour distances can vary between 10 and 90 μm (Fig. 2.5D). As mentioned above, such positional noise in a raster reduces visual resolution (French *et al.*, 1977) and quite likely abolishes the $\sqrt{2}$ advantage. The separate homogeneous ON- and OFF-mosaics fulfil the requirement of regular spacing.

In the following, therefore, the spatial resolution of the ON- and OFF-systems will be tested independently. About 55% of all ganglion cells in the cat retina, including the central area, are beta cells (Illing & Wässle, 1981). This gives a peak density of about 5500 cells mm^2 (Hughes, 1975). If one approximates the beta cell mosaic as a hexagonal lattice, the intercell spacing (a) can be calculated from the density (D) by the formula $a = (2/D\sqrt{3})^{\frac{1}{2}}$ and the resolution limit (λ) would be $\lambda = a\sqrt{3}$ (Snyder & Miller, 1977). For both ON- and OFF-beta cells, the intercell distance in the central area would be about 20 μm, which gives a grating resolving power of 0.16 degrees or 6–7 cycles/degree. The smallest RFC diameters of brisk-sustained (X) cells measured with physiological methods were 0.26 degrees (Cleland, Harding & Tulunay-Keesey, 1979) and 0.29 degrees (Peichl & Wässle, 1979). They could resolve, respectively, 9.5 and 6 cycles/degree. The grating resolution found behaviourly is 5–10 cycles/degree (Smith, 1936; Berkley & Watkins, 1973; Campbell, Maffei & Piccolino, 1973; Jacobson, Franklin & McDonald, 1976; Mitchell *et al.*, 1976; Timmey & Mitchell, 1979). This agreement between the calculated and measured visual acuity closely supports the assumption that beta cells are involved in the resolution of fine detail by the cat's visual system.

Orientation specificity of cortical neurons

The rather coarse arrangement of the retinal ganglion cells must be reflected in the spatial sensitivity of the visual cortex. In the following, the restraints placed by the mosaic of retinal ganglion cells will be considered in terms of the orientation specificity of simple cerebral cortical cells. Since the RFCs of other ganglion cells are too large to provide the limitation for optimal spatial sensitivity, only the beta cells are considered to be the raster from which simple cells select their orientation.

According to the model proposed by Hubel & Wiesel (1962), centres of retinal ganglion cells arrayed on a straight line in the retina relay into the lateral geniculate, those axons then converge onto 'simple cells'. Depending on the distance from the central area, the widths of simple cell discharge centres range between 0.3 and 1.5 degrees with a mean of 0.8 degrees (Kato, Bishop & Orban, 1978). These data fall within the range of the RFCs of brisk-sustained (X) cells. Length summation curves of simple cells, especially in layer IV of the visual cortex (Bullier, Mustari & Henry, 1982), indicate that only a few cells need to be aligned to give an optimal orientation. Summation data (Henry, Goodwin & Bishop, 1978; Mustari, Bullier & Henry, 1982) indicate that the RFCs, and therefore the dendritic fields, of the ganglion cells along the long axes of simple cells, must overlap. As shown in the upper left corner of Fig. 2.6A, two close but non-neighbouring beta cells will define an orientation. However, such an arrangement must be ruled out since it leaves a long gap between the receptive fields of these two retinal units which disagrees with the direct cortical measurements.

The upper right hand corner of Fig. 2.6A illustrates a second possibility in which beta cells, together with neighbours with overlapping dendritic fields, provide the input for a cortical cell. This possibility is shown diagrammatically in the lower right of Fig. 2.6A. A straight line connects the central cell with all its neighbours within a circle of radius 2r, where r is the radius of the dendritic field. On average, 12 different lines of orientation can be constructed around every beta cell. This would provide 30 degrees of resolution in the orientation domain. The sharpness of orientation-tuning curves of simple cells in the cat striate cortex (Henry, Dreher & Bishop, 1974), taken as width at half height, is a measure of the orientation specificity of the cells. For 40 simple cells, the range was from 16 to 62 degrees with a mean of 34 degrees, in good agreement with the present estimate of 30 degrees resolution.

It has to be emphasized that this construction of the cortical orientation specificity assumes convergence solely from geniculate afferents, and therefore might only hold for simple cells of lamina IV. The coarseness of the beta-cell mosaic could be overcome in the cortex by an interpolation procedure, where the position of every physical point on the retina could be calculated from the relative responses of surrounding beta cells (Barlow, 1979, 1981). Such a calculation would require a more complex wiring pattern, which would increase the receptive field sizes and one could imagine that such operations could take place outside layer IV.

There is recent evidence that the Hubel & Wiesel (1962) model might need qualification. Inhibitory sidebands are considered to be a major source of orientation specificity (Bishop, Coombs & Henry 1973; Creutzfeldt, Kuhnt & Benevento, 1974; Sillito *et al.*, 1980), but such models also have to use the retinal mosaic for constructing the different orientation of simple cell receptive fields. In Fig. 2.6B, dendritic fields of cells forming the inhibitory sidebands as proposed by Henry & Bishop (1972) are inserted. The excitatory input would be provided by one or several beta cells in between the two sidebands. It is obvious that this model builds up orientation specificity by *avoiding inhibition* along the optimal orientation. Thus, the cells along the optimal orientation (indicated by the two lines) are taken out of the inhibitory pool, and their mosaic defines the number of possible orientations. As

Fig. 2.6. Modelling of the simple cell orientation specificity from the beta-cell mosaic. Each dot represents the cell-body position of a single ON-beta cell labelled after HRP injection into the lateral geniculate (eccentricity 3.5 mm, temporal retina). The inserted dendritic fields are taken from a Golgi-stained beta cell of comparable eccentricity. (A) Modelling of simple cell receptive field according to the scheme proposed by Hubel & Wiesel (1962), where excitatory convergence of the cells define the orientation specificity. Details are given in the text. (B) Construction of orientation specificity by inhibitory convergence of the cells drawn onto a single cortical cell. The excitatory input of the cell would be provided by one or several ganglion cells from between the two lines.

A

B

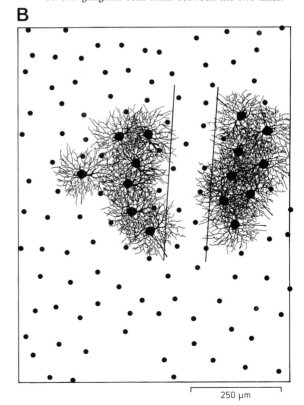

250 μm

in the Hubel & Wiesel (1962) model, the coarse arrangement of retinal ganglion cells sets the limits of the orientation tuning by inhibitory sidebands.

Beta-cell disorder, aliasing and noise

The mosaic of beta cells shows neither square nor hexagonal symmetry; only the distances between the cells and not the angles seem to be regular (Wässle & Riemann, 1978; Wässle *et al.*, 1981*a*). This apparent 'imperfection' has the advantage of preventing spatial interferences known as aliasing, but it has the disadvantage of introducing spatial noise. It was recently shown for extrafoveal monkey cones, which have the same general pattern of arrangement as beta cells (Wässle & Riemann, 1978), that their array is optimized: they introduce minimal noise for spatial frequencies below their resolution limit, while frequencies above this limit are not converted into moire patterns, but are scattered into broadband noise (Yellott, 1982, 1983). In the human fovea, the cones are spherically packed in a hexagonal array, with the result that aliasing might occur (Williams & Collier, 1983).

ON–OFF cooperation

So far, the mosaics of ON- and OFF-beta cells have been considered as independent sampling points. Marr & Hildreth (1980) have shown that ON- and OFF-centre X cells with spatially separated receptive fields might cooperate to detect zero crossings in the second derivative of the light intensity and thus signalling edges of the light distribution. It is very tempting to analyse the mosaic of ON- and OFF-beta cells of Fig. 2.5A along these lines. One might calculate the optimum distance of ON–OFF pairs for detecting zero crossings and thus predict spatial convergence of the ON- and OFF-channel in the cortex. Marr & Ullman (1981) have proposed an algorithm to calculate directionality of a moving edge by adding a Y cell input to the above-mentioned detector of zero crossings. Again, there must be an optimal spatial arrangement and receptive field size of the underlying X and Y cells to perform such an operation. This chapter contains the grain the brain can use for visual information processing.

Acknowledgement

The author would like to thank B. B. Boycott and L. Peichl for the enjoyable cooperation during the last 10 years in London, Konstanz, Tübingen and Frankfurt.

References

Barlow, H. B. (1964) The physical limits of visual discrimination. In *Photophysiology*, vol. 2, ed. A. C. Giese, pp. 163–202. New York: Academic Press

Barlow, H. B. (1979) Reconstructing the visual image in space and time. *Nature, Lond.*, **279**, 189–90

Barlow, H. B. (1981) Critical limiting factors in the design of the eye and visual cortex. *Proc. R. Soc. Lond., Biol. Sci.*, **212**, 1–34

Berkley, M. A. & Watkins, D. W. (1973) Grating resolution and refraction of the cat estimated from evoked cerebral potentials. *Vision Res.*, **13**, 403–15

Bishop, P. O., Coombs, J. S. & Henry, G. H. (1973) Receptive fields of simple cells in the cat striate cortex. *J. Physiol. (Lond.)*, **231**, 31–60

Bonds, A. B. (1974) Optical quality of the living cat eye. *J. Physiol. (Lond.)*, **243**, 777–95

Boycott, B. B. & Kolb, H. (1973) The connections between bipolar cells and photoreceptors in the retina of the domestic cat. *J. Comp. Neurol.* **148**, 91–114

Boycott, B. B. & Wässle, H. (1974) The morphological types of ganglion cells of the domestic cat's retina. *J. Physiol. (Lond.)*, **240**, 397–419

Brown, J. E. & Major, D. (1966) Cat retinal ganglion cell dendritic fields. *Exp. Neurol.* **15**, 70–8

Bullier, J., Mustari, M. J. & Henry, G. H. (1982) Receptive-field transformations between LGN neurons and S-cells of cat striate cortex. *Neurophysiol.*, **47**, 417–38

Campbell, F. W. & Green, D. G. (1965) Optical and retinal factors affecting visual resolution. *J. Physiol. (Lond.)*, **181**, 576–93

Campbell, F. W., Maffei, L. & Piccolino, M. (1973) The contrast sensitivity of the cat. *J. Physiol. (Lond.)*, **229**, 719–31

Cleland, B. G., Harding, T. H. & Tulunay-Keesey, U. (1979) Visual resolution and receptive field size: examination of two kinds of cat retinal ganglion cell. *Science*, **205**, 1015–17

Cleland, B. G., Levick, W. R. & Wässle, H. (1975) Physiological identification of a morphological class of cat retinal ganglion cells. *J. Physiol. (Lond.)*, **248**, 151–71

Creutzfeldt, O. D., Kuhnt, U. U. & Benevento, L. A. (1974) An intracellular analysis of visual cortical neurones to moving stimuli: responses in a co-operative neuronal network. *Exp. Brain Res.*, **21**, 251–74

Creutzfeldt, O. D., Sakmann, B., Scheich, H. & Korn, A. (1970) Sensitivity distribution and spatial summation within receptive field centre of retinal on-centre ganglion cells and transfer function of the retina. *J. Neurophysiol.*, **33**, 654–71

Dowling, J. E. & Boycott, B. B. (1966) Organization of the primate retina: electron microscopy. *Proc. R. Soc., Biol. Sci.*, **160**, 80–111

French, A. S., Snyder, A. W. & Stavenga, D. G. (1977) Image degradation by an irregular retinal mosaic. *Biol. Cybernetics,* 27, 229–33

Gallego, A. (1954) Conexiones transversales retinianas. *Ann. Inst. Farmacol. Esp.,* 3, 31–9

Henry, G. H. & Bishop, P. O. (1972) Striate neurons: receptive field organization. *Invest. Ophthalmol. Vis. Sci.,* 11, 357–68

Henry, G. H., Dreher, B. & Bishop, P. O. (1974) Orientation specificity of cells in cat striate cortex. *J. Neurophysiol.,* 8, 1394–409

Henry, G. H., Goodwin, A. W. & Bishop, P. O. (1978) Spatial summation of responses in receptive fields of single cells in cat striate cortex. *Exp. Brain Res.,* 32, 245–66

Hubel, D. H. & Wiesel, T. N. (1962) Receptive fields, binocular interaction and functional architecture in the cat's visual cortex. *J. Physiol. (Lond.),* 160, 106–54

Hughes, A. (1975) A quantitative analysis of the cat retinal ganglion cell topography. *J. Comp. Neurol.,* 163, 107–28

Hughes, A. (1980) Cat retina and the sampling theorem: the relation of transient and sustained brisk-unit cut-off frequency to alpha and beta-mode cell density. *Exp. Brain Res.,* 40, 250–7

Illing, R.-B. & Wässle, H. (1981) The retinal projection to the thalamus in the cat: a quantitative investigation and a comparison with the retinotectal pathway. *J. Comp. Neurol.,* 202, 265–85

Jacobson, S. G., Franklin, K. B. J. & McDonald, W. I. (1976) Visual acuity of the cat. *Vision Res.,* 16, 1141–3

Kato, H., Bishop, P. O. & Orban, G. O. (1978) Hypercomplex and simple/complex cell classification in cat striate cortex. *J. Neurophysiol.,* 41, 1071–95

Kawamura, S., Fukushima, N. & Hattori, S. (1979) Topographical origin and ganglion cell type of the retinopulvinar projection in the cat. *Brain Res.,* 173, 419–29

Kolb, H. (1970) Organization of the outer plexiform layer of the primate retina: electron microscopy of Golgi-impregnated cells. *Phil. Trans. R. Soc., Biol. Sci.,* 258, 261–83

Kolb, H., Nelson, R. & Mariani, A. (1981) Amacrine cells, bipolar cells and ganglion cells of the cat retina: a Golgi study. *Vision Res.,* 21, 1081–114

Kuffler, S. W. (1953) Discharge patterns and functional organization of mammalian retina. *J. Neurophysiol.,* 16, 37–68

Leventhal, A. G., Rodieck, R. W. & Dreher, B. (1981) Retinal ganglion cell classes in the old world monkey; morphology and central projections. *Science,* 213, 1139–42

Levick, W. R. (1975) Form and function of cat retinal ganglion cells. *Nature,* 254, 659–62

Marr, D. & Hildreth, E. (1980) Theory of edge detection. *Proc. R. Soc., Biol. Sci.,* 207, 187–217

Marr, D. & Ullman, S. (1981) Directional selectivity and its use in early visual processing. *Proc. R. Soc., Biol. Sci.,* 211, 151–80

Mitchell, D. E., Giffin, F., Wilkinson, F., Anderson, P. & Smith, M. L. (1976) Visual resolution in young kittens. *Vision Res.,* 16, 363–6

Mustari, M. J., Bullier, J. & Henry, G. H. (1982) Comparison of response properties of three types of monosynaptic S-cell in cat striate cortex. *J. Neurophysiol.,* 47, 439–54

Nelson, R., Famiglietti, E. V., Jr & Kolb, H. (1978) Intracellular staining reveals different levels of stratification for on- and off-center ganglion cells in cat retina. *J. Neurophysiol.,* 41, 472–83

Peichl, L. & Wässle, H. (1979) Size, scatter and coverage of ganglion cell receptive field centres in the cat retina. *J. Physiol. (Lond.),* 291, 117–41

Peichl, L. & Wässle, H. (1981) Morphological identification of on- and off-centre brisk transient (Y) cells in the cat retina. [With an appendix: Neurofibrillar staining of cat retinae by B. B. Boycott and L. Peichl.] *Proc. R. Soc. Lond., Biol. Sci.,* 212, 139–56

Peichl, L. & Wässle, H. (1983) The structural correlate of the receptive field centre of α ganglion cells in the cat retina. *J. Physiol. (Lond.),* 341, 309–24

Robson, J. G. & Enroth-Cugell, C. (1978) Light distribution in the cat's retinal image. *Vision Res.,* 18, 159–73

Saito, H.-A. (1983) Morphology of physiologically identified X-, Y- and W-type retinal ganglion cells of the cat. *J. Comp. Neurol.,* 221, 279–88

Shannon, C. E. (1949) Communication in the presence of noise. *Proc. Inst. Radio Engineers, New York,* 37, 10–21

Sillito, A. M., Kemp, J. A., Milson, J. A. & Berardi, N. (1980) A re-evaluation of the mechanism underlying simple cell orientation selectivity. *Brain Res.,* 194, 517–20

Smith, K. U. (1936) Visual discrimination in the cat. IV. The visual acuity of the cat in relation to stimulus distance. *J. Gen. Physiol.,* 49, 297–313

Snyder, A. W. & Miller, W. H. (1977) Photoreceptor diameter and spacing for highest resolving power. *J. Optical Soc. America,* 67, 696–8

Steinberg, R. H., Reid, M. & Lacy, P. L. (1973) The distribution of rods and cones in the retina of the cat (*Felis domesticus*). *J. Comp. Neurol.,* 148, 229–48

Stone, J. & Clarke, R. (1980) Correlation between soma size and dendritic morphology in cat retinal ganglion cells. Evidence for further variation in the γ-cell class. *J. Comp. Neurol.,* 192, 211–18

Sur, M., Humphrey, A. L. & Sherman, S. M. (1982) Monocular deprivation affects X- and Y-cell retinogeniculate terminations in cats. *Nature,* 300, 183–5

Timmey, B. & Mitchell, D. E. (1979) Behavioural recovery from visual deprivation: comments on the critical period. In *Developmental neurobiology of vision,* ed. R. D. Freeman, pp. 149–60. New York & London: Plenum Press

Wässle, H. (1971) Optical quality of the cat eye. *Vision Res.,* 11, 995–1006

Wässle, H., Boycott, B. B. & Illing, R.-B. (1981a) Morphology and mosaic of on- and off-beta cells in the cat retina and some functional considerations. *Proc. R. Soc. Lond., Biol. Sci.,* 212, 177–95

Wässle, H., Levick, W. R. & Cleland, B. G. (1975) The distribution of the alpha type of ganglion cells in the cat's retina. *J. Comp. Neurol.,* 159, 419–38

Wässle, H., Peichl, L. & Boycott, B. B. (1978) Topography of horizontal cells in the retina of the domestic cat. *Proc. R. Soc. Lond., Biol. Sci.,* 203, 269–91

Wässle, H., Peichl, L. & Boycott, B. B. (1981*b*) Morphology and topography of on- and off-alpha cells in the cat retina. *Proc. R. Soc. Lond., Biol. Sci.*, **212**, 157–75

Wässle, H., Peichl, L. & Boycott, B. B. (1983) A spatial analysis of ON- and OFF-ganglion cells in the cat retina. *Vision Res.*, **23**, 1151–60

Wässle, H. & Riemann, H. J. (1978) The mosaic of nerve cells in the mammalian retina. *Proc. R. Soc. Lond., Biol. Sci.*, **200**, 441–61

Williams, D. R. & Collier, R. (1983) Consequences of spatial sampling by a human photoreceptor mosaic. *Science*, **221**, 385–7

Yellott, J. I., Jr (1982) Spectral analysis of spatial sampling by photoreceptors: topological disorder prevents aliasing. *Vision Res.*, **22**, 1205–10

Yellott, J. I., Jr (1983) Spectral consequences of photoreceptor sampling in the rhesus retina. *Science*, **221**, 382–5

3

Sampling of information space by retinal ganglion cells

W. R. LEVICK

Bottleneck at the optic nerve

The optic nerve constitutes a major information bottleneck in the visual system. There are about 900 times as many photoreceptors in the cat (150 million, based on the densities of Steinberg, Reid & Lacy (1973) and the average retinal area of Hughes (1975)) as there are ganglion cells and thus optic nerve fibres. Neglecting the minor complication of functional interphotoreceptor connections, the photoreceptor array has the potential for representing the light distribution of the retinal image in far greater detail than the ganglion cell array. The number of cells in the intermediate neural arrays is less than that of photoreceptors but still considerably greater than the number of ganglion cells (170 000 (Hughes, 1981)).

The evidence of diminished numbers from photoreceptors to ganglion cells is not in itself sufficient to sustain the argument for the existence of an *information* bottleneck. It might be the case, for example, that the ocular optics were not of sufficient quality to produce better than a seriously degraded retinal image. If this were the case, the image would be greatly oversampled by the receptor matrix and it would make good economic sense to reduce the number of elements representing the image in the ganglion cell array much closer to that required by Shannon's sampling theorem. This would be sufficient to preserve all of the image information available to the nervous system and allow full reconstruction of the degraded image. Under these circumstances, the 'bottleneck' would be assigned to poor-quality optics rather than to diminished numbers of ganglion cells.

The evidence on optical quality of the cat's eye points in the other direction, to the effect that the retinal image is surprisingly good. Robson & Enroth-Cugell (1978) illustrate a line-spread function for a 2 mm pupil which corresponded to a spatial frequency cut-off of around 20 cycles/degree. Taking half this period as the spacing of a hexagonal matched sampling array and using 143° as the angular extent of the retina (Hughes, 1976), assumed to be a spherical segment and discounting peripheral aberrations, the retina would hold about 13 million such sampling points. Thus there are many more spatial degrees of freedom in the retinal image than can be represented by the ganglion cell array. So it is appropriate to speak of an information bottleneck at the optic nerve.

There may also be a bottleneck in terms of temporal factors. It is to do with the fact that the nature of signalling in the optic nerve is by all-or-none impulses. In the preganglionic arrays the representation of visual signals is by continuous variation of membrane potential (Werblin & Dowling, 1969). The impulse-transmission type of system conveys no additional information beyond the timing of the impulse and is effectively blanked out for the fraction of time occupied by each impulse and its refractory period. The non-impulsive system is not limited in this way but it is difficult to state how much better it is. The answer would depend upon the temporal bandwidth and noise level of membranes embedded in cells of complex topology. Furthermore, the non-impulsive mode, coupled with the distributed mingling of input and output especially on the dendritic trees of amacrine cells, raises the possibility of a degree of independence of signal transmission at different positions within an individual neuron. Thus, single neurons operating in this fashion may represent multiple information channels in the preganglionic network.

Why is there a bottleneck? An obvious possibility is that it is a necessary consequence and an undesired side effect of the design of a mobile eye. If other factors are unchanged, a 100-fold increase in the number of optic nerve fibres would require an expansion of the optic nerve to a great cord about 10 mm in diameter, which would be hardly suitable for wide-ranging, speedy and accurate ocular positioning. Of course, this situation could be improved by trade-off against other factors. For example, more fibres could be squeezed in without increasing the overall size of the optic nerve by scaling the fibre calibre spectrum towards smaller diameters, but then other problems (larger absolute conduction-time differences; greater conduction-time variabilities), possibly less tractable, would emerge.

For the purposes of the discussion that follows, the optic nerve bottleneck will be regarded as an unavoidable constraint placed on the organization of the visual system. From this viewpoint, the spatial organization of ganglion cells can be considered as a particular solution to the problem of compressing a large number of degrees of freedom in the retinal image to a much smaller number of degrees of freedom in the optic nerve.

Representation of the visual world

The myriads of successive spatial intensity distributions which constitute the natural visual world possess regularities which, although difficult to express in full generality, nevertheless provide the opportunities for developing strategies to compress the messages of the visual system. The scene is populated by features characterized by groupings of regions having similar or related attributes. These groupings have a degree of both durability and coherent development as the relative positions of visual system and feature change in the real world. In terms of representation, having a neural signal correspond with a feature would amount to a compression if the intensity distribution implied by the feature occupied a significant spatial extent. Thus, one signal may serve where otherwise several would have been required.

What constitutes a 'feature' may be a very subtle matter. The term is used in the more general sense given by its dictionary definition. According to the *Oxford English Dictionary*, such a meaning is: 'distinctive or characteristic part of a thing, part that arrests attention'. So 'feature' is made to depend upon the existence of some 'thing' which has a special significance. Its meaning in the present context depends both on the constitution of the visual world and on the use made of it by the possessor of the visual system under consideration.

The occupation of different ecological niches by different animals implies important differences in the visual world they must interpret in the struggle for survival (Hughes, 1977). The size of the eyes, the positions in which they are carried in the head, and the way in which they are transported and manipulated during exploration are key factors determining how the visual scene is presented to the visual system. All of these factors are relevant to the consideration of what might constitute a 'feature'.

In the context of the ganglion cell layer, representing the visual scene in terms of features implies at least three things. First, there should be a specific pattern of connectivity through the dendritic tree and preganglionic networks which in some sense matches the structure of the feature in the outside intensity array. Secondly, the sensitivity of the cell representing the feature should be spread out in visual space. The recognition that a neural element has a spatial distribution of sensitivity underlies the concept of the receptive field which has played an important role in the description and interpretation of sensory organization. It is with this concept that much of this essay is concerned. The third implication comes from an appreciation of the possibility that more than one type of feature may refer to the same place in the visual scene. To give an example it is necessary to anticipate the later discussion a little. A small detail lighter than the background may be moving in a particular direction in the field. A cell representing the feature: 'local patch lighter than surroundings' will provide its report, as well as other cells representing: 'contrast edge moving within a certain range of directions'. The same place in the visual scene may be indicated by all of the cells reporting. Thus it comes about that receptive fields overlap and different representations are carried in parallel. These notions are further developed in the next section.

Receptive field overlap and parallel pathways

When cat ganglion cells were regarded as falling into just two complementary classes (ON-centre and OFF-centre (Kuffler, 1952, 1953)), which mapped the visual field in fine detail centrally and progressively more coarsely peripherally (Wiesel, 1960), the demonstration of an extraordinary amount of overlap of receptive field centres (RFCs) (60- to 1000-fold, Rodieck & Stone, 1965; 35-fold, Fischer, 1973) was something of a puzzle. A system of large, homogeneous, partially shifted overlapping receptive fields would seem an odd way to compress visual information into the optic nerve bottleneck.

The key advance which greatly altered attitudes to the puzzle of partially shifted overlap was the recognition of qualitative diversity of receptive fields. Depending on how one counts them, there are at least 13 different classes (Cleland & Levick, 1974a,b; Stone & Fukuda, 1974a) and the diversity is a local property throughout the retina, as far as can be determined on the limited data from the classes encountered uncommonly. Although there is no unanimity about the relative proportions in different regions (area centralis, visual streak, retina generally (Rowe & Stone, 1976; Hughes, 1981)), there is agreement that all classes are nevertheless represented in all localities.

If different classes of ganglion cells are making qualitatively different kinds of assessments of the retinal image, it obviously makes sense of overlap since a report is required from each class for each part of the retinal image. Furthermore, there is no necessity for the receptive fields of all cells serving a locality to have the same size: it would depend on the type of assessment being made. Cells responsive to the low end of the spatial frequency spectrum would have large fields whereas those reporting on the high end would have small fields (Enroth-Cugell & Robson, 1966; Thibos & Levick, 1983). So a cell with a large field could potentially overlap a large number of cells with small fields. The extent to which this happens depends upon the coverage of the retina by the different classes.

The notion of coverage expresses the idea that the receptive fields of a particular class of cell taken together cover the retina with some particular multiplicity (Cleland, Levick & Wässle, 1975). Coverage is the product of the local density of a particular class of ganglion cell by the local average area of the RFCs of that class. A simple way of appreciating the concept is to suppose a pin pushed through the array of receptive field centres at a point. Coverage at that point is the average number of RFCs transfixed by the pin.

The same concept is equally applicable to the spread of dendritic trees of ganglion cells, although some unstated assumptions are usually implied in the application. The point is that the dendrites are very thin structures and the area literally covered by their profiles is only a small fraction of the area of the region over which they ramify. The implied assumption is that the elements making synapses on the dendrites already have some spatial spread over which they themselves have picked up their visual excitation. This has the effect of extending the effective coverage of dendrites to include the empty spaces between them. Something like this is almost certainly occurring, because careful search for a receptive field microstructure which might correlate with the detailed arrangement of dendrites has not been successful (Peichl & Wässle, 1983). These considerations form the basis for representing dendritic trees by the area enclosed by some profile matching the furthest extents of the dendrites. Such profiles may then be treated like receptive field centres in calculating the corresponding coverage.

The intriguing result that has emerged is that coverage within particular classes of cells is rather constant over the retina despite substantial changes in cell density (Peichl & Wässle, 1979). It suggests that RFC area is regulated in accordance with local cell density and a simple rule for achieving the regulation has recently been proposed (Wässle, Peichl & Boycott, 1981, 1983): during development, dendrites grow out from their cell bodies at the same time and rate, and stop growing when they reach dendrites of neighbouring cells of the same class. This works out nicely for ON-centre and OFF-centre brisk-transient (Y) cells, operating as independent classes. The rule generates a mosaic of domains, the shapes of which agree rather well with the dendritic fields of the corresponding cells. The coverage is about 1.4 for each of these two classes. For the other commonly encountered classes, the ON- and OFF-centre brisk-sustained (X) cells, the coverage is about 3, so some modification of the simple rule would be needed. What this all amounts to is that the puzzle of a high degree of overlap largely disappears when ganglion cells are first subdivided into their basic functional classes and these are then regarded as independent organizational sets. Each set forms a space-filling mosaic with only minor overlap. If the design goals of the system were completeness and economy, a process of this kind would certainly achieve it.

The foregoing considerations lead naturally to the notion of 'parallel pathways'. In some ways, the choice of the expression has been unfortunate. It was introduced at a time when comparisons of brain and computer organizations were used to emphasize the serial nature of computer processing whereas the brain conducted its operations in parallel. Nowadays it is more fashionable, as well as more accurate, to speak of the multiplicity of 'instruction paths' and 'data paths' and to describe the classical computer as a 'single-instruction single-data-path machine' whereas the brain would be likened to a 'multiple-instruction multiple-data-path' system. The parallelism of 'parallel pathways' is not really the antithesis of serialism

as applied to the processing.

The term 'parallel pathways' also figured prominently in discussions of the nature of visual processing in the cortex. On the one hand was a view of specific types of operations being performed sequentially by hierarchical sets of neurons to generate progressively more elaborate receptive fields (simple, complex, hypercomplex (Hubel & Wiesel, 1962, 1965)). An alternative view was that visual information reached the different sets of cortical neurons in parallel from functionally different classes of subcortical neurons (Stone, Dreher & Leventhal, 1979). Again, the parallelism is not really in contraposition to the sequentialism of hierarchical organization.

Nor does the parallelism merely refer to the patently obvious fact that the visual system's elements are arranged in parallel: for example, the receptors are arrayed in parallel side by side as are also the ganglion cells and their axons carrying the visual signals up the optic nerve. More is implied than mere topographic juxtaposition. Instead, the expression 'parallel pathways' encompasses the notion that for every small patch of visual field there exists a number of spatially distinct neural channels conveying functionally different information to higher levels. The idea is thus the conjunction of multiplicity of analysis of visual events centred on a single visual field location.

Classes of receptive fields

There are at least 13 distinct classes of cat retinal ganglion cells. Yet the view still abounds that there are just three. The popularity of a three-way classification may well be related to the existence of three conduction groups in the cat optic pathway (Bishop, Clare & Landau, 1969) and the relation of those groups to particular functional classes (Cleland, Dubin & Levick, 1971; Stone & Hoffmann, 1972; Cleland & Levick, 1974a,b; Fukuda & Stone, 1974; Stone & Fukuda, 1974a) and to particular morphological classes (Boycott & Wässle, 1974). The reduction of functional classes to three certainly eases the burden on the mind but it seriously handicaps a proper understanding of functional organization. For example, the three-way classification forces together such functionally disparate classes as blue–yellow colour-coded cells with direction-selective cells. The relation of classifications to conduction velocity groups was recently reviewed (Levick & Thibos, 1983).

Eight of the classes have the concentric centre-surround organization described by Kuffler (1952, 1953). At first glance, it might seem unnecessary and deleterious from the point of view of the optic nerve bottleneck to provide so much apparent duplication. There are reasons to believe otherwise.

First, half of the concentric units are of the ON-centre type and half are OFF-centre. The information carried by their respective signals is not really duplicated because it would not be possible to reconstruct centrally the signals of, say, the OFF-centre cells serving the same region of the visual field. Although the ON-centre classes signal local darkening by a reduction in the ongoing discharge, in all cases the discriminable gradations of reduced discharge are much fewer and coarser than the gradations of increased discharge generated by the corresponding OFF-centre cells. Furthermore, in the case of the sustained classes, the speed with which a signal of sudden local darkening can be extracted from the decrement in discharge of ON-centre cells is not as great as when the increment in discharge of the corresponding OFF-centre cells is used. This is a limitation inherent in the pulse-frequency-modulation coding used by these classes of ganglion cells. One must wait longer to be sure that a significantly long gap is present in a decremental response. The extra impulses of an incremental response necessarily create multiple short intervals in the discharge which surpass levels of significance all the sooner.

Of course, the problem of reconstructing the signals of OFF-centre cells from the discharges of ON-centre cells could be largely avoided by boosting the unstimulated maintained discharge up to say 400/s, about half the maximum possible. Then, the asymmetry in the capacities of incremental and decremental responses would be largely removed. Presumably the metabolic and cellular cost of defending the local environment within the optic nerve against such an avalanche of ionic fluxes would be prohibitive.

The concentric classes also differ in the scale of their spatial organization. In each local region of retina, the brisk-sustained units have the smallest RFCs and the brisk-transient units the largest. Although the responses of a set of small-centred fields having at least unity coverage could not generally be reconstructed unambiguously from a set of large-centred fields having only unity coverage, the converse possibility certainly needs to be considered. If it proved to be a simple matter in the higher visual centres to reproduce the

signals of large-centred units by combination of signals from a sufficient number of small-centred units, there would be no need to make available precious space in the optic nerve for the large-centred units. This viewpoint has hardly been considered in the past, so most of the experimental results on the different classes have not been focussed on the specific issues that the question raises. The discussion that follows is therefore necessarily rather speculative.

A useful starting point is to consider only the component of response at the same temporal frequency as the stimulus and use it heuristically as the basis for comparing spatial frequency tuning curves. The spatial modulation transfer functions of all the concentric classes have a band-pass characteristic with roughly the same bandwidth regardless of where the high-frequency cut-off is located (Thibos & Levick, 1983). This means that the small-centred units have substantially reduced responses for low spatial frequencies where the large-centred units respond strongly. The question is: would it be a simple matter centrally to reconstruct responses of high signal-to-noise ratio in large-centred units by suitably combining responses of low signal-to-noise ratio from the available number of small-centred units? Without actually answering the question it may be useful to mention some of the factors that need to be resolved experimentally.

In principle, improvement by a factor k would require the combination of in-phase signals from at least k^2 independent contributors. The available number of such contributors is deducible from the size of the dendritic tree of alpha cells (Boycott & Wässle, 1974), the relation between brisk-transient RFC size and alpha dendritic tree size (Peichl & Wässle, 1983) and the number of beta cells encompassed by the area of the receptive field centre at any particular retinal locus (Hughes, 1981). Since about 60 beta cells are encompassed by an alpha dendritic tree at 4 mm from the centre of the area centralis, the opportunity for achievement of the required signal-to-noise ratio must be regarded as substantial. The spatial weighting function for combining the activities of a set of small-centred units would be given, under linear conditions, by the inverse Fourier transform of the ratio of the spatial frequency tuning curves of the large-centred and small-centred units. The square of the noise after combination would depend on the sum of the squares of the weighting coefficients of this function. A complication is that the impulse trains of ganglion cells with overlapping receptive field centres are not independent (Mastronarde, 1983). The signal-to-noise ratio of the reconstruction would also be impaired by any source of noise originating in the mechanism of the small-centred units, for example, by representing continuous signals in the inner plexiform neuropil with a train of discrete impulses in the axon of a small-centred unit.

An interesting challenge to reconstructibility is posed by a characteristic response of brisk-transient (Y) cells (Hochstein & Shapley, 1976): the second harmonic response to phase reversal of a stationary fine grating pattern. In the typical situation, there are two transitions in each temporal cycle of the stimulus; in the first, all the light bars become dark and all the dark bars become light; in the second, the original state is restored. Brisk-transient cells (and other types of Y-cells) yield an excitatory response at both transitions, but brisk-sustained cells (and other types of X-cells) are excited at one or other transition but not at both. One problem confronting the central reconstruction of the brisk-transient responses from the responses of a set of brisk-sustained cells is that the latter show equivalent decrements in response at transitions opposite to those where they show excitation. Hence, forming a simple superposition of the responses of a spatially distributed set fails to generate the second harmonic response, since equivalent decrements balance increments at both the transitions. Nevertheless, the reconstruction can be rescued by first passing the signals to be combined through a threshold non-linearity to remove enough of the average maintained discharge to nullify the decremental phase of responses.

That is not the end of the problem. The characteristic second harmonic response, previously considered to be an attribute confined to the locality of the classical centre-surround receptive field, has recently (Derrington, Lennie & Wright, 1979; Krüger 1980) been brought into coincidence with the periphery effect (McIlwain, 1964), now usually referred to as the shift-effect (Krüger & Fischer, 1973; Fischer, Krüger & Droll, 1975). The shift-effect corresponds to a second-harmonic response to an alternating grating at distances beyond the borders of the classical receptive field of 50° or more on all sides. This implies an extension of the reconstruction considered above to cover a large fraction of the whole retina. When it is recalled that the nasal and temporal contingents of brisk-sustained units on which the reconstruction is to be based actually project to opposite sides of the brain, the impossibility becomes obvious. On a less-grand scale, the different decussation patterns of brisk-sustained and brisk-transient cells (Stone & Fukuda, 1974*b*; Kirk *et al.*, 1976) similarly negate the reconstruction near the zero vertical meridian where brisk-transient and the subjacent brisk-sustained units project to opposite sides of the brain.

It is instructive to consider the penalty implicit in supplying the brisk-transient array at the retinal rather than the central level. The RFCs are large and the coverage (4–7, Cleland *et al.*, 1975; 3–6, Peichl & Wässle, 1979) is less than that of the brisk-sustained cells (7–20, Peichl & Wässle, 1979). The net effect is that only some 6000–7000 ganglion cells are required. How many preganglionic cells must be dedicated to the function and how much thicker the inner plexiform layer would then be are still quite unknown, but the guess is that this part of the cost would be minimal. The surprise is the cost in terms of optic nerve cross-section. The axons of brisk-transient cells are the largest (mean diameter about 8 μm (Hughes & Wässle, 1976; Hughes, 1981)) whereas those of brisk-sustained cells are intermediate (modal diameter about 3.5 μm). So 6500 brisk-transient axons occupy as much space as about 34000 brisk-sustained axons. This a considerable cost since there are only about 80000 of the latter in total (Hughes, 1981). The only obvious advantage of the large diameter is the two- to three-fold increase in conduction velocity. It seems a high price to gain, say, 2–3 ms of conduction speed-up. A more subtle and possibly much more significant advantage is the precision in signalling temporal events. Unpublished observations indicate that the trial-to-trial variation of antidromic conduction latency between a site of suprathreshold electrical stimulation in the optic tract and the ganglion cell is decidedly smaller in the brisk-transient classes than in any of the others. In the past, not much significance has been attached to the accuracy of conduction timing because the temporal resolution of the visual system assessed by reference to Bloch's law and the flicker fusion frequency (see Brindley, 1970) was considered much poorer. Nevertheless, more recent psychophysical observations (Burr & Ross, 1979) seem to require a timing accuracy some 30-fold better ($\sim$ 200 μsec). It would be interesting if the brisk-transient system provided the neural substrate; however, dispersion of conduction time is only one component of an as yet incomplete picture.

A discussion of the central reconstructability of the responses of the sluggish-sustained and sluggish-transient classes of cells would probably follow similar lines to the foregoing, perhaps with special emphasis on the temporal modulation transfer functions. However, there is still insufficient parametric information on these classes to make the attempt informative.

The non-concentric classes

Five of the 13 classes in the cat fall into a grouping conveniently referred to as 'non-concentric'. The meaning of this designation is that their receptive field structures differ radically from the antagonistic centre-surround arrangement first described in the cat by Kuffler (1952, 1953) and later called concentric by others (Hubel & Wiesel, 1959; Barlow, Hill & Levick, 1964; Kozak, Rodieck & Bishop, 1965). 'Non-concentric' is thus synonymous with 'non-Kufflerian', and the term should not be taken literally to imply the absence of concentricity in the arrangement of receptive field components. In fact, the local-edge detectors and direction-selective units have what is called a silent inhibitory surround (Levick, 1972), arranged as an annulus concentric with the part of the receptive field yielding excitatory responses. The interactions, however, are not of the mutually antagonistic kind and the fields cannot be described as ON-centre or OFF-centre.

Collectively, the non-concentric units form a subset of W-cells. In the original description they were the almost exclusive occupants of the W-cell class (Stone & Hoffmann, 1972), but later (Stone & Fukuda, 1974a) the slowly conducting concentric classes (sluggish-sustained and sluggish-transient cells) were bundled together with the non-concentric classes to constitute the full set of W-cells. The merits of these groupings are considered elsewhere (Rowe & Stone, 1977; Levick & Thibos, 1983).

Far less is known about the non-concentric classes because they are relatively rarely encountered in microelectrode explorations of the retina. At least two factors determine rarity of encounter: (1) the small size of the cell bodies, certainly in the case of the small-centred local-edge detectors and direction-selective units; and (2) the genuinely low proportion in the population, probably in the case of the blue–yellow colour units, uniformity detectors and edge-inhibitory OFF-centre units. Since these have larger receptive fields, fewer would be required for unity coverage.

Another factor handicapping an exposition of the non-concentric classes is the rather undifferentiated linkage between physiological classes and morphological classes. The current view is that W-cells (i.e. the sluggish-sustained and sluggish-transient members of the concentric classes plus all of the non-concentric classes) correspond to the gamma cells of Boycott & Wässle (1974). But what is sorely needed is a structural marker for each individual functional class. The differentiation into morphological subclasses (Kolb, Nelson & Mariani, 1981) and the pursuit of functional-structural identification by intracellular marking (Saito, 1983) are important steps on the path.

In what follows, it is taken for granted that the rarely encountered classes are not merely artefacts generated by the experimental procedures, nor are they dismissed as vestigial remnants of development or isolated atrophied examples of exaggerated variation. Instead it is supposed that each identified class represents a system which, independently of the others, maps the retina with at least unity coverage.

The question immediately arises: are there enough ganglion cells for such an arrangement? A direct answer cannot be given because it depends on the density of cells, in their separate classes, as a function of retinal location, and the information is not available. Nevertheless, some idea may be gained by making certain assumptions. Let the estimates be based on (1) such measures of RFC size as are available; (2) a similarly scaled form of the function of centre size with eccentricity as holds for the well-studied brisk-transient and brisk-sustained classes (Cleland, Harding and Tulunay-Keesey, 1979); and (3) a coverage similar to that of the brisk-transient classes. The approach is to express the measures relative to the parameters for the ON-centre or OFF-centre brisk-transient class and make use of the facts that the coverage for each of these is about 3 (Cleland *et al.*, 1975) and their total number is about 6000 (Wässle, Levick & Cleland, 1975).

For the four sluggish concentric classes the scatter of measured RFC diameters is rather large at each eccentricity (Cleland & Levick, 1974a; Stone & Fukuda, 1974a), but it seems reasonable to take an average of about 0.77 of the brisk-transient RFC diameter. The local density of one of the classes thus amounts to about 1.67 times the density of the ON-centre or OFF-centre brisk-transient density and the total number to about 5000. Four classes therefore account for 20 000 ganglion cells. Similar calculations for the non-concentric classes are summarized, along with results for the others, in Table 3.1. Special considerations arise with three of the classes and these merit specific discussion.

With the direction-selective cells, the preferred directions differ from one unit to another. Such variation is expected because unambiguous signalling of the direction of movement requires it. To provide unambiguous signalling locally requires local variation of preferred directions, and to be practical and efficient in relation to higher-level requirements to combine directional signals globally, some ordering of directional preferences would be necessary. So systematic local grouping of preferred directions is also expected. But existing data do not yet enable the number of local groups to be recognized. There may be three, as in the case of the ON-direction-selective ganglion cells in the rabbit (Oyster & Barlow, 1967) for which Simpson, Soodak & Hess (1979) have revealed a beautiful relation to the vestibulo-ocular control system: the directions preferred correspond with the motions of the retinal image associated with head rotations about the axes of the three pairs of semicircular canals (Simpson, 1984). Alternatively, the number of groups might be four, as

with the rabbit ON–OFF direction-selective cells (Barlow *et al.*, 1964), where the preferred directions appear to align with the pulls of the four rectus muscles of the eye (Oyster, 1968). In Table 3.1, four subgroups of direction-selective ganglion cells are assumed.

Similarly, there appear to be at least two subgroups of colour-coded ganglion cells. Measurement of area-threshold curves (Cleland & Levick, 1974b) with differently coloured light revealed that the receptive field was composite, with sometimes the blue-sensitive process having the smaller area and sometimes the antagonistic minus-blue-sensitive process having the smaller area. Two subgroups are assumed in Table 3.1.

The complication presented by the edge-inhibitory OFF-centre units is associated with the details of the receptive field structure. The field is a composition of three distinct processes. There is an innermost patch within which the appearance of a contrasting border leads to potent inhibitory effects. Concentric with this and occupying a larger area is a region for which the effective stimulus for excitation is local dimming. The third component, again concentric with the other two, is an outermost annular zone which behaves like an antagonistic surround to the second

Table 3.1. *Provisional assignments to ganglion cell classes*

Receptive field class	Relative RFC diameter	Coverage	Relative density	[a]Total number
ON-brisk-sustained[b]	0.39[c]	6[d]	13.33	40 000[e]
OFF-brisk-sustained[b]	0.39[c]	6[d]	13.33	40 000[e]
ON-brisk-transient	1.00	3[d]	1.00	3000[f]
OFF-brisk-transient	1.00	3[d]	1.00	3000[f]
ON-sluggish-sustained	0.77	3	1.67	5000
OFF-sluggish-sustained	0.77	3	1.67	5000
ON-sluggish-transient	0.77	3	1.67	5000
OFF-sluggish-transient	0.77	3	1.67	5000
Local-edge detector	0.61	3	2.67	8000
Direction-selective	1.00	12	4.00	12 000
Colour-coded	1.55	6	0.83	2500
Edge-inhibitory OFF-centre	1.41	3	0.50	1500
Uniformity detector	1.73	3	0.33	1000
			Subtotal	131 000
			Total count, histological	170 000[e]
			Balance for assignment	39 000
			(increased coverage; more classes)	

[a] Rounded arbitrarily   [b] At about 2 mm from the centre of the area centralis   [c] Cleland *et al.* (1979)
[d] Peichl & Wässle (1979)   [e] Hughes (1981)   [f] Wässle *et al.* (1975)

OFF-centre region. The functional meaning of this arrangement is unclear and this leads to uncertainty as to which component, the edge-inhibitory part or the OFF-centre part, is to be considered as the spatial element of the corresponding mosaic of such cells. In Table 3.1, the size of the OFF-centre region is used.

The outcome of the accounting in Table 3.1 is that all of the currently known classes of ganglion cells could be fitted together in a common retinal plan based on similar cell density functions with respect to eccentricity and a coverage factor not less than 3, and yet the arrangement would not exceed the total number of ganglion cells available. Indeed it would not strain credibility to suppose the existence of more classes than have been currently accepted. There are in fact some candidates.

Stone & Fukuda (1974a) described ON-centre direction-selective cells, which differed from the ON–OFF type in yielding a receptive field map with just ON-responses. Whether these and other differences are sufficient to justify segregating them is open to question. Certainly the differences are not as prominent as between the ON–OFF and the ON-type of direction-selective cells in the rabbit retina, where differences in velocity tuning are prominent (Oyster, 1968; Oyster, Takahashi & Collewijn, 1972). In the rabbit, there is good evidence that the ON-direction-selective ganglion cells send their axons to the terminal nuclei of the accessory optic system (Simpson *et al.*, 1979; Oyster *et al.*, 1980). It is therefore significant that neurons of the medial terminal nucleus of the accessory optic tract in the cat are direction-selective and have large fields (Grasse & Cynader, 1982) just as in the rabbit (Simpson *et al.*, 1979). They receive direct input from retinal ganglion cells as shown by retrograde labelling with horseradish peroxidase (Farmer & Rodieck, 1982) and anterograde transport of radio-active amino acid (Grasse & Cynader, 1982). So it is just possible that there may be a special class of direction-selective ganglion cells, distinct from the ON–OFF type, to feed the accessory optic system. As many as 1000 ganglion cells could be back-filled from the medial terminal nucleus and, even if a like number was required for both of the other terminal stations of this pathway, the number of extra ganglion cells to be added to Table 3.1 would still leave the accumulation well short of the histological total.

That there may be additional classes not previously recognized is suggested by the following argument. Loosely speaking, the ganglion cell classes can be grouped in such a way that the members of a set are complementary with respect to some parameter of interest. For example, the eight classes of concentric receptive fields can be viewed as four sets where each set contains the ON-centre and OFF-centre varieties of the particular type. For example, the set of sluggish-sustained cells contains ON-centre and OFF-centre varieties; the members of each pair in the set are complementary with respect to signalling the contrast of some detail of the retinal image over a continuous range including both decrements and increments. Such groupings are also evident among the non-concentric classes. Local-edge detectors may be considered as complementary to the uniformity detectors with respect to signalling the density of texture in a patch of visual field. Similarly, direction-selective units having different preferred directions make up a mutually complementary group with respect to the signalling of local direction of movement. Even the blue colour-coded units can be considered as the composite of complementary pairs signalling the distribution of blueness between object and background, because of the different relative sizes of the blue-sensitive component of their receptive fields.

However, in the published descriptions there seems to be no obvious complement for the edge-inhibitory OFF-centre units. It is therefore of interest that a possible candidate was recently encountered (Harrison & Levick, unpublished). The unit gave a sustained elevation of discharge when a stationary fine square-wave grating was left exposed on the receptive field, regardless of the precise spatial position of the grating. For large uniform spots, the unit responded like a fuzzy ON-centre concentric type. These properties complement those of the edge-inhibitory OFF-centre units.

Concluding remarks

The notion of reconstructability, commonly employed in connection with the performance of sampling arrays in representing images, is here used to focus attention on the question: why are there so many different classes of receptive fields of retinal ganglion cells? By asking to what extent the responses of one class of cells can be reproduced accurately and reliably by operating on the responses of another class, one is seeking the dimensions along which the performances of the various classes are well separated. Since central reconstructability is also a function of the impulse-coded form of transmission in the optic nerve and also of the partitioning of channels to different sides of the brain and to different central nuclei, an approach is thus opened to the problem of evaluating alternative design strategies in terms of their organizational costs.

References

Barlow, H. B., Hill, R. M. & Levick, W. R. (1964) Retinal ganglion cells responding selectively to direction and speed of image motion in the rabbit. *J. Physiol. (Lond.)*, **173**, 377–407

Bishop, G. H., Clare, M. H. & Landau, W. M. (1969) Further analysis of fiber groups in the optic tract of the cat. *Exp. Neurol.*, **24**, 386–99

Boycott, B. B. & Wässle, H. (1974) The morphological types of ganglion cells of the domestic cat's retina. *J. Physiol. (Lond.)*, **240**, 397–419

Brindley, G. S. (1970) *Physiology of the Retina and Visual Pathway*, 2nd edn. London: Arnold

Burr, D. C. & Ross, J. (1979) How does binocular delay give information about depth? *Vision Res.*, **19**, 523–32

Cleland, B. G., Dubin, M. W. & Levick, W. R. (1971) Sustained and transient neurones in the cat's retina and lateral geniculate nucleus. *J. Physiol. (Lond.)*, **217**, 473–96

Cleland, B. G., Harding, T. H. & Tulunay-Keesey, U. (1979) Visual resolution and receptive field size: examination of two kinds of cat retinal ganglion cell. *Science*, **205**, 1015–17

Cleland, B. G. & Levick, W. R. (1974*a*) Brisk and sluggish concentrically organized ganglion cells in the cat's retina. *J. Physiol. (Lond.)*, **240**, 421–56

Cleland, B. G. & Levick, W. R. (1974*b*) Properties of rarely encountered types of ganglion cell in the cat's retina and an overall classification. *J. Physiol. (Lond.)*, **240**, 457–92

Cleland, B. G., Levick, W. R. & Wässle, H. (1975) Physiological identification of a morphological class of cat retinal ganglion cell. *J. Physiol. (Lond.)*, **248**, 151–71

Derrington, A. M., Lennie, P. & Wright, M. J. (1979) The mechanism of peripherally evoked responses in retinal ganglion cells. *J. Physiol. (Lond.)*, **289**, 299–310

Enroth-Cugell, C. & Robson J. G. (1966) The contrast sensitivity of retinal ganglion cells of the cat. *J. Physiol. (Lond.)*, **187**, 517–52

Farmer, S. G. & Rodieck, R. W. (1982) Ganglion cells of the cat accessory optic system: morphology and retinal topography. *J. Comp. Neurol.*, **205**, 190–8

Fischer, B. (1973) Overlap of receptive field centers and representation of the visual field in the cat's optic tract. *Vision Res.*, **13**, 2113–20

Fischer, B., Krüger, J. & Droll, W. (1975) Quantitative aspects of the shift-effect in cat retinal ganglion cells. *Brain Res.*, **83**, 391–403

Fukuda, Y. & Stone, J. (1974) Retinal distribution and central projections of Y-, X-, and W-cells of the cat's retina. *J. Neurophysiol.*, **37**, 749–72

Grasse, K. L. & Cynader, M. S. (1982) Electrophysiology of medial terminal nucleus of accessory optic system in the cat. *J. Neurophysiol.*, **48**, 490–504

Hochstein, S. & Shapley, R. M. (1976) Quantitative analysis of retinal ganglion cell classifications. *J. Physiol (Lond.)*, **262**, 237–64

Hubel, D. H. & Wiesel, T. N. (1959) Receptive fields of single neurones in the cat's striate cortex. *J. Physiol. (Lond.)*, **148**, 574–91

Hubel, D. H. & Wiesel, T. N. (1962) Receptive fields, binocular interaction and functional architecture in the cat's visual cortex. *J. Physiol. (Lond.)*, **160**, 106–54

Hubel, D. H. & Wiesel, T. N. (1965) Receptive fields and functional architecture in two nonstriate visual areas (18 and 19) of the cat. *J. Neurophysiol.*, **28**, 229–89

Hughes, A. (1975) A quantitative analysis of the cat retinal ganglion cell topography. *J. Comp. Neurol.*, **163**, 107–28

Hughes, A. (1976) A supplement to the cat schematic eye. *Vision Res.*, **16**, 149–54

Hughes, A. (1977) The topography of vision in mammals of contrasting life style: comparative optics and retinal organization. In *Handbook of Sensory Physiology*, vol. VII/5, *The Visual System in Vertebrates*, ed. F. Crescitelli, pp. 613–756. Berlin, Heidelberg & New York: Springer

Hughes, A. (1981) Population magnitudes and distribution of the major modal classes of cat retinal ganglion cell as estimated from HRP filling and a systematic survey of the soma diameter spectra for classical neurones. *J. Comp. Neurol.*, **197**, 303–39

Hughes, A. & Wässle, H. (1976) The cat optic nerve fibre total count and diameter spectrum. *J. Comp. Neurol.*, **169**, 171–84

Kirk, D. L., Levick, W. R., Cleland, B. G. & Wässle, H. (1976) Crossed and uncrossed representation of the visual field by brisk-sustained and brisk-transient cat retinal ganglion cells. *Vision Res.*, **16**, 225–31

Kolb, H., Nelson, R. & Mariani, A. (1981) Amacrine cells, bipolar cells and ganglion cells of the cat retina: a Golgi study. *Vision Res.*, **21**, 1081–114

Kozak, W., Rodieck, R. W. & Bishop, P. O. (1965) Responses of single units in lateral geniculate nucleus of cat to moving visual patterns. *J. Neurophysiol.*, **28**, 19–47

Krüger, J. (1980) The shift-effect enhances X- and suppresses Y-type response characteristics of cat retinal ganglion cells. *Brain Res.*, **201**, 71–84

Krüger, J. & Fischer, B. (1973) Strong periphery effect in cat retinal ganglion cells. Excitatory responses in ON- and OFF-center neurones to single grid displacements. *Exp. Brain Res.*, **18**, 316–18

Kuffler, S. W. (1952) Neurons in the retina: organization, inhibition, and excitation problems. *Cold Spring Harbor Symp. Quant. Biol.*, **17**, 281–92

Kuffler, S. W. (1953) Discharge patterns and functional organization of mammalian retina. *J. Neurophysiol.*, **16**, 37–68

Levick, W. R. (1972) Receptive fields of retinal ganglion cells. In *Handbook of Sensory Physiology*, vol. VII/2, *Physiology of Photoreceptor Organs*, ed. M. G. F. Fuortes, pp. 531–66. Berlin: Springer

Levick, W. R. & Thibos, L. N. (1983) Receptive fields of cat ganglion cells: classification and construction. *Prog. ret. Res.*, 2, 267–319

McIlwain, J. T. (1964) Receptive fields of optic tract axons and lateral geniculate cells: peripheral extent and barbiturate sensitivity. *J. Neurophysiol.*, 27, 1154–73

Mastronarde, D. N. (1983) Correlated firing of cat retinal ganglion cells. I. Spontaneously active inputs to X- and Y-cells. *J. Neurophysiol.*, 49, 303–24

Oyster, C. W. (1968) The analysis of image motion by the rabbit retina. *J. Physiol (Lond.)*, 199, 613–35

Oyster, C. W. & Barlow, H. B. (1967) Direction-selective units in rabbit retina: distribution of preferred directions. *Science*, 155, 841–2

Oyster, C. W., Simpson, J. I., Takahashi, E. S. & Soodak, R. E. (1980) Retinal ganglion cells projecting to the rabbit accessory optic system. *J. Comp. Neurol.*, 190, 49–61

Oyster, C. W., Takahashi, E. & Collewijn, H. (1972) Direction-selective retinal ganglion cells and control of optokinetic nystagmus in the rabbit. *Vision Res.*, 12, 183–93

Peichl, L. & Wässle, H. (1979) Size, scatter and coverage of ganglion cell receptive field centres in the cat retina. *J. Physiol. (Lond.)*, 291, 117–41

Peichl, L. & Wässle, H. (1983) The structural correlate of the receptive field centre of alpha ganglion cells in the cat retina. *J. Physiol. (Lond.)*, 341, 309–24

Robson, J. G. & Enroth-Cugell, C. (1978) Light distribution in the cat's retinal image. *Vision Res.*, 18, 159–73

Rodieck, R. W. & Stone, J. (1965) Analysis of receptive fields of cat retinal ganglion cells. *J. Neurophysiol.*, 28, 833–49

Rowe, M. H. & Stone, J. (1976) Properties of ganglion cells in the visual streak of the cat's retina. *J. Comp. Neurol.*, 169, 99–125

Rowe, M. H. & Stone, J. (1977) Naming of neurones. *Brain Behav. Evol.*, 14, 185–216

Saito, H. (1983) Morphology of physiologically identified X-, Y-, and W-type retinal ganglion cells of the cat. *J. Comp. Neurol.*, 221, 279–88

Simpson, J. I. (1984) The accessory optic system. *Ann. Rev. Neurosci.*, 7, 13–41

Simpson, J. I., Soodak, R. E. & Hess, R. (1979) The accessory optic system and its relation to the vestibulocerebellum. *Prog. Brain Res.*, 50, 715–24

Steinberg, R. H., Reid, M. & Lacy, P. L. (1973) The distribution of rods and cones in the retina of the cat (*Felis domesticus*). *J. Comp. Neurol.*, 148, 229–48

Stone, J., Dreher, B. & Leventhal, A. (1979) Hierarchical and parallel mechanisms in the organization of visual cortex. *Brain Res. Rev.*, 1, 345–94

Stone, J. & Fukuda, Y. (1974a) Properties of cat retinal ganglion cells: a comparison of W-cells with X- and Y-cells. *J. Neurophysiol.*, 37, 722–48

Stone, J. & Fukuda, Y. (1947b) The naso-temporal division of the cat's retina re-examined in terms of Y-, X- and W-cells. *J. Comp. Neurol.*, 155, 377–94

Stone, J. & Hoffmann, K. -P. (1972) Very slow-conducting ganglion cells in the cat's retina: a major, new functional type? *Brain Res.*, 43, 610–16

Thibos, L. N. & Levick, W. R. (1983) Spatial frequency characteristics of brisk and sluggish ganglion cells of the cat's retina. *Exp. Brain Res.*, 51, 16–22

Wässle, H., Levick, W. R. & Cleland, B. G. (1975) The distribution of the *alpha* type of ganglion cells in the cat's retina. *J. Comp. Neurol.*, 159, 419–37

Wässle, H., Peichl, L. & Boycott, B. B. (1981) Dendritic territories of cat retinal ganglion cells. *Nature (Lond.)* 292, 344–5

Wässle, H., Peichl, L. & Boycott, B. B. (1983) Mosaics and territories of cat retinal ganglion cells. *Prog. Brain Res.*, 58, 183–90

Werblin, F. S. & Dowling, J. E. (1969) Organization of the retina of the Mudpuppy, *Necturus Maculosus*. II. Intracellular recording. *J. Neurophysiol.*, 32, 339–54

Wiesel, T. N. (1960) Receptive fields of ganglion cells in the cat's retina *J. Physiol. (Lond.)*, 53, 583–94

4

Parallel and serial pathways in the inner plexiform layer of the mammalian retina

DAVID I. VANEY

Introduction

There are six classes of neurons in the mammalian retina and their cell bodies are packed in three well-defined layers that are separated by two synaptic regions (Cajal, 1892; Polyak, 1941; Boycott & Dowling, 1969). The rod and cone receptors in the outer nuclear layer send their axons to the outer plexiform layer where they contact the dendrites of bipolar and horizontal cells, whose perikarya are located in the inner nuclear layer. The bipolar cells feed the visual information to the inner plexiform layer where their axons make synaptic contact with the dendrites of ganglion cells and the processes of amacrine cells. The perikarya of the great majority of amacrine cells are situated at the inner border of the inner nuclear layer but several types are located in the ganglion cell layer and account for 35–80% of the neurons in that region (Hughes & Vaney, 1980; Hughes & Wieniawa-Narkiewicz, 1980; Vaney, 1980; Perry, 1981). Inter-plexiform cells, like the amacrine cells, are both pre- and postsynaptic to other neurons in the inner plexiform layer but, in addition, they send processes to the outer plexiform layer that are exclusively presynaptic (Boycott et al., 1975; Dowling & Ehinger, 1975; Kolb & West, 1977) (Fig. 4.1).

The inner plexiform layer in primate retina is 30–60 μm thick and contains 15–40 million synapses per mm^2 of retina (Dubin, 1970). In rhesus monkey and human retinae, the peak density of cones and thus of midget bipolars is about 150000/mm^2 (Steinberg, Reid & Lacy, 1973; Mariani, Kolb & Nelson, 1984),

44

indicating that in the fovea there are about 270 inner plexiform synapses for each midget bipolar/midget ganglion cell pair. In peripheral primate retina and in the retinae of other mammals, there is significant convergence in both the outer and inner plexiform layers with the result that in the area centralis of cat retina there are about 1300 inner plexiform synapses for each ganglion cell. When it is considered that only a minority of these synapses are on the direct bipolar pathway to ganglion cells (Dubin, 1970), it becomes apparent that the amacrine cell pathways can play a

Fig. 4.1. Semithin transverse section of cat retina showing the three nuclear layers and the two plexiform layers: ONL, outer nuclear layer; OPL, outer plexiform layer; INL, inner nuclear layer; IPL, inner plexiform layer; GCL, ganglion cell layer. (Scale bar: 10 μm.) (Micrograph courtesy of R. O. L. Wong.)

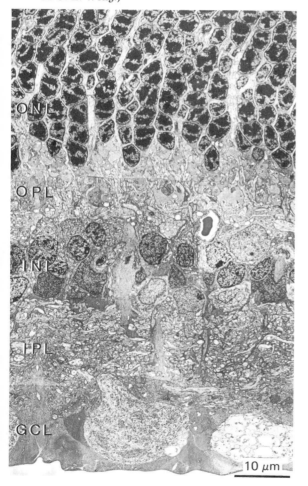

substantial role in shaping the receptive field properties of retinal ganglion cells.

This review describes the anatomy, physiology and pharmacology of the synaptic pathways in the inner plexiform layer of selected mammalian retinae, principally cat and rabbit. The bipolar and amacrine pathways are considered separately, although inevitably there is considerable overlap between the two sections.

Bipolar cell pathways

ON and OFF pathways

All vertebrate photoreceptors are depolarized in the dark and hyperpolarized in the light: they are excited tonically by OFF-stimuli. The receptive field centres of retinal ganglion cells, however, are excited by either ON- or OFF-stimuli. In cat and rabbit retinae, the four or five types of concentric ganglion cells occur in two forms, one ON-centre and the other OFF-centre (Cleland & Levick, 1974a; Stone & Fukuda, 1974; Caldwell & Daw, 1978a; Vaney, Levick & Thibos, 1981). In each case the centre is antagonized by a surround response of the opposite polarity (Kuffler, 1953). The organization of visual information into parallel ON- and OFF-channels appears to be generated early in the process and is preserved throughout the visual system.

Intracellular dye injection of physiologically characterized ganglion cells in cat and rabbit retinae reveals that the ON–OFF dichotomy is correlated with the level of dendritic stratification in the inner plexiform layer (Nelson, Famiglietti & Kolb, 1978; Bloomfield, 1981; Saito, 1983). Ganglion cells with OFF-centre receptive fields branch near the inner nuclear layer (sublamina a), while ON-centre cells branch lower in the inner plexiform layer (sublamina b). The dendrites of ON–OFF ganglion cells either branch in both sublaminae or straddle the sublaminar border (Amthor, Oyster & Takahashi, 1984).

In cat retina, brisk-responding types of concentric ganglion cells are encountered more often than sluggish-responding types and, therefore, they have been studied in more detail. The brisk-transient (Y) cells correspond to the alpha morphological type and the brisk-sustained (X) cells correspond to the beta type (Boycott & Wässle, 1974; Cleland, Levick & Wässle, 1975; Saito, 1983). Both alpha and beta ganglion cells occur in paramorphic pairs, with the OFF-cells branching in sublamina a and the ON-cells branching in sublamina b (Famiglietti & Kolb, 1976; Wässle,

Boycott & Illing, 1981*a*; Wässle, Peichl & Boycott, 1981*b*). For both morphological types, the ON- and the OFF-populations form regular cell mosaics that are independent of each other.

The synaptic input to alpha and beta cells is concentrated on those dendrites contained in the plane of arborization: the cell body and proximal primary dendrites have few synapses (Stevens, McGuire & Sterling, 1980). Beta cells receive about 70% of their input from cone bipolar cells, whereas alpha cells receive much of their input from amacrine cells (Kolb, 1979).

Cone pathways

Like the ganglion cell dendrites, the axon terminals of cone bipolar cells are broadly stratified within the inner plexiform layer. In monkey retina, the flat and invaginating types of midget bipolars branch near the inner nuclear and ganglion cell layers, respectively (Kolb, Boycott & Dowling, 1969). The two levels of branching correspond to the dendritic strata of midget ganglion cells (Polyak, 1941), suggesting that the ON-and OFF-responses of the ganglion cells are determined primarily by the laminar organization of cone bipolar terminals (Gouras, 1971).

Similarly, in cat retina the axon terminals of most cone bipolars are confined to either sublamina a, where OFF-centre ganglion cells branch, or to sublamina b, where ON-centre ganglion cells branch (Boycott & Kolb, 1973). When other morphological features, such as the branching pattern and diameter of the dendritic tree, are also taken into consideration, at least eight types of cone bipolar cells can be distinguished (Famiglietti, 1981*a*; Kolb, Nelson & Mariani, 1981; McGuire, Stevens & Sterling, 1984). These include four narrow-field types, two in each sublamina, that contact beta ganglion cells directly and which may be analogous to the midget bipolars in primate retina (Sterling, 1983; Kolb & Nelson, 1984).

In cat retina, however, the interactions between narrow-field bipolar cells and narrow-field ganglion cells are more complex than the simple excitation postulated for the midget system in monkey retina. The narrow-field bipolar cells that branch in sublamina b of cat retina can either be ON-centre (depolarizing to light) or OFF-centre (hyperpolarizing; Nelson *et al.*, 1981; Nelson & Kolb, 1983), suggesting that one type (CBb1) is excitatory and the other type (CBb2) is inhibitory to ON-centre ganglion cells (Sterling, 1983;

Kolb & Nelson, 1984). This hypothesis is supported by evidence that the CBb2 bipolar accumulates the inhibitory transmitter glycine (McGuire *et al.*, 1984). The two types of narrow-field cone bipolars in sublamina a also appear to form a complementary pair (CBa1 and CBa2), with both cells making multiple synapses onto OFF-centre ganglion cells. The CBa2 bipolar accumulates glycine but it is not known whether it responds antagonistically to the OFF-centre CBa1.

These findings suggest that the cone bipolar inputs to beta ganglion cells are arranged in a 'push–pull' fashion: excitation from one bipolar type may be accompanied by withdrawal of inhibition from the other, and *vice versa* (Sterling, 1983). As a result the dynamic range of the response would be increased and the time constant shortened. In addition to the stratified narrow-field bipolars, each beta cell receives input from several multistratified and wide-field bipolars, and their dendritic fields overlap extensively, indicating that all types of cone bipolar cells are present in each patch of cat retina (McGuire *et al.*, 1984). (Figure 2).

The density of both types of narrow-field bipolars that branch in sublamina b can be estimated from published data. The ON-centre, CBb1 bipolars form a regular array with an intercell distance of $10 \pm 1 \ \mu$m in the area centralis (Freed & Sterling, 1983), indicating a density of 10 000–11 500 cells/mm^2. The OFF-centre, CBb2 bipolars make invaginating contacts with about five cones in the area centralis and each terminal cluster occupies the majority of triads in each cone pedicle (Boycott & Kolb, 1973; Nelson & Kolb, 1983). The cone density reaches a peak of 26 000 mm^2 in the area centralis (Steinberg *et al.*, 1973) and, assuming that all spectral types of cones contact the CBb2 bipolars, this indicates a maximum density of about 5200 cells/mm^2.

There are about 5600 beta ganglion cells/mm^2 in the peak area centralis, 48% of which are ON-centre cells (Hughes, 1981; Wässle *et al.*, 1981*a*). It seems plausible therefore that there are about four CBb1 and two CBb2 bipolar cells for each ON-centre beta cell. Serial reconstruction confirms that the bipolar input to beta ganglion cells is convergent rather than distributed: each of three CBb1 bipolars contacts a central beta cell at least 50 times, ending on dendrites of all orders (Freed & Sterling, 1983). A CBb2 bipolar probably makes as many synapses onto beta ganglion cells as each CBb1 bipolar, but the number of CBb1 synapses on each beta cell is twice that of the CBb2 synapses (McGuire *et al.*, 1984). This provides independent evidence that there are twice as many CBb1 bipolars as CBb2 bipolars.

The CBb1 input to alpha ganglion cells in central cat retina is quite different from the input to beta ganglion cells. Each of 100 CBb1 bipolars makes only about four synapses that are restricted to alpha-cell dendrites of a single order, and contacts from one bipolar are not located between those of another (Freed & Sterling, 1983). The input from CBb2 bipolars is correspondingly sparse (McGuire *et al.*, 1984) and serial reconstruction of the primary dendrites of an ON-centre alpha cell indicates that bipolar cells only provide about 20% of the total synaptic input (Kolb, 1979).

The receptive field centre of each beta ganglion cell is shaped by the envelope of excitatory bipolars converging on the cell. Consequently the centre

Fig. 4.2. Wiring diagram of the cone bipolar input to beta ganglion cells in cat retina. ON-centre beta cells branch in sublamina b of the IPL and probably receive excitatory input from ON-centre CBb1 cells and inhibitory input from OFF-centre CBb2 cells. Similarly, OFF-centre beta cells receive input from a complementary pair of bipolar cells branching in sublamina a. (After Sterling (1983) and Kolb & Nelson (1984).)

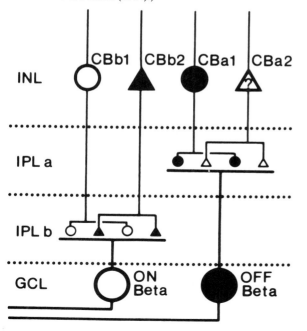

diameter of a brisk-sustained (X) cell is about twice the dendritic field diameter of the corresponding beta cell (Boycott & Wässle, 1974; Cleland, Harding & Tulunay-Keesey, 1979; Peichl & Wässle, 1979). It is not clear, however, whether the antagonistic surround of a concentric ganglion cell is derived from the same bipolar input as the centre component. The CBb1 bipolar has an antagonistic OFF-surround but it is not thought to be due to horizontal cell interaction as in lower vertebrates (Werblin & Dowling, 1969; Nelson *et al.*, 1981). The complementary CBb2 bipolar does not have a surround, nor does the CBa1 bipolar which is excitatory to OFF-centre ganglion cells (Nelson & Kolb, 1983). This suggests that the centre and surround components of OFF-centre, brisk-sustained cells may be mediated by separate inputs from ON- and OFF-bipolars.

Alternatively, the antagonistic surround may result from lateral interactions mediated by amacrine cells. The surrounds of brisk-transient (Y) cells are probably of this type, as the extent of the surround greatly exceeds the envelope of bipolar cells contacting the alpha-cell dendritic tree (Peichl & Wässle, 1983). In cat retina, the bistratified A8 amacrine receives input from bipolar and amacrine terminals in both sublaminae and is presynaptic in sublamina a to OFF-centre ganglion cells. The A8 amacrine cell has a transient OFF-centre and a strong ON-surround and its cone-dominated receptive field resembles that of brisk-transient ganglion cells (Nelson & Kolb, 1983).

Contrary to earlier reports (Famiglietti & Kolb, 1976; Kolb, 1979), there is no correlation in cat retina between the type of synaptic contacts a bipolar cell makes with the cone pedicles and its axonal stratification in the inner plexiform layer: cone bipolars making invaginating, semi-invaginating or basal (flat) contacts are found in each sublamina (Nelson & Kolb, 1983). Furthermore, the CBb2 bipolar makes both invaginating and basal contacts, as does the B4 bipolar in grey squirrel retina (West, 1978).

Rod pathways

The rod bipolar cells in cat retina, like the rod receptors they contact, are depolarized by OFF-stimuli (Nelson *et al.*, 1976). Thus, unlike the cone pathway, the information in the rod pathway is not split into separate ON- and OFF-channels at the outer plexiform layer. Moreover, the rod bipolars are a single morphological type: they terminate deep in sublamina b amongst the dendrites of ON-centre ganglion cells and make almost no synapses in sublamina a (Boycott & Kolb, 1973; Kolb & Famiglietti, 1974; Kolb, 1979). Each rod bipolar in central cat retina makes about 29 ribbon contacts in sublamina b (McGuire *et al.*, 1984) and, like those of cone bipolar cells, each is presynaptic to two processes (Kidd, 1962; Dowling & Boycott, 1966). One component of the dyad is always a dendrite of an AII amacrine cell and the other is typically an amacrine varicosity that makes reciprocal contact with the rod bipolar (Kolb & Famiglietti, 1974; Kolb, 1979). The output is thus directed almost exclusively to amacrine cells, with only 1% of the ribbon synapses contacting alpha ganglion cells (Raviola & Raviola, 1967; Freed & Sterling, 1983; McGuire *et al.*, 1984).

The AII amacrine in cat retina has been studied in more detail than any other retinal amacrine cell. Although originally described from Golgi-stained material (Kolb & Famiglietti, 1974; Famiglietti & Kolb, 1975), the population of AII amacrine cells can be selectively stained with the fluorescent dye DAPI or by [³H]-glycine accumulation (Pourcho, 1980; Vaney, 1985). The AII amacrines range in density from 500 cells/mm² in the far periphery to 5000 cells/mm² within 150 μm of the peak area centralis (Sterling, 1983; Vaney, 1985), and the increasing cell density is compensated by a reduction in the dendritic field diameter from 95 μm to 18 μm (Vaney, 1985).

The AII amacrine cells have a vertically organized, narrow-field, bistratified dendritic tree. Generally a single stout dendrite descends from the cell body and arborizes into a conical field in sublamina b. The AII cell bodies are ringed with short, extremely fine stalks that terminate in large varicosities (or 'lobular appendages') in sublamina a and the proximal inner nuclear layer. Amacrine cells with a similar morphology have also been described in monkey, dog and rat retinae (Cajal, 1892; Boycott & Dowling, 1969; Perry & Walker, 1980). The AII dendrites in sublamina b receive multiple input from up to 30 rod bipolars and from unidentified amacrine cells; they also make small gap junctions with each other and extensive gap junctions with cone bipolar axons in sublamina b, particularly those of CBb1 bipolar cells (Famiglietti & Kolb, 1975; Kolb, 1979; Sterling, 1983). The lobular appendages in sublamina a are presynaptic to cone bipolar, amacrine and ganglion cell processes. They provide about 36% of the amacrine input to CBa1 bipolars and, in return, receive about 64% of the CBa1

output directed to amacrine cells (McGuire *et al.*, 1984). (Figs. 4.3 and 4.4).

The AII amacrine cells respond to an ON-stimulus with a pronounced transient followed by a sustained depolarization that is strongly rod-dominated; in some cases a hyperpolarizing (OFF) surround can also be elicited (Nelson *et al.*, 1976; Nelson, 1982). The ON-centre AII amacrines can therefore feed rod signals into ON-centre bipolars through the gap junctions in sublamina b, and into OFF-centre bipolars and ganglion cells through the glycinergic (sign-inverting) synapses in sublamina a (Famiglietti & Kolb, 1975). It is appropriate therefore that the CBb1 cone bipolar is rod-dominated under mesopic conditions and that its antagonistic surround may be derived from that of the AII amacrine (Kolb & Nelson, 1983). Under these conditions, the CBb1 bipolar is functioning as the fourth element in the chain of rod interneurons and this

Fig. 4.3. An AII amacrine cell in cat retina injected with Lucifer yellow and viewed in flat-mount under fluorescence illumination. In the upper micrograph the focus is on the lobular appendages at the INL–IPL border; in the lower micrograph the focus is on the processes in sublamina b.

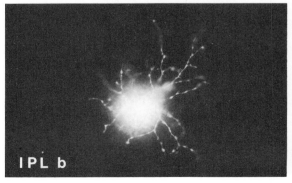

challenges the traditional assumption that bipolar cells are vertical, rather than lateral elements.

Although the CBb1 bipolar is an extreme case, there is considerable mixing of rod and cone signals in all types of cone bipolars, much of which occurs in the outer plexiform layer (Nelson, 1977). There are small gap junctions between the basal processes of cone

Fig. 4.4. An AII amacrine cell and its synaptic connections partially reconstructed from electron micrographs of 189 serial sections. The processes in sublamina b receive input from rod bipolars (□) and unidentified amacrine cells (■) and have gap junctions (=) with ON-centre cone bipolars. The lobular appendages in sublamina a and the cell body receive input from dopaminergic amacrines (●) and OFF-centre cone bipolars (○); they provide output to the same bipolars and to OFF-centre ganglion cells (▲). (Reproduced with permission from Sterling (1983).)

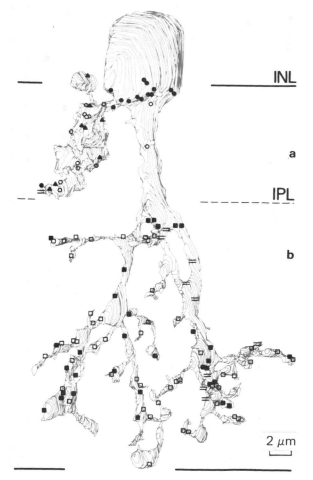

INL

a

IPL

b

2 μm

pedicles and neighbouring rod spherules (Raviola & Gilula, 1975; Kolb, 1977) and the cones themselves receive about 50% of their input from rods. It is difficult to judge, therefore, whether the rod component of amacrine cells with mixed signals is mediated by rod bipolars or cone bipolars. Of the four types of amacrine cells known to be paired with the AII amacrines at rod bipolar dyads (A6, A8, A13, A17), the first three receive a mixed input from both rod and cone bipolars and the A8 amacrine, in particular, is cone-dominated (Kolb & Nelson, 1981, 1983). The A6, A8 and A13 amacrines respond to an ON-stimulus with a sustained hyper-polarization and this contrasts curiously with the depolarization of the rod-dominated A17 and AII amacrines. It is not clear if the rod bipolar is both excitatory and inhibitory at the ribbon synapse or whether the three amacrines with mixed bipolar input are swamped by cone bipolar signals.

The A17 amacrine is a wide-field cell with fine beaded dendrites ramifying throughout the inner plexiform layer (Kolb *et al.*, 1981). Each varicosity in sublamina b is postsynaptic to a ribbon synapse and locally presynaptic to the same rod bipolar, while the varicosities in sublamina a are pre- and postsynaptic to other amacrine cells. In contrast, the AII, A6 and A8 amacrines make direct synapses with the dendrites of ON-centre ganglion cells and the A13 amacrine contacts OFF-centre cells (Kolb & Nelson, 1983). The single OFF-centre signal transmitted by rod bipolars to the inner retina is therefore split into parallel ON- and OFF-pathways that have different routes via specific amacrine cells to the retinal ganglion cells.

Amacrine cell pathways

The amacrine cells play a major role in the information flow to the retinal ganglion cells. In central cat retina, the rod bipolars synapse almost exclusively onto amacrine cells and about 45% of the identified output of cone bipolars also goes to amacrines (Kolb, 1979; McGuire *et al.*, 1984). It is often said that amacrine cells mediate complex interactions in the retina, but we know little about how this is done. Microelectrode penetration of amacrine cells is difficult, particularly in mammalian retinae, and the responses recorded in the cell body may not be representative of the activity in the dendrites.

The processes of amacrine cells possess both axonal and dendritic properties in that they generate tetrodotoxin-sensitive spikes superimposed on summed, postsynaptic potentials (Miller, 1979). The close apposition of excitatory and inhibitory inputs provides scope for very local interactions that may be highly non-linear in character and isolated from activity elsewhere in the dendritic field. A passive dendritic tree may then perform hundreds of independent analogue operations on its synaptic inputs without requiring any threshold (spiking) mechanism (Poggio & Torre, 1981; Koch, Poggio & Torre, 1982, 1983).

There may be over 20 morphological types of amacrine cells in each mammalian retina (Cajal, 1892; Boycott & Dowling, 1969; West, 1976; Perry & Walker, 1980; Kolb *et al.*, 1981) and half as many pharmacological types (Ehinger, 1982; Karten & Brecha, 1983). Intuitively this provides scope for each type to serve a specific function or to influence a limited number of ganglion cell classes. This however is not the case: the effects of pharmacologically isolated amacrine cells appear surprisingly general and diffuse. Nevertheless, pharmacological dissection provides a broad framework to which other studies can be linked (Kolb & Nelson, 1984). Of the dozen or so amines, amino acids and neuropeptides that have been identified as putative neurotransmitters in amacrine cells, only glycine, dopamine, acetylcholine and γ-aminobutyric acid (GABA) have been soundly established in mammalian retinae (Ehinger, 1982). The pathways served by these four transmitters are described here to illustrate the diversity and complexity of amacrine cell function.

The dopaminergic pathway

In most mammalian retinae there are amacrine cells that contain, synthesize and degrade dopamine, take it up from the extracellular space and release it upon light stimulation: it is likely therefore that dopamine is the transmitter of these cells (Ehinger, 1976). The number of dopaminergic neurons is very small however. In rhesus monkey retina, for example, there are only 7500 dopaminergic amacrine cells. Their density distribution parallels that of the rod photoreceptors and peaks in a ring of 3 mm radius centred on the fovea (Mariani *et al.*, 1984).

In rabbit retina, about 0.1% of the 8 000 000 amacrine cells are dopaminergic. They range in density from 13 cells/mm² in the periphery to 23 cells/mm² on the visual streak and have wide-field, sparsely branching dendrites that achieve multiple coverage of the retina (Brecha, Oyster & Takahashi, 1984). As in

other mammals, most of the dopaminergic processes form a dense plexus at the border of the inner nuclear and inner plexiform layers (stratum one), although some processes are sparsely distributed in strata three and five (Ehinger, 1982) (Fig. 4.5).

In rabbit and other mammalian retinae, the dopaminergic neurons are interamacrine cells that do not make synaptic contact with bipolar cells or ganglion cells (Dowling & Ehinger, 1978; Holmgren, 1982; Pourcho, 1982). They receive input from other dopaminergic amacrines on their dendritic varicosities and from unlabelled amacrine cells on the intervaricose segments. Their output is directed to the cell bodies and processes of unlabelled amacrine cells and, less frequently, to the labelled processes of other dopaminergic cells. The number of output synapses is far greater than the number of input synapses.

In cat retina, the dopaminergic processes encircle the glycinergic, AII amacrine cells near their descending dendrites and are presynaptic to both the cell bodies and the lobular appendages (Törk & Stone,

Fig. 4.5. Montage of the dopaminergic and the cholinergic amacrine cells in a peripheral field of rabbit retina, as shown by tyrosine hydroxylase immunoreactivity and neurofibrillar staining, respectively. Although the dendritic trees of the four dopaminergic cells (stippled) are understained, the characteristic branching pattern of their proximal dendrites is apparent. There are approximately equal numbers of cholinergic cells in the inner nuclear (outlined) and ganglion cell layers (filled) and the dendritic tree of a Lucifer-filled cholinergic cell is superimposed on one of the cell bodies to illustrate the substantial dendritic coverage of the cholinergic amacrines. Field: 580 μm × 430 μm. (Dopaminergic cells courtesy of N. C. Brecha, C. W. Oyster and E. S. Takahashi.)

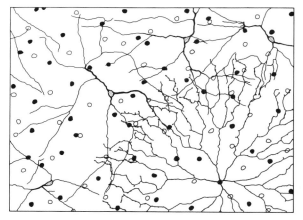

1979; Pourcho, 1982). There are numerous processes, however, that do not form junctional rings, and some may be presynaptic to A17 amacrine cells (Kolb & Nelson, 1984). Both the AII and A17 cells are integral to the rod pathways and this indicates, together with the evidence from rhesus monkey, that the dopaminergic pathways are convergent on rod-dominated circuits. The cells that are presynaptic to the dopaminergic amacrines have not been identified in cat, but evidence from rat retina indicates that there are GABA receptors on the dopaminergic neurons and the effects of GABAergic drugs on dopamine turnover suggests that the dopaminergic amacrines are tonically inhibited by GABA (Kamp & Morgan, 1981; Marshburn & Iuvone, 1981).

The responses of all types of concentric ganglion cells in cat retina to stimulation by a central spot or a surround annulus are reduced by iontophoretically applied dopamine. The surround response is reduced more than the centre response, however, shifting the centre-surround balance in favour of the centre. The consequent reduction in surround antagonism can give rise during whole-field stimulation to a centre-dominated response that is greater than the predrug level. Dopamine also reduces the spontaneous activity of all types of ganglion cells (Straschill & Perwein, 1969; Thier & Alder, 1984).

The low density of dopaminergic neurons, their interamacrine position and the general inhibitory effects of dopamine on the responses of retinal ganglion cells are all consistent with the idea that dopamine has a metabotropic influence, whereby postsynaptic cells are sensitized by dopamine to give a modified response to subsequent, more discrete ionotropic input (McGeer, Eccles & McGeer, 1978). Dopaminergic inhibition can be elicited under both scotopic and mesopic conditions and may be mediated in part by the rod-dominated, AII amacrine cells (Pourcho, 1982; Thier & Alder, 1984).

Inhibition of both ON-centre and OFF-centre ganglion cells by iontophoretically applied dopamine is difficult to reconcile with the microcircuitry of the AII amacrines however (Famiglietti & Kolb, 1975; Kolb, 1979; Sterling, 1983). If dopamine depolarizes the AII cells, this would inhibit OFF-centre ganglion cells through increased glycine release at conventional synapses, but excite ON cone bipolars in sublamina b through their gap junctions with the AII cells. The metabotropic effects of dopamine are probably not straightforward, however, and dopamine may reduce the electrical coupling through the gap junctions of AII amacrines (Mariani *et al.*, 1984; Vaney, 1985), just as it regulates the gap junctions of horizontal cells in teleost retina (Teranishi, Negishi & Kato, 1983).

Dual cholinergic pathways

There is substantial evidence that acetylcholine is a retinal neurotransmitter (Neal, 1983). It is synthesized and released by two populations of amacrine cells in rabbit retina, one in the inner nuclear layer and the other in the ganglion cell layer (Masland & Mills, 1979; Hayden, Mills & Masland, 1980). The two populations of cells form independent, regular arrays that range in density from 850 cells/mm² on the peak visual streak to 150 cells/mm² in the far periphery (Masland, 1983; Vaney, Peichl & Boycott, 1981). The symmetrical distribution of acetylcholine-synthesizing cells in the inner retina corresponds to the localization of the cholinergic markers acetylcholinesterase and choline acetyltransferase (Nichols & Koelle, 1967; Ross & McDougal, 1976). (Fig. 4.6).

In the inner plexiform layer, the synthesized acetylcholine is confined to two discrete bands, one in sublamina a and the other in sublamina b (Masland & Mills, 1979). The band in sublamina a is formed by the unistratified dendrites of those cholinergic amacrines with their cell bodies in the inner nuclear layer, while the band in sublamina b is formed by the dendrites of

Fig. 4.6. The cholinergic amacrines in the inner nuclear layer comprise only 2% of the amacrine cells in rabbit retina and their uniform mosaic within the matrix of other cell bodies is shown in this drawing of a neurofibrillar-stained flat-mount. Field: 300 μm × 200 μm. (Reproduced with permission from Vaney, Peichl & Boycott (1981*b*).)

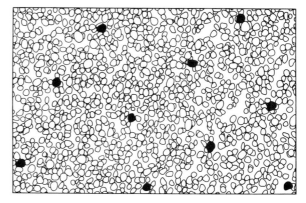

the displaced cholinergic amacrines. In rabbit retina both types of cholinergic cells have a distinctive dendritic morphology that resembles a 'starburst' firework, with the distal dendrites containing numerous preterminal and terminal varicosities (Famiglietti, 1983*a*; Vaney, 1984). Mirror-symmetrical amacrine cells with a starburst dendritic morphology have also been described in cat, rat and ground squirrel retinae (West, 1976; Perry, 1979; Famiglietti, 1981*b*).

Intracellular recordings from the cell bodies of cholinergic amacrines in rabbit retina indicate that the type b cells respond at light onset with transient excitation followed by a sustained depolarization, whereas the type-a cholinergic cells respond to diffuse illumination with a sustained hyperpolarization (Bloomfield, 1981). The cholinergic neurons thus form independent ON and OFF populations, with the sign of the response determined by the level of dendritic stratification. In agreement with this, the release of acetylcholine from rabbit retina is increased 10-fold at both light onset and light offset, whereas the resting release is insensitive to a gradual change in the background illumination (Massey & Neal, 1979; Masland, 1982*a*; Massey & Redburn, 1982). Part of the resting release is calcium-independent and this implies that there is non-vesicular as well as vesicular release of acetylcholine (Masland, 1982*b*).

Electron microscopy of Golgi-stained cholinergic cells shows that they receive cone bipolar and amacrine cell input over the whole dendritic tree. Their output is, however, largely confined to the distal varicose zone and impinges on ganglion cell dendrites in dense clusters (Famiglietti, 1983*b*). Some of the presynaptic amacrine cells may be GABAergic, since GABA antagonists cause a marked increase in the resting release of acetylcholine and also potentiate the light-evoked release (Massey & Neal, 1979; Massey & Redburn, 1982). The cholinergic amacrine cells are not part of an obligatory pathway because the postsynaptic ganglion cells also receive a significant input from bipolar cells and from other amacrine cells (Masland & Ames, 1976; Famiglietti, 1983*b*).

The evidence thus indicates that the cholinergic cells in rabbit retina are sign-conserving interneurons that are tonically driven by cone bipolar cells and GABAergic amacrine cells. The cholinergic cells in turn modulate the activity of retinal ganglion cells by both a light-evoked, transient release and a light-independent, tonic release of acetylcholine.

Acetylcholine affects the receptive field properties of rabbit retinal ganglion cells in two ways (Masland & Ames, 1976; Ariel & Daw, 1982*a*). First, cholinergic agonists increase the spontaneous activity of nearly all ganglion cell types, with those cells having high maintained firing rates showing large increases. This tonic modulation of ganglion cell activity, which acts primarily on the brisk-concentric cells (Caldwell & Daw, 1978*a*; Vaney *et al.*, 1981*a*), does not affect the basic organization of the receptive field, such as the balance between centre and surround. Secondly, cholinergic agonists enhance the light-evoked responses of the sluggish-concentric and non-concentric cells, indicating that these cells receive significant excitatory input from cholinergic amacrine cells. The input to direction-selective ganglion cells is particularly strong as only they are sufficiently depolarized by cholinergic agonists to produce a depolarizing block (Masland & Ames, 1976; Ariel & Daw, 1982*a*). The role of the cholinergic cells in the generation of direction selectivity is reviewed in a later section.

Multiple GABAergic pathways

Although GABA and related drugs have pronounced effects on the receptive field properties of retinal ganglion cells, information about the GABAergic pathways in the retina is incomplete and confusing. This reflects the fact that GABAergic neurons are not a homogeneous population but comprise a number of distinct morphological types. Cellular identification by autoradiography of accumulated GABA and by immunocytochemical localization by the GABA-synthesizing enzyme (GADase) indicates that most GABAergic neurons in mammalian retinae are amacrine cells. Although their dendrites are distributed throughout the inner plexiform layer, the GAD-positive terminals are concentrated in four or five narrow bands in rats and rabbits (Ehinger & Falck, 1971; Brandon, Lam & Wu, 1979; Pourcho, 1980; Vaughn *et al.*, 1981).

The information from cat and rat retinae indicates that the GABAergic amacrine cells function at two levels: they feed activity forward to the ganglion cells and also provide a significant recurrent input to the bipolar cells. In rat retina, rod and cone bipolar cells provide over 80% of the synaptic input to the GAD-positive terminals with the remaining input coming from other labelled amacrine cells. The synaptic output from the GAD-positive terminals is greatest to bipolar cells, followed by unlabelled amacrine cells, ganglion cells, and labelled amacrine cells (Vaughn, *et al.*, 1981). These results suggest that the GABAergic amacrines are the first link in multiple parallel pathways from bipolar cells to ganglion cells, with one-third of the pathways involving serial synapses between pharmacologically distinct amacrine cells.

Some 25–35% of the neurons in the amacrine layer of cat retina accumulate labelled GABA and its analogues (Nakamura, McGuire & Sterling. 1980; Pourcho, 1980). Autoradiography of these cells combined with Golgi-staining indicates that they include at least five types of conventional amacrine cells, a displaced amacrine cell and an interplexiform cell (Pourcho & Goebel, 1983). All but one of the amacrine cells are characterized by numerous dendritic varicosities that may function to isolate electrically adjacent segments of a dendrite (Ellias & Stevens, 1980). (Fig. 4.7).

The synaptic connections of three of the GABA-accumulating amacrines in cat retina (A10, A13, A17) have been examined and each type makes numerous reciprocal contacts with rod and/or cone bipolar cells (Kolb & Nelson, 1981; Pourcho & Goebel, 1983). Moreover, it is claimed that all amacrine processes that are reciprocal to the rod bipolar accumulate GABA (Freed & Sterling, 1982): this has been confirmed for the A13 and A17 amacrines but has not been tested for the A6 and A8 cells. There is a paradox, however, that 75% of the amacrine processes that are reciprocal to rod bipolar terminals also accumulate the indoleamine

5,6-dihydroxytryptamine (Holmgren-Taylor, 1982), and the synaptic connections of the indoleamine-accumulating cells are similar to those of the A8 and A17 amacrines (Kolb & Nelson, 1984). It is possible, therefore, that the indoleamine-accumulating amacrines in cat retina are GABAergic but, first, it will be necessary to establish which of the GABA-accumulating amacrines also contain the GABA-synthesizing enzyme.

In cat retina, GABA antagonists applied systemically alter the centre–surround balance of brisk-transient ganglion cells in favour of the centre but have little effect on the responses of brisk-sustained cells. In addition, centre size decreases for ON-centre brisk-transient cells but increases for the OFF-centre cells (Kirby & Enroth-Cugell, 1976; Kirby & Schweitzer-Tong, 1981*a*,*b*). These effects only occur, however, when the ganglion cells are rod-driven. Under mesopic and photopic conditions, GABA antagonists have little effect on the linear components of the receptive field but strongly affect the non-linear responses: the contrast sensitivity of the non-linear subunits is reduced, as is the shift response and inhibition by a radial grating (Frishman & Linsenmeier, 1982).

In rabbit retina, as in cat, the basic organization of concentric ganglion cells is preserved when GABAergic drugs are applied. GABAergic antagonists increase the centre responses of brisk-transient and sluggish-transient ganglion cells but do not affect the centre-surround balance of brisk-sustained and sluggish-sustained cells. In addition, inhibition by a radial grating is abolished for OFF-centre brisk-transient cells but not for the ON-centre cells (Caldwell & Daw, 1978*b*). Rabbit retinal ganglion cells with complex receptive fields and pronounced non-linear characteristics (Barlow, Hill & Levick, 1964; Levick, 1967) are dramatically affected by GABAergic antagonists: their

Fig. 4.7. Diagram of neurons in cat retina that accumulate the GABA analogue muscimol, showing the laminar distribution of their processes: A2–A19, conventional amacrines; dA, displaced amacrine; IPC, interplexiform cell. (Scale bar: 50 μm.) (Reproduced with permission from Pourcho & Goebel (1983).)

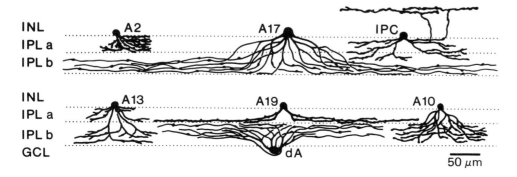

INL   A2   A17   IPC
IPL a
IPL b

INL   A13   A19   A10
IPL a
IPL b
GCL   dA

50 μm

receptive field properties are much simplified and their velocity tuning is broadened. The antagonists abolish the direction specificity of both types of direction-selective ganglion cells and eliminate the orientation specificity of orientation-selective cells (Wyatt & Daw, 1976; Caldwell, Daw & Wyatt, 1978).

These physiological experiments support the concept, originally based on anatomical evidence, that the temporal processing of complex visual information is largely mediated by amacrine cell inhibition in the inner plexiform layer (Dowling, 1968; Dubin, 1970). Pharmacological dissection provides information about the synaptic pathways in the retina but does not reveal how the excitatory and inhibitory elements interact to produce the complex receptive field properties of the ganglion cells. Nevertheless, the established micro-circuits not only constrain the possible interactions but also suggest likely mechanisms. In the following section, neuronal circuitry that may underlie direction selectivity is examined.

Direction selectivity

Retinal ganglion cells with direction-selective receptive fields have been recorded in rabbits, cats, goats, rats and ground squirrels (Barlow *et al.*, 1964; Michael, 1966; Hughes & Whitteridge, 1973; Cleland & Levick, 1974*b*; Stone & Fukuda, 1974; Hughes, 1980). The cells respond maximally to stimuli moving in the preferred direction whereas movement in the opposite (null) direction inhibits the maintained discharge. The receptive field anisotropy is not apparent in the responses to stationary, flashed stimuli nor in the gross morphology of the dendritic tree (Barlow *et al.*, 1964; Amthor *et al.*, 1984). (Fig. 4.8).

Two-spot experiments on direction-selective ganglion cells indicate that each part of the receptive field receives an excitatory input from overlying receptors and an inhibitory input through a delayed pathway from a larger area that excludes the preferred direction. For null-direction movement, multiplicative interaction between the pathways results in the excitatory input being vetoed by the slow inhibitory input from the adjacent region (Barlow & Levick, 1965; Wyatt & Daw, 1975). It is not known whether the inhibitory interactions that underlie direction selectivity take place presynaptically (Dowling, 1970; Miller, 1979) or at the level of the ganglion cell membrane (Torre & Poggio, 1978; Koch *et al.*, 1982), but a recent theoretical analysis suggests that a wide range of primitive non-linear operations can be implemented by passive interactions between synaptic inputs (Poggio & Torre, 1981).

Intracellular injection of direction-selective cells in rabbit retina confirms that they have a dendritic morphology that is well suited for neural interactions that are both local and non-linear: the small calibre, highly branched dendrites give rise to over 200 terminal branches that are distributed throughout the bistratified dendritic tree (Amthor *et al.*, 1984). Postsynaptic inhibition at the base of a terminal dendrite could shunt excitation on the distal part and the inhibition would be localized in subunits of the dendritic field (Koch *et al.*, 1982, 1983). Under these

Fig. 4.8. Drawing of the dendritic tree of an ON–OFF direction-selective ganglion cell in rabbit retina filled with horseradish peroxidase; the dendritic ramification in sublamina a has been drawn separately from that in sublamina b. (Scale bar: 50 μm.) (Reproduced with permission from Amthor, Oyster & Takahashi (1984).)

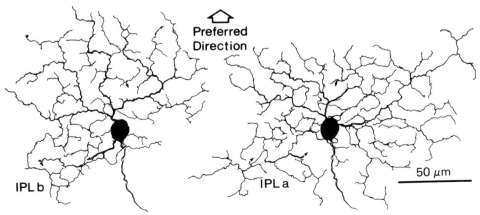

Preferred Direction

IPL b   IPL a   50 μm

conditions the spatial resolution of direction selectivity is much finer than the size of the receptive field (Barlow & Levick, 1965). Intracellular recordings from direction-selective cells in rabbit retina are consistent with this scheme: stimulus movement in the null or preferred directions evokes net inhibition or net excitation, respectively, and fluctuations in the postsynaptic potentials suggest that there is interaction between the two inputs (Dacheux, 1977).

Pharmacological experiments on direction-selective ganglion cells in rabbit retina indicate that the excitatory and inhibitory conductance changes may be mediated by acetylcholine and GABA, respectively (Wyatt & Daw, 1976; Ariel & Daw, 1982b). Cholinergic amacrine cells in both sublaminae provide a substantial direct input to the direction-selective cells and, if potentiated by physostigmine, the cholinergic excitation overcomes the null-direction inhibition. Ganglion cells that are postsynaptic to the cholinergic amacrines also receive a direct input from cone bipolar cells, however, and about half of the light-evoked excitation of direction-selective cells is mediated by transmitter(s) other than acetylcholine (Masland & Ames, 1976; Ariel & Daw, 1982b; Famiglietti, 1983b). (Fig. 4.9).

GABA antagonists applied systemically to direction-selective ganglion cells abolish or reduce those receptive field properties, such as direction selectivity and size specificity, that are associated with lateral inhibition (Wyatt & Daw, 1976; Caldwell et al., 1978). The spatial extent of the GABAergic inhibition is asymmetric to, and larger than, the extent of the cholinergic excitation (Ariel & Daw, 1982b). Pharmacological and anatomical evidence indicates that GABA acts directly on the ganglion cell membrane (as well as providing a tonic input to the cholinergic amacrines). Alternative schemes, with the GABAergic inhibition presynaptic to the excitatory input, would require the release of the excitatory transmitters to be direction sensitive. This does not seem to be the case because the cholinergic inputs in the null and preferred directions are symmetrically potentiated by physostigmine (Ariel & Daw, 1982b).

Parallel pathways and complexity

Many retinal workers believe that the simplicity of organization makes the vertebrate retina a model system for examining local circuit interactions in the brain (Dowling, 1979). For the same reason many cortical workers regard the retina as an inappropriate model of higher CNS function. There are only about 50 types of nerve cells in the three nuclear layers of cat retina (Kolb et al., 1981) whereas, in cat striate cortex, there may be that many cell types in layer IVab alone (Solnick, Davis & Sterling, unpublished). Nevertheless there are some important similarities between retina and cortex. In both systems information flow is essentially hierarchical, but imposed on the serial pathways there is a complex web of parallel and recurrent pathways. Increased recognition of the complexity of retinal function has outstripped the

Fig. 4.9. Wiring diagram of the synaptic pathways that may underlie direction selectivity in rabbit retinal ganglion cells. In each sublamina, proximal inhibition from a GABAergic amacrine cell is presumed to veto distal excitation from cholinergic amacrine cells and cone bipolar cells. The asymmetric inhibitory input is delayed relative to the excitatory input and is thus only effective for null direction movement.

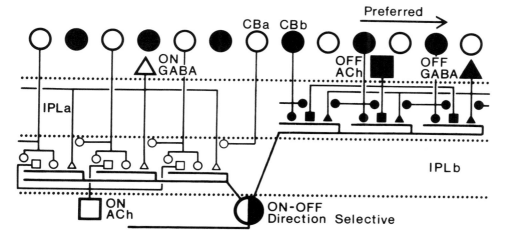

description of new microcircuits, so the wiring diagram of the retina is now perceived to be less complete than was thought 10 years ago.

It does appear, however, that each retinal neuron makes limited connections with specific types of cells and the same is probably true of cortical neurons. The heterogeneity of amacrine structure and function is clearly related to the diversity of ganglion cell types but there is no simple correlation between the two. For example, the bipolar and amacrine cells presynaptic to direction-selective ganglion cells are also integral components of other pathways (Ariel & Daw, 1982b). Moreover, pharmacological experiments indicate that each receptive field property is influenced by a variety of amacrine types and it is often difficult to differentiate the nature of one amacrine input from that of another.

The response pathways in the inner plexiform layer thus appear curiously redundant with present physiological methods. Considering, however, that the receptive field organization of ganglion cells is designed to reduce redundancy by compressing the representation of the visual scene (Barlow, 1961), it seems probable that the parallel pathways within the retina subserve orthogonality rather than redundancy and that their physiology is more subtle than hitherto described. Thus the ON- and OFF-pathways would be differentiated not only by the sign of their response, but also by the sum of their spatial, temporal, spectral and adaptive characteristics (Levick & Thibos, 1983). It is possible that the nature and variety of the intrinsic retinal circuits can be elicited using existing technology, providing a multidisciplinary approach is adopted that uses the many tools of modern neuroscience in combination.

Acknowledgements

I thank Anthea Hills for typing the paper and H. Kolb, W. R. Levick, C. W. Oyster, P. Sterling, H. Wässle and their colleagues for access to unpublished manuscripts.

References

Amthor, F. R., Oyster, C. W. & Takahashi, E. S. (1984) Morphology of on–off direction-selective ganglion cells in the rabbit retina. *Brain Res.*, **298**, 187–90

Ariel, M. & Daw, N. W. (1982a) Effects of cholinergic drugs on receptive field properties of rabbit retinal ganglion cells. *J. Physiol.*, **324**, 135–60

Ariel, M. & Daw, N. W. (1982b) Pharmacological analysis of directionally sensitive rabbit retinal ganglion cells. *J. Physiol.*, **324**, 161–85

Barlow, H. B. (1961) Possible principles underlying the transformations of sensory messages. In *Sensory communication*, ed. W. Rosenblith, pp. 217–34. Cambridge, Mass.: M.I.T. Press

Barlow, H. B., Hill, R. M. & Levick, W. R. (1964) Retinal ganglion cells responding selectively to direction and speed of image motion in the rabbit. *J. Physiol.*, **173**, 377–407

Barlow, H. B. & Levick, W. R. (1965) The mechanism of directionally selective units in rabbit's retina. *J. Physiol*, **178**, 477–504

Bloomfield, S. A. (1981) *A physiological–morphological study of neuronal pathways in the rabbit retina.* Ph.D. Thesis, Washington University, St. Louis

Boycott, B. B. & Dowling, J. E. (1969) Organization of the primate retina: light microscopy. *Phil. Trans. R. Soc. Lond., B.*, **255**, 109–76

Boycott, B. B., Dowling, J. E., Fisher, S. K., Kolb. H. & Laties, A. M. (1975) Interplexiform cells of the mammalian retina and their comparison with catecholamine-containing retinal cells. *Proc. R. Soc. Lond., B.*, **191**, 353–68

Boycott, B. B. & Kolb, H. (1973) The connections between bipolar cells and photoreceptors in the retina of the domestic cat. *J. Comp. Neurol.*, **148**, 91–114

Boycott, B. B. & Wässle, H. (1974) The morphological types of ganglion cells of the domestic cat's retina. *J. Physiol.*, **240**, 397–419

Brandon, C., Lam, D. M. K. & Wu, J. Y. (1979) The gamma-aminobutyric acid system in rabbit retina: localisation by immunocytochemistry and autoradiography. *Proc. Natl Acad. Sci. USA*, **76**, 3557–61

Brecha, N. C., Oyster, C. W. & Takahashi, E. S. (1984) Identification and characterization of tyrosine hydroxylase immunoreactive amacrine cells. *Invest. Ophthalmol.*, **25**, 66–70

Caldwell, J. H. & Daw, N. W. (1978a) New properties of rabbit retinal ganglion cells. *J. Physiol.*, **276**, 257–76

Caldwell, J. H. & Daw, N. W. (1978b) Effects of picrotoxin and strychnine on rabbit retinal ganglion cells: changes in centre surround receptive fields. *J. Physiol.*, **276**, 299–310

Caldwell, J. H., Daw, N. W. & Wyatt, H. J. (1978) Effects of picrotoxin and strychnine on rabbit retinal ganglion cells: lateral interactions for cells with more complex receptive fields. *J. Physiol.*, **276**,. 277–98

Cajal, S. R. (1892) La retiné des vertébrés. *La Cellule*, **9**, 17–257

Cleland, B. G., Harding, T. H. & Tulunay-Keesey, U. (1979) Visual resolution and receptive field size: examination of two kinds of cat retinal ganglion cell. *Science*, **205**, 1015–17

Cleland, B. G. & Levick, W. R. (1974a) Brisk and sluggish concentrically organized ganglion cells in the cat's retina. *J. Physiol.*, **240**, 421–56

Cleland, B. G. & Levick, W. R. (1974b) Properties of rarely encountered types of ganglion cells in the cat's retina and an overall classification. *J. Physiol.*, **240**, 457–92

Cleland, B. G., Levick, W. R. & Wässle, H. (1975) Physiological identification of a morphological class of cat retinal ganglion cells. *J. Physiol.*, **248**, 151–71

Daucheux, R. F. (1977) *A physiological study of ontological formation of synaptic interactions in the rabbit retina.* Ph.D. Thesis, S.U.N.Y., Buffalo

Dowling, J. E. (1968) Synaptic organization of the frog retina: an electron microscopic analysis comparing the retinas of frogs and primates. *Proc. R. Soc. Lond., B.,* **170**, 205–28

Dowling, J. E. (1970) Organization of vertebrate retinas. *Invest. Ophthalmol.,* **9**, 655–80

Dowling, J. E. (1979) Information processing by local circuits: the vertebrate retina as a model system. In: *The neurosciences, Fourth study program,* ed. F. O. Schmitt & F. G. Worden, pp. 163–81. Cambridge, Mass.: M.I.T. Press

Dowling, J. E. & Boycott, B. B. (1966) Organization of the primate retina: electron microscopy. *Proc. R. Soc. Lond. B.,* **166**, 80–111

Dowling, J. E. & Ehinger, B. (1975) Synaptic organization of the amine-containing interplexiform cells of the goldfish and *Cebus* monkey retinas. *Science,* **188**, 270–3

Dowling, J. E. & Ehinger, B. (1978) Synaptic organization of the dopaminergic neurons in the rabbit retina. *J. Comp. Neurol.,* **180**, 203–20

Dubin, M. W. (1970) The inner plexiform layer of the vertebrate retina: a quantitative and comparative electron microscopic analysis. *J. Comp. Neurol.,* **140**, 479–506

Ehinger, B. (1976) Biogenic monoamines as transmitters in the retina. In *Transmitters in the visual process,* ed. S. L. Bonting, pp. 145–63. Oxford: Pergamon Press

Ehinger, B. (1982) Neurotransmitter systems in the retina. *Retina,* **2**, 305–21

Ehinger, B. & Falck, B. (1971) Autoradiography of some suspected neurotransmitter substances: GABA, glycine, glutamic acid, histamine, dopamine, and L-dopa. *Brain Res.,* **33**, 157–72

Ellias, S. A. and Stevens, J. K. (1980) The dendritic varicosity: a mechanism for electricity isolating the dendrites of cat retinal amacrine cells? *Brain Res.,* **196**, 365–72

Famiglietti, E. V. (1981*a*) Functional architecture of cone bipolar cells in mammalian retina. *Vision Res.,* **21**, 1559–63

Famiglietti, E. V. (1981*b*) Displaced amacrine cells of the retina. *Soc. Neurosci. Abstr.,* **7**, 620

Famiglietti, E. V. (1983*a*) 'Starburst' amacrine cells and cholinergic neurons: mirror-symmetric ON and OFF amacrine cells of rabbit retina. *Brain Res.,* **261**, 138–44

Famiglietti, E. V. (1983*b*) On and off pathways through amacrine cells in mammalian retina: the synaptic connections of 'starburst' amacrine cells. *Vision Res.,* **23**, 1265–79

Famiglietti, E. V. & Kolb, H. (1975) A bistratified amacrine cell and synaptic circuitry in the inner plexiform layer of the retina. *Brain Res.,* **84**, 293–300

Famiglietti, E. V. & Kolb, H. (1976) Structural basis for ON- and OFF-center responses in retinal ganglion cells. *Science,* **194**, 193–5

Freed, M. & Sterling, P. (1982) Amacrines reciprocal to the rod bipolar in cat retina are GABA-accumulating. *Soc. Neurosci. Abstr.,* **8**, 46

Freed, M. A. & Sterling, P. (1983) Spatial distribution of input from depolarizing cone bipolars to dendritic tree of on-center alpha ganglion cell. *Soc. Neurosci. Abstr.,* **9**,

Frishman, L. J. & Linsenmeier, R. A. (1982) Effects of picrotoxin and strychnine on non-linear responses of Y-type cat retinal ganglion cells. *J. Physiol.,* **324**, 347–63

Gouras, P. (1971) The function of the midget cell system in primate color vision. *Vision Res.,* Suppl. 3, 397–410

Hayden, S. A., Mills, J. W. & Masland, R. H. (1980) Acetylcholine synthesis by displaced amacrine cells. *Science,* **210**, 435–7

Holmgren, I. (1982) Synaptic organization of the dopaminergic neurons in the retina of the *Cynomolgus* monkey. *Invest. Ophthalmol.* **22**, 8–24

Holmgren-Taylor, I. (1982) Electron microscopical observations on the indoleamine-accumulating neurons and their synaptic connections in the retina of the cat. *J. Comp. Neurol.,* **208**, 144–56

Hughes, A. (1980) Directional units in the rat optic nerve. *Brain Res.,* **202**, 196–200

Hughes, A. (1981) Population magnitudes and distribution of the major modal classes of cat retinal ganglion cell as estimated from HRP filling and a systematic survey of the soma diameter spectra for classical neurones. *J. Comp. Neurol.,* **197**, 303–39

Hughes, A. & Vaney, D. I. (1980) Coronate cells: displaced amacrines of the rabbit retina? *J. Comp. Neurol.,* **189**, 169–89

Hughes, A. & Whitteridge, D. (1973) The receptive fields and topographical organization of goat retinal ganglion cells. *Vision Res.,* **13**, 1101–14

Hughes, A. & Wieniawa-Narkiewicz, E. (1980) A newly identified population of presumptive microneurones in the cat retinal ganglion cell layer. *Nature,* **284**, 468–70

Kamp, C. W. & Morgan, W. W. (1981) GABA antagonists enhance dopamine turnover in rat retina *in vivo. Eur. J. Pharmacol.,* **69**, 273–9

Karten, H. J. & Brecha, N. (1983) Localization of neuroactive substances in the vertebrate retina: evidence for lamination in the inner plexiform layer. *Vision Res.,* **23**, 1197–205

Kidd, M. (1962) Electron microscopy of the inner plexiform layer of the retina in the cat and the pigeon. *J. Anat. (Lond).,* **96**, 179–88

Kirby, A. W. & Enroth-Cugell, C. (1976) The involvement of gamma-aminobutyric acid in the organization of cat retinal ganglion cell receptive fields. *J. Gen. Physiol.,* **68**, 465–84

Kirby, A. W. & Schweitzer-Tong, D. E. (1981*a*) GABA-antagonists and spatial summation in Y-type cat retinal ganglion cells. *J. Physiol.,* **312**, 335–44

Kirby, A. W. & Schweitzer-Tong, D.E. (1981*b*) GABA-antagonists alter spatial summation in receptive field centres of rod- but not cone–driven cat retinal ganglion Y-cells. *J. Physiol.,* **320**, 303–8

Koch, C., Poggio, T. & Torre, V. (1982) Retinal ganglion cells: a functional interpretation of dendritic morphology. *Phil. Trans. R. Soc. Lond., B,* **298**, 227–64

Koch, C., Poggio, T. & Torre, V. (1983) Nonlinear interactions in a dendritic tree: localization, timing, and role in information processing. *Proc. Natl Acad. Sci. USA,* **80**, 2799–802

Kolb, H. (1977) The organization of the outer plexiform layer in the retina of the cat: electron microscopic observations. *J. Neurocytol.,* **6**, 131–53

Kolb, H. (1979) The inner plexiform layer in the retina of the cat: electron microscopic observations. *J. Neurocytol.,* **8**, 295–329

Kolb, H., Boycott, B. B. & Dowling, J. E. (1969) A second type of midget bipolar cell in the primate retina. *Phil. Trans. R. Soc. Lond.*, *B*, **255**, 177–84

Kolb, H. & Famiglietti, E. V. (1974) Rod and cone pathways in the inner plexiform layer of cat retina. *Science*, **186**, 47–9

Kolb, H. & Nelson, R. (1981) Amacrine cells of the cat retina. *Vision Res.*, **21**, 1625–33

Kolb, H. & Nelson, R. (1983) Rod pathways in the retina of the cat. *Vision Res.*, **23**, 301–12

Kolb, H. & Nelson, R. (1984) Neural architecture of the cat retina. In *Progress in retinal research, vol. 3*, ed. N. N. Osborne & G. J. Chader, pp. 21–60. Oxford: Pergamon Press

Kolb, H., Nelson, R. & Mariani, A. (1981) Amacrine cells, bipolar cells, and ganglion cells of the cat retina: a Golgi study. *Vision Res.*, **21**, 108i–114

Kolb, H. & West, R. W. (1977) Synaptic connections of the interplexiform cell in the retina of the cat. *J. Neurocytol.*, **6**, 155–70

Kuffler, S. W. (1953) Discharge patterns and functional organization of mammalian retina. *J. Neurophysiol.*, **16**, 37–68

Levick, W. R. (1967) Receptive fields and trigger features of ganglion cells in the visual streak of the rabbit's retina. *J. Physiol.*, **188**, 285–307

Levick, W. R. & Thibos, L. N. (1983) Receptive fields of cat ganglion cells: classification and construction. In *Progress in retinal research, vol. 2*, ed. N. N. Osborne & G. J. Chader, pp. 267–319. Oxford: Pergamon Press

McGeer, P. L., Eccles, J. C. & McGeer, E. G. (1978) *Molecular neurobiology of the mammalian brain*. New York: Plenum Press

McGuire, B. A., Stevens, J. K. & Sterling, P. (1984) Microcircuitry of bipolar cells in cat retina. *J. Neurosci.* **4**, 2920–38.

Mariani, A. P., Kolb. H. & Nelson, R. (1984) Dopamine-containing amacrine cells of rhesus monkey retina parallel rods in spatial distribution. *Brain. Res.*, **322**, 1–7

Marshburn, P. B. & Iuvone, P. M. (1981) The role of GABA in the regulation of the dopamine/tyrosine hydroxylase-containing neurons of the rat retina. *Brain Res.*, **214**, 335–47

Masland, R. H. (1982*a*) High resolution measurement of acetylcholine release by the rabbit retina. *Invest. Ophthalmol.*, Suppl. 22, 82

Masland, R. H. (1982*b*) Evidence for two kinds of release by an identified retinal neuron. *Soc. Neurosci. Abstr.*, **8**, 519

Masland, R. H. (1983) Direct fluorescent labeling of amacrine cell subpopulations. *Invest. Ophthalmol.*, Suppl. 24, 260

Masland, R. H. & Ames, A. (1976) Responses to acetylcholine of ganglion cells in an isolated mammalian retina. *J. Neurophysiol.*, **39**, 1220–35

Masland, R. H. & Mills, J. W. (1979) Autoradiographic identification of acetylcholine in the rabbit retina. *J. Cell Biol.*, **83**, 159–78

Massey, S. C. & Neal, M. J. (1979) The light evoked release of acetylcholine from the rabbit retina *in vivo* and its inhibition by gamma-aminobutyric acid. *J. Neurochem.*, **32**, 1327–9

Massey, S. C. & Redburn, D. A. (1982) A tonic gamma-aminobutyric acid-mediated inhibition of cholinergic amacrine cells in rabbit retina. *J. Neurosci.*, **2**, 1633–43

Michael, C. R. (1966) Receptive fields of directionally selective units in the optic nerve of the ground squirrel. *Science*, **152**, 1092–5

Miller, R. F. (1979) The neuronal basis of ganglion-cell receptive-field organization and the physiology of amacrine cells. In *The neurosciences, Fourth study program*, ed. F. O. Schmitt & F. G. Worden, pp. 227–45. Cambridge, Mass.: MIT Press

Nakamura, Y., McGuire, B. A. & Sterling, P. (1980) Interplexiform cell in cat retina: identification by uptake of gamma-[³H]aminobutyric acid and serial reconstruction. *Proc. Natl Acad. Sci. USA*, **77**, 658–61

Neal, M. J. (1983) Cholinergic mechanisms in the vertebrate retina. In *Progress in retinal research, vol. 2*, ed. N. N. Osborne & G. J. Chader, pp. 191–212. Oxford: Pergamon Press

Nelson, R. (1977) Cat cones have rod input: a comparison of the response properties of cones and horizontal cell bodies in the retina of the cat. *J. Comp. Neurol.*, **172**, 109–35

Nelson, R. (1982) AII amacrine cells quicken time course of rod signals in the cat retina. *J. Neurophysiol.*, **47**, 928–47

Nelson, R., Famiglietti, E. V. & Kolb, H. (1978) Intracellular staining reveals different levels of stratification for on- and off-center ganglion cells in cat retina. *J. Neurophysiol.*, **41**, 472–83

Nelson, R. & Kolb, H. (1983) Synaptic patterns and response properties of bipolar and ganglion cells in the cat retina. *Vision Res.*, **23**, 1183–95

Nelson, R., Kolb, H., Famiglietti, E. V. & Gouras, P. (1976) Neural responses in the rod and cone systems of the cat retina: intracellular records and procion stains. *Invest. Ophthalmol.*, **15**, 946–53

Nelson, R., Kolb, H., Robinson, M. M. & Mariani, A. P. (1981) Neural circuitry of the cat retina: cone pathways to ganglion cells. *Vision Res.*, **21**, 1527–36

Nichols, C. W. & Koelle, G. B. (1967) Acetylcholinesterase: method for demonstration in amacrine cells of rabbit retina. *Science*, **155**, 477–8

Peichl, L. & Wässle, H. (1979) Size, scatter and coverage of ganglion cell receptive field centres in the cat retina. *J. Physiol.*, **291**, 117–41

Peichl, L. & Wässle, H. (1983) The structural correlate of the receptive field centre of alpha ganglion cells in the cat retina. *J. Physiol.*, **341**, 309–24

Perry, V. H. (1979) The ganglion cell layer of the retina of the rat: a Golgi study. *Proc. R. Soc. Lond. B.* **204**, 363–75

Perry, V. H. (1981) Evidence for an amacrine cell system in the ganglion cell layer of the rat retina. *Neurosci.*, **6**, 931–44

Perry, V. H. & Walker, M. (1980) Amacrine cells, displaced amacrine cells and interplexiform cells in the retina of the rat. *Proc. R. Soc. Lond.*, *B*, **208**, 415–31

Poggio, T. & Torre, V. (1981) A theory of synaptic interactions. In *Theoretical approaches in neurobiology*, ed. W. Reichardt & T. Poggio, pp. 28–38. Cambridge, Mass.: M.I.T. Press

Polyak, S. L. (1941) *The retina*. Chicago: University Press

Pourcho, R. G. (1980) Uptake of [³H]glycine and [³H]GABA by amacrine cells in the cat retina. *Brain Res.*, **198**, 333–46

Pourcho, R. G. (1982) Dopaminergic amacrine cells in the cat retina. *Brain Res.*, **252**, 101–9

Pourcho, R. G. & Goebel, D. J. (1983) Neuronal subpopulations in cat retina which accumulate the GABA agonist, [³H]muscimol: a combined Golgi and autoradiographic study. *J. Comp. Neurol.*, **219**, 25–35

Raviola, E. & Gilula, N. B. (1975) Intramembrane organization of specialized contacts in the outer plexiform layer of the retina. A freeze fracture study in monkey and rabbit. *J. Cell Biol.*, **65**, 192–222

Raviola, G. & Raviola, E. (1967) Light and electron microscopic observations on the inner plexiform layer of the rabbit retina. *Am. J. Anat.*, **120**, 403–26

Ross, C. D. & McDougal, D. B. (1976) The distribution of choline acetyltransferase activity in vertebrate retina. *J. Neurochem.*, **26**, 521–6

Saito, H. A. (1983) Morphology of physiologically identified X-, Y-, and W-type retinal ganglion cells of the cat. *J. Comp. Neurol.*, **221**, 279–88

Steinberg, R. H., Reid, M. & Lacey. P. L. (1973) The distribution of rods and cones in the retina of the cat (*Felis domesticus*). *J. Comp. Neurol.*, **148**, 229–48

Sterling, P. (1983) Microcircuitry of the cat retina. *Ann. Rev. Neurosci.*, **6**, 149–85

Stevens, J. K., McGuire, B. A. & Sterling, P. (1980) Toward a functional architecture of the retina: serial reconstruction of adjacent ganglion cells. *Science*, **207**, 317–19

Stone, J. & Fukuda, Y. (1974) Properties of cat retinal ganglion cells: a comparison of W-cells with X- and Y-cells. *J. Neurophysiol.*, **37**, 722–48

Straschill, M. & Perwein, J. (1969) The inhibition of retinal ganglion cells by catecholamines and gamma-aminobutyric acid. *Pflügers Arch.*, **312**, 45–54

Teranishi, T., Negishi, K. & Kato, S. (1983) Dopamine modulates S-potential amplitude and dye-coupling between external horizontal cells in carp retina. *Nature*, **301**, 243–6

Thier, P. & Alder, V. (1984) Action of iontophoretically applied dopamine on cat retinal ganglion cells. *Brain Res.*, **292**, 109–21.

Törk, I. & Stone, J. (1979) Morphology of catecholamine-containing amacrine cells in the cat's retina, as seen in retinal whole mounts. *Brain Res.*, **169**, 261–73

Torre, V. & Poggio, T. (1978) A synaptic mechanism possibly underlying directional selectivity to motion. *Proc. R. Soc. Lond.*, B, **202**, 409–16

Vaney, D. I. (1980) A quantitative comparison between the ganglion cell populations and axonal outflows of the visual streak and periphery of the rabbit retina. *J. Comp. Neurol.*, **189**, 215–33

Vaney, D. I. (1984) 'Coronate' amacrine cells in the rabbit retina have the 'starburst' dendritic morphology. *Proc. R. Soc. Lond.*, B, **220**, 501–8

Vaney, D. I. (1985) The morphology and topographic distribution of AII amacrine cells in the cat retina. *Proc. R. Soc. Lond.*, B, **224**, 475–88.

Vaney, D. I., Levick, W. R. & Thibos, L. N. (1981a) Rabbit retinal ganglion cells: receptive field classification and axonal conduction properties. *Exp. Brain Res.*, **44**, 27–33

Vaney, D. I., Peichl, L. & Boycott, B. B. (1981b) Matching populations of amacrine cells in the inner nuclear and ganglion cell layers of the rabbit retina. *J. Comp. Neurol.*, **199**, 373–91

Vaughn, J. E., Famiglietti, E. V., Barber, R. P., Saito, K., Roberts, E. & Ribak, C. E. (1981) GABAergic amacrine cells in rat retina: immunocytochemical identification and synaptic connectivity. *J. Comp. Neurol.*, **197**, 113–27

Wässle, H., Boycott, B. B. & Illing, R. -B. (1981a) Morphology and mosaic of on- and off-beta cells in the cat retina and some functional considerations. *Proc. R. Soc. Lond.*, B, **212**, 177–95

Wässle, H., Peichl, L. & Boycott, B. B. (1981b) Morphology and topography of on- and off-alpha cells in the cat retina. *Proc. R. Soc. Lond.*, B, **212**, 157–75

Werblin, F. S. & Dowling, J. E. (1969) Organization of the retina of the mudpuppy, *Necturus maculosus*. II. Intracellular recording. *J. Neurophysiol.*, **32**, 339–55

West, R. W. (1976) Light and electron microscopy of the ground squirrel retina: functional considerations. *J. Comp. Neurol.*, **168**, 355–78

West, R. W. (1978) Bipolar and horizontal cells of the gray squirrel retina: Golgi morphology and receptor connections. *Vision Res.*, **18**, 129–36

Wyatt, H. J. & Daw, N. W. (1975) Directionally sensitive ganglion cells in the rabbit retina: specificity for stimulus direction, size, and speed. *J. Neurophysiol.*, **38**, 613–26

Wyatt, H. J. & Daw, N. W. (1976) Specific effects of neurotransmitter antagonists on ganglion cells in rabbit retina. *Science*, **191**, 204–5

5

The schematic eye comes of age

AUSTIN HUGHES

Thomas Young, 'On the mechanism of the eye'

The majority of visual physiologists are indifferent to schematic eyes. This attitude has its origins, I believe, not only in the inadequacies of past models but also in a very narrow conception of their potential role as prosaic tools rather than as a means of gaining insight into the comparative design strategies of a sophisticated organ. Surprisingly, as long ago as 1801, the broader problems of ocular organization were considered in that magnificent paper by Thomas Young, 'On the mechanism of the eye' (Young, 1801).

Young was concerned to justify teleologically the limited field of 'perfect vision' which extends in man for only a degree or two from the visual axis. In part, he attributed this falling off in performance to what he regarded as the unavoidable oblique optical aberrations. But, in addition, he regarded it as inconsistent with the 'economy of Nature' to endow a greater area of retina with increased sensibility. The optic nerve is already large, susceptible to injury and inflammation; it would be inadvisable to increase the sensible area of retina and so enlarge it. He regarded the disadvantage of this situation as small; motion of the eyes suffices to extend the field of acute vision to some 110 degrees.

However, given that the retina limits performance, he noted that it is,

> of such a form as to receive the most perfect image on every part of its surface, that the state of each refracted pencil will admit; and the

varying density of the crystalline renders that state more capable of delineating such a picture, than any other imagineable contrivance could have done.

Retinal magnification is noted as decreasing in the periphery. His conclusions were attained by ray tracing and estimation of the role of the crystalline lens in optimizing the form of the surface of minimum confusion.

Young's reflections delineated the major challenges to physiological optics. He clearly appreciated the distinction between the optimal image shell and the form of the retinal surface. Retinal image quality could be modulated by varying either.

1. He saw the retina as, in a sense, 'matched' to the optics in the degree of its sensibility.
2. Most importantly, he regarded the design considerations underlying both central and peripheral imagery as important.
3. He recognized the potential role of the gradient index lens in optimizing the match between retinal sensibility and optical quality in the peripheral field.

What a collection of richly absorbing problems arise just from these considerations! Given that there is optimization in ocular design, how is it achieved? How is normal vision maintained during growth of the eye without disturbing such a balance? What goes wrong in the development of refractive errors? Are the eyes of all vertebrate species built according to the same design?

However, the pace set by Young was not maintained. Human physiological optics became primarily concerned with normal and abnormal central vision. The human fovea is so situated as to permit its imagery to be treated as a problem in axial optics alone; quite apart from the intrinsic difficulties of their analysis, peripheral vision and optical performance were regarded as of little consequence. Such an attitude is still common. I have to hand the reports on my recent grant application for work on peripheral optics. It was the view of one referee that, a striking piece of evidence for the biological insignificance of peripheral image quality in man, is the minimal handicap experienced by diabetics who have had peripheral retinal ablation in both eyes. Tell that to the marines!

Comparative optics; a universal 'bauplan' for the mammalian optical system?

I was not indoctrinated with this anthropomorphic attitude because I was fortunate to begin my work in vision at Edinburgh under the supervision of David Whitteridge. His experience of a wide range of experimental species, his enthusiasm for the works of Johnson (1901), Walls (1942), Rochon-Duvigneaud (1943) and Duke-Elder (1958) and emphasis on adaptive function, ensured a sound comparative base to the thinking of those in his group. We could never have thought only in terms of human axial vision. This was relatively unusual at a time when cats were regarded by many as little people in furry suits.

It was not long before I was intrigued by Whitteridge's work on the factors which determine mapping of visual space into the central nervous system and had taken the visual system of the rabbit for my own. Thus, even before my introduction to schematic eyes, I was concerned with peripheral optics and had begun to wonder if the peripheral compression in cortical maps of the visual field was in part determined by a change in the magnification of the optical image.

Is there an obvious optical reason why such peripheral compression occurs and cannot readily be avoided? The ill-remembered remnants of Nelkon's sixth-form optics text indeed suggested to me that the effective focal length of a simple lens would fall off inversely with eccentricity from the optic axis. A consequence would be reduced magnification in the surface of circles of least confusion.

One concomitant of such simple optics is that image quality also reduces with eccentricity from the optic axis because oblique astigmatism increases the size of the blur circle, the minimum image of a point source. This design might be qualitatively well suited to man, whose visual acuity, ganglion cell and photoreceptor density also fall off inversely with eccentricity. Would similar optical properties obtain with the more complex optics of the real eye? I was not aware, at that time, of Young's sophisticated demonstration that it was plausible. Was such a design inevitable? Did a system like this represent a universal 'bauplan' for the mammalian optical system?

An obvious obstacle to the latter possibility was immediately apparent in the familiar rabbit eye. The central area in this animal is a band of high photoreceptor and ganglion cell density, the visual streak, which ensures 180 degrees of nearly uniform resolution along the image of the horizon in each eye (Hughes, 1971). If, in accord with a universal 'bauplan', optical quality falls off with increasing eccentricity in the eyes of all species and were to be just adequate to serve the central streak on-axis in the rabbit, then it must be less than adequate to serve the peripheral ends of the uniform streak. With such organization the optical quality would have to be much better than necessary on-axis, so that its fall off just brings it to adequacy on the peripheral streak.

Now it could no doubt be objected that the attainment of unneeded optical quality on the optical axis is not in keeping with the 'economy of Nature'. However, if only one design of optical system is available to the eye then it could be the only means of achieving the necessary quality of image over a wide angle.

A goat eye organized in this manner would be a great strain on credibility. This animal has an area of best vision second, among mammals, only to the primates but it is set at an angle of some 58 degrees to the optic axis (Hughes & Whitteridge, 1973). If optical quality falls off in similar fashion to man from the optical axis of the goat eye then for this peripheral *area* to have the necessary image quality (2 min resolution) the axial image must be 12 times better than that of man or six times better than that of an eagle.

A more attractive possibility is that the optical sophistication of the eye is such that the distribution of image quality can be modulated by the form of cornea, lens, index gradient and retinal shape in accord with the needs of each species. But how do we determine what is possible within the bounds of the physiological constraints?

It may seem extraordinary to many people that a sophisticated, comparative, wide-angle analysis of visual optics which can answer such questions is not a commonplace. The science of geometrical optics pre-dates Newton and, of all physiological systems, the optics of the eye would appear most readily susceptible to a conventional physical analysis. Indeed, quantitative studies of physiological optics have been made for hundreds of years; Newton himself established the cornea as elliptical in the 1660s. Limited quantitative schematic eyes date at least from Young (1801) and there was extensive work on this subject in the last century. However there are sound reasons why such an analysis has not been achievable until recently.

Information theory; the matching of optical and neural image quality

Before tackling the technical difficulties which have hindered advances in this area, it is worthwhile diverting to a modern consideration of the idea of 'matching' optics and retina which is implicit in Young's thoughts. He and the discussion above have only been concerned with ensuring that the optical image serving a given retinal location is of adequate quality for the local neural matrix. The possibility that optics might be detrimental if too good has not been introduced; such a limitation would require much more sophisticated ocular design than is commonly envisaged.

Sampling theorem and aliasing

Any 'image', whether optical or neural, can be treated as if comprised of the sum of a set of independent sinusoidal gratings with differing frequencies, phases and amplitudes; these are its Fourier components (Bracewell, 1965). If the highest Fourier component frequency in an image can be relayed by an information channel then the image can be transmitted without loss of information. Description of optical, retinal or visual system performance in terms of the highest frequency of grating which it can faithfully transmit is thus convenient because it defines the upper bound on the complexity and quality on image transmission through the system.

Helmholtz long ago pointed out that at least one relatively unstimulated sampling element must be interposed between each of those activated by bright bars if a sampling array is to resolve a grating pattern (von Helmholtz, 1856). A grating image so matched to the sampling array is said to be at the cut-off frequency of the array, f_c (Fig. 5.1) because it represents the highest frequency which the system can accurately resolve and transmit. A grating of lower frequency than this is oversampled but nevertheless generates a unique set of output values which define the original stimulus with no loss of information and permit its accurate reconstruction, $f_c/2$ (Fig. 5.1) A grating of higher frequency than f_c, however, is undersampled and does not generate a unique array output. The original stimulus thus cannot be reconstructed and masquerades as a spurious lower frequency signal. Corruption of information in this way is known as 'aliasing' and is readily understood in its simplest form by considering a grating of frequency $3f_c$ sampled by an array at the rate $2f_c$ (Fig. 5.1). The array output wrongly suggests the stimulus to be of the cut-off frequency, f_c, rather than $3f_c$ and, although its presence is detected, it is not resolved.

These ideas now have a formal basis in the sampling theorem (Cauchy, 1841; Nyquist, 1919; Gabor, 1946; Shannon & Weaver, 1949). Given an array of n elements per degree, the image Fourier component of highest frequency, f_c, which can be relayed by the array without loss of information is $f_c = n/2$. If a sampling array is to be free from spurious imagery then it must not be presented with significant signal above the array cut-off frequency, determined from its density.

Information theory thus suggests that the most economical transfer of optical information to the brain will occur when the cut-off frequency of the optical image on the retina is 'matched' to the cut-off-frequency of the corresponding 'cone array'. An image which is of poorer quality than the resolution of the sampling array is wasteful of sampling cells. An image which is

too good to be resolved by the sampling elements is 'costly' to organize and results in the masquerading of its high-frequency Fourier components as low frequencies in the array output. The potential for irreversible corruption after reconstitution is thus introduced.

The eye forms a communication channel in which visual information is fed from the optics to the retina and thence to the optic nerve. The continuous form of the retinal brightness distribution does not mean that it contains an infinite amount of information. A point source external to the eye forms an image of finite width, the pointspread or blurring function, which limits the highest grating frequency the optics can reproduce. Its properties are determined by diffraction and the various aberrations of the eye. All the information in the optical image may thus be acquired by the retina if the photoreceptor array can resolve the highest Fourier component transmitted by the optical elements of the eye, $f_c(o)$. This requires a photoreceptor frequency of $2f_c(o)$ and suggests the simplest matching strategy.

However, other schemes are possible. As the cut-off frequency of the optical system is approached, the transfer of contrast from the object to the image falls off rapidly (Campbell & Green, 1965). Contrast in the optical image at the sampling array cut-off frequency can be massively improved, with increased signal-to-noise ratio in the neural image, if the optical cut-off frequency is raised by only a factor of 2. Such undersampling of the optical image by the photoreceptor array inevitably contaminates its output with energy from spatial frequencies above the array cut-off frequency, so that any subsequent attempt at reconstitution must result in aliasing. It may be that the level of potential aliasing so introduced is tolerable in normal existence with substantial gain in performance at the array cut-off frequency. In such instances we would expect $f_c(\text{optics}) = n.f_c(\text{array})$ where n is a factor, which may vary from one part of the retina to another, defining the tolerance of potential aliasing. In areas of high resolution and cone density the finite width of the cone entrance apertures sets an upper bound on their modulation by high spatial frequencies which would protect the system from massive aliasing. This would provide no protection in areas with widely spaced cones.

Alternatively, aliasing is only a nuisance if the central visual mechanisms employ an interpolation between sampling points in some form of central reconstruction of the original brightness distribution in

Fig. 5.1. The nature of aliasing. Three sinusoidal grating patterns of differing frequencies are illustrated in cross-section on the left of the figure. One, of frequency f_c, is optimally matched to the sampling array symbolized by the row of arrowheads at frequency $2f_c$. The two values sampled by the array in each grating cycle suffice for reconstruction of the original signal, f_c, with no information loss, as demonstrated on the right-hand side. Frequencies, such as $f_c/2$, lower than the critical value, f_c, defined by sampling theory for the array are oversampled but reconstructable. Frequencies such as $3f_c$, which are higher than the critical value f_c, are undersampled and alias at lower frequencies, f_c in this instance, which becomes apparent when the output of the array is reconstituted. (Courtesy of *Experimental Brain Research*, Berlin: Springer.)

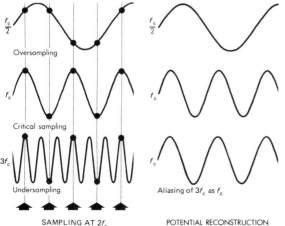

the image. In a system where such interpolation is not effected then it need not be detrimental. The eyes of fast-moving insects (Snyder, 1979) enjoy higher image contrast at the expense of tolerating aliasing because of its advantage in detection tasks. The same may be true of the cat alpha cell (Y,BT) system, which has been thought to be involved in a motion detecting system rather than interpolative pattern recognition.

If the information of the retinal image is acquired by a 'matched' array of photoreceptors without significant undersampling then there is of course no problem in organizing neural 'filters' to permit its processing through a variety of parallel channels (Bracewell, 1965). At subsequent levels we might then encounter coextensive systems with different functions comprising matched arrays in which the elements transmit signals closely related to the array cut-off and others which undersample. The pattern recognition system could be free of significant aliasing while other systems involved in different tasks would be precluded from receiving high spatial frequencies by an alternate route. For instance, the demonstrated behavioural grating resolution of the rabbit is substantially smaller than its optokinetic displacement threshold.

Our working hypothesis is thus that image quality will be 'matched' in some way to the local cone array at photopic levels; the array cut-off frequencies may be expected to vary with eccentricity in a similar fashion to image quality (Snyder, Bossomaier & Hughes, 1985). Such an arrangement permits any form of subsequent differential neural filtering.

Aliasing in the primate foveal region

It may be thought that the concept of aliasing is rather abstruse and unconvincing. Indeed, it has already been shown that its importance depends on what the system is designed to do with visual information. However, there have been recent demonstrations that it can occur and that the match between the optics and photoreceptor array is a consideration in ocular design.

Blue mechanism

De Monasterio, Schein & McCrane (1981) and McCrane *et al.* (1983) found that the blue cone array can be stained in a great variety of species by means of procion yellow and other dyes. The lattice is not perfect nor is it random. More important, however, is its relatively low density compared to that of the surrounding red and green cones in the perifoveal region. The human eye is known from psychophysical studies to have a correspondingly low resolving power for blue stimuli. Williams & Collier (1983) have demonstrated that, when the blue mechanism is isolated by viewing a violet grating on a yellow background, observers can resolve grating bars up to some 10 cycles/degree, yet detect the presence of field 'mottling' up to some 20 cycles/degree.

In the same region of retina the green mechanism has a capability of resolving gratings of up to more than 40 cycles/degree. Optical image quality should therefore be adequate to resolve the violet grating at frequencies above 10 cycles/degree. Note that the effects of chromatic aberration were eliminated in these experiments.

It was therefore concluded that the observers resolved the blue grating up to a spatial frequency of 10 cycles/degree, but for higher frequencies the blue cone array is too coarse to permit reconstitution so that aliasing occurs and is manifested psychophysically by a blotchy appearance of the grating. These experiments were carried out with natural optics.

It will be noted that under normal conditions the fovea focusses for the red and green content of the image and that the macular filtering reduces the blue content. The chromatic aberration of the eye would produce an out-of-focus blue image and reduce the susceptibility of the blue receptor array to aliasing.

Receptor array regularity

Yellot (1983) has argued that the receptor array in the primate retina is optimally organized to minimize the effects of aliasing. He assumes that, other than in the fovea, image quality is substantially above the resolving capability of the cone array. From array patterns determined on tangential sections through monkey outer segments he concludes that the degree of irregularity at each eccentricity is sufficient to prevent 'Moiré' aliasing in which, say, an 80 cycles/degree grating on a cone array capable of 60 cycles/degree resolution would produce a 20 cycles/degree subjective aliased grating. Rather, he claims, the irregular sampling would spread the aliasing energy over a broad band of frequencies so as to be less conspicuous.

Miller & Bernard (1983) have pointed out that the cone entrance aperture is the inner segment and that its diameter is such as to set an upper cut-off frequency of about 150 cycles/degree. It is inner segment regularity which must be considered because this is the region of the 'effective image shell' for receptor stimulation (Hughes, 1977b). Indeed, using Miller's preparation, Hirsch & Hylton (1984a) showed that the monkey fovea has a highly regular hexagonal array of inner segments with a positional correlation length of at least 130 receptors. It is also probable that the peripheral inner segment array is much more regular than suggested by Yellot's (1983) outer segment data. There is psychophysical evidence of a similar regular hexagonal array in man from the orientation dependence of visual hyperacuity tasks (Hirsch & Hylton, 1984b). It is readily apparent that central vision, at least, is not protected from 'Moiré' aliasing by irregular receptor arrays.

Both Miller & Bernard (1983) and Hirsch & Hylton (1984a) argue that formation of an image at above the array cut-off frequency in man should result in aliasing of both the Moiré and broad-band kind. Recently (Williams, 1984) has demonstrated that interference gratings, which 'short-circuit' the natural optics, are detectable up to some 150 cycles/degree in the human fovea and that low-frequency aliased patterns are quite visible under these conditions of viewing.

It is commonly assumed that the optics which serve the fovea are such as would preclude the presentation of such high spatial frequencies at the photoreceptor array. However, von Helmholtz (1856) and Byram (1944) described detection at high intensity and for stimuli of greater than 60 cycles/degree spatial frequency with natural optics.

Yellot (1984) now accepts these points but continues to regard non-foveolar vision as protected from aliasing by cone irregularity. The irregularity of the inner segment arrays still require validation and quantification. Indeed, Bossomaier, Snyder & Hughes (1985) argue that irregularity always has a penalty (French, Snyder & Stavenga, 1977) and that there would in general be an evolutionary drive towards regular sampling arrays. In information theory terms, the broad-band aliasing favoured by Yellot (1983) is just as detrimental as Moiré aliasing. To emphasize the Moiré effects is to assume a neural homunculus.

'Matching of optics and the receptor array'

The preceeding arguments have implicit the classic concept that optical image quality is predetermined in some way and that the receptor array is chosen to fit it. The investigations into optical imagery described below rather imply that optical imagery may equally vary in its intraocular distribution and could readily match itself to different retinal arrays. But how would the optical and neural 'images' to be related in quality?

Snyder *et al.*, (1985) argue from information theory and other models that, because of the decline in the modulation transfer function of the real optic system near to the cut-off frequency, it is best to have local image quality significantly better than the resolving power of the local receptor array. This ensures good contrast near the cut-off frequency of the array and reduces susceptibility to noise.

The model predicts the real situation in a variety of species. The major problem with current data lies in the interpretation of the organization of the human fovea where the optics and receptor array appear to be closely matched in resolving power (Green, 1970). It is possible that the optical quality is held down to avoid aliasing in red and green resolution systems. If the optics are bypassed then aliasing can readily be demonstrated with interference gratings exceeding the array cut-off frequency (Williams, 1984).

In humans, some recent results suggest that the peripheral retina does undersample information in the image (Levi & Klein, 1985). Studies of peripheral perception now suggest that aliasing occurs and is perceptible (Thibos & Walsh, 1985) substantiating the argument above that image quality there exceeds the array resolution.

The all-cone retina of the garter snake fits the theoretical expectations well. Its image cut-off frequency is some three times greater than the cut-off of the cone array. In general, it is expected that optical quality would be about twice as good as the receptor array. This is well established to be a feature in the organization of invertebrate eyes (Wehner, 1981).

Significance of the marginal aberrations of the eye

In certain species, such as the rat (Hughes, 1979a,b), the expansion of the pupil at lower light intensities is associated with the introduction of major optical aberrations and a reduction of image quality. At low light intensities quantum limitations and the finite integration time of the eye (Rose, 1942) set bounds on physically achievable spatial resolution. It is feasible that the aberrations are specifically introduced to degrade the image in order to maximize matching of the optics to the resolution system under these conditions.

Observation of receptors in the living eye

Another immediate implication of our argument (Snyder *et al.*, 1985) is based on the optical principle of reciprocity. If a grating of twice the receptor array cut-off frequency can be imaged in the eye, then the receptors themselves, if contrast is high enough, should be visible from without the eye and through the natural optics.

One approach is to employ transretinal illumination to cause the receptors to act as waveguides in directing light to the exit pupil. Jagger (1985) has used this method to demonstrate that adjacent photo-

Fig. 5.2. (A) The first photograph of vertebrate photoreceptors visualized directly through the optics of the living eye – that of a garter snake. (Courtesy of Mike Land and Allan Snyder.) (B) A light microphotograph of cone inner segments in the retina of a garter snake. (Courtesy of Miss Rachel Wong.)

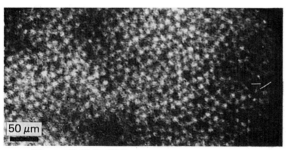

50 μm

A

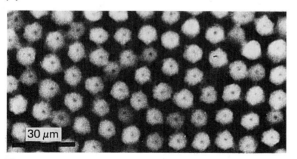

30 μm

B

receptors can be resolved and that the optics of the toad are better than equal to the resolution of the retina.

Land & Snyder (1985) employed a simple ophthalmoscopic technique to obtain the first photograph of the photoreceptor array in the living vertebrate eye, Fig. 5.2*a*. This work has interesting implications for potential developments in clinical ophthalmoscopy.

Organization of the cat retina

The concepts of sampling may also be applied at the ganglion cell level, albeit less readily. If the photoreceptor array is matched to optics, it would be possible for some ganglion cells to reveal matched, and others unmatched, properties depending on the nature of their connections. If the photoreceptors are not matched to the optics then all ganglion cells are precluded from a matched performance. Once aliasing enters the system it cannot be eliminated.

The organization of separate channels with different sampling strategies is best understood in the cat retina (Hughes, 1981*a*) which has many physiological classes of ganglion cell. These grossly comprise the BT (brisk-transient), BS (brisk-sustained), sluggish and rarely encountered units (Cleland & Levick, 1974), comparable to the X, Y and W cells of Stone & Fukuda (1974). The equivalent anatomical classes are now unequivocally demonstrated (Saito, 1983) to be represented by the alpha, beta and gamma cells of Boycott & Wässle (1974).

Given density estimates for the anatomical cell classes (Hughes, 1981*b*), it is then possible to compare the actual physiological cut-off frequency of the corresponding single unit classes to grating stimuli (Cleland, Harding & Tulunay-Keesey, 1979) with their theoretical array cut-off frequencies, calculated from Shannon's Sampling Theorem (Shannon & Weaver, 1949) in the manner described above.

Unfortunately, these cells are rectifiers and cannot be regarded as simple apertures. It is necessary to decide whether ON and OFF cells are to be regarded as independent samplers. The calculation may assume that each independent sampler consists of one ON and one OFF cell or it can be assumed that the two classes form a single array which operates according to a more sophisticated sampling theorem (see Bracewell (1965) for derivative sampling). It is also necessary to assess the regularity of the array. The topic is discussed in much more detail by Hughes (1985).

Comparison of the BT (Y) cut-off frequency with

the equivalent alpha cell density profile over a range of retinal eccentricity (Hughes, 1981*a*) shows that the cut-off frequency of the cellular array is much lower than the cut-off frequency for signals passing through individual cells over the entire range of studied retinal eccentricity and regardless of the assumptions made for the calculation. Undersampling and aliasing must therefore result upon reconstitution; however, these cells are commonly regarded as more likely to be involved in the detection of fast movements.

By contrast, Hughes (1981*a*) shows that the physiological cut-off spatial frequency of BS (X) cells could be consistent, at all retinal locations tested, with the sampling matrix cut-off frequency as calculated from the corresponding beta cell array. This class of cell is thought, for a variety of reasons, to be the substrate of the grating resolution and pattern recognition mechanisms. It is the class which, *a priori*, might be expected to avoid aliasing and ensure the possibility of reconstruction by interpolation.

The potential for matching thus exists in the BS system. If the array were hexagonal and the ON and OFF cells are totally independent samplers, then 1:1 matching occur between the channels and the array. Alternatively, if one ON and one OFF cell is required for each independent channel then some aliasing must occur because the array cut-off would be lower than that of the individual cells.

It has been noted above that some degree of aliasing may be a normal feature of such systems (Snyder *et al.*, 1985) in order to ensure maximum information capture. We cannot yet tell whether the array is tolerant of some aliasing or not. More accurate information is necessary before the actual organization of the cat area centralis can be determined, however, the available results indicate its performance to be within the bounds set by information theory.

The failure of cat peripheral BS (X) units to respond to spatial frequencies above their calculated density matrix cut-off frequency, suggests that either the optical image cut-off frequency is matched to this array of ganglion cells or, if higher, does not greatly exceed that of the local cone array. If it did, then ineradicable aliasing would be introduced at the receptor level and BS (X) responses would be reported for spatial frequencies above their array cut-off frequency. This argument is particularly interesting in the light of Robson & Enroth-Cugell's (1978) claim that the cat optical image contains frequencies of up to

20–30 cycles/degree. The central retinal image could thus be optimally matched to the 20 cycles/degree cut-off of the cone array (Steinberg, Read & Lacey, 1973; Hughes, 1977*b*) but not to the anatomical beta-cell array which cannot relay more than 10 cycles/degree. Neural filters could match the minimally aliased output of the photoreceptors to the ganglion cells without penalty in the form of information loss by further aliasing. In man there are also indications that the foveal optical image quality does not massively exceed the photoreceptor array cut-off frequency (Green, 1970; Williams, 1984).

Implications for behavioural acuity performance

It follows from the above considerations that we may expect aliasing to be behaviourally demonstrable in some species; it has long been known in invertebrates (Goetz, 1970). Detection and resolution of gratings must therefore be distinguished in behavioural grating resolution tests. It is not enough to require discrimination of a grating from an intensity matched grey field. It is minimally essential to require discrimination between gratings at different orientations and the same spatial frequency and should ideally incorporate a measure of absolute frequency perception.

In principle, there are numerous further complications because sensitivity at different orientations may differ for optical or neural reasons or aliased images might guide such discriminations. However, there is not space to enter into discussion of these matters.

The need for sophisticated considerations is, however, emphasized by results from experiments with falconiform birds. It has been demonstrated that a falcon can detect gratings of 2.6 times the cut-off frequency for man (Fox, Lehmkuhle & Westendorf, 1976). Although the image quality of their eyes is high (Shlaer, 1972), their central photoreceptor density is very similar to that of man's fovea (Snyder & Miller, 1978) and could only perform 2.6 times better with conventional optics if the eye were 2.6 times longer; however, the experimental animal's eyes were of similar size to those of man.

It might be concluded from the above discussion that the birds were detecting the presence of an aliased grating rather than resolving it. Snyder & Miller (1978) put forward the alternate suggestion that the foveal pit forms a concave lens which turns the eagle eye into a

Galilean telescope with an effective focal length longer than the eye. The evidence in support of such an arrangement has been strengthened (Hirsch, 1982), although indirectly and not without question (Dvorak, Mark & Reymond, 1983; Hirsch, 1983).

Although unconfirmed, this theory is a warning to those who feel the vertebrate eye is of simple and self-evident design. The spectacular variety of invertebrate eyes described by Land (1981) may not be immediately apparent in the vertebrates but similar evolutionary processes have been operating on their eyes and appropriate analysis will indubitably reveal analogous elegance of adaptation.

The topography of ocular image quality

Within recent years it has become apparent that retinal topography is highly labile between species. In particular, the distribution of neurons in the retinal ganglion cell layer is varied and forms circular areas, streaks, anakatabatic regions and other specializations (Hughes, 1977*b*; 1981*c*) which have been related to the lifestyles of their owners. The topography of these maps can be quite complex. If the optical and neural image quality are locally related, does it follow that optical image quality must also achieve such complex distributions? It is hard to see how that might be done.

Of course, it could be suggested that the pattern of photoreceptor distribution tends to be more radially symmetrical, less variant between species and more readily matched by a standard distribution of retinal image quality. Neural filtering might then optimally tailor the topography of receptor plane 'image' quality to the more complex 'bespoke' form of that in the ganglion cell layer. However, it remains, as pointed out above, that the human and rabbit cone distributions alone differ sufficiently to require substantial differential specialization in the optical apparatus if image quality were to be matched to each of them in some constant fashion across the retina. Does the rabbit have an optical visual streak?

Unfortunately we do not know how the optical pointspread function varies across the rabbit eye, but Hughes & Vaney (1979) demonstrated that oblique astigmatism is absent at up to 80 degrees of horizontal eccentricity, as might be expected if retinal image quality remained high on the peripheral visual streak. We do know that the optics of the toad are quite uniform with eccentricity so that such organization is

achievable in an eye with a cornea and lens (Jagger, in press). In man, with his restricted area of high-resolution vision, oblique astigmatism is prominent at much smaller eccentricities (Le Grand, 1967). Such species differences are in accord with our expectations.

Systematic experimental investigation of off-axis image quality has primarily been concerned with the degree to which optical factors limit visual acuity rather than with the absolute quality of the imagery *per se*. Robson & Enroth-Cugell (1978) did not address the matter in the cat but the double pass method of Bonds (1974) suggested that image quality changed little at medium eccentricities. Although image quality may be quite closely matched to receptor density at the human fovea, we have no clear idea as to how it varies with eccentricity. Frisen & Glansholm (1975) suggest that peripheral optical factors limit visual acuity which implies substantial decline. Jennings & Charman (1981) report a small change in central to peripheral retinal image quality. However, they used a fully dilated pupil, introduced a spectacle lens in the path, employed a double pass technique and are of little help in discussion of diurnal optical quality. Recent psychophysical work does suggest that aliasing is perceptible in the periphery with natural optics (Thibos & Walsh, 1985) and that the image is therefore undersampled (Levi & Klein, 1985) and is superior to the retinal 'grain' in resolving power.

By now, it must be apparent that ocular simulation, a sophisticated schematic eye, offers a powerful means for comparative exploration of the theoretically indicated relations between optical and neural apparatus. It is not enough simply to expect the retinal image quality to exceed locally the resolution of the cone array. There is good reason to regard the properties of the optical image and the corresponding retinal array as optimized in a subtle and related fashion which has evolved in terms of the role of that region of visual field in the lifestyle of a given species. The nature of these optimizations has major implications with respect to the coding of visual information (Hughes, 1981*a*) and sets important bounds on the potential performance of the central visual system in such tasks as hyperacuity judgements (Westheimer, 1979; Geisler, 1984). This requires consideration of the entire image space of the eye in the context of lifestyle and retinal organization. The route for achieving this is not an easy one.

Optics

If we are to consider any of the fascinating optical problems which abound in visual physiology then we need some of the basic theory of geometrical optics. There are three approaches to geometrical optics; simple and approximate, extraordinarily complicated and less approximate, or amazingly tedious, simple and completely accurate. These represent the analytical first-order, third-order and ray-tracing approaches, respectively. But the mathematically naïve physiologist need not fear; understanding of past work in physiological optics mainly requires simple first-order approximations. Future work will be done by ray-tracing, but digital computers will eliminate the tedium. The third-order analytical treatment has not been employed for schematic eye work.

There are several derivations for the simple optical equations. The geometrical ray diagrams of school (Fincham, 1959; Davson, 1962) do not generate functional insight; the wavefront method and vergences may be unfamiliar (Fincham, 1959; Davson, 1962). Instead we approach by Fermat's principle (Feynman, Leighton & Sands, 1966). Enunciated in 1650, it states that 'light takes the path which requires the least time'.

Consider a single refracting interface between air and glass (Fig. 5.3). If this surface is to form a perfect image of a point source O on the optic axis at a corresponding point O′ in image space, then rays from O along all paths through the surface must be brought to O′. For this to happen, according to Fermat's principle, the light must take the same time for each route.

The velocity of light varies inversely with the refractive index of the medium it is passing through. By making the glass surface convex to the air, the proportion of the path in the medium of high refractive index is increased and the time of axial passage from O to O′ increased. The path OPO′ is obviously longer than the path OQO′ but the surface shape may be so adjusted that the time for the passage OQO′ is matched to that for OPO′ and to paths involving other incident heights.

The surface required for perfect imagery at large apertures is aspherical. However, it often suffices to treat rays close to the optic axis so that only small angles are involved. The necessary mathematics is simplified [Note 1] by the assumption that radian angle $i = \sin i$. The required form of the refracting surface is then found to be spherical [Note 2] and its properties are described by the *fundamental paraxial equation*

$$1/f = n'/f' = (n'-n)/r = F \text{ (in dioptres) (5.1)}$$

In practice, rays incident at angles of up to about 10 degrees are satisfactorily treated by this approximation, which results in what is known as *paraxial optics* and forms the substrate for most of the eye-modelling literature currently available.

This simple equation (5.1) can already be employed to model an aphakic eye, one from which the lens has been removed after injury or the development of a cataract. The radius of curvature of the human cornea is 7.7 mm (Gullstrand, 1924); neglecting the thin cornea we take the refractive index as that of aqueous and vitreous at 1.336 whence, from eqn 5.1 above, the power of the surface, F, is 43.64 D and the posterior focal length is $1.336/43.64 = 30.61$ mm. The retinal photoreceptors lie about 24 mm behind the vertex of the cornea; parallel incoming rays would have to be brought to a focus at this point for normal vision, which indicates a required power of $1.336/0.024 = 55.67$ D, some 12.03 D more than that available with the cornea alone. A 12.03 D contact lens would make the eye emmetropic.

But the normal eye is made up of a sequence of refracting surfaces separated by different media and distances. Gauss (1841) was the first to employ the paraxial optics to enable the performance of a sequence of surfaces arranged on a common optic axis to be

Fig. 5.3. A single refracting surface between air and glass. If this surface is to form a perfect image, O′, of a point source O on the optic axis, then rays from O along all paths through the surface must be brought to O′. According to Fermat's Principle, for this to happen the light must take the same time to travel along each path. The path OQO′ is obviously shorter than OPO′ but by making the surface convex it can be arranged that the proportion of time spent in glass on OQO′ is longer than on OPO′ so as to increase the passage time to match that for OPO′. The necessary surface is aspherical.

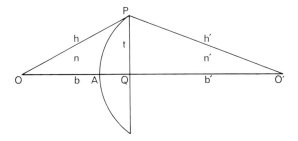

expressed in unified fashion by employing two pairs of cardinal points, the focal and principal points, along with the refractive index of the first and last media to describe the performance of the entire system.

Consider any optical system consisting of a series of refracting surfaces. An axial distant object will have its image, of given size, formed at the axial focal point of the system, F'. The Gaussian trick is to place a thin lens on the optic axis at a position H', which will form an image at F' which is the same size as that produced by the original system. The distance H'F' is known as the *equivalent focal length* of the original system and the position H' is its second principal point. The same considerations for rays passing in the opposite direction through the system defines the primary focal point, F, and the primary focal length HF by employing the same thin lens now situated at the primary principal point, H. A thin lens in one position cannot simulate the original system. For incident rays, a system of lenses is thus regarded as an equivalent thin lens situated at the first encountered principal point and for emergent rays like the same thin lens placed at the subsequent principal point.

It is the cardinal points of the equivalent thin lens which describe the behaviour of the original system and these can be substituted in the usual thin lens formulae to study the system's behaviour.

If the refractive index of the first and last media of the system is not the same, as in the eye, then a ray which enters the first principal point does not leave the second principal point at the same angle to the optic axis. Under these conditions that property belongs to an additional pair of nodal points whose important physiological role was emphasized by Listing (1845). The separation of the posterior nodal point from the posterior focal point, the posterior nodal distance (PND), defines both image scale and curvature of the retinal image.

The derivation of cardinal points from anatomical knowledge of the eye was first carried out by Moser (1844) but with erroneous results. These were corrected by Listing (1845) whose model eye was adopted by von Helmholtz. The procedure of derivation is tedious but straightforward and consists of deriving an equivalent lens for the cornea surfaces, an equivalent lens for the two lens surfaces and then combining them to produce an equivalent lens for the optical system of the whole eye [Note 3].

This approach has obvious limitations even when carried out in accord with the assumptions made. For instance, it deals only with paraxial ray bundles so cannot treat an eye with a finite pupil. It deals only with axial imagery and so can tell us nothing about what happens in even slightly displaced regions of the peripheral image and finally it makes no provision for aspheric surfaces or non-homogeneous media.

Off-axis imagery

You might ask whether it is possible to extend some form of simplified analytical treatment to marginal rays, i.e. finite apertures, and to oblique imagery, if not at the level of accuracy that permits quantitative prediction, at least adequate to permit intuitive grasp of the possibilities open in the evolution of ocular performance. After all, it is inconvenient to remain incapable of even generalized quantitative statements about ecological optics. Unfortunately all the useful analytic solutions are exceeding complex.

Between 1852 and 1856, Seidel (1856) published a series of papers in the *Astronomische Nachrichten* in which he retained terms in the expansion of the sine and cosine expressions of eqns iii and iv [Note 1] up to the third order. By this means he introduced the effects of aperture and field of view into consideration but the approach is still valid only for a vanishingly small region near the optic axis; its extension may lead to great inaccuracy as the higher-order terms become more significant.

Oblique imagery

It will now be apparent that, within the context of standard paraxial optics, back in 1965, I was not going to obtain a model of the overall workings of the optical system of the eye. It was, of course, only ignorance which had led to that expectation; Seidel's extension to third-order treatment permits a little obliquity and marginality to be considered but even now I cannot follow his treatment without a feeling of being hopelessly out of my depth.

How does the physiologist get some insight into what the optics can do in the periphery? The answer lies in ray tracing, but this will not give a handy analytical expression.

The simplest intuitive model for the ocular off-axis image surface is to consider a centrally stopped thin lens with rays passing through it obliquely from a source at a great distance. Such oblique rays have angles of incidence which differ in the horizontal and vertical planes so that a ray bundle comes to two separated line foci, rather than to a point focus, for an object point anywhere other than on the optic axis. These are known as the meridional, or tangential, and sagittal images (the meridional plane contains the optic axis and the off-axis image point; the sagittal plane is perpendicular to it). Between these two foci the image forms a circle of minimum diameter, the circle of least confusion. This aberration is called oblique astigmatism.

As a first approximation, therefore, we might expect the eye to evolve with the retinal photoreceptor entrance apertures in the surface formed by the circles of least confusion. The form of the surfaces containing the meridional and sagittal line images for rays of increasing obliquity is defined by the Coddington equations for a thin lens with a stop. For an object at infinity and unit radius of curvature of the lens,

$$V_m = 1/\{(2/\cos i)\{(n.\cos i'/\cos i)-1\}\} \quad (5.2)$$

$$V_s = 1/2 . \cos i . \{(n.\cos i'/\cos i)-1\} \quad (5.3)$$

where V_m and V_s are the distances from the optical centre of the lens to the meridional and sagittal foci, respectively, and n is the refractive index of the lens (Strong, 1958). The resultant form of the astigmatic and best focus image surfaces is shown in Fig. 5.4.

The increasing separation of the meridional and sagittal image surfaces with increasing eccentricity

Fig. 5.4. Meridional and sagittal image surfaces for a thin lens, demonstrating their increasing separation with increasing eccentricity of the source image at infinity. The surface of least confusion lies between these two surfaces and the magnitude of the circle of least confusion must increase with eccentricity.

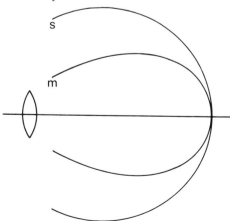

means that the minimum size of the circle of least confusion would increase and optical resolution would fall off radially with increasing eccentricity. This resembles the arrangement in the human eye. But is it of physiological significance?

The presence of two optical elements, a cornea and a lens, the use of aspherical surfaces and the introduction of a gradient of refractive index through the crystalline lens means that the optimal image surface of a real eye could be of almost any arbitrary form. Only a very sophisticated design results in its apparent similarity to the performance of the simple thin lens.

Imagine trying to obtain a simple model of the observed properties of rabbit optics. Image quality would need to be identical for rays entering at any eccentricity from the optical axis; oblique astigmatism must be eliminated. A satisfactory model is beyond the capabilities of the thin lens and paraxial optics.

Gaussian paraxial schematic eyes

The homogeneous model lens

It is a major assumption of the Gaussian treatment that the media composing the system are homogeneous. In the case of the crystalline lens this is patently untrue. As long ago as 1759, in his *Treatise on the Eye*, Porterfield (1759) discussed in detail the difference in optical density of the cortex and core of the crystalline lens and correctly connected its organization with the potential of reducing spherical aberration. Young (1801) also regarded the gradient of refractive index through the lens as playing a role in matching the image plane to the locus of the peripheral retina.

The inventor of the kaleidoscope, David Brewster, used polarized light to demonstrate the variation of optical density in the intact crystalline lenses of a variety of species (Brewster, 1816). The form of the axial distribution of refractive index was concluded to be approximately parabolic by Mathiessen (1877).

In order to accommodate a treatment of physiological optics within the framework of the Gaussian system, a common procedure from the time of Listing has been to consider the index gradient of the crystalline lens to be replaced by a homogeneous lens of the same shape and equivalent paraxial back vertex power (the distance from the second principal focus to the rear vertex of the lens – employed because the location of the second principle point of the thick lens cannot be observed directly). This can be achieved by setting the refractive index of the homogeneous lens at an appropriate value unrelated to any found in the real lens. This fictitious refractive index is called the 'total' index of the lens (Senff, 1846).

The procedure for calculating the total index is tedious if a programmable calculator is not employed. The crystalline lens is supported in saline and a narrow beam of parallel light arranged to pass along its optic axis. Either a travelling microscope or photography is employed to determine the distance from the back of the lens to the paraxial focus of the narrow beam. Using the equations ix to xi [Note 3] and the known front and rear radii of curvature of the lens, it is possible to employ repetitive approximation to determine the value of total index which, substituted into eqn 5.4 (Davson, 1962), gives rise to a value for the back vertex power, V_p, equal to that measured

$$V_p = F_{1,2}/\{1-(A_1A_2/T)F_2\} \qquad (5.4)$$

where A_1A_2 is the thickness of the lens, T is the total index, $F_{1,2}$ is the power of the lens and F_2 the power of its second surface.

It is most important that those who employ paraxial schematic eyes as a tool, but are unfamiliar with their derivation, realize that the total index, T, is a fiction. It is not related to the real refractive indices of the lens and must be calculated for each lens considered. Massof & Chang (1972) invalidly revised Block's rat schematic eye because of a misunderstanding as to the nature of the total index and in refereeing it is common to find errors in this matter.

This procedure provides only a paraxial fit for the lens. It is thus essential that the light beam employed to measure the back vertex distance is very narrow and axial if the resultant model is to represent the paraxial state of the eye with any accuracy. The homogeneous lens cannot replicate the performance of the crystalline lens for rays significantly displaced from the optic axis. Thus, the common schematic eye is limited not only because of its paraxial assumptions but also by its inability to incorporate the performance of the crystalline lens operating at finite apertures.

The core model lens

Gullstrand (1908; see also von Helmholtz, 1856) noted that the principal points of a homogeneous lens model are separated by a greater distance than those of

the corresponding crystalline lens so that the power of the model is greater. In large eyes, such as that of the cat, the resultant error in the refractive state of the eye is small, some 0.5 D (Vakkur & Bishop, 1963). But in a small eye an error of up to 4 D may be introduced by this means. Hughes (1979b) attempted to reduce this effect by substituting a two-shell model for the rat crystalline lens. The core of this model was intended to approximate more closely the real spheroidal core of high refractive index which can be extracted from the rat crystalline lens.

Gullstrand (1908; 1924) employed an alternative model, the *equivalent core lens*. To derive this he approximated the human axial refractive index distribution by a series expansion. After indirect determination of the constants, the power and principal points of the lens were derived by integration. These values were then employed to determine a unique model containing a core of the same refractive index as the lens centre suspended in a cortex of refractive index the same as that at the lens surface. Gullstrand claims for this equivalent core model that, 'so far as the imagery laws of the first order are concerned the entire lens system then has precisely the optical characteristics of the actual crystalline lens.' The model thus remains a paraxial approximation, although its principal points occupy positions exactly corresponding to those of the crystalline lens. Its core is a fictitious entity and has no relation to the anatomical core demonstrable in many vertebrate lenses. This approach has not been adopted by other workers.

Conventional schematic eyes

If paraxial optics are so limited in explanatory power, why bother constructing conventional schematic eyes which are based on them? There is a variety of reasons. According to Vakkur & Bishop (1963), 'A schematic eye is a self-consistent mathematical model of the optical system of the average eye. Its development in man forms the science of physiological optics, and the design of spectacles and the use of optical instruments in ophthalmology have largely depended upon it.'

The efficacy of the paraxial model for such purposes is related to the small displacement of the human fovea from the optic axis. Listing (1845) was the first to produce a human schematic eye by means of the Gaussian approximations. The approach was developed by von Helmholtz (1856) and culminated in Gullstrand's schematic eye (Gullstrand, 1908; 1924) which is in almost universal use today for computations on human optics (Duke-Elder, 1958; Fincham, 1959; Davson, 1962).

After the early work of Mathiessen (1877–1893) and Hirschberg (1882), very little quantitative comparative optical research took place until recently. Clearly, the inadequacies of the Gaussian approach for dealing with the fundamental problems of wide-angle ocular design and with non-homogeneous lenses, inhibited the basic interest in development of paraxial eye models. Nor was there a practical call for means of calculating dog and cat spectacles in everyday life.

With the development of extensive investigation of the retina and central visual system the situation has changed. Comparative neuro-ophthalmology in effect began with the seminal papers on cat optics by Vakkur & Bishop (1963) and Vakkur, Bishop & Kozak (1963) which were devoted to the practical needs of providing a quantitative optical framework for neurophysiological studies.

As well as the schematic eye for the cat, we now also have versions for monkey (Vakkur, 1967), rabbit (Hughes, 1972), rat (Hermann, 1958; Block, 1969; Massof & Chang, 1972; Hughes, 1979b; Campbell & Hughes, 1981; Chaudhuri, Hallett & Parker, 1983), mouse (Remtulla, Hallett & Chaudhuri, 1984), bat (Suthers & Wallis, 1970), oppossum (Oswaldo-Cruz, Hokoc & Sousa, 1979), pigeon (Marshall, Mellerio & Palmer, 1973), owl (Martin, 1982), frog (DuPont & de Groot, 1976), gecko and iguana (Citron & Pinto, 1973), goldfish (Charman & Tucker, 1973) and lemon shark (Hueter, 1980).

It is, of course, much more straightforward in principle to determine directly the cardinal points of the equivalent system than to calculate them by means of the conventional schematic eye. From a knowledge of the refractive error of the eye, the refractive index of the vitreous and the posterior nodal distance, it is easy to obtain the posterior cardinal points of the system. The PND is obtained either by measurements of the transclerally viewed image of an object of known angular dimensions (Vakkur *et al.*, 1963) or from the angular dimensions of fundus features of known size (Hughes, 1976).

This approach was employed by Vakkur & Bishop (1963) for one treatment of the cat eye. But the general schematic eye paper contains more than just the development of cardinal points. It is a source for a complete metrological treatment of the globe and its components, their average values and change with weight or age, measurements of the visual fields of the eye, assessment of the projections of visual landmarks into visual space, and so on. The majority of workers thus compute the paraxial eye from tabulated average measurements of surface position, curvature and refractive index, employing the direct assessment of the cardinal points as a confirmation for the calculation rather than as prime data.

The information in published schematic eyes is thus far more extensive and useful than that of the equivalent paraxial system derived from the cardinal points alone and this pragmatic base further explains the large amount of work in this area, in spite of the limitations inherent in the optical treatment. Data on corneal and ocular dimensions can be employed for contact lens design. At the time the axial results are obtained, measurements of the PND and peripheral retinal magnification factor (mm of retina/degree) can be made off-axis to estimate their variation (Vakkur *et al.*, 1963; Hughes, 1976, 1979b) – something that the paraxial theoretical treatment cannot deal with. These data may subsequently be combined with those from retinal flat mounts to establish the location and projection of otherwise undistinguishable specialized retinal areas.

These considerations suggest that there may be good reason to distinguish between schematic eyes, obtained by computation, and model eyes, which incorporate much important information either outside the scope of the paraxial treatment or of a more general nature.

Ray-trace schematic eyes
Gaussian inadequacies

Consider the patent inadequacies of the conventional schematic eye which arise out of its paraxial derivation: (1) it is effective for infinitesimal apertures alone; (2) it cannot deal with significantly oblique imagery; and (3) surfaces are assumed spherical and coaxial. These difficulties may all be avoided by employing ray-tracing rather than unifying formulae. Snell's Law, eqn i [Note 1], completely specifies the path of a given ray at any interface between the media of an optical system. It is thus possible to employ trigonometrical or geometrical construction to follow rays through a system such as the eye without paraxial approximation. This approach is general in its power but it can be a demanding and tedious exercise in three-dimensional trigonometry and certainly does not result in convenient analytical forms. The procedures can be systematized but only the advent of the digital computer has brought this method out of the hands of highly specialized practitioners.

Lotmar (1971) based a ray-trace model on the data of the Gullstrand schematic eye improved by the use of an accurate corneal profile (Bonnet, 1964). Lotmar's aim was to deal with finite apertures and wide-angle imagery. The peripheral flattening of the cornea did reduce spherical aberration and an approximation to the level of experimental values was gained by making the rear lens surface aspherical. But his was an entirely pragmatic approach which still retained a homogeneous lens. The design parameters for the Gullstrand lens are strictly developed for paraxial use alone. Drasdo & Fowler (1974) used an ellipsoidal cornea with a homogeneous spherical lens based on the paraxial model of Cowan (1928) to study the non-linear mapping of the visual field onto the retina by means of ray tracing. Neither study provides strong experimental justification for the models and computational simplicity was emphasized more than accuracy of modelling.

The problem of the lens refractive index gradient

Although these approaches abandon the paraxial model in favour of ray tracing, thus freeing themselves from the three assumptions above, they fail to deal with two further assumptions of the Gaussian approximation: (1) it assumes the media are homogeneous and ray paths rectilinear; and (2) it takes no account of the evidence for gradients of chromatic dispersion through the non-homogeneous crystalline lens. Unfortunately, the ray-tracing procedure provides no simple means for the non-mathematical physiologist to drop these assumptions in order to derive schematic eyes incorporating a more realistic lens model with a continuous gradient of refractive index from its surface to its core.

Two types of model have been employed: (1) approximation of the continuous gradient by means of a series of thin shells; and (2) an analytical treatment of the ray path through the lens medium. Both approaches now rely extensively on the use of digital computers.

Wide-angle shell model eyes

In order to obtain a closer approximation to the refractive index gradient of the real crystalline lens, Lotmar (1971) made preliminary ray traces through a model in which the continuous lens was replaced by a nest of seven shells arranged such that each shell, itself uniform, was of higher refractive index than that outside it. The first extensive study by this technique, through a 2000-shell model for the rabbit crystalline lens. Some problems obviously remained because the focal length of this model was not in agreement with that measured for the rabbit schematic eye (Hughes, 1972) and confirmed by Ludlam & Twarowski (1973). Nakao *et al.* (1970) have described a similar shell model lens which has been developed into a full schematic eye for man.

Pomerantzeff, Govignon & Schepens (1971) and Pomerantzeff (1985) also described a wide-angle model human eye based on a lens of 98 shells. The model lens parameters were fitted by relaxation procedures to a curve of spherical aberration measured from the living eye. This model employed some 96 parameters to define the index gradients and aspherical isoindicial surfaces. The approach ensured modelling of axial spherical aberration but gave no guarantee of valid modelling of off-axis imagery. Recently, this model has been employed to explore oblique astigmatism (Wang, Pomerantzeff & Panktratov, 1983). The surface of best imagery is assumed to define the retinal entrance aperture surface, the effective image shell of Hughes (1977*b*), for an emmetropic eye and no attempt is made to fit the model with an experimentally determined retinal surface. The model, however, suggests a pattern consistent with clinical observation. The relatively arbitrary nature of this model's design means that the degree to which it accurately predicts features other than spherical aberration is unclear. Its primary use has been in the design of wide-angle optical instrumentation where the model's success is regarded as justifying its inherent assumptions.

Analytical ray-trace model eyes

It is not certain that such shell models are satisfactory for studying oblique imagery. Consider a continuous gradient of refractive index from one side of a slab of glass to the other; a ray which initially runs parallel to the surface will be bent towards the surface of higher refractive index. If we consider the block substituted by a shell model with a sequence of thin plates of different refractive index, arranged with the plate of highest refractive index at the bottom, then the same ray would remain undeviated and the model is clearly invalid.

But if such models are to be avoided it is necessary to solve the differential equations which describe the path of each ray through the non-homogeneous medium. It has already been pointed out that the gradient of refractive index through the crystalline lens was well known and its possible role in controlling aberrations was understood in the eighteenth century. Young (1801) explored the influence of the gradient on the matching of the image shell to the form of the retina and presented calculations based on the assumption of a *spherical* lens whose refractive index falls off as the distance from the lens centre increases. This amenable analysis is little use for the aerial eye; however, it does provide insight into the basic principles as they apply to aquatic eye design and has a long history.

The 'fish lens' eye

Maxwell (1854) employed fish and ox eyes for the demonstration of the principles of optics to his students and became intrigued by the design of the crystalline lens. He showed that perfect imagery could be achieved with a radially symmetrical parabolic distribution, like that of the fish lens, only if the object and image lay within the medium (Maxwell, 1858); a fish-eye universe. The relative analytical accessibility of such models has maintained interest in them.

Mathiessen (1877) was struck by the short focal length of the fish lens and its apparently fixed ratio of 2.5–2.6 (Mathiessen's ratio) to the lens radius in a variety of species. He concluded that the refractive index distribution in the lens must be parabolic. Luneberg (1944) demonstrated that an aplanatic spherical lens could exist without the continuity with the environment implicit in Maxwell's solution (1854). More recently Fletcher, Murphy & Young (1954) made an analytic demonstration of the absence of spherical aberration in the imagery of a spherical lens with a radially symmetrical gradient index (GRIN) distribution.

Pumphrey (1961) discussed these results in relation to the spherical fish lens which operates in the effective absence of corneal refraction, when the eye is immersed in sea water, and might achieve aberration-free imagery; the rays entering over its entire aperture

would be brought to one focal point if it contained a spherically symmetrical, parabolic distribution of refractive index which fell from a value of 1.52 at the centre to the value of the medium at the surface.

A *homogeneous* lens of refractive index 1.52 would have a focal length of about 2.8 r compared to the 2.5 r of a gradient index lens of the same core refractive index. The refractive index gradient thus enables an increase in the aperture of the eye, the F-number is conversely reduced and the retinal image is brighter because of increased light capture. In addition, it eliminates the terrible spherical aberration of the homogeneous lens. An increase in the refractive index to the biologically unachievable value of 1.67 would reduce the focal length of the homogeneous lens to that of the gradient index lens but not reduce the spherical aberration.

Not only does a spherical gradient index lens permit control of spherical aberration but it also produces a spherical image surface at which imagery of oblique incidence is also perfect. To match this the retina must thus take on the spherical form found in the fish eye.

It will be clear that the fish eye appears to be remarkably straightforward to analyse when compared to the vertebrate eye, although it is far from elementary in its properties. Pumphrey (1961) suggests that the 'relative freedom from aberration of the human eye was all that remained of an original beauty and symmetry only possible for aquatic animals – a reminder of the sacrifices which had to be made for the purchase of a terrestial existence'.

However, these accounts may have been too strongly influenced by theoretical considerations. The work of Sroczynski (1975a, 1975b, 1976a, 1976b, 1977) has shown that the ratio of focal length to lens radius is not constant across the lens and that significant spherical aberration is present. He obtains values for different species in the range 2.2 to 2.4 for paraxial rays. According to Sivak & Kreuzer (1983), the perch and goldfish show massive spherical aberration of the lens, only the rock bass appears compensated. More striking is the fact that the aberrations differ from those of land animals in being positive aberrations rather than negative.

Pumphrey (1961) emphasized the possibility that the fish eye was organized to minimize chromatic aberration. However, Sivak (1974) found chromatic aberration in a variety of fish lenses to range from 8%

to 15% of the focal length. Sroczynski (1976*b*; 1978) confirms this with reports of Mathiessen's ratio ranging from 2.35 ($\lambda = 640$ nm) to 2.26 ($\lambda = 435$ nm).

Even the spherical shape of the retina appears not to be attained in practice. The peripheral tapetal plates of certain fish eyes are not tangential to the retinal surface but, because of its deviation from a sphere, are rather organized perpendicular to the rays emerging from the exit pupil (Gruber & Cohen, 1978).

Such deviations from ideal organization might be seen as shortfalls in performance. However, it seems more likely that our view of the fish eye has been oversimplified because of its consequent analytical approachability. We must employ a sophisticated analysis similar to that for aerial eyes if we are to come to a teleological basis for understanding the aquatic eye's apparent 'imperfections'.

The means for analysing non-homogeneous lens behaviour

Thus, whether interested in fish, reptiles, birds or mammals we ultimately need to develop a wide-angle, finite-aperture, eye simulation with a non-homogeneous lens.

This was begun a long time ago by Mathiessen (1877) who assumed a paraboloidal indicial function for the crystalline lens and employed it in the derivation of the focal length and principle point positions by means of a definite integral. The approach was, however, confined to the accurate derivation of the cardinal points of the system. In similar fashion, Blaker (1980) has more recently employed Moore's (1971) paraxial ray-trace method for gradient index lenses, itself based on the work of Sands (1970), to obtain a more accurate version of Gullstrand's model eye from modern data. No attempt was made to extend the approach to off-axis and oblique imagery.

The approach is analytical because it provides an equation which describes the true path of the rays through the ocular media. This contrasts to the approximation method of employing shells. It is not clear whether the analytical approach has been employed outside the Canberra group within recent years. Fitzke (1981) has developed a wide-angle ray-trace eye model with aspherical surfaces and incorporates the zonal discontinuities of the human lens as well as a continuous refractive index gradient. However, the approximations involved in his non-paraxial technique for deriving the ray paths have not been presented, so that computational basis of the representational model has not been identified.

The analytical model eye of the Canberra group
Axial and marginal ray imagery

Our attack on the development of an analytical model eye began with the arrival of Melanie Campbell to undertake doctoral studies in my laboratory. She brought mathematical and computing skills which we employed to set up a gradient index model eye which could deal with finite pupils and spherical aberration in the relatively simple, spherical eye of the rat. This was to be done by two independent approaches. Using the refractive index distributions of Phillipson (1969), and our own anatomical data we would develop an analytical model of the lens and from it derive a whole-eye model. In addition, and quite independently, we would study the aberrations of the real lens by laser ray tracing to test the predictions of the model computations.

The outcome (Campbell & Hughes, 1981) was most satisfactory. Using spherical approximations for corneal and lens surfaces, the calculated ray paths through the gradient index lens model were found to be in excellent agreement with those measured for the crystalline lens with axial and marginal rays incident over the entire extent of its aperture. The core and homogeneous lens models were found to be quite inadequate for modelling non-paraxial behaviour.

When the model lens was combined with a spherical cornea, it was found that the positive aberrations of the cornea were more than compensated by the negative aberrations of the lens to leave slight overall negative aberration. The model eyes were slightly myopic, but the range of experimental error is consistent with their being emmetropic as demonstrated experimentally by Hughes (1977*b*). The aberrations were of considerable magnitude in the marginal pupil and theoretical calculation confirmed the observation of Hughes (1977*a*), that the rat eye becomes substantially hypermetropic as the pupil dilates outside of the scotopic light intensity range. It could be argued that this is a means for matching the image quality of the eye at low intensities to the resolving power of the rod system.

Error of retinoscopy

The discrepancy between the photopic emmetropia of the rat eye and the finding of some 9.5 D of retinoscopic hypermetropia is explicable as an artefact of retinoscopy (Sheard, 1922; Glickstein & Millodot, 1970) resulting from the observer employing reflections from the retinal surface (Hughes, 1977a). The evidence for this view is assessed by Hughes (1977b). However, it has been postulated that chromatic aberration is involved (Charman & Jennings, 1976; Nuboer & Van Genderen Takken, 1978). This claim was strongly challenged by Millodot & Sivak (1978) and Millodot & O'Leary (1980) who showed an erroneous calculation in the work of Nuboer and Van Genderen Takken (1978). Their view was confirmed by Campbell (1986b) who has recently employed the gradient index model eye to examine the magnitude of chromatic aberration in the rat. She finds it much too small to explain the artefact of retinoscopy, a result in agreement with the experimental findings of Hughes (1979a).

The presence of a specular reflection, presumed to be from the retinal surface, and a diffuse reflection has been confirmed in man with polarization retinoscopy by O'Leary & Millodot (1978). In young eyes the former predominates to produce the artefact of retinoscopy but in older eyes the specular reflection is reduced and it disappears after the age of 55 (Millodot & O'Leary, 1978).

Remtulla *et al.* (1984) regard the question of origin of the artefact as open in the mouse, for which 10 D would be predicted and 20 D has been observed, and in which a calculated 50–70 D chromatic aberration exists to complicate the issue. Given that genuine marginal ametropia may exist when the pupil dilates, in order to match optical quality to low light intensity performance, it becomes necessary to refract with a small pupil and this can be subject to great error with small animals. In addition, it would be wise to consider only feral specimens because other strains are not necessarily subject to strong visual selection pressures.

Some technical considerations

It can be seen from the above discussions that the ability conferred by computer programs to generate more complex and accurate models by no means results in automatic insight into visual function.

Aspherical surfaces

For the generation of Gaussian model eyes it has been convenient to approximate the refracting and imaging surfaces by spherical curves. But, of course, it has long been known that the human cornea has an aspherical surface (Bonnet, 1964) which is concluded to be involved in control of spherical aberration. As long as schematic eyes were confined to consideration of paraxial rays this feature could not be modelled and errors were small. Now the situation has changed.

As soon as we wish to come to a full understanding of eye performance through the use of computer models, then we are concerned with a range of pupil sizes and thus with the true form of the refracting surfaces. However, measurements are available for a variety of surfaces which establish aspherical parameters (e.g. Parker, 1972; Howcroft & Parker, 1977). Some workers have already begun to employ aspherical data for standard homogeneous eyes (Chaudhuri, *et al.* 1983). As model eyes, compendia of useful information, these are valuable, but as schematic eyes they are subject to so many limitations on their usefulness that they are not helpful in generating further insight into eye design. However, this approach points the way for those who wish to model wide-angle performance accurately (Campbell, in press, *a*). Sands (in press) reports on the poorer performance of the spherical model compared to the aspherical model of the rat eye.

Marginal aberrations

It has been known for some time that the human eye has little spherical aberration and El Hage & Berny (1973) concluded that the cornea introduces substantial positive aberration which is compensated for by the negative aberration of the lens. More recently, Millodot & Sivak (1979) suggest that the cornea and lens of man each have their aberration independently minimized and do not systematically compensate for one another.

Campbell & Hughes (1981) employed their gradient-index model rat eye to study its image properties over a range of pupil sizes. One of their interesting findings is that, whatever the situation in man, the spherical aberration of the rat cornea is balanced by the negative aberrations of the lens.

The possibility that different species tolerate patterns of spherical aberration determined by the balance between their corneal and lens aberrations is commonly accepted simply to reflect their varying reliance on 'good' vision. Whether more sophisticated considerations are involved in the choice of differing balances must await further investigation. It is conceivable that the depth of field of an eye can be set this way at the cost of a penalty in image contrast. The phenomenon of 'pseudoaccommodation' or increased depth of field in aphakia has already been attributed simply to the increased spherical aberration of the eye (El Hage & Berny, 1973).

Of course, it is not necessary to employ computer simulation to begin such comparative studies. Sivak & Kreuzer (1983) have used laser ray tracing (Campbell & Hughes, 1981) to examine the aberrations of lenses from a wide range of animals. Lenses from different species were found to possess negative, positive or negligible spherical aberration. The interpretation of such results is difficult because the variation of accommodative mechanisms brings into question the representative state of the excised lens. Nonetheless, the results suggest that comparative computer modelling will be a rewarding area, especially when conjoined with complementary information about the cornea.

It is interesting to note that Sivak & Kreuzer (1983) found evidence for spherical aberration with marginal aberration in man which was similar to that encountered in my own laboratory (Hughes & Campbell, unpublished) but conflicted with the findings of Millodot & Sivak (1979).

Chromatic aberrations

The refractive index of optical materials varies as a function of wavelength. The measured optical properties of a system are thus a function of the light with which it is examined. In addition, conclusions about the performance of an eye must be drawn with reference to the normal spectral environment of the species in question and its ability to sense the wavelengths involved. Optics are usually analysed without reference to the infrared or ultraviolet component of the spectrum. However, it is now established that ultraviolet light is detected not only by insects but also by a variety of birds (Chen, Collins & Goldsmith, 1984). This opens up a whole new series of neglected questions. What is the quality of the ultraviolet image and how is it employed in these species?

Until recently, the chromatic aberrations of the eye have been either measured directly or estimated from the assumption that the media of the optical apparatus have the dispersion of salt water (Le Grand, 1967). Having calculated a homogeneous schematic eye, it is possible to compact this to a 'reduced' form comprising one equivalent refracting surface. The variation in focal length, or longitudinal chromatic aberration, is then determined as a function of wavelength on the assumption that the reduced eye is full of sea water of tabulated dispersive power (Hughes, 1979*c*; note that the formula used in this paper is an approximation, introducing little error, rather than incorrect as noted by Mandelman & Sivak, 1983). The results obtained using this formula for man and rat are in quite good agreement with measurement.

Palmer & Sivak (1981) and Sivak & Mandelman (1982), however, have employed a monochromator to reassess the chromatic dispersion of the crystalline lens. It was found that the dispersion substantially increased over that of water in the blue/violet range of the spectrum so that prior computations would underestimate chromatic aberration.

Mandelman & Sivak (1983) have since employed averaged lens sample dispersions to calculate chromatic aberration for homogeneous lens eye models. The new calculation and the greater dispersions considerably increase estimates of chromatic aberration over that of water models, especially for man.

Unfortunately their model remains invalid. Campbell (in press, *b*) has pointed out that real dispersions cannot be used for a homogeneous lens model because this is as invalid as employing real refractive indices to represent the 'effective' or total index of such models. The 'effective' dispersion of a homogeneous lens must be larger than any dispersion in the real lens in order to replicate the chromatic dispersion of the real refractive index gradient. Calculations for the rat employing central dispersions in a homogeneous model indicate 1.8–4.5 D of chromatic aberration compared to 3.4–10.4 D for a gradient index lens model.

Hughes (1979*a*) employed the measured back vertex chromatic difference of the rat lens to compute the 'effective' chromatic indices for the equivalent homogeneous lens and incorporated the result in a model eye to obtain the paraxial chromatic difference in refraction of 5.4 D.

In a lens-dominated eye, the dispersive properties of the lens require careful measurement, not an easy task, if an accurate model is to result. Cornea-dominated eyes will be better approximated by the water models.

It will be no mean feat to develop a model which can satisfactorily handle white light imagery to predict image quality.

A further complexity has been introduced by Millodot & Newton (1976) who describe a change in the chromatic aberration of the eye with age. This is suggested to have its origin in the ocular media.

Individual variation

A major problem in dealing with schematic eyes is to decide on the necessary magnitude of the sampled population and its composition. Mandell (1971) illustrated the range of corneal shape and Howcroft & Parker (1977) displayed a wide range of human lens shapes. Measures of human optical aberrations suggest a corresponding range of values consistent with normal function. The range of variability appears to be less in laboratory animal populations.

Given the earlier discussion of developmental correlations between ocular elements leading to emmetropia then it must be decided whether it is necessary to compute complete schematic eyes each based on data from one eye alone (e.g. Vakkur & Bishop, 1963), or whether correlations between individual elements of each eye are to be ignored by employing pooled data (Hughes, 1979b). With the complex data sets required for gradient index eyes, only the latter course would appear to be feasible at present.

Current information also draws attention to the need for paying attention to the variation of the ocular parameters with age. Data for the more complex schematic eyes which are employed for subtle investigations must be derived from animals of similar strain and age group. Indeed, the techniques now available are admirably suited to a comparative longitudinal study of the coordinated development of the optical system from birth.

Radial symmetry

It may be felt that the whole subject is becoming unduly complex, even at this level of qualitative discussion. Unfortunately, we have hardly touched on some of the potentially most important areas for modulating image quality. The whole of the treatment has so far assumed that the modelled system is rotationally symmetric about one axis.

There is evidence that the rabbit cornea has different radii of curvature in the vertical and horizontal axes (Prince, 1964) and cats may have a similar corneal astigmatism (Levick, personal communication). Such asymmetry might be significant in controlling image quality. Certainly, the retina of the cat eye has major asymmetry in its curvature during development and evidence of this is present in the adult eye (unpublished observations). Nobody has even begun the theoretical treatment of such matters.

Indeed it may be felt that it is better to examine the overall performance of differing visual systems by mapping image quality directly with single or double pass methods rather than by attempting analytical computations.

In man some progress has been made along these lines by the study of wavefront aberrations employing subjective techniques whose results can indicate image quality in any cross-section of the optical system and for high orders of aberration. Smirnov (1961) and Van den Brink (1962) both report enormous variation and asymmetry in the aberrations measured across the pupil by this method. Howland & Howland (1977) describe a very powerful and rapid subjective technique which enables high orders of aberration to be measured. They report that spherical aberration often appears dominant only in one axis and that third order, coma, aberrations predominate. Their polynomial descriptions of the aberrations permit them to compute the modulation transfer function for the optics at any orientation (Hopkins, 1962). The range in form of the MTF curve for 55 subjects is extraordinarily varied and these analyses open up a completely new approach to the comparative study of optical performance.

More recently, Walsh, Charman & Howland (1984) have employed a similar but objective technique to examine monochromatic aberrations in man. They confirm the importance of coma, rather than spherical aberration, as the primary source of image degradation but the variation between individuals is less marked than in the results of Howland & Howland (1977). Whether a radially symmetrical schematic eye can suffice as an ideal model from which individual eyes deviate in random fashion, or whether there are universal deviations from symmetry, remains to be seen. The schematic eyes with high-order aberrations will play an important role in the analysis of this area.

Apart from introducing new approaches, we have not yet even considered the most interesting potential of the relatively conventional, new gradient index model eyes – that of permitting the analysis of off-axis optics, a relatively unexplored area of increasing interest which has been conveniently summarized by Ronchi & Cetica (1974).

Wide-angle imagery of the analytical eye models

During the period in which the gradient index model eye was being developed we began a collaboration with Peter Sands of the Canberra CSIRO. He is a specialist in computer modelling who also has an extensive record of work on inhomogeneous lens design (see Sands, 1970). He developed a powerful interactive computer program which enabled the modelling and exploration of gradient index eye designs to move on to the study of off-axis imagery (Sands, in press).

This program has been employed to develop a wide-angle model eye for the rat (Campbell, Hughes & Sands, 1982; Sands, in press; Campbell, in press, *b*; Campbell & Hughes, unpublished; Campbell & Sands, unpublished). Off-axis aberrations were explored for models with spherical or aspherical refracting surfaces. The spherical model was not as good as the aspherical model for predicting peripheral imagery beyond some 50 degrees eccentricity. However, the entrance pupils of the photoreceptors, the outer limiting membrane, are found to be well matched to the surface of best focus for nearly 70 degrees' eccentricity in the rat eye modelled by aspherical surfaces. Retinal magnification varies little over this range.

These models are all computed directly from anatomical or physiological measurements and gain support from the consistency between the results of the two approaches. No attempt is made to optimize their predictions. However, although the models have an organization which is functionally sensible, it is not clear whether the peripheral fall off in image quality and refractive state, which arises from the separation of the outer limiting membrane from the image surface of minimum confusion, is an artefact of experimental error of measurement or whether it reflects the situation in the living eye.

A particularly important property of the computer-based gradient index model eye is its provision of a means for exploring sensitivity to parameter variation. It is thus possible to examine the properties of models which fall within the range of error of the parameter measurements (Sands, 1985). This permits the experimenter (1) for the first time to ask questions such as 'could the eye design be optimal in some respect?', and (2) by exploratory computation to study

the trade-offs inherent in the evolution of ocular designs. By employing such relaxation procedures, Sands (1985) was able to establish an optimized rat eye whose parameter values were within the range of experimental error and which had uniform refractive state and image quality as a function of some 70 degrees of hemi-eccentricity.

The potential for such a design was foreshadowed in the results of Hughes & Vaney (1978) which showed the rabbit eye to lack significant oblique astigmatism over a range of hemi-eccentricity of some 80 degrees. This result is comprehensible in terms of the presence of an extended area centralis, or visual streak, in the rabbit retina. The rat retina has no visual streak but is relatively uniform with respect to cone and ganglion cell density distributions. Although the rat's visual pole and visual axis lie at a lesser angle to the optic axis than those in the rabbit, the 58-degree eccentricity is substantial (Hughes, 1979*b*), and the theoretical demonstration of the eye's potentially uniform image quality is in keeping with its retinal organizaiton.

These are most import findings. The absence of substantial oblique astigmatism in these two species contrasts well with the presumed progressive increase of this aberration to a substantial value in the periphery of the human eye (Millodot, 1981). The results comprise the beginning of an answer to my question of some 20 years ago: 'Is there a universal Bauplan for the imagery of the mammalian eye?' It looks as if the answer will be 'no'.

The majority of available results support the view that human image quality, unlike that of the rat, falls off substantially away from the visual axis. This has long been assumed because of calculations made by Le Grand (1967) which indicated high degrees of oblique astigmatism based on models with spherical refracting surfaces. In fact, the 80 D of astigmatism at 50 degrees' eccentricity in such models is reduced to about 15 D by the real optical system, but this remains a substantial aberration. The psychophysical results of Frisen & Glansholm (1975) do not establish absolute values for image quality but indicate that it is sufficiently low at eccentricities up to 40 degrees to substantially reduce the local resolving power of the retinal apparatus.

The question is not closed, however. Although the preliminary measurements of Röhler (1962) suggested a rapid fall off in the image quality of man away from the optic axis, Jennings & Charman (1981) argued that human optical quality varies little with eccentricity. Unfortunately, they employed a dilated pupil, as well as spectacle lenses, so that even foveal image quality was massively reduced from its known value at the optimal pupil sizes. It is not acceptable to assume, as they did, that the distribution of image quality with eccentricity would have the same form for large and small pupils.

We have only just begun the computational exploration of this problem for the human eye and have no contribution to the discussion, but the theoretical results of Wang, Pomerantzeff & Pankratov (1983) certainly support the view that there is a substantial reduction in human image quality as eccentricity increases.

The finding of relatively uniform image quality across the eyes of some species thus enables us to conclude, in contrast to Young (1801), that when we find image quality falling off as a function of eccentricity we are not observing an obligatory consequence of optical limitations on ocular design but rather may be examining a specific adaptation with functional implications.

When we consider the results of Ferree & Rand (1933) and of Rempt, Hoogerheide & Hoogenboom (1976), which show that the retina lies in very different relation to the optimal image shell in some five classes of human eye type, we are faced with a challenge to this view. The pattern of image quality variation across the eye may differ substantially in different people. Does this variability simply result from the absence, in the modern environment, of selection pressure to optimize human peripheral vision? Is the eye with little oblique astigmatism the norm? If so, how can it attain a fall off in its image quality to match that of its photoreceptor distribution? Perhaps coma dominates to ensure the desired pattern of image topography in such eyes. Maybe all the identified distributions achieve similar image quality distributions in spite of the differences in their oblique astigmatism.

Man apart, the fascinating problem which remains is the extent to which the topography of image quality varies between species. How complex will its distribution maps turn out to be; to what extent will they resemble receptor distributions? It is only by the continuing dialogue between experimental synthesis and theoretical analysis that we will get the answers.

Comparative use of computer-based analytical eye models

An important outcome of Sands' (1985) relaxation studies was a demonstration that the predicted image surface is very sensitive to parameter variation within the range of error of the original measurements. The message is that only the most accurate modelling of physiological optical systems can be of benefit to a sophisticated comparative study of eye design. For all its interest, the comparative study of hypothetical eye designs by means of paraxial, homogeneous models (Martin, 1983) is of little practical use. Martin himself (1983) notes that the gradient index models reveal the exquisite subtlety of the optical design of the vertebrate lens. He concludes that 'because of the complexity of the measures and computational techniques used, and because of intra- and interspecific variations of gradients, the possibility of general solutions, or of solutions which permit rapid interspecific comparisons of optical mechanisms seem remote at this time'. His conclusion is already outdated. The computer techniques described above make such methods available now.

An unfortunate block to the development of extensive comparative studies on ocular design is the unavailability of reliable refractive index distributions for a large number of species. The majority of techniques available require destruction of the lens and are very susceptible to distortion of the sections. Campbell (1984) has provided a relatively straightforward, non-destructive means of determining refractive index distribution in the intact crystalline lens. The method simply requires ray path determination for laser beams passing through the lens immersed in special media and a back computation to the refractive index distribution; the method is based on techniques employed for optical fibres (Barrell & Pask, 1978) and the theory is presented by our colleague Colin Pask (Pask, in press) of the ANU Department of Applied Mathematics, RSPhys.S.

Applied significance of sophisticated model human eyes
The normal eye has highly optimized design

It is important to emphasize that the normal eye appears to be beautifully organized to optimize its aberrations. The results of Stenström (1948) and Sorsby *et al.* (1957) suggest that there is a negative correlation between the components of the ocular dioptrics which ensures that the system approaches axial emmetropia more commonly than random errors would suggest. However, note the comment of Carroll (1980) that leptokurtosis of ametropia distribution does not necessarily arise from negative correlation between the optical components of the eye. The aberrations of cornea and lens may offset one another to establish a specific pattern of image degradation as the pupil dilates. In the periphery, as indicated above, the aspherical surfaces and lens index gradient may commonly ensure that oblique astigmatism is minimized and are possibly employed to match image quality to retinal resolution as a function of eccentricity.

Similar mechanisms appear to operate during development (Sorsby *et al.*, 1961). Disease-induced image degradation appears to break down these controls (Rabin, Van Sluyters & Malach, 1981; Nathan, Keily, Crewther & Crewther, 1985) but the mechanisms are only just being analysed now that primate models of myopia are available (Criswell & Goss, 1983).

Conventional contact and implant lenses degrade vision when the pupil is large

It is an outcome of the mathematical problems associated with accurate modelling of physiological optics that ophthalmic lens design has become almost entirely based on paraxial computations and spherical surfaces. It takes little imagination to consider the potential effect of replacing the evolutionarily optimized corneal surface form by a contact lens with a spherical surface chosen by paraxial design. Millodot (1969) has demonstrated that such lenses lead to more rapid reduction in visual acuity as the pupil dilates under low light conditions than would occur with spectacle lenses. Kerns (1974) describes pragmatic lens designs which may reduce this effect. The soft lens does not suffer from the defect, as it must preserve the original corneal aspherical form to some degree (Millodot, 1975).

Considering the sophisticated nature of the crystalline lens's design it is a theoretically outrageous act to replace it with a simple homogeneous, planoconvex plastic lens implant with surfaces defined by Gaussian paraxial optics (Ridley, 1951; Binkhorst, 1975; Jalie, 1978; Miller, 1979). However, this is now done some 700000 times per year in the USA without much complaint from the patients about image quality (Wigton, 1978). High Snellen acuities are attained in many such pseudophakic eyes (see Shirley, 1984); characteristically some 80% of patients attain better than 6/12 vision and, when those with pre-existing pathology are excluded, the figure may rise to 95% (McConnell, Duvall & Cullen, 1983; Choyce, 1984). Sophisticated assessment of lens performance by means of contrast sensitivity function tests does not yet appear to have been carried out. It would seem likely that the combination of uncompensated corneal spherical aberration (but see C. E. Campbell, 1981) and implant aberration might be significant enough at large pupil apertures to reduce acuity in similar fashion to a hard contact lens.

The reduced incidence of complications after intraocular lens surgery (Nordlohne, 1974; Apple *et al.*, 1984) means that the use of these lenses will increase rapidly and that there is extensive interest in their improvement (Choyce, 1977). The optical quality is currently chosen to be high enough not to limit acuity (Olson, Kolodner & Kaufman, 1979) and the major considerations relate to mechanical and surgical improvements (Anis, 1984). Great care is required to ensure that all the implications of changing an implant design have been considered; many of these would not be obvious to those not involved in the entire spectrum of surgical considerations (e.g. see Hoffer, 1983).

As yet, there is little experimental data to support the need for implementing theoretically apparent optical improvements. Rosenblum & Hendrickson (1981) described methods for assessing the quality of individual lenses and discussed the optimal choice of lens form to obtain good paraxial image quality with minimized spherical aberration and coma in the pseudophakic eye. Holladay *et al.* (1983) discussed possible improvements, such as using biconvex lenses with a front to-rear surface power ratio of about 1:3 (Wang & Pomerantzeff, 1982) and spectral absorption matched to the real lens to protect the retina from excessive ultraviolet exposure. The majority of such theoretical analyses employ a simple Gaussian approach. A theoretical ray-trace analysis of the pseudophakic eye (Jagger & Hughes, unpublished) demonstrates that an aplanatic implant could significantly improve image quality for large pupil diameters.

Off-axis vision can be vital

Current ophthalmic practice pays no attention to peripheral field refraction. By turning the head or eye it is easy to compensate for peripheral optical defects.

But the peripheral field has tasks other than providing for resolution; it is the visual 'early warning system' for man. Any peripheral refractive errors not only impair resolution (Millodot, Johnson, Lamont & Leibowitz, 1975) but also reduce the contrast sensitivity of the eye for detecting peripheral point sources or movement. Unfortunately these latter deficits cannot be compensated for by eye or head movements; the whole field should be surveyed in parallel.

Many people with what is commonly regarded as 'normal vision', i.e. corrected for central vision, have high levels of peripheral oblique astigmatism. It has been demonstrated that the threshold for motion detection at 50 degrees' eccentricity can be halved by elimination of oblique astigmatism (Leibowitz, Johnson & Isabella, 1972). Such errors may have a major effect on behavioural performance and survival under conditions of low illumination or need to detect the rapid movement of small images. An obvious instance is the pilot for whom such considerations may determine life or death. Visual contact remains a vital consideration in aerial combat even with today's radar and computer equipped 'superfighters' (Kinnucan, 1984).

Spectacles preclude whole-field correction

The role of the modern best-form periscopic spectacle lens precludes specific correction of peripheral vision. Its design is aimed at providing optimal correction for ametropic foveal vision over a wide range of eye positions. Such a lens aids the peripheral field only to the extent that its refractive state is the same as that at the fovea, but this is so for only a small proportion of eyes.

In general, ametropia varies with eccentricity in different, but characteristic, fashions in the majority of hypermetropic, emmetropic and myopic eyes (Millodot, 1981) so that off-axis retina may be under- or overcorrected when served by the best-form lens with the central prescription. Even 'normal' eyes with emmetropic central refractions may be ametropic off-axis.

It is probably fortunate that the region of peripheral field where such errors would be greatest is not served by the spectacle lens.

More sophisticated correction may be possible

Unlike spectacle lenses, the contact and implant lens influence the optics of the entire field of view of the eye. However, their selection is based only on paraxial considerations. The outcome is that the majority of current contact and implant lenses, like spectacle lenses, not only fail to correct peripheral vision but in addition must inevitably degrade it, as they do axial vision at wider pupil apertures. Such detrimental effects should be in principle avoidable. Whether it is feasible or desirable, at present, to produce such lenses requires investigation. For instance, simple manipulation of the form of the plastic implant lens cannot improve the massive oblique astigmatism present at high eccentricities (Jagger & Hughes, unpublished). It might be necessary to employ a gradient index implant lens to achieve this aim. Such a lens would have many degrees of freedom if it were given aspherical surfaces.

In addition, there are many new means of axial optical correction under investigation which also influence peripheral vision. The technology of spin-casting contact lenses has opened up the independent control of asphericity of the convex and concave surfaces of hard and possibly soft contact lenses (Neefe, 1984). Bifocal contact lenses are now in use (Andrasko, 1984). The 'alternating' design requires movement of lens zones with respect to the visual axis but the 'simultaneous' type has deliberate 'aberrations' which increase depth of field of the presbyopic eye by means of pseudoaccommodation.

An area of increasing significance is that of keratorefractive surgery. Ophthalmic surgeons appear to be accepting the viability of the technique of radial keratotomy (Fyodorov & Durneu, 1979) for the correction of myopia: 24 months after surgery some 87% of patients are reported to have uncorrected visual acuities of 20/20 to 20/40 and methods of reducing glare have been established (Smith & Kutro, 1984). More dramatic is the technique of keratomileusis which involves removal of the corneal vertex, its reshaping on a computer controlled cryolathe, and replacement. The majority of these new approaches to ametropia show the classic concern with paraxial vision, pay little attention to marginal rays and totally neglect peripheral image quality.

We referred above to the view that peripheral vision is not important to man and that peripheral correction is immaterial. The ability to adjust to sensory distortion (Kohler, 1964) and loss is remarkable but is not to be taken as signifying the redundancy of the normal apparatus. It was a classic experiment of Canella (1936*a*, *b*) which demonstrated that monocular kestrels, cats and frogs appear to be as capable as binocular animals in the conduct of their everyday life. Is one eye then to be simply regarded as a spare in case the other is mislaid? That would be very hard to accept now that the complexity of the neural mechanisms of stereopsis has been established. Two eyes must be of vital adaptive significance. Similarly we require the correct tests to lay bare the importance of peripheral vision but, of course, most of us could manage in our everyday life were it deficient.

In future years we can expect an increasing awareness of subtle visual defects induced by these corrective procedures. The advent of computers may soon make it no more arduous to provide good periscopic correction than it is at present to deal with central vision alone. Why should someone with an implant lens, for instance, ultimately not expect a performance equal to the original, or even better?

Acknowledgements

The author wishes to note that The National Health and Medical Research Council, The Department of Science and Technology, The Institute of Advanced Studies of the Australian National University and the Commonwealth Scientific and Industrial Research Organisation have provided financial support for the research programme of his group. N.V.R.I. is an Affiliate Institution of Melbourne University.

Notes

1. A less intuitive geometrical derivation is based on application of Snell's Law across the refracting surface. It states, at the interface between two media of refractive index n and n', that

$$\sin i = (n'/n) \cdot \sin i' \qquad (i)$$

where i is the angle of incidence of the ray to the normal to the surface and i' is the angle of its refraction. The paraxial derivation assumes the angles dealt with are small enough for the sine of the angle and its magnitude in radians to be equal:

$$i = (n'/n) \cdot i' \qquad (ii)$$

The expansions of sines and cosines

$$\sin x = x - x^3/3! + x5/5! - x^7/7! + \ldots \qquad (iii)$$

$$\cos x = 1 - x^2/2! + x^4/4! - x^6/6! + \ldots \qquad (iv)$$

have thus been reduced to the first term only. This is why paraxial optics is also referred to as 'first-order theory'. The Seidel theory includes the second term of the expansion for greater accuracy slightly away from the axis and is thus 'third-order theory'.

2. First consider the triangle base b, height t, and hypotenuse h (see Fig. 5.3). How much longer, d, is the hypotenuse than the base? From Pythagoras we have $t^2 = h^2 - b^2$ or $(h-b)(h+b) = t^2$. But $(h-b) = $ d and $(h+b) \approx 2h$. *Thus*

$$d \approx t^2/2h \qquad (v)$$

Now consider a single refracting surface between, say, air and glass. In air the velocity of light is 1 and in the glass it is lower, $1/n$ where n is the refractive index of the glass. Imagine that the surface is plane. From (i) the path OPO' (Fig. 5.3) is longer than OQO' by $d + d' = (PQ^2/2OP) + (PQ^2/2PO')$ and the excess time for the light to travel the path is $nd + n'd'$ or, because $n = 1$, $(PQ^2/2OP) + n'(PQ^2/2PO')$.

For perfect imagery, the light arising at a point O, whatever its path through the surface, must be reassembled at a point O'. Fermat's principle then indicates that the time for it to cover any one of those paths must be the same. If this is to be true in Fig. 5.3, a delay must be introduced into the axial path OQO' which eliminates the difference in time for light to traverse it relative to other paths. This can be attained by making the surface spherical so that the bulge in the glass introduces an appropriate segment AQ such that the extra delay is $(n'-1) \cdot AQ$. From eqn v, AQ is PQ/2R so we now have a relation defining the necessary radius of curvature, r, for perfect imagery

$$(PQ^2/2OP) + (n'PQ^2/2PO') = (n-1)PQ^2/2r$$

thus

$$(1/OP) + (n'/PO') = (n'-1)/r$$

As PQ tends to 0, AQ tends to 0 so that we have OP = OA = u, the object distance from the lens vertex, and PO' = AO = v, the image distance

$$1/u + n'/v = (n'-1)/r \qquad (vi)$$

If u is at infinity then $1/u = 0$, the image is at the focal point and v is the posterior focal length of the surface, f', so we have the classic formula

$$1/f = n'/f' = (n'-1)/r = F \quad \text{(in diopters)} \qquad (viii)$$

where f is the anterior focal length of the surface and F its power in diopters.

The approximations made to obtain eqn i [Note 1] limit the validity of these equations to rays only infinitesimally displaced from the optic axis or at very small angles to it; the treatment is thus known as paraxial optics and it deals with spherical surfaces.

3. For example, the power F_1 and F_2 of the anterior and posterior surfaces of the cornea, respectively, can be derived from eqn viii [Note 2]. These surfaces at A_1 and A_2 are separated by medium of refractive index n_2 which

establishes a reduced distance $c = A_1 A_2 / n_a$ (Davson, 1962; Southall, 1964). The combined surfaces then have a refractive power of

$$F_{1,2} = F_1 + F_2 - c \cdot F_1 \cdot F_2 \qquad \text{(ix)}$$

The principal point positions are determined as

$$A_1 H_{1,2} = (n_1 \cdot A_1 A_2 \cdot F_2)/F_{1,2} \qquad \text{(x)}$$

$$A_2 H_{1,2}' = -(n_3 \cdot A_1 A_2 \cdot F_1)/F_{1,2} \qquad \text{(xi)}$$

A similar procedure is then carried out for the lens surfaces and for the combination of lens and cornea systems. The details of the calculations are available in Southall (1964), Davson (1962) and in many schematic eye papers (Hughes, 1972, 1977b). The outcome is a set of values for the cardinal points of the system and a unified description of its paraxial performance. In the so-called 'reduced' model eye of Listing the whole optical system of the eye is replaced, after further simplifying assumptions, by a single fictitious surface of appropriate power.

References

Andrasko, G. J. (1984) Bifocal soft lenses: a comparison. *Contact Lens Forum*, **Sept.**, 53–65

Anis, A. Y. (1984) Guidelines for the selection of an anterior chamber lens: a personal viewpoint. *CLAO J.*, **10**, 213–17

Apple, D. J., Mamalis, N., Loftfield, K., Googe, J. M., Novak, L. C., Kavka-Van Norman, D., Brady, S. E. & Olson, R. J. (1984) Complications of intraocular lenses. A historical and histopathological review. *Surv. Ophthalmol.*, **29**, 1–54

Barrell, K. F. & Pask, C. (1978) Nondestructive index profile measurement of non-circular optical fibre preforms. *Opt. Commun.*, **27**, 230–4

Binkhorst, R. D. (1975) The optical design of intraocular lens implants. *Ophthalmic Surg.*, **6**, 17–31

Blaker, W. J. (1980) Toward an adaptive model of the human eye. *J. Opt. Soc. Am.*, **70**, 220–3

Block, M. T. (1969) A note on the refraction and image formation of the rat's eye. *Vision Res.*, **9**, 705–11

Bonds, A. B. (1974) Optical quality of the living cat eye. *J. Physiol.*, **24**, 777–95

Bonnet, R. (1964) *La Topographie Cornéenne*. Paris: Desroches

Bossomaier, T. R. J., Snyder, A. W. & Hughes, A. (1985) Irregularity and aliasing: Solution? *Vision Res.*, **25**, 145–7

Boycott, B. B. & Wässle, H. (1974) The morphological types of ganglion cell of the domestic cat's retina. *J. Physiol.*, **240**, 397–419

Bracewell, R. (1965) *The Fourier Transform and its Applications*. New York: McGraw-Hill

Bracey, R. J. (1960) *The Technique of Optical Instrument Design*. London: English Universities Press Ltd

Brewster, D. (1816) On the structure of the crystalline lens in fishes and quadrupeds as ascertained by its action on polarised light. *Phil. Trans.*, **106**, 311–17

Byram, A. M. (1944) The physical and photochemical basis of visual resolving power. *J. Opt. Soc. Am.*, **34**, 718–38

Campbell, C. E. (1981) The effect of spherical aberration of

contact lens to the wearer. *Am. J. Optom. Physiol. Optic.*, **58**, 212–17

Campbell, F. W. & Green, D. G. (1965) Optical and retinal factors affecting visual resolution. *J. Physiol.*, **181**, 576–93

Campbell, M. C. W. (1984) Measurement of refractive index in an intact crystalline lens. *Vision Res.*, **24**, 409–15

Campbell, M. C. W. (in press, *a*) A full-field gradient-refractive index eye model. In *Modelling the Eye with Gradient Index Optics*, ed. A. Hughes. Cambridge: Cambridge University Press.

Cambell, M. C. W. (in press, *b*) A wavelength dependent gradient refractive index model of the rat eye predicts chromatic aberration. In *Modelling the Eye with Gradient Index Optics*, ed. A. Hughes, Cambridge: Cambridge University Press.

Campbell, M. C. W. & Hughes, A. (1981) An analytic, gradient index schematic lens and eye for the rat which predicts aberrations for finite pupils. *Vision Res.*, **21**, 1129–48

Campbell, M. C. W., Hughes, A. & Sands, P. J. (1982) Anatomically based refractive index model of the rat eye predicting image quality across the retina. *J. Opt. Soc. Am.*, **72**, 1110

Canella, F. (1936*a*) Quelques recherches sur la vision monoculair. *C.R. Soc. Biol.*, **122**, 1221–4

Canella, F. (1936*b*) Les problémes du chiasma et de la vision binoculaire. Quelques recherches sur la vision monoculaire. *J. Psychol. Norm. Path.*, **33**, 696–711

Carroll, J. P. (1980) Geometrical optics and statistical analysis of refractive error. *Am. J. Optom. Physiol. Optic.*, **57**, 367–71

Cauchy, A. D. (1841) Memoire sur diverses formules d'analyse. *C.R. Acad. Sci. (Paris)*, **12**, 283–98

Charman, W. N. & Jennings, J. A. M. (1976) Objective measurements of the longitudinal chromatic aberration of the human eye. *Vision Res.*, **16**, 999–1005

Charman, W. N. & Tucker, J. (1973) The optical system of the goldfish eye. *Vision Res.*, **13**, 1–8

Chaudhuri, A., Hallett, P. E. & Parker, J. A. (1983) Aspheric curvatures, refractive indices and chromatic aberration for the rat eye. *Vision Res.*, **23**, 1351–63

Chen, de Mao, Collins, J. S. & Goldsmith, T. H. (1984) The ultraviolet receptor of bird retinas. *Science*, **225**, 337–40

Choyce, D. P. (1977) The theoretical ideal for an artificial lens implant to correct aphakia. *Trans. Ophthal. Soc. UK*, **97**, 94–5

Choyce, D. P. (1984) The first 1,000 Mark IX implants in the practice of The Regional Eye Centre, Southend General Hospital, Essex, England: 1978–1983. *CLAO J.*, **10**, 218–21

Citron, M. C. & Pinto, L. H. (1973) Retinal image: larger and more illuminous for a nocturnal than for a diurnal lizard. *Vision Res.*, **13**, 873–6

Cleland, B. G., Harding, T. H. & Tulunay-Keesey, U. (1979) Visual resolution and receptive field size: examination of two kinds of cat retinal ganglion cell. *Science*, **205**, 1015–171

Cleland, B. G. & Levick, W. R. L. (1974) Properties of rarely encountered types of ganglion cells in the cat's retina and an overall classification *J. Physiol.*, **240**, 457–92

Conrady, A. E. (1929) *Applied Optics and Optical Design* Oxford:

Oxford University Press. (Reprint: New York, Dover, 1957)

Cowan, A. (1928) *Ophthalmic Optics*. Philadelphia: Davies

Criswell, M. H. & Goss, D. A. (1983) Myopia development in non-human primates. *Am. J. Optom. Physiol. Optic.*, **60**, 250–68

Davson, H. (1962) *The Eye*, 2nd edn. London: Academic Press

DeMonasterio, F. M., Schein, S. J. & McCrane, E. P. (1981) Staining of blue-sensitive cones of the Macaque retina by a fluorescent dye. *Science*, **213**, 1278–81

Duke-Elder, W. S. (1958) *System of Ophthalmology*, Vol. 1, *The Eye in Evolution*. London: Henry Kimpton

Duke-Elder, W. S. (1970) *System of Ophthalmology*, Vol. 5, *Ophthalmic Optics and Refraction*. London: Henry Kimpton

Drasdo, N. & Fowler, C. W. (1974) Non-linear projection of the retinal image in a wide-angle schematic eye. *Br. J. Ophthalmol.* **58**, 709–14

DuPont, J. & de Groot, P. J. (1976) A schematic dioptric apparatus for the frog's eye. *Vision Res.*, **16**, 803–10

Dvorak, D., Mark, R. & Reymond, L. (1983) Factors underlying falcon grating acuity. *Nature*, **303**, 729–30

El Hage, S. & Berny, F. (1973) Contribution of the crystalline lens to the spherical aberration of the eye. *J. Opt. Soc. Am.*, **63**, 205–9

Ferree, C. E. & Rand, G. (1933) Interpretation of refractive conditions in the peripheral field of vision. *Arch. Ophthal.*, **9**, 925–38

Feynman, R., Leighton, R. & Sands, M. (1966) The Feynman Lectures on Physics. Reading, Mass.: Addison Wesley

Fincham, W. H. A. (1959) *Optics*. London: Hatton Press

Fitzke, F. W. (1981) Optical properties of the eye. *Invest. Ophthalmol.*, **20**, suppl., 144

Fletcher, A., Murphy, T. & Young, A. (1954) Solutions of two optical problems. *Proc. R. Soc. A.*, **223**, 216–25

Fox, R., Lehmkuhle, S. & Westendorf, D. H.(1976) Falcon visual acuity. *Science*, **192**, 263–5

French, A. S., Snyder, A. W. & Stavenga, D. G. (1977) Image degradation by an irregular retinal mosaic. *Biol. Cybernet.*, **27**, 229–33

Frisen, L. & Glansholm, A. (1975) Optical and neural resolution in peripheral vision. *Invest. Ophthalmol.*, **14**, 528–36

Fyodorov, S. N. & Durnev, V. V. (1979) Operation of dosaged dissection of corneal circular ligament in cases of myopia of mild degree. *Ann. Ophthalmol.*, **11**, 1885–90

Gabor, D. (1946) Theory of communication. *J.I.E.E.E.*, **93**, 429–57

Gauss, J. K. F. (1841) *Dioptrische Untersuchungen*. Göttingen

Geisler, W. S. (1984) Physical limits of acuity and hyperacuity. *J. Opt. Soc. Am. A*, **1**, 77–782

Glickstein, M. & Millodot, M. (1970) Retinoscopy and eye size. *Science*, **168**, 605–6

Goetz, K. G. (1970) Die optischen Übertragungseigenschaften der Komplexaugen von *Drosophila*. *Kybernetic*, **2**, 215–21

Green, D. G. (1970) Regional variation in the visual acuity for interference fringes on the retina. *J. Physiol.*, **207**, 351–6

Gruber, S. H. & Cohen, J. L. (1978) Visual System of the Elasmobranchs: State of the Art 1960–1975. In *Sensory Biology of Sharks, Skates and Rays*, ed. S. H. Gruber. Arlington: Office of Naval Research, Dept of the Navy

Gullstrand, A. (1908) Die optische Abbildung in heterogenen Medien und die Dioptrik der Kristallinse des Menschen. *K. Vet. Handl.* **43**, 1–58

Gullstrand, A. (1924) Appendix. In *Helmholtz' Physiological Optics*, 3rd edn (1909). Reprint: New York: Dover (1962) of translation by Southall, J. P. C. for Am. Opt. Soc. (1924)

Helmholtz, H. von (1856–1866) *Handbuch der Physiologischen Optik*, ed. A. Gullstrand, J. Kries & W. Nagel, 3rd edn (1909). Reprint: New York: Dover (1962), of translation by Southall J. P. C. for Am. Opt. Soc. (1924)

Hermann, G. (1958) Beiträge zur Physiologie des Rattenauges. *Z. Tierpsychol.*, **15**, 462–518

Hirsch, J. (1982) Falcon visual sensitivity to grating contrast. *Nature*, **300**, 57–8

Hirsch, J. (1983) Factors underlying falcon grating acuity, reply. *Nature*, **303**, 729–30

Hirsch, J. & Hylton, R. (1984*a*) Quality of the primate photoreceptor lattice and limits of spatial vision. *Vision Res.*, **24**, 347–55

Hirsch, J. & Hylton, R. (1984*b*) Orientation dependence of visual hyperacuity contains a component with hexagonal symmetry. *J. Opt. Soc. Am. A*, **1**, 300–8

Hirschberg, J. (1882) Zur Dioptrik und Ophthalmoskopie der Fisch und Amphibienaugen. *Arch Anat. Physiol. Lpz.*, **6**, 493–526

Hoffer, K. J. (1983) The ideal IOL. *CLAO J.*, **9**, 204–5

Holladay, J. T., Bishop, J. E., Prager, T. C. & Blaker, J. W. (1983) The ideal intraocular lens. *CLAO J.*, **9**, 15–19

Hopkins, H. H. (1962) The application of frequency response techniques in optics. *Proc. Phys. Soc.*, **79**, 889–919

Howcroft, M. J. & Parker, J. A. (1977) Aspheric curvature for the human lens. *Vision Res.*, **17**, 1217–23

Howland, H. C. & Howland, H. C. (1977) A subjective method for the measurement of monochromatic aberrations of the eye. *J. Opt. Soc. Am.*, **67**, 1508–17

Hueter, G. (1980) *Physiological Optics of the Juvenile Lemon Shark*. Tech. Rep., Office Naval Res., 145 pp. Miami: University of Miami

Hughes, A. (1971) Topographical relationships between the anatomy and physiology of the rabbit visual system. *Doc. ophthal.*, **30**, 33–159

Hughes, A. (1972) A schematic eye for the rabbit. *Vision Res.*, **12**, 123–38

Hughes, A. (1976) A supplement to the cat schematic eye. *Vision Res.*, **16**, 149–54

Hughes, A. (1977*a*) The refractive state of the rat eye. *Vision Res.*, **17**, 927–39

Hughes, A. (1977*b*) The topography of vision in mammals of contrasting life style: comparative optics and retinal organisation. In *Handbook of Sensory Physiology*, vol. 7/5, ed. F. Crescitelli, pp. 613–756. Berlin: Springer

Hughes, A. (1979*a*) The artefact of retinoscopy in the rat and rabbit eye has its origin at the retina/vitreous interface rather than in longitudinal chromatic aberration. *Vision Res.*, **19**, 1293–4

Hughes, A. (1979*b*) A schematic eye for the rat. *Vision Res.*, **19**, 56–588

Hughes, A. (1981*a*) Cat retina and the sampling theorem; the relation of transient and sustained brisk-unit cut-off frequency to alpha- and beta-mode cell density. *Exp. Brain Res.*, 42, 196–202

Hughes, A. (1981*b*) Population magnitudes and distribution of the major modal classes of cat retinal ganglion cell as estimated from HRP filling and a systematic survey of the soma diameter spectra from classical neurones. *J. Comp. Neurol.*, 197, 303–39

Hughes, A. (1981*c*) One brush tailed possum can browse as much pasture as 0.06 sheep which may indicate why this 'arboreal' animal has a visual streak; some comments on the terrain theory. *Vision Res.*, 21, 957–8

Hughes, A. (1985) New perspectives in retinal organisation. In *Progress in Retinal Research*, vol. 4, ed. N. N. Osborne & G. J. Chader, pp. 243–313. Oxford: Pergamon Press

Hughes, A. & Vaney, D. I. (1978) The refractive state of the rabbit eye: variation with eccentricity and correction for oblique astigmatism. *Vision Res.*, 18, 1351–5

Hughes, A. & Whitteridge, D. (1973) The receptive fields and topographical organisation of goat retinal ganglion cells. *Vision Res.*, 13, 1101–14

Jagger, W. S. (1985) Visibility of photoreceptors in the intact cane toad eye. *Vision Res.*, 25, 729–31

Jagger, W. S. (in press) Optical performance of the cane toad eye. *Vision Res.*

Jalie, M. (1978) The design of intra-ocular lenses. *Br. J. Physiol. Optics*, 32, 1–21

Jennings, J. A. M. & Charman, W. N. (1981) Off-axis image quality in the human eye. *Vision Res.*, 21, 445–55

Johnson, G. L. (1901) contributions to the comparative anatomy of the mammalian eye, chiefly based on ophthalmoscopic examination. *Phil. Trans. B.*, 194, 1–82

Kerns, R. L. (1974) Clinical evaluation of the merits of an aspheric front surface contact lens for patients manifesting residual astigmatism. *Am. J. Optom. Physiol. Optics*, 51, 750–7

Kinnucan, P. (1984) Superfighters. *High Tech.*, 4, 36–48

Kohler, I. (1964) The formation and transformation of the visual world. *Psychol. Issues*, 3, 28–46, 116–33

Land, M. F. (1981) Optics and Vision in Invertebrates. In *Comparative Physiology and Evolution of Vision in Invertebrates, B. Invertebrate Visual Centers and Behaviour, I.* ed. H. Autrum, pp. 471–592. Berlin: Springer Verlag

Land, M. F. & Snyder, A. W. (1985) Cone mosaic observed directly through natural pupil of live vertebrate. *Vision Res.* 25, 1519–23

Le Grand, Y. (1967) *Form and Space Vision.* Bloomington, Ind.: Indiana University Press

Leibowitz, H., Johnson, C. & Isabella, E. (1972) Peripheral motion detection and refractive error. *Science*, 177, 1207–8

Levi, D. M. & Klein, S. (1985) Sampling of spatial information in central and peripheral retina. In *Proc. O.S.A., Optics News*, 11, 115

Listing, J. B. (1845) *Beitrag zur Physiologischen Optik.* Göttingen

Lotmar, W. (1971) Theoretical eye model with aspherics. *J. Opt. Soc. Am.*, 61, 1522–9

Ludlam, W. M. & Twarowski, C. J. (1973) Ocular-dioptric component changes in the growing rabbit. *J. Opt. Soc. Am.*, 63, 95–8

Luneberg, R. K. (1944) *Mathematical Theory of Optics.* Providence, R. I.: Brown University Graduate School

McConnell, J. M. S., Duvall, J. & Cullen, J. F. (1983) Posterior chamber intraocular lens implantation. *Trans. ophthal. Soc. UK*, 103, 532–6

McCrane, E. P., DeMonasterio, F. M., Schein, S. J. & Carusa, R. C. (1983) Non-fluorescent dye staining of primate blue cones. *Invest. Ophthalmol. Vis. Sci.*, 24, 1449–55

Mandell, R. B. (1971) Mathematical model of the corneal contour. *Br. J. Physiol. Optics*, 26, 183–97

Mandelman, T. & Sivak, J. G. (1983) Longitudinal chromatic aberration of the vertebrate eye. *Vision Res.*, 23, 1555–9

Marshall, J., Mellerio, J. & Palmer, D. A. (1973) A schematic eye for the pigeon. *Vision Res.*, 13, 2449–53

Martin, G. R. (1982) An owl's eye: schematic optics and visual performance in *Strix aluco* L. *J. Comp. Physiol.*, 145, 341–9

Martin, G. R. (1983) Schematic eye models. In *Progress in Sensory Physiology*, vol. 4, ed. D. Ottoson, pp. 147–98. Berlin: Springer-Verlag

Massof, R. W. & Chang, F. W. (1972) A revision of the rat schematic eye. *Vision Res.*, 12, 793–6

Mathiessen, L. (1877) *Grundriss der Dioptrik geschichteter Linsensysteme.* Leipzig.

Mathiessen, L. (1879) Die Differentialgleichungen der Dioptrik der geschichteten Krystallinse. *Pflügers Arch. Ges. Physiol.*, 19, 480–562

Mathiessen, L. (1882) Über die Beziehungen welche zwischen dem Brechungsindex des Kerncentrums der Krystallinse und der Dimensionen des Auges bestehen. *Pflügers Arch. Ges. Physiol.*, 27, 510–23

Mathiessen, L. (1886*a*) Beiträge zur Dioptrik der Krystallinse, I. *Z. Vergl. Augenheilk.*, 4, 1–39

Mathiessen, L. (1886*b*) Über den physikalisch-optiken Bau des Auges der Cetacean und der Fische. *Pflügers Arch. Ges. Physiol.*, 38, 521–8

Mathiessen, L. (1886*c*) Über den physikalisch-optischen Bau des Auges der Vögel. *Pflügers Arch. Ges. Physiol.*, 38, 521–8

Mathiessen, L. (1887*a*) Beiträge zur Dioptrik der Kristallinse, II & III. *Z. Vergl. Augenheilk.*, 5, 21–44, 97–126

Mathiessen, L. (1887*b*) Über den physikalisch-optischen Bau des Auges von *Corvus alcesmos. Pflügers Arch. Ges. Physiol.*, 40, 314–23

Mathiessen, L. (1893) Über den physikalisch-optischen Bau des auges von Knölwal (*Megaptera boops*, Fabr.) und Finwal (*Balaenoptera musculus* Comp.). *Z. vergl. Augenheilk.*, 7, 77–101

Maxwell, J. C. (1854) Solutions of problems. *Cambridge and Dublin Mathematical Journal* 8, 188. In *The Scientific Papers of J. C. Maxwell*, vol. I, (1890), ed. W. D. Niven, pp. 74–9. London: Cambridge University Press

Maxwell, J. C. (1858) On the general laws of optical instruments. *Quarterly Journal of Pure and Applied Mathematics*, VII. In *The Scientific Papers of J. C. Maxwell*, vol. I, (1890), ed. W. D. Niven, pp. 271–85. London: Cambridge University Press

Miller, D. (1979) On understanding the optics of intraocular lenses. *Surv. Ophthalmol.*, 24, 39–44

Miller, W. H. & Bernard, G. D. (1983) Averaging over the foveal receptor aperture curtails aliasing. *Vision Res.*, 23, 1365–9

Millodot, M. (1969) Variation of visual acuity with contact lenses. *Arch. Opthalmol.*, 82, 461–5

Millodot, M. (1975) Variation of visual acuity with soft contact lenses: a function of luminance. *Am. J. Optom. Physiol. Optics*, 52, 541–4

Millodot, M. (1981) Effect of ametropia on peripheral refraction. *Am. J. Optom. Physiol. Opt.*, 58, 691–5

Millodot, M., Johnson, C. A., Lamont, A. & Leibowitz, H. W. (1975) Effect of dioptrics on peripheral visual acuity. *Vision Res.*, 15, 1357–62

Millodot, M. & Newton, I. A. (1976) A possible change of refractive index with age and its relevance to chromatic aberration. *Albrecht v. Graefes. Arch. Ophthal.*, 201, 159–67

Millodot, M. & O'Leary, D. J. (1978) The discrepancy between retinoscopic and subjective measurements: effect of age. *Am. J. Optom. Phys. Optic.*, 55, 309–16

Millodot, M. & O'Leary (1980) On the artefact of retinoscopy and chromatic aberration. *Am. J. Opt. Phys. Opt.*, 57, 822–4

Millodot, M. & Sivak, J. (1978) Hypermetropia of small animals and chromatic aberration. *Vision Res.*, 18, 125–6

Millodot, M. & Sivak, J. (1979) Contribution of the cornea and lens to the spherical aberration of the eye. *Vision Res.*, 19, 685–7

Moore, D. T. (1971) Design of singlets with continuously varying indices of refraction. *J. Opt. Soc. Am.*, 61, 886–94

Moser, L. (1844) Ueber das Auge. *Repert. Physik.*, V, 240–390

Nakao, S. N., Fujimoto, S., Nagata, R. & Iwata, K. (1968) Model of refractive index distribution in the rabbit crystalline lens. *J. Opt. Soc. Am.*, 58, 1125–30

Nakao, S. N., Mine, K., Nishioka, K. & Kamiya, S. (1970) New schematic eye and its clinical applications. *Abstracts 21st Int. cong. Ophthalmol. Mexico*, E 102

Nathan, J., Keily, P. M., Crewther, S. G. & Crewther, D. P. (1985) Disease-induced visual image degradation and spherical refractive errors in children. *Am. J. Optom. Physiol. Optics*, 62, in press.

Neefe, C. W. (1984) The optics of spin-cast lenses. *Contact Lens Forum*, Sept., 23–9

Nordlohne, M. E. (1974) The intraocular implant lens, development and results with special reference to the Binkhorst lens. *Doc. Ophthalmol*, 38, 1–270

Nuboer, J. F. W. & Van Genderen Takken, H. (1978) The artifact of retinoscopy. *Vision Res.*, 18, 1091–6

Nyquist, H. (1919) Certain topics in telegraph transmission theory. *Trans. AIEE*, 47, 617–44

O'Leary, D. & Millodot, M. (1978) The discrepancy between retinoscopic and subjective refraction: effect of light polarisation. *Am. J. Optom. Phys. Optic.*, 55, 553–6

Olson, R. J., Kolodner, H. & Kaufman, H. E. (1979) The optical quality of currently manufactured intraocular lenses. *Am. J. Ophthalmol.*, 88, 548–51

Oswaldo-Cruz, E., Hokoc, J. N. & Sousa, A. P. B. (1979) A schematic eye for the opossum. *Vision Res.*, 19, 263–78

Palmer, D. A. & Sivak, J. (1981) Crystalline lens dispersion. *J. Opt. Soc. Am.*, 71, 780–2

Parker, J. A. (1972) Aspheric optics of the human lens. *Can. J. Ophthalmol.*, 7, 168–75

Pask, C. (in press) The theory of non-destructive lens index distribution measurement. In *Modelling the Eye with Gradient Index Optics*, ed. A. Hughes. Cambridge: Cambridge University Press

Phillipson, B. (1969) Distribution of protein within the normal rat lens. *Invest. Ophthal. Vis. Sci.*, 8, 258–70

Pomerantzeff, O. (in press) A wide angle optical model of the human eye. In *Modelling the eye with Gradient Index Optics*, ed. A. Hughes. Cambridge: Cambridge University Press

Pomerantzeff, O., Govignon, J. & Schapens, C. L. (1971) Wide-angle optical model of the human eye. *Ann. Ophthalmol.*, 3, 815–19

Porterfield, W. (1759) *A Treatise on the Eye, the Manner and Phenomena of Vision*. Edinburgh: Hamilton and Balfour

Prince, J. H. (1964) *The Rabbit Eye in Research*. Springfield, Ill.: Thomas

Pumphrey, R. J. (1961) Concerning vision. In *The Cell and the Organism*, ed. J. A. Ramsay & V. B. Wigglesworth, pp. 193–208. London: Cambridge University Press

Rabin, J., Van Sluyters, R. C. & Malach, R. (1981) Emmetropization: a vision-dependent phenomenon. *Invest. Ophthalmol. Vis. Sci.*, 20, 561–4

Rempt, F., Hoogerheide, J. & Hoogenboom, W. P. H. (1976) Influence of correction of peripheral refractive errors on peripheral static vision. *Ophthalmologica*, 173, 128–35

Remtulla, S. & Hallett, P. E. (1985) A schematic eye for the mouse and comparisons with the rat. *Vision Res.*, 25, 21–31

Ridley, H. (1951) Intra-ocular acrylic lenses. *Trans. Ophthal. Soc. UK*, 71, 617–21

Rivamonte, A. (1977) The under-corrected lens of the frog eye (*Rana esculenta*) could yield comparable aerial and underwater vision. *Vision Res.*, 17, 1237–8

Robson, J. G. & Enroth-Cugell, C. (1978) Light distribution in the cat's retinal image. *Vision Res.*, 18, 159–73

Rochon-Duvigneaud, A. (1943) *Les Yeux et la Vision des Vertèbres*. Paris: Masson

Röhler, R. (1962) Die Abbildungseigenschaften der Augenmedien. *Vision Res.*, 2, 391–492

Ronchi, L. & Cetica, M. (1974) Off-axis aberrations of the human eye. An annotated bibliography. *Fond. G. Ronchi. At.*, 29, 803–10

Rose, A. (1942) The relative sensitivities of television pick-up tubes, photographic film and the human eye. *Proc. IRE*, 30, 293–300

Rosenblum, W. M. & Hendrickson, P. (1981) Edge-trace analysis of ocular implant lenses. *Cont. Intraoc. Lens Med. J.*, 7, 345–7

Rosenblum, V. M. & Shealy, D. L. (1979) Caustic analysis of intraocular lens implants in humans. *Cont. Intraoc. Lens Med. J.*, 5, 136–40

Saito, H. A. (1983) Morphology of physiologically identified X-, Y- and W-type retinal ganglion cells of the cat. *J. Comp. Neurol.*, 221, 279–88

Sands, P. J. (1970) Third order aberrations of inhomogeneous lenses. *J. Opt. Soc. Am.*, 60, 1436–43

Sands, P. J. (in press) Modelling the geometrical optics of eyes in *Modelling the Eye with Gradient Index Optics.*, ed. A. Hughes. Cambridge: Cambridge University Press.

Seidel, L. (1856) *Astronomische Nachr.*, **43**, 289–321

Senff, R. (1846) Sehen. In *Handwörterbuch der Physiologie*, Band III, ed. R. Wagners.

Shannon, C. E. & Weaver, W. (1949) *The Mathematical Theory of Communication*. Illinois: University of Illinois Press

Sheard, C. (1922) The comparative value of various methods and practices of skiametry. *Am. J. Physiol. Optics*, **3**, 177–208

Shirley, S. Y. (1984) Complications of intracapsular cataract extraction with anterior chamber implants. *CLAO J.*, **10**, 140–2

Shlaer, R. (1972) An eagles eye: quality of the retinal image. *Science*, **176**, 920–2

Sivak, J. G. (1974) The refractive error of the fish eye. *Vision Res.*, **14**, 209–13

Sivak, J. G. & Kreuzer, R. O. (1983) Spherical aberration of the crystalline lens. *Vision Res.*, **23**, 59–70

Sivak, J. G. & Mandelman, T. (1982) Chromatic dispersion of the ocular media. *Vision Res.*, **22**, 997–1003

Smirnov, H. S. (1961) Measurement of wave aberration in the human eye. *Biophysics*, **6**, 52–66

Smith, R. S. & Kutro, J. (1984) Computer analysis of radial keratotomy. *CLAO J.*, **10**, 241–8

Snyder, A. W. (1979) the Physics of Vision in Compound Eyes. In *Handbook of Sensory Physiology*, vol. VII/6A., ed. Autrum, pp. 225–314. Berlin: Springer Verlag

Snyder, A. W., Bossomaier, T. R. J. & Hughes, A. (1985) Optical image quality and the cone mosaic. *Science*, in press

Snyder, A. W. & Miller, W. H. (1978) Telephoto lens system of falconiform eyes. *Nature*, **275**, 127–9

Sorsby, A., Benjamin, B., Davey, J. B., Sheridan, M. & Tanner, J. M. (1957) Emmetropia and its aberrations. *MRC Report*, No. 293

Sorsby, A., Benjamin, B., Davey, J. B., Sheridan, M. & Tanner, J. M. (1961) Refraction and its components during the growth of the eye from the age of three. *MRC Report*, No. 301

Southall, J. P. C. (1964) *Mirrors, Prisms and Lenses*. New York: Dover.

Sroczynski, S. (1975*a*) Die sphärische Aberration der Augenlinse der Regenbogenforelle (*Salmo gairdneri* Rich.). *Zool. Jb. Physiol.*, **79**, 204–12

Sroczynski, S. (1975*b*) Die spärische Aberration der Augenlinse des Hechts (*Esox lucius* L.). *Zool. Jb. Physiol.*, **79**, 547–58

Sroczynski, S. (1976*a*) Untersuchungen über die Wachstumsgesetzmassigkeiten des Sehorgans des Hechts (*Esox lucius* L.). *Arch. Fisch Wiss.*, **26**, 137–50

Sroczynski, S. (1976*b*) Die chromatische Aberration der Augenlinse der Regenbogenforelle (*Salmo gairdneri* Rich.) *Zool. Jb. Physiol.*, **80**, 432–50

Sroczynski, S. (1977) Spherical aberration of crystalline lens in the roach, *Rutilus rutilus* L. *J. Comp. Physiol. A*. **121**, 135–44.

Sroczynski, S. (1978) Die chromatische Aberration der Augenlinse der Bachforelle (*Salmo trutta fario* L.). *Zool. Jb. Physiol.*, **82**, 113–33

Steinberg, R. H., Reid, M. & Lacey, P. L. (1973) The distribution of rods and cones in the retina of the cat. *J. Comp. Neurol.*, **148**, 229–48

Stenström, S. (1948) Investigation of the variation and correlation of the optical elements of human eyes. *Am. J. Optom.*, **25**, 218–32.

Stone, J. & Fukuda, Y. (1974) Properties of cat retinal ganglion cells: A comparison of W-cells with X- and Y-cells. *J. Neurophysiol.*, **37**, 722–48

Strong, J. (1958) *Concepts of classical Optics*. San Francisco: Freeman

Suthers, R. A. & Wallis, N. E. (1970) Optics of the eyes of echo locating bats. *Vision Res.*, **10**, 1165–73

Thibos, L. N. & Walsh, D. J. (1985) Detection of high frequency gratings in the periphery. *Proc. O.S.A. Optics News*, **11**, 117

Vakkur, G. J. (1967) *Studies on the Optics and Neurophysiology of Vision*. M.D. Thesis, University of Sydney

Vakkur, G. J. & Bishop, P. O. (1963) The schematic eye in the cat. *Vision Res.*, **3**, 357–81

Vakkur, G. J., Bishop, P. O. & Kozak, W. (1963) Visual optics in the cat, including posterior nodal distance and retinal landmarks. *Vision Res.*, **3**, 289–314

Van den Brink, G. (1962) Measurements of the geometrical aberrations of the eye. *Vision Res.*, **2**, 233–4

Walls, G. L. (1942) *The Vertebrate Eye and its Adaptive Radiation*. New York: Hefner

Walsh, G., Charman, W. N. & Howland, H. C. (1984) Objective technique for the determination of monochromatic aberrations of the human eye. *J. Opt. Soc. Am. A*, **1**, 987–92

Wang, G. J. & Pomerantzeff, O. (1982) Obtaining a high quality retinal image with a biconvex intraocular lens. *Am. J. Ophthalmol.*, **94**, 87–90

Wang, G. J., Pomerantzeff, O. & Pankratov, M. M. (1983) Astigmatism of oblique incidence in the human model eye. *Vision Res.*, **23**, 123–31

Wehner, R. (1981) Spatial vision in arthropods. In *Handbook of Sensory Physiology. Comparative Physiology and Evolution of Vision in Invertebrates. C, Invertebrate Visual Centres and Behaviour, II*, ed. H. Autrum, pp. 287–616. Berlin: Springer

Westheimer, G. (1979) The spatial sense of the eye. *Invest. Ophthalmol.*, **18**, 893–912

Whitteridge, D. W. (1973) Projection of the optic pathways to the visual cortex. In *Handbook of Sensory Physiology, VII/3B*, ed. R. Jung, pp. 247–68. Berlin: Springer

Wigton, J. S. (1978) Review of the literature on intraocular lenses 1976–1977. *Am. J. Optom. Physiol. Optics*, **55**, 780–91

Williams, D. R. (1984) Aliasing in human fovea. *Vision Res.*, **25**, 195–206.

Williams, D. R. & Collier, R. (1983) Consequences of spatial sampling by a human photoreceptor mosaic. *Science*, **221**, 385–7

Yellot, J. I. (1983) Spectral consequences of photoreceptor sampling in the rhesus retina. *Science*, **22**, 382–5

Yellot, J. I. (1984) Image sampling properties of photoreceptors: a reply to Miller and Bernard. *Vision Res.*, **24**, 281–2

Young, T. (1801) On the mechanism of the eye. *Phil. Trans. R. Soc.*, B, **92**, 23–88

6

Control of intensity in a raster-based colour exchange stimulator for studies on vision

R. W. RODIECK and D. M. McLEAN

Photoreceptors, like photodiodes, respond to the number or rate of photons they effectively catch. William Rushton recognized the analytic power of this relation, which he termed 'univariance' (Naka & Rushton, 1966). It implies that, even when two lights have different spectral compositions, a photoreceptor can be made unaware of a substitution of one light for the other. All that is necessary is for the intensity of one of the lights to be adjusted so that the photoreceptor catches the same number of photons from either light.

Rodieck & Rushton (1976a,b) investigated the interactions of rods and cones at the level of cat ganglion cells, using an optical stimulator that neatly exchanged two lights of different spectral composition. When the two lights were made equal for the rods, the response to their exchange was then due entirely to the cones.

Thus, when there are two spectrally distinct photoreceptors contributing to a response, it is possible to silence one of them by adjusting the intensity of one of two spectrally distinct lights. When the number of photoreceptors rises to three, an additional light is required. Then, by adjusting the intensities of two of them, two of the cone types can be silenced, so that any response to exchange is due entirely to the third. Estévez & Spekreijse (1982) and Rodieck (1983) have discussed how this can be done.

When exchange stimulation is used to observe the response of a single photoreceptor type, the critical factor is not the stimulation of the type that is isolated

90

but the balance of the lights that makes the other photoreceptors unaware of the exchange. This places strong constraints on the control of intensity. Here we assess whether a visual stimulator, based upon a color monitor, can meet these requirements. The stimulator has been described by Rodieck (1983) and consists of a digital image processor and an RGB color monitor.

Fig. 6.1 shows the intensity control circuit for one of the three spectral channels. Within the image processor, the digital signal is transformed to a voltage level by a digital-to-analog converter (D/A), and sent to the color monitor. We rebuilt this circuitry to allow the full eight bits of digital information to specify 256 different intensity levels, and made use of an additional line to transmit the synchronization signal. The voltage is amplified within the monitor and applied to the cathode of the display tube, where it controls the beam current to one of the three phosphors.

Fig. 6.2 (upper) shows measured screen radiance, plotted as a function of the digital value presented to the D/A. It is apparent that the relation is highly non-linear. As discussed elsewhere (Rodieck, 1983), this non-linearity originates in the relation between the grid-to-cathode voltage (E) and the beam current (I) of the display tube, and has the form $I \sim E^{\gamma}$; the factor γ was close to 2.0 for each of the three beams of the display tube we tested.

The fact that the output intensity is related non-linearly to input voltage is not in itself important, provided that this relation is known. The computer that controls the image processor can be programmed to select the digital value presented to the D/A that will produce the desired intensity. Originally, we determined

the value of γ from three intensity measurements and used this value in the above equation to predict screen radiance for each of the digital inputs. However, certain digital inputs yielded measured intensities that consistently deviated from the calculated value. These deviations, which result from imperfections in the D/A of the image processor, are seen as slight kinks in the curve of intensity *versus* input value, indicated by arrows in Fig. 6.2 (upper). They are more evident in the derivative of this curve, shown in Fig. 6.2 (lower). Because of them, we now measure and store within the computer a table of the screen radiance of each beam at each of the 256 digital input levels.

A consequence of the non-linearity between input voltage and screen radiance is that equal steps of the input do not produce equal steps in screen radiance. This situation is summarized in Fig. 6.3, in which change in intensity per unit change of the digital input to the D/A is plotted as a function of relative intensity, assuming a value of 2.0 for γ. The upper line shows the percentage change, expressed relative to the intensity at that point, while the dashed line below it shows the percentage change expressed relative to maximum intensity. For intensity values greater than 60% of the maximum, the smallest intensity change, relative to that value, is less than 1%.

Fig. 6.2. Screen intensity *versus* digital input value. The upper graph shows the measured screen intensity generated by each of the 256 digital input values. The arrows indicate the points at which the measured intensity deviated from a smooth curve. These kinks, which are caused by imperfections in the D/A within the image processor, are seen more clearly in the lower graph, which plots the change in intensity per unit increase in the digital input value.

Fig. 6.1. Schematic showing one of three identical intensity-control circuits of a visual stimulator, consisting of an image processor and color monitor. Eight bits of digital information are converted within the image processor to a voltage which, when amplified within the color monitor, is applied to one of the cathodes of the color tube to control beam intensity. The photodiode is used for measurement and calibration.

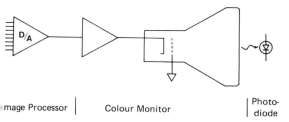

| Image Processor | Colour Monitor | Photo-diode |

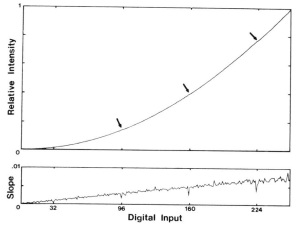

A resolution of 1 % for each of the three beams is sufficient for our studies on single visual neurons. The critical issue is then whether the values in the intensity table are reliable. There are three factors that can compromise reliability. First, the screen radiance may drift with time. Secondly, the measured values may be inaccurate because of non-linearities or noise in the measuring apparatus. Thirdly, the R(red), G(green) and B(blue) channels may interact, so that a change in the intensity of one beam would influence the intensity of the other beams.

In order to assess these factors, it is first necessary to determine some bounds for the degree of non-linearity and drift of the apparatus used to determine the relative screen radiance. This apparatus is shown schematically in Fig. 6.4. The photodetector is a PIN photodiode, whose properties are summarized by Wyszecki & Stiles (1982). Light falling on the photodiode produces a current that is converted to a voltage by an operational amplifier. This signal is smoothed by a second operational amplifier, and converted to a digital value by the analog-to-digital converter (A/D). The converter has a 12 bit resolution

(i.e. 0.025 % of the full-scale value); over the smaller voltage range that we presented, the resolution for single measurements was about 0.1 %.

The computer was programmed to cause the A/D to sample the smoothed signal from the photodiode at the same point on each frame, so as to eliminate the residual ripple in the output of the smoothing circuit. We then observed a slow and periodic change in the measured intensity, having an amplitude of 1 % of the steady level, and a frequency of about 0.3 Hz. This was traced to a beating of the mains frequency with the field frequency of the color monitor, presumably caused by a small amount of 'hum' on the R, G and B signal lines. Since all three beams were influenced in the same way, the effect was a small and slow intensity change that would not disturb the intensity ratios between the three beams, and thus would not affect the exchange stimulation. However, this fluctuation was about as large as the minimum intensity change the

Fig. 6.3. Intensity step size *versus* relative intensity. Each unit step in the digital input produces a corresponding step in screen intensity. The size of the intensity step varies according to a power–law relation between input voltage and screen radiance. Here, the intensity step size is shown as a function of intensity, assuming a value of 2.0 for the exponent of the power–law relation. The upper solid line shows the step as a percentage of the intensity at that point; the dashed line, as a percentage of the maximum intensity.

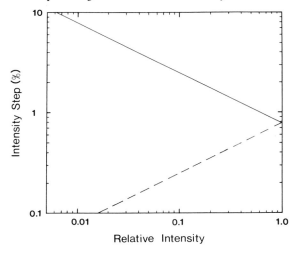

Fig. 6.4. Schematic diagram of the measuring apparatus. The photodiode (P) (United Detector Technology PIN-10DF) had a filter before it to produce a flat energy spectrum ($\pm 5\%$). The photodiode/filter combination had a responsivity of 0.5 A/W, and an active area of 1 cm^2. It was placed directly against the screen, so that its output could be used to calculate the local screen radiance. The current generated by the incident light was converted to a voltage by an operational amplifier having a transimpedance of 10 V/A (UDT Model 101A). The rise time of this combination (10 μs) was short compared to the time constants of the phosphors (about 200 μs), so that the voltage signal followed the beam each time it crossed the screen. Since the height of the active area of the photodiode (1 cm) was small compared to the height of the screen (18 cm), the response had a short duty cycle, appearing for only a small part of the period of each field. To avoid the complexities of sampling the voltage during this period, and to match the output voltage to the dynamic range of the analog-to-digital converter, an intermediate low-pass amplifier was used, which had a time constant of 300 ms and a low-pass gain of 4. Both amplifiers were powered by batteries, to avoid ground-loop noise.

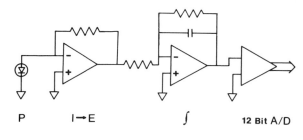

apparatus could produce (Fig. 6.3). When the intensity table was constructed, this fluctuation was cancelled by averaging the measured values across a single cycle of the beat frequency, using a circuit that detected the zero crossings of this subharmonic.

Fig. 6.5 shows the drift in the ouput of the measuring system when the light source was a 15 W bulb, driven by a regulated DC power supply. As shown in this figure, the combined drift of source and detector was less than 0.15% which was sufficient for our purposes.

Linearity of the measuring system was evaluated by a superposition test, using the apparatus shown in Fig. 6.6. Light from the 15 W bulb (L) was collimated (lenses not shown), attenuated by a neutral wedge (W), and split into two beams (I_1 and I_2). Two masks (M_1 and M_2) determined whether either of the two beams,

or both, were recombined and imaged on the photodiode (P). If the measuring apparatus is linear, then the measure of both beams together ($I_{1,2}$) should equal the sum of the measures of each beam separately ($I_1 + I_2$). The neutral wedge allowed the linearity of the photodiode to be tested over different parts of its operating range. Over the range from the equivalent of maximum screen radiance to 10% of that value, the deviation from linearity, based on the measure for non-linearity devised by Sanders (1962), was less than 0.17%.

To summarize, the measuring apparatus is sufficiently linear to obtain reliable values of relative screen radiance, and sufficiently stable to detect drifts in these values with time.

Fig. 6.7 shows the variation in screen radiance of two of the three beams with time. Shortly before recording, the measuring apparatus was turned on, and the initial variations in measured intensity presumably reflect 'warm-up' drift in this apparatus. Subsequently, the measured intensity of either beam was observed to vary by as much as 0.7%.

From the point of exchange stimulation, however, the critical issue is not the degree to which the intensities of the individual beams vary, but the degree

Fig. 6.5. Drift in the output of the intensity measuring system, in response to a 15 W bulb driven by a regulated DC power supply.

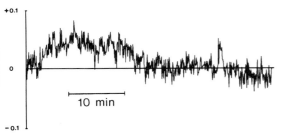

Fig. 6.6. Apparatus used to test the linearity of the intensity measuring system. Light from a collimated light source (L) is attenuated by a neutral wedge (W) and split into two paths, I_1 and I_2. Each beam is gated by a mask (M_1 and M_2), and then recombined and imaged on the photodiode (P). If the system is linear, then the measured value to the two beams presented together, ($I_{1,2}$), will equal the sum of the two measured separately ($I_1 + I_2$).

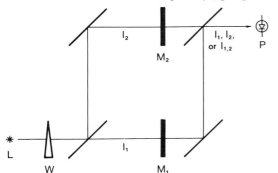

Fig. 6.7. The top two records show the change in screen radiance of the red and green beams of the color monitor, expressed as a percentage change. The screen was alternately made red, then green, allowing the intensities of the two beams to be recorded over the same time interval. The measuring apparatus had been turned on shortly before these records were compiled. The ratio of the two intensity records is shown at the bottom.

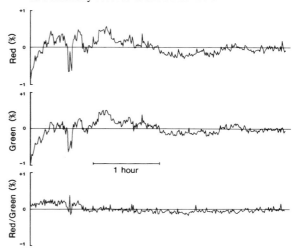

to which the ratio of these intensities vary. For if the ratio remains constant, the effect is equivalent to a small and slow intensity change, which would not disturb the relative contributions of the two lights in the exchange stimulus. In fact, the intensities of the beams tended to vary together, so that their ratio, shown in the lower record, varied by no more than about 0.33%.

The final intensity factor to be evaluated is the independence of the three beams. We checked for independence by arranging the display for one beam to be a small spot on the monitor screen, centered on the receptive field of the photodiode. The output of the photodiode was recorded when the spot was turned on and then off, and the difference between these readings provided a measure of the intensity of the spot. Another beam was arranged to cover most of the screen. This background beam was set to a given intensity, and remained steady at that value while the measurement of spot intensity was made. If the beams are independent, then variation in the intensity of one of them should not influence the intensity change produced by the other. Fig. 6.8 shows the percentage change in the measured spot intensity, plotted as a function of the intensity of the background beam. The maximum change illustrated here (1%) was typical of that observed for other combinations. In this situation the background beam also covered the region of the spot, but similar results were observed when the background beam formed an annulus around the spot.

To summarize, the important intensity factors concerned with exchange stimulation are: drift in the intensity ratios of the three beams, which was found to vary by no more than 0.33%; interactions between beam intensities, which was found to be less than 1%; and intensity resolution, which was about 1% in midrange. These factors are small enough to allow a wide range of visual investigations.

Acknowledgements

Supported by NIH grants EY-02923, EY-01730, and the E.K. Bishop Foundation.

References

Estévez, O. & Spekreijse, H. (1982) The 'silent substitution' method in visual research. *Visual Res.*, **22**, 681–91

Naka, K.-I. & Rushton, W. A. H. (1966) An attempt to analyse colour vision by electrophysiology. *J. Physiol.*, **185**, 556–86

Rodieck, R. W. (1983) Raster-based colour stimulators. In *Colour Vision: Physiology and Psychophysics*, ed. J. D. Mollon & L. T. Sharpe, pp. 133–44. London: Academic Press

Rodieck, R. W. & Rushton, W. A. H. (1976a) Isolation of rod and cone contributions to cat ganglion cells by a method of light exchange. *J. Physiol.*, **254**, 759–73

Rodieck, R. W. & Rushton, W. A. H. (1976b) Cancellation of rod signals by cones and cone signals by rods in the cat retina. *J. Physiol.*, **254**, 775–85

Sanders, C. L. (1962) A photocell linearity tester. *Appl. Optics*, **1**, 207–11

Wyszecki, G. & Stiles, W. S. (1982) *Color Science: Concepts and Methods, Quantitative Data and Formulae*, 2nd edn. New York: John Wiley

Fig. 6.8. Test for intensity interactions between beams. A spot, generated by a fixed digital input to the red channel (Fig. 6.1), was presented on the screen, centered on the receptive area of the photodiode. The relative radiance of this spot was determined by taking the difference between the measured intensity while the spot was on and while it was off. A steady green background covered the screen, including the region of the spot, during the measurement of spot intensity. The diagram plots the variation in the relative radiance of the spot for different intensity values of the green background.

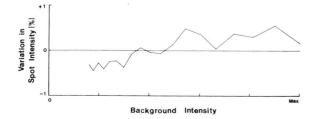

II

Retinogeniculate connections

When Peter Bishop built his first amplifiers they were put to work on a study of synaptic transmission through the cat's lateral geniculate nucleus (LGN) in the early 1950s. At that time attention was focussed on establishing a 'circuit diagram' for this part of the thalamus, with the appropriate connections between the retinal inputs, the relay cells projecting on to the cortex and the inhibitory interneurons. The LGN was found to conform to the general picture of thalamic relays at that time, with some special features such as a very high safety factor for transmission and a marked sensitivity to serotonin antagonists such as LSD. It was also noticed that the retinal inputs to the LGN were in three different groups of conduction velocity. The functional significance of these different fibre types required another 20 years for its elucidation, in the form of the X, Y, W story which unfolded in large part because of the efforts of two of Bishop's teams in Canberra.

This thread linking the early and later work on the LGN is represented in this Section by W. Burke (Chapter 7) who worked on the early LGN circuit diagrams and conduction fibre groups with Peter Bishop and now presents more recent work on the LGN guided by the discovery of the X and Y subclasses.

Burke's chapter is concerned with the function of optic nerve fibre groups in the cat, a topic examined by P. O. Bishop early in his career. By applying a pressure block which selectively knocks out the fast conducting y fibres, Burke and his co-workers have been able to differentiate some of the functional effects of the X and Y systems and to show that visual acuity is probably determined by the X system and the Y system subserves some aspects of fast motion detection.

Cleland (Chapter 8) belonged to one of the Bishop teams who put the X/Y subdivision on the map by correlating the two different receptive-field properties with two conduction velocity groups and by showing that the functional division is maintained after synaptic transmission through the lateral geniculate nucleus.

Cleland's chapter is a review of the circuitry of the dorsal LGN of the cat, with particular emphasis on the role of X and Y cells and interneurons and relay cells. Some of this circuitry has been proposed from the elegant dual recording experiments of Cleland and his co-workers, in which the recordings are from lateral geniculate neurons and the retinal ganglion cells which provide their excitatory input. Supplementing this data are the estimates, taken from the literature, of the numbers of X and Y cells in the retina and the LGN.

Orban's chapter (9) examines the responses of neurons in the primary visual cortex to moving bars and is thus most closely allied with P. O. Bishop's recent work, in which Bishop has also analysed the responses of striate visual cortical neurons to moving targets, and examined the receptive-field properties of different classes of striate neurons.

7

The function of optic nerve fibre groups in the cat studied by means of selective block

W. BURKE

One of Peter Bishop's important early papers was a thorough description of the timecourse of excitation and inhibition in the dorsal lateral geniculate nucleus (LGNd) of the cat following electrical stimulation of the optic nerve (Bishop & Davis, 1960). He was not, however, a pioneer in this area, previous papers being by G. H. Bishop & O'Leary (1940), Marshall & Talbot (1940, 1941, 1942), Talbot & Marshall (1941), Marshall (1949) and Vastola (1959). Bishop & Davis (1960) used the classical two-shock technique for testing the excitability of the LGNd neurons but made a very careful correction for presynaptic excitability changes. Unless this is done, the changes in postsynaptic excitability, at least in the 20–30 ms after the first stimulus, cannot be accurately described. When this correction is made there is a supernormality at the LGNd synapses lasting about 10 ms, following a brief refractory period, and this is then followed by a deep and prolonged inhibition extending to beyond 1 s (Fig. 7.1).

Subsequent work has tended to ignore the supernormality and attention has concentrated on the inhibitory phase. Research has been mainly concerned with the mechanism of the inhibition but several authors have also commented on its possible function. The role of the inhibition is likely to become clearer as we learn more about its details.

Inhibitory mechanisms in the LGNd

Eccles and his colleagues in a series of papers (Andersen & Eccles, 1962; Andersen, Brooks & Eccles, 1964; Andersen, Eccles & Sears, 1964; Andersen & Sears, 1964; Andersen *et al.*, 1964*b*) on the ventrobasal nucleus of the thalamus in the cat proposed a model for inhibition essentially similar to that involving the spinal motor neuron and the Renshaw cell (Eccles, Fatt & Koketsu, 1954). Evidence that this model could be adapted for the LGNd in the rat was provided by Ann Sefton and me (Sefton & Burke, 1965, 1966; Burke & Sefton, 1966*a,b,c*). The evidence is as follows:

Fig. 7.1. Recovery of responsiveness of geniculate synapses following a maximal conditioning shock applied to the contralateral optic nerve. (*a*) Recovery of presynaptic spike (t₁). (*b*) Relation between amplitudes of presynaptic and postsynaptic spikes in response to increasing stimulus strength; open circle shows the mean unconditioned test response during recovery cycle. (*c*), (*d*) Recovery of postsynaptic spike (r_1) uncorrected (*c*) and corrected (*d*) for presynaptic excitability changes. Time scale (abscissa) applies to (*a*), (*c*), (*d*). Ordinate: amplitude of conditioned test response as a percentage of the unconditioned test response (*a*), (*c*), (*d*). (From Bishop & Davis, 1960.)

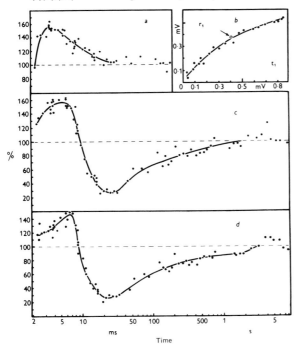

1. Two types of cell can be distinguished in the LGNd of the rat. One type (P cell) responds to an electrical stimulus to the optic nerve with only one or two action potentials whereas the other (I cell) responds with a long burst of spikes (about 10). P cells can be activated from the visual cortex at a short latency consistent with being relay cells. I cells have a longer latency both to optic nerve and to cortical stimulation but they have a low threshold to optic nerve stimulation; they are probably interneurons.

2. Inhibition is deepest when the I cells are discharging and begins to decrease when the I cell discharge ceases.

3. The P cells exhibit a phase of hyperpolarization during the inhibitory phase, which appears to be composed of a succession of inhibitory post-synaptic potentials (IPSPs) at a frequency corresponding to the I cell spike discharge.

4. I cells cannot be re-excited from the optic nerve during the period of P cell inhibition, but they can be re-excited from the visual cortex during this time.

The model which emerged from these data is shown in Fig. 7.2. There is both anatomical and physiological evidence for the existence of short-axon cells in the LGNd both in the rat (Grossman,

Fig. 7.2. Model for inhibition in the LGNd proposed by Burke & Sefton. Excitatory optic tract impulses activate P cells (relay cells) which excite I cells through collaterals. I cells project back on to P cells to inhibit them (modified from Burke & Sefton, 1966*b*).

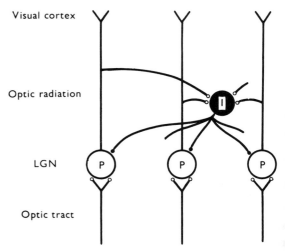

Visual cortex

Optic radiation

LGN

Optic tract

Lieberman & Webster, 1973; Lieberman, 1973; Brauer et al., 1974; Lieberman & Webster, 1974; Hunt et al., 1982) and in the cat (O'Leary, 1940; Tömböl, 1967, 1969; Guillery, 1969, 1971; Famiglietti & Peters, 1972; Dubin & Cleland, 1977; Lin, Kratz & Sherman, 1977; Ahlsén & Lindström, 1978; LeVay & Ferster, 1979; Geisert, 1980; Lindström, 1982) and there is also evidence for collaterals of the relay cells terminating within the nucleus (Tello, 1904; Cajal, 1911; O'Leary, 1940; Szentágothai, 1963). More recent work has, however, modified this model in two significant ways.

Many, if not all, of the I cells described by Burke & Sefton (1966a) lie not in the LGNd but in an adjacent region, part of the reticular nucleus of the thalamus (Sumitomo, Kurioka & Iwama, 1975; Sumitomo, Iwama & Nakamura, 1976; Sumitomo, Nakamura & Iwama, 1976; Hale & Sefton, 1977; French, Sefton & Mackay-Sim, 1979). The homologous region in the cat appears to be the perigeniculate nucleus. There is now anatomical evidence that LGNd relay cells give branches to these regions in both the rat (Lieberman & Ohara, 1981) and the cat (Ahlsén, Lindström & Sybirska, 1978; Friedlander, Lin & Sherman, 1979; Ahlsén & Lindström, 1982a), and that the cells in these regions project back to the LGNd in the rat (Ohara, Sefton & Lieberman, 1980) and in the cat (Ahlsén & Lindström, 1978) and are inhibitory to the relay cells there, at least in the rat (Sumitomo et al., 1975; Sumitomo et al., 1976a; French et al., 1979).

The second way in which the model is inadequate is that it is now very probable that some, and possibly all, intrinsic interneurons in the LGNd receive a direct innervation from the retina. The evidence is:

1. The occurrence of cells activated at a reliable short latency from the retina but not antidromically activated from the visual cortex when adjacent relay cells are so activated (Dubin & Cleland, 1977; Lindström, 1983).

2. The occurrences of inhibitory postsynaptic potentials (IPSPs) in relay cells, of latency too short to be generated via a recurrent inhibitory circuit and of a different waveform to the IPSPs produced by cortical stimulation (fast rising phase and short duration *versus* slow, rippling rising phase and long duration) (Lindström, 1982).

The picture is further complicated by the existence of mutual inhibition between the perigeniculate neurons in the cat (Ahlsén & Lindström, 1982b) and

by the fact that the perigeniculate cells make contact not only with the relay cells of the LGNd but also with intrinsic interneurons there (Ahlsén, Lindström & Lo, 1982). In the rat LGNd there is also the possibility of presynaptic inhibition (Kelly, Godfraind & Maruyama, 1979). In summary, there are at least two powerful inhibitory systems in the LGNd. A further insight into the role of inhibition in the LGNd may come from a consideration of the way in which different classes of relay cells in the LGNd participate in inhibition.

Classes of LGNd relay cells and their roles in inhibition

Optic nerve fibres in the cat are classified into three groups based on conduction velocity. The t_1 fibres have conduction velocities more than about 28 m/s, t_2 fibres have conduction velocities in the region 14–28 m/s, while t_3 fibres have conduction velocities below 14 m/s (for a list of references see Stone, Dreher & Leventhal (1979) and Levick & Thibos (1983)). At the present time there is good reason to think that the t_1 fibres arise from alpha cells in the retina, t_2 fibres from beta cells, and t_3 fibres from gamma cells (Boycott & Wässle, 1974; Cleland & Levick, 1974).

The correlation with physiological properties is more controversial. Enroth-Cugell & Robson (1966) classified retinal ganglion cells as X or Y. X cells showed linear spatial summation in their receptive fields whereas Y cells showed non-linear summation. Cleland, Dubin & Levick (1971b) identified the Y cells with their transient cells having axons in the t_1 group and X cells with their sustained cells with axons in the t_2 group. Subsequently these two groups were referred to as brisk-transient and brisk-sustained cells, respectively (Cleland & Levick, 1974). Although it was logically necessary that retinal ganglion cells possessing t_3 axons must also have either linear or non-linear receptive field properties (and this has been confirmed by Levick & Thibos, 1980), nevertheless the labels 'Y' and 'X' are now firmly attached to cells supplying the t_1 and t_2 axons, respectively, and to the LGNd cells which they innervate. This usage may not be the most desirable but it is now too firmly entrenched to make it worthwhile to try to change it.

The work about to be described was based entirely on measurements of latencies and therefore the terms t_1, t_2 and t_3 are appropriate. However, since the emphasis of the experiments is on the functional importance of these different groups it is necessary to relate these terms to a physiological classification. Accordingly the following equivalencies are used: $t_1 \equiv$ brisk-transient $\equiv$ Y; and $t_2 \equiv$ brisk-sustained $\equiv$ X.

The Y ganglion cells correspond to the morphological alpha cells of Boycott & Wässle (1974), and their axons have the fastest conduction velocity, corresponding to the t_1 fibres of Bishop & McLeod (1954). They respond to a sudden change of illumination or movement with a transient burst of firing but do not respond well or at all to a maintained stimulus. They respond to fast movements rather better than do the X cells (Cleland, Dubin & Levick, 1971b) and it has been suggested that they may function as 'movement detectors' (Ikeda & Wright, 1972). The X ganglion cells correspond to the morphological beta cells of Boycott & Wässle (1974), they are highly concentrated in the area centralis of the retina, and have smaller receptive fields than do Y cells. Their axons have intermediate conduction velocities corresponding to the t_2 cells of Bishop & McLeod (1954). They respond with a sustained discharge to a maintained photic stimulus and hence may function as 'pattern detectors'. The gamma cells of Boycott & Wässle (1974), corresponding to the slowest fibres in the optic nerve (t_3; Bishop, Clare & Landau, 1969), are a mixed group and probably serve different functions. However, they appear to project to specific groups of cells in the LGNd and other visual nuclei which are separate from the Y and X cells (see Rodieck (1979) for a fuller account of these pathways).

Although the three groups maintain almost completely separate pathways, at least up to the visual cortex, there is considerable inhibitory interaction at LGNd level, at least between the Y and X systems. It is generally accepted that activation of the Y system inhibits both Y and X cells (Hoffmann, Stone & Sherman, 1972; Rodieck, 1973; Singer & Bedworth, 1973). However, it was unclear whether the X system could also exert inhibition either on X cells or on Y cells. Rodieck (1973) considered that the X system did not mediate any inhibition whereas Hoffmann *et al.* (1972) and Singer & Bedworth (1973) thought it did. All three groups of workers reached their conclusions by observing the effect of increasing the strength of the electrical stimulus so as to include X fibres, and comparing this effect with that obtained in the absence of X-fibre stimulation. These experiments were not ideal because any effect due to the X fibres was superimposed on the effect due to the Y fibres. Also, if the stimulus is applied at some distance from the

LGNd, the Y fibres, by virtue of their faster conduction velocity, can inhibit the X cells even before the impulses in the stimulated X fibres arrive at the synapses.

Selective block of Y fibres

An attempt has recently been made to circumvent this difficulty by selectively blocking the Y fibres (Martin, Burne & Burke, 1980, 1981; Burke, Burne & Martin, 1985). The technique depends on the well-known observation that large nerve fibres are more susceptible to pressure block than are small fibres (Gasser & Erlanger, 1929). A theoretical explanation to account for this fact has been provided by MacGregor, Sharpless & Luttges (1975).

In a cat anaesthetized with pentobarbitone, one optic nerve is exposed by a dorsal approach. The part of the skull overlying the frontal sinus is removed as is the thin bone overlying the orbital cavity. The contents of the orbital cavity covering the nerve are removed or retracted. It is necessary to pull the eyeball forward for a good view of the nerve. We found that stretching the optic nerve in this way does not cause any deleterious effects apart from occasionally temporarily restricting the blood supply. A bipolar insulated stainless steel electrode is inserted into the optic nerve just behind the eyeball. The pressure blocking device is applied further back along the nerve. In the original design it consisted of a narrow rigid tube attached to a small balloon which lay within a cut-away section of a brass tube 4 mm long and 5 mm in diameter. This hemi-cylinder was hooked around the nerve so that inflation of the balloon compressed the nerve against the wall of the cylinder. Other electrodes are placed in the intracranial portion of the optic nerve, in the optic tract, LGNd and visual cortex as required. Fig. 7.3 shows a diagrammatic view.

Block of optic nerve fibres is monitored by stimulating behind the eyeball and recording in the nerve or tract, or *vice versa*. There are, of course, an infinite number of combinations of magnitude and duration of pressure that would produce any desired effect. However, application of pressure cuts off the blood supply to the retina and part of the nerve and so should not be maintained indefinitely. On release of pressure there is always some recovery of response. It is necessary, therefore, to proceed by trial and error and to increase either the magnitude or the duration of pressure until a satisfactory result has been obtained.

In practice, we used pressures of 70–200 kPa applied for periods of 10 s to 2 m. The magnitude of these pressures is not of any significance because the balloon does not closely envelop the nerve when at zero pressure.

Fig. 7.4 shows the response recorded from the optic tract to stimulation of the optic nerve behind the eyeball. The t_1 (Y) response appears at shorter latency than the t_2 (X) response. The amplitudes of these responses before, during and after the application of pressure were measured and are shown in the graph. The t_1 response was completely abolished and did not recover within 17 min. In this and other experiments the response was regularly monitored to ensure that there was no recovery of the t_1 response. It was essential, of course, that there should be a complete block of the Y system, otherwise any effects obtained could not be ascribed unequivocally to the smaller fibres. In our initial experiments we recorded from several sites in the optic tracts in order to be sure that the block was satisfactory not merely in one place. We always found good agreement between the levels of block achieved in different sites. In subsequent experiments we always recorded from three or four places before accepting that the block was satisfactory. We found that to produce a complete block of the t_1 response it was necessary to block partly or completely the t_2 response during the application of pressure. However, the t_2 response always made a good recovery, although not always a complete one (Fig. 7.4).

Inhibitory effects of t_2 and t_3 fibres

Test shocks were applied either to the t_1-blocked nerve (stimulating t_2 and t_3 fibres) or to the optic tract (stimulating all types of fibres). A conditioning shock was given to the blocked nerve at various intervals prior to the test shock. The field response due to the conditioned test shock was compared to that due to the unconditioned test shock. These experiments, performed in five cats, showed that the smaller optic fibres could exert an inhibitory effect on Y and X LGNd cells. The main effect is probably due to stimulation of the t_2

Fig. 7.3. Diagram to show the positions of the implanted electrodes and the arrangement of retrobulbar electrode and pressure device in the cat. 1, retrobulbar bipolar optic nerve electrode. 2, pressure cuff. 3, intracranial optic nerve bipolar electrode. 4, optic tract electrode. 5, lateral geniculate nucleus. Each electrode could be used for either stimulation or recording. (From Burke, Burne & Martin, 1985.)

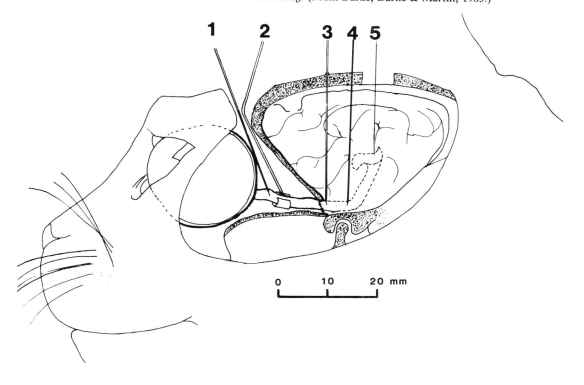

fibres, although, it would have been impossible to avoid stimulating some t_3 fibres. Therefore some caution must be expressed in interpreting the results. Fig 7.5 shows the time-course of the inhibitory effect on the r_1 and r_2 responses (respectively, the postsynaptic Y and X responses in the LGNd) in one experiment. In general, the inhibitory effect of stimulating all fibres was much greater than that due to stimulation of only the t_2 and t_3 fibres. It is difficult to make an accurate comparison but it is clear that the inhibition due to the t_1 fibres far exceeds that due to the t_2 or t_3 fibres. Nevertheless, the smaller fibres can exert an inhibition, in agreement with the conclusions of Hoffmann *et al.* (1972) and Singer & Bedworth (1973).

We have made only a few observations on single fibres but these have confirmed the inhibitory effect exerted on X LGNd cells by the X fibre input. The sample did not include a Y cell. A full single-unit study is planned.

Functional implications of inhibition due to X cells

If X cells are to be regarded as important in high-resolution vision, it seems probable that they would exert a strong lateral inhibition in order to enhance contours. Although this may occur at retinal level there is good evidence that it is a very powerful effect at LGNd level. Hubel & Wiesel (1961) demonstrated that the threshold illumination of certain LGNd cells was greatly increased when the photic stimulus exceeded the receptive field centre although the threshold of the retinal ganglion cells supplying the cells was little affected. That these cells were X cells may be inferred from the results of Fukuda & Stone (1976) showing that Y LGNd cells tended not to show this effect.

We do not know the pathway for this inhibition but we do know that X LGNd cells send axonal branches to the perigeniculate nucleus (Ahlsén *et al.*, 1978; Friedlander *et al.*, 1979), and there is also some

Fig. 7.4. Acute pressure block of t_1 optic nerve fibres. The optic nerve was stimulated behind the eyeball and the response recorded in the optic tract. After a few control responses a pressure of 200 kPa was applied for 10 s. Representative responses are shown on the right, the letters indicating by reference to the graph the times at which they were obtained. Both t_1 and t_2 responses were initially abolished but the t_2 response recovered well, whereas the t_1 response remained depressed (from Burke *et al.*, 1985).

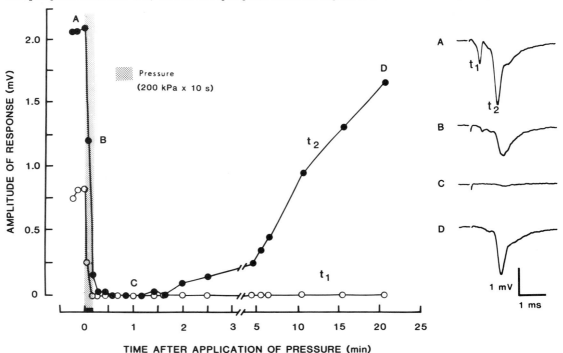

evidence that these branches supply different cells to those supplied by Y cells (Ahlsén & Lindström, 1982*a*). However, because of their proximity to the relay cells, one might expect the intrinsic interneurons to be more appropriate for mediating lateral inhibition.

Long-term effects of selective block

By implanting electrodes in the optic nerves (intracranial segment), optic tracts, LGNd or optic radiation and visual cortex, and by performing the block on one optic nerve using sterilization procedures and aseptic technique, we were able to study the long-term effects of the block. For these experiments we were careful to avoid injuring the extra-ocular muscles. We also retained as much of the skull as possible, including the supra-orbital bony ridge. These cats looked completely normal and appeared to use their eyes in a perfectly normal way. We found that if we used a degree of pressure just sufficient to abolish the entire response $(t_1 + t_2)$ even for 10 s (cf. Fig. 7.4), we could produce a complete loss of t_1 fibres from that nerve.

The time-course of this loss is shown in Fig. 7.6. In this experiment the electrodes had been implanted several weeks before the block in order to establish a steady baseline. The responses were obtained by stimulation of the intracranial optic nerve and recording from the contralateral optic tract (and also *vice versa*). We interpret the loss of the t_1 response after about 4 days as indicating anterograde Wallerian degeneration of the t_1 fibres (Kumarasinghe *et al.*, 1982) and this interpretation was confirmed by histological examination (Westland, Freeman & Burke, 1983; see below). In some of these experiments the loss of the t_1 response was not complete and in others, in addition to the total loss of the t_1 response, there was also some decrease in the t_2 response. This preparation is an ideal one for studying the behavioural consequences of a loss of the Y system.

Effect of loss of Y system on acuity

If it is assumed that a major role of the X system is acuity, then it would be expected that loss of the Y system would have little effect on acuity. Using the

Fig. 7.5. Graph of amplitudes of postsynaptic responses r_1 and r_2 *versus* interval after conditioning stimulus applied only to the t_2 fibres. The r_2 responses are expressed as a percentage of the unconditioned response (right ordinate). Because the r_1 responses were superimposed on the t_2 responses the results are expressed as a change in the amplitude of the combined response (left ordinate). Up to an interval of 20 ms the conditioning stimulus was a single shock: from 30 ms onwards the conditioning stimulus consisted of a train of six shocks at 3.2-ms intervals. Asterisks indicate a significant difference from the control value ($P < 0.05$) (From Burke *et al.*, 1985).

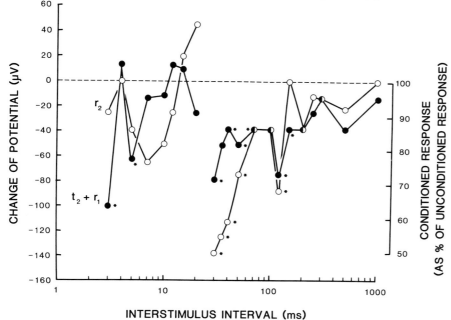

chronic preparation described above, we determined the acuity of four cats using the Mitchell jumping apparatus (Mitchell, Giffin & Timney, 1977). In each cat only one optic nerve was blocked (with 100% loss of the t_1 response). Each eye was separately tested, the other eye being occluded either by an opaque contact lens or a special shield.

The cats were trained to discriminate a square-

Table 7.1. *Estimates of acuity. (Values in cycles/degree of visual angle; N = cat tested via normal optic nerve; B = cat tested via blocked optic nerve; for criteria see text)*

Animal	KK 10		KK 12		KK 14		KK 17	
	N	B	N	B	N	B	N	B
'Passing' value	4.0	4.0	4.2	4.1	4.4	4.4	3.2	3.2
'Failing' value	4.6	4.3	4.6	4.5	4.7	4.6	3.5	3.5

Fig. 7.6. Loss of response of t_1 fibres following pressure block. Stimulation and recording were both posterior to the block (electrodes 3 and 4 in Fig. 7.3). Pressure block at 0 h did not affect the t_2 response but the t_1 response declined to zero between 80 and 114 h.

wave black–white grating (contrast 0.5–0.8) from a grey surface of equal average luminance (140 cd/m²). The spatial frequency of the grating or the height from which the cat jumped were increased until the cat failed to discriminate. We noted: (1) the highest spatial frequency (measured at the level of the cat's eye) which the cat selected correctly at least 7 times out of 10 in three successive sessions (this is equivalent to a probability of about 0.5% or less that the result was due to chance); and (2) the lowest spatial frequency which the cat selected 6 times or less out of 10 in three successive sessions. These 'passing' and 'failing' values are shown in Table 7.1. The acuity values obtained for the normal eye are somewhat below published values (e.g. 6.9 cycles/degree; Timney, Mitchell & Giffin, 1978) but this may be because all the animals were fully adult at the time of training; higher values are obtainable if training is commenced when the animals are young. There was no significant difference between the two eyes of any of the four animals examined.

Thus, visual acuity is unaffected by the loss of the Y system and is probably determined by the X system (Hughes, 1981). In one cat (KK-12), the block had been rather excessive and the t_2 response had decreased to about 45%, yet the acuity was not affected. Probably

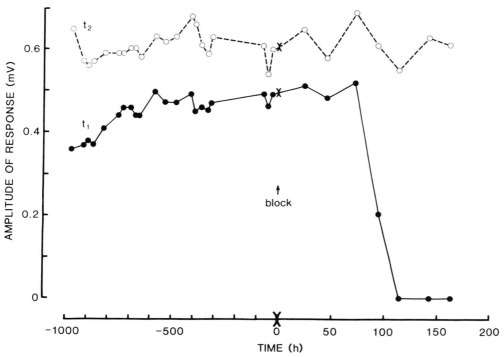

the lost X fibres were the larger ones and came from the periphery of the retina (Stone & Fukuda, 1974). If acuity is dependent mainly on the area centralis, the loss of these fibres might not be important.

Anatomical effects of block

A mild degree of pressure produces an ischaemic block from which recovery is fairly rapid. High pressures will produce a rupture of the axon leading to Wallerian degeneration. Intermediate pressures cause a demyelination block without axonal degeneration (Holmes, 1906; Denny-Brown & Brenner, 1944; Ochoa, Fowler & Gilliatt, 1972). Naturally, there is an overlap of effects depending on the size of the fibres. Although the classical view is that large fibres are more susceptible to pressure block than are small fibres, it has recently been shown by Battista & Alban (1983) that if the length of nerve over which the pressure is applied is very narrow, for example by the use of a ligature, then unmyelinated C fibres are more susceptible than the large myelinated A fibres.

In our original acute experiments, pressure would have been applied over a length of nerve of about 3 mm. For the acuity experiments on chronic cats, we wished to reduce the size of the pressure device and in place of the 4 mm long brass cylindrical section through which the optic nerve ran, we substituted a 1 mm diameter rod, so that the nerve was compressed against this rod. In a few cats, prepared by this method,

Fig. 7.7. Electron micrographs of optic nerve from normal side (A) and blocked side (B) approximately one week after block. The sections were taken posterior to the region of block and close to the optic chiasm. B shows one large and one small degenerating fibre (arrows).

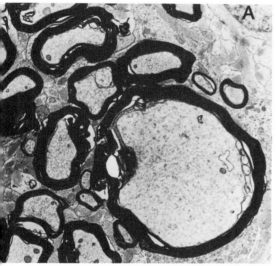

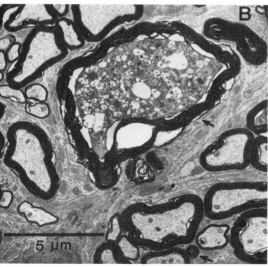

Fig. 7.8. Histograms of diameters of retinal ganglion cells over 20 μm from normal retina and from retina attached to blocked nerve, 2 years after pressure block. Each count was taken from an area of 17.5 mm² in the nasal retina, centred about 9 mm above the horizontal line through the area centralis and about 9 mm nasal to the vertical line through area centralis. There is a loss of all cells more than 30 μm in diameter in the retina attached to the blocked nerve.

LOSS OF LARGE RETINAL GANGLION CELLS

(2 YEARS AFTER PRESSURE BLOCK OF OPTIC NERVE)

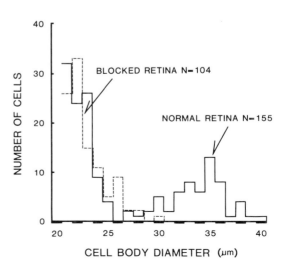

which were not used in the behavioural studies, the optic nerve was examined histologically about 1 week after the block. While there was extensive injury to the largest myelinated fibres, there were also signs of degeneration in some of the smallest myelinated fibres. Examples of this degeneration are shown in Fig. 7.7. We also examined the retina of a cat that had been blocked about 2 years earlier. There was a virtually total disappearance of the large (alpha) cells (Fig. 7.8).

Fast motion detection

Because of the superior ability of the Y retinal ganglion cells to detect fast movement (Cleland *et al.*, 1971*b*), we anticipated that a loss of the Y system would produce significant behavioural effects. We measured the cat's ability to discriminate fast motion by using a modification of the Mitchell jumping apparatus (Mitchell *et al.*, 1977). Each of the two platforms on to which the cat could jump was painted the same off-white colour. A light spot generated by a projector was reflected off a mirror attached to a galvanometer. The galvanometer rotated the mirror so that the spot moved backwards and forwards across the platforms (Fig. 7.9). The shutter of the projector was triggered so that it opened only when the spot was moving in one direction. The background illumination, the size and intensity of the spot, the direction and velocity of movement could all be independently varied. However,

because this created a very large number of possible combinations, we restricted our conditions by fixing the background illumination at a low mesopic level (0.09 cd/m² on the platforms) and the spot size (about 11.5 degrees when the cat's eyes were 35 cm from the platform, the platform having a width of 98 degrees at this distance), and varied only the spot intensity, velocity and direction of movement.

On any given day, only one spot intensity was used. The normal eye and the blocked eye were tested on alternate days, using the same spot intensity. Direction of movement (left-to-right or right-to-left) was varied in pseudorandom fashion. The cat was rewarded with a small piece of meat if it jumped to the side towards which the spot was moving. The usual training strategies were employed. The criterion for discriminating a particular velocity was a correct jump seven times or more in ten trials in three successive sessions, combined with less than seven correct jumps in three successive sessions to a velocity not more than one-third greater. This 'failing' velocity ranged from 11 to 32% above the 'passing' velocity (mean 19.6%).

The results for one cat are shown in Fig. 7.10. Two other cats are at present being tested and show similar results at low contrasts. With strong contrast between spot and background (ratio of spot luminance to background luminance of 35.8), all three cats could detect a velocity of 6260 degrees/s using the normal

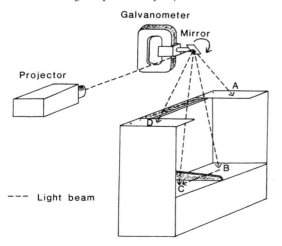

Fig. 7.9. Diagram to show method of projecting a moving light spot on to the platform of a Mitchell jumping apparatus. The spot is made to move from B to C or C to B. At A and D the spot is prevented from reaching the platform by adjustable baffles.

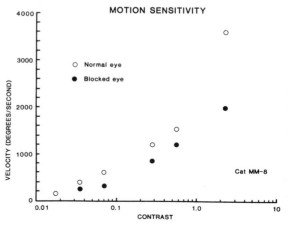

Fig. 7.10. Graph to show the fastest velocity detected using the normal eye only (open circles) and the eye attached to the blocked nerve only (closed circles) as a function of contrast (spot/background) in one cat. At a contrast of 17.9 this cat could detect 6260 degrees/s using either eye.

eye. However, one cat could detect this velocity with a contrast of 17.9 and one cat even at a contrast of 1.12. This velocity (6260 degrees/s) is the limit of our present equipment (for a viewing distance of 35 cm). Doubtless, higher velocities could be discriminated provided the contrast was adequate. One can see intuitively that there must be an upper limit, whatever the contrast, because eventually there would not be a sufficient time difference across the retina. Although these are very high velocities, they are similar to those recently obtained in man, albeit under slightly different conditions (Burr & Ross, 1982).

Fig. 7.10 shows that when this cat was using its normal eye, the velocities it could detect were 28–88% greater than when using its blocked eye (mean 60%). However, this cat could also detect a velocity of 6260 degrees/s at a contrast of 17.9 when using its blocked eye. For all three cats and for all contrast levels (18 pairs of points, to date), the performance using the normal eye was superior to that using the blocked eye and in all 18 pairs the 'failing' velocity for the blocked eye was less than or (in 4 cases) equal to the 'passing' velocity for the normal eye. Thus there is a highly significant difference between the two eyes.

The results show that in the discrimination of fast velocity the Y system is not unique and the X system can discriminate very high velocities provided the contrast is great enough. For the cat illustrated in Fig. 7.10, the X system required the contrast to be increased about 2.4 times to see the same velocity as the intact eye. It is possible that the t_3 fibres may play some role although their sensitivity to fast movement is very limited (Cleland & Levick, 1974).

Although this result surprised us, in view of the general belief that X cells do not respond well to fast velocities, in fact there is probably no discrepancy between these behavioural results and the velocity sensitivities of X and Y cells. Cleland et al. (1971b) did not use velocities greater than about 200 degrees/s and reported that, provided the contrast was appropriate to the centre of the receptive field, X retinal ganglion cells often gave responses to fast velocities. Cohen, Winters & Hamasaki (1980) reported that 5 of 14 Y LGNd cells and 2 of 25 X cells responded at velocities of 1000 degrees/s. Orban, Kennedy & Maes (1981) found that many cells in the visual cortex would respond at velocities of 700 degrees/s. Cohen et al. (1980) summed 25 responses and Orban et al. (1981) up to 40 responses for their analyses. These workers were primarily interested in peak sensitivities and did not aim to explore the upper velocity limits.

For an animal to detect a particular velocity of movement, it is obviously necessary that certain cells at all levels of the visual system must *each* alter their response in some way. The response of a single retinal ganglion cell (for example) in one trial might be undetectable electrophysiologically. The brain presumably detects the change by virtue of an adequate amount of convergence within the central visual pathways. It is a prediction from our results that, provided contrast is great enough and the discharges in a sufficient number of trials are summed, certain retinal ganglion cells, both X and Y cells, will show a response to velocities in excess of 6000 degrees/s.

Acknowledgements

The experiments based on the selective blocking technique described here were performed in collaboration with John Burne, Lynne Cottee, Brian Freeman, John Garvey, Kim Hamilton, Lorraine Kerr, Rischeh Kumurasinghe, Chris Kyriacou, Paul Martin, Marianna Milosavljevic, Colette Van Hees and Kornelius Westland, to all of whom I am deeply indebted. I am grateful to Bill Levick and Brian Cleland for valuable suggestions and discussion. The work was supported by the National Health and Medical Research Council of Australia.

References

Ahlsén, G. & Lindström, S. (1978) Axonal branching of functionally identified neurones in the lateral geniculate body. *Neurosci. Lett.*, Suppl. 1, S156

Ahlsén, G. & Lindström, S. (1982a) Excitation of perigeniculate neurones via axon collaterals of principal cells. *Brain Res.*, 236, 477–81

Ahlsén, G. & Lindström, S. (1982b) Mutual inhibition between perigeniculate neurones. *Brain Res.*, 236, 482–6

Ahlsén, G., Lindström, S. & Lo, F.-S. (1982) Interactions between inhibitory interneurones in the lateral geniculate nucleus of the cat. *J. Physiol.*, 328, 38–9

Ahlsén, G., Lindström, S. & Sybirska, E. (1978) Subcortical axon collaterals of principal cells in the lateral geniculate body of the cat. *Brain Res.*, 156, 106–9

Andersen, P., Brooks, C.McC. & Eccles, J. C. (1964a) Electrical responses of the ventro-basal nucleus of the thalamus. *Progr. Brain Res.*, 5, 100–13

Andersen, P., Brooks, C.McC., Eccles, J. C. & Sears, T. A. (1964b) The ventro-basal nucleus of the thalamus: potential fields synaptic transmission and excitability of both presynaptic and post-synaptic components. *J. Physiol.*, 174, 348–69

Andersen, P. & Eccles, J. C. (1962) Inhibitory phasing of neuronal discharge. *Nature*, 196, 645–7

Andersen, P., Eccles, J. C. & Sears, T. A. (1964*c*) The ventro-basal complex of the thalamus: types of cells, their responses and their functional organization. *J. Physiol.*, **174**, 370–99

Andersen, P. & Sears, T. A. (1964) The role of inhibition in the phasing of spontaneous thalamo-cortical discharge. *J. Physiol.*, **173**, 459–80

Battista, A. F. & Alban, E. (1983) Effect of graded ligature compression on nerve conduction. *Exp. Neurol.*, **80**, 186–94

Bishop, G. H., Clare, M. H. & Landau, W. M. (1969) Further analysis of fiber groups in the optic tract of cat. *Exp. Neurol.*, **24**, 386–99

Bishop, G. H. & O'Leary, J. S. (1940) Electrical activity of the lateral geniculate of cats following optic nerve stimuli. *J. Neurophysiol.*, **3**, 308–22

Bishop, P. O. & Davis, R. (1960) The recovery of responsiveness of the sensory synapses in the lateral geniculate nucleus. *J. Physiol.*, **150**, 214–38

Bishop, P. O. & McLeod, J. G. (1954) Nature of potentials associated with synaptic transmission in lateral geniculate of cat. *J. Neurophysiol.*, **17**, 387–414

Boycott, B. B. & Wässle, H. (1974) The morphological types of ganglion cells of the domestic cat's retina. *J. Physiol.*, **240**, 397–419

Brauer, K., Winkelmann, E., Marx, I. & David, H. (1974) Licht- und elektronenmikroskopische Untersuchungen an Axonen und Dendriten in der Pars dorsalis des Corpus geniculatum laterale (Cgl d) der Albinoratte *Z. Mikrosk.-Anat. Forsch.*, **88**, 596–626

Burke, W., Burne, J. A. & Martin, P. (1985) Selective block of Y optic nerve fibres in the cat and the occurrence of inhibition in the dorsal lateral geniculate nucleus. *J. Physiol*, **364**, 81–92

Burke, W. & Sefton, A. J. (1966*a*) Discharge patterns of principal cells and interneurones in lateral geniculate nucleus of rat. *J. Physiol.*, **187**, 201–12

Burke, W. & Sefton, A. J. (1966*b*) Recovery of responsiveness of cells of lateral geniculate nucleus of rat. *J. Physiol.*, **187**, 213–29

Burke, W. & Sefton, A. J. (1966*c*) Inhibitory mechanisms in lateral geniculate nucleus of rat.*J. Physiol.*, **187**, 231–46

Burr, D. C. & Ross, J. (1982) Contrast sensitivity at high velocities. *Vision Res.*, **22**, 479–84

Cajal, S. R. (1911) *Histologie du système nerveux de l'homme et des vertébrés. Vol. II.* Madrid: Consejo Superior de Investigaciones Cientificas. (1955 edition)

Cleland, B. G., Dubin, M. W. & Levick, W. R. (1971*a*) Simultaneous recording of input and output of lateral geniculate neurones. *Nature New Biol.*, **231**, 191–2

Cleland, B. G., Dubin, M. W. & Levick, W. R. (1971*b*) Sustained and transient neurones in the cat's retina and lateral geniculate nucleus. *J. Physiol.*, **217**, 473–96

Cleland, B. G. & Levick, W. R. (1974) Brisk and sluggish concentrically organized ganglion cells in the cat's retina. *J. Physiol.*, **240**, 421–56

Cohen, H. I., Winters, R. W. & Hamasaki, D. I, (1980) Response of X and Y cat retinal ganglion cells to moving stimuli. *Exp. Brain Res.*, **38**, 299–303

Denny-Brown, D. & Brenner, C. (1944) Paralysis of nerve induced by direct pressure and by tourniquet. *Arch. Neurol. Chicago*, **51**, 1–26

Dubin, M. W. & Cleland, B. G. (1977) Organization of visual inputs to interneurons of lateral geniculate nucleus of the cat. *J. Neurophysiol.*, **40**, 410–27

Eccles, J. C., Fatt, P. & Koketsu, K. (1954) Cholinergic and inhibitory synapses in a pathway from motor-axon collaterals to motoneurones. *J. Physiol.*, **126**, 524–62

Enroth-Cugell, C. & Robson, J. G. (1966) The contrast sensitivity of retinal ganglion cells of the cat. *J. Physiol.*, **187**, 517–52

Famiglietti, E. V., Jr & Peters, A. (1972) The synaptic glomerulus and the intrinsic neuron in the dorsal lateral geniculate nucleus of the cat. *J. Comp. Neurol.*, **144**, 285–333

French, C., Sefton, A. J. & Mackay-Sim, A. (1979) Abolition of two-shock inhibition of geniculo-cortical relay cells by a lesion in the thalamic reticular nucleus of the rat. *Proc. Aust. Physiol. Pharmacol. Soc.*, **10**, 119P

Friedlander, M. J., Lin, C. S. & Sherman, S. M. (1979) Structure of physiologically identified X and Y cells in the cat's lateral geniculate nucleus. *Science*, **204**, 1114–17

Fukuda, Y. & Stone, J. (1976) Evidence of differential inhibitory influences on X- and Y-type relay cells in the cat's lateral geniculate nucleus. *Brain Res.*, **113**, 188–96

Gasser, H. S. & Erlanger, J. (1929) The role of fiber size in the establishment of a nerve block by pressure or cocaine. *Am. J. Physiol.*, **88**, 581–91

Geisert, E. E., Jr (1980) Cortical projections of the lateral geniculate nucleus in the cat. *J. Comp. Neurol.*, **190**, 793–812

Grossman, A., Lieberman, A. R. & Webster, K. (1973) A Golgi study of the rat dorsal lateral geniculate nucleus. *J. Comp. Neurol.*, **150**, 441–66

Guillery, R. W. (1969) The organization of synaptic interconnections in the laminae of the dorsal lateral geniculate nucleus of the cat. *Z. Zellforsch-mikr. Anat.*, **96** 1–38

Guillery, R. W. (1971) Patterns of synaptic interconnections in the dorsal lateral geniculate nucleus of cat and monkey: a brief review. *Vision Res.*, Suppl. 3, 211–27

Hale, P. T. & Sefton, A. J. (1977) Inhibition of the transfer of visual information through lateral geniculate nucleus. *Proc. Aust. Physiol. Pharmacol. Soc.*, **8**, 100P

Hoffmann, K.-P., Stone, J. & Sherman, S. M. (1972) Relay of receptive-field properties in dorsal geniculate nucleus of the cat. *J. Neurophysiol.*, **35**, 518–31

Holmes, G. (1906) On the relation between loss of function and structural change in focal lesions of the central nervous system, with special reference to secondary degeneration. *Brain*, **29**, 514–23

Hubel, D. H. & Wiesel, T. N. (1961) Integrative action in the cat's lateral geniculate body. *J. Physiol.*, **155**, 385–98

Hughes, A. (1981) Cat retina and the sampling theorem: the relation of transient and sustained brisk-unit cut-off frequency to α and β-mode cell density. *Exp. Brain Res.*, **42**, 196–202

Hunt, S. P., Lieberman, A. R., Ohara, P. T. & Wu, J. Y. (1982) Interneurones in the dorsal lateral geniculate nucleus (LG of the adult rat are GABAergic: evidence from immunocytochemistry. *J. Physiol.*, 332, 61P

Ikeda, H. & Wright, M. J. (1972) Receptive field organization of 'sustained' and 'transient' retinal ganglion cells which subserve different functional roles. *J. Physiol.*, 227, 769–800

Kelly, J. S., Godfraind, J. M. & Maruyama, S. (1979) The presence and nature of inhibition in small slices of the dorsal lateral geniculate nucleus of cat and rat incubated in vitro. *Brain Res.*, 168, 388–92

Kumarasinghe, R., Kyriacou, C., Westland, K. C. & Burke, W. (1982) Failure of axonal conduction in the optic nerve and of synaptic transmission in the LGN of the cat after selective block of fast optic fibres. *Neurosci. Lett.*, 8, S62

LeVay, S. & Ferster, D. (1979) Proportion of interneurons in the cat's lateral geniculate nucleus. *Brain Res.*, 164, 304–8

Levick, W. R. & Thibos, L. N. (1980) X/Y analysis of sluggish-concentric retinal ganglion cells of the cat. *Brain Res.*, 74, 156–60

Levick, W. R. & Thibos, L. N. (1983) Receptive fields of cat ganglion cells: classification and construction. *Prog. Retinal Res.*, 2, 267–319

Lieberman, A. R. (1973) Neurons with presynaptic perikarya and presynaptic dendrites in the rat lateral geniculate nucleus. *Brain Res.*, 59, 35–59

Lieberman, A. R. & Ohara, P. T. (1981) Functionally relevant anatomical features of synaptic organization in the reticular nucleus of the thalamus of the rat. *J. Physiol.*, 313, 42–3P

Lieberman, A. R. & Webster, K. E. (1974) Aspects of the synaptic organization of intrinsic neurons in the dorsal lateral geniculate nucleus. An ultrastructural study of the normal and of the experimentally deafferented nucleus in the rat. *J. Neurocytol.*, 3, 677–710

Lin, C. S., Kratz, K. E. & Sherman, S. M. (1977) Percentage of relay cells in the cat's lateral geniculate nucleus. *Brain Res.*, 131, 167–73

Lindström, S. (1982) Synaptic organization of inhibitory pathways to principal cells in the lateral geniculate nucleus of the cat. *Brain Res.*, 234, 447–53

Lindström, S. (1983) Interneurones in the lateral geniculate nucleus with monosynaptic excitation from retinal ganglion cells. *Acta Physiol. Scand.*, 119, 101–3

MacGregor, R. J., Sharpless, S. K. & Luttges, M. W. (1975) A pressure vessel model for nerve compression. *J. Neurol. Sci.*, 24, 299–304

Marshall, W. H. (1949) Excitability cycle and interaction in geniculate-striate system of cat. *J. Neurophysiol.*, 12, 277–88

Marshall, W. H. & Talbot, S. A. (1940) Recovery cycle of the lateral geniculate of the nembutalized cat. *Am. J. Physiol.*, 129, P417–18

Marshall, W. H. & Talbot, S. A. (1941) Relation of the excitability cycle of the geniculate-striate system to certain problems of monocular and binocular vision. *Am. J. Physiol.*, 133, 378–9

Marshall, W. H. & Talbot, S. A. (1942) Recent evidence for neural mechanisms in vision leading to a general theory of sensory acuity. In *Biological Symposia, Vol. VII, Visual mechanisms*, ed. H. Klüver, pp. 117–64. Lancaster, Pa: Cattell Press

Martin, P. R., Burne, J. A. & Burke, W. (1980) Selective block of the larger fibres in the optic nerve of the cat. *Proc. Aust. Physiol. Pharmacol. Soc.*, 11, 146P

Martin, P. R., Burne, J. A. & Burke, W. (1981) Inhibition in the lateral geniculate nucleus produced by the slower optic nerve fibres. *Aust. Neurosci. Proc.*, 1, 94P

Mitchell, D. E., Giffin, F. & Timney, B. (1977) A behavioural technique for the rapid assessment of the visual capabilities of kittens. *Perception*, 6, 181–93

Ochoa, J., Fowler, T. J. & Gilliatt, R. W. (1972) Anatomical changes in peripheral nerves compressed by a pneumatic tourniquet. *J. Anat. (Lond.)*, 113, 433–55

Ohara, P. T., Sefton, A. J. & Lieberman, A. R. (1980) Mode of termination of afferents from the thalamic reticular nucleus in the dorsal lateral geniculate nucleus of the rat. *Brain Res.*, 197, 503–6

O'Leary, J. L. (1940) A structural analysis of the lateral geniculate nucleus of the cat. *J. Comp. Neurol.*, 73, 405–30

Orban, G. A., Kennedy, H & Maes, H. (1981) Velocity sensitivity of areas 17 and 18 of the cat. *Acta Psychol.*, 48, 303–9

Rodieck, R. W. (1973) Inhibition in cat lateral geniculate nucleus, role of Y system. *Proc. Aust. Physiol. Pharmacol. Soc.*, 4, 167

Rodieck, R. W. (1979) Visual pathways. *Ann. Rev. Neurosci.*, 2, 193–225

Sefton, A. J. & Burke, W. (1965) Reverberatory inhibitory circuits in the lateral geniculate nucleus of the rat. *Nature*, 205, 1325–6

Sefton, A. J. & Burke, W. (1966) Mechanism of recurrent inhibition in the lateral geniculate nucleus of the rat. *Nature*, 211, 1276–8

Singer, W. & Bedworth, N. (1973) Inhibitory interaction between X and Y units in the cat lateral geniculate nucleus. *Brain Res.*, 49, 291–307

Stone, J., Dreher, B. & Leventhal, A. (1979) Hierarchical and parallel mechanisms in the organization of visual cortex. *Brain Res. Revs.*, 1, 345–94

Stone, J. & Fukuda, Y. (1974) The naso-temporal division of the cat's retina re-examined in terms of Y-, X- and W-cells. *J. Comp. Neurol.*, 155, 377–94

Sumitomo, I., Iwama, K. & Nakamura, M. (1976a) Optic nerve innervation of so-called interneurons of the rat lateral geniculate body. *Tohoku J. Exp. Med.*, 119, 149–58

Sumitomo, I., Kurioka, Y. & Iwama, K. (1975) Location and function of so-called interneurons of rat lateral geniculate body. *Proc. Jap. Acad.*, 51, 74–9

Sumitomo, I., Nakamura, M. & Iwama, K. (1976b) Location and function of the so-called interneurons of rat lateral geniculate body. *Exp. Neurol.*, 51, 110–23

Szentágothai, J. (1963) The structure of the synapse in the lateral geniculate body. *Acta Anat.*, 55, 166–85

Talbot, S. A. & Marshall, W. H. (1941) Binocular interaction and excitability cycles in cat and monkey. *Am. J. Physiol.*, 133, P467–8

Tello, F. (1940) Disposición macroscópica y estructura del cuerpo geniculado externo. *Trab. Lab. Invest. Biol., Univ. Madrid*, 3, 39–62

Timney, B., Mitchell, D. E. & Giffin, F. (1978) The development of vision in cats after extended periods of dark-rearing. *Exp. Brain Res.*, 31, 547–60

Tömböl, T. (1967) Short neurons and their synaptic relations in the specific thalamic nuclei. *Brain Res.*, 3, 307–26

Tömböl, T. (1969) Terminal arborization in specific afferents in the specific thalamic nuclei. *Acta Morphol. Acad. Sci. Hung.*, 17, 273–84

Vastola, E. F. (1959) After-positivity in lateral geniculate body. *J. Neurophysiol.*, 22, 258–72

Westland, K. C., Freeman, B. & Burke, W. (1983) Distribution of X and Y optic nerve terminals in the lateral geniculate nucleus. *Neurosci. Lett.*, 11, S80

8

The dorsal lateral geniculate nucleus of the cat

B. G. CLELAND

The shape that is so readily recognized today as the lateral geniculate nucleus (LGN) of the cat was first described and illustrated by Tello (1904), who carried out a detailed histological study of this nucleus. He described it as having, in parasagittal section, the form of a comma whose head is located rostral and inferior and whose tail is located caudal and superior. Three cellular layers can be clearly seen in Nissl-stained parasagittal sections and these were labelled A, A1 and B by Thuma (1928). Following enucleation of one eye, cellular atrophy was observed in the central layer (A1) on the ipsilateral side and in the two outer layers (A and B) on the contralateral side (Minkowski, 1920), clearly demonstrating separation of the projections from each eye.

In Nissl-stained sections three types of cells can be easily recognized in the dorsal layers (A and A1) and these were described as large, medium and small by Thuma (1928). The cells in layer B are rather elongated although they otherwise have a similar morphology to the medium-sized cells distinguished in the dorsal layers. Fibrous laminae, with only a few scattered large cells among them, separate the three cellular layers. Due to their distinctive nature, Thuma (1928) designated them the central interlaminar nuclei, although he had some doubts as to whether they were unique and should be accepted as a distinct nucleus.

In contrast to the three layers defined by Thuma (1928), Rioch (1929) proposed a system of four layers. These were the laminae principalis anterior, principalis posterior and parvocellularis, which were equivalent to layers A, A1 and B, respectively, and in addition a lamina magnocellularis, which was proposed to be between layers A1 and B. In effect, Rioch (1929) emphasized one of Thuma's central interlaminar nuclei and raised it to the status of a full lamina.

By placing small lesions in the retina of one eye and observing the resulting degeneration in the LGN, Overbosch (1927) was able to demonstrate a clear retinotopic organization within the nucleus. More detailed studies of the retinotopic organization, together with the evidence that fibres from each eye end in separate laminae, have since been reported by Hayhow (1958), Laties & Sprague (1966) and Stone & Hansen (1966), with physiological confirmation coming from studies by Hubel & Wiesel (1961) and Bishop *et al.* (1962). This line of investigation was largely brought to a conclusion with the detailed map of the retinotopic organization presented by Sanderson (1971).

In anatomical studies of the LGN, Guillery (1970) and Hickey & Guillery (1974) extended the observations made by Hayhow (1958), indicating that the B lamina is more complex than originally believed, and demonstrated four rather than one lamina ventral to layer A1. Two of these laminae, layers C and C2, receive a contralateral innervation, while layer C1 receives an ipsilateral input. An additonal layer, C3, which is the most ventral layer of the dorsal LGN, has also been observed but this appears to receive no direct afferents from either eye. While the two A layers have a closely matched appearance, this is not the case with the ventral layers. Layer C is the thickest of the ventral layers and contains both large and medium-sized cells, with the large cells concentrating dorsally and the medium-sized cells ventrally within the same lamina. Neither layer C1 nor layer C2 contains the larger cells found in layer C.

In Golgi-stained material, Guillery (1966) distinguished four morphological classes of cells in the dorsal LGN. The largest cells (class 1) have perikarya with diameters in the range 25–40 μm and dendrites that follow a relatively straight course in a plane perpendicular to the plane of the laminae. The dendrites cross interlaminar borders freely and are moderately well covered with fine spines. The second class of cells (class 2) have medium-sized perikarya (15–30 μm in diameter) with fewer principal dendrites than the large cells. The dendrites are also shorter, more sinuous and show less orientation. The character-

istic feature of these cells is the clusters of large, grape-like appendages close to the branching points of the dendrites. There are fine spines on the more peripheral segments. Both of these classes of cells are observed throughout laminae A and A1 and the central interlaminar nucleus, which would presumably include the upper portion of lamina C. The smaller, class 3 cells (10–20 μm in diameter) are seen only rarely and have few dendrites with relatively limited branching. There are characteristic appendages to the dendrites and these have long slender stalks with extensive branching and end in swollen terminal portions 1–2 μm in diameter. These cells occur throughout the geniculate. The fourth class of cells, which are only found in the C laminae, are medium-sized multipolar cells with five to ten main dendrites. They appear somewhat flattened in a plane parallel to the lamina. Relatively few of these cells have dendritic appendages (Guillery, 1966). The limitation of this classification is that Guillery found it only fully covered about 60% of the cells and that there were many cells with intermediate properties. More recently a similar observation has been made by Meyer & Albus (1981), who studied the morphology of the geniculate relay cells after injections of horseradish peroxidase (HRP) into Areas 17 and 18 of the visual cortex. Despite these reservations, LeVay & Ferster (1979) were able to carry out a quantitative study of the A laminae, allocating all cells to one of the three classes, and their findings are presented in Table 8.1.

According to LeVay & Ferster (1977), most class 1 cells and all class 2 cells project to Area 17 of the visual cortex, while a few large class 1 cells project to Area 18. (Contrary to the observations of LeVay & Ferster, Geisert (1980) found that most of the lamina A and A1 cells which project to Area 18 have branches which also project to Area 17). Class 4 cells project to Areas 17 and

Table 8.1. *Percentage of the three cell classes observed in the A and A1 laminae by LeVay & Ferster (1979). (Numbers in brackets represent X or Y cells as a percentage of relay cells in lamina.)*

Cell class	Lamina		
	A	A1	
1	15 (19)	26 (34)	Presumed Y relay cells
2	64 (81)	50 (66)	Presumed X relay cell
3	21	24	Presumed interneuron

18 with branching axons, while class 3 cells do not project to the visual cortex and would appear to serve the role of interneurons (Guillery, 1966). Meyer & Albus (1981), however, report that some class 3 cells project to Areas 17 and 18, so that this class should not be regarded as exclusively representing interneurons.

In an electron microscopic study, Guillery (1969) distinguished four types of axon terminals within the LGN. Two of these contain vesicles with round profiles and can be distinguished from each other by their size, appearance of their contents and the types of contact they make. The larger axons are believed to be retinal in origin and are commonly found within the encapsulated synaptic zones which form around the grape-like dendritic appendages in the A laminae. The smaller axons are believed to be of cortical origin and are mostly found within interstitial zones which lie between the encapsulated zones. The other two axon types contain many flattened or irregular vesicles and are regarded as two types of intrageniculate fibre. They are commonly found in both encapsulated and interstitial zones, generally in close association with the presumed retinal and cortical axons.

More recently, Friedlander *et al.* (1981) studied the geniculate by filling individual cells with HRP and have put forward a rather different scheme for the classification of geniculate cells. By correlating structure and function they have presented a morphological description that fits in with the physiological classification of X and Y cells (see below). They found that Y cells had large cell bodies with large axons and dendritic trees which were radially symmetrical and which freely crossed laminar boundaries. The individual dendrites were large, fairly straight and had a small number of simple appendages. In contrast, X cells had small cell bodies with thinner axons and dendritic trees which were asymmetrically elongated along projection lines. The dendrites were thin, sinuous and had many complex appendages. Lin, Kratz & Sherman (1977) have challenged the very existence of a separate interneuron class and proposed that the axons of all geniculate neurons probably leave the nucleus. To account for possible interneuron action they have suggested that such a function may be performed by axonal branches of relay cells.

Following the work of Lin *et al.* (1977), LeVay & Ferster (1979) repeated their earlier study, again observing cells which they considered to be interneurons and these represented about 20–25% of the cells in the

geniculate. They suggested that the failure of Lin *et al.* to observe interneurons resulted from the use of thick sections in which unlabelled cells, presumed interneurons, would be difficult to recognize.

Medial interlaminar nucleus

As well as the laminar structure described above, the dorsal LGN also includes a smaller medial structure, the media interlaminar nucleus (Thuma, 1928; Hayhow, 1958). This nucleus receives inputs from both eyes and these tend to terminate in separate regions of the nucleus (Hayhow, 1958; Laties & Sprague, 1966; Stone & Hansen, 1966: Garey & Powell, 1968), although an actual lamination has only recently become clear through the use of HRP (Guillery *et al.*, 1980; Rowe & Dreher, 1982). Layer 1, which is the most dorsomedial of the three layers, receives input from the contralateral nasal retina, layer 2 from the ipsilateral temporal retina, and layer 3 from the contralateral temporal retina. The arrangement of layers 1 and 2 is such that they appear as mirror reversals of layers A and A1 which they join at the representation of the retinal midline. Layer 3 occupies a rather different position with regard to layers C and C2, acting more as an extension of these layers into the temporal retina. Hence this layer should probably be considered part of the C laminae rather than part of the medial interlaminar nucleus (Mason, 1975; Guillery *et al.*, 1980).

Thalamic reticular complex and perigeniculate nucleus

The thalamic reticular complex is a thin neuronal shell which covers the lateral and anterior aspects of the dorsal thalamus. It receives afferent input from collaterals of thalamic axons projecting to the cortex and from collaterals of cortical axons projecting to the thalamus. There is some suggestion of spatial segregation of the regions receiving the two different inputs. The axons of the thalamic reticular complex project back to the specific thalamic nuclei from which they receive their excitation (Scheibel & Scheibel, 1966; Jones, 1975). Immediately dorsal to the LGN lies a specialized portion of the reticular complex, the perigeniculate nucleus (Thuma, 1928), which is about 200–300 μm thick and separated from lamina A by a cell-free zone about 100 μm wide (Schmielau, 1979; Ahlsén, Lindström & Lo, 1982). Above the perigeniculate, the reticular nucleus is 300–500 μm wide and is separated from the perigeniculate nucleus by a zone containing scattered cells. This zone varies in width from 200 μm to 1 mm. The perigeniculate nucleus clearly receives input from the collaterals of geniculate relay cells (Ahlsén, Lindström & Sybirska, 1978; Ferster & LeVay, 1978; Friedlander et al., 1981; Meyer and Albus, 1981), but the morphological evidence for other visual inputs, especially from the visual cortex, to both the perigeniculate nucleus and the adjacent reticular complex still needs some clarification.

Receptive field properties

In general, cells in the LGN have receptive field properties that are similar to those of their retinal afferents, but with the addition of an inhibitory region surrounding the normal excitatory field (Hubel & Wiesel, 1961). The excitatory input for a relay cell comes from one eye only, either ipsilateral or contralateral, but this is not the case for the inhibitory input which has been shown to come from the corresponding position in the visual field of the non-excitatory eye (Sanderson, Darian-Smith & Bishop, 1969; Singer, 1970; Sanderson, Bishop & Darian-Smith, 1971; Rodieck and Dreher, 1979) as well as from the excitatory eye (Hubel & Wiesel, 1961; Cleland, Dubin & Levick, 1971b). As in the retina, the majority of cells have centre/surround properties and the action of the inhibition is to oppose a response whether generated by the centre or the surround. The main excitatory input comes from either one or two ganglion cells, although there may be as many as six ganglion cells contributing varying amounts of excitation (Cleland, Dubin & Levick, 1971a,b). In general, these cells have the same functional properties as their retinal inputs, so that the classes of cells observed in the geniculate are representative of those observed in the retina. (The one class not so far observed represents cells with direction-selective properties.) Hence those tests which are applicable in the retina for the separation of receptive fields into classes are still largely applicable in the geniculate, whether it be on the basis of linearity/non-linearity (Enroth-Cugell & Robson, 1966; So & Shapley 1979) or the battery of tests (Cleland et al., 1971b; Hoffmann, Stone & Sherman, 1972). In a small proportion of the geniculate cells, however, there is convergence of X and Y inputs onto a single cell.

The most widely distributed of the different classes of cells is the Y (brisk-transient) cell which is found in high proportions in layers A, A1 (Cleland et al., 1971b, 1976; Hoffmann et al., 1972; Wilson, Rowe & Stone, 1976; Friedlander et al., 1981) and C (Cleland et al., 1976; Wilson et al., 1976; Friedlander et al., 1981) and in the medial interlaminar nucleus (Mason, 1975; Kratz, Webb & Sherman, 1978; Dreher & Sefton, 1979). In many cases it would appear that a given retinal Y cell may project to several of these areas through axonal branching, e.g. layers A, C and medial interlaminar nucleus (Bowling & Michael, 1980). Cells in the X (brisk-sustained) class are mainly found in the A layers (Cleland et al., 1971b,1976; Hoffmann et al., 1972; Wilson et al., 1976; Bowling & Michael, 1980; Friedlander et al., 1981) but also to a lesser extent in the C layers (Cleland et al., 1976; Wilson et al., 1976; Sur & Sherman, 1982) and in the medial interlaminar nucleus (Dreher & Sefton, 1979; Bowling & Michael, 1980). The sluggish concentric cells have been found to a limited extent in the A layers (Cleland et al., 1976) but are principally observed in the C layers (Cleland et al., 1975, 1976; Wilson & Stone, 1975; Wilson et al., 1976; Friedlander et al., 1981: Stanford, Friedlander & Sherman, 1981) and the medial interlaminar nucleus (Dreher & Sefton, 1979). Sluggish non-concentric cells have been reported in the C layers (Cleland et al., 1975, 1976; Wilson & Stone, 1975; Wilson et al., 1976) but have been observed in such small numbers overall that it must be considered a possibility that they also occur in other regions of the geniculate, especially the medial interlaminar nucleus.

The majority of studies of geniculate cells have used flashing spots to determine receptive field properties. However, when moving bars were used as stimuli a clear orientation preference was noted for a proportion of the cells (Daniels, Norman & Pettigrew, 1977; Vidyasagar & Urbas, 1982). This was especially the case if maximal inhibition was produced through use of an elongated bar. If the cat had been stripe-reared or if lesions had been placed in Areas 17 and 18, biases for particular orientations were observed, leading to the conclusion that there was a definite cortical contribution to the inhibitory properties of relay cells.

The conduction velocity of the geniculate axons is closely related to the conduction velocity of the incoming retinal afferents, so that Y cells with their high-velocity afferents conduct information to the visual cortex at high velocity and can be antidromically stimulated from both Areas 17 and 18. By contrast, X cells with medium-velocity afferents have medium-velocity axons and project exclusively to Area 17 (Stone & Dreher, 1973). These properties have led to the association of Y cells with the class 1 cells of Guillery and of X cells with the class 2 cells.

The proportions of X and Y cells recorded in the geniculate are shown in Table 8.2 for a sample of 269 cells from the study of Cleland et al. (1976): similar proportions were also obtained by Wilson et al. (1976). From Table 8.2, the proportion of Y cells is much higher in layer A1 than in layer A. However, if layers A and C are combined so that the proportions from the contralateral eye are compared with those from the ipsilateral eye, the figures are much closer to one another, although the proportions of Y cells are still greater in the lamina receiving information from the ipsilateral eye. Considering the differing crossed/uncrossed projection from each eye for X and Y cells (Kirk et al., 1976; Stone & Fukuda, 1974), one would expect a greater proportion of Y cells than X cells crossing in the chiasma and hence a greater proportion of Y cells in the combined A and C laminae. This expectation is the reverse of what is observed.

The Y cells appear to have larger cell bodies than X cells and therefore constitute a larger target for the electrode, the probability of recording from a given Y cell should be higher than that for a given X cell. Thus it might be expected that the proportion of recorded Y cells would be higher than the true proportion. In practice it will also be affected by electrode size and configuration, the type of electrode and even the way the electrode is advanced by the individual investigator. Rather surprisingly, the actual proportions determined histologically by LeVay & Ferster (1979) are similar to the recorded proportions shown in Table 8.2 for layers A and A1 (Cleland et al., 1976), and are almost identical to the results of Wilson et al. (1976). Therefore the still

higher proportion of histologically determined Y cells (35%) reported by Friedlander et al. (1981) seems especially surprising.

Madarasz et al. (1978) estimated from Nissl-stained sections that the number of neurons in layer A is approximately 280 000 with the number in layer A1 being approximately 208 000. Using the proportions of cells determined by LeVay & Ferster (1979), this would give a total of 283 000 X cells in the geniculate which would compare with a total of 88 000 such cells in the retina, assuming 55% of the population have X properties (Illing & Wässle, 1981) and also assuming a total ganglion cell count of 160 000 (Illing & Wässle, 1981; Williams, Bastiani & Chalupa, 1983). Thus the number of X cells in the geniculate is probably 3.2 times greater than the number of X cells in the retina. A similar calculation for Y cells is complicated by the presence of a significant number of such cells in layer C. However, if layer A1 were considered by itself, then there would be a total of 54 000 Y relay cells and this would compare to 1700 Y cells in the retina projecting to the ipsilateral geniculate (Illing & Wässle, 1981). Thus the number of Y cells in lamina A1 of the geniculate would be 32 times greater than in the retina, a rather surprising increase especially in view of the fact that there appears to be so little difference between the receptive field properties of ganglion cells and geniculate neurons.

In the retina of the cat the ON- and OFF-beta cells each form an almost hexagonal array (Wässle, Boycott & Illing, 1981). These cells are generally believed to be the morphological equivalent of retinal X cells. Thus it might be argued that the receptive fields of ON- and OFF-centre X cells likewise form a hexagonal array. Such an arrangement is shown in Fig. 8.1. Superimposed on this is a pattern of relay cells with the geniculate density four times greater than the ganglion-cell density. One-quarter of the relay cells (R_1) are centred on ganglion cells, indicating that they obtain their principal excitation from a single ganglion cell, while three-quarters of the relay cells (R_2) lie midway between two ganglion cells, indicating that their principal excitation comes equally from the two ganglion cells. From examples where the ganglion cell input has been extracted on the basis of a large prepotential recorded at the same time as an associated relay cell, it can be seen that for an optimal stimulus the geniculate response is about 80% of the ganglion cell response (Cleland, Lee & Vidyasagar, 1982) or

Table 8.2. *Numbers of X and Y cells recorded in the geniculate by Cleland* et al. (*1976*). (*Numbers in brackets represent percentage of cells in each column.*)

Cell type	Lamina			Projection	
	A	A1	C	Contra (A+C)	Ipsi (A1)
X	103 (79)	61 (57)	3	106 (66)	61 (57)
Y	28 (21)	47 (43)	27	55 (34)	47 (43)
Total	131	108	30	161	108

where sensitivity has been measured that the ganglion cell and geniculate have equal sensitivity (Hubel & Wiesel, 1961). The importance of this is that there is very little loss in sensitivity for an optimum stimulus as information is relayed through the geniculate. In Fig. 8.1, the R_1 geniculate cells could be expected to have an 80% correlation with the underlying ganglion cell for a well-focussed, optimal image (i.e. 80% of the ganglion cell spikes will be followed after a conduction delay by a spike in the relay cell), and the R_2 geniculate cells a 40% correlation with each of the adjoining ganglion cells. Assuming independence between the firing of neighbouring geniculate cells, it can be seen that each R_1 geniculate cell will have a 40% correlation with each adjacent R_2 cell, and neighbouring R_2 cells will have a 16% correlation. The model seen in Fig. 8.1 can then be regarded as a 'correlation' map of the visual field. It is not essential that the cells should be in the exact positions shown, as reasonable variation should have only a second-order effect. However, if the correlation has any functional significance it would be optimized by having a clear, well-focussed image in order to produce crisp responses. A similar 'correlation' map will exist for both ON-and OFF-centre Y cells,

and as some geniculate cells have a mixed X/Y input this will lock the X and Y maps together, but independently for both the ON- and OFF-centre subtypes. (The geniculate is not the first level where correlation has been observed. Mastronarde (1983*a,b,c*) has reported correlation between the maintained activity of adjacent ganglion cells, but it is much smaller than that proposed here and it is uncertain yet to what extent it is reflected in the maintained activity of geniculate cells.)

As sharp focus is necessary to produce a crisp response and hence high correlation, it is interesting to speculate on the importance of correlated firing of geniculate neurons for normal development of the visual cortex. Many of the changes seen in visual deprivation may be due to an effective reduction, or loss, of this correlated firing.

Circuitry of the LGN

The first model of the LGN was developed for the rat by Burke & Sefton (1966*a,b*) on the basis of the responses of cells to single shocks of the optic nerve and visual cortex. They proposed a subdivision of the cells into two distinct groups. These were P (principal or relay) cells, which received afferents from retinal ganglion cells and projected to the visual cortex, and I cells or interneurons, which received recurrent collaterals from P cells and then projected back onto P cells. The ON/OFF cells subsequently reported in the geniculate of the cat by Dreher & Sanderson (1973) were generally considered to represent these interneurons. Two distinct groups of cells that could be orthodromically stimulated from the visual cortex were described as potential interneurons by Cleland *et al.* (1976). The cells of one group, which were only observed just above the first recorded relay cells, had ON/OFF receptive fields, often with binocular excitation, while the cells of the second group had classical ON- and OFF-centre receptive fields and were only encountered within the main layers of the geniculate. The cells of the latter group could also be distinguished from relay cells by the variability of their latency to electrical stimulation of the retina or optic chiasm.

The detailed circuitry of these interneurons has been proposed on the basis of simultaneous recording from an interneuron within the geniculate, a relay cell and a ganglion cell that was excitatory to both (Cleland & Dubin, 1976; Dubin & Cleland 1977). From this and

Fig. 8.1. Correlations between relay cells in the LGN.

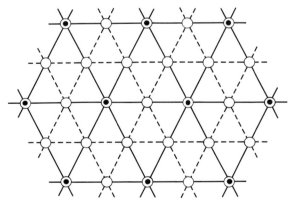

●	Centre of ganglion cell receptive field
○	Centre of relay cell receptive field
◉	80% correlation between cells
○—○	40% correlation between cells
○- -○	16% correlation between cells

an analysis of the ON/OFF cells in the perigeniculate, an extension of the Burke & Sefton model was proposed (Fig. 8.2). This consists of intrageniculate interneurons which resemble relay cells in being directly excited by a small number of X or Y retinal afferents which are either ON-centre or OFF-centre. These interneurons provide feed-forward inhibition from the retina onto relay cells and feedback inhibition from the visual cortex. There are also perigeniculate interneurons which behave in the manner described by Burke & Sefton (1966a) and are thus excited by recurrent collaterals of geniculate relay cells onto which they provide feedback inhibition. Those perigeniculate cells studied had clear Y cell input, but the model was minimal in that it was not possible to rule out a contribution from X cells nor direct excitation from the visual cortex. A similar model has also been proposed for the rat geniculate (Sumitomo, Nakamura & Iwama, 1976; Sumitomo & Iwama, 1977).

This detailed circuitry of the geniculate has been further extended by Lindström, Ahlsén and co-workers using intracellular recording techniques to resolve both excitatory postsynaptic potential (EPSP) and inhibitory postsynaptic potential (IPSP) activity. They were able to demonstrate that relay cells are excited directly from the visual cortex but not in the ready manner that

Fig. 8.2. Model of connections to relay cells in the LGN.

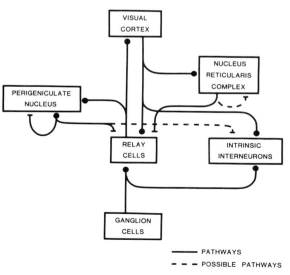

PATHWAYS
- - - POSSIBLE PATHWAYS

CIRCUITRY OF VISUAL PATHWAYS RELATING TO
DORSAL LATERAL GENICULATE NUCLEUS

intrageniculate interneurons can be excited with a single shock (Ahlsén, Grant & Lindström, 1982). Some perigeniculate cells were observed which appeared to receive input from X and not Y cells and hence, presumably, the X and Y cell pathways remain separate in the perigeniculate neucleus (Ahlsén & Lindström, 1982b). There is no evidence for direct cortical excitation of these cells and it is argued that the visual inhibition that is observed comes from neighbouring perigeniculate cells (Ahlsén & Lindström, 1982a).

Cells in the thalamic reticular nucleus, where it overlies the LGN, do not appear to give a visual response even though they receive excitation from the visual cortex (Ahlsén et al., 1982b). It is proposed that similar to the situation in other thalamic nuclei (Scheibel & Scheibel, 1966; Jones, 1975), these reticular neurons project to the LGN, where they inhibit relay cells.

While the X and Y pathways generally remain separate for excitation, the inhibitory pathways would appear not to be so specific (Hoffmann, et al., 1972; Singer & Bedworth, 1973). This is probably most clearly argued in situations where the afferents to relay cells are electrically stimulated from the retina. By adjusting the stimulus strengh, Y relay cells can be made to follow 100% of the time. However, this is not possible for many X cells where the rate of following may actually decrease with increasing stimulus strength. In this case it is proposed that the faster-conducting Y cells excite interneurons which then inhibit the X cells before excitation arrives from the slower-conducting X afferents. This does not, however, fit in with the observations of Lindström (1982) who found the earliest IPSPs to follow the initial EPSP in both X and Y cells. Thus there are clearly details still to be resolved, especially with regard to inhibition onto intrageniculate interneurons and inhibition between thalamic reticular neurons. There is also the whole situation of non-visual inputs.

Schmielau (1979) produced a rather different appraisal of the inhibitory circuitry, suggesting that perigeniculate neurons receive direct retinal input, while thalamic reticular neurons receive excitation from the collaterals of relay cells. There is, however, no evidence for the termination of retinal afferents in the perigeniculate nucleus (Bowling & Michael, 1980). Ahlsén et al. (1982b) also could find no evidence to support these claims and have suggested that Schmielau's results may have come from a differing evaluation of the borders of the LGN, perigeniculate nucleus and thalamic reticular nucleus. They pointed out that at times they recorded intrageniculate inter-neurons before the first relay cells were observed, but the depth of the electrode was such that the recording was always well within the LGN.

References

Ahlsén, G., Grant, K. & Lindström, S. (1982*a*) Monosynaptic excitation of principal cells in the lateral geniculate nucleus by corticofugal fibres. *Brain Res.*, **234**, 454–8

Ahlsén, G. & Lindström, S. (1982*a*) Excitation of perigeniculate neurones via axon collaterals of principal cells. *Brain Res.*, **236**, 477–81

Ahlsén, G. & Lindström, S. (1982*b*) Mutual inhibition between perigeniculate neurones. *Brain Res.*, **236**, 482–6

Ahlsén, G., Lindström, S. & Lo, F.-S. (1982*b*) Functional distinction of perigeniculate and thalamic reticular neurons in the cat. *Exp. Brain Res.*, **46**, 118–26

Ahlsén, G., Lindström, S. & Sybirska, E. (1978) Subcortical axon collaterals of principal cells in the lateral geniculate body of the cat. *Brain Res.*, **156**, 106–9

Bishop, P. O., Kozak, W., Levick, W. R. & Vakkur, G. J. (1962) The determination of the projection of the visual field on to the lateral geniculate nucleus in the cat. *J. Physiol.*, **163**, 503–39

Bowling, D. B. & Michael, C. R. (1980) Projection patterns of single physiologically characterized optic tract fibres in cat. *Nature*, **286**, 899–902

Burke, W. & Sefton, A. J. (1966*a*). Discharge patterns of principal cells and interneurones in lateral geniculate nucleus of rat. *J. Physiol.*, **187**, 201–12

Burke, W. & Sefton, A. J. (1966*b*) Recovery of responsiveness of cells of lateral geniculate nucleus of rat. *J. Physiol.*, **187**, 213–29

Cleland, B. G. & Dubin, M. W. (1976) The intrinsic connectivity of the LGN of the cat. *Exp. Brain Res.*, Suppl. 1, 493–6

Cleland, B. G., Dubin, M. W. & Levick, W. R. (1971*a*) Simultaneous recording of input and output of lateral geniculate neurones. *Nature, New Biol.*, **231**, 191–2

Cleland, B. G., Dubin, M. W. & Levick, W. R. (1971*b*) Sustained and transient neurones in the cat's retina and lateral geniculate nucleus. *J. Physiol.*, **227**, 473–96

Cleland, B. G., Lee, B. B. & Vidyasagar, T. R. (1982) Response of neurons in the cat's lateral geniculate nucleus to moving bars of different length. *J. Neuosci.*, **3**, 108–16

Cleland, B. G., Levick, W. R., Morstyn, R. & Wagner, H. G. (1976) Lateral geniculate relay of slowly conducting retinal afferents to cat visual cortex. *J. Physiol.*, **255**, 299–320

Cleland, B. G., Morstyn, R., Wagner, H. G. & Levick, W. R. (1975) Long-latency retinal input to lateral geniculate neurones of the cat. *Brain Res.*, **91**, 306–10

Daniels, J. D., Norman, J. L. & Pettigrew, J. D. (1977) Biases for oriented moving bars in lateral geniculate nucleus neurons of normal and stripe-reared cats. *Exp. Brain Res.*, **29**, 155–72

Dreher, B. & Sanderson, K. J. (1973). Receptive field analysis:

responses to moving visual contours by single lateral geniculate neurones in the cat. *J. Physiol.*, **234**, 95–118

Dreher, B. & Sefton, A.J. (1979) Properties of neurons in cat's dorsal lateral geniculate nucleus: a comparison between medial interlaminar and laminated parts of the nucleus. *J. Comp. Neurol.*, **183**, 47–64

Dubin, M. W. & Cleland, B. G. (1977) Organization of visual inputs to interneurons of lateral geniculate nucleus of the cat. *J. Neurophysiol.*, **40**, 410–27

Enroth-Cugell, C. & Robson, J. G. (1966) The contrast sensitivity of retinal ganglion cells of the cat. *J. Physiol.*, **187**, 516–52

Ferster, D. & LeVay, S. (1978) The axonal arborization of lateral geniculate neurons in the striate cortex of the cat. *J. Comp. Neurol.*, **182**, 923–44

Friedlander, M. J., Lin, C.-S., Stanford, L. R. & Sherman, S. M. (1981) Morphology of functionally identified neurons in the lateral geniculate nucleus of the cat. *J. Neurophysiol.*, **46**, 80–129

Garey, L. J. & Powell, T. P. S. (1968) The projection of the retina in the cat. *J. Anat. (Lond.)*, **102**, 189–222

Geisert, E. E., Jr., (1980) Cortical projection of the lateral geniculate nucleus in the cat. *J. Comp. Neurol.*, **190**, 793–812

Guillery, R. W. (1966) A study of Golgi preparations from the dorsal lateral geniculate nucleus of the adult cat. *J. Comp. Neurol.*, **238**, 21–50

Guillery, R. W. (1969) The organization of synaptic interconnections in the laminae of the dorsal lateral geniculate nucleus of the cat. *Z. Zellforsch.*, **96**, 1–38

Guillery, R. W. (1970) The laminar distribution of retinal fibers in the dorsal lateral geniculate nucleus of the cat: a new interpretation: *J. Comp. Neurol.*, **138**, 339–68

Guillery, R. W., Geisert, E. E., Jr., Polley, E. H. & Mason, C. A. (1980) An analysis of the retinal afferents to the cat's medial interlaminar nucleus and to its rostral thalamic extension, the 'geniculate wing'. *J. Comp. Neurol.*, **194**, 117–42

Hayhow, W. R. (1958) The cytoarchitecture of the lateral geniculate body in the cat in relation to the distribution of crossed and uncrossed optic fibers. *J. Comp. Neurol.*, **110**, 1–64

Hickey, T. L. & Guillery, R. W. (1974) An autoradiographic study of retinogeniculate pathways in the cat and the fox. *J. Comp. Neurol.*, **156**, 239–54

Hoffmann, K.-P., Stone, J. & Sherman, S. M. (1972) Relay of receptive-field properties in dorsal lateral geniculate nucleus of the cat. *J. Neurophysiol.*, **35**, 518–31

Hubel, D. H. & Wiesel, T. N. (1961) Integrative action in the cat's lateral geniculate body. *J. Physiol.*, **155**, 385–98

Illing, R.-B. and Wässle, H. (1981) The retinal projection to the thalamus in the cat: a quantitative investigation and a comparison with the retinotectal pathway. *J. Comp. Neurol.*, **202**, 265–85

Jones, E. G. (1975) Some aspects of the organization of the thalamic reticular complex. *J. Comp. Neurol.*, **162**, 285–308

Kirk, D. L., Levick, W. R., Cleland, B.G. & Wässle, H. (1976) Crossed and uncrossed representation of the visual field by brisk-sustained and brisk-transient cat retinal ganglion cells. *Vision Res.*, 16, 225–31

Kratz, K. E., Webb, S. V. & Sherman, S.M. (1978) Electrophysiological classification of X- and Y-cells in dorsal lateral geniculate nucleus of the cat. *Vision Res.*, 18, 489–92

Laties, A. M. & Sprague, J. M. (1966) The projection of optic fibers to the visual centers in the cat. *J. Comp. Neurol.*, 127, 35–70

LeVay, S. & Ferster, D. (1977) Relay cell classes in the lateral geniculate nucleus of the cat and the effects of visual deprivation. *J. Comp. Neurol.*, 172, 563–84

LeVay, S. & Ferster, D. (1979) Proportions of interneurons in the cat's lateral geniculate nucleus. *Brain Res.*, 164, 304–8

Lin, C.-S., Kratz, K. E. & Sherman, S. M. (1977) Percentage of relay cells in the cat's lateral geniculate nucleus. *Brain Res.*, 131, 167–73

Lindström, S. (1982) Synaptic organization of inhibitory pathways to principal cells in the lateral geniculate nucleus of the cat. *Brain Res.*, 234, 447–53

Madarasz, M., Gerle, J., Hajdu, F., Somogyi, G. & Tömböl, T. (1978) Quantitative histological studies on the lateral geniculate nucleus in the cat. II. Cell numbers and densities in the several layers. *J. Hirnforsch.*, 19, 159–64

Mason, R. (1975) Cell properties in the medial interlaminar nucleus of the cat's lateral geniculate complex in relation to the transient/sustained classification. *Exp. Brain Res.*, 22, 327–9

Mastronarde, D. N. (1983a) Correlated firing of cat retinal ganglion cells. I. Spontaneously active inputs to X- and Y-cells. *J. Neurophysiol.*, 49, 303–24

Mastronarde, D. N. (1983b) Correlated firing of cat retinal ganglion cells. II. Responses of X- and Y-cells to single quantal events. *J. Neurophysiol.*, 49, 325–49

Mastronarde, D. N. (1983c) Interactions between ganglion cells in cat retina. *J. Neurophysiol.*, 49, 350–65

Meyer, G. & Albus, K. (1981) Topography and cortical projections of morphologically identified neurons in the visual thalamus of the cat. *J. Comp. Neurol.*, 201, 353–74

Minkowski, M. (1920). Uber den Verlauf, die Endigung und die Zentrale Represantation von gekreuzten und ungekreuzten Sehnervenfasern bei einigen Saugetieren und beim Menschen. *Schweiz. Arch. Neurol. Psychiat.*, 6, 201–52

Overbosch, J. F. A. (1927). *Experimenteel-anatomische Onderzoekingen over de Projectie der Retine in het centrale Zenuwstelsel.* Dissertation. Amsterdam: H. J. Paris

Rioch, D.McK. (1929) Studies on the diencephalon of carnivora. I. The nuclear configuration of the thalamus, epithalamus and hypothalamus of the dog and cat. *J. Comp. Neurol.*, 49, 1–118

Rodieck, R. W. & Dreher, B. (1979). Visual suppression from nondominant eye in the lateral geniculate nucleus: a comparison of cat and monkey. *Exp. Brain Res.*, 35, 465–77

Rowe, M. H. & Dreher, B. (1982). Retinal W-cell projections to the medial interlaminar nucleus in the cat: implication for ganglion cell classification. *J. Comp. Neurol.*, 204, 117–33

Sanderson, K. J. (1971) The projection of the visual field to lateral geniculate and medial interlaminar nuclei in the cat. *J. Comp. Neurol.*, 143, 101–18

Sanderson, K. J., Bishop, P. O. & Darian-Smith, E. (1971) The properties of the binocular receptive fields of lateral geniculate neurons. *Exp. Brain Res.*, 13, 178–207

Sanderson, K. J., Darian-Smith, I. & Bishop, P. O. (1969) Binocular corresponding receptive fields of single units in the cat lateral geniculate nucleus. *Vision Res.*, 9, 1297–303

Scheibel, M. E. & Scheibel, A. B. (1966) The organization of the nucleus reticularis thalami: a Golgi study. *Brain Res.*, 1, 43–62

Schmielau, F. (1979) Integration of visual and nonvisual information in nucleus reticularis thalami of the cat. In *Developmental neurobiology of vision*, ed. R. D. Freeman, pp. 205–26. New York: Plenum Press

Singer, W. (1970) Inhibitory binocular interaction in the lateral geniculate body of the cat. *Brain Res.*, 18, 165–70

Singer, W. & Bedworth, N. (1973) Inhibitory interaction between X and Y units in the cat lateral geniculate nucleus. *Brain Res.*, 49, 291–307

So, Y. T. & Shapley, R. M. (1979) Spatial properties of X and Y cells in the lateral geniculate nucleus of the cat and conduction velocities of their inputs. *Exp. Brain Res.*, 36, 533–50

Stanford, L. R., Friedlander, M. J. & Sherman, S. M. (1981) Morphology of physiologically identified W-cells in the C laminae of the cat's lateral geniculate nucleus. *J. Neurosci.*, 1, 578–84

Stone, J. & Dreher, B. (1973) Projection of X- and Y-cells of the cat's lateral geniculate nucleus to areas 17 and 18 of the visual cortex. *J. Neurophysiol.*, 36, 450–7

Stone, J. & Fukuda, Y. (1974) The naso-temporal division of the cat's retina re-examined in terms of Y-, X- and W- cells. *J. Comp. Neurol.*, 155, 377–94

Stone, J. & Hansen, S. M. (1966) The projection of the cat's retina on the lateral geniculate nucleus. *J. Comp. Neurol.*, 126, 601–24

Sumitomo, I. & Iwama, K. (1977) Some properties of intrinsic neurons of the dorsal lateral geniculate nucleus of the rat. *Jap. J. Physiol.*, 27, 717–30

Sumitomo, I., Nakamura, M. & Iwama, K. (1976) Location and function of so-called interneurons of rat lateral geniculate body. *Exp. Neurol.*, 51, 110–23

Sur, M. & Sherman, S. M. (1982) Linear and non-linear W-cells in C-laminae of the cat's lateral geniculate nucleus. *J. Neurophysiol.*, 47, 869–84

Tello, F. (1904). Disposicion macroscopica y estructura del cuerpo geniculado externo. *Trab. Lab. Invest. biol., Univ. Madrid*, 3, 39–62

Thuma, B. D. (1928) Studies on the diencephalon of the cat. I. The cyto-architecture of the corpus geniculatum laterale. *J. Comp. Neurol.*, 46, 173–99

Vidyasagar, T. R. & Urbas, J. V. (1982) Orientation sensitivity of cat LGN neurones with and without inputs from visual cortical areas 17 and 18. *Exp. Brain Res.*, 46, 157–69

Wässle, H., Boycott, B. B. & Illing, R.-B. (1981) Morphology and mosaic of on- and off-beta cells in the cat retina and some functional considerations. *Proc. R. Soc. Lond., B*, 212, 177–95

Williams, R. W., Bastiani, M. J. & Chalupa, L. M. (1983) Loss of axons in the cat optic nerve following fetal unilateral enucleation: an electron microscopic analysis. *J. Neurosci.*, 3, 133–44

Wilson, P. D., Rowe, M. H. & Stone, J. (1976) Properties of relay cells in the cat's lateral geniculate nucleus: a comparison of W-cells with X- and Y-cells. *J. Neurophysiol.*, 39, 1193–209

Wilson, P. D. & Stone, J. (1975) Evidence of W-cell input to the cat's visual cortex via the C-laminae of the lateral geniculate nucleus. *Brain Res.*, 92, 472–8

9

Processing of moving images in the geniculocortical pathway

GUY A. ORBAN

Do moving images have a special nature?

Since Hubel & Wiesel noted in 1962 that moving stimuli were often more potent stimuli than stationary flashed stimuli, it has been a longstanding debate whether or not moving images are special compared to stationary ones. The answer to that question turns out to be complex.

Moving images as retinal input

Moving images are special in the sense that for higher mammals, including man, they are the most common, if not the only, natural retinal input. Since the eyes are mobile in the orbit and the head itself can be moving, movement of retinal images will occur in different stimulus situations and hence give rise to different perceptions. These situations are listed in Table 9.1 for the simplified case of an observer facing a single visual object. The picture would be even more complicated if one was to consider the many different objects that make up our visual world.

Table 9.1. *Stimulus situations from moving retinal images*

Situation	Retinal input	Perception
(1) Stationary stimulus – fixation – no body movement	Slowly moving image	Stationary object; subject stands still
(2) Moving stimulus – fixation – no body movement	Moving image	Moving object; subject stands still
(3) Stationary stimulus – saccade – no body movement	Fast-moving image	Stationary object; subject stands still
(4) Moving stimulus – pursuit – no body movement	Slowly moving image	Moving image; subject stands still
(5) Stationary stimulus – no eye movement – body movement	Moving image	Object stationary; subject moving

It is clear that in situations (3) to (5) of Table 9.1, neuronal centres outside the visual system will be active: either those eliciting the eye or body movements, or those receiving input from receptors activated by these movements, such as extraocular proprioceptors and vestibular receptors as well as skin, joint and muscle proprioceptors. It is still largely unknown to what extent the extraretinal signals are necessary for an adequate perception in these situations, as the visual system on its own has some cues about the situation. During saccades (situation 3), the input to the visual system will be a very fast movement of the whole retinal image, typical of this situation. During pursuit (situation 4), the visual input will be a small, almost stationary, object on a relatively slowly moving visual background, while body movement (situation 5) will generate relatively smooth velocity gradients over the visual field (flow patterns). The distinction between situations (1) and (2), however, has necessarily to be made by the visual system itself. We will take up this question later.

Moving *versus* stationary standing images in the awake, non-paralysed animal

Moving images are clearly special in the sense that in behaving animals they will excite a certain group of cells that are unresponsive to motionless stimuli. Such cells have been described in the awake monkey by Wurtz (1969) and in the awake cat by Noda *et al.* (1971). Wurtz (1969) distinguished between adapting and non-adapting responses of Area 17 cells depending on whether the responses to the onset of a stationary stimulus faded or continued as long as the stimulus was present. Adapting units responded well to moving stimuli, were usually optimally driven by speeds of 8–12 degrees/s and were often direction-selective. Noda *et al.* (1971) recorded, probably, both adapting and non-adapting responses from Areas 17 and 18 of the cat and distinguished four groups of cells: (1) neurons responding to medium fast target motion (12–100 degrees/s) and not to the motionless grating; (2) neurons with orientation-specific, continuous responses to motionless gratings, which mostly did not respond to target motion (over 12 degrees/s); (3) neurons which clearly responded to saccadic eye movements in the presence of patterned targets (and hence also to very fast movement of these targets) but did not respond or responded little to the motionless grating or to slower target motion (below 80–100 degrees/s); and (4) units

which responded only to fine irregular movements within their receptive area. The latter units were probably end-stopped cells with little response to a large grating (48 degrees × 56 degrees). It is plain that a moving stimulus was special for the adapting units of Wurtz and for group 1 and 3 cells of Noda *et al.*

Moving *versus* stationary flashed images in paralysed, lightly anaesthetized animals

Since Hubel & Wiesel's claim that moving stimuli can be more potent than stationary flashed stimuli, it has been frequently reported that in paralysed and lightly anaesthetized cats some cortical cells (about 15% in area 17) do not respond to stationary flashed stimuli (Pettigrew, Nikara & Bishop, 1968; Singer, Tretter & Cynader, 1975; Kato, Bishop & Orban, 1978; for review see Orban, 1984). All these reports were based on qualitative testing on a plotting table while listening to the units' discharges. These findings seem, however, to be due more to limitation of the human ear than to the properties of the cortical cells (Duysens & Orban, 1981). We have measured quantitatively, with a multihistogram technique, the response of cortical cells to stationary flashed stimuli presented in different positions in the receptive field and to the same slit moving through the receptive field at optimal velocity. We have computed a stationary-to-moving ratio, comparing the responses to the stationary slit flashed in the optimal position and to the same slit moving at optimal speed (as determined by a velocity–response curve). The outcome for 173 Area 17 cells is shown in Fig. 9.1A for both the preferred and non-preferred directions. In the preferred direction, the average stationary-to-moving ratio is close to 1, demonstrating that, on average, responses of cortical cells to flashed and moving slits are equally strong. In fact, when tested quantitatively, all cells, including those seemingly unresponsive to flashed slits on the plotting table, did respond to stationary flashed stimuli, although for a few cells the flash duration had to be adjusted.

The slits used for quantitative testing were of high contrast (2.6 cd/m² on a 0.005 cd/m² background), while on the plotting table lower contrasts were used. Hence one could argue that the difference in stimulus strength explains the discrepancy between qualitative and quantitative observations. Indeed the contrast– response relation for stationary flashed

Fig. 9. 1. Distribution of the stationary-to-moving ratio of cortical cells for the preferred (upper histograms of each pair) and the non-preferred (lower histograms) directions. (A) Whole sample of Area 17 cells recorded over a wide range of eccentricity (0–40 degrees). (B) Direction-selective and asymmetric cells. (C) Non-direction-selective cells. A stationary-to-moving ratio smaller than 0.5 in the preferred direction (stippled areas) is taken as an indication of facilitatory sequential effects (F) and a stationary-to-moving ratio over 2.5 in the non-preferred direction (hatched areas) is taken as indication of inhibitory sequential effects (I). See text for definition of stationary-to-moving ratio, and of direction-selective, asymmetric and non-selective cells. Arrows indicate median values.

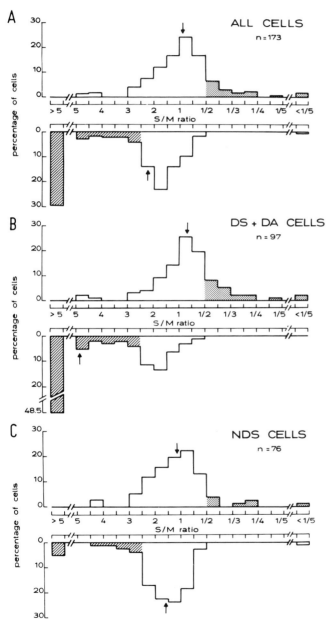

stimuli may be shifted towards higher contrasts compared to that for moving stimuli. Contrast–response relations of cortical cells show saturation at high contrasts (Albrecht & Hamilton, 1982). Therefore the use of high-contrast stimuli could yield equally strong responses to both moving and stationary flashed stimuli, while near-threshold contrasts would reveal a difference in sensitivity for both stimulation modes. We have made a number of controls at low contrast which indicate that the difference in stimulus intensity does not explain the discrepancy between our hand-plotting results and quantitative measurements. It seems rather that the human ear is very bad at picking out extremely tonic responses and, even more so, short phasic

responses, especially when the cell has some spontaneous activity as do C cells. This probably explains why it is commonly believed that many complex cells only respond to moving stimuli (Orban, 1984).

The stationary-to-moving ratio is also interesting from another point of view. It indicates the degree to which the response at the most sensitive point of the receptive field (or at least of the main subregion) is modified by stimulation of preceding (and following) loci in the receptive field. In other words, it measures the effects of sequential stimulation of different loci typical for a moving stimulus. This will be true to the extent that other differences between both stimulation modes, notably in local effects (see later) as stimulus

Fig. 9.2. Peristimulus–time histograms, representing the average responses to stop–go–stop movement in the preferred direction at different speeds (indicated on the left hand side) for representative examples of the four velocity types. The duration of the movement is indicated by the horizontal black bar below each histogram. The time scale is different for each velocity in order to obtain histograms of equal length: the duration of movement was the same for the cells in (A) and (B) and is indicated between the sets of histograms for both cells; the same holds for the cells in (C) and (D). All four cells were recorded

in Area 17. The velocity low-pass cell was an S cell with a single ON subregion (eccentricity 2.8 degrees, deep layer IV). The velocity-tuned cell (optimum 48 degrees/s) was an HS cell with a single OFF subregion (eccentricity 26 degrees, deep layer IV). The velocity broad-band cell was a C cell (eccentricity 7 degrees, layer V). The velocity high-pass cell was an A cell (eccentricity 24.5 degrees, superficial layer IV). See Orban & Kennedy (1981) and Orban (1984a) for the receptive field classification criteria.

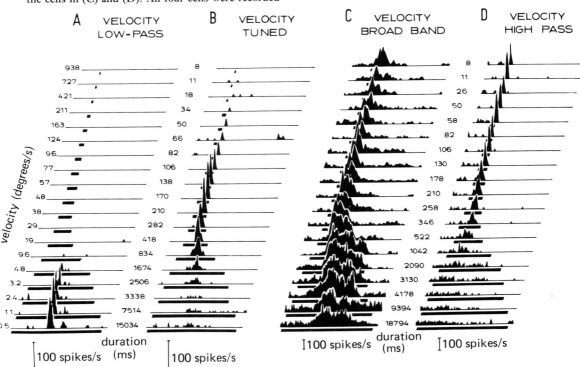

duration, energy or rise time, do not affect the visual neurons. Small stationary-to-moving ratios correspond to instances where the moving slit is a more potent stimulus than a stationary flashed one and thus indicate facilitation by the sequential stimulation; this occurs chiefly in the preferred direction. Large stationary-to-moving ratios occur when the moving slit is a less potent stimulus than a stationary flashed one and hence indicate inhibition by the sequential stimulation: this occurs chiefly in the non-preferred direction. That these two effects are of importance for direction selectivity of cortical cells is shown by the comparison of stationary-to-moving ratios of direction-selective and non-selective cells (cf. Fig. 9.1B and C). This point will be taken up later.

The fact that cortical cells, on average, respond as well to moving as to stationary flashed slits has the important consequence that one cannot use the comparison of these two responses to distinguish between cells involved in vision of stationary or moving objects (i.e. distinguishing between situation (1) or (2) of Table 9.1). As argued by Orban, Kennedy & Maes (1981*b,c*), the degree to which cells respond to very slow movement (0.3 degrees/s), of the order of magnitude of slow drifts during fixation, is a much better indicator. This has been quantified as the response to slow movement (Orban *et al.*, 1981*b*). Cells with little response to slow movement, most notably velocity-tuned cells (Fig. 9.2), will be involved in vision of moving objects. Vision of stationary objects will involve cells with a strong response to slow movement. Both in cats and monkeys, a particular group of cortical cells, specialize for slow speeds (Orban, *et al.*, 1981*b*; Orban, Kennedy & Bullier, 1983). These cells, labelled velocity low-pass cells, respond optimally to speeds of 0.3 degrees/s and their response falls to half the maximum for speeds between 2 and 20 degrees/s (Fig. 9.2). In both species they occur chiefly in Area 17 subserving central vision, the region where the cells have the highest spatial resolution (see Orban, 1984, for review). In this respect it is worth mentioning that human acuity is optimal for stationary stimuli and stimuli moving at speeds below 3 degrees/s and falls off rapidly with increasing velocity (Westheimer & McKee, 1975).

Moving stimuli in the spatiotemporal frequency domain

Moving images are not special in another way: a continuously illuminated moving stimulus (smooth motion) is undistinguishable from a stroboscopically illuminated moving stimulus (stroboscopic motion) under circumstances of high flash rates. This is predicted by theoretical analysis (Watson & Ahumada, 1983) and verified by experimental observation (Cremieux, Orban & Duysens, 1984). Indeed, cortical cells have a spatial and temporal frequency selectivity (Movshon, Thompson & Tolhurst, 1978). The representation of a stroboscopically illuminated moving bar in the spatiotemporal frequency plane is a set of lines (including one through the origin) with slope $1/v$ (v is the bar velocity) and separated on the abscissa (representing temporal frequency) by a distance ω (ω is the strobe rate). A continuously illuminated moving slit (smooth motion) corresponds in the frequency domain to a single line with slope $1/v$ through the origin. Both situations will be undistinguishable when the replicas of the line through the origin fall outside the spatiotemporal frequency sensitivity of the neuron which is centred near the origin. This will occur when ω is large (i.e. the time between flashes is short). All geniculate and cortical cells we have tested responded equally smoothly to stroboscopic motion as to smooth motion for strobe rates over 30–35 Hz (Orban *et al.*, 1982) This corresponds to the observation of Levick & Zacks (1970) that under low background illumination levels, comparable to ours, the temporal resolution of retinal ganglion cells (tested with two light pulse stimuli) is of the order of 35 ms. Hence there is a first temporal integration in the retinocortical pathway at the ganglion cell level (or before). While this integration is important for the understanding of our perception of movies or television, it also shows that for cells in the retinocortical pathway a moving stimulus is only the limiting case of a sequence of briefly flashed stationary stimuli. This temporal integration in the retina has a further consequence: single flashes will have different effects on central visual neurons depending on whether the temporal interval between the flashes exceeds the temporal resolution of the ganglion cells or not. In the former case, each flash will act in isolation and one will have to consider the energy, duration and rise time of the flash. In the latter case, the sequence of flashes acts as a whole and one will have to consider the 'equivalent' energy, duration and rise time of the whole sequence. As a final remark, one should notice that this spatiotemporal frequency approach can predict the transition from a single smooth motion response to multiple flash responses, but that this linear description cannot account for changes with strobe rate of direction selectivity and velocity tuning, which both result from non-linear sequential interactions.

Prediction of responses to moving stimuli from the response plane

The response plane is the quantification of the original cortical receptive field plots obtained by Hubel & Wiesel (1962) with stationary flashed stimuli. The response plane is a set of stacked poststimulus-time histograms corresponding to the average response to slits flashed ON and OFF in different positions in the receptive field (see Figs 9.6B, 9.9B and 9.10B for examples). It allows one to outline the subregions from which the cell is activated by turning the light slit ON and OFF. The outline of these excitatory subregions allows classification of cortical cells according to receptive field structure. The two most common types are S-family cells having non-overlapping ON and OFF subregions and narrow receptive fields and C-family cells having overlapping ON and OFF subregions and wide receptive fields (Henry, Harvey & Lund, 1979; Orban & Kennedy, 1981; Duysens *et al.*, 1982*b*; Orban, 1984). A major improvement (Palmer & Davis, 1981) is to combine the presentation of the flashing stimulus with a conditioning stimulus (Henry, Bishop & Coombs, 1969). This allows the outlining of the inhibitory subregions as well as of the excitatory subregions. From the response plane, both qualitative and quantitative predictions of responses to moving stimuli can be made. Moving stimuli include not only narrow light and dark bars but also light and dark edges, introduced as elementary stimuli by Bishop, Coombs & Henry (1971).

In contrast to the initial belief that little correlation existed (Bishop *et al.*, 1971), there is rather good qualitative correlation between excitatory ON and OFF subregions and moving light-and dark-edge discharge regions (Goodwin, Henry & Bishop, 1975; Emerson & Gerstein, 1977*a*; Kulikowski, Bishop & Kato, 1981; Palmer & Davis, 1981; Bullier, Mustari & Henry, 1982). This qualitative correspondence can be summarized using the spatial sensitivity profile of the ON and OFF excitatory subregions. The responses to light

and dark edges correspond to integration of the stationary spatial sensitivity profile: their locations correspond to those of ON and OFF excitatory subregions, respectively, and their strengths (heights of the response peaks) to the integral of the area under the spatial sensitivity profile of the ON or OFF region. The responses to light and dark bars correspond to differentiation of the stationary spatial sensitivity profile. The cell responds to a light slit when the profile turns up (under the convention that ON sensitivity is plotted upward and the OFF sensitivity downward), i.e. by leaving an OFF area or entering an ON area. Conversely, under the same convention a cell will respond to a dark bar when the profile turns down, i.e. by entering an OFF region or leaving an ON region. For both light slits and dark bars the response strength depends on the slope of the profile. Despite claims to the contrary by Kulikowski *et al.* (1981), examination of a large number ($n = 257$) of Area 17 cells by Palmer & Davis (1981) has shown that when the qualitative correlation fails, it is usually due to a response to a moving slit or edge that is missing.

Quantitative predictions derived from the response plane are usually made for responses to slowly moving light slits. These quantitative predictions test for linear superposition of the responses elicited by static stimuli in different loci. In other words, one assumes that moving a slit is equivalent to flashing the slit ON in a sequence of locations, e.g. a slit moving at 1 degree/s will be equated to a sequence of flashing the slit ON for 250 ms at locations separated by 0.25 degrees.

If one derives a response plane for this static stimulation and mathematically superimposes the response at different locations (taking into account the time to reach the next position before adding the responses), one can predict the response to the moving stimulus. According to Palmer & Davis (1981), the predictions from the response plane are much better when both excitatory and inhibitory responses elicited at a given locus are considered, and more discrepancies could be removed if the depth of inhibition could be evaluated more precisely. When the prediction from the response plane fits the observation, it means that the cell behaves linearly, in the sense that the different loci give an independent contribution to the activity of the cell and that the resultant activity is only the (algebraic) sum of these individual contributions. Failure of the prediction shows that the cell behaves non-linearly, in the sense that the contributions of the different loci are non-independent but interact with each other and hence that the resultant activity is no longer the sum of the contribution of the different loci. In terms of information processing in the dendritic tree, the first instance will only occur when the signals (excitatory postsynaptic potentials (EPSPs) and inhibitory post-synaptic potentials (IPSPs)) elicited in the dendritic branches reach the soma unaltered and when the inhibition acts in a subtractive manner. The non-linear situation will arise either when there are facilitatory or inhibitory interactions at the level of the dendrites, modifying the signals reaching the soma, or when the inhibition at the soma acts divisively.

Palmer & Davis (1981) further showed that the major exceptions to the quantitative predictions from the response plane are strong direction selectivity of S cells and direction selectivity (of any degree) of C cells. This fits with the observation of Emerson & Coleman (1981) that for Area 17 cells, whether simple or complex, the strength of response to a moving slit (in peak firing rate) can be almost perfectly predicted from linear superposition when one considers the preferred direction, while in the non-preferred direction the predicted strength is double that actually observed. One should clearly distinguish the ratio between predicted and observed strength of Emerson & Coleman (1981) from our stationary-to-moving ratio. The stationary-to-moving ratio tests (by deviation from one) for any effect, either linear or non-linear, of sequential activation of different loci. In contrast, the ratio of Emerson & Coleman (1981) only tests (by deviation from one) for non-linear effects of sequential activation of different loci. Indeed the linear superposition of the spatially distinct static responses allows for one type of sequential effect, namely the synergy of leaving one subregion and entering an antagonistic one. This mechanism was initally put forward by Hubel & Wiesel (1962) as explanation of direction selectivity in simple cells and has been labelled the 'superposition' mechanism by Emerson & Gerstein (1977a). It occurs in S cells with asymmetric receptive fields (including either one ON and one OFF subregion or one subregion flanked by two opposite subregions of unequal strength) and implies that direction preference is correlated with stimulus contrast, i.e. for a light bar the preferred direction will correspond to entering the ON region after leaving the OFF zone, while for a dark bar the preferred direction will be the reverse one. This

explains why in S cells only extreme direction selectivity, larger than that which superposition mechanism can generate, violates the linear super-position hypothesis (Palmer & Davis, 1981). In our stationary-to-moving ratio terminology, the (linear) superposition mechanism of direction selectivity could account for a number of the facilitatory influences observed in the preferred direction of direction-selective or asymmetrical cells (Fig. 9.1). According to Emerson & Gerstein (1977b), the two major types of non-linear sequential interactions contributing to direction selectivity are forward inhibition in the non-preferred direction and forward facilitation in the preferred direction.

It should be noted that all of these studies of linear superposition were done on Area 17 cells with central receptive fields. As will be shown in the next paragraph, these cells exhibit relatively little velocity and direction selectivity and one would thus expect less sequential interactions, especially non-linear ones, in these cells.

Selectivities of cortical cells tested with moving stimuli

When tested with moving stimuli, cortical cells can display selectivity for two important parameters: direction and velocity of movement. The shapes of velocity–response curves are highly dependent on the response measure used (Pettigrew *et al.*, 1968; Orban *et al.*, 1981b; Orban, 1984) and in the initial studies it was not clear whether one should use number of spikes or firing rate (Pettigrew *et al.*, 1968; Emerson & Gerstein, 1977a). It gradually became clear that number of spikes could not be used, since it depends on stimulus duration which decreases with increasing velocity. Moreover, Movshon (1975) showed that contrast sensitivity at different velocities was correlated with peak firing rate and not with number of spikes. Some authors (Movshon, 1975; Goodwin & Henry, 1978) have argued that peak and mean firing rate change in parallel with velocity, since the shape of the response peaks changes little with velocity. Our own experiments, in which we could plot the same data with both measurements, clearly showed that this assumption is not verified and hence one has to choose either peak or average firing rate. Some authors have chosen average firing rate, averaged over a minimum duration of 200 ms (Maunsell & Van Essen, 1983), which depresses the responses to the extreme velocities and accentuates the tuning. We believe that the measure that makes the least assumptions is the peak firing rate measured with a short (8 ms) bin width. Since response duration of cortical cells is never shorter than 20 ms, this sampling with a 8 ms bin width will not distort the cortical response even to very fast motion. We also average over an equal number of presentations for each velocity, generally 10–20 presentations. If the neurons' response is weak or if it is unclear to which type the velocity–response curve of the neuron belongs, we average over more than 10 presentations.

We have systematically tested neurons through the dominant eye with optimal, narrow slits of high contrast, moving at 19 or 20 velocities ranging between a minimum of 0.5 or 0.3 and a maximum of 700 or 900 degrees/s. We use the multihistogram method pioneered by the Canberra group (Henry, Bishop & Dreher, 1974) with interleaving in random order. We (Orban *et al.*, 1981*b*; Duysens *et al.*, 1982*a*; Cremieux *et al.*, 1984; Duysens, Orban & Cremieux, 1984) have tested over 450 cortical cells in Areas 17, 18 and 19 over different eccentricities. We (Orban, Hoffmann & Duysens, 1981*a*) have also recorded from 92 lateral geniculate nucleus (LGN) cells at different eccentricities. Velocity-response curves are obtained by fitting a spline function through the data points plotting the maximum firing rate as a function of velocity. The velocity sensitivity of the cell is characterized by its velocity-response curve in the preferred direction. Most of these velocity-response curves can be fitted into one of the four types illustrated (Fig. 9.2) by the peristimulus-time histograms from which the curves are computed: (1) velocity low-pass cells have responses exceeding 50% of the maximum response at the lowest velocity (0.5 degrees/s) and their response drops to less than half the maximum response at velocities below 20 degrees/s; (2) velocity-tuned cells have a small ratio (less than 50) between the upper and lower velocities at which the response equals half the maximum response; (3) velocity high-pass cells have a response less than 33% of the maximum to the slowest movement (0.5 degrees/s), a maximum response to movements of 100–200 degrees/s and a good response (40–100% of maximum) to the fastest velocity; and (4) all the remaining cells are considered as velocity broad band. The latter cells could well display specificities when tested with stimuli other than a light slit moving in a frontal plane. Except for velocity high-pass cells, these velocity types have also been observed in V1 and V2 of the monkey (Orban, *et al.*, 1983).

Henry *et al.*, (1973) made clear the distinction between orientation, axis and direction of movement. Provided that one follows the convention that the axis of movement is orthogonal to the orientation, a cortical cell tested with an elongated moving stimulus such as a moving slit, will display an axial selectivity reflecting its orientation selectivity which is brought into play both by moving and stationary stimuli. Some cells have in addition a preference for one direction of movement on the optimal axis; this is referred to as direction selectivity. When one tests these cells with a slit moving in all possible directions, a directional tuning is measured which results both from the orientation- and direction-selective mechanisms. Using a non-oriented stimulus such as a spot or a random noise field, one also obtains a directional tuning. This tuning is probably only due to the direction-selective mechanism(s) which hence has an axial-sensitivity profile. Complex cells respond equally well to slits and two-dimensional visual noise (Hammond and MacKay, 1975; Orban, 1975) and in about two out of three complex cells, directional tunings obtained with both stimuli have similar optima, but the width of tuning is wider for noise stimuli (Hammond, 1978).

For elongated moving stimuli one usually considers just direction selectivity. There is a general agreement to use an index for direction selectivity at a given velocity, the direction index, DI = 1 − response in non-preferred direction/response in preferred direction (Kato *et al.*, 1978), where all responses are net responses (i.e. the mean spontaneous activity is subtracted). There is some discussion as to whether this is a full description of the direction-selective properties of cortical cells. Emerson & Gerstein (1977*b*) and Albus (1980) have suggested that one should measure the direction indices for both a moving light slit and a moving dark bar. These authors considered a cell to be direction-selective only when the direction index was large for both stimuli and when the preferred direction was the same for both stimuli. Cells for which the preferred direction switched with contrast sign were considered to be direction-asymmetric. We have not yet explored this invariance systematically. We (Orban, Kennedy & Maes, 1981*c*), as others (Movshon, 1975), were however able to show that the direction index depends on velocity. Hence we characterize a cell by its mean direction index (MDI) which is a weighted average of the direction indices obtained at 19 or 20 velocities, the responses in the preferred direction at each velocity being used as weighting factors. The MDI ranges from 0 (no selectivity) to 100 (complete direction selectivity). We define (Fig. 9.3) cells with MDI > 66 as direction-selective (DS) and those with MDI < 50 as non-direction-selective (NDS); those in between are considered to be direction-asymmetric (DA). The response in the preferred direction of direction-selective cells is, on average, three times larger than that in the non-preferred direction. This is a more stringent definition than one should think, since we use maximum firing rate and not average firing rate as a response measure. None of 92 LGN cells tested qualified as direction-selective (Fig. 9.3).

Our systematic study of the primary cortical areas and of the LGN of the cat over a wide range of eccentricities has shown that velocity selectivity and direction selectivity are strongly dependent on cortical area and eccentricity (see Orban, 1984, for review). Velocity low-pass cells dominate in Area 17 receiving central projections and are absent in the LGN and in Area 18. Velocity-tuned cells occur mainly in Area 18 subserving central vision. In Areas 17 and 19 and in the LGN they constitute less than 15% of the sample. Velocity high-pass cells occur mainly in Area 18 subserving peripheral vision, in the LGN and, to a lesser extent, in Areas 17 and 19 subserving peripheral vision. Direction-selective cells occur mainly in area 18 subserving central vision, not at all in the LGN, and are almost absent in Area 19. It should be noted that velocity-tuned and velocity high-pass cells do hardly respond to slowly moving stimuli and they will therefore only be active in situation (2) of Table 9.1. They probably correspond to the adapting units of Wurtz (1969) and to group 1 (velocity-tuned) and 3 (velocity high-pass) cells of Noda *et al.* (1971). In contrast, velocity low-pass cells will be strongly active in situation (1), i.e. fixation of a stationary object. They may correspond to the non-adapting cells of Wurtz (1969) and group 2 of Noda *et al.* (1971) (see above). The importance of velocity-tuned cells for motion perception is further underscored by their direction selectivity and the distribution of their optima which fits the range over which cats (Vandenbussche, Orban & Maes, unpublished) and human subjects (McKee, 1981; Orban, 1985; Orban & Maes, 1984; Orban *et al.*, 1984) can make fine velocity discriminations.

Fig. 9.3. Distributions of mean direction index (MDI) for a sample of cortical cells with the receptive field uncovered (A), for the same cortical cells with the receptive field masked except for a 0.3-degree wide window placed over the most sensitive part of the receptive field (B), and a sample of LGN cells (C). Sample sizes are indicated above the histograms. The dark areas indicate direction-selective cells; open areas indicate non-direction-selective cells; and stippled areas show direction-asymmetric cells (see text for definition). Notice the similarity of distributions in B and C.

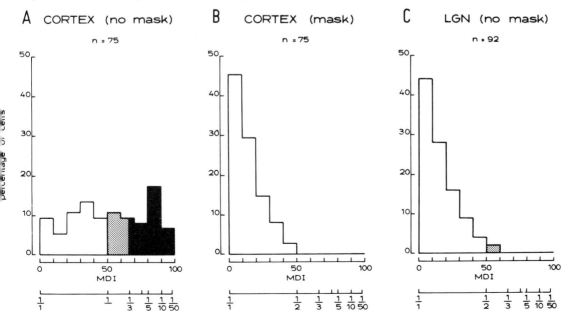

Mechanisms underlying direction and velocity selectivity

Movement of a light slit over a receptive field induces a number of changes which depend either on the velocity and/or the direction of the moving slit. They can be divided into two broad classes, local and sequential effects (Table 9.2). Local effects only depend on slit velocity and hence can only contribute to velocity selectivity. Sequential effects depend on both direction and velocity of movement and hence can both determine direction selectivity and contribute to velocity selectivity.

The distinction between sequential and local effects depends solely on whether or not the response to stimulation of a given locus is dependent on stimulation of preceding or following loci. It has no spatial connotation in the sense that the local effects can extend over a region of space that is wider than the range over which sequential effects operate. Indeed, in cortical cells with non-overlapping ON and OFF subregions, many of the local effects can summate over the whole subregion, while the sequential effects can occur either between subregions or within a subregion. That local effects occur over a small region of space rather than a point is due to the limitations imposed on spatial sampling by the cone and retinal ganglion cell mosaics (see Chapter 2). That sequential effects can occur within a subregion is likely to be due to the overlap of the receptive field centres (RFCs) of retinal ganglion cells projecting onto a cortical cell. Local effects will depend on the width of the region over which they extend, e.g. the width of the subregion of a cortical cell with non-overlapping subregions or the diameter of the RFC of a LGN cell. Indeed, effects (1) to (3) of Table 9.2 depend on stimulus duration which is given by the width of the locus divided by velocity. Effects (4) and (5) depend much less on the width of the locus, although they require some minimal width of the stimulus and the locus. Inasmuch as they depend on the spatial extent over which they occur, these local temporal effects can be considered as being spatiotemporal. The sequential effects are genuinely spatiotemporal since successive loci have to be reached successively. The latter effects will escape exploration by single slit tests such as response planes. The only way to test these sequential effects is by using multiple slit experiments, in which at least two slits are pre-

Table 9.2. *Local and sequential effects of changes in direction and velocity of a moving light slit*

Effect of moving light slit	Change with increasing velocity
Local effects	
(1) Illumination of the locus for a given duration	Duration decreases
(2) The locus is exposed to a given light energy (luminance × duration × stimulus area)	Light energy decreases
(3) After being illuminated for some duration, the locus is darkened	ON–OFF alternation time decreases
(4) The luminance over the locus increases with a given rise time	Rise time decreases
(5) The luminance over the locus decreases in some time	The decay time decreases
Sequential effects	
(6) Several spatially distinct loci are illuminated in turn (Onset sequence)	The time between light Onset in 2 successive loci decreases
(7) Light is offset sequentially in different loci (Offset sequence)	The time between light Offset in 2 successive loci decreases
(8) After light goes OFF at one locus, it goes ON in the next locus	The time between light OFF and light ON in successive loci decreases

ented (ON or OFF) in an asynchronous fashion: temporally coincident stimuli will probe spatial interactions rather than spatiotemporal ones.

It is worth noting that in addition to the eight effects produced by a moving light slit (those for a moving dark bar would be symmetrical), a drifting sinusoidal luminance grating will produce two additional local rate effects: (1) the locus will be illuminated a number of times per second; and (2) a number of times per second the locus will be darkened. It has been shown that Area 18 cells at least are very sensitive to these local rate effects (Dinse & von Seelen, 1981). This analysis of the many different effects that moving slits and gratings can exert on cells in the visual pathway has important consequences on the choice of stimuli used to test cortical cells. First, it is not possible to use a drifting grating to evaluate the temporal frequency tuning of visual cells and especially cortical cells, since such stimuli could evoke sequential effects in addition to the local effects. Recent work of Bisti *et al.* (1984) in the cat has shown that the responses of Area 18 cells to drifting gratings depend in a complex way on temporal frequency and velocity. One can rely only on a stationary slit modulated sinusoidally in time to measure the temporal frequency tuning of a cortical cell (a stationary grating modulated sinusoidally in time is equivalent to the sum of two gratings drifting in opposite directions and would not be adequate). Secondly, there is no simple way by which one can predict the velocity sensitivity of a visual cell from its temporal and spatial frequency tuning. Indeed the temporal frequency tuning does not capture the sequential effects that can contribute to velocity sensitivity and, on the other hand, temporal frequency tuning depends on local rate effects which do not contribute to velocity sensitivity. In addition it is possible that, due to its continuous presence over the receptive field, a grating, unlike a moving slit, induces adaptation of visual cells (Vautin & Berkley, 1977). It seems therefore that at least for the study of velocity coding (estimation of the magnitude of the velocity vector) in the visual system, drifting gratings, although mathematically simple, are too complicated in practice, while moving slits, although mathematically complicated, are much more adequate in practice. In this respect it is worth noting that human velocity discrimination truly depends on velocity and not on temporal frequency (McKee, Silverman & Nakayama, 1984).

The sequential effects (6) and (7) in Table 9.2 are symmetrical and seem to occur mainly within ON or OFF subregions. It seems that forward facilitation and inhibition, as direction-selective mechanisms, depend on this type of sequential effect within subregions (Emerson & Gerstein, 1977b; Movshon *et al.*, 1978; Ganz, 1984). This fits with the observation that direction selectivity is not abolished by intraocular injection of APB (DL-2-amino-4-phosphonobutyric acid) which silences the ON pathway (Schiller, 1982). The available data suggest that the asymmetrical effect (8) in Table 9.2 has, in contrast, to occur between subregions, although Emerson & Gerstein (1977b) have alluded to the possibility that forward inhibition and facilitation within a subregion may depend on this type of interaction. The 'superposition' direction-selective mechanisms (see above) are an effect of type (8). It is likely that in addition to this linear interaction, non-linear interactions also occur between subregions of opposite polarity (Citron & Emerson, 1983).

That ONset or OFFset sequence effects occur within the subregion of a cortical cell can be difficult to understand, especially if the cell is a first-order cortical neuron and the subregion is small (and apparently receives input from only one row of LGN afferents) or if no sign of inhibition within the subregion can be brought forward, as in C cells. An elegant solution to this problem has been suggested by the theoretical work on passive dendritic trees by Koch, Poggio & Torre (1983). Although this study intended to explain retinal ganglion cell direction selectivity it may apply to cortical cells as well. Koch *et al.* (1983) have shown that inhibitory synaptic inputs can veto excitatory inputs (i.e. producing non-linear inhibitory interactions) provided that (1) the peak conductance changes are large, (2) inhibition lies on the direct path from the excitation to the soma, and (3) the time courses of excitation and inhibition overlap substantially. With respect to condition (2), it is interesting to note that Davis & Sterling (1979) have shown that geniculate afferents end much more distally on layer IV stellate cells than the flat vesicle asymmetrical terminals which are supposedly inhibitory in nature. Due to local forbidding mechanisms, such as the one suggested by Koch *et al.* (1983), as well as other mechanisms (e.g. presynaptic inhibition; Levick, personal communication), the local effects impinging on the dendritic tree do not necessarily have a reflection at the level of the cell soma. It should be stressed that these non-linear interactions will escape detection in the classical receptive field studies (producing response planes), even those using conditioning stimuli. Two-slit experiments with sequential presentation (Emerson & Gerstein, 1977*b*; Ganz, 1984) can reveal such interactions.

The importance of sequential effects for velocity and direction selectivity can be asessed by departure of the stationary-to-moving ratio from 1 and also by masking experiments in which most of the receptive field, except a small part (the most sensitive one) of a subregion, is covered. The stationary-to-moving ratio shows that sequential effects are more prominent in direction-selective cells than in non-direction-selective

cells (Fig. 9.1). The masking experiments show th sequential effects are of paramount importance f direction selectivity: under our stimulus condition leaving only a 0.3 degree wide strip of the receptive fiel unmasked completely wipes out direction selectivity cortical cells (Fig. 9.3). More work is required evaluate the contribution of the different sequenti effects to direction selectivity of the different cortic cells. It can already be concluded from data (Fig. 9. and from Sillito's iontophoretic experiments (197 with the γ-aminobutyric acid (GABA) antagoni bicuculline, that forward inhibition must be the maj mechanism for direction selectivity in most cells. Fi

Fig. 9.4. Duration-sensitive cells. (A) Peri-stimulus-time histograms, showing the average response to a light slit flashed ON for different durations at the most sensitive part of the ON subregion. The cell was a velocity low-pass cell (velocity-response curve shown in C), recorded in layer VI of Area 17. Its S receptive field consisted of a single ON subregion and was located 2.7 degrees from the fixation point. The durations of the ON presentation are indicated by the horizontal black bars below the histograms. (B) Proportion of

duration-sensitive cells in Area 17 as a function of their upper cut-off velocity (i.e. the upper velocity at which the response equals half the maximum response). (C) Comparison of velocity-response curves in preferred direction of a velocity low-pass cell in unmasked (full line) and masked condition (dashed line) with the prediction (stippled line) made from the duration–response curve (same cell as in A). The cell had no spontaneous activity. (D) Distribution of upper cut-off velocities of LG. cells.

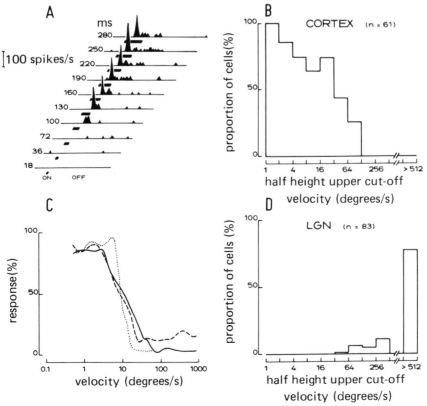

Fig. 9.5. Properties of velocity low-pass cells. (A) Comparison of velocity-response curves obtained at two stimulus strengths: 2.6 cd/m² (full line – the standard strength) and 0.26 cd/m² (dashed line). (B) Comparison of velocity-response curves obtained with light slit (full line) and light edge (dashed line). The arrows indicate the corresponding significance levels, i.e. the mean spontaneous activity plus twice its standard deviation (P = 0.025) (notice that both response and spontaneous activity are measured in peak firing rate): full arrows relate to curves in full lines, open arrows to stippled lines.

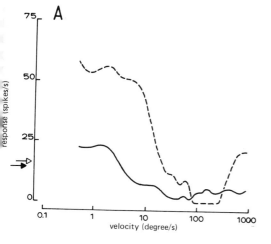

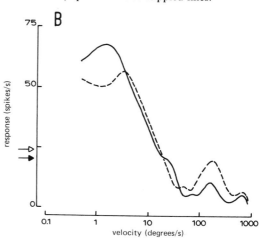

Fig. 9.6. Invariance of duration thresholds of a duration-sensitive cell to changes in stimulus strength. (A) Response plane (standard luminance 2.6 cd/m²). (B), (C), (D) Average responses to a light slit flashed ON for different durations at different stimulus strengths: 2.6 cd/m² (B); 0.26 cd/m² (C) and 0.03 cd/m² (D). Same conventions as in Fig. 9.4 (A). The cell was an Area 17 HS cell (eccentricity 8 degrees, layer VI). Notice

that for the two lower strengths the threshold duration was the same (about 100 ms). At the higher strength the threshold did not decrease as one would expect if the light energy was critical, but rather increased to 400 ms. This was fairly typically for strong stimuli presented to HS cells, as was the appearance of phasic OFF bursts (peak latency of 300 ms) at some positions in the ON subregion.

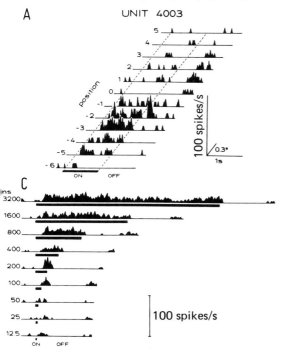

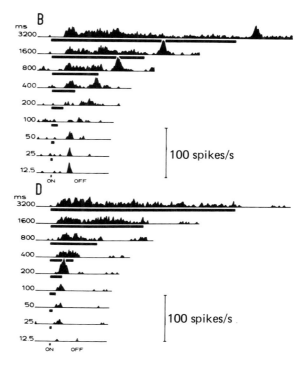

9.4 to 9.12 show that most effects listed in Table 9.2 occur in cortical cells and hence can contribute to velocity selectivity. The aim of our more recent experiments has been to evaluate the relative contribution of these different effects to the velocity selectivity of the different cortical cell types.

Fig. 9.4 shows the influence of duration of illumination on cortical cells. We have been able to show that many cortical cells require a critical duration of ON stimulation before discharging (Duysens *et al.*, 1984). This is demonstrated by flashing light slits ON for different durations in the most sensitive part of the ON subregion (determined by a response plane). For example, the cell in Fig. 9.4A required a minimum ON duration (threshold) of 72–100 ms. Duration-sensitive cells, as the one in Fig. 9.4A, usually have low cut-off velocities (velocity at which the response drops to half the maximum response) as shown in Fig. 9.4B. This distribution is the mirror image of that of the LGN cells (Fig. 9.4D). The upper cut-off velocity of duration-sensitive cells was little affected by masking and could effectively be predicted from the duration–response curve and the width of the subregion (Fig. 9.4C). Duration-sensitive cells make up almost the entire velocity low-pass population, with the exception of a few C cells, and a sizeable proportion of slow velocity-tuned and velocity broad-band cells, but none of the velocity high-pass cells was duration-sensitive. That duration of illumination, rather than light energy, at least for medium and high luminances, is critical for duration-sensitive cells, is shown by the observation that the upper velocity cut-off of these cells hardly changes with decreasing the luminance 10-fold (Fig. 9.5A). Also the duration threshold changes little with changes in stimulus strength (Fig. 9.6). Fig. 9.5 further shows that the upper cut-off velocity is little affected by changing from a narrow (0.3-degree) light slit to a light edge. This shows that the critical factor for moving stimuli is not exactly the time of illumination but what we have called the activation time, i.e. the time the light edge takes to cross the ON subregion. What seems to be important for duration-sensitive cells is the duration of activation by a positive or negative luminance step. This activation time for a moving slit will depend on receptive field width. Supposing that the critical duration is on average equal for S cells receiving X and Y input, one would expect S cells receiving Y input to have slightly higher upper cut-off velocities than S cells receiving X input, since Y-driven S cells

have larger subregions than X-driven S cells. This is indeed what has been observed by Mustari, Bullier & Henry (1982).

We also have evidence that for other cortical cells, chiefly velocity high-pass cells, light energy is critical

Fig. 9.7. Comparison of velocity-response curves (A) and duration-response curves (B) of a duration-insensitive cell at different stimulus strengths. The cell was an Area 18 C cell recorded in layer V (eccentricity 24 degrees). The stimulus strengths were 0.26 cd/m² (full line), 0.03 cd/m² (stippled line) and 0.007 cd/m² (dashed line). Arrows indicate the significance levels ($P = 0.025$). In the top panel the full arrow refers to the curve in full line, the open arrow to the two other curves. Notice that the duration sensitivity does predict the velocity upper cut-off but not the shape of the velocity–response curve. This is in all likelihood due to the rise-time effects (effect 4 in Table 9.2).

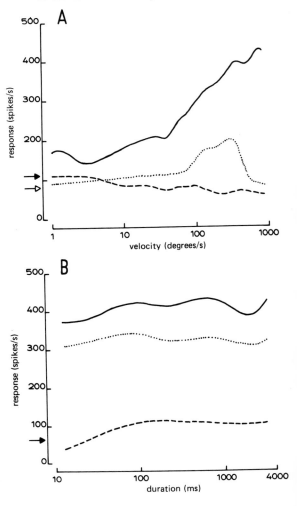

o set the upper cut-off velocity. It has been shown that he preferred velocity and the upper cut-off velocity of etinal ganglion cells (Cleland, 1980) and, to a lesser xtent, of geniculate cells (Frishman, Schweitzer-Tong & Goldstein, 1983) depend on light energy. Since light nergy (luminance × duration × stimulus area) depends n duration, and hence, for the same velocity, on eceptive field width, this energy sensitivity may xplain why, when tested with low stimulus luminances, etinal and geniculate Y cells, with a larger receptive ield than X cells, are sensitive to faster velocities than X cells. Provided that luminance and duration can be nterchanged, the high-stimulus luminance used in our ystematic study of velocity sensitivity of cortical and geniculate cells (see above), will tend to decrease the ontribution of the duration and hence of receptive field vidth. This explains why, under our experimental onditions, the difference in velocity preference

between X and Y geniculate cells is much smaller (Orban *et al.*, 1983) than that reported for lower stimulus strengths (Frishman *et al.*, 1983). The cortical cell of Fig. 9.7 behaved exactly in the way just described for geniculate cells. This cell had a high-pass curve at high and medium stimulus luminance (0.26 cd/m²) and no critical duration was observed (Fig. 9.7B). Decreasing the stimulus luminance to 0.03 cd/m² reduced the response level, but all velocities still yielded a significant response and so did all durations tested. Further reduction of the stimulus strength, however, introduced a velocity upper cut-off and also a threshold duration. This strongly suggests that in this type of cell the light energy determines the range of effective velocities and at high stimulus strength this range can be quite large.

While effects (1) and (2) set the upper limit of the range of effective velocities and hence determine the

Fig. 9.8. Influence of rise time of luminance increments on a cortical cell (same cell as in Fig. 9.7). (A) Average response to luminance increments over the most sensitive part of the ON subregion for different rise times (from 3430 ms to 5 ms). (B) Average response to a positive luminance pulse 3200 ms long (i.e. light slit flashed ON in the same position for 3200 ms). The shutter has an opening time of 3 ms. The changes in luminance are indicated above each peristimulus-time histogram.

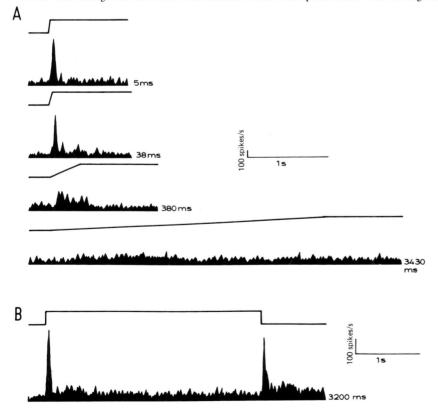

descending leg of velocity-response curves, effect (4) probably contributes to the ascending leg of velocity-response curves. Fig. 9.8 shows that cortical cells are sensitive to the rise time of a stationary flashed ON stimulus. In Fig. 9.8(A), a cortical velocity high-pass cell of the C family was subjected to a positive luminance increase of fixed amplitude but with different rise times ranging from 3430 ms to 5 ms: while the cell was hardly affected by the slowest increase in luminance, the medium fast one (rise time 38 ms) was almost as effective as a step in luminance (i.e. the slit flashed ON) as shown in Fig. 9.8(B). Our masking experiments, as well as our experiments with stroboscopically illumin-

ated moving slits, suggest that the ascending leg of the velocity-response curve of LGN cells and cortical velocity high-pass cells of the C family depends on the rise time of the luminance increase. Effect (5) probably contributed in a similar way for cells dominated by OFF input.

The last local effect, the ON–OFF alternation time, can only be important for OFF dominant cells responding to a light slit. Indeed, responses to moving slits originate from the ON subregion in most cells having both ON and OFF subregions in their receptive field. For OFF cells the duration of the preceding ON can be critical, i.e. it has to exceed a minimum duration; this can explain small upper cut-off velocities of dominant OFF cells tested with light slits. A more subtle effect is shown in Fig. 9.9(B) where the OFF response is optimal for intermediate durations of the preceding ON duration. This correlated to some extent with the preference of this cell for medium velocities (Fig. 9.9A).

As has been mentioned before, sequential effects which are of paramount importance for direction selectivity can also contribute to velocity sensitivity. Masking experiments have shown that sequential effects determine the ascending leg of velocity-tuned cells of any receptive field type and of velocity high-pass cells of the S and A families, mainly by facilitation between subregions (i.e. effects of type 8 in Table 9.2) but perhaps also by forward facilitation (effects of type 6 or 7). Fig. 9.10 shows that superposition (a linear effect of type 8) can also affect the velocity-response curve of other S cells. The cell has an asymmetric receptive field typical for S cells displaying superposition (i.e. the OFF subregion precedes the ON subregion in the preferred direction), however the two subregions had different duration requirements: the ON subregion had a duration threshold of about 25 ms (effect 1), while the OFF region required that the preceding ON stimulation lasted at least 600 ms (effect 3). When tested with a moving light slit, the velocity-response curve was similar for both directions in the medium and fast velocity range. This can be taken as indication that the descending leg of both curves was set by the threshold duration of the ON subregion. In the slow velocity range, however, the backward direction (going from positive to negative positions in Fig. 9.10(B)) was clearly preferred, and the difference between both directions faded at a velocity corresponding to the threshold duration of the OFF

Fig. 9.9. Influence of the preceding ON duration on the OFF response. Data from an HS cell with a simple OFF subregion recorded in superficial layer IV of Area 17 (eccentricity 4.5 degrees). (A) Velocity–response curve in the preferred direction (full line) compared with the predictions made from the responses to different Onset durations shown in B (same conventions as in Fig. 9.4). The arrow indicates the significance level ($P = 0.025$). (B) Peristimulus-time histograms, representing the average response of the cell to a slit flashed ON for different durations in the centre of the OFF subregion. The cell responded better to intermediate velocities, but failed to qualify as velocity-tuned, since the tuning width was too wide, hence the cell was classified as velocity broad-band.

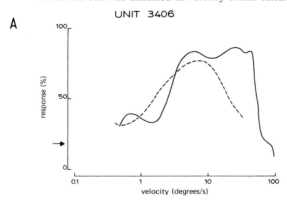

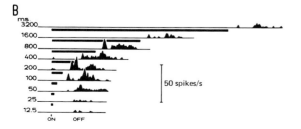

ubregion. Only slow stimuli could trigger the OFF egion and facilitate the subsequent ON response.

Stroboscopic motion is another way to remove equential effects, since at low strobe rates the temporal nd spatial requirements of the sequential effects will e violated. One should note, however, that decreasing trobe rate also alters the local effects. As long as the trobe interval is short enough to allow temporal ntegration of the light pulses, it decreases the apparent' luminance and hence the 'apparent' light nergy (effect 2). Outside the temporal integrating apacities of the cells, it reduces the illumination luration (effects 1 and 3) and, most importantly, it lecreases the rise time (effect 4). Figs 9.11 and 9.12 how the effect of decreasing strobe rate on an Area 17 relocity high-pass cell with an S-receptive field. The ell had an asymmetric receptive field typical for uperposition facilitations. Indeed, the preferred lirection (velocity-response curves shown in Fig. .11(A)) corresponded to the OFF–ON sequence (E).

Hence the rise in the velocity-response curve was probably due to superposition facilitation. This facilitation amplified at medium and fast velocities the direction selectivity generated by inhibition in the non-preferred direction (Fig. 9.11(C) and Fig. 12). With decreasing strobe rate the ascending leg of the velocity-response curve became flat (Fig. 9.11B) and the difference in response between both directions faded (Fig. 9.11D). It should be noted that the drop in response to fast velocities in the preferred direction observed under stroboscopic illumination is probably due to the reduction in light energy, since such a reduction also influenced the upper end of the velocity-response curves obtained with a continuously illuminated slit (Fig. 9.11A). The changes in direction- and velocity-selectivity induced by low strobe rates are further illustrated in Fig. 9.12. This figure shows the peristimulus–time histograms obtained for the stop-motion-stop of the light slit in opposite directions at two velocities (a slow one and a near-optimal one) and

Fig. 9.10. Influence of duration threshold on linear superposition facilitation. Data from an Area 17 S cell recorded in deep layer III of a strobe-reared animal (eccentricity receptive field 7 degrees). (A) Velocity-response curves for both directions of movement: the full line corresponds to movement from negative to positive positions in the response

plane shown in (B) and the dashed line represents the response in the opposite direction. The arrow indicates the significance level ($P = 0.025$). (C), (D) Responses to Onset presentations of different durations in the ON subregion (C) and the OFF subregion (D).

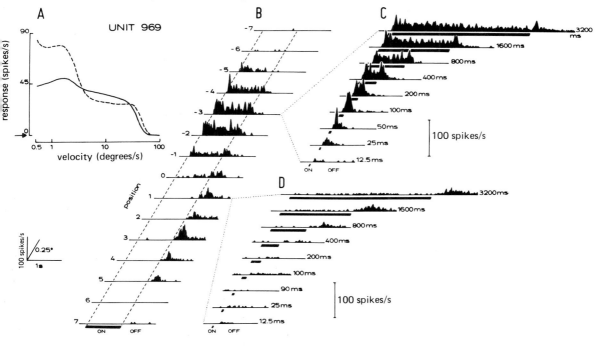

six different strobe rates. The histograms have been plotted base to base so that the parts corresponding to the motion in opposite directions (marked by vertical stippled lines) are in register. Comparison of responses to motion in opposite directions shows that there was no response at all in the non-preferred direction at high strobe rate, but that as the strobe rate decreased responses appeared at 8 Hz for 21 degrees and 4 Hz for 1.2 degrees/s. This may have been due to the loss of the forward inhibition (a type 6 effect). Change in rise time at low rates could also have contributed to the loss of direction selectivity. This is unlikely, at least at the low velocity (1.2 degrees/s), since at 8 Hz the temporal integration capacities had been violated and responses to single flashes appeared and yet the response was still direction-selective. The direction selectivity was only lost at 4 Hz, indicating that the interval between flashes exceeded the spatiotemporal limits of the forward inhibition in the non-preferred direction. Comparison of the responses in the preferred direction at both speeds shows that at high strobe rates there was a much stronger response at 21 degrees/s than at 1.2 degrees/s and that this difference vanished at 4–8 Hz. This was probably due to loss of the superposition facilitation in

Fig. 9.11. Influence of strobe rate on velocity characteristics and direction selectivity. Data from an Area 18 HS cell recorded in layer VI at eccentricity of 5 degrees. (A)–(D) Velocity-response curves in preferred (A and B) and non-preferred directions (C and D) for smooth (A and C) and stroboscopic motions (B and D). For stroboscopic motion, the apparent velocity equals the ratio between the spatial and the temporal interval between flashes. (E) Response plane: the preferred direction corresponds to movement from the negative to positive positions. The arrows through the bars indicate the movement traverse used (6 degrees) for the stroboscopic motion in (A) and (B). For smooth motion, a larger (10 degrees) traverse was used. The smooth motion in (A) and (C) was tested at two stimulus strengths: 2.6 cd/m² (high) and 0.26 cd/m² (low). The mean direction index for each condition was, respectively, 79 and 73. Only strobe rates of 50 and 2 Hz are illustrated in (B) and (D). The velocity–response curves for the different strobe rates were obtained in an interleaved fashion; those for smooth motion in separate tests. The arrows indicate the significance levels ($P = 0.025$).

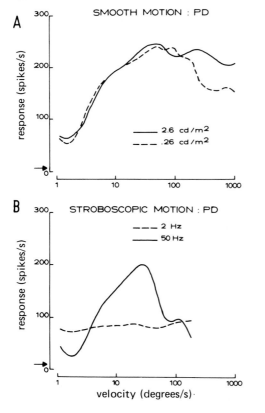

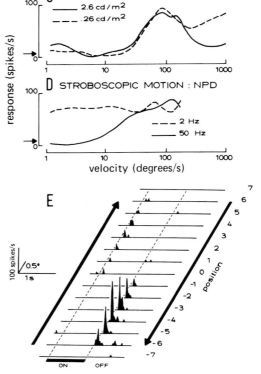

the preferred direction. Indeed, for the testing with stroboscopic stimuli the traverse of movement was reduced to 6 degrees (rather than 10 degrees for the smooth movement) in order to save time (even then the test interleaving six strobe rates and 10 velocities took over 3 h). Because of this shorter traverse, the slit stopped in the OFF subregion at the end of the backward motion (from positive to negative positions in Fig. 9.1(E); lower histograms of each pair) and then disappeared after the last strobe flash. This disappearance produced an OFF response visible both at 1.2 degrees/s and 21 degrees/s for rates between 50 and 15 Hz (arrows in Fig. 9.12). This OFF response faded at low rate (below 15 Hz) and so did the difference in response between the two velocities. This suggests that

the rise in the velocity-response curve of the preferred direction indeed resulted from (non-linear) superposition facilitation. That the superposition only operated at fast velocities was probably due to the phasic character of the OFF subregion. Since the OFF response decayed quickly, it could only facilitate an ON response elicited just after the OFF response.

In conclusion, it can be stated that the descending legs of cortical velocity-response curves are determined mainly by local effects: duration for most velocity low-pass and slow velocity-tuned and velocity broad-band cells, and light energy for the velocity high-pass and most fast velocity-tuned and velocity broad-band cells. The rise in velocity-response curves of velocity-tuned cells and of S and A family velocity

Fig. 9.12. Peristimulus-time histograms, representing the average response to opposite directions of motion at two apparent velocities and six strobe rates, for the same cell as in Fig. 9.11. The duration of the movement is indicated by the interval between the two vertical stippled lines and the horizontal black bars. The histograms are drawn base-to-base, with the parts corresponding to motion in register (activity was also measured while the stimulus stood still before and after the motion).

Notice that in the preferred direction (upper histograms) at 21 degrees/s and high strobe rates (50–28 Hz), the response was more than twice as large as that to a single strobe flash (visible in histograms for 2 and 4 Hz strobe rate), indicating clear facilitation in the preferred direction, and that in the non-preferred direction (lower histograms) at high strobe rates there was no response at all, indicating inhibition in the non-preferred direction.

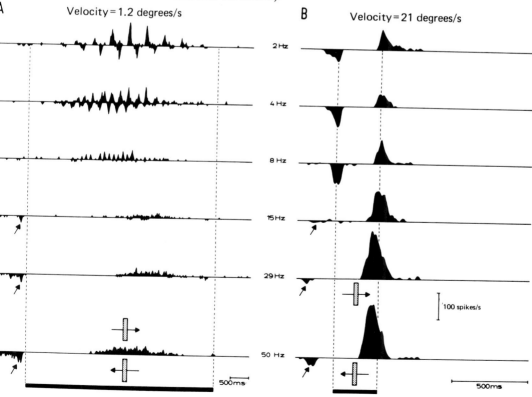

A   Velocity = 1.2 degrees/s

B   Velocity = 21 degrees/s

2 Hz

4 Hz

8 Hz

15 Hz

29 Hz

50 Hz

100 spikes/s

500 ms

500 ms

high-pass cells is due to sequential effects, mainly to facilitation occurring between subregions of S and A family cells and probably within subregions of C family cells. It should be noted that in cortical cells with strong facilitation in the preferred direction, as in velocity-tuned cells, the upper cut-off could also result from the spatiotemporal limitations of the facilitation. The ascending leg of velocity high-pass curves of C family cells is probably due to rise time effects. Since velocity-response curves of velocity low-pass cells and of most velocity broad-band cells have no ascending leg but rather an initial plateau, due to their strong temporal low-pass filtering, cortical velocity-response curves are largely understood (except perhaps those of some C family velocity low-pass cells). More work is however needed to find out the relative contribution of the different sequential effects (superposition, forward inhibition in the non-preferred direction and forward facilitation in the preferred direction) to the direction selectivity of different cortical cells. It seems however that forward inhibition in the non-preferred direction is of major importance in this respect.

Acknowledgements

Most of the work from this laboratory reviewed here was supported by the national research council of Belgium (NFWO) and the research council of the Katholieke Universiteit, Leuven. It was a pleasure to collaborate in these studies with H. Kennedy, H. Maes, J. Duysens, H. van der Glas, K.-P. Hoffmann and J. Cremieux. Thanks are due to J. Duysens for his comments on the manuscript. The drawings of G. Vanparrijs, the photographs of J. Warmoeskerken, and the computerized typework of Y. Celis are gladly acknowledged.

References

Albrecht, D. G. & Hamilton, D. B. (1982) Striate cortex of monkey and cat: contrast response function. *J. Neurophysiol.*, **48**, 217–37

Albus, K. (1980) The detection of movement direction and effects of contrast reversal in the cat's striate cortex. *Vision Res.*, **20**, 289–93

Bishop, P. O., Coombs, J. S. & Henry, G. H. (1971) Responses to visual contours: spatio-temporal aspects of excitation in the receptive fields of simple striate neurones. *J. Physiol. (Lond.)*, **219**, 625–57

Bisti, S., Carmignoto, G., Galli, L. & Maffei, L. (1984) Spatial response characteristics of neurones of cat area 18 at different speeds of the visual stimulus. *J. Physiol. (Lond.)*, **349**, 23P

Bullier, J., Mustari, M. J. & Henry, G. H. (1982) Transformation of receptive field properties between LGN neurons and

first-order S cells in cat striate cortex. *J. Neurophysiol.*, **47**, 417–38

Citron, M. C. & Emerson, R. C. (1983) White noise analysis of cortical directional selectivity in cat. *Brain Res.*, **279**, 271–7

Cleland, B. G. (1980) A comparison of the properties of the brisk-sustained (X) and brisk-transient (Y) cells in the retina of the cat. *Exp. Brain Res.*, **41**, A1–A2

Cremieux, J., Orban, G. A. & Duysens, J. (1984) Responses of cat visual cortical cells to continuously and stroboscopically illuminated moving light slits compared. *Vision Res.*, **24**, 449–57

Davis, T. L. & Sterling, P. (1979) Microcircuitry of cat visual cortex: classification of neurons in layer IV of area 17, and identification of the patterns of lateral geniculate input. *J. comp. Neurol.*, **188**, 599–628

Dinse, H. R. O. & von Seelen, W. (1981) On the function of cell systems in area 18. Part I. *Biol. Cybern.*, **41**, 47–57

Duysens, J. & Orban, G. A. (1981) Is stimulus movement of particular importance in the functioning of cat visual cortex? *Brain Res.*, **220**, 184–7

Duysens, J., Orban, G. A. & Cremieux, J. (1984) Functional basis for the preference for slow movement in area 17 of the cat. *Vision Res.*, **24**, 17–24

Duysens, J., Orban, G. A., van der Glas, H. W. & de Zegher, F. E. (1982a) Functional properties of area 19 as compared to area 17 of the cat. *Brain Res.*, **231**, 279–91

Duysens, J., Orban, G. A., van der Glas, H. W. & Maes, H. (1982b) Receptive field structure of area 19 as compared to area 17 of the cat. *Brain Res.*, **231**, 293–308

Emerson, R. C. & Coleman, L. (1981) Does image movement have a special nature for neurons in the cat's striate cortex? *Invest. Ophthalmol. Visual Sci.*, **20**, 766–83

Emerson, R. C. & Gerstein, G. L. (1977a) Simple striate neurons in the cat. I. Comparison of responses to moving and stationary stimuli. *J. Neurophysiol.*, **40**, 119–35

Emerson, R. C. & Gerstein, G. L. (1977b) Simple striate neurons in the cat. II. Mechanisms underlying directional asymmetry and directional selectivity. *J. Neurophysiol.*, **40**, 136–55

Frishman, L. J., Schweitzer-Tong, D. E. & Goldstein, E. B. (1983) Velocity tuning of cells in the dorsal lateral geniculate nucleus and retina of the cat. *J. Neurophysiol.*, **50**, 1393–413

Ganz, L. (1984) Visual cortical mechanisms responsible for direction selectivity. *Vision Res.*, **24**, 3–11

Goodwin, A. W. & Henry, G. H. (1978) The influence of stimulus velocity on the responses of single neurons in the striate cortex. *J. Physiol. (Lond.)*, **277**, 467–82

Goodwin, A. W., Henry, G. H. & Bishop, P. O. (1975) Direction selectivity of simple striate cells: properties and mechanisms. *J. Neurophysiol.*, **38**, 1500–23

Hammond, P. (1978) Directional tuning of complex cells in area 17 of the feline visual cortex. *J. Physiol. (Lond.)*, **285**, 479–91

Hammond, P. & MacKay, D. M. (1975). Differential responses of cat visual cortical cells to textured stimuli. *Exp. Brain Res.*, **22**, 427–30

Henry, G. H., Bishop, P. O. & Coombs, J. S. (1969) Inhibitory and subliminal excitatory receptive fields of simple units in cat striate cortex. *Vision Res.*, 9, 1289–96

Henry, G. H., Bishop, P. O. & Dreher, B. (1974) Orientation, axis and direction as stimulus parameters for striate cells. *Vision Res.*, **14**, 767–77

Henry, G. H., Bishop, P. O., Tupper, R. M. & Dreher, B. (1973) Orientation specificity and response variability of cells in the striate cortex. *Vision Res.*, **13**, 1771–9

Henry, G. H., Harvey, A. R. & Lund, J. S. (1979) The afferent connections and laminar distribution of cells in the cat striate cortex. *J. Comp. Neurol.*, **187**, 725–44

Hubel, D. H. & Wiesel, T. N. (1962) Receptive fields, binocular interaction and functional architecture in the cat's visual cortex. *J. Physiol. (Lond.)*, **160**, 106–54

Kato, H., Bishop, P. O. & Orban, G. A. (1978) Hypercomplex and the simple/complex cell classifications in cat striate cortex. *J. Neurophysiol.*, **41**, 1071–95

Koch, C., Poggio, T. & Torre, V. (1983) Nonlinear interactions in a dendritic tree: localization, timing, and role in information processing. *Proc. Natl. Acad. Sci. USA*, **80**, 2799–802

Kulikowski, J. J., Bishop, P. O. & Kato, H. (1981) Spatial arrangements of responses by cells in the cat visual cortex to light and dark bars and edges. *Exp. Brain Res.*, **44**, 371–85

Levick, W. R. & Zacks, J. L. (1970) Responses of cat retinal ganglion cells to brief flashes of light. *J. Physiol. (Lond.)*, **96**, 677–700

Maunsell, J. H. R. & Van Essen, D. C. (1983) Functional properties of neurons in middle temporal visual area of the macaque monkey. I. Selectivity for stimulus direction, speed, and orientation. *J. Neurophysiol.*, **49**, 1127–47

McKee, S. P. (1981) A local mechanism for differential velocity detection. *Vision Res.*, **21**, 491–500

McKee, S. P., Silverman, G. & Nakayama, K. (1984) Velocity descrimination is not affected by random variations in temporal frequency. *Suppl. Invest. Ophthalmol. Visual Sci.*, **25**, 13

Movshon, J. A. (1975) The velocity tuning of single units in cat striate cortex. *J. Physiol. (Lond.)*, **249**, 445–68

Movshon, J. A., Thompson, I. D. & Tolhurst, D. J. (1978) Spatial and temporal contrast sensitivity of neurones in areas 17 and 18 of the cat's visual cortex. *J. Physiol. (Lond.)*, **283**, 101–20

Mustari, M. J., Bullier, J. & Henry, G. H. (1982) Comparison of response properties of three types of monosynaptic S-cell in cat striate cortex. *J. Neurophysiol.*, **47**, 439–54

Noda, H., Freeman, R. B. Jr, Gies, B. & Creutzfeldt, O. D. (1971) Neuronal responses in the visual cortex of awake cats to stationary and moving targets. *Exp. Brain Res.*, **12**, 389–405

Orban, G. A. (1975) Movement-sensitive neurones in the peripheral projections of area 18 of the cat. *Brain Res.*, **85**, 181–2

Orban, G. A. (1984) Neuronal operations in the visual cortex. In *Studies of Brain Function. Vol. 11*, ed. H. B. Barlow, T. H. Bullock, E. Florey, O. J. Grüsser & A. Peters, 367 pp. Berlin: Springer-Verlag

Orban, G. A. (1985) Velocity tuned cortical cells and human velocity discrimination. In *Brain Mechanisms of Spatial Vision*, ed. D. J. Ingle, M. Jeannerod & D. N. Lee, pp. 371–88. La Haye: Martinus Nijhof (in press)

Orban, G. A., Cremieux, J., Duysens, J. & Maes, H. (1982) Response of cat's geniculate and visual cortical cells to stroboscopically illuminated moving light bars. *Soc. Neurosci. Abstr.*, **8**, 678

Orban, G. A., De Wolf, J. & Maes, H. (1984) Factors influencing velocity coding in the human visual system. *Vision Res.*, **24**, 33–9

Orban, G. A., Hoffmann, K. -P. & Duysens, J. (1981*a*) Influence of stimulus velocity on LGN neurons. *Soc. Neurosci. Abstr.* **7**, 24

Orban, G. A. & Kennedy, H. (1981) The influence of eccentricity on receptive field types and orientation selectivity in areas 17 and 18 of the cat. *Brain Res.*, **208**, 203–8

Orban, G. A., Kennedy, H. & Bullier, J. (1983) Influence of stimulus velocity on monkey striate neurons: changes with eccentricity. *Soc. Neurosci. Abstr.*, **9**, 477

Orban, G. A., Kennedy, H. & Maes, H. (1981*b*) Response to movement of neurons in areas 17 and 18 of the cat: velocity sensitivity. *J. Neurophysiol.*, **45**, 1043–58

Orban, G. A., Kennedy, H. & Maes, H. (1981*c*) Response to movement of neurons in areas 17 and 18 of the cat: direction selectivity. *J. Neurophysiol.*, **45**, 1059–73

Orban, G. A. & Maes, H. (1984) The velocity dependence of human direction and velocity discrimination: changes with eccentricity. *Suppl. Invest. Ophthalmol. Visual Sci.*, **25**, 71

Palmer, L. A. & Davis, T. L. (1981) Comparison of responses to moving and stationary stimuli in cat striate cortex. *J. Neurophysiol.*, **46**, 227–95

Pettigrew, J. D., Nikara, T. & Bishop, P. O. (1968) Responses to moving slits by single units in cat striate cortex. *Exp. Brain Res.*, **6**, 373–90

Schiller, P. H. (1982) Central connections of the retinal ON and OFF pathways. *Nature*, **297**, 580–3

Sillito, A. M. (1977) Inhibitory processes underlying the directional specificity of simple, complex and hypercomplex cells in the cat's visual cortex. *J. Physiol. (Lond.)*, **271**, 699–720

Singer, W., Tretter, F. & Cynader, M. (1975) Organization of cat striate cortex: a correlation of receptive-field properties with afferent and efferent connections. *J. Neurophysiol.*, **38**, 1080–98

Vautin, R. G. & Berkley, M. A. (1977) Responses of single cells in cat visual cortex to prolonged stimulus movement: neural correlates of visual aftereffects. *J. Neurophysiol.*, **40** 1051–65

Watson, A. B. & Ahumada, A. J., Jr (1983) A look at motion in the frequency domain. *Nasa tech. mem.* 84352.

Westheimer, G. & McKee, S. P. (1975) Visual acuity in the presence of retinal-image motion. *J. Opt. Soc. Am.*, **65**, 847

Wurtz, R. H. (1969) Visual receptive fields of striate cortex neurons in awake monkeys. *J. Neurophysiol.*, **32**, 727–42

III

Visual development

Despite the presence in Australia of marsupials ideal for developmental work and the big impact of Wiesel & Hubel's discovery of developmental plasticity in the mammalian visual pathway, visual development has not been a focus of Peter Bishop's laboratory. Nevertheless many of his students have gone on to contribute much to the current ferment in developmental visual neuroscience. This section deals with two of the most important areas in visual development: (1) the role of neuronal cell death; and (2) developmental plasticity and the effects of deprivation on binocular connections in the geniculocortical pathway.

The first area is the subject of intense research at the moment. Although massive neuronal death during development was first described for spinal motoneurons, the phenomenon is now known to be widespread in the nervous system during ontogeny. The retina is a very convenient place in which to study this phenomenon and some Australian research groups have led the way in demonstrations that dying retinal ganglion cells can be 'rescued' by growth factors (as yet unidentified) released from target tissue in the central visual pathway. The high degree of understanding of the different morphological and functional subclasses of retinal ganglion cells and their various patterns of central connections will enable these studies on cell death to be taken considerably further than in other systems. Ann Sefton (chapter 10) presents current work in this area on the rat, a favourite species on account of its immaturity at birth. Stone & Rapaport (chapter 11) examine the possibility that the phenomenon of cell death might play a role in the development of the steep centroperipheral gradient of ganglion cell density which is characteristic of the cat retina. Although they do find evidence of cell death in the developing retina, in the absence of definitive markers for immature ganglion cells, they found no evidence that differential cell death contributes to the centroperipheral specialization.

David and Sheila Crewther (chapter 12) represent the second area, with a behavioural and physiological account of the developmental disorder of binocular vision called 'amblyopia' in cats suffering from different models of this disorder.

10

The regulation of cell numbers in the developing visual system

ANN JERVIE SEFTON

Introduction

In this chapter I shall first summarize features of the development of the visual system in the rat, and then discuss a number of interesting problems which are common to rats and other mammals. Since the retina is developed at a distance from the rest of the brain, the major questions relate to the mechanisms by which the axons of retinal ganglion cells are initially directed towards their targets, and the means by which precise retinotopic maps are established in each of the visual target nuclei. It is possible to rephrase these questions. Since each retinal ganglion cell is unique by virtue of its type, retinal location and projections, how is this uniqueness established, so that the visual system of adult animals within one species is so strikingly similar?

Development in the visual system of the rat

In the rat, the optic vesicle is formed by embryonic day 12 (E12), invaginating an outpocket of the forebrain (Kuwabara & Wiedman, 1974). From the inner layer arise the precursors of retinal cells, of which the ganglion cells are the first to translocate from the ventricular layer and to differentiate (Morest, 1970). By E14–15, several layers of ganglion cells can clearly be distinguished in the centre of the retina. As seen in Fig. 10.1(A) and (B), their axons will travel through channels present on the retinal surface to invade the optic stalk (Horsburgh & Sefton, in press), as in the mouse (Silver and Sidman, 1980). At the stage when

Fig. 10.1. Micrographs of the developing optic stalk, developing optic nerve, and adult optic nerve: (A), (B) light micrographs; (D)–(F) electron micrographs. (A) The developing eye and optic stalk at E13.5, showing the retina which contains channels through which will course the optic axons. Section 1 μm thick, stained with Azur II. (B) The eye and optic stalk at E14, before optic axons have left the retina and entered the stalk. Section 6 μm thick, silver stain. (C) A bundle of axons in the optic stalk at E15.5, at the stage when there are 400 axons in the stalk, lying amongst glial precursors. Note the variation in sizes of the axons. (D) Tightly packed axons in the optic nerve at birth, when maximum numbers of axons are present. A dark tongue of glial tissue separates two fascicles. (E) Optic nerve at 5 days after birth (P5) showing two myelinating profiles. (F) Adult optic nerve; virtually all axons myelinated.

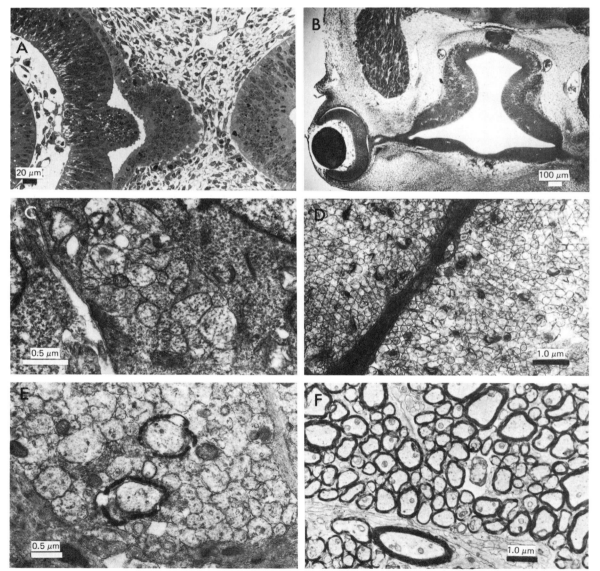

400 axons and growth cones have reached the distal stalk, they are grouped into seven bundles (Sefton & Lam, 1985), one of which is illustrated in Fig. 10.1(C). Within 0.5 days, the number of axons has increased to 10 000 and the separate bundles have coalesced into a single group, tightly packed into fascicles. The proportion of growth cones, initially high, has decreased. At birth, the non-myelinated axons are packed closely together in fascicles, separated by glial processes, see Fig. 10.1(D).

Amongst central nuclei, the major target of retinal ganglion cell axons is the superior colliculus (Linden & Perry, 1983*a*), the cells of which are generated between E12 and E18 (Bruckner, Mares & Biesold, 1976; Altman & Bayer, 1981). The other major target, the dorsal lateral geniculate nucleus, is generated between E12 and E16 (Bruckner *et al*, 1976; Lund & Mustari, 1977; McAllister & Das, 1977; Altman & Bayer, 1979). Both these nuclei are innervated by axons of retinal ganglion cells between E17 and about the time of birth (Lund & Bunt, 1976; Bunt, Lund & Land, 1983). Dendritic and synaptic development of their constituent cells accelerates at about the end of the first postnatal week, and again at the time of eye opening (colliculus: Lund & Lund, 1971; Labriola & Laemle, 1977; geniculate: Parnavelas *et al.*, 1977).

Cell death and axonal loss

One of the most striking events occurring during normal development is the phenomenon of cell death (see Cowan, 1979). As seen in Fig. 10.2, the maximum number of optic axons (242 000) is found at the time of birth (P0) in the albino rat (Sefton & Lam, 1985) and about 60% of these axons are lost in the first postnatal week (Lam, Sefton & Bennett, 1982). Similar losses of optic axons during development have been quantified postnatally in the hooded rat (Perry, Henderson & Linden, 1983), pre- and postnatally in the cat (Ng & Stone, 1982; Williams, Bastiani & Chalupa, 1983), and during incubation in the chick (Rager and Rager, 1978). Because of a possible overlap of generation and degeneration, however, the total number actually lost is impossible to estimate. Throughout the period during which axons are lost, there are few signs of damage, degeneration or disruption in the nerve. Dark degeneration, obvious glial reaction and phagocytosis are not apparent, (see, for example, Fig. 10.1D). Even in embryos, occasional clear profiles appear swollen,

Fig. 10.2. Graph redrawn from Sefton & Lam (1985) and Potts *et al.* (1982), illustrating the timecourse of development and loss of axons from the optic nerve of the rat (black circles), the number of retinal ganglion cells (black triangles), and the number of axons in the remaining optic nerve after enucleation of one eye at birth (open squares). Note that the peak number of axons is reached at birth, and that there is a 60% loss of axons from the nerve during the the first postnatal week. From P1 to P5, the number of axons present after enucleation is similar to the number of retinal ganglion cells, that is, less than the number of axons normally present.

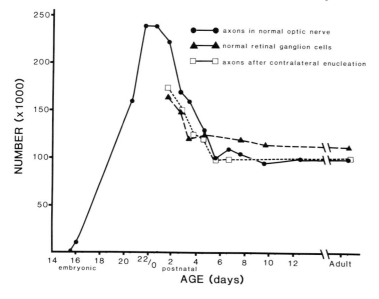

axons in normal optic nerve
normal retinal ganglion cells
axons after contralateral enucleation

but these features cannot with certainty be distinguished from perfusion artefacts. Throughout the period of axonal loss, there is no change in the mean diameter of the optic nerve, as the three-fold loss is compensated for by a three-fold increase in the area of individual axons (Fig. 10.3). Thus the massive reduction in the number of axons occurs in the absence of unequivocal morphological evidence of degeneration and phagocytosis. This form of apparently autophagic degeneration seen during normal development, as well as in other situations, has been called apoptosis (see Weedon, Searle & Kerr, 1979). It is not clear to what extent this form of degeneration seen during normal development resembles that seen after axotomy (Sefton & Lam, 1985).

Myelination (as illustrated in Fig. 10.1E) begins on P5 when the axonal loss is complete; by adult life all optic axons are myelinated (Fig. 10.1F). It is therefore possible that myelination is only initiated when an axon is assured of survival. In addition, because non-myelinated and unmyelinated nerves and pathways degenerate rapidly, and the debris is removed in hours or days compared with weeks or months in myelinated nerves (Pecci-Saavedra, Mascitti & Lloret, 1973; Cook, Ghetti & Wisniewski, 1974; Bignami *et al.*, 1981; Miller & Oberdorfer, 1981), it seems likely that myelination will be shown to follow the period of cell death occurring in other systems and in other species.

Although the loss of optic axons might represent only the withdrawal of branches which fail to project far into the nerve, there is an associated loss of retinal ganglion cells which project to central nuclei in the rat (Jeffery & Perry, 1981; Potts *et al.*, 1982; Dreher, Potts & Bennett, 1983; Martin, Sefton & Dreher, 1983; Perry *et al.*, 1983) and hamster (Insausti, Blakemore & Cowan, 1984). During the first postnatal week, degenerating cells have been identified in the inner retinal layer of the rat (Cunningham, Mohler & Giordano, 1982; Perry *et al.*, 1983) and hamster (Sengelaub & Finlay, 1981, 1982), but it is not known what fraction are ganglion cells. Elsewhere in the visual system of rodents during early postnatal life, cells in the lateral geniculate nucleus (LGN) die (mouse: Heumann & Rabinowitz, 1980), and degenerating cells have been reported in the retinorecipient layers of the superior colliculus (rat: Giordano, Murray & Cunningham, 1980; hamster: Finlay, Berg & Sengelaub, 1982). Thus amongst mammals, neuronal death occurs during the normal development of the visual system, as has previously been reported for the motor system (Romanes, 1946; Hutchinson, Davey & Bennett, 1981). Similarly, cell death has been reported in the avian

Fig. 10.3. Graph to illustrate the relation between the density of axons, mean area of each axon and the area of the optic nerve during development. Note that the density increases until birth, then falls to adult values. This is accompanied by an increase in the mean area of axons from birth. However, the mean area of the whole nerve does not change during the period of loss of axons (birth to P5).

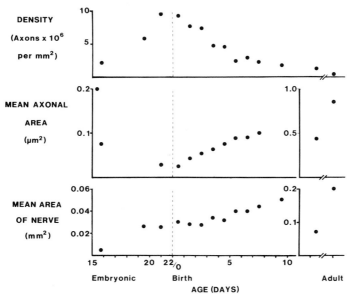

(chick) visual system (isthmo–optic nucleus: Clarke, Rogers & Cowan, 1976; retinal ganglion layer: Hughes & McLoon, 1979; Rager, 1980; optic nerve axons: Rager & Rager, 1978), as well as in the avian motor system (Hamburger, 1975; Chu-Wang & Oppenheim, 1978).

Which axons are lost?

In the optic nerve of the rat, there are no obvious functional or morphological criteria which distinguish the optic axons which are lost. The optic nerve at birth conducts impulses (Foster, Connors & Waxman, 1982), although it is not clear whether all axons within the nerve are able to do so. However, optic axons, including those that die, transport horseradish peroxidase (Potts *et al.*, 1982; Dreher *et al.*, 1983; Martin *et al.*, 1983; Perry *et al.*, 1983). Furthermore, as a result of various manipulations (described later), their survival can be enhanced; thus some which would otherwise have died are capable of surviving and conducting impulses (Fukuda, Sumitomo & Hsiao, 1983; Shirokawa, Fukuda & Sugimoto, 1983). In consequence, it seems unlikely that the death of neurons is restricted to a non-functioning class, or is predetermined genetically.

The group of axons which are lost includes a grossly misrouted contingent. On any one day from birth to P5, there is a discrepancy between the number of optic axons counted by Lam *et al.* (1982) and the number of centrally projecting ganglion cells labelled by Potts *et al.* (1982) (see Fig. 10.2). However, one day after the removal of one eye at birth, all axons have degenerated in the stump ipsilaterally, and disrupted axons are present in the contralateral (remaining) nerve. The number of axons in the contralateral nerve from P2 onwards is similar to the number of retinal ganglion cells. The axons which degenerate after neonatal enucleation are therefore considered to be the retinoretinal axons described by Bunt & Lund (1981) and representing about 40% of the 80000 axons normally lost from the optic nerve between P0 and P5 (Sefton & Lam, 1985). The somata of ganglion cells which project to the opposite eye are distributed across the retina (Bunt & Lund, 1981) and their axons are found throughout the optic nerve (Sefton & Lam, 1985).

The remaining axons which are lost project centrally (Potts *et al.*, 1982; Dreher *et al.*, 1983; Perry *et al.*, 1983), some ipsilaterally (Jeffery & Perry, 1981; Martin *et al.*, 1983). Whether any of these die because they fail

to reach appropriate visual nuclei is not known, but optic axons inappropriately invade the inferior colliculus in the neonatal period (Kato, 1983), although these might only be branches of axons which project to the neighbouring superior colliculus. It seems likely that many of the axons which later die have innervated inappropriate regions of 'correct' targets, since during the period of axonal loss there is a reduction of an initially diffuse innervation of visual nuclei to the more restricted adult pattern. For example, at birth in the rat, terminals of axons arising in one retina are distributed throughout the contralateral lateral posterior nucleus (Perry & Cowey, 1982) and retinorecipient layers of the ipsilateral superior colliculus (Land & Lund, 1979; Laemle & Labriola, 1982). By contrast, in the adult rat, terminals of retinal origin within these nuclei are more restricted in their distribution. The reorganization, which seems to be associated with a period of plasticity, occurs within the first postnatal week. Although the ipsilateral terminals are diffusely distributed in the superior colliculus, the great majority of their parent somata are located in the lower temporal retina (Martin *et al.*, 1983), the region which in the adult rat contains cells which project ipsilaterally (Cowey & Perry, 1979). More recently, similar results have been reported in another rodent, the hamster (Insausti *et al*, 1984). Thus, death amongst the population of cells which projects ipsilaterally is not restricted to those lying outside the 'appropriate' retinal region.

As seen in Fig. 10.2, the elimination of grossly misrouted optic axons (for example, those projecting to the opposite eye) occurs at the same time as the loss of centrally projecting axons (Bunt & Lund, 1981; Sefton & Lam, 1985). This simultaneous loss is in sharp contrast with motoneurons, which appear to be withdrawn in an orderly sequence with those supplying inappropriate muscle groups being eliminated first, followed by those invading inappropriate muscles, then by those invading inappropriate parts of the 'correct' muscle (Bennett, 1981).

Role of competition

Following removal of one eye in the kitten (Williams *et al.*, 1983) and monkey (Rakic & Riley, 1983) before the period of cell death, fewer axons are lost from the remaining optic nerve than in the normal animal. This result demonstrates that the removal of terminals from visual centres contralateral to the enucleated eye provides enhanced opportunities for the survival of ipsilaterally projecting terminals from the remaining eye. Such an interpretation is consistent with similar conclusions reached by many authors who had reported that prenatal or neonatal enucleation resulted in an enhanced distribution of ipsilateral terminals to the superior colliculus of the rat (Lund & Lund, 1971, 1973, 1976; Lund & Miller, 1975; Lund, Land & Boles, 1980; Jen & Lund, 1981).

The wider ipsilateral terminal field after enucleation is associated with an increased survival of retinal ganglion cells in the rat (Jeffery & Perry, 1981) and hamster (Insausti *et al.*, 1984), together with a reduction in the normally occurring degeneration in the ganglion cell layer in the hamster (Sengelaub and Finlay, 1981). However, no increased survival was demonstrated amongst axons in the remaining optic nerve in the rat (Lam *et al.*, 1982). As pointed out by Sefton & Lam (1985), the increased survival represents only about 1 % of the total number of axons in the optic nerve (calculated from Jeffery & Perry, 1981; Jen & Lund, 1981; Lam *et al.*, 1982; Potts *et al.*, 1982), a value too small to be resolved. The results and conclusions of Shirokawa *et al.* (1983), who reported massive changes in the number of axons in the optic tract following neonatal enucleation, are in conflict with those of Jen & Lund (1981) and Sefton & Lam (1985). However, the number of axons in the optic tract normally exceeds that in the nerve (Lam, unpublished), because of the presence of additional axons such as those passing from one parabigeminal nucleus to the contralateral superior colliculus and LGN, themselves affected by enucleation (Stevenson & Lund, 1982*b*).

Elimination of a large number of competing terminals by removing the opposite eye appears to stabilize the widespread distribution which is already present in the ipsilateral superior colliculus at birth (Land & Lund, 1979; Laemle & Labriola, 1982; Bunt *et al.*, 1983; Martin *et al.*, 1983). Similarly, neonatal tectal damage, alone or associated with visual cortical ablation, stabilizes the widespread retinal projection in the lateral posterior nucleus, presumably because of the removal of competition (Perry & Cowey, 1981).

The effects of enucleation in the rat are greatest when the eye is removed prenatally (Lund & Miller, 1975) or at birth, when the number of axons is greatest (Sefton & Lam, 1985). Enucleation after P5 does not substantially change the distribution of retinofugal terminals (Lund & Lund, 1971, 1973, 1976; Lund & Miller, 1975; Lund *et al.*, 1980), suggesting that

significant anatomical modification of the primary visual pathways is possible only before or during the period of cell death and axonal loss. Thus the small physiological changes reported by Sakai & Yagi (1981) to follow enucleations performed between P5 and P30 must involve mechanisms (such as terminal sprouting) other than enhancing the survival of cells which would otherwise die.

The effect of enucleation is restricted to one side of the brain. Prenatal or neonatal removal of one eye effectively deafferents one superior colliculus by removing over 98% of its afferent input. Although ipsilaterally projecting optic axons arising from the remaining eye can take advantage of the increased space, their contralaterally projecting counterparts do not do so, even though optic axons can make synaptic connections, even with non-visual targets (Schneider, 1973; Baisinger, Lund & Miller, 1977; Perry & Cowey 1979*b*). However, optic axons after crossing at the chiasm do not readily recross the midline (Lund & Lund, 1976), except in the presence of intertectal damage (Schneider, 1973; Jen & Lund, 1979; So, 1979). Thus, anatomical barriers might prevent optic axons from taking advantage of the largely deafferented tectum ipsilateral to the remaining eye.

If competition plays a role in regulating cell numbers during normal development, further questions arise. Where is it occurring? Competition could be occurring between dendrites for afferents, or between axonal terminals for target tissue. In either case, what is being competed for? Several suggestions, including synaptic sites, terminal space, or trophic molecules, will be considered below.

Afferent influences on developing neuronal populations

In the visual system of the chick, the normal development of the optic tectum (Rager, 1980) and isthmo-optic nucleus (Cowan & Clarke, 1976) depends on their receiving an afferent input, although the interpretation of the latter observations is not straightforward. In mammals, too, the orderly development of many visual nuclei depends on their receiving a normal innervation. Thus, after neonatal enucleation primary visual targets shrink (rat: Tsang, 1936) associated with a loss of their constituent cells and an increase in the number of degenerating cells within them (rat: Cunningham, Huddleston & Murray, 1979 Lund *et al.*, 1980; hamster: Finlay *et al.*, 1982; mouse

DeLong & Sidman, 1962; Heumann & Rabinowitz, 1980). In the primate, prenatal enucleation disrupts development of the normal lamination in the dorsal LGN (Rakic, 1981). Similar observations have been made in the auditory systems of the chick (Parks, 1979) and mouse (Trune, 1982). On the other hand, decortication in the neonatal rat leads to degeneration of cells in the LGN and an increase in the number of cells found in the nucleus of the optic tract and superior colliculus, perhaps because they are innervated by optic axons deprived of normal targets (Cunningham *et al.*, 1979).

In addition, there is a correlation between the time at which the number of optic axons stabilizes (P5) and the onset of rapid morphological development (of dendritic complexity and synapses) in primary visual nuclei of the rat (dorsal LGN: Karlsson, 1967; Raedler & Sievers, 1975; Parnavelas *et al.*, 1977; superior colliculus: Lund & Lund, 1972; Labriola & Laemle, 1977; Giordano *et al.*, 1980). However, because the state of development of the targets of cells in the retinorecipient layers of the superior colliculus is unknown, it cannot be concluded that it is only the afferent supply which is crucial for their development.

Although it has been suggested that retinal ganglion cells require an afferent input for their survival (Linden & Perry, 1982), this seems unlikely. Synapses in the inner plexiform layer do not develop during the first five postnatal days (Wiedman & Kuwubara, 1968; Raedler & Sievers, 1975; Sefton, 1984) during the period of cell death. The earliest synapses appear to be those in which amacrine cells form the presynaptic elements (Morest, 1970; Sefton, 1984), but the postsynaptic elements cannot be identified at early stages.

However, although not responsible for ensuring the survival of retinal ganglion cells, retinal afferents may play a crucial role in their growth and development. Thus, Perry & Linden (1982) presented evidence that dendritic morphology can be modified by the removal at birth of adjacent ganglion cells, suggesting that this is due to competition between ganglion cell afferents. If afferent competition plays such a role, then similar lesions made in the second postnatal week, when amacrine synapses are developing and bipolar synapses begin to be formed, should prove equally effective in modifying ganglion cell dendrites. On the other hand, it seems equally plausible that dendrites extending from one cell might inhibit the development of the dendrites

of adjacent cells of the same subtype. Across the retina, although the dendrites of different classes overlap extensively, those of one type and class remain separated to a greater extent, such that, for example, ON- and OFF-centre alpha and beta cells in the cat (Wässle, Peichl & Boycott, 1983) and Type I cells in the rat are evenly distributed across the retina (Dreher *et al.*, 1985). If such a competitive mechanism is operating between dendrites of the same type, then the dendritic spread of each class of neuron would be expected to be smaller in situations where there is a significantly enhanced survival of retinal ganglion cells, and perhaps greater when cell death is increased.

Role of targets in regulating cell numbers

The sequence of degeneration of retinal ganglion cells and their axons is not known. In the albino rat, axons have withdrawn from the optic nerve by P5 (Lam *et al.*, 1982), even though the number of cells degenerating in the inner retinal layer was reported to be maximal on P6 and P7 (Cunningham *et al.*, 1982). Similarly, in the chick, significant axonal loss precedes the appearance of degeneration of retinal cells (Rager & Rager, 1978; Hughes & McLoon, 1979; Rager, 1980). If these degenerating retinal cells include ganglion cells, then the withdrawal of its axon precedes death of the soma. Such a sequence of events might be initiated (1) if an axon did not reach its 'appropriate' target, perhaps by failing to match some specification; (2) if it were actively rejected from some 'inappropriate' region; or (3) if it were unsuccessful in competition for general synaptic space, for specific synaptic sites, or for trophic molecules.

The importance of the presence of the target has been demonstrated in the chick, both in the development of the isthmo-optic nucleus (Cowan & Clarke, 1976) and in the effects of tectal damage prior to innervation in increasing degeneration in the retinal ganglion cell layer (Hughes & LaVelle, 1975; Hughes & McLoon, 1979). Although in the rat tectal lesions at birth reduced the number (Dreher *et al.*, 1983) and density (Perry & Cowey, 1979*a*, 1981) of retinal ganglion cells, the lesions were made after the tectum had been innervated (Bunt *et al.*, 1983). Thus, in the rat, the increased cell death may reflect no more than the rapid and complete retrograde degeneration of parent somata noted by Miller & Oberdorfer (1981) after section of optic axons in the retina; axotomy in infant nervous systems often causes more profound effects than in adults (Lieberman, 1974). Such an interpretation is consistent with the observation that tectal lesions made at P5 have more profound effects than those made at birth (Perry & Cowey, 1981). However, in other parts of the visual system, the removal of a target before it is innervated profoundly affects its major afferents. For example, after neonatal removal of the visual cortex before it is innervated (Lund & Mustari, 1977), cells in the LGN degenerate and disappear (Cunningham *et al.*, 1979). Tectal lesions made at birth reduced the number of neurons in the parabigeminal nucleus, which projects extensively to the tectum (Linden & Perry, 1983*b*). However, the interpretation of this result is complicated by possible afferent influences, since the tectum also projects extensively to the parabigeminal nucleus (Stevenson & Lund, 1982*a*), and by the possibility of retrograde degeneration following axonal damage, since the time at which these projections develop is not known.

One test of the dependence of axons on their targets involves the provision of additional target tissue; for example, grafting a supernumary limb increases the survival of motoneurons (Hollyday & Hamburger, 1976). In the visual system of the rat, additional target tissue has been provided by the grafting of an additional superior colliculus (Harvey & Lund, 1981; Lund & Harvey, 1981). Provided that it is possible for significant numbers of optic axons to invade this additional tissue, more retinal ganglion cells would be expected to survive if the presence of the target were crucial.

Role of trophic factors

Developing cells could be influenced by trophic molecules secreted either by targets or by afferents; such molecules could be absorbed after diffusion for some distance or exchanged by contact between cell membranes. Molecules secreted by targets have been frequently considered, and they could influence invading terminals in a number of different ways. First, they could play a chemotropic role, attracting axons from a distance and directing their growth into the target nucleus. Secondly, having invaded, terminals could be directed to specific parts of the target. Thirdly, once appropriate connections had been made, trophic molecules could play a role in sustaining the parent neuron. Either neurons or glia could serve as the source of such molecules.

Since normal mechanical factors are disrupted, chemical attraction may operate to guide axons arising in the ganglion cell layer of transplanted, disassociated retinae to appropriate visual target nuclei (McLoon, Lund & McLoon, 1982) and host neurons into transplanted tecta (Harvey & Lund, 1981; Lund & Harvey, 1981). Similarly, *in vitro*, retinae removed from mice on E14 will extend axons specifically into adjacent tectal explants (Smalheiser, Crain & Bornstein, 1981*a*; Smalheiser, Peterson & Crain, 1981*b*). In contrast, tectal transplants *in vivo* do not appear to be invaded extensively by optic axons (Harvey & Lund, 1981) although this may be due to mechanical barriers. If chemotropism plays any role in attracting axons to their normal targets, then the retinoretinal pathway may result from the attraction of the retina itself (or other ocular tissues) for axons arising in the opposite retina. If trophic factors are important in generating specific retinotopic maps within a target, it seems likely that several would need to be secreted. Because of differential effects of half-tecta on half-retinae (Smalheiser, 1981), at least two factors appear to be present affecting either the affinity of the axons for their targets or their survival (or both). Alternatively, a target, or a subregion within it, might actively reject inappropriate terminals.

How could specific connections be initiated? have observed in developing retinae that synaptic vesicles are present in many terminals well before postsynaptic densities have developed. Therefore terminals may secrete their specific transmitter(s) as they grope towards their targets; each subset of retinal ganglion cells may have a different transmitter. Obviously, a viable synaptic connection will be possible only if the target neuron is able to generate the appropriate receptor(s), thus perhaps accounting for one sort of specificity. In the retina of the cat, ON- and OFF-centre mechanisms for X- and Y-ganglion cells appear to utilize different transmitters (Saito, 1983), implying that each subclass generates different receptors.

In tissue culture, the presence of tectum enhances the survival of identified retinal ganglion cells of neonatal rats (McCaffery, Bennett & Dreher, 1982). However, the interpretation of these *in vitro* experiments is complicated, since the axons of the retinal ganglion cells have been severed. Degeneration of retinal ganglion cells is rapid and complete in rats after section of their axons (Miller & Oberdorfer, 1981), so the enhanced survival of the cultured cells may reflect improved regeneration rather than mechanisms related to normal growth and survival. By using retinae from mice taken on E14, before the axons invade the stalk, Smalheiser *et al.* (1981*a*,*b*) demonstrated that cells in the retinal ganglion layer can invade the tectal tissue, providing evidence that developing, as well as regenerating, retinal ganglion cells are supported by the presence of target tissue.

How could non-synaptic trophic interactions be mediated? One possibility is that trophic protein molecules are secreted from, and taken up by, coated pits and vesicles, which play a role in the transfer of proteins between cells elsewhere in the body (see Goldstein, Anderson & Brown, 1979), as well as the uptake of foreign proteins such as horseradish peroxidase (Trojanowski & Gonatas, 1983). I have observed that coated vesicles and pits in various stages of development (see Fig. 10.4) are present in the inner plexiform layer of the retina of neonatal and juvenile rats at frequencies between five and ten times those found in adults, and they are also present in large numbers in the developing superior colliculus. Although it is not possible to distinguish between

secretion and uptake, it seems reasonable to suggest that these coated pits and vesicles could represent morphological evidence for the transfer of trophic substances. Although Altman (1971) similarly found them to be present in large numbers in developing cerebellum, he suggested that coated pits might flatten out at the surface of a cell to create the substrate for the development of dense postsynaptic specializations. However, in the retina, coated pits and vesicles are present in large numbers in all types of profiles, including presynaptic elements, and are present at birth, a week before any synaptic development has occurred.

Conclusion

It is not yet possible to define the developmental mechanisms underlying the specificity of neuronal connections in the visual system. However, several factors seem to play a role. These include the sequence of axonal outgrowth from the retina, and the ready fasciculation of axons as they enter the optic stalk; such mechanisms can account for the maintenance of topography in the optic nerve and, ultimately, at least to some extent, in the target nuclei. Some sort of chemotropism may play a role in attracting some axons rather than others, although mechanical barriers may define routes for particular sets of axons. The time of arrival of particular axons in relation to the degree of maturity of cells within a nucleus may also be important. The development of specific connections may depend on the matching of an axon, which secretes a particular transmitter, to a neuron which is capable of expressing the appropriate surface receptor(s). Once terminals are securely lodged in the appropriate region, trophic molecules necessary for their survival may be exchanged.

Acknowledgements

The results reported here were obtained with the considerable assistance of Kit Lam, Paul Martin, Gwynn Horsburgh, Mick Branley and Kerrie Nichol, to whom I express my thanks. I gratefully acknowledge the valuable discussions held with Bogdan Dreher and Max Bennett, the help of the Electron Microscope Unit, University of Sydney, the photographic assistance of the Department of Anatomy, University of Sydney, and technical assistance from Judy Furby, Yuri Ryuntyu and Gayle Nisbett.

Fig. 10.4. Electron micrographs to illustrate coated pits and vesicle in the developing visual system of the rat: (A) retina, P12; (B), (C) superior colliculus, P4. (A) The coated vesicle is contained within a terminal which, in adjacent sections, was seen to contain synaptic vesicles. (B) Coated pit in soma. (C) Coated pit (arrowed) in presynaptic terminal.

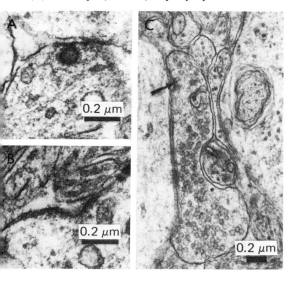

References

Altman, J. (1971) Coated vesicles and synaptogenesis. A developmental study in the cerebellar cortex of the rat. *Brain Res.*, 30, 311–22

Altman, J. & Bayer, S. A. (1979) Development of the diencephalon in the rat. IV Quantitative study of the time of origin of neurons and the internuclear chronological gradients in the thalamus. *J. Comp. Neurol.*, 188, 455–72

Altman, J. & Bayer, S. A. (1981) Time of origin of neurons of the rat superior colliculus in relation to other components of the visual and visuomotor pathways. *Exp. Brain Res.*, 42, 424–34

Baisinger, J., Lund, R. D. & Miller, B. (1977) Aberrant retinothalamic projections resulting from unilateral tectal lesions made in fetal and neonatal rats. *Exp. Neurol.*, 54, 369–82

Bennett, M. R. (1981) The development of neuromuscular synapses. *Proc. Aust. Physiol. Pharmacol. Soc.*, 12, 41–58

Bignami, M. D., Dahl, D., Nguyen, B. T. & Crosby, C. J. (1981) The fate of axonal debris in Wallerian degeneration of rat optic and sciatic nerves. *J. Neuropath. Exp. Neurol.*, 40, 537–50

Bruckner, G., Mares, V. & Biesold, D. (1976) Neurogenesis in the visual system of the rat. An autoradiographic investigation. *J. Comp. Neurol.*, 166, 245–56

Bunt, S. & Lund, R. D. (1981) Development of a transient retino-retinal pathway in hooded rats. *Brain Res.*, 211, 339–404

Bunt, S., Lund, R. D. & Land, P. (1983) Prenatal development of the optic projection in albino and hooded rats. *Dev. Brain Res.*, 6, 149–68

Chu-Wang, I. -W. & Oppenheim, R. W. (1978) A quantitative and qualitative analysis of degeneration in the ventral root, including evidence for axon outgrowth and limb innervation prior to cell death. *J. Comp. Neurol.*, 177, 59–86

Clarke, P. G. H., Rogers, L. A. & Cowan, W. M. (1976) The time of origin and the pattern of survival of neurons in the isthmo-optic nucleus in the chick. *J. Comp. Neurol.*, 167, 125–42

Cook, R. D., Ghetti, B. & Wisniewski, H. M. (1974) The pattern of Wallerian degeneration in the optic nerve of newborn kittens: an ultrastructural study. *Brain Res.*, 75, 261–75

Cowan, W. M. (1979) Selection and control in neurogenesis. In *The Neurosciences, Fourth Study Program* ed. F. O. Schmitt & F. G. Worden, pp. 59–79. Cambridge: MIT Press

Cowan, W. M. & Clarke, P. G. H. (1976) The development of the isthmo-optic nucleus. *Brain Behav. Evol.*, 13, 345–75

Cowey, A. & Perry, V. H. (1979) The projection of the temporal retina in rats, studied by retrograde transport of horseradish peroxidase. *Exp. Brain Res.*, 35, 457–64

Cunningham, T. J., Huddleston, C. & Murray, M. (1979) Modification of neuron numbers in the visual system of the rat. *J. Comp. Neurol.*, 184, 423–34

Cunningham, T., Mohler, M. & Giordano, D. (1982) Naturally occurring neuron death in the ganglion cell layer of the neonatal rat: morphology and evidence for regional correspondence with neuron death in superior colliculus. *Dev. Brain Res.*, 2, 203–15

DeLong, G. R. & Sidman, R. L. (1962) Effects of eye removal at birth on histogenesis of the mouse superior colliculus: an autoradiographic analysis with tritiated thymidine. *J. Comp. Neurol.*, 118, 205–23

Dreher, B., Potts, R. A. & Bennett, M. R. (1983) Evidence that the early postnatal reduction in the number of rat retinal ganglion cells is due to a wave of ganglion cell death. *Neurosci. Lett.*, 36, 255–60

Dreher, B., Potts, R. A., Ni, S. Y. K. & Bennett, M. R. (1984) The development of heterogeneities in distribution and soma sizes of rat retinal ganglion cells. In *Development of Visual Pathways in Mammals*, ed. J. Stone, B. Dreher, & D. H. Rapaport, pp. 39–58. New York: Alan R. Liss

Dreher, B., Sefton, A. J., Ni, S. Y. K. & Nisbett, G. (1985) The morphology, number, distribution and central projections of Class I retinal ganglion cells in albino and hooded rats. *Brain. Behav. Evol.* 26, 10–48

Finlay, B. L., Berg, A. T. & Sengelaub, D. R. (1982) Cell death in the mammalian visual system during normal development: II. Superior colliculus. *J. Comp. Neurol.*, 204, 318–24

Foster, R., Connors, W. & Waxman, S. (1982) Rat optic nerve: electrophysiological, pharmacological and anatomical studies during development. *Dev. Brain Res.*, 3, 371–86

Fukuda, Y., Sumitomo, I. & Hsiao, C.-F. (1983) Effects of neonatal enucleation on excitatory and inhibitory organizations of the albino rat lateral geniculate nucleus. *J. Neurophysiol.*, 50, 46–60

Giordano, D. L. & Cunningham, T. J. (1982) Optic afferents, neuron maturation, and neuron survival in the rat superior colliculus. *Dev. Brain Res.*, 4, 365–8

Giordano, D. L., Murray, M. & Cunningham, T. J. (1980) Naturally occurring neuron death in the optic layers of superior colliculus of the postnatal rat. *J. Neurocytol.*, 9, 603–14

Goldstein, J. L., Anderson, R. G. W. & Brown, M. S. (1979) Coated pits, coated vesicles, and receptor-mediated endocytosis. *Nature (Lond.)*, 279, 679–85

Hamburger, V. (1975) Cell death in the development of the lateral motor column of the chick embryo. *J. Comp. Neurol.*, 160, 535–46

Harvey, A. R. & Lund, R. D. (1981) Transplantation of tectal tissue in rats. II. Distribution of host neurons which project to transplants. *J. Comp. Neurol.*, 202, 505–20

Heumann, D. & Rabinowitz, T. (1980) Postnatal development of the dorsal lateral geniculate nucleus in the normal and enucleated albino mouse. *Exp. Brain Res.*, 38, 75–85

Hollyday, M. & Hamburger, V. (1976) Reduction of the naturally occurring motor neuron loss by enlargement of the periphery. *J. Comp. Neurol.*, 170, 311–20

Horsburgh, G. & Sefton, A. J. (In press) The early development of optic nerve and chiasm in embryonic rat. *J. Comp. Neurol*

Hughes, W. F. & LaVelle, A. (1975) Effects of early tectal lesions on development in retinal ganglion cell layer of chick embryos. *J. Comp. Neurol.*, 163, 265–84

Hughes, W. F. & McLoon, S. C. (1979) Ganglion cell death during normal retinal development in the chick: comparisons with cell death induced by early target field destruction. *Exp. Neurol.*, **66**, 587–601

Hutchinson, I., Davey, D. F. & Bennett, M. R. (1981) The loss of ventral root axons accompanies the early decrease of polyneuronal innervation in developing rat muscle. *Proc. Aust. Physiol. Pharmacol. Soc.*, **12**, 54P

Insausti, R., Blakemore, C. & Cowan, W. M. (1984) Ganglion cell death during development of ipsilateral retino-collicular projection in golden hamster. *Nature*, **308**, 362–5

Jeffery, G. & Perry, V. H. (1981) Evidence for ganglion cell death during development of the ipsilateral retinal projection in the rat. *Dev. Brain Res.*, **2**, 176–80

Jen, L.-S. & Lund, R. D. (1979) Intertectal crossing of optic axons after tectal fusion in neonatal rats. *Brain Res.*, **178**, 99–105

Jen, L.-S. & Lund, R. D. (1981) Experimentally induced enlargement of the uncrossed retinotectal pathway in rats. *Brain Res.*, **211**, 37–57

Karlsson, U. (1967). Observations on the postnatal development of neuronal structures in the lateral geniculate nucleus of the rat by electron microscopy. *J. Ultrastr. Res.*, **17**, 158–75

Kato. T. (1983) Transient retinal fibers to the inferior colliculus in the newborn albino rat. *Neurosci. Lett.*, **37**, 7–9

Kuwabara, T. & Wiedman, T. A. (1974) Development of the prenatal rat retina. *Invest. Ophthalmol.*, **13**, 725–39

Labriola, A. & Laemle, L. (1977) Cellular morphology in the visual layer of the developing rat superior colliculus. *Exp. Neurol.*, **55**, 247–68

Laemle, L. & Labriola, A. (1982) Retinocollicular projections in the neonatal rat: an anatomical basis for plasticity. *Dev. Brain Res.*, **3**, 317–22

Lam, K., Sefton, A. J. & Bennett, M. R. (1982) Loss of axons from the optic nerve of the rat during early postnatal development. *Dev. Brain Res.*, **3**, 487–91

Land, P. W. & Lund, R. D. (1979) Development of the rat's uncrossed retinotectal pathway and its relation to plasticity studies. *Science, N. Y.*, **205**, 698–700

Lieberman, A. R. (1974) Some factors affecting retrograde neuronal responses to axonal lesions. In *Essays on the Nervous System*, ed. R. Bellairs & E. G. Gray, pp. 71–105. Oxford: Clarendon

Linden, R. & Perry, V. H. (1982) Ganglion cell death within the developing retina: a regulatory role for retinal dendrites? *Neuroscience*, **11**, 2813–27

Linden, R. & Perry, V. H. (1983*a*) Massive retinotectal projection in rats. *Brain Res.*, **272**, 145–9

Linden, R. & Perry, V. H. (1983*b*) Retrograde and anterograde-transneuronal degeneration in the parabigeminal nucleus following tectal lesions in developing rats. *J. Comp. Neurol.*, **218**,. 270–81

Lund, R. D. & Bunt, A. H. (1976) Prenatal development of central optic pathways in albino rats. *J. Comp. Neurol.*, **165**, 247–64

Lund, R. D. & Harvey, A. (1981) Transplantation of tectal tissue in rats. I. Organization of transplants and patterns of distribution of host afferents within them. *J. Comp. Neurol.*, **201**, 191–209

Lund, R. D., Land, P. W. & Boles, J. (1980) Normal and abnormal uncrossed retinotectal pathways in rats: an HRP study in adults. *J. Comp. Neurol.*, **189**, 711–20

Lund, R. D. & Lund, J. S. (1971) Modifications of synaptic patterns in the superior colliculus of the rat during development and following deafferentation. *Vision Res.*, Suppl. 3, 281–98

Lund, R. D. & Lund, J. S. (1972) Development of synpatic patterns in the superior colliculus of the rat. *Brain Res.*, **42**, 1–20

Lund, R. D. & Lund, J. S. (1973) Reorganization of the retinotectal pathway in rats after neonatal retinal lesions. *Exp. Neurol.*, **40**, 377–90

Lund, R. D. & Lund, J. S. (1976) Plasticity in the developing visual system: the effects of retinal lesions made in young rats. *J. Comp. Neurol.*, **169**, 133–54

Lund, R. D. & Miller, B. F. (1975) Secondary effects of fetal eye damage in rats on intact central optic projections. *Brain Res.*, **92**, 279–89

Lund, R. D. & Mustari, M. J. (1977) Development of the geniculocortical pathway in rats. *J. Comp. Neurol.*, **173**, 289–306

McAllister, J. P. & Das, G. D. (1977) Neurogenesis in the epithalamus, dorsal thalamus and ventral thalamus of the rat. An autoradiographic and cytological study. *J. Comp. Neurol.*, **172**, 647–86

McCaffery, C. A., Bennett, M. R. & Dreher, B. (1982) The survival of neonatal rat retinal ganglion cells in vitro is enhanced in the presence of appropriate parts of the brain. *Exp. Brain Res.*, **48**, 377–86

McLoon, L. K., Lund, R. D. & McLoon, S. C. (1982) Transplantation of reaggregates of embryonic neural retinae to neonatal rat brain: differentiation and formation of connections. *J. Comp. Neurol.*, **205**, 179–89

Martin, P. R., Sefton, A. J. & Dreher, B. (1983) The retinal location and fate of ganglion cells which project to the ipsilateral superior colliculus in neonatal albino and hooded rats. *Neurosci. Lett.*, **41**, 219–26

Miller, N. M. & Oberdorfer, M. (1981) Neuronal and neuroglial responses following retinal lesions in the neonatal rats. *J. Comp. Neurol.*, **202**, 493–504

Morest, D. K. (1970) The pattern of neurogenesis in the retina of the rat. *Z. Anat. Entwickl.-Gesch.*, **131**, 45–67

Mustari, M. J., Lund, R. D. & Graubard, K. (1979) Histogenesis of the superior colliculus of the albino rat: a tritiated thymidine study. *Brain Res.*, **164**, 39–52

Ng, A. Y. K. & Stone, J. (1982) The optic nerve of the cat: appearance and loss of axons during normal development. *Dev. Brain Res.*, **5**, 263–71

Parks, T. N. (1979) Afferent influences on the development of the brain stem auditory nuclei of the chicken. I. Otocyst ablation. *J. Comp. Neurol.*, **183**, 665–78

Parnavelas, J. G., Mounty, E. J., Bradford, R. & Lieberman, A. R. (1977) The postnatal development of neurons in the dorsal lateral geniculate nucleus of the rat: a Golgi study. *J. Comp. Neurol.*, **171**, 481–500

Pecci-Saavedra, J., Mascitti, T. A. & Lloret, L. P. (1973) Increased rate of anterograde degeneration in the visual pathway of kittens. *Brain Res.*, **50**, 265–74

Perry, V. H. & Cowey, A. (1979a) The effects of unilateral cortical and tectal lesions on retinal ganglion cells in rats. *Exp. Brain Res.*, **35**, 85–95

Perry, V. H. & Cowey, A. (1979b) Changes in the retino-fugal pathways following cortical and tectal lesions in neonatal and adult rats, *Exp. Brain Res.*, **35**, 97–108

Perry, V. H. & Cowey, A. (1981) Degeneration and re-organization following neonatal tectal lesions in rats. In *Functional Recovery from Brain Damage*, ed. M. W. Hoff & G. Mohn, pp. 335–47. Amsterdam: Elsevier/North Holland Biochemical Press

Perry, V. H. & Cowey, A. (1982) A sensitive period for ganglion cell degeneration and the formation of aberrant retino-fugal connections following tectal lesions in rats. *Neuroscience*, 7, 583–94

Perry, V. H., Henderson, Z. & Linden, R. (1983) Postnatal changes in retinal ganglion cell and optic axon populations in the pigmented rat, *J. Comp. Neurol.*, **219**, 356–68

Perry, V. H. & Linden, R. (1982) Evidence for dendritic competition in the developing retina of the rat. *Nature (Lond.)*, **297**, 683–5

Potts, R. A., Dreher, B. & Bennett, M. R. (1982) The loss of ganglion cells in the developing retina of the rat. *Dev. Brain Res.*, **3**, 481–6

Raedler, A. & Sievers, J. (1975) The development of the visual system of the albino rat. *Adv. Anat. Embryol. Cell Biol.*, **50**, 1–87

Rager, G. (1980) Development of the retinotectal projection in the chicken. *Adv. Anat. Embryol. Cell Biol.*, **63**, 1–92

Rager, G. & Rager, U. (1978) Systems-matching by degeneration. I. A quantitative electron microscopic study of the generation and degeneration of retinal ganglion cells in the chicken. *Exp. Brain Res.*, **33**, 65–78

Rakic, P. (1981) Development of visual centers in the primate brain depends upon binocular competition before birth. *Science, N. Y.*, **214**, 928–31

Rakic, P. & Riley, K. P. (1983) Regulation of axon number in primate optic nerve by prenatal binocular competition. *Nature (Lond.)*, **305**, 135–7

Romanes, G. J. (1946) Motor localization and the effects of nerve injury on the ventral horn cells of the spinal cord. *J. Anat. (Lond.)*, **80**, 117–31

Saito, H.-A. (1983) Pharmacological and morphological differences between X- and Y-type ganglion cells in the cat's retina. *Vision Res.*, **23**, 1298–308

Sakai, M. & Yagi, F. (1981) Evoked potential in the lateral geniculate body as modified by enucleation of one eye in the albino rat. *Brain Res.*, **210**, 91–102

Schneider, G. E. (1973) Early lesions of superior colliculus: factors affecting the formation of abnormal retinal projections. *Brain Behav. Evol.*, **8**, 73–109

Sefton, A. J. (1983) Survival of retinal ganglion cells in the rat is not regulated by afferent synaptic inputs. *Neurosci. Lett.*, Suppl. 15, S59

Sefton, A. J. & Lam, K. (1985) Quantitative and morphological studies on developing optic axons in the normal and enucleated albino rat. *Exp. Brain Res*, **57**, 107–17

Sengelaub, D. R. & Finlay, B. L. (1981) Early removal of one eye reduces normally occurring cell death in the remaining eye. *Science, N. Y.*, **212**, 573–4

Sengelaub, D. R. & Finlay, B. L. (1982) Cell death in the mammalian visual system during normal development: I. Retinal ganglion cells. *J. Comp. Neurol.*, **204**, 311–17

Silver, J. & Sidman, R. L. (1980) A mechanism for the guidance and topographic patterning of retinal ganglion cell axons. *J. Comp. Neurol.*, **189**, 101–11

Shirokawa, T., Fukuda,. Y. & Sugimoto, T. (1983) Bilateral reorganization of the rat optic tract following enucleation of one eye at birth. *Exp. Brain Res.*, **51**, 172–8

Smalheiser, N. R. (1981) Positional specificity tests in co-cultures of retinal and tectal explants. *Brain Res.*, **213**, 493–9

Smalheiser, N. R., Crain, S. M. & Bornstein, M. B. (1981a) Development of ganglion cells and their axons in organized cultures of fetal mouse retinal explants. *Brain Res.*, **204**, 159–78

Smalheiser, N. R., Peterson, E. R. & Crain, S. M. (1981b) Neurites from mouse retina and dorsal root ganglion explants show specific behaviour within co-cultured tectum or spinal cord. *Brain Res.*, **208**, 499–505

So, K. -F. (1979) Development of abnormal recrossing retinotectal projections after superior colliculus lesions in newborn Syrian hamsters. *J. Comp. Neurol.*, **186**, 241–58

Stevenson, J. A. & Lund, R. D. (1982a) Alterations of the crossed parabigeminotectal projection induced by neonatal eye removal in rats. *J. Comp. Neurol.*, **207**, 191–202

Stevenson, J. A. & Lund, R. D. (1982b) A crossed parabigemino-lateral geniculate projection in rats blinded at birth. *Exp. Brain Res.*, **45**, 95–100

Trojanowski, J. Q & Gonatas, N. K. (1983) A morphometric study of the endocytosis of wheat germ agglutinin-horseradish peroxidase conjugates by retinal ganglion cells in the rat. *Brain Res*, **272**, 201–10

Trune, D. R. (1982) Influence of neonatal cochlear removal on the development of mouse cochlear nucleus. I. Number, size and density of its neurons. *J. Comp. Neurol.*, **209**, 425–34

Tsang, Y.-C. (1936) Visual centers in blinded rats. *J. Comp. Neurol.*, **66**, 211–61

Wässle, H., Peichl, L. & Boycott, B. B. (1983) A spatial analysis of on- and off-ganglion cells in the cat retina. *Vision Res.*, **23**, 1151–60

Weedon, D., Searle, J. & Kerr, J. F. R. (1979) Apoptosis. Its nature and implications for dermatopathology. *Am. J. Dermatopathol.*, **1**, 133–44

Weidman, T. A. & Kuwabara, T. (1968) Postnatal development of the rat retina. *Arch. Ophthalmol.*, **79**, 470–84

Williams, R. W., Bastiani, M. J. & Chalupa, L. M. (1983) Loss of axons in the cat optic nerve following fetal unilateral enucleation: an electron microscopic analysis. *J. Neurosci.*, **3**, 133–44

11

The role of cell death in shaping the ganglion cell population of the adult cat retina

JONATHAN STONE and
DAVID H. RAPAPORT

Introduction

The regional specializations of mammalian retina, such as the visual streak and the fovea or area centralis, are familiar to the visual scientist. They have been studied intensively in a number of species, and their importance in visual function has been established. The fovea centralis of human retina, for example, contains far higher concentrations of ganglion cells and cones than the peripheral retina, and is capable of correspondingly higher spatial and chromatic resolution. In addition, because the high degree of specialization established at the fovea is maintained over only a small patch of retina, less than 1% of the retina by area, we have developed a complex set of precise eye movements to keep the image of objects of interest on the fovea, indeed on the fovea in both eyes. Moreover, large proportions of the visual centres of the brain are dedicated to the processing of information coming from the small foveal region of retina. In short, our understanding of spatial and chromatic acuity, eye movements, interocular alignment and the retinotopic organization of the visual centres of the brain has required the continuing study of the regional specializations of the retina.

To the embryologist, these specializations pose an equal but less familiar challenge. The degree of regional specialization achieved in some mammalian retinae is very high; around the human and rhesus monkey foveas, for example, ganglion cells and cones concentrate at densities 200–300 times greater than in peripheral retina, and in the cat the gradient between the area centralis and the periphery is 50–100:1. These gradients in cell concentration, and the corresponding specializations of neural circuitry, are far higher than are found within any other functionally homogeneous region of the CNS. In the striate cortex, for example, no gradients in cell concentrations, morphology or circuitry appear to exist between areas representing central and peripheral regions of the retina. The foveal region of the retina is represented by a large number of cortical neurons, but they are present in the same concentration as cells representing more peripheral retina, and simply occupy a large proportion of Area 17. Given the optics of the eye, however, retinal neurons cannot be allowed to spread; the large number of ganglion cells needed at the fovea or area centralis must be located close to the receptors they subserve. The developmental mechanisms which produce the remarkable gradients in cell concentration are little studied and only in recent years has work begun to establish the role of classic parameters of neuro-embryology – tissue growth, synapse formation and the genesis, movement, death and differentiation of cells – in the 'topogenesis' of the retina. The observations described here concern one of these parameters, cell death, and its role in the formation of the regional specializations of the retina of the cat.

Cell death is an ubiquitous phenomenon in the developing nervous system, and its occurrence in the ganglion cell layer of mammalian retina has recently been documented (Cunningham, Mohler & Giordano, 1981; Sengelaub & Finlay, 1981, 1982; Jeffery & Perry, 1982; Dreher, Potts & Bennett, 1983). In the studies described here, we have investigated the role of cell death in determining the numbers and distribution of ganglion cells in the adult retina of the cat.

The adult population of ganglion cells in the cat's retina numbers between 100 000 and 160 000 (Stone, 1965, 1978; see also Hughes, 1975, 1981), with considerable variation between individual animals. The ganglion cells are born between the 21st and 36th days of embryonic life (E21 to E36; Kliot & Shatz, 1982; Walsh *et al.*, 1983). Axons have entered the optic nerve by E30 (Anker, 1977; Shatz, 1983), and reach the lateral geniculate nucleus (LGN) and superior colliculus in the early E30s (Anker, 1977; Williams & Chalupa, 1982; Shatz, 1983). Ganglion cell precursors begin to form a layer separate from the neuroblast layer at approximately E30, but until E50 these precursors differ from adult ganglion cells in (at least) three ways:

1. The precursors are morphologically immature. Although many have axons, their somas are small ($< 7 \mu$m in diameter) with little cytoplasm, and show none of the size differentiation which characterizes adult ganglion cells (Rapaport & Stone, 1983b). Their nuclei lack nucleoli, but contain scattered chromatin granules.
2. The number of precursors is several times higher than the number of ganglion cells in the adult. Their number at E47 was estimated at over 800 000 by Stone *et al.* (1982) and large numbers of axons have been reported in the optic nerve in the E40s by Ng & Stone, (1982; $> 400 000$) and by Williams, Bastiani & Chalupa (1983; 328 000).
3. The precursors appear to be uniformly distributed over the retina, at a density (10 000–20 000/mm^2) higher than the peak density of ganglion cells found in the adult at the area centralis (Stone *et al.*, 1982).

The maturation of ganglion cells involves (at least) three distinct processes: the morphological differentiation of individual cells; a reduction in their numbers; and the 'sculpting' of the regional variations of ganglion cell density found in the adult from the apparently uniform distribution of precursors. In this chapter we discuss the contribution of cell death to the latter two processes.

Mechanisms which might shape the adult ganglion cell population

Reduction of numbers

Two mechanisms have been discussed in the literature which might reduce the high foetal population of ganglion cells to adult levels. These are the death of precursors and the transformation of precursors into amacrine cells, an idea developed by Hinds & Hinds (1978, 1983). The fate of precursors might in either case be determined by their success in establishing synapses on neurons in visual centres of the brain, from which they may receive a trophic molecule needed for their survival (McCaffery, Bennett & Dreher, 1982), or at least for their survival as ganglion cells. The fate of precursors may also be determined by a competition among their dendrites for afferent synapses from amacrine and bipolar cells (Linden & Perry, 1982).

Variations in ganglion cell density

Several mechanisms have been proposed to account for the development of the variations of ganglion-cell density across the retina which are found in the adult. The density of ganglion cells is, for example, 50–100 times higher at the area centralis than at the edge of the retina, and 5–10 times higher along the ridge of the visual streak than at the retinal edge. Mechanisms considered include:

A 'hot-spot' of ganglion-cell generation at the area centralis, and extending along the visual streak.

Transretinal migration of ganglion cells to form the central and streak concentration.

Differential retinal growth, the growth of retinal area after the birth of ganglion cells occurring predominantly in the periphery.

Differential cell death, the death of ganglion-cell precursors occurring predominantly in the retinal periphery.

Differential maturation, the transformation of precursors into amacrines occurring predominantly in the periphery.

Narrowing down the possibilities

The case against a 'hot-spot'

Ganglion cells are born by E36 (Kliot & Shatz, 1982; Walsh et al., 1983). The ganglion cell layer first becomes apparent between E21 and E25, and in Nissl-stained material the distribution of cells in the layer appears uniform until about E50, when the area centralis and visual streak begin to appear (Stone et al., 1982; Rapaport & Stone, 1983b). Moreover, Rapaport & Stone (1983a) showed that dividing cells at the ventricular surface are uniformly distributed over the retina between E29 (the youngest age studied) and E50, when cytogenesis starts to cease. We have subsequently observed that uniformity at E25 and E21, i.e. during the period of ganglion cell birth. In terms either of cell density in the ganglion cell layer or of mitotic activity during the birth of ganglion cells, there is little evidence of a 'hot-spot' of ganglion-cell generation.

Lia, Williams & Chalupa (1983) used the retrograde transport of horseradish peroxidase from injections into the visual centres of the brain to identify ganglion cells and estimate their distribution in the fetal retina. They reported a 2:1 gradient in ganglion-cell concentration between the area centralis and peripheral retina at E34, thus confirming that at this age the distribution of ganglion cells is much more uniform than in the adult. Even this gradient may result from the circumstance that the latest ganglion cells generated are born in the early E30s and are located in the retinal periphery. The axons of many of them may not have reached the brain by E34. Their results do indicate, however, that a gradient in the density of cells which have sent an axon in the optic nerve may go undetected in Nissl-stained material.

The case against transretinal migration

Because the density of ganglion-cell precursors is high between E33 and E50, greater than the peak of ganglion-cell density found in the adult (Stone et al., 1982; Rapaport & Stone, 1984), there is no need for transretinal migration to explain the concentration of ganglion cells at the area centralis or visual streak. (Migration of this sort does appear, however, to be involved in the formation of the concentration of cones at the primate fovea (Hendrickson & Kupfer, 1976)).

The inadequacy of differential growth

Mastronarde, Thibeault & Dubin (1980) presented evidence that the redistribution of ganglion cells which occurs postnatally is produced by a gradient in retinal growth, growth being least at the area centralis and greatest at the edge of the retina. By such a mechanism the density of ganglion cells at the area centralis may remain close to foetal values, while the peripheral cells are progressively spread as the retina expands. Our own results (Stone et al., 1982) generally confirm this idea, but indicate that growth can account for only part of the observed gradients. In particular we noted that between E47 and E57 a sharp (18:1) centroperipheral gradient of ganglion cell density appears from a uniform precursor population. The area of retina increases two- to three-fold during this period. Making a reasonable assumption of the gradient of retinal growth, for example that growth varies linearly with distance from the area centralis, this growth can account for the appearance of only about a 2:1 gradient.

Two of the mechanisms set out above remain to be investigated: differential maturation; and the parameter examined below, differential cell death.

Tempo and topography of cell death assessed from pyknotic cells

Tempo

Pyknotic cells (Fig. 11.1) occur commonly in the ganglion cell layer of developing retina. At E33, near the end of the period of ganglion cell birth, few were apparent; they occurred at an average density of only $10/mm^2$ in a retina only 9 mm^2 in area (Figs 11.2A and

11.3). By E38, however, pyknotic cells occurred at an average density of 325/mm² in all areas of the retina, and their density remained higher than 30/mm² until after birth (Figs 11.2A and 11.3), after which density declined. Expressed as the number of pyknotic cells occurring per retina (Fig. 11.2B), the rate of cell death generally increases from E33 until just before birth, after which numbers decline to low levels.

In making these estimates, we counted as pyknotic only cells which were clearly degenerative, with their nuclear chromatin condensed to one or a few dense spheres, or into dark clumps located around the rim of the nucleus (Fig. 11.1). We also sought, by careful focussing with a 100× objective, to count *only pyknotic cells in the ganglion cell layer.* A qualification to the interpretation of these results is that we could not identify whether a particular degenerating cell had been a ganglion, amacrine, glial or vascular cell. Nevertheless, the results provide support for the idea that cell death occurs among ganglion cell precursors and contributes to the reduction of the high numbers of precursors towards adult values. A surprising result was that pyknotic cells appeared very early, well before E47 when the precursor population was estimated at over 800 000 (Stone *et al.*, 1982). The overproduction of precursors may therefore be even greater than that estimate indicates.

Topography

At all ages examined (E33, E38, E47, E48, E52, E61, P2, P7, P30 adult), the distribution of pyknotic cells across the retina appeared approximately uniform. There was always variation in the count of pyknotic cells between sampling areas (Figs 11.3, 11.4 and 11.5),

but there was no evidence of a tendency for pyknotic cells to be more frequent in any particular region of retina. If the death of precursors were to contribute to the development of ganglion cell gradients, pyknotic cells should concentrate in the retinal periphery. They do not, and these results go against the hypothesis that cell death contributes to the development of adult topography.

Fig. 11.2. Tempo of cell death in the developing retina of the cat. (A) Mean density of pyknotic cells as a function of age. (B) The data points show the total number of pyknotic cells in the cat retina, as a function of age, estimated as the product of mean density and the area of the retina. The dashed line shows the change in axon numbers in the developing nerve, as determined by Ng & Stone (1982). (C) The area of the cat's retina, as a function of age.

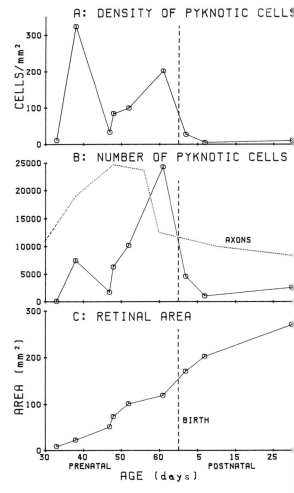

Fig. 11.1. Pyknotic cells (arrowed) in the ganglion cell layer of the wholemounted retina of an E62 foetus. (A) Area centralis. (B) Peripheral retina.

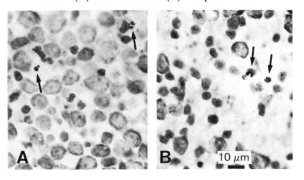

Fig. 11.3. Maps of the distribution of pyknotic cells in the foetal cat retina at ages E33, E38 and E48. In each map the open circle represents the optic disc, and the arrow points to the disc across the area centralis. Each number represents the number of pyknotic cells counted in an area of 0.01 mm². At these ages the cells in the ganglion cell layer are distributed uniformly across the retina at high density.

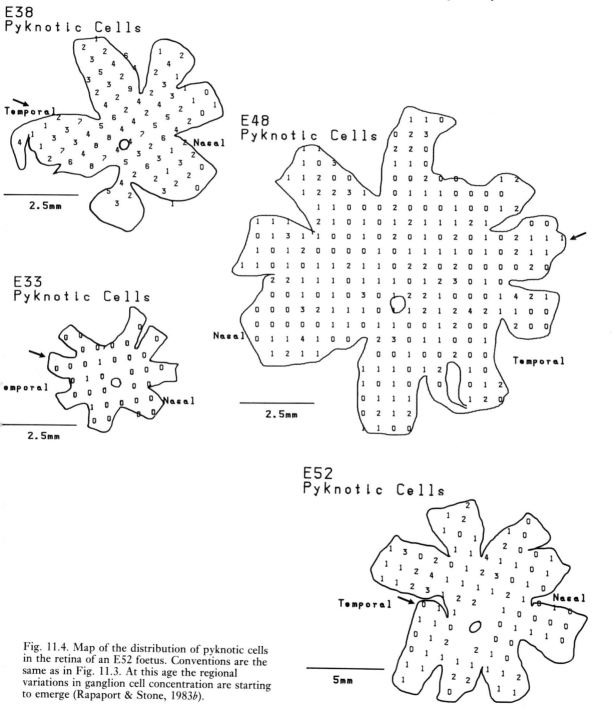

Fig. 11.4. Map of the distribution of pyknotic cells in the retina of an E52 foetus. Conventions are the same as in Fig. 11.3. At this age the regional variations in ganglion cell concentration are starting to emerge (Rapaport & Stone, 1983*b*).

Fig. 11.5. Maps of the distribution of pyknotic cells in the retinae of kittens at P30 and P2. Conventions are the same as in Fig. 11.3, except that the area centralis is represented by a black dot. At these ages the topography of the ganglion cells is well developed.

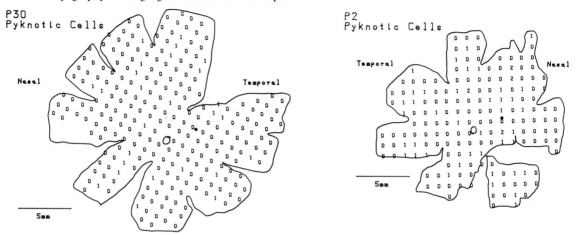

Fig. 11.6. The photomicrographs show the appearance of the ganglion cell layer in two kittens at P30. The kittens were littermates, and the left eye of one had been removed at E42. (A) Area centralis in the retinas of the unoperated animal. (B) Area centralis of the surviving retina of the operated animal. (C) Peripheral retina, 3 mm above the area centralis in the unoperated animal. (D) Peripheral retina, 3 mm above the area centralis in the operated animal.

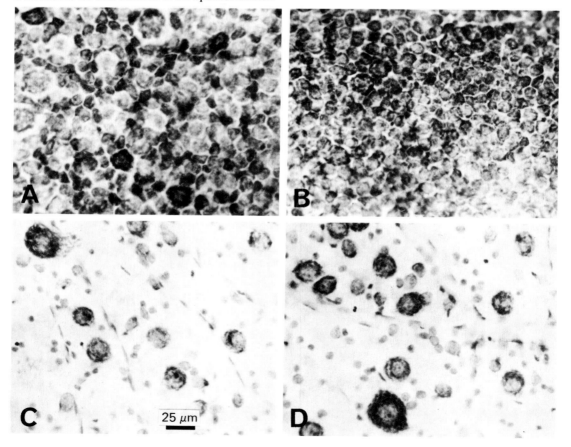

Evidence from early unilateral enucleation

Several workers have suggested that ganglion cells die or transform into amacrine cells because they fail to make synapses on neurons in the visual centres of the brain. That failure may result from competition among an excess of optic axons for available synaptic space, and relief of the competition should result in increased survival of ganglion cells. Competition can be relieved by removal of one eye, thereby freeing all of the central neurons normally contacted by the enucleated eye to accept afferents from the surviving eye. There is evidence that after unilateral enucleation most neurons in the LGN survive, and are contacted by the remaining eye (Rakic, 1979, in the monkey; Williams & Chalupa, 1982, in the cat). Moreover, Sengelaub & Finlay (1981) working with the hamster, and Jeffery & Perry (1982) with the rat, have reported increased survival of ganglion cells in one retina after early removal of the other; and Williams *et al.* (1983) with the cat and Rakic & Riley (1983) with the monkey have

reported increased numbers of axons in the optic nerve of one eye after early removal of the other. In the rodents the effect was small, the increase amounting to less than 10% of the normal ganglion-cell population. In the cat the increase was approximately 25%, and in the monkey the increase was as high as 35%.

We have examined the right retinae of a 30-day-old kitten whose left eye was removed at E42, and of an unoperated littermate at the same age. To inspection, the ganglion cell layers of the two retinae appeared very similar (Fig. 11.6). Counts of ganglion-cell density, however, showed a consistently higher density in the experimental retina. Fig. 11.7, for example, shows the density of ganglion cells in the two retinae, along a vertical transect across the visual streak 5 mm nasal to

Fig. 11.8. Graph of the density of ganglion cells as a function of distance superior to the area centralis. The counts were made in the two retinae also used for Fig. 11.6 and 11.7. The circles represent values from the enucleated animal, the crosses represent values from the normal littermate.

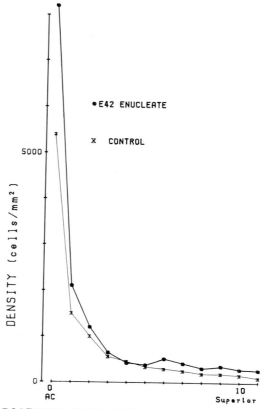

Fig. 11.7. Graph of the density of ganglion cells along a vertical transect across the visual streak, 5 mm nasal to the area centralis. The counts were made in the two retinae also used for Figs 11.6 and 11.8. The circles represent values from the enucleated animal, the crosses represent values from the normal littermate.

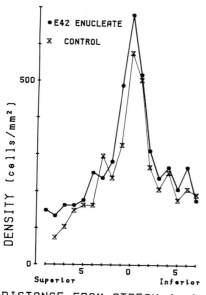

the optic disc. At all points except two, density was higher in the enucleate, by an average of 21%. Fig. 11.8 shows similar data taken along an axis extending from the area centralis into the superior retina. Density was higher in the enucleate at all points except one, the mean increase being 56%. The increased survival is not strongly topographic; ganglion-cell density was higher in both peripheral and central areas of retina.

These observations support the view that the death (or perhaps the transformation) of ganglion-cell precursors occurs partly in response to competition between axons from the two eyes, and contributes to determining the adult ganglion-cell population. Like the observation on the topography of pyknotic cells, however, they go against the view that cell death contributes to the topography of the adult ganglion cell layer.

Conclusion

The two sets of observations presented here – of pyknotic cells in the ganglion cell layer and of increased survival of ganglion cells following unilateral enucleation in the foetus – confirm the occurrence of considerable cell death in the developing retina of the cat. Without knowledge of the speed of pyknotic changes, however, it is difficult to estimate how many cells in the ganglion cell layer die during development. With between 1500 and 25 000 dying cells present in the ganglion cell layer throughout the last half of gestation, however, the number is presumably in the tens or hundreds of thousands, and would make a substantial contribution to the reduction of the high foetal ganglion cell population to adult levels.

As to the causes of cell death, the present results confirm that competition for central synapses plays a role, since reduction of that competition by unilateral enucleation increases ganglion-cell survival in the remaining eye. Can competition for synaptic space account fully for observed levels of ganglion-cell death? Counts of cells in the foetal ganglion cell layer (Stone *et al.*, 1982) and of axons in the foetal optic nerve (Ng & Stone, 1982; Williams *et al.*, 1983) indicate that the number of ganglion cells which die during normal development may be several times the adult population. The 21–56% increase in survival observed here is considerably higher than reported in earlier studies in rodents (Sengelaub & Finlay, 1981; Jeffery & Perry, 1982), yet it may have resulted from the 'rescue' of only a small fraction of the ganglion cells which normally

die. Previously we have argued that events in the ganglion cell layer, including cell death, may be determined by factors in more peripheral layers (Rapaport & Stone, 1982, 1984), including the availability of afferent synapses for the ganglion cells (Linden & Perry, 1982) and properties of the mitotic cells at the ventricular surface. The present results emphasize the need to test that suggestion.

We were surprised when our results indicated that the death of ganglion cell precursors is not topographically organized. So many features of the adult and developing retina are topographically arranged that the exceptions are notable; and at least in one other species, the hamster (Sengelaub & Finlay, 1982), cell death has been reported to be more prominent in the periphery. The clear implication of the present findings is that in the cat cell death does not contribute significantly to the development of the gradients of ganglion cell concentrations which characterize the regional specializations of the adult retina. In searching for the developmental mechanism responsible for this gradient, therefore, attention will need to be given to the one idea remaining from earlier work – that ganglion cells in peripheral retina transform into amacrine cells – as well as to weaknesses in the analyses reviewed and developed above. One such weakness, for example, is our lack of knowledge of cell death. For how long does a cell appear pyknotic. How many cells in the ganglion cell layer die? What precipitates most of the cell death? As those parameters are established, the role of cell death in the topogenesis of the retina will be clarified

References

Anker, R. L. (1977) The prenatal development of some of the visual pathways in the cat. *J. Comp. Neurol.*, **173**, 185–203

Cunningham, T. J., Mohler, I. M. & Giordano, D. L. (1981) Naturally occurring neuron death in the ganglion cell layer of the neonatal rat: morphology and evidence for regional correspondence with neuron death in superior colliculus. *Dev. Brain Res.*, **2**, 203–17

Dreher, B., Potts, R. A. & Bennett, M. R. (1983) Evidence that the early postnatal reduction in the number of rat retinal ganglion cells is due to a wave of ganglion cell death. *Neurosci. Lett.*, **36**, 255–60

Hendrickson, A & Kupfer, C. (1976) The histogenesis of the fovea in the macaque monkey. *Invest. Ophthalmol.*, **15**, 746–56

Hinds, J. W. & Hinds, P. L. (1978) The development of amacrine cells in the mouse retina: an electron microscopic, serial section analysis. *J. Comp. Neurol.*, **179**, 277–300

Hinds, J. W. & Hinds, P. L. (1983) Development of retinal

amacrine cells in the mouse embryo: evidence for two modes of formation. *J. Comp. Neurol.*, 213, 1–23

Hughes, A. (1975) A quantitative analysis of the cat retinal ganglion cell topography. *J. Comp. Neurol.*, 163, 107–28

Hughes, A. (1981) Population magnitudes and distribution of the major modal classes of cat retinal ganglion cell as estimated from HRP filling and a systematic survey of the soma diameter spectra for classical neurones. *J. Comp. Neurol.*, 197, 303–40

Jeffery, G. & Perry, V. H. (1982) Evidence for ganglion cell death during development of the ipsilateral retinal projection in the rat. *Dev. Brain Res.*, 2, 176–80

Kliot, M. & Shatz, C. J. (1982) Genesis of different ganglion cell types in the cat. *Soc. Neurosci. Abstr.*, 8, 815

Lia, B., Williams, R. W. & Chalupa, L. M. (1983) Early development of retinal specialization: the distribution and decussation patterns of ganglion cells in the prenatal cat demonstrated by retrograde peroxidase labelling. *Soc. Neurosci. Abstr.*, 9, 702

Linden, R. & Perry, V. H. (1982) Ganglion cell death within the developing retina: a regulatory role for dendrites. *Neuroscience*, 7, 2813–27

McCaffery, C. A., Bennett, M. R. & Dreher, B. (1982) The survival of neonatal rat retinal ganglion cells in vitro is enhanced in the presence of appropriate parts of the brain. *Exp. Brain Res.*, 48, 377–86

Mastronarde, D. N., Thibeault, M. A. & Dubin, M. W. (1980) How ganglion cells redistribute during postnatal growth of the cat's retina. *Invest. Ophthalmol.*, 19, (Suppl.) 70

Ng, A. Y. K. & Stone, J. (1982) The optic nerve of the cat: appearance and loss of axons during normal development. *Dev. Brain Res.*, 5, 263–71

Rakic, P. (1979) Genesis of visual connections in the rhesus monkey, In *Developmental Neurobiology of Vision*, ed. R. D. Freeman, pp. 249–69. New York: Plenum Press

Rakic, P. & Riley, K. P. (1983) Overproduction and elimination of retinal axons in the fetal rhesus monkey. *Science*, 219, 1441–4

Rapaport, D. H. & Stone, J. (1982) The site of commencement of maturation in mammalian retina: observations in the cat. *Dev. Brain Res.*, 5, 273–9

Rapaport, D. H. & Stone, J. (1983*a*) The topography of cytogenesis in the retina of the cat. *J. Neurosci.*, 3, 1824–34

Rapaport, D. H. & Stone, J. (1983*b*) Time course of morphological differentiation of cat retinal ganglion cells: influences on cell size. *J. Comp. Neurol.*, 221, 42–52

Rapaport, D. H. & Stone, J. (1984) The area centralis of mammalian retina: focal point for function and development of the visual system. *Neuroscience*, 11, 289–301

Sengelaub, D. R. & Finlay, B. L. (1981) Early removal of one eye reduces normally occurring cell death in the remaining eye. *Science*, 213, 573–4

Sengelaub, D. R. & Finlay, B. L. (1982) Cell death in the visual system during normal development: I. Retinal ganglion cells. *J. Comp. Neurol.*, 204, 311–17

Shatz, C. J. (1983) The prenatal development of the cat's retinogeniculate pathway. *J. Neurosci.*, 3, 482–99

Stone, J. (1965) A quantitative analysis of the distribution of ganglion cells in the cat's retina. *J. Comp. Neurol.*, 124, 337–52

Stone, J. (1978) The number and distribution of ganglion cells in the cat's retina. *J. Comp. Neurol.*, 180, 753–72

Stone, J., Rapaport, D. H., Williams, R. W. & Chalupa, L. (1982) Uniformity of cell distribution in the ganglion cell layer of prenatal cat retina: implications for mechanisms of retinal development. *Dev. Brain Res.*, 2, 231–42

Walsh, C., Polley, E. H., Hickey, T. L. & Guillery, R. W. (1983) Generation of cat retinal ganglion cells in relation to central pathways. *Nature*, 302, 611–13

Williams, R. W., Bastiani, M. J. & Chalupa, L. M. (1983) Loss of axons in the cat optic nerve following fetal unilateral enucleation: an electron microscopic analysis. *J. Neurosci.*, 3 133–44

Williams, R. W. & Chalupa, L. M. (1982) Prenatal development of retinocollicular projections in the cat: an anterograde tracer transport study. *J. Neurosci.*, 2, 604–22

Williams, R. W. & Chalupa, L. M. (1982) The effects of prenatal unilateral enucleation upon the functional organization of the cat's lateral geniculate nucleus. *Soc. Neurosci. Abstr.*, 8, 223

12

Abnormal binocular vision

D. P. CREWTHER and
S. G. CREWTHER

Normal binocular vision

Practically all vertebrate species possess two eyes, but it is only in the higher vertebrates (with a few notable exceptions) and in particular in the mammals that the eyes are used in a cooperative fashion to produce more than just a probabilistic sum of two images. The convergence of afferent information from small patches of both retinae on to a single neuron in the visual cortex enables retinal disparities to be coded in addition to the normal spatial relations that result from an ordered topographic mapping from retina to brain. Normal binocular vision has two requirements: first, that each eye has normal monocular function; and secondly, that the afferent fibres combine during development to allow for the correct binocular interaction.

Normal functioning of the visual system can be most easily expressed in terms of visual acuity. Such measurement requires the determination of a threshold, the boundary between discriminating or not discriminating, that is, seeing or not seeing. The factors that affect this discrimination are physical, anatomical and physiological. The physical quality of an image projected on to the retina is reduced by the diffraction caused by passage of the light through a restricted aperture – the pupil. Diffraction effects dominate light spread for pupil sizes of less than 2 mm (Westheimer 1981). For larger pupil sizes, optical aberrations of the various refractive elements in the eye become important. Apart from the effects of some scatter and absorption caused by the passage of light through the

16

optic media, the other major physical influence on acuity is the refractive state of the eye. If the stimulus is sufficiently delineated and the subject possesses normal accommodative powers, this will not normally contribute to light spread. From an anatomical viewpoint, the resolution of patterns is limited by the mosaic of retinal receptors, which is of finite grain. In the foveal region of the human retina, the cone spacing is approximately 0.5 min of arc (Polyak, 1941). Resolving power is also affected by the way in which connections are made between the various neural elements of the retina. This is especially apparent in the periphery, where single retinal ganglion cells achieve summation over several degrees of visual angle, encompassing the inputs from many rod photoreceptors.

Depending on the type of resolution task, the threshold measurements will differ markedly. Resolution tasks are generally of three types. The first requires the detection of a single feature such as a speck on a wall. This is called the 'minimum visible' criterion and is really a contrast threshold measurement, as the line spread will always produce a finite width image whose contrast will depend on the object contrast. Under normal photopic levels of illumination, the minimum visible subtence of a dark line on a light background is approximately 1 s of arc. The second type of resolution requires the determination of the presence of more than one identifying feature (such as in the classical Rayleigh criterion for resolution), or the discrimination between more than one identifying target. This is called the 'minimum resolvable' criterion and results in a value of about 30 s of arc (Westheimer, 1981). The third type requires the determination of the relative location of two or more features with respect to each other and is called a hyperacuity task involving the 'minimum discriminable criterion'. Normal threshold values for hyperacuity tasks vary from 2 to 10 s of arc. For the threshold values to be so small, complex neural processing must be involved, but the mechanisms at a cellular level are basically unknown, except perhaps in the case of stereoacuity (Nikara, Bishop & Pettigrew, 1968; Poggio & Fischer, 1977).

Normal binocular vision implies binocular single vision, that is, the two slightly different images on the two eyes are neurally fused so that only one image is perceived. Obviously, binocular vision requires overlap of the visual fields of the two eyes, often entailing

frontally placed eyes, such as in carnivores and primates, and a partial decussation of the optic nerve fibres (or in the case of some birds, via the supraoptic decussation connecting the opticus pricipalis thalami with the visual Wulst (Pettigrew & Konishi, 1976)) so that information from corresponding points of the two retinae can synapse on single cells in the cortex.

The idea of a horopter – the locus of points seen as single with the two eyes – appeared early in the seventeenth century, but the existence of a partial decussation of visual information from one eye to the two brain hemispheres was not established anatomically until the work of von Gudden (Duke-Elder & Wybar, 1960). The idea that disparate retinal images would provide a cue for binocular depth perception was due to Wheatstone (1838). Panum (1858) proposed that for any point on one retina, there is a small area on the other over which fusion of the two monocular inputs can be maintained. This fusional area would allow for single binocular images for more than just visual points on the surface comprising the horopter. Mitchell (1966) showed that the extent of Panum's area is about 15 min of arc for central vision but rapidly increases with eccentricity.

All discussion of stereo performance and the different nature of binocular viewing compared with monocular viewing, presupposes that at some stage in the visual system the afferents from the two eyes synapse on single cells and that the stimulation of one eye's afferents can influence the firing of the cell due to stimulation of the other eye's afferents. In man, we can assume that this is indeed the case based on extensive studies in various mammalian species. In the cat, approximately 80% of cells in area 17, the primary visual cortex, are binocular in the sense that they can be excited by either eye separately (Hubel & Wiesel, 1962). However, practically all striate neurons show binocular influence when properly controlled binocular stimulation is employed (Bishop, Henry & Smith, 1971). In the monkey, the proportion of binocularly driven cells in the corresponding visual cortical area, V1, is somewhat lower (Hubel & Wiesel, 1970), due to segregation by eye of geniculocortical afferents in the layer IV of V1.

The development of visual acuity in young children was for a long time ignored because of the problems of subject communication. However, the advent of three techniques for measuring acuity in preverbal subjects has made the study of the postnatal development of vision accessible. These techniques are preferential looking, optokinetic nystagmus and visually evoked potentials (Banks & Salapatek, 1978; Dobson & Teller, 1978; Gwiazda *et al.*, 1978). Letter acuity (minimum resolvable) improves from about 6/90 (Snellen) neonatally to 6/6 by about 6 months of age. Similar studies in the macaque monkey have shown that adult acuity levels are reached at about 6 weeks of age (Teller *et al.*, 1978).

Although one might expect the development of stereopsis to be preceeded by the development of monocular visual acuity, the precocious development of the oculomotor system allows even the newborn infant to perform some coordinated eye-scanning movements (Haith, 1978). Stereopsis appears to undergo striking development between about 15 and 20 weeks in children (Fox *et al.*, 1980).

Abnormal binocular vision

Abnormalities of binocular vision can occur in adults due to either disease or accidents; however, if the eyes can be realigned or vision through each eye can be reinstated, then binocular vision can be achieved again. Thus we will concentrate on children's vision where permanent deficits do arise if the clarity of visual information to the eye is interrupted or interfered with during formative years, or if there is a misalignment of the eyes or a motor imbalance in the control of movement of one or both eyes. To date, comparatively few complete experimental analyses of the various types of abnormal binocular vision have been attempted. However, most developmental disorders of vision are characterized by a loss of visual acuity (usually only in one eye) and some anomalies in the ability of the two eyes to fuse and maintain stability of the two images, resulting in a lowered stereoacuity. Together, these losses in acuity have been referred to clinically as 'amblyopia'.

Amblyopia is characterized by the lack of an apparent pathology within the eye and the inability to improve visual performance by optical correction. This, together with psychophysical evidence, has led to the conclusion that the loss of acuity involves a central neural abnormality. Classical indicators of retinal function such as increment thresholds and dark adaptation are normal in amblyopic patients (Wald & Burian, 1944; Duke-Elder & Wybar, 1973; Burian & von Noorden, 1974). However, spatial perception, form vision and binocular function are usually disrupted –

these are typical indicators of higher-order (cortical) visual processing. The only evidence for retinal involvement in amblyopia comes from pattern electro-retinograms which have been found to be abnormal in many amblyopes (Arden *et al.*, 1980; Vaegen, Arden & Fells, 1983).

There are five main clinical groupings of amblyopia. *Deprivation amblyopia* arises after deprivation of all form vision due to cataract, ptosis, corneal opacification or haemangioma in early childhood. Three months of deprivation results in acuities of the order of Snellen 6/60 or worse through the deprived eye and in binocularity deficits (Taylor *et al.*, 1979). *Anisometropic* (unequal refraction without eye misalignment) is the most common cause of amblyopia (65% – Schapiro (1971)), although this figure may include a population with microtropia – a strabismus so small as to escape the usual clinical scrutiny. Anisometropia may result from genetic factors controlling eye growth or corneal development, or it may follow various types of disease states such as congenital cataract, optic atrophy maculopathies and retrolental fibroplasia. The effect of unequal refraction on retinal image quality include both magnification differences and high spatial frequency filtering for the blurred image (Campbell & Green, 1965). *Ametropic amblyopia*, due to a large (equal) bilateral refractive error, is characterized by lowered monocular acuities in both eyes, but subjects may retain some stereopsis if the ametropia is not accompanied by a strabismus. *Meridional amblyopia* is due to an uncorrected, cylindrical refractive error in early life. Although blur in the defocussed meridian selectively attenuates high spatial frequencies, the contrast thresholds are increased uniformly down to very low spatial frequencies (Freeman & Thibos, 1975). This presumably reflects a loss of cortical cells which code for orientations of that meridian. *Strabismic amblyopia* results from an early misalignment of the eyes. The misalignment usually has one of two outcomes: either one eye fixates for a time and then fixation is transferred to the other eye; or fixation in one eye is suppressed. The latter situation, that of suppression during development, invariably results in a monocular loss of visual acuity in that eye. The former situation, alternation of fixation, is usually accompanied by normal monocular acuity; however binocular functions such as stereopsis are usually disrupted.

Sophisticated psychophysical experiments

which investigate more than just the threshold visual performance, point to different neural bases for the various human amblyopias. While all types of amblyopia share the feature of large threshold losses, anisometropic amblyopes appear to be relatively more handicapped in their suprathreshold contrast processing than are strabismic amblyopes (Hess & Bradley, 1980; Hess, Campbell & Zimmern, 1980; Hess, Bradley & Piotrowski, 1983).

A similar distinction between strabismic and anisometropic amblyopes has been drawn by Levi & Klein (1982) using a hyperacuity task. While grating acuities for amblyopes do not predict their considerably worse letter acuities, there is a good correlation between Snellen acuities and vernier hyperacuities. Using such a hyperacuity task, Levi and Klein showed that anisometropic amblyopes demonstrate hyperacuity, even at spatial frequencies approaching their grating resolution, while strabismic amblyopes demonstrate a loss of hyperacuity at high spatial frequencies. Thus strabismic amblyopes suffer relatively more spatial confusion.

Stereoblindness does not have a direct correlation with amblyopia. For example, subjects with strabismus who alternately fixate with one eye followed by the other, can have normal monocular acuity but abnormal stereoacuity. Typically, stereoblind people perform visual tasks no better with two eyes than with one eye alone (Lema & Blake, 1977). This reflects an absence of binocular summation which in normal people results in approximately 40% improvement in binocular over monocular performance (Campbell & Green, 1965; Blake & Fox, 1973). It is likely that one of the chief causes in humans of stereoblindness is a paucity of binocular neurons. This conclusion is based on poor interocular transfer of various after-effects (Movshon, Chambers & Blakemore, 1972; Mitchell & Ware, 1974), and on the superior utrocular discrimination possessed by stereoblind subjects (Blake & Cormack, 1979).

Animal models

The use of animal models for the study of human developmental disorders of vision derives from the observation of close similarities in the functional architecture and the neuronal properties of homologous visual areas in marsupials, eutherian mammals and birds. The neural substrate subserving binocular vision has many similarities in most vertebrate species including man. A knowledge of the neural substrate of a visual abnormality enables better design of treatments and the testing of such treatments on a physiological and anatomical basis as well as using animal behaviour testing to gauge overall visual performance.

The first and most extensively modelled visual abnormality is that of visual deprivation. In man, this occurs through cataract, cloudy media, ptosis or extreme refractive error. In cats and monkeys (as well as other species), these conditions have been modelled by eyelid suture, dark-rearing, atropine penalization and artificially induced refractive errors. Wiesel & Hubel, (1963a) showed that unilateral eyelid closure in young kittens resulted in practically no binocularly driven cells in striate cortex, and with only about 5% of cells responding to stimulation through the deprived eye. Behaviourally, there was a drastic loss of vision through the deprived eye (Dews & Wiesel, 1970). Subsequent studies into the time course of susceptibility to lid closure (Hubel & Wiesel, 1970), led to the definition of a critical period for plastic changes in the visual system, a concept central to the treatment of visual disorders. The effects of deprivation on visual acuity are generally consistent with the cortical neurophysiology. Monocular deprivation through eyelid closure in rhesus monkeys causes an amblyopia in the deprived eye (von Noorden, Dowling & Ferguson, 1970), and contrast threshold studies (Harwerth et al., 1981) have shown that there is reduced sensitivity at all spatial frequencies.

Acuity losses following form deprivation without light deprivation have been shown with anisometropia induced with chronic atropinization (Ikeda & Tremain, 1978; Crewther, Crewther & Cleland, 1985) or through lens defocus (Crewther et al., 1985). Binocularly deprived animals have a large population of visually unresponsive cells (Baker, Grigg & von Noorden, 1974; Crawford et al., 1975) and many cells with immature properties, characterized by receptive fields which are less sharply defined and less selectively tuned for orientation. Such properties would suggest lowered acuities through both eyes, and this is borne out in animals as close to the human as the chimpanzee (Riesen & Zilbert, 1975).

Meridional amblyopia has been modelled by raising animals in visual environments selective for a particular orientation. Hirsch & Spinelli (1970) raised kittens wearing goggles through which one eye viewed horizontal contours and the other eye viewed vertical contours, while Blakemore & Cooper (1970) raised kittens in a drum containing stripes of one orientation. When tested behaviourally, the cats displayed an amblyopia when presented with the non-exposed orientation (Muir & Mitchell, 1973, 1975). This acuity loss is accompanied by a lack of cells in the striate cortex which are responsive to the deprived orientation (Blasdel *et al.*, 1977) and by modification of dendritic arborizations, so that the dendritic fields become elongated along axes which are either parallel to (stellate cells) or orthogonal to (pyramidal cells) the projection of the orientation seen on the cortex (Spencer, 1974; Tieman, Butterfield & Hirsch, 1981).

As strabismus often results in the development of an amblyopia, it is often categorized as a form of visual deprivation. However, there appear to be good reasons for treating it separately. Strabismus in animals has been modelled both by surgical intervention and by the use of ophthalmic prisms held in goggles in front of the eyes. Hubel & Wiesel (1965) showed that raising kittens with a divergent strabismus produced by section of the medial rectus muscle of one eye led to a situation where the usual proportion of binocularly driven cells in the visual cortex was drastically reduced, while there were approximately equal proportions of cells driven separately by the two eyes. In the so-called higher visual areas such as the lateral suprasylvian areas in the cat (e.g. the posterolateral lateral suprasylvian sulcus (PLLS) and posteromedial LS (PMLS), there are marked differences between the effects of visual deprivation by lid closure (Spear & Tong, 1980) and those of surgical strabismus, which while causing a loss of binocular cells in striate cortex, has little effect on the percentage of binocular cells in the cortical areas PLLS and PMLS (von Grunau, 1982). While Hubel and Wiesel's results immediately suggested a mechanism for the loss of stereopsis in strabismic persons, they did not establish the monocular resolving power of their cats. Subsequent behavioural evaluation of the visual acuity of strabismic cats has highlighted the sensitivity of the results to the experimental preparation. Jacobson & Ikeda (1979) found that esotropic cats, in which the strabismus was produced by removal of the lateral rectus muscle and sometimes the superior oblique muscle, were amblyopic; by using a cover test, Ikeda & Tremain, (1979) suggested that other esotropic cats, in which the strabismus was created by section of the muscle, were not amblyopic. On the other hand, Cleland *et al.*, (1982) found that both esotropic and exotropic strabismic cats (strabismus created by muscle

section) were highly amblyopic through the deviating eye. Strabismus in monkeys (produced by muscle extirpation) is accompanied by reduced visual acuity through the deviating eye for esotropia (von Noorden & Dowling, 1970), while for exotropia, equal acuity was observed.

Mechanisms for amblyopia

Most of the theories of the breakdown of normal binocular vision have resulted from experiments in which the normal visual environment of developing animals has been altered in some way. There are striking results (described below) when one eye is allowed a competitive advantage over the other during development. These results have led to the idea of a dynamic fight for synaptic terminal space between the visual afferents. This is commonly referred to as binocular competition. Cats appear blind when first exposed to visual stimuli through an eye deprived of form vision by eyelid suture from soon after birth (Wiesel & Hubel, 1963a; Ganz & Fitch, 1968; Dews & Wiesel, 1970; Giffin & Mitchell, 1978). These behavioural observations were supported by physiological findings that almost all cells in the striate visual cortex of the cat (Wiesel & Hubel, 1963a) and monkey (Hubel, Wiesel & Le Vay, 1976) are solely excited by stimulation through the visually experienced eye, with the exception of some cells with strongly monocular inputs in layer IV (Shatz & Stryker, 1978), which presumably are less susceptible to capture by terminals from the visually experienced eye. The behaviourally measured loss of visual acuity in visually deprived animals (Giffin & Mitchell, 1978) has also been observed at the single cell level in the striate cortex of cats raised with an artificial anisometropia (Eggers & Blakemore, 1978), where cells driven by the defocussed eye were less able to resolve gratings of high spatial frequency.

The degradation of the retinal image itself has some effect on the developing visual system, as evidenced by binocular deprivation and dark-rearing experiments (Regal *et al.*, 1976). Changes in cortical neuronal density and dendritic morphology have been observed. There is an increase in neuronal density of the visual cortex in both the mouse (Gyllensten, 1959) and the cat (Cragg, 1975). Valverde (1967) observed a reduction in the linear density of spines in the apical dendrites of pyramidal cells of mice reared in darkness. Concurrent with the morphological changes, a retardation in the development of neural connectivity has also been observed (Swindale, 1981). In contrast to the changes observed in the cortex, due to the lack of ordered afferent activity, visually deprived retinal ganglion cells appear to function normally. This was shown by recording from cat optic nerve fibres (Sherman & Stone, 1973; Kratz, Mangel, Lehmkuhle & Sherman, 1979) with only chance encounters with fibres from the area centralis, and by direct recording from the area centralis region of the deprived eye (Cleland, Mitchell, Crewther & Crewther, 1980), where brisk-sustained (X) cells which resolved drifting gratings of spatial frequency in excess of 10 cycles/degree were found. Similarly, direct optical defocus produced by rearing kittens with high-powered contact lenses has no effect on retinal ganglion cell acuities (Crewther *et al.*, 1985). Thus, the lowered ganglion cell acuities observed in retinal ganglion cells of cats reared with a strabismus caused by surgical removal of extraocular muscle tissue (Ikeda & Tremain, 1979) may not be due to a degraded retinal image. The resolving properties of retinal ganglion cells from totally dark-reared animals have not been reported and it is possible that in the light of a study of excitation of kitten cortical cells due to stimulation through closed eyelids (Spear, Tong & Langsetmo, 1978), different results might obtain.

The rather simple principle entailed in binocular competition seems to explain well the afferent terminal distribution in primary visual cortex. Using only a competitive theory, Swindale (1980) successfully produced a computer model for the formation of ocular dominance stripes. If the visual cortex were isolated, with only a purely visual input, then arguments based on binocular competition alone would possibly be sufficient to explain all observed phenomena. However, recent reports (Mitchell and Murphy, 1984) indicate that while the behaviourally measured acuity of a kitten subjected to a period of reverse occlusion follows the ocular dominance shifts (the acuity of the initially deprived eye improves with a period of reverse occlusion while the acuity of the initially experienced eye drops), the restoration of equal binocular vision to the animal results in an amblyopia developing rapidly in the initially deprived eye, while the initially experienced eye reasserts dominance.

The situation is made complex by two features. First, the cells in the visual cortex of young animals are not in a steady state but are undergoing maturation, and

this is generally accompanied by a decreasing ability to reflect the afferent balance from the two eyes. Secondly, there are a myriad of connections: between cortical areas, with thalamic and midbrain regions, and to areas such as the visual claustrum (Olson & Graybiel, 1980; Le Vay & Sherk, 1981). Different temporal maturation of these extracortical areas could play a role, especially in situations such as the reverse occlusion described above. For example, the circuitry that controls target fixation may lose its plasticity at an earlier stage than the primary visual cortex. Thus, even at the end of a period of reverse occlusion, when visual cortex is totally dominated by the initially deprived eye, the fixation reflex may be controlled by the initially open eye. One might expect this controlling eye to increase its influence to produce an ocular dominance shift in striate cortex towards the initially open eye. Although this suggestion is hypothetical, it is known that different properties of cortical cells have different critical periods. For instance, the sensitivity to monocular deprivation continues (Hubel & Wiesel, 1970; Blakemore, Van Sluyters & Movshon, 1976) after the period of susceptibility to directional rearing has ended (Daw & Wyatt, 1976; Berman & Daw, 1977; Cynader & Mitchell, 1980).

Visual synergy

Lack of visual synergy has been promoted as the cause of the breakdown of binocularity in strabismic cats (Hubel & Wiesel, 1965). Corresponding points on the two retinae no longer cause stimulation of the same cortical cell, and thus the input from one eye onto a cortical neuron will be uncorrelated with that from the other eye. If there is a competition between the visual afferents, then the most binocular cells seem to be the most easily changed (in terms of ocular dominance). This idea is reinforced by the protection against plastic changes observed in strabismic kittens that are subsequently monocularly deprived (Mustari & Cynader, 1981). Presumably, if the striate cortex is already monocular as a result of the early strabismus, the ocular dominance of cells driven by the eye that is later occluded is more difficult to change.

The relation between lack of visual synergy and the development of amblyopia is not at all clear. There is no apparent difference in image quality on the retina of the deviating eye compared with that of the normal eye. Changes in retinal response could occur, however, if there was an oculomotor abnormality, such as a

Fig. 12.1. Final training curves on a Mitchell jumping stand to establish a behavioural measure of visual acuity of the cat shown in Fig. 9.1. Performance on a two-choice discrimination between a black and white grating and a grey field was evaluated for both eyes open (binocular–*a*) and for the left eye (*b*) and right eye (*c*) alone. Criterion performance on a block of trials at one spatial frequency is indicated by a filled circle, while performance below criterion is indicated by an open circle. In each case, the acuity was of the order of 5.5 cycles/degree.

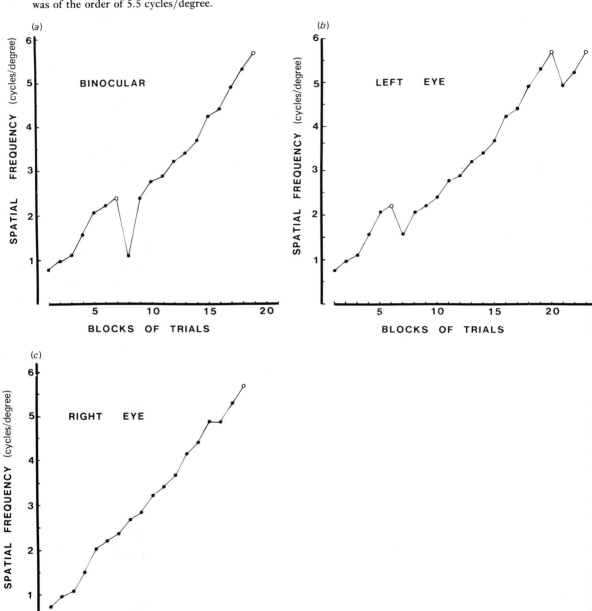

nystagmus, associated with the strabismus. Yet, while monkeys which have been made esotropic by extirpation of the lateral rectus muscle become amblyopic, similar monkeys made exotropic by extirpation of the medial rectus muscle have equal acuities in the two eyes (von Noorden & Dowling, 1970). One might expect that the clearest indication of the effects of visual asynchrony would come from experiments in which there was interference only with the visual image, such as rearing animals with optically induced squint. Such a procedure leads to a loss of binocular cells in the striate visual cortex (Smith *et al.*, 1979; Crawford & von Noorden, 1980; Van Sluyters & Levitt, 1980). It also leads in some cases to a behavioural amblyopia (Bennett *et al.*, 1980). The cause of this amblyopia is unlikely to be due to a degraded retinal image with subsequent competitive advantage to one eye. It is more likely to be due to some oculomotor change, such as a fixation dominance, to compensate for the altered eye alignment. In this regard it is interesting that Bennett *et al.* noted esotropic misalignments in most of their kittens after the period of prism-rearing.

Although amblyopia is normally associated with loss of fixation in one eye, our ideas may be changed by bilaterally strabismic cats, which appear to have no central fixation with either eye, yet possess normal acuity through both eyes (Fig. 12.1).

Extraretinal factors

Without question, visual experience is important for the maturation of the perceptual capacities, at least in those animals that are born functionally immature. However, recent reports have shown that rather than visual stimulation alone, development requires concomitant oculomotor activity. Six hours of exposure to a normal visual environment was sufficient for the development of refined orientation tuning in a 6-week-old dark-reared kitten, but only if eye movements were present throughout the visual exposure (Buisseret, Gary-Bobo & Imbert, 1978). The role of the eye movements in permitting neural change has been recently refined. Freezing of neuronal plasticity has been observed in freely behaving animals in which afferent information about eye movements has been blocked by bilateral section of the ophthalmic branch of the trigeminal nerve a few days before the delayed visual experience was presented (Buisseret & Gary-Bobo, 1979; Trotter, Gary-Bobo & Buisseret, 1981; Trotter, Fregnac & Buisseret, 1983). The inference that

the recovery of visual function is dependent, in part, on the afferent input from the proprioceptive fibres, receives support from a similar experiment in which the maxillary branch of the trigeminal nerve was cut and recovery of orientation tuning was observed (Trotter *et al.*, 1981). Similarly, unilateral section of the ophthalmic branch of the trigeminal nerve inhibits plastic changes under conditions of delayed monocular experience, irrespective of the laterality of the section to the eye receiving visual experience.

Section of the ophthalmic branch of the fifth nerve also has effects on the ocular dominance distribution when performed in combination with certain other rearing procedures. For example, a considerable reduction in the percentage of binocular cells was observed for kittens raised in a normal visual environment with unilateral section of the ophthalmic branch of the trigeminal nerve (Imbert & Fregnac, 1983). While such a result may be explained away by a microtropia (eye misalignment of very small magnitude), reduction in binocularity also occurs if the kittens with unilateral trigeminal section are raised in total darkness (Trotter, Fregnac & Buisseret, 1983). The latter result is reminiscent of Maffei and Bisti's (1976) study of the cortical effects of strabismus with bilateral visual occlusion or dark-rearing. Their result, of a loss in binocularity, was not confirmed by Van Sluyters & Levitt (1980). While the difference in results may be due in some part to differences in environment (dark-rearing *versus* lid closure), it may also involve the period of time for which the strabismus was maintained before recording, since three strabismic kittens recorded in our laboratory showed a reduction in binocularity after a long period of dark-rearing (Fig. 12.2).

Not all strabismic cats lose binocular connectivity in the visual cortex. Cats raised with large, chronic, bilateral cyclotorsions of the eyes show binocularity in Area 17 within the normal range (75% binocular) (Crewther *et al.*, 1981). Such cats possess an extremely large strabismus: the visual axes may be misaligned by as much as 60 degrees.

On the effects of symmetrical, bilateral strabismus, results in the literature are divided. Bisti & Maffei (1979) found a maintenance of binocularity in some cases. Singer, von Grunau & Rauschecker (1979) thought that maintenance of binocularity was associated with abnormal oculomotor behaviour. Singer (1979) suggests that maintenance is associated with a marked impairment of excitatory transmission, since the expected ocular dominance shift towards the open eye does not occur if this eye's effectiveness in driving cortical neurons is reduced by rotation (Singer, Tretter & Yinon, 1982). It is possible that the lack of ocular dominance shift in both these experiments could be the result of the immobilization's effect in reducing stimulation of the extraocular afferents. In this way, it would mimic the effects of section of the trigeminal, which, as already described, appears to protect the cortex from plastic change (Buisseret & Singer, 1983). In cats where the rotated eye has normal mobility (Crewther *et al.*, 1981), no protection is afforded by the rotation. For example, monocular deprivation of one eye combined with rotation of the other leads to a corte. totally dominated by the rotated eye (Fig. 12.3).

The gating role of extraocular afferents has als been shown in the postcritical period reversal of th effects of ocular dominance. While enucleation of th visually experienced eye of a cat raised through th critical period under conditions of monocular depriva tion causes the appearance of a large number of cell responsive to stimulation through the deprived ey (Kratz, Spear & Smith, 1976), removal of the retina afferents of the visually experienced eye by pressur blinding does not cause responses to be elicited throug! the deprived eye (Blakemore & Hillman, 1977; Harri & Stryker, 1977). Extraocular afferents were shown t play a modulatory or gating role (Crewther, Crewthe & Pettigrew, 1978). Visual responses were recorde due to stimulation of the deprived eye only when bot the retinal and extraocular afferents from the visuall experienced eye were removed. This finding als indicated that the proprioceptive afferents play a rol in maintaining the *status quo* during adult life.

Buisseret & Maffei (1977) showed directly tha units in cat visual cortex were affected by signals fror the extraocular muscle proprioceptors. Effects o passive eye movement in conjunction with visua stimulation have also been found in the superio colliculus (Rose & Abrahams, 1975; Donaldson & Long, 1980) and in the lateral geniculate nucleus (Donaldson & Dixon, 1981).

The anatomy of the proprioceptive pathway i not yet well established. The nature of extraocula muscle receptors varies from species to species. Muscl spindles are present in primates and ungulates but no in cats. In cats , the pallisade endings which surroun the junction of the muscle with tendon are the mos likely candidates for the stretch receptor (Alvarado Mallart & Pinçon-Raymond, 1979). The pathway take by the sensory fibres may not be consistent. Either the leave the motor nerves to join the ophthalmic brancl of the trigeminal (Batani & Buisseret, 1974; Batin Buisseret & Buisseret-Delmas, 1975), or they enter th brain with the oculomotor nerves (Bach-y-Rita & Murata, 1964). Although the trigeminal pathways t the mesencephalic nucleus of the fifth nerve (Fillenz 1955; Alvarado-Mallart *et al.*, 1975) and to th cerebellum (Baker, Precht & Llinas, 1972; Batini *et al* 1975) are well established, the anatomical pathways t the visual cortex are still unknown.

Changes in cortical plasticity without inter

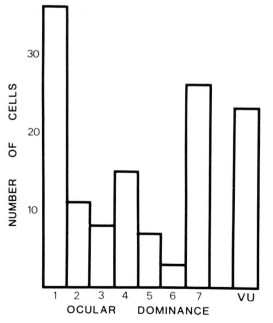

Fig. 12.2. Ocular dominance histogram of the combined data from three cats which were made strabismic by section of the medial rectus muscle of one eye at 1 week of age. The kittens were immediately placed with their mothers in a totally dark, light-locked room, until recorded at 4 months of age. Recordings of 129 units were made from Area 17 of the visual cortex contralateral to the strabismic eye. The convention of Hubel & Wiesel (1962) was used for the ocular dominance groups 1–7, while the group VU comprises visually unresponsive cells. The percentage of binocularly driven cells (groups 2–6) (41%) is approximately midway between the percentages of binocular cells reported for normal cats and dark-reared cats on the one hand, and the percentages reported for strabismic cats on the other (20%).

ference to the primary visual pathway, i.e. without visual deprivation *per se*, have been observed under two other general descriptions: interference with the reticular activating system, and depletion of endogenous catecholamines. Singer (1982) showed that unilateral destruction of the nuclear complex of the medial thalamus leads to an impairment of experience-dependent modifications in the visual cortex ipsilateral to the lesion. Also, stimulation studies suggest that the mesencephalic reticular formation has an ascending projection which contributes to the gating of cortical changes (Singer & Rauschecker, 1982).

It has been proposed that neuronal plasticity early in life is, in part, controlled by the catecholaminergic system (Kasamatsu & Pettigrew, 1976, 1979; Kasamatsu, Pettigrew & Ary, 1979). This control was first expressed by protecting the neurons of Area 17 from the effects of monocular deprivation by depleting the cortex of catecholamines by an intraventricular injection of 6-hydroxydopamine (6-OHDA) in young kittens, prior to a period of monocular deprivation. Rather more refined injections, given intracortically via a mini-pump, have since added to the evidence that the

effect of the 6-OHDA is to block the neural plasticity to environmental change that would normally be observed in a young kitten. Catecholaminergic fibres project to practically all parts of the central nervous system (Moore & Bloom, 1979). However, the cell bodies are quite few in number and are restricted to a few brainstem nuclei. One of the chief sources of catecholamines is the locus coeruleus. Injection of 6-OHDA into this structure causes a widespread depletion of endogenous noradrenaline. However, plasticity in the visual cortex can be restored by the intracortical microperfusion of noradrenaline (Pettigrew & Kasamatsu, 1978; Kasamatsu *et al.*, 1979).

Since its projections are so diffuse, the locus coeruleus has been suggested as a source for global reinforcement in learning (Crow, 1968; Anlezark, Crow & Greenway, 1973). In the context of visual plasticity during development, the locus coeruleus could cause reinforcing signals which would be distributed over the whole cortex and which would cause an incoming visual signal to be validated, perhaps by lowering thresholds for cortical transmission, or by opening reverberatory circuits, for example in the geniculocortical loop. The

Fig. 12.3. A comparison of the cortical effects of rearing kittens with and without monocular deprivation (MD) when the other eye has been surgically rotated. Recordings were made from Area 17 of both hemispheres. The ocular dominance histogram on the left (9 kittens, 263 cells) shows that the normal (unrotated) eye has a tendency to dominate over the rotated eye, while there is a reasonably large percentage of binocular cells (groups 2–6) (54%) amongst the visually responsive cells. On the other hand, when the unrotated eye is deprived of visual experience by lid suture, almost total domination of visual cortex by the rotated eye is found (2 kittens, 61 cells). There is no evidence of suppressed plasticity as reported by Singer *et al.* (1982). (Data redrawn from Crewther *et al.*, 1981.)

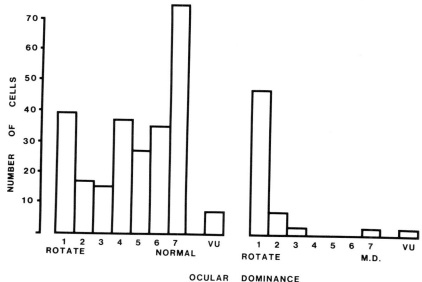

cause of the coeruleus reinforcement could well be provided by the proprioceptive signals (the mesencephalic nucleus of the trigeminal nerve is situated adjacent to the locus coeruleus in many species), or by motor corollary discharge, or by visual afferents. The last alternative is not attractive, however, as it begs the question of the choice of stimuli that should provide reinforcement.

Conclusions

In this chapter we have concentrated on the aspects of abnormal acuity and abnormal binocularity which are associated with amblyopia. However, there is an equally large amount that could be written on the neuromuscular anomalies that accompany many cases of amblyopia (see Burian & von Noorden, (1974) for review). Part of the problem is that there are so many different motor strategies that may result from a visual imbalance, and there are similarly many places in the oculomotor system where a breakdown during development will lead to a visual abnormality. For example, bilaterally strabismic cats may adopt the strategy of turning their heads to look directly forward in preference to moving either eye. However, if one eye is occluded during behavioural testing sessions, such cats can move the uncovered eye to increase their field of view, indicating that the eyes are not immobilized.

Perhaps the best investigated eye movement property in any of the animal models of abnormal vision is that of optokinetic nystagmus (OKN), and it is encouraging that the horizontal asymmetries observed for patients with amblyopia are similar to the disturbed eye movements of kittens which have been monocularly deprived (Hoffmann, 1979) or reared with a strabismus (Cynader & Harris, 1980). However, the reduced gain of the temporal OKN is more likely to be associated with disturbance of the inputs to the nucleus of the optic tract (Harris & Cynader, 1981), than to be directly associated with a loss of visual acuity through one or both eyes.

Although the subclasses of amblyopia have not yet been specifically modelled, it now appears fairly certain that therapies designed to improve acuity will need to take into account the oculomotor balance as well as the visual. Both clinical observations and experimental studies using animal models indicate that for amblyopia treatment to be successful, it must be initiated very early in childhood. Although a strabismus is often detected early enough, the visual

recovery rate after surgical correction of the misalignment is surprisingly low. For young anisometropes, the problem is often not diagnosed until school age, i.e. 4–6 years of age. A solution to this situation would be the widespread refractive assessment of infant's vision.

Clinical treatment procedures are yet to benefit from the wealth of insight obtained from the animal models. The treatments in use have remained virtually unchanged over the past 200 years. Patching, or constant occlusion, of the eye with better sight is the established therapy, and many months are often required before the maximum achievable acuity is attained (often at the expense of the vision in the occluded eye). The time needed for patching is directly related to the time of onset of deprivation. The effects of treatment are variable, and the gains in terms of acuity are often not sustained, reminiscent of the behaviour when binocular experience is restored to reverse occluded cats (Mitchell & Murphy, 1984). Secondly, such treatments based on monocular stimulation of the deprived or strabismic eye is that the recovery of binocular vision is very seldom achieved. Lastly, such an extreme and prolonged treatment is often socially and psychologically distressing to the child. Minimal occlusion goes some way to avoiding these problems, and has been shown in cats to be more effective per unit time than massed periods of occlusion in promoting recovery from the effects of monocular deprivation (Crewther, Crewther & Mitchell, 1983). A recent adaptation of this method is the introduction of brief periods of occlusion under conditions of heightened motivation. A rotating grating device (Banks, Campbell & Hood, 1977) was designed to stimulate maximally the majority of visual cortical cells by subjecting the child to different spatial frequency, high-contrast, square-wave gratings presented at all orientations. The results of testing indicated that vision in an amblyoptic eye could be significantly improved by weekly periods of concentrated visuomotor tasks against the background rotating gratings (Campbell *et al.*, 1978). However, recent clinical trials of this method (Keith *et al.*, 1980) have suggested that although the technique has resulted in significant improvements in acuity, the visual grating stimulus is not important but that motivated minimal occlusion is a successful treatment regimen. The key to future treatments of amblyopia, and possible cure, appears to lie in the understanding of the oculomotor involvement in the suppression or arrest of development of vision during childhood, and in understanding the circuitry involved in attention and motivation, in relation to plastic changes.

Acknowledgement

We should like to thank Don Mitchell for the behavioural results on bilateral strabismic cats shown in Fig. 9.1.

References

Alvarado-Mallart, R. M., Batini, C., Buisseret, P., Gueritaud, J. P. & Horcholle-Bosavit, G. (1975) Mesencephalic projections of the rectus lateralis muscle afferents in the cat. *Arch. Ital. Biol.*, 113, 1–20

Alvarado-Mallart, R. M. & Pinçon-Raymond, M. (1979) The palisade endings of cat extraocular muscles: a light and electron microscope study. *Tissue Cell*, 11, 567–84

Anlezark, G. M., Crow, T. J. & Greenway, A. P. (1973) Impaired learning and decreased cortical norepinephrine after bilateral locus coeruleus lesions. *Science*, 181, 682–4

Arden, G. B., Vaegan, Hogg, C. R., Powell, D. J. & Carter, R. M. (1980) Pattern ERGs are abnormal in many amblyopes. *Trans. Ophthalmol. Soc. UK*, 100, 453

Bach-y-Rita, P. & Murata, K. (1964) Extraocular proprioceptive responses in the VI nerve of the cat. *Q. J. Exp. Physiol.*, 49, 408–16

Baker, F. H., Grigg, P. & von Noorden, G. K. (1974) Effects of visual deprivation and strabismus on the response of neurons in the visual cortex of the monkey, including studies on the striate and prestriate cortex in the normal animal. *Brain Res.*, 66, 185

Baker, R., Precht, W. & Llinas, R. (1972) Mossy and climbing fibre projections of extraocular muscles afferents to the cerebellum. *Brain Res.*, 38, 440–5

Banks, M. S. & Salapatek, P. (1978) Acuity and contrast sensitivity in 1, 2, and 3 month old human infants. *Invest. Ophthalmol. Vis. Sci.*, 17, 361–5

Banks, R. V., Campbell, F. W. & Hood, C. (1977) A neurophysiological approach to the treatment of amblyopia. *J. Physiol.*, 275, 16–17P

Batini, C. & Buisseret, P. (1974) Sensory peripheral pathway from extrinsic eye muscles. *Arch. Ital. Biol.*, 112, 18–32

Batini, C., Buisseret, P. & Buisseret-Delmas, C. (1975) Trigeminal pathways of the extrinsic eye muscle afferents in cat. *Brain Res.*, 85, 74–8

Bennett, M. J., Smith, E. L., Harwerth, R. L. & Crawford, M. L. J. (1980) Ocular dominance, eye alignment and visual acuity in kittens reared with an optically induced squint. *Brain Res.*, 193, 33

Berman, N. & Daw, N. W. (1977) Comparison of the critical periods for monocular deprivation and directional deprivation in cats. *J. Physiol.*, 265, 249–59

Bishop, P. O., Henry, G. H. & Smith, C. J. (1971) Binocular interaction fields of single units in the cat striate cortex. *J. Physiol.*, 216, 39–68

Bisti, S. & Maffei, L. (1979) Binocular interactions in bilateral strabismic kittens. *Brain Res.*, 170, 359–61

Blake, R. & Cormack, R. H. (1979) Psychophysical evidence for a monocular visual cortex in stereoblind humans. *Science*, 203, 274–5

Blake, R. & Fox, R. (1973) The psychophysical inquiry into binocular summation. *Percept. Psychophys.*, 14, 161–85

Blakemore, C. & Cooper, G. F. (1970) Development of the brain depends on the visual environment. *Nature*, 228, 477

Blakemore, C. & Hillman, P. (1977) An attempt to assess the effects of monocular deprivation and strabismus on synaptic efficacy in the kitten's visual cortex. *Exp. Brain Res.*, 30, 187–202

Blakemore, C., Van Sluyters, R. C. & Movshon, J. A. (1976) Synaptic competition in the kitten's visual cortex. *Cold Spring Harbor Symp. Quant. Biol.*, 40, 601–9

Blasdel, G. G., Mitchell, D. E., Muir, D. W. & Pettigrew J. D. (1977) A physiological and behavioural study in cats of the effect of early visual experience with contours of a single orientation. *J. Physiol.*, 265, 615–36

Buisseret, P. & Gary-Bobo, E. (1979) Development of visual cortical orientation specificity after dark-rearing: role of extraocular proprioception. *Neurosci. Lett.*, 13, 259–63

Buisseret, P., Gary-Bobo, E. & Imbert, M. (1978) Ocular motility and recovery of orientational properties of visual cortical neurons in dark-reared kittens. *Nature*, 272, 816

Buisseret, P. & Maffei, L. (1977) Extraocular proprioceptive projections to the visual cortex. *Exp. Brain Res.*, 28, 421–5

Buisseret, P. & Singer, W. (1983) Proprioceptive signals from extraocular muscles gate experience dependent modifications of receptive fields in the kitten visual cortex. *Exp Brain Res.*, 51, 443–50

Burian, H. M. & von Noorden, G. K. (1974) *Binocular Vision and Ocular Motility*. St Louis: CV Mosby

Campbell, F. W. & Green, D. G. (1965) Monocular versus binocular acuity. *Nature*, 208, 192–3

Campbell, F. W., Hess, R. S., Watson, P. G. & Banks, R. V. (1978) Preliminary results of a physiologically based treatment of amblyopia. *Br. J. Ophthalmol.*, 62, 748–55

Cleland, B. G., Crewther, D. P., Crewther, S. G. & Mitchell, D. E. (1982) Normality of spatial resolution of retinal ganglion cells in cats with strabismic amblyopia. *J. Physiol.*, 326, 235–49

Cleland, B. G., Mitchell, D. E., Crewther, S. G. & Crewther, D. P. (1980) Visual resolution of retinal ganglion cells in monocularly-deprived cats. *Brain Res.*, 192, 261–6

Cragg, B. G. (1975) The development of synapses in kitten visual cortex during visual deprivation. *Exp. Neurol.*, 46, 445–51

Crawford, M. L. J., Blake, R., Cool, S. J. & von Noorden, G. K. (1975) Physiological consequences in unilateral and bilateral eye closure in macaque monkeys: some further observations. *Brain Res.*, 84, 150–4

Crawford, M. L. J. & von Noorden, G. K. (1980) Optically induced concomitant strabismus in monkeys. *Invest. Ophthalmol. Vis. Sci.*, 19, 1105–9

Crewther, D. P., Crewther, S. G. & Cleland, B. G. (1985) Is the retina sensitive to the effects of prolonged blur? *Exp. Brain Res.*, 58, 427–34

Crewther, S. G., Crewther, D. P. & Mitchell, D. E. (1983) The effects of short-term occlusion therapy on reversal of the anatomical and physiological effects of monocular deprivation in the lateral geniculate nucleus and visual cortex of kittens. *Exp. Brain Res.*, 51, 206–16

Crewther, D. P., Crewther, S. G. & Pettigrew, J. D. (1978) A role for extraocular afferents in post-critical period reversal of monocular deprivation. *J. Physiol.*, 282, 181–95

Crewther, S. G., Crewther, D. P., Peck, C. K. & Pettigrew, J. D. (1980) Visual cortical effects of rearing kittens with monocular or binocular cyclotorsion. *J. Neurophysiol.*, 44, 97–117

Crow, T. J. (1968) Cortical synapses and reinforcement: a hypothesis. *Nature*, 219, 736–7

Cynader, M. & Harris, L. (1980) Eye movements in strabismic cats. *Nature*, 286, 64–5

Cynader, M. & Mitchell, D. E. (1980) Prolonged sensitivity to monocular deprivation in dark-reared cats. *J. Neurophysiol.*, 43, 1026–40

Daw, N. & Wyatt, H. J. (1976) Kittens reared in a unidirectional environment: evidence for a critical period. *J. Physiol.*, 257, 155–70

Dews, P. B. & Wiesel, T. N. (1970) Consequences of monocular deprivation on visual behaviour in kittens. *J. Physiol.*, 206, 437–55

Dobson, V. & Teller, D. Y. (1978) Visual acuity in the human infant. *Vision Res.*, 18, 1469–84

Donaldson, I. M. L. & Dixon, R. A. (1981) Excitation of units in the lateral geniculate and contiguous nuclei by stretch of the extraocular muscles. *Exp. Brain Res.*, 38, 245–55

Donaldson, I. M. L. & Long, A. C. (1980) Interaction between extraocular proprioceptive and visual signals in the superior colliculus of the cat. *J. Physiol.*, 298, 85–110

Duke-Elder, S. & Wybar, K. (1960) The anatomy of the visual system. In *System of Ophthalmology*, Vol. 2, ed. S. Duke-Elder p. 588. London: Henry Kimpton

Duke-Elder, S. & Wybar, K. (1973) Ocular motility and strabismus. In *System of Ophthalmology*, Vol. 6, S. Duke-Elder, p. 301. London: Henry Kimpton

Eggers, H. M. & Blakemore, C. (1978) Physiological basis of anisometropic amblyopia. *Science*, 201, 264–7

Fillenz, M. (1955) Responses in the brainstem of the cat to stretch of extrinsic ocular muscles. *J. Physiol.*, 128, 182–99

Fox, R., Aslin, R. N., Shea, S. L. & Dumais, S. T. (1980) Stereopsis in human infants. *Science*, 207, 323–4

Freeman, R. D. & Bonds, A. B. (1979) Cortical plasticity in monocularly deprived immobilized kittens depends on eye movement. *Science*, 206, 1093–5

Freeman, R. D & Thibos, L. N. (1975) Visual evoked responses in humans with abnormal visual experience. *J. Physiol.*, 247, 711–24

Ganz, L. & Fitch, M. (1968) The effect of visual deprivation on perceptual behaviour. *Exp. Neurol.*, 22, 639–60

Giffin, F. & Mitchell, D. E. (1978) The rate of recovery of vision after early monocular deprivation in kittens. *J. Physiol.*, 274, 511–37

Grunau, M. W. von (1982) Comparison of the effects of induced strabismus on binocularity in area 17 and LS area in the cat. *Brain Res.*, 246, 325–32

Gwiazda, J., Brill, S., Mohindra, I. & Held, T. (1978) Infant visual acuity and its meridional variations. *Vision Res.*, 18, 1557–64

Gyllensten, L. (1959) Postnatal development of the visual cortex in darkness (mice). *Acta Morphol, Neerl. Scand.*, 2, 331–45

Haith, M. M. (1978) Visual competence in early infancy. In *Handbook of Sensory Physiology*, Vol. 8, ed. R. Held *et al.*, pp. 311–56. Berlin: Springer-Verlag

Harris, L. & Cynader, M. (1981) The eye movements of the dark-reared cat. *Exp. Brain Res.*, 44, 41–56

Harris, W. A. & Stryker, M. P. (1977) Attempts to reverse the effects of monocular deprivation in the adult visual cortex. *Neurosci. Abstr.*, 1785, 562

Harwerth, R. S., Crawford, M. L. J., Smith, E. L. & Bolz, R. L. (1981) Behavioural studies of deprivation amblyopia in monkeys. *Vision Res.*, 21, 779–89

Hess, R. & Bradley, A. (1980) Contrast perception above threshold is only minimally impaired in human amblyopia. *Nature*, 287, 463–4

Hess, R., Bradley, A. & Piotrowski, L. (1983) Contrast-coding in amblyopia. I. Differences in the neural basis of human amblyopia. *Proc. R. Soc. Lond., Ser. B*, 217, 309–30

Hess, R., Campbell, F. W. & Zimmern, R. (1980) Differences in the neural basis of human amblyopias: the effect of mean luminance. *Vision Res.*, 20, 295–305

Hirsch, H. V. B. & Spinelli, D. N. (1970) Visual experience modifies distribution of horizontally and vertically oriented receptive fields in cats. *Science*, 168, 869

Hoffmann, K-P. (1979) Optokinetic nystagmus and single-cell responses in the nucleus tractus opticus after early monocular deprivation in the cat. In *Developmental Neurobiology of Vision*, ed. R. D. Freeman, pp. 63–72. New York: Plenum Press

Hubel, D. H. & Wiesel, T. N. (1962) Receptive fields, binocular interaction and functional architecture in the cat's visual cortex. *J. Physiol.*, 160, 106–54

Hubel, D. H. & Wiesel, T. N. (1965) Binocular interaction in striate cortex of kittens reared with artificial squint. *J. Neurophysiol.*, 28, 1041–59

Hubel, D. H. & Wiesel, T. N. (1970) The period of susceptibility to the physiological effects of unilateral eye closure in kittens. *J. Physiol.*, 206, 419–36

Hubel, D. H., Wiesel, T. N. & Le Vay, S. (1976) Functional architecture of area 17 in normal and monocularly deprived macaque monkeys. *Cold Spring Harbor Symp. Quant. Biol.*, 40, 581–9

Ikeda, H. & Tremain, K. E. (1978) Amblyopia resulting from penalization: neurophysiological studies of kittens reared with atropinization of one or both eyes. *Br. J. Ophthalmol.* 62, 21–8

Ikeda, H. & Tremain, K. E. (1979) Amblyopia occurs in retinal ganglion cells in cats reared with convergent squint without alternating fixation. *Exp. Brain Res.*, 35, 559–82

Imbert, M. & Fregnac, Y. (1983) Specification of cortical neurone by visuomotor experience. In *Prog. Brain Res.*, 58, 427–36

Jacobson, S. G. & Ikeda, H. (1979) Behavioural studies of spatial vision in cats reared with convergent squint: is amblyopia due to arrest of development? *Exp. Brain Res.*, 34, 11–26

Kasamatsu, T. & Pettigrew, J. D. (1976) Depletion of brain catecholamines: failure of ocular dominance shift after monocular occlusion in kittens. *Science*, **194**, 206–9

Kasamatsu, T. & Pettigrew, J. D. (1979) Preservation of binocularity after monocular deprivation in the striate cortex of kittens treated with 6-hydroxydopamine. *J. Comp. Nuerol.*, **185**, 139–61

Kasamatsu, T., Pettigrew, J. D. & Ary, M. (1979) Restoration of visual cortical plasticity by local microperfusion of norepinephrine. *J. Comp. Neurol.*, **185**, 163–82

Keith, C. G., Howell, E. R., Mitchell, D. E. & Smith, S. (1980) A clinical trial of the use of rotating grating patterns in the treatment of amblyopia. *Br. J. Ophthalmol.*, **64**, 597–606

Kratz, K. E., Mangel, S. C., Lehmkuhle, S. & Sherman, S. M. (1979) An effect of early monocular lid suture upon the development of X-cells in the cat's lateral geniculate nucleus. *Brain Res.*, **172**, 545–51

Kratz, K. E., Spear, P. D. & Smith, D. C. (1976) Postcritical-period reversal of the effects of monocular deprivation on striate cortex cells in the cat. *J. Neurophysiol.*, **39**, 501–11

Le Vay, S. & Sherk, H. (1981) The visual claustrum of the cat. I. Structure and connections. *J. Neurosci.*, **1**, 956–80

Lema, S. A. & Blake, R. (1977) Binocular summation in normal and stereoblind humans. *Vision Res.*, **17**, 691–5

Levi, D. M. & Klein, S. (1982) Hyperacuity and amblyopia. *Nature*, **298**, 268–70

Levitt, F. B. & Van Sluyters, R. C. (1982) The sensitive period for strabismus in the kitten. *Dev. Brain Res.*, **3**, 323–7

Maffei, L. & Bisti, S. (1976) Binocular interaction in strabismic kittens deprived of vision. *Science*, **191**, 579–80

Mitchell, D. E. (1966) Retinal disparity and diplopia. *Vision Res.*, **6**, 441–51

Mitchell, D. E. & Murphy, K. (1984) The effectiveness of reverse occlusion as a means of promoting visual recovery in monocularly deprived kittens. In *Development of Visual Pathways in Mammals*, ed. J. Stone, B. Dreher & D. H. Rapaport, pp. 381–92. New York: A. R. Liss.

Mitchell, D. E. & Ware, C. (1974) Interocular transfer of a visual aftereffect in normal and stereoblind humans. *J. Physiol.*, **236**, 707–21

Moore, R. Y. & Bloom, F. E. (1979) Central catecholamine neuron systems: anatomy and physiology of the norepinephrine and epinephrine systems. *Ann. Rev. Neurosci.*, **2**, 113–68

Movshon, J. A., Chambers, B. E. I. & Blakemore, C. (1972) Interocular transfer in normal humans, and those who lack stereopsis. *Perception*, **1**, 483–90

Muir, D. W. & Mitchell, D. E. (1973) Visual resolution and experience; acuity deficits in cats following early selective visual deprivation. *Science*, **180**, 420–2

Muir, D. W. & Mitchell, D. E. (1975) Behavioural deficits in cats following early selected visual exposure to contours of a single orientation. *Brain Res.*, **85**, 459–77

Mustari, M. & Cynader, M. (1981) Prior strabismus protects kitten cortical neurons from the effects of monocular deprivation. *Brain Res.*, **211**, 165–70

Nikara, T., Bishop, P. O. & Pettigrew, J. D. (1968) Analysis of retinal correspondence by studying receptive fields of binocular single units in cat striate cortex. *Exp. Brain Res.*, **6**, 353–72

Noorden, G. K. von & Dowling, J. E. (1970) Experimental amblyopia in monkeys. II. Behavioural studies in strabismic amblyopia. *Arch. Ophthalmol.*, **84**, 215–20

Noorden, G. K. von, Dowling, J. E. & Ferguson, D. C. (1970) Experimental amblyopia in monkeys. I. Behavioural studies of stimulus deprivation amblyopia. *Arch. Ophthalmol.*, **84**, 206–14

Olson, C. R. & Graybiel, A. M. (1980) Sensory maps in the claustrum of the cat. *Nature*, **288**, 479–81

Panum, P. L. (1858) *Untersuchungen über das Sehen mit zwei Augen*. Kiel: Schwering

Pettigrew, J. D. & Kasamatsu, T. (1978) Local perfusion of noradrenaline maintains cortical plasticity. *Nature*, **271**, 761–3

Pettigrew, J. D. & Konishi, M. (1976) Neurons selective for orientation and binocular disparity in the visual Wulst of the barn owl (*Tyto alba*). *Science*, **193**, 675–8

Poggio, G. F. & Fischer, B. (1977) Binocular interaction and depth sensitivity in striate and prestriate cortex of behaving monkeys. *J. Neurophysiol.*, **40**, 1392

Polyak, S. (1941) *The Retina*. Chicago: University of Chicago Press

Regal, D. M., Boothe, R., Teller, D. Y. & Sackett, G. P. (1976) Visual acuity and visual responsiveness in dark-reared monkeys. (*Macaca nemestrina*). *Vision Res.*, **16**, 523–30

Riesen, A. H. & Zilbert, D. E. (1975) Behavioural consequence of variations in early sensory environments. In *Developmental Neuropsychology of Sensory Deprivation*, ed. A. H. Riesen, pp. 21–52. New York: Academic Press

Rose, P. K. & Abrahams, V. C. (1975) The effect of passive eye movement on unit discharge in the superior colliculus of the cat. *Brain Res.*, **97**, 95–106

Schipiro, M. (1971) *Amblyopia*. Radnor, Pa: Chilton

Shatz, C. J. & Stryker, M. P. (1978) Ocular dominance in layer IV of the cat's visual cortex and the effects of monocular deprivation. *J. Physiol.*, **282**, 267

Sherman, S. M. & Stone, J. (1973) Physiological normality of the retina in visually deprived cats. *Brain Res.*, **60**, 224–30

Singer, W. (1979) Evidence for central control of developmental plasticity in the striate cortex of kittens. *Dev. Neurobiol. Vision*, **27**, 135–47

Singer, W. (1982) The role of attention in developmental plasticity. *Hum. Neurobiol.*, **1**, 41–3

Singer, W., von Grunau, M. & Rauschecker, J. P. (1979) Requirements for the disruption of binocularity in the visual cortex of strabismic kittens. *Brain Res.*, **171**, 536–40

Singer, W. & Rauschecker, J. P. (1982) Central core control of developmental plasticity in the kitten visual cortex. II. Electrical activation of mesencephalic and diencephalic projections. *Exp. Brain Res.*, **47**, 223–33

Singer, W., Tretter, F. & Yinon, U. (1982) Central gating of developmental plasticity in kitten visual cortex. *J. Physiol.*, **324**, 221–37

Smith, E. L., Bennett, M. J., Harwerth, R. L. & Crawford, M. L. J. (1979) Binocularity in kittens reared with an optically induced squint. *Science*, **204**, 875

Spear, P. D. & Tong, L. (1980) Effects of monocular deprivation on neurons in cat's lateral suprasylvian visual area. I. Comparison of binocular and monocular segments. *J. Neurophysiol.*, 44, 568

Spear, P. D., Tong, L. & Langsetmo, A. (1978) Striate cortex neurons of binocularly deprived kittens respond to visual stimuli through closed eyelids. *Brain Res.*, 155, 141–6

Spencer, R. F. (1974) *Influence of selective visual experience upon the postnatal maturation of the visual cortex of the cat.* Thesis, University of Rochester, New York

Swindale, N. V. (1980) A model for the formation of ocular dominance stripes. *Proc. R. Soc. Lond., Ser. B*, 208, 243–64

Swindale, N. V. (1981) Absence of ocular dominance patches in dark-reared cats. *Nature*, 290, 332–3

Taylor, D., Vaegan, Morris, J. A., Rodgers, J. E. & Warland, J. (1979) Amblyopia in bilateral infantile and juvenile cataract: relationship to timing of treatment. *Trans. Ophthalmol. Soc. UK*, 99, 170–5

Teller, D. Y., Regal, D. M., Videen, T. O. & Pulos, E. (1978) Development of acuity in infant monkeys (*Macaca nemestrina*) during the early prenatal weeks. *Vision Res.*, 18, 561–6

Tieman, S. B., Butterfield, K. & Hirsch, H. V. B. (1981) Stripe-rearing modifies dendritic morphology of cells in visual cortex of the cat. *Soc. Neurosci. Abstr.*, 7, 141

Trotter, Y., Fregnac, Y. & Buisseret, P. (1983) Synergie de la vision et de la proprioception extroculaire dans les mechanismes de plasticité fonctionelle du cortex visual primaire du chaton. *C. R. Acad. Sci. Ser. C*, 296, 665–8

Trotter, Y., Gary-Bobo, E. & Buisseret, P. (1981) Recovery of orientation selectivity in kitten primary visual cortex is slowed down by bilateral section of ophthalmic trigeminal afferents. *Dev. Brain Res.*, 1, 450–4

Vaegan, Arden, G. B. & Fells, P. (1983) Amblyopia: some possible relations between experimental models and clinical experience. In *Paediatric Ophthalmology*, ed. K. Wybar & D. Taylor, pp. New York: Marcel Decker

Valverde, F. (1967) Apical dendrite spines of the visual cortex and light deprivation in the mouse. *Exp. Brain Res.*, 3, 337–52

Van Sluyters, R. C. & Levitt, F. B. (1980) Experimental strabismus in the kitten. *J. Neurophysiol.*, 43, 686–99

Wald, G. & Burian, H. M. (1944) The dissociation of form, vision and light perception in strabismic amblyopia. *Am. J. Ophthalmol.*, 27, 950–62

Westheimer, G. (1981) Visual acuity. In *Adler's Physiology of the Eye*, ed. A. Moses, pp. 530–44. St Louis: CV Mosby

Wheatstone, C. (1838) Contributions to the physiology of vision. I. On some remarkable and hitherto unobserved, phenomena of binocular vision. *Phil. Trans. R. Soc. Lond.*, 128, 371

Wiesel, T. N. & Hubel, D. H. (1963a) Single cell responses in striate cortex of kittens deprived of vision in one eye. *J. Neurophysiol.*, 26, 1003–17

Wiesel, T. N. & Hubel, D. H. (1963b) Effects of visual deprivation on morphology and physiology of cells in the cat's lateral geniculate body. *J. Neurophysiol.*, 26, 978–83

Wiesel, T. N. & Hubel, D. H. (1965) Extent of recovery from the effects of visual deprivation in kittens. *J. Neurophysiol.*, 28, 1060–72

IV

Comparative visual physiology

Evolution can be an important guide to the functional significance, or otherwise, of some puzzling morphological feature. For this reason, comparative neuroscience provides an inexhaustible supply of clues to the function of the brain. Nowhere is this more evident than in visual neuroscience. For example, the powerful idea of parallel visual processing originated with the work of Herrick and his school of comparative neurology on anurans in the early part of the century, was fostered by Karten's work on the avian brain in the 1960s, and now blossoms with the realization that primates too have just as many multiple, parallel paths from eye to brain (Kaas: Chapter 21). There is little doubt that some of the functionally most important subpathways in the primate's visual system (for example, those for dual-opponent colour processing, or visuomotor stabilization) can also be the least evident in physiological recordings because of small size or scarcity. A postulate for their presence based on comparisons with other species can therefore be the decisive factor in their eventual discovery.

Comparative insights have likewise played an important role in the understanding of the lateral geniculate nucleus (LGN) whose evolution is described by Sanderson (Chapter 13). The paradox of LGN function concerns the way it appears to go to great lengths, in evolutionary terms, to sandwich closely together layers representing each eye, yet at the same time keeps inputs to those layers from each eye functionally and anatomically segregated. The whole structure seems to be designed for binocular interaction but at the same time appears to avoid interaction between the eyes at all cost. The solution to the riddle is provided by two findings on the LGN, binocular inhibition and parallel visual processing, the first due wholly to Peter Bishop's own determined search (as described in Sanderson's personal note), and the second due in large part to the concerted efforts of his disciples (as described in Chapters 2, 3, 7, 8, 17, 18, 20). Binocular inhibition allows a limited form of interaction between adjacent layers representing different eyes but the same subclass of visual neuron. The excitatory retinal inputs are therefore segregrated into different layers according to eye and functional subclass so that inhibitory interactions between the layers can be restricted to the pair of layers subserving the same submodality for each eye. Presumably, at some time in the future the same scholars who have elucidated some of the parallel submodalities within the cat and monkey visual pathways will tell us what submodalities are subserved by the 11 different layers Sanderson describes in the LGN of the kangaroo!

Just as evolution can illuminate form and function, so we can try to reverse the process in an attempt to reconstruct an evolutionary tree from the various forms currently present. This is a most dangerous game, because one must choose morphological features which are the least likely to be subject to strong selection pressures and therefore not of vital functional significance. These last aspects would tend to make this enterprise rather unattractive to scholars interested in studying aspects of the brain which *are* o vital functional significance, but Johnson (Chapter 14 shows how a phylogenetic tree can be constructed fo mammals using some brain traits as the basis fo classification.

Pettigrew (Chapter 15) gives an account of hi discovery of an example of convergent evolution a intriguing as the simple eye invented independently by cephalopod molluscs and vertebrates: the processing o binocular information by the owl and the cat. Althoug they do it with completely different underlyin structures, both owl and cat achieve a remarkabl similar end result for the analysis of stereoscopi information. The evolutionary origins for this conver gence have important ramifications for our under standing of the way in which the geniculostriat pathway is organized, such as the way in which th elaboration of highly specialized receptive field types i delayed in this pathway (in contrast to the pathway from eye to midbrain) until after information has com together from both eyes at the cortical level.

Organizational differences between pathway within the visual system can be further highlighted b a comparison with the auditory pathways. The recen discovery of separate channels for time and intensity i the auditory system suggests that some powerfu concepts developed for vision such as paralle processing, may be even more apposite for audition Aitkin (Chapter 16) examines the auditory system wit the visual system in mind.

13

Evolution of the lateral geniculate nucleus

K. J. SANDERSON

The dorsal lateral geniculate nucleus (LGNd) is a well-defined nucleus in the thalamus of mammals which relays information from the retina to the visual cortex. It is of course not a pure relay nucleus, since it receives input from many regions besides the retina, and input–output studies (Cleland, Dubin & Levick, 1971) and receptive field studies show that some analysis of visual information occurs in the LGNd. In mammals, the LGNd may be highly laminated or have little lamination; the significance and diversity of lamination patterns have been of considerable interest to vision scientists for many years. This chapter explores the range of LGNd lamination which occurs in different groups of mammals and describes some of the possible functions of lamination. In addition there is a brief description of the 'lateral geniculate body' in non-mammalian vertebrates. This chapter covers the LGNd literature from 1900 to 1982; however, the citation of references includes only key references, some original descriptions of LGNd organization and recent autoradiographic studies.

Non-mammalian vertebrates

In fishes, amphibians, reptiles and birds, there is a direct retinal projection to one or more regions in the thalamus and some of these thalamic visual nuclei relay information to a telencephalic visual area. In birds and reptiles, the retinothalamotelencephalic pathway is usually considered homologous to the retinogeniculo-cortical pathway of mammals and thus one or all of the thalamic relay nuclei are sometimes called the dorsal lateral geniculate body.

In birds, the 'lateral geniculate body' is a complex of four or five cell masses in the dorsal thalamus (Karten *et al.*, 1973); all of these receive input from the contralateral eye and probably all send projections to the visual wulst with projections from the individual nuclei being ipsilateral, contralateral or bilateral to the wulst (for a recent summary see Nixdorf & Bischof, 1982). In reptiles also, there appear to be a number of thalamic nuclei which receive direct retinal input, mostly from the contralateral eye, with a small component from the ipsilateral eye. The name 'dorsal lateral geniculate body' has been used for only one of these nuclei, which projects to the visual cortex in turtles (Hall *et al.*, 1977) but apparently not in lizards and snakes. Further results are needed to clarify the details of the retinothalamotelencephalic visual pathway in reptiles.

Organization of the LGNd in mammals

Monotremes

In the echidna and platypus, two thalamic nuclei receive retinal input and these have been designated LGNa and LGNb by Campbell & Hayhow (1971, 1972). These nuclei receive a contralateral retinal input and a much smaller ipsilateral retinal input. Retrograde cell degeneration studies (Welker & Lende, 1980) and horseradish peroxidase studies (Neylon, personal communication) in the echidna show that LGNb projects to the visual cortex.

Marsupials

About 250 species of living marsupials are known (Kirsch & Calaby, 1977) and most belong to two orders: Diprotodonta (Australian kangaroos, possums, wombats and koala) and Polyprotodonta (Australian carnivorous marsupials: dasyurids and bandicoots; and American didelphid opossums). LGNd cytoarchitecture has been examined in only about 30 species but representatives of most marsupial families are included in this number. The marsupial LGNd may be non-laminated or may contain two, three, four or more cell layers. A variety of patterns of retinogeniculate input also occurs in the 19 species which have been examined.

Cytoarchitecture of the LGNd

In most marsupials, the LGNd consists of an α and a β segment. The α segment is adjacent to the optic tract and contains one or more layers of densely packed cells. The β segment forms the medial part of the nucleus, is usually not subdivided into cell layers, and has a sparser cell density. Exceptions are the Tasmanian devil and bandicoots, in which no separation of the LGNd into α and β segments is apparent (Sanderson, Pearson & Haight, 1979). In other marsupials, the number of cell layers in the α segment ranges from one in the Tasmanian forest wombat to four in some Australian possums. Where a cell layer accommodates retinal input from both eyes and the retinal input is segregated, sublaminae may form (Sanderson & Pearson, 1981; Sanderson, Haight & Pettigrew, 1984) and increase the number of cell layers in the binocular part of the nucleus. In general, the cell layers are not separated by interlaminar zones, so the identification of cell layers is more difficult in marsupials than in carnivores or primates. Fig. 13.1 shows the LGNd in the marsupial mouse *Sminthopsis murina*.

Retinogeniculate organization

In the polyprotodont marsupials, zones of binocular interaction are a common feature of LGNd

---

Fig. 13.1. The LGNd in the marsupial mouse *Sminthopsis murina*. Frontal section, thionin stain. Note the two layers of single cells in the α segment adjacent to the optic tract. Photography by J. R. Haight from the comparative neurology collection by J. I. Johnson.

Fig. 13.2. Contralateral retinal input to the top half of the LGNd in the southern native cat *Dasyurus viverrinus*. Frontal section, [³H]-leucine autoradiography. The ipsilateral retinal input (not illustrated) overlaps the contralateral input substantially.

Fig. 13.3. Contralateral retinal input to the LGNd in the Tammar wallaby *Macropus eugenii*. Frontal section, [³H]-leucine autoradiography. Note the four contralateral terminal bands and the gaps in the label corresponding to the ipsilateral input. Photography by J. D. Pettigrew.

Fig. 13.4. Contralateral retinal input to the LGNd in the brushtailed possum *Trichosurus vulpecula*. Frontal section, [³H]-proline autoradiography. Note the three contralateral terminal bands and the gaps in the label corresponding to the ipsilateral input (cf. Sanderson, Pearson & Dixon, 1978).

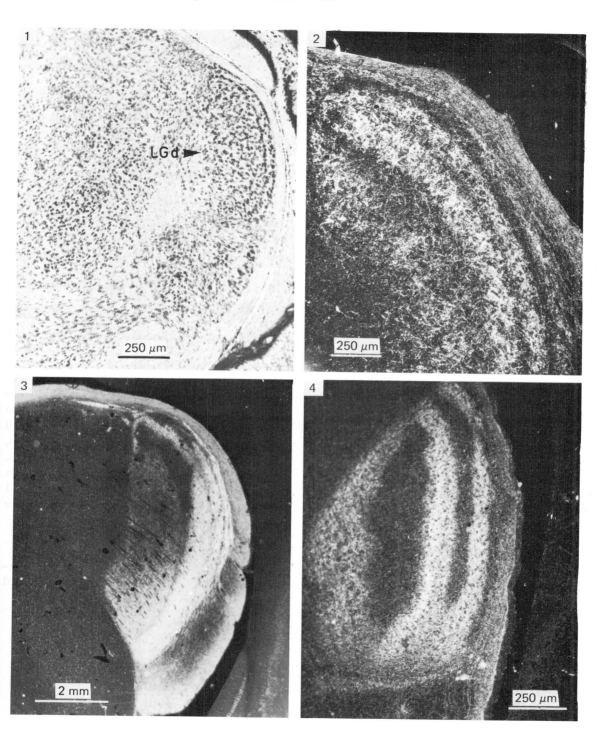

organization. In the three Australian dasyurid marsupials and three American opossums which have been studied (Lent, Cavalcante & Rocha-Miranda, 1976; Royce, Ward & Harting, 1976; Sanderson & Pearson, 1977; Sanderson *et al.*, 1979), the LGNd has areas which receive retinal input from both eyes, as well as areas receiving only a contralateral retinal input and, in some cases, other areas receiving only an ipsilateral retinal input. Fig. 13.2 shows the contralateral retinal input to the LGNd in the southern native cat *Dasyurus viverrinus*. In the bandicoots, the retinal input from the two eyes appears to be segregated in the LGNd. The extensive binocular interaction which is seen in the LGNd in some polyprotodonts has not been described in placental mammals and is possibly a primitive arrangement, as it seems that binocular integration in the visual system may be occurring at the LGNd in some polyprotodonts, whereas it occurs at the cortex in other mammals.

In the diprotodont marsupials, the retinal input to the LGNd consists typically of alternating inputs from the two eyes, with occasional regions where two contralateral or ipsilateral layers are side by side. Where the cell layers in the LGNd can be readily identified, it is apparent that each of the terminal bands of retinal input lies within a cell layer or sublayer. The total number of terminal bands varies, from five in the ringtailed possum, two ipsilateral and three contralateral (Pearson, Sanderson & Wells, 1976), to 11 in the parma wallaby, six ipsilateral, and five contralateral, and there are generally more ipsilateral than contralateral terminal bands. Figs 13.3 and 13.4 show the contralateral terminal bands in the LGNd of a Tammar wallaby, with four, and a brushtailed possum, with three.

Placental mammals
There are 17 orders of living placental mammals with a total of about 4000 species (Walker, 1975). The cytoarchitecture of the LGNd has been studied in about 120 species and retinogeniculate organization in about 50 species. No information is available for three of the 17 placental orders: Tubulidentata (aardvark), Proboscidaea (elephants), and Sirenia (manatees and dugongs). The general principles of organization of the LGNd in mammals are well known; a detailed exposition of these principles is provided in the review by Kaas, Guillery & Allman (1972).

Insectivora, e.g. moles, shrews, hedgehogs. LGNd cytoarchitecture has been studied in 12 species and retionogeniculate connections in five species. The LGNd is not subdivided into separate cell laminae, however the ipsilateral and contralateral retinal inputs to the LGNd are segregated; in hedgehogs there are usually two patches of ipsilateral input and in moles the retinal input to the LGNd is probably only from the contralateral eye (Campbell, 1972).

Dermoptera – gliding lemurs. LGNd cytoarchitecture has been studied in one of the two species, *Cynocephalus volans*. The LGNd has six well-defined cell layers, separated by interlaminar zones (Kaas *et al.*, 1978).

Chiroptera – bats. There are two suborders: the Megachiroptera, or fruit bats, which have large eyes and a well-developed visual system; and the Microchiroptera, or insectivorous bats, which are mostly nocturnal echolocators with small eyes and poorly developed visual systems. In the three microchiropterans which have been studied, the LGNd is undifferentiated into layers and receives a retinal input predominantly from the contralateral eye (Crowle, 1974; Pentney & Cotter, 1976; Suthers & Braford, 1980). In *Pteropus* and *Rousettus*, the megachiropterans which have been studied, the LGNd consists of three well-defined cell laminae, together with a superficial (s) layer adjacent to the optic tract, and a medial (m) cell region. In *Pteropus*, each of the three cell layers receives a contralateral retinal input to the outer part of the layer and an ipsilateral input to the inner part (Crowle, 1974; Pentney & Cotter, 1981). There is an ipsilateral input to the s layer also. In *Rousettus*, the retinogeniculate organization is slightly different: the first and third cell layers receive segregated inputs from the two eyes and the superficial layer receives an ipsilateral input as in *Pteropus*; however, the second cell layer receives only a contralateral retinal input and the medial cell region receives an input from both eyes (Suthers & Braford, 1980; Cotter, 1981).

Primates, e.g. lemurs, monkeys, apes, man. The organization of the LGNd in primates has been studied extensively: cytoarchitecture has been described in about 40 species and retinogeniculate connections in about 20 species, including man. A comprehensive account of LGNd organization in primates has recently been provided by Kaas *et al.* (1978). Most classifications recognize two suborders of primate: Prosimii – tree shrews, lemurs, lorises and tarsiers; and Anthropoidea – monkeys, apes and man. Most of the primates have a laminated LGNd with four, six or more layers of cells, each of which receives input from the contralateral or the ipsilateral eye, but not from both eyes.

Tree shrews. In the pen-tailed tree shrew *Ptilocerus lowii*, the LGNd is rather simple and consists of one or two cell layers (Simmons, 1979). In the tree shrews *Tupaia glis* and *T. minor*, the LGNd has five or six well-defined cell layers separated by interlaminar zones. Layers 1 (the most medial) and 5 of the LGNd receive an ipsilateral retinal input and layers 2, 3, 4 and 5 (the most lateral) receive a contralateral retinal input in *T. glis* (Casagrande & Harting, 1975; Hubel, 1975). The LGNd layers do not appear to be paired as they are in most other primates (Kaas *et al.*, 1978). Note that the convention for numbering of layers is the reverse of that used for the LGNd in other primates.

Tarsiers. The LGNd consists of four layers, two parvocellular layers and two magnocellular layers adjacent to the optic tract (Kaas *et al.*, 1978).

Lemurs. The LGNd consists of six layers, two magnocellular and four parvocellular (see Simmons, 1980) for a summary). The pattern of retinogeniculate connections has been studied in one species, the dwarf lemur *Cheirogaleus* (Hassler, 1966). Layers 1 (the most lateral, next to the optic tract), 5 and 6 receive a contralateral retinal input and layers 2, 3 and 4 receive an ipsilateral retinal input.

Lorises and galagos. The LGNd is similar to that of lemurs: there are usually six complete layers and an incomplete superficial (s) layer next to the optic tract. Of the six complete layers, two are magnocellular (layers 1 and 2), two are parvocellular (layers 3 and 6) and two (layers 4 and 5) have smaller cells and have been termed 'koniocellular' layers by Kaas *et al.* (1978). The retinogeniculate connections are similar to those of lemurs, but in addition there may be an ipsilateral projection to the superficial (s) layer (Kaas *et al.*, 1978). At least 90% of the neurons in the LGNd, including those in the superficial layer, are relay neurons which project to the visual cortex (Norden, 1979). In addition to their obvious size differences, the cells in the magnocellular, parvocellular and koniocellular layers have distinctive receptive field properties. Thus Norton & Casagrande (1982) found that the magnocellular layers of the galago LGNd contained many y-like cells, the parvocellular layers contained many x-like cells and

the koniocellular layers contained a mixed population of cells, some of which were w-like cells.

New World monkeys. The LGNd in New World monkeys consists of one or two superficial (s) layers adjacent to the optic tract, two magnocellular layers, and a parvocellular region which may appear cytoarchitectonically as a single parvocellular mass, as two parvocellular layers, or as four parvocellular layers (Kaas *et al.*, 1978). Autoradiographic studies (Tigges & O'Steen, 1974; Kaas *et al.*, 1978; Spatz, 1978) show that the ipsilateral eye projects to one s layer and one magnocellular layer, and the contralateral eye projects also to one s layer and one magnocellular layer. The parvocellular region receives an input from the two eyes which varies between species and also within species (Spatz, 1978) but which may be interpreted as a single ipsilateral parvocellular layer and a single contralateral parvocellular layer, together with variable leaflets which Kaas *et al.* (1978) regard as being derived from one or other of the basic two parvocellular layers.

Most of the neurons in the parvocellular and magnocellular layers are relay neurons which project to the visual cortex (Norden & Kaas, 1978) and many of the neurons in the s layers also project to the visual cortex. The parvocellular neurons have been classified physiologically as x cells and the magnocellular neurons as y cells (Sherman *et al.*, 1976).

Old World monkeys. Most investigators have regarded the LGNd in the macaques and the vervet monkey *Cercopithecus aethiops* as being composed of two magnocellular layers, four parvocellular layers, and an s layer lying between the magnocellular layers and the optic tract. It is clear, however, that there are not four complete parvocellular layers (cf. Malpeli & Baker, 1975) and Kaas *et al.* (1978) have proposed that the LGNd in these animals should be considered as having two magnocellular layers and two parvocellular layers which partly subdivide to form four leaflets and sometimes further subleaflets. The lamination of the LGNd in the baboon *Papio ursinus* is apparently even more complex, since Campos-Ortega & Hayhow (1970) have observed nine laminae. Autoradiographic studies (Hendrickson, Wilson & Toyne, 1970; Rakic, 1976; Hubel, Wiesel & LeVay, 1977; Kaas *et al.*, 1978; Gross & Hickey, 1980; Nakagawa *et al.*, 1980) have revealed a segregated input from the two eyes to the s layer, a contralateral input to one magnocellular layer and two parvocellular leaflets, and an ipsilateral input to one magnocellular layer and two parvocellular leaflets. In the commonly used number terminology, layers 1 (magnocellular), 4 and 6 (parvocellular) are contralaterally innervated, and layers 2 (magnocellular), 3 and 5 (parvocellular) are ipsilaterally innervated. Most of the cells in all layers are relay neurons (Norden & Kaas, 1978) and there is a segregation of functional properties in different layers of the LGNd, as is the case in a number of other species. The parvocellular layers contain x-like cells which may be colour-coded and the magnocellular layers contain y-like cells which are not colour-coded (Dreher, Fukada & Rodieck, 1976). In addition, cells in the dorsal pair of parvocellular layers or leaflets are predominantly ON centre and those in the ventral pair are mostly OFF centre (Schiller & Malpeli, 1978).

Apes and man. The LGNd in gibbons has four layers, two magnocellular and two parvocellular, with an ipsilateral retinal input to one magnocellular and one parvocellular layer, and a contralateral retinal input to one magnocellular and one parvocellular layer (Chacko, 1955; Kanagasuntheram, Krishnamurti & Wong, 1969). In the related siamang there is a partial subdivision of the magnocellular layers into four, in parts of the the LGNd (Kanagasuntheram *et al.*, 1969).

In the orangutan and chimpanzee, the LGNd has six layers in parts of the nucleus (Balado & Franke, 1937) and the autoradiographic studies (Tigges, Bos & Tigges, 1977) show a similar arrangement of retinal projections to that seen in the macaque. The retinal input to the human LGNd is again similar, but there is a lot of variation in the cytoarchitectonic structure of the LGNd, significantly more than is seen in cats or monkeys (Hickey & Guillery, 1979). There are two magnocellular layers and there may be two, four or six parvocellular leaflets in different parts of the nucleus, with only two parvocellular layers containing a complete representation of the visual field. Some human LGNd shows a subdivision of the parvocellular layers into four and others into six leaflets.

Edentata – sloths, anteaters, armadillos. LGNd cytoarchitecture has been examined in three species, the anteaters *Tamandua tetradactyla* and *Myrmecophaga jubata* (Kaelber, 1966) and the armadillo *Dasypus novemcinctus* (Papez, 1932). The nucleus is not subdivided into cellular laminae.

Lagomorpha – pikas, rabbits, hares. In rabbits and hares the LGNd is subdivided into α and β segments (Rose & Malis, 1965; Giolli & Guthrie, 1969; Brauer Schober & Winkelmann, 1978). The α segment is large and densely packed with cells; the small β segment is located medially and has sparsely packed cells. There is a small ipsilateral retinal input to a medial part of the α segment and a contralateral retinal input to the remainder of the α segment and to the β segment (Sanderson, 1975; Takahashi, Hickey & Oyster, 1977).

Rodentia, e.g. squirrels, cavies, mice, rats, gophers beavers, hamsters, porcupines. There are over 200 species of living rodents including some 570 different *Rattus* species. LGNd organization has been studied in 14 rodent species.

Squirrels. The LGNd is often subdivided into three cell laminae (Kaas *et al.*, 1972; Abplanalp, 1974). In the grey squirrel *Sciureus carolinensis*, lamina 1 (the most medial) receives contralateral retinal input, lamina receives ipsilateral retinal input, and lamina 3 (nearest the optic tract) has three layers of retinal terminals There is a contralateral retinal input to 3a and 3c and an ipsilateral retinal input to 3b (Kaas *et al*, 1972 Weber, Casagrande & Harting, 1977). In the ground squirrels of the genus *Citellus*, laminae 1 and 3 receive contralateral retinal input, and lamina 2 ipsilateral retinal input (Tigges, 1970; Abplanalp, 1974). In the flying squirrel there is again a slight variation in the pattern of retinogeniculate input: laminae 0, 2 and receive ipsilateral retinal input, and laminae 1 and receive contralateral retinal input (Tigges, 1970).

Hamsters and guinea-pigs. The LGNd has α and segments similar to those described in the rabbit (Giolli & Creel, 1973; Frost, So and Schneider, 1979). There is a small ipsilateral projection to the medial edge of the α segment and a contralateral retinal projection the remainder of the α segment and to the β segment

Rats and mice. The LGNd in the commonly used laboratory rat (*Rattus norvegicus*) and mouse (*Mus musculus*) has no obvious cytoarchitectonic subdivision In some Australian rodents, *Notomys alexis* and *Pseudomys australis*, the LGNd is subdivided into two parts which are similar to the α and β segments described in other rodents (Mayner, Pearson Sanderson, 1980.) Autoradiographic studies show that the retinogeniculate organization is relatively simple with an ipsilateral retinal input to the medial edge of the nucleus and a contralateral retinal input to the remainder of the nucleus (Hickey & Spear, 1976 LeVail, Nixon & Sidman, 1978; Mayner *et al.*, 1980

Compoint-Monmignaut, 1983). Receptive field studies reveal that relay neurons in the rat LGNd fall into two classes, y-like cells and w-like cells (Fukuda *et al.*, 1979; Hale, Sefton & Dreher, 1979).

Cetacea – whales and dolphins. The LGNd in the dolphin *Tursiops truncatus* is unlike that of other mammals. It does not have a distinct lamination but consists of irregular layers of cells separated by fibre bundles (Kruger, 1959) and probably receives only a contralateral retinal input (Jacobs, Morgane & McFarland, 1975). Zvorykin (1976) has described a somewhat different cytoarchitectonic organization in the LGNd of the porpoise *Phocoena phocoena*, which apparently has a much smaller LGNd.

Carnivora e.g. dogs, cats, raccoons, weasels, hyenas, pandas, mongooses, bears. There are seven families in this order of about 250 species, and all of them have LGNds with a well-defined pattern of lamination, similar to that observed in the cat.

Felidae – cats. Cytoarchitectonically, it is possible to recognize the magnocellular layers A and A_1, layer C, which has a dorsal magnocellular part (Cm) and a ventral parvocellular part (Cp), and the parvocellular layers C_1, C_2 and C_3. Well-defined interlaminar zones separate A from A_1 and A_1 from C. Contralateral retinal fibres terminate in laminae A, C and C_2 and ipsilateral retinal fibres in laminae A_1 and C_1 (Guillery 1970; Hickey & Guillery, 1974). The proportion of interneurons in the LGNd appears to be 20–25% (LeVay & Ferster, 1979) and they are located both within the nucleus and above it, in the perigeniculate nucleus (Dubin & Cleland, 1977). The principal or relay cells of the LGNd divide into a number of different classes: x or brisk-sustained, y or brisk-transient and w, a physiologically fixed population comprising four to six different receptive field types (Cleland *et al.*, 1976; Wilson, Rowe & Stone 1976). The x and y cells are located primarily in layers A and A_1 and magnocellular C (Cm), and w cells are located in the parvocellular C layers (Cp, C_1, C_2). The x, y and w cells each project to different layers of striate cortex (Leventhal, 1979). In addition to its laminated part, the LGNd in carnivores has a medial interlaminar nucleus (MIN) which receives input from both eyes (Guillery *et al.*, 1980) and contains y cells and w cells (Rowe & Dreher, 1982) which project to different layers of the visual cortex (Leventhal, 1979).

The LGNd in the 'big cats' – lion, puma,

leopard, tiger – and a smaller cat, the jaguarundi, has a very similar organization (Kaas, Guillery & Allman, 1973; Kaelber, Yarmat & Afifi, 1982).

Canidae – dogs. The LGNd organization is very similar to that observed in cats, except that the layers may be more sharply defined (Kaas *et al.*, 1972; Sanderson, 1974). Autoradiographic studies (Hickey & Guillery, 1974) show a contralateral input to laminae A, C and C_2, and an ipsilateral retinal input to laminae A_1 and C_1 in the fox. Our observations show that the LGNd in the coyote *Canis latrans* is similar to that of other canids.

Procyonidae – raccoons, coatimundis, kinkajous, pandas. Studies of the LGNd in four procyonids (raccoon, coatimundi, kinkajou and ringtailed cat: Sanderson, 1974) show slightly more variation than is observed in the cat family. In the Procyonidae, the LGNd usually shows some rotation from the orientation found in the cat, so that the layers tend to run vertically. In addition there is a splitting or duplication of laminae A and A_1 which is faintly suggested in the raccoon, quite obvious in the coatimundi and not present in the kinkajou or ringtailed cat.

Mustelidae – weasels. LGNd cytoarchitecture has been observed in the skunk and weasel (Sanderson, 1974; Brauer *et al.*, 1978) and retinogeniculate projections in the ferret and mink (Sanderson, 1974; Guillery and Oberdorfer, 1977; Linden, Guillery & Cucchiaro, 1981). In the skunk, the LGNd is small and has single A and A_1 laminae. In the ferret, there is a suggestion of a subdivision of laminae A and A_1, and in the mink and weasel, the subdivision of laminae A and A_1 is well defined. In the mink, the anterior leaflets of laminae A and A_1 contain neurons with ON-centre receptive fields, and the posterior leaflets contain neurons with OFF-centre receptive fields (LeVay & McConnell, 1982). In the mustelids, the LGNd tends to have layers which run vertically, as in the procyonids.

Ursidae (bears), Viverridae (e.g. civets, mongooses) and Hyaenidae (hyenas). Our observations of the LGNd in a polar bear, a yellow mongoose (Fig. 13.5), and a spotted hyaena suggest that the organization of the LGNd in these orders is probably similar to that seen in other carnivores. In the polar bear LGNd there is an obvious subdivision of one of the A laminae, as is observed in some procyonids and mustelids.

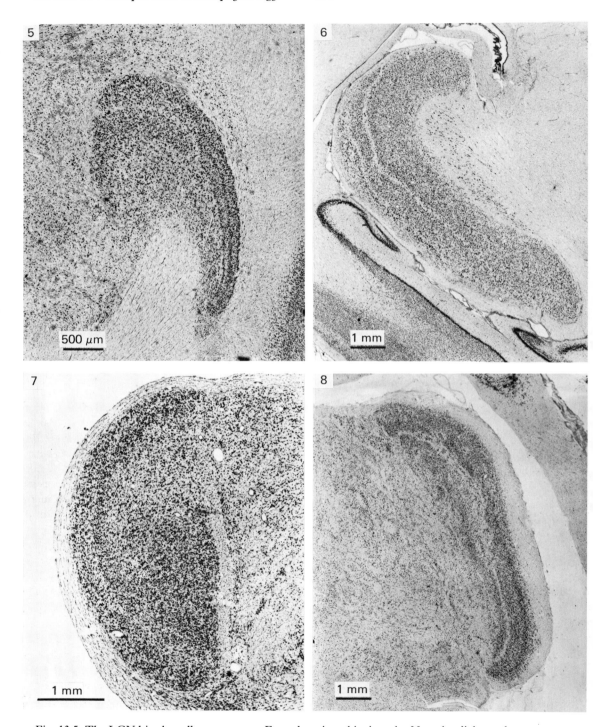

Fig. 13.5. The LGNd in the yellow mongoose. Frontal section, thionin stain. Note the slight tendency to a subdivision of lamina A, the most dorsal lamina, into two. Photographic assistance from T. P. Stewart, from the University of Wisconsin, Department of Neurophysiology Comparative Neurology Collection (also Figs 13.6 and 13.8).

Fig. 13.6. The LGNd in the fur seal. Frontal section, thionin stain.

Fig. 13.7. The LGNd in the hyrax *Procavia capensis*. Frontal section, thionin stain. Photography by J. R. Haight.

Fig. 13.8. The LGNd in the llama. Frontal section, thionin stain.

Pinnipedia – seals, sea lions, walruses. Our observations of the LGNd in a fur seal (Fig. 13.6) show that it is very similar to the LGNd in cats and confirm Kornyey's early (1927) notes.

Hyracoidea – hyraxes. The LGNd is large and contains three well-defined cell laminae: two relatively thin outer layers and a broad inner layer (Ibrahim & Shanklin, 1941) (Fig. 13.7).

Perissodactyla – odd-toed ungulates (hoofed mammals) – horses; and Artiodactyla – even-toed ungulates – pigs, sheep, cattle, goats, deer, camels. The LGNd is very similar in these two orders. LGNd organization has been examined in both normal cytoarchitectonic studies and fibre degeneration studies. The LGNd consists of a least three layers and a medial interlaminar nucleus and its structure is more complex in wild forms (deer, llama: Kanaseki, Sakamoto & Sakai, 1957) than in domesticated animals (pig, sheep, goat, cattle, camel: Solnitzky, 1938; Rose, 1942; Kanagasuntheram, Kanan & Krishnamurti, 1970). In the domestic species the LGNd has two contralateral layers sandwiching an ipsilateral layer (Nichterlein & Goldby, 1944; Cummings & de LaHunta, 1969; Campos-Ortega, 1970; Karamanlidis & Magras, 1972, 1974). Fig. 13.8 shows the cytoarchitecture of the LGNd in a llama.

Pholidota – pangolins. Our observations of the LGNd in the African tree pangolin *Manis tricuspis* are that the LGNd is not obviously laminated in this species. (The material was kindly provided by Drs H. Stephan and J. Nelson.)

Function of lamination in the LGNd

A variety of evidence (presented above) shows that the lamination of the LGNd serves to segregate (1) cells receiving input from the two eyes and (2) cells receiving input from different classes of retinal ganglion cells. Sometimes the segregation is into x, y and w functional classes, sometimes into ON-centre and OFF-centre receptive field types and sometimes into colour-sensitive and broad-band types.

Some other suggestions for the function of lamination are that interlaminar zones may be important for corticogeniculate interactions (Guillery, 1979) and that the formation of leaflets in the primate LGNd may occur in order to limit laminar thickness

so that binocular interaction can occur (Kaas *et al.*, 1978).

Correlations of laminar structure with the behaviour of an animal have been sought, mostly in support of the hypothesis that animals with good eye–hand coordination or good depth perception or arboreal lifestyle or predatory habits should have laminated LGNds. Correlations which do appear to exist are that carnivorous placental mammals and primates, both of which require good binocular depth perception, do indeed have highly laminated LGNds, as do some flying bats and gliding squirrels, lemurs and possums. However, well-laminated LGNds can be found also in kangaroos and deer and llamas, herbivores living both in open plains and in forests. An interesting contrast appears in the LGNd of two aquatic carnivorous mammals – seals, which have a highly laminated LGNd, and dolphins, whose LGNd probably has little lamination, though this may be derived secondarily as a result of the totally crossed optic pathway.

In summary, there is strong evidence that lamination of the LGNd separates different classes of retinal input in the LGNd. The argument that laminar pattern is related to the lifestyle of the animal can be made but does not seem to fit all cases.

Evolutionary patterns of LGNd structure

A number of authors have proposed schemes suggesting how the LGNd of different classes of mammals might be derived (for example Brauer *et al.*, 1978; Kaas *et al.*, 1978; Suthers & Braford, 1980), starting with a knowledge of patterns of LGNd lamination in existing mammals and schemes of evolutionary relationships of mammals, and assuming a fairly simple LGNd in prototype mammals. We have attempted a similar synthesis here for all mammals, with considerable borrowing from the previous syntheses.

The first marsupials and placental mammals appear in the fossil record in the late Cretaceous period, more than 65 million years ago. The ancestral-marsupials probably resembled the modern didelphid opossums and the ancestral placental mammals were probably small nocturnal insectivores with large, laterally directed eyes (Romer, 1966; Cartmill, 1972). From our knowledge of living insectivores and opossums, we can speculate that the early mammals probably also had a fairly simple plan of LGNd organization, with no distinct cell laminae and a single patch of ipsilateral retinal input to the nucleus, either superimposed on the contralateral retinal input (marsupials) or surrounded by the contralateral retinal input (placental mammals).

The subsequent marsupial radiation produced two major patterns of LGNd organization: the poly-protodont pattern, with significant overlap of the retinogeniculate fibres from the two eyes, and the diprotodont pattern, in which the inputs from the two eyes segregate into separate cell laminae and sub-laminae. In both the polyprotodonts and diprotodonts there are species which have well-laminated LGNds with several cell layers in the α division of the nucleus; probably the most extensive lamination of the LGNd is seen in some of the Australian possums, in which there may be four well-differentiated cell layers in the α division. The number of terminal bands of retinal input varies widely among the different species with up to 11 bands of retinal input in some kangaroo species. In the marsupials, it is difficult to find much correlation between the lamination of the LGNd and the lifestyle of the animal, since highly laminated LGNds are to be found in most diprotodonts, including the plains-dwelling hairy-nosed wombat, the kangaroos of the plains, open forest and thick scrub, and the tree-dwelling possums. The diprotodonts are generally believed to be more advanced than the polyprotodonts (Kirsch, Johnson & Switzer, 1983). There is no evidence for any nocturnal/diurnal dichotomy of LGNd structure in marsupials, since most of the marsupials studied thus far are nocturnal or crepuscular forms; only a few diurnal species are known – the numbat or banded anteater, and the rock wallabies - and the LGNds of these have not been examined.

The radiation of placental mammals produced a wide range of patterns of LGNd organization also, including (1) forms with a primitive LGNd organiza-tion with no differentiation into separate cell layers and only one or two patches of ipsilateral input – insecti-vores, most rodents, lagomorphs, microchiropteran bats, edentates and cetaceans – and (2) forms with a laminated LGNd, usually with several cell layers for the retinal inputs of each eye – primates, gliding lemurs, megachiropteran bats, squirrels, carnivores, seals, hyraxes and ungulates. The species with primitively organized LGNds possibly include some in which the primitive condition is a secondary state arising from the loss of a laminated LGNd; cetaceans may be a case in point. However, the evolutionary tree of placental mammals cannot be drawn with enough certainty to verify such speculations (Kirsch *et al.* 1983). In a similar fashion, it seems possible that there has been independent evolution of lamination in the LGNd several times, since it occurs in one group of rodents – the squirrels, in primates and their close relatives megachiropteran bats and flying lemurs, and in the ferungulates – ungulates, hyraxes, carnivores and seals. The acquisition of a laminated LGNd in placental mammals is usually associated with life in the trees (squirrels, flying lemurs, fruit bats, primates) or as a predator (carnivores and seals), but the function of the lamination is probably best understood as a means of segregating different classes of information from the ganglion cells in the retina. Within the primates, Hassler (1966) could see some evidence for a distinction between nocturnal and diurnal forms: the magnocellular layers were relatively larger in the nocturnal species and the parvocellular layers were relatively larger in the diurnal species. However, it should be noted that some diurnal primates may have small parvocellular layers (the titi monkey: Kaas *et al.* 1972) or large magnocellular layers (the siamang: Kanagasuntheram *et al.*, 1969); thus there is not a consistent day/night dichotomy of LGNd organization in primates.

Acknowledgements

This work has been made possible by grants from Australian Research Grants Scheme and the Flinders University Research Budget to K. J. Sanderson, and by grants from the U.S.P.H.S. to Wally Welker and Clinton Woolsey, who collected much of the comparative brain collection at the University of Wisconsin, Madison. I am indebted to my colleagues of the last 12 years who have assisted with my comparative studies of the lateral geniculate nucleus: Ray Guillery, Jon Kaas, John Allman, Lyn Pearson, Pam Stainer, John Haight, Jack Pettigrew, Sheila Crewther and David Crewther. Expert photographic assistance has been provided over a number of years by Terry Stewart at the University of Wisconsin and by Reg Brook and the staff of the photographic unit at Flinders University, and secretarial assistance was provided by Jo Hill and Joy Noakes. Finally I acknowledge the help of Professor P.O. Bishop who initiated my work on the lateral geniculate nucleus.

References

Abplanalp, P. (1974) Topography of retinal efferent connections in sciurids. *Brain, Behav. Evol.*, 9, 333–75

Balado, M. & Franke, E. (1937) Das corpus geniculatum externum. Berlin: Springer Verlag

Brauer, K., Schober, W. & Winkelmann, E. (1978) Phylogenetical changes and functional specializations in the dorsal lateral geniculate nucleus (dLGN) of the mammals. *J. Hirnforsch.*, 19, 177–87

Campbell, C. B. G. (1972) Evolutionary patterns in mammalian diencephalic visual nuclei and their fiber connections. *Brain, Behav. Evol.*, 6, 218–36

Campbell, C. B. G. & Hayhow, W. R. (1971) Primary optic pathways in the echidna *Tachyglossus aculeatus*: an experimental degeneration study. *J. Comp Neurol.*, 143, 119–36

Campbell, C. B. G. & Hayhow, W. R. (1972) Primary optic pathways in the duckbill platypus *Ornithorhynchus anatinus*: an experimental degeneration study. *J. Comp. Neurol.*, 145, 195–208

Campos-Ortega, J. A. (1970) The distribution of retinal fibers in the brain of the pig. *Brain Res.*, 19, 306–12

Campos-Ortega, J. A. & Hayhow, W. R. (1970) A new lamination pattern in the lateral geniculate nucleus of primates. *Brain Res.*, 20, 335–9

Cartmill, M. (1972) Arboreal adaptations and the origin of the order primates. In *Functional and Evolutional Biology of Primates*, Ed. R. Tuttle, pp. 97–122. Chicago: Aldine-Atherton

Casagrande, V. A. & Harting, J. K. (1975) Transneuronal transport of tritiated fucose and proline in the visual pathways of tree shrew *Tupaia glis*. *Brain Res.*, 96, 367–72

Chacko, L. W. (1955) The lateral geniculate body in gibbon (*Hylobates hoolock*). *J. Anat. Soc. India*, 4, 69–81

Cleland, B. G., Dubin, M. W. & Levick, W. R. (1971) Simultaneous recording of input and output of lateral geniculate neurons. *Nature*, 23, 191–2

Cleland, B. G., Levick, W. R., Morstyn, R. & Wagner, H. G. (1976) Lateral geniculate relay of slowly conducting retinal afferents to cat visual cortex. *J. Physiol.*, 255, 299–320

Compoint-Monmignaut, C. (1983) Organisation du système visuel primaire de deux rongeurs *Arvicola terrestris* et *Meriones shawi*. *J. Hirnforsch.*, 24, 43–55

Cotter, J. R. (1981) Retinofugal projections of an echolocating megachiropteran. *Am. J. Anat.*, 160, 159–74

Crowle, P. K. (1974) *Experimental investigation of retinofugal connections to the diencephalon and midbrain of Chiroptera*. Ph.D. thesis, Indiana University, Bloomington

Cummings, J. F. & de Lahunta, A. (1969) An experimental study of the retinal projections in the horse and sheep. *Ann. N.Y. Acad. Sci.*, 167, 293–318

Dreher, B., Fukada, Y. & Rodieck, R. W. (1976) Identification, classification and anatomical segregation of cells with x-like and y-like properties in the lateral geniculate nucleus of old-world primates. *J. Physiol.*, 258, 433–52

Dubin, M. W. & Cleland, B. G. (1977) Organization of visual inputs to interneurons of lateral geniculate nucleus of the cat. *J. Neurophysiol.*, 40, 410–27

Frost, D. O., So K-F. & Schneider, G. E. (1979) Postnatal development of retinal projections in Syrian hamsters: a study using autoradiographic and anterograde degeneration techniques. *Neuroscience*, 4, 1649–77

Fukuda, Y., Sumitomo, I., Sugitani, M. & Iwama, K. (1979) Receptive-field properties of cells in the dorsal part of the albino rat's lateral geniculate nucleus. *Jap. J. Physiol.*, 29, 283–307

Giolli, R. A. & Creel, D. J. (1973) The primary optic projections in pigmented and albino guinea pigs: an experimental degeneration study. *Brain Res.*, 55, 25–39

Giolli, R. A. & Guthrie, M. D. (1969) The primary optic projections in the rabbit. An experimental degeneration study. *J. Comp. Neurol.*, 136, 99–126

Gross, K. J. & Hickey, T. L. (1980) Abnormal laminar patterns in the lateral geniculate nucleus of an albino monkey. *Brain Res.*, 190, 231–7

Guillery, R. W. (1970) The laminar distribution of retinal fibers in the dorsal lateral geniculate nucleus of the cat: a new interpretation. *J. Comp. Neurol.*, 138, 339–68

Guillery, R. W. (1979) A speculative essay on geniculate lamination and its development. *Prog. Brain Res.*, 51, 403–18

Guillery, R. W., Geisert, E. E., Polley, E. H. & Mason, C. A. (1980) An analysis of the retinal afferents to the cat's medial interlaminar nucleus and to its rostral thalamic extension, the 'geniculate wing'. *J. Comp. Neurol.*, 194, 117–42

Guillery, R. W. & Oberdorfer, M. D. (1977) Study of fine and course retino-fugal axons terminating in the geniculate C laminae and in the medial interlaminar nucleus of the mink. *J. Comp. Neurol.*, 176, 515–26

Hale, P. T., Sefton, A. J. & Dreher, B. (1979) A correlation of receptive field properties with conduction velocity of cells in the rat's retino-geniculo-cortical pathway. *Exp. Brain Res.*, 35, 425–42

Hall, J. A., Foster, R. E., Ebner, F. F. & Hall, W. C. (1977) Visual cortex in a reptile, the turtle (*Pseudemys scripta* and *Chrysemys picta*). *Brain Res.*, 130, 197–216

Hassler, R. (1966) Comparative anatomy of the central visual systems in day- and night-active primates. In *Evolution of the Forebrain*, Ed. R. Hassler & H. Stephan, pp. 419–34. Stuttgart: Thieme

Hendrickson, A., Wilson, M. E. & Toyne, M. J. (1970) The distribution of optic nerve fibers in *Macaca mulatta*. *Brain Res.*, 23, 425–7

Hickey, T. L. & Guillery, R. W. (1974) An autoradiographic study of retinogeniculate pathways in the cat and the fox. *J. Comp. Neurol.*, 156, 239–54

Hickey, T. L. & Guillery, R. W. (1979) Variability of laminar patterns in the human lateral geniculate nucleus. *J. Comp. Neurol.*, 183, 221–46

Hickey, T. L. & Spear, P. D. (1976) Retinogeniculate projections in hooded and albino rats: an autoradiographic study. *Exp. Brain Res.*, 24, 523–9

Hubel, D. H. (1975) An autoradiographic study of the retino-cortical projections in the tree shrew (*Tupaia glis*). *Brain Res.* 96, 41–50

Hubel, D. H., Wiesel, T. N. & LeVay, S. (1977) Plasticity of ocular dominance columns in monkey striate cortex. *Phil. Trans. R. Soc. Lond. B.*, **278**, 377–409

Ibrahim, M. & Shanklin, W. M. (1941) The diencephalon of the coney, *Hyrax syriaca. J. Comp. Neurol.*, **75**, 427–85

Jacobs, M. S., Morgane, P. J. & McFarland, W. L. (1975) Degeneration of visual pathways in the bottlenose dolphin. *Brain Res.*, **88**, 346–52

Kaas, J. H., Guillery, R. W. & Allman, J. M. (1972) Some principles of organization in the dorsal lateral geniculate nucleus. *Brain, Behav. Evol.*, **6**, 253–99

Kaas, J. H., Guillery, R. W. & Allman, J. M. (1973) Discontinuities in the dorsal lateral geniculate nucleus corresponding to the optic disc: a comparative study. *J. Comp. Neurol.*, **147**, 163–80

Kaas, J. H., Huerta, M. F., Weber, J. T. & Harting, J. K. (1978) Patterns of retinal terminations and laminar organization of the lateral geniculate nucleus of primates. *J. Comp. Neurol.*, **182**, 517–54

Kaelber, W. W. (1966) Nuclear configuration of the diencephalon of *Tamandua tetradactyla* and *Myrmecophaga jubata. J. Comp. Neurol.*, **128**, 133–70

Kaelber, W. W., Yarmat, A. J. & Afifi, A. K. (1982) A comparison of the diencephalic and subcortical telencephalic areas of the brain of *Felis jaguarondi* and *Felis domestica. J. Hirnforsch.*, **23**, 709–19

Kanagasuntheram, R., Kanan, C. V. & Krishnamurti, A. (1970) Nuclear configuration of the diencephalon of *Camelus dromedarius. Acta Anat.*, **75**, 301–18

Kanagasuntheram, R., Krishnamurti, A. & Wong, W. C. (1969) Observations on the lamination of the lateral geniculate body in some primates. *Brain Res.*, **14**, 623–31

Kanaseki, T., Sakamoto, K. & Sakai, T. (1957) [The comparative anatomical study of the lateral geniculate body in ungulates.] *Repts Second Sect. Anat. Dep. Sch. Med., Tokushima Univ.*, **3**, 79–110. (In Japanese)

Karamanlidis, A. N. & Magras, J. (1972) Retinal projections in domestic ungulates. I. The retinal projections in the sheep and the pig. *Brain Res.*, **44**, 127–45

Karamanlidis, A. N. & Magras, T. (1974) Retinal projections in domestic ungulates. II. The retinal projections in the horse and the ox. *Brain Res.*, **66**, 209–25

Karten, H. J., Hodos, W., Nauta, W. J. H. & Revzin, A. M. (1973) Neural connections of the 'Visual Wulst' of the avian telencephalon. Experimental studies in the pigeon (*Columbia livia*) and owl (*Speotyto cunicularia*). *J. Comp. Neurol.*, **150**, 253–78

Kirsch, J. A. W. & Calaby, J. H. (1977) The species of living marsupials. In *The Biology of Marsupials* Ed. B. Stonehouse & D. Gilmore, pp. 9–26. London: Macmillan

Kirsch, J. A. W., Johnson, J. I. & Switzer, R. C. (1983) Phylogeny through brain traits: the mammalian family tree. *Brain Behav. Evol.*, **22**, 70–4

Kornyey, S. (1927) Zur vergleichenden Morphologie des lateralen Kniehöckers der Säugetiere. *Arb. Neurol. Inst. Wien*, **30**, 93–120

Kruger, L. (1959) The thalamus of the dolphin (*Tursiops truncatus*) and comparison with other mammals. *J. Comp. Neurol.*, **111**, 133–94

LeVail, J. H., Nixon, R. A. & Sidman, R. L. (1978) Genetic control of retinal ganglion cell projections. *J. Comp. Neurol.*, **182**, 399–422

Le Gros Clark, W. E. (1930) The thalamus of *Tarsius. J. Anat.*, **64**, 371–414

Lent, R., Cavalcante, L. A. & Rocha-Miranda, C. E. (1976) Retinofugal projections in the opossum. Anterograde degeneration and radioautographic study. *Brain Res.*, **107**, 9–26

LeVay, S. & Ferster, D. (1979) Proportion of interneurons in the cat's lateral geniculate nucleus. *Brain Res.*, **164**, 304–8

LeVay, S. & McConnell, S. K. (1982) ON and OFF layers in the lateral geniculate nucleus of the mink. *Nature*, **300**, 350–1.

Leventhal, A. G. (1979) Evidence that the different classes of relay cells of the cat's lateral geniculate nucleus terminate in different layers of the striate cortex. *Exp. Brain Res.*, **37**, 349–72

Linden, D. C., Guillery, R. W. & Cucchiaro, J. (1981) The dorsal lateral geniculate nucleus of the normal ferret and its postnatal development. *J. Comp. Neurol.*, **203**, 189–211

Malpeli, J. G. & Baker, F. H. (1975) The representation of the visual field in the lateral geniculate nucleus of *Macaca mulatta. J. Comp. Neurol.*, **161**, 569–94

Mayner, L., Pearson, L. J. & Sanderson, K. J. (1980) An autoradiographic study of visual pathways of the Australian rodents *Notomys alexis, Pseudomys australis* and *Rattus villosissiumus. Aust. J. Zool.*, **28**, 381–93

Nakagawa, S., Konishi, N, Yasui, M. & Kanagawa, R. (1980) The termination of retinofugal fibers in the crab-eating monkey (*Macaca irus*): an autoradiographic study. *Folia Psychiatr. Neurol. Japonica*, **34**, 135–45

Nichterlein, O. E. & Goldby, F. (1944) An experimental study of optic connexions in the sheep. *J. Anat.*, **78**, 59–67

Nixdorf, B. E. & Bischof, H-J. (1982) Afferent connections of the ectostriatum and visual wulst in the zebra finch (*Taeniopygia guttata castanotis* Gould) – an HRP study. *Brain Res.*, **248**, 9–17

Norden, J. J. (1979) Some aspects of the organization of the lateral geniculate nucleus in *Galago senegalensis* revealed by using horseradish peroxidase to label relay neurons. *Brain Res.*, **174**, 193–206

Norden, J. J. & Kaas, J. H. (1978) The identification of relay neurons in the dorsal lateral geniculate nucleus of monkeys using horseradish peroxidase. *J. Comp. Neurol.*, **182**, 707–26

Norton, T. T. & Casagrande, V. A. (1982) Laminar organisation of receptive-field properties in lateral geniculate nucleus of bush baby (*Galago crassicaudatus*). *J. Neurophysiol.*, **47**, 715–41

Papez, J. W. (1932) The thalamic nuclei of the nine-banded armadillo (*Tatusia novemcinta*). *J. Comp. Neurol.*, **56**, 49–104

Pearson, L. J., Sanderson, K. J. & Wells, R. T. (1976) Retinal projections in the ringtailed possum, *Pseudocheirus peregrinus. J. Comp. Neurol.*, **170**, 227–40

Pentney, R.P. & Cotter, J. R. (1976) Retinofugal projections in an echolocating bat. *Brain Res.*, **115**, 479–84

Pentney, R.P. & Cotter, J. R. (1981) Organization of the

retinofugal fibers to the dorsal lateral geniculate nucleus of *Pteropus giganteus. Exp. Brain Res.*, **41**, 427–30

Rakic, P. (1976) Prenatal genesis of connections subserving ocular dominance in the rhesus monkey. *Nature*, **261**, 467–71

Romer, A. S. (1966) *Vertebrate Palaeontology.* (3rd edn) Chicago: University of Chicago Press

Rose, J. E. (1942) The thalamus of the sheep: cellular and fibrous structure and comparison with pig, rabbit and cat. *J. Comp. Neurol.*, **77**, 469–523

Rose, J. E. & Malis, L. I. (1965) Geniculo-striate connections in the rabbit. II. Cytoarchitectonic structure of the striate region and of the dorsal lateral geniculate body; organization of the geniculo-striate projections. *J. Comp. Neurol.*, **125**, 121–40

Rowe, M. H. & Dreher, B. (1982) Retinal w-cell projections to the medial interlaminar nucleus in the cat: implications for ganglion cell classification. *J. Comp. Neurol.*, **204**, 117–33

Royce, G. J., Ward, J. P. & Harting, J. K. (1976) Retinofugal pathways in two marsupials. *J. Comp. Neurol.*, **170**, 391–414

Sanderson, K. J. (1974) Lamination of the dorsal lateral geniculate nucleus in carnivores of the weasel (*Mustelidae*), raccoon (*Procyonidae*) and fox (*Canidae*) families. *J. Comp. Neurol.*, **153**, 239–66

Sanderson, K. J. (1975) Retinogeniculate projections in the rabbits of the albino allelomorphic series. *J. Comp. Neurol.*, **159**, 15–28

Sanderson, K. J., Haight, J. R. & Pettigrew, J. D. (1984) The dorsal lateral geniculate nucleus of macropodid marsupials: cytoarchitecture and retinal projections. *J. Comp. Neurol.*, **224**, 85–106

Sanderson, K. J. & Pearson, L. J. (1977) Retinal projections in the native cat, *Dasyurus viverrinus. J. Comp. Neurol.*, **174**, 347–58

Sanderson, K. J. & Pearson, L. J. (1981) Retinal projections in the hairy-nosed wombat, *Lasiorhinus latifrons* (Marsupialia: Vombatidae). *Aust. J. Zool.*, **29**, 473–81

Sanderson, K. J., Pearson, L. J. & Dixon, P. G. (1978) Altered retinal projections in brushtailed possum, *Trichosurus vulpecula*, following removal of one eye. *J. Comp. Neurol.*, **180**, 841–68

Sanderson, K. J., Pearson, L. J. & Haight, J. R. (1979) Retinal projections in the Tasmanian devil, *Sarcophilus harrisii. J. Comp. Neurol.*, **188**, 335–46

Schiller, P. H. & Malpeli, J. G. (1978) Functional specificity of lateral geniculate nucleus laminae of the rhesus monkey. *J. Neurophysiol.*, **41**, 788–97

Sherman, S. M., Wilson, J. R., Kaas, J. H. & Webb, S. V. (1976)

x and y cells in the dorsal lateral geniculate nucleus of the owl monkey (*Aotus trivirgatus*). *Science*, **192**, 475–6

Simmons, R. M. T. (1979) The diencephalon of *Ptilocercus lowii* (pen tailed tree shrew). *J. Hirnforsch.*, **20**, 69–92

Simmons, R. M. T. (1980) The morphology of the diencephalon in the Prosimii II. The Lemuroidea and Lorisoidea. Part I. Thalamus and metathalamus. *J. Hirnforsch.*, **21**, 449–91

Solnitsky, O. (1938) The thalamic nuclei of *Sus scrofa. J. Comp. Neurol.*, **69**, 121–69

Spatz, W. B. (1978) The retino-geniculo-cortical pathway in *Callithrix*. I. Intraspecific variations in the lamination pattern of the lateral geniculate nucleus. *Exp. Brain Res.*, **33**, 551–63

Suthers, R. A. & Braford, M. R. (1980) Visual systems and the evolutionary relationships of the Chiroptera. In *Proceedings of the Fifth International Bat Research Conference.*, ed. D. E. Wilson & A. L. Gardner, pp. 331–46. Lubbock, Texas: Texas Technical University Press

Takahashi, E. S., Hickey, T. L. & Oyster, C. W. (1977) Retinogeniculate projections in the rabbit: an autoradiographic study. *J. Comp. Neurol.*, **175**, 1–12

Tigges, J. (1970) Retinal projections to subcortical optic nuclei in diurnal and nocturnal squirrels. *Brain Behav. Evol.*, **3**, 121–34

Tigges, J., Bos, J. & Tigges, M. (1977) An autoradiographic investigation of the subcortical visual system in chimpanzee. *J. Comp. Neurol.*, **172**, 367–80

Tigges, J. & O'Steen, K. (1974) Termination of retinofugal fibers in squirrel monkey: a re-investigation using autoradiographic methods. *Brain Res.*, **79**, 489–95

Walker, E. P. (1975) *Mammals of the World.* (3rd edn.) Baltimore: Johns Hopkins

Weber, J. T., Casagrande, V. A. & Harting, J. K. (1977) Transneuronal transport of ^{3}H proline within the visual system of the grey squirrel. *Brain Res.*, **129**, 346–52

Welker, W. I. & Lende, R. A. (1980) Thalamocortical relationships in echidna (*Tachyglossus aculeatus*). In *Comparative Neurology of the Telencephalon*, ed. S. O. E. Ebbesson, pp. 449–81. New York: Plenum Press

Wilson, P. D., Rowe, M. H. & Stone, J. (1976) Properties of relay cells in cat's lateral geniculate nucleus: a comparison of w-cells with x- and y-cells. *J. Neurophysiol.*, **39**, 1193–209

Zvorykin, V. P. (1976) [Specific features of the structural organization of the external geniculate body in the dolphin *Phocaena phocaena* as compared with other dolphins.] *Arkh. Anat. Gistol. Embriol.*, **71**, 58–66. (In Russian, figure legends and abstract in English)

14

Mammalian evolution as seen in visual and other neural systems

JOHN IRWIN JOHNSON

I came to Peter Bishop's department in Sydney to collect brains of Australian mammals and to learn what was going on in visual neuroscience. Largely through Peter's efforts I was able to assemble the world's most extensive collection of monotreme and marsupial neural specimens. He was also indirectly responsible for my meeting John Kirsch. In recent years Dr Kirsch and I, in collaboration with my student Dr Robert Switzer, have used this collection, along with the great collection of placental mammal brains gathered at the University of Wisconsin by Drs Wally Welker and Clinton Woolsey, to examine the record of mammalian evolution as revealed by diversification of brain characteristics.

Phylogeny through brain traits

We first identify a trait that occurs in at least two different states in mammals. These states we then classify as primitive or derived, a risky procedure no matter how it is done (Switzer, Johnson & Kirsch, 1980; Johnson, Kirsch & Switzer, 1982; Kirsch, 1983). One of the more effective (but far from foolproof) ways to do this is the procedure known as out-group comparison. As an example I shall use here a visual trait, that of oil droplets in the cones of retinal receptors. (I am indebted to Robert Rodieck for acquainting me with this character; meeting him was another very fruitful result of my stay in Peter's department.)

Some mammals (most of them Australian) have oil droplets at the distal end of the inner segment of some of their cone receptors (Fig. 14.1). Most other mammals do not. Here is a variation existing within the mammalian nervous system: two contrasting states with possible phylogenetic implications. Which of the two states represents the earlier, primitive condition, and which the later, more evolved, derived condition? We see the cone oil droplets in most of the *out-group* reptiles, wherefrom we believe, from other data, that mammals evolved; and we see them in the birds as well, the other great group of reptilian descendants. This suggests that the presence of cones with oil droplets is the primitive condition, and their absence is the derived state. This belief is strengthened when we find that those mammals with oil droplets also possess other primitive characters, both neural and non-neural, and that those without oil droplets have large numbers of derived traits which is evidence of phylogenetic relationship (all are on the same branch of the evolutionary tree). All the specimens without oil

Fig. 14.1. Varieties of cones bearing oil droplets. Those from pigeon, platypus and Australian native cat are reproduced from O'Day (1938); those from *Marmosa* are reproduced from Walls (1939). Only Australian marsupials show twin cones, where both members of a double-cone pair possess oil droplets. Monotremes and American marsupials, like birds and reptiles, have oil droplets in single cones and in one member of a double-cone pair. Placental mammals have no oil droplets in their cones.

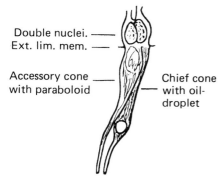

Double nuclei. ——
Ext. lim. mem. ——

Accessory cone
with paraboloid

Chief cone
with oil-
droplet

Double-cone of the pigeon

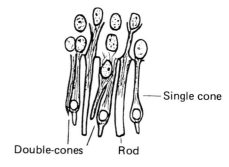

—— Single cone

Double-cones   Rod

Visual cells of the platypus

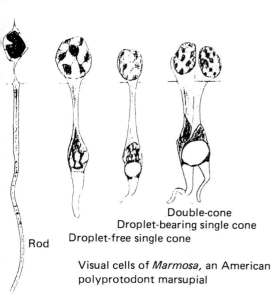

Double-cone
Droplet-bearing single cone
Droplet-free single cone
Rod

Visual cells of *Marmosa*, an American
polyprotodont marsupial

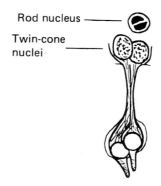

Rod nucleus ——

Twin-cone
nuclei

Twin-cones of the Australian
native cat

Fig. 14.2. (A) Phylogenetic tree constructed from the distribution of primitive derived states of 12 brain characters as seen in 38 mammalian species, reproduced from Kirsch & Johnson (1983). (Three of the original set of 15 characters were not used for this tree, in order to enhance the number of usable species; using all 15 characters with fewer species yielded a less-detailed tree that was consonant with this one.) (B) The same tree with species grouped into orders and subclasses.

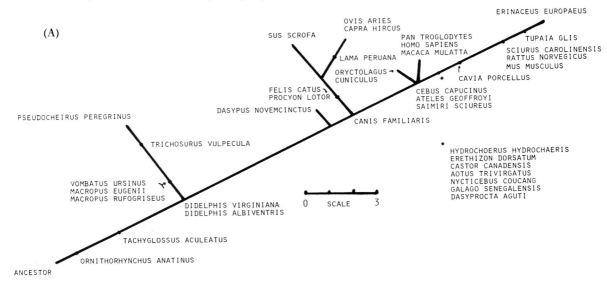

TREE 1-A: 12 CHARACTERS, 38 TAXA; C = .64

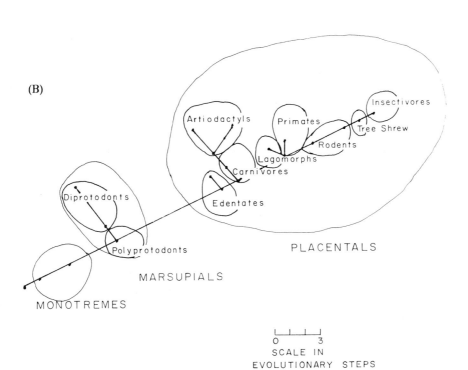

droplets also possess, for example, a corpus callosum (Abbie, 1937, 1939: Johnson, 1977), and a medial positioning of the central nucleus of the inferior olive (Watson & Herron, 1977; Johnson, Switzer & Kirsch, 1982), and they nurture their young through a long-lasting placenta.

Through a similar series of observations and arguments, we have assembled, thus far, 15 brain traits whose variable states we can classify as primitive or derived; each thereby gives us possible clues about mammalian descent and relationships. By assigning numerical values to these primitive and derived states (some traits have several progressive states of derivation), we can apply computerized programs to cross-correlate the conclusions drawn from each of the separate traits. I will not go further into the method here as John Kirsch has published the details (Kirsch, 1983). The result obtained is a phylogenetic tree designed according to the shared derived states of all the traits, considered in relation to one another (Kirsch & Johnson, 1983; Kirsch, Johnson & Switzer, 1983).

Fig. 14.2A shows our 'best' tree, and the species used in its construction. For those not familiar with the individual species, I have relabeled the tree with the more familiar names of the orders to which the species belong (Fig. 14.2B). The brain traits alone tell us, (as do the osteological, reproductive system, blood group, DNA, and other evidence), that there are three large groupings of mammals, the monotremes, marsupials and placentals. The monotremes are close to the reptiles and to the stem of mammalian evolution. The marsupials include a less-derived polyprotodont group found in America and Australia, and a more-derived diprotodont group that is exclusively Australian. The edentates are close to the stem of the large placental grouping, which then bifurcates. On one branch we find carnivores and the still more derived artiodactyls; on the other are the lagomorph rabbits, the primates, rodents and insectivores. We have discussed at length in other publications the implications of this tree for mammalian phylogeny and systematics (Kirsch & Johnson, 1983; Kirsch *et al.*, 1983).

Brain traits through phylogeny

In addition to its interest for speculations about mammalian interrelationships, this resulting tree provides a framework for speculation about the course of evolution of each of the individual traits, and that is what I now wish to do here.

Four of our 15 traits are visual. I learned about each of them from my experience in Peter Bishop's department. Two of them were discovered in the Sydney department. Having built our tree from the data of 15 traits, let us now look at the progress of evolution of each of these four visual traits as our tree says they happened.

Loss of retinal oil droplets

To look at the first of our evolving visual traits, on this tree, I will return to our first example, oil droplets in retinal cones. Using the tree generated from the cross-correlation of the 15 characters, we can locate on the tree that place in evolutionary history where the change occurred (Fig. 14.3). Oil droplets in retinal cones appear to have been lost by the ancestor of the placental radiation. Science advances by the generation of new questions, and those generated here are most interesting. What induced the change in this ancestor? How does the visual function in the ancestor, and all of us his descendants, differ from those of our 'older' relatives? This is another statement of the basic question, what is the role of the oil droplets? How is it we function well without them? What do we have that the collateral lines do not have and that enables this oil-dropletless function?

Twinning of dropleted cones

Another of our traits with variable states concerns the cone oil droplets. In birds, reptiles, monotremes and American polyprotodont marsupials, oil droplets frequently occur in one member of what O'Day (1936, 1938) has termed 'double cones' (Fig. 14.1). Kevin O'Day worked in the Anatomy Department at the University of Melbourne; in the 1930s he reported the existence of this variable character state: In Australian diprotodont marsupials, both members of a pair of double cones possess oil droplets. These he termed 'twin cones', alike in every respect. More interesting in the taxonomist's view, he found an Australian polyprotodont, the native cat or quoll, with twin cones. This is the one shred of evidence available, to our knowledge, that the Australian polyprotodonts and diprotodonts share a common ancestry separate from that of American polyprotodonts. This means that, on the tree, the separation within the polyprotodonts marked by the appearance of twin cones (arrow in Fig. 14.4) also denotes the separation between American and Australian marsupials.

Fig. 14.3. Location on the phylogenetic tree where
oil droplets in retinal cones were lost: where the
shift from the primitive state (solid line) to the
derived state (dashed line) occurred.

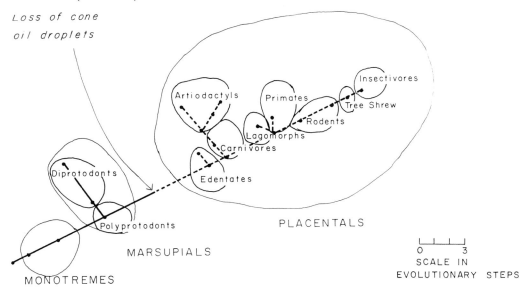

Fig. 14.4. The derived state (dashed line) of
doubling of oil droplets to form twin cones was
introduced in the Australian members of the
polyprotodont marsupials, and provides evidence
that they are ancestors of the diprotodonts.

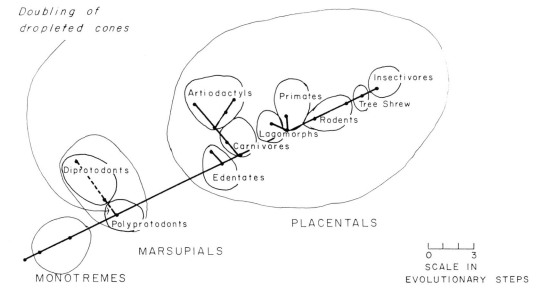

Visually, what does the twinning of dropleted cones mean? What is the function of doubling of cones, where only one has the droplet? What is there about Australia that fostered the twinning of dropleted cones? Is this a minor genetic change with little functional significance?

Superficial tectal gray

Our third visual character with variable states was reported by Campbell & Hayhow (1971, 1972) from their work in the Sydney department. They traced the degeneration of retinofugal fibers following removal of

Fig. 14.5. Optic tecta of anterior colliculi of four species, as seen in transverse sections stained with iron haematoxylin to show myelinated fibres. (A) Reptile, diamond-backed python *Morelia spilotes*. (B) Monotreme, duckbill platypus *Ornithorhynchus anatinus*. (C) Monotreme, spiny anteater *Tachyglossus aculeatus*. (D) Marsupial, gray kangaroo *Macropus fuliginosus*. Arrows show the entering fibers of the optic tract, located on the dorsalmost aspect of the tectum in the reptiles and monotremes, but subjacent to a superficial gray layer in the therian mammals (both marsupials and placentals).

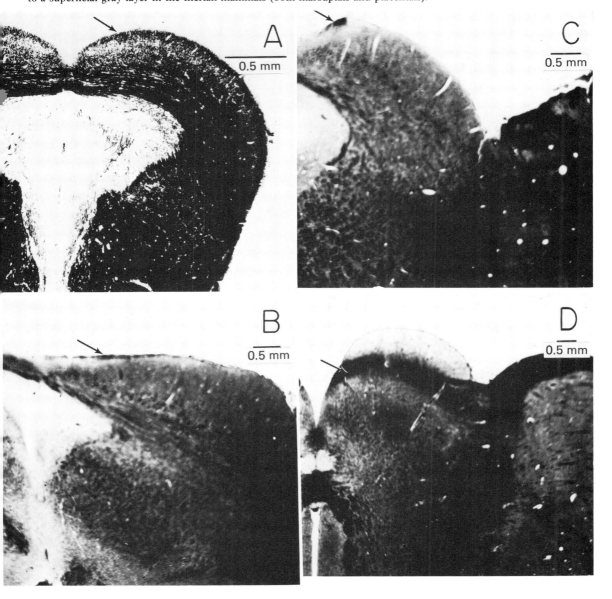

the eye in two species of monotremes, the platypus and the spiny anteater (echidna). One thing they found was that the retinofugal fibers occupy the dorsalmost aspect of the tectum of the anterior colliculus (Fig. 14.5), from whence they descend into more ventral tectal layers to terminate.

This is similar to what is seen in amphibians, reptiles and birds. Among the marsupial and placental mammals, a contrasting state is seen; they possess a superficial layer of gray matter capping the tectum (Fig. 14.5D); the optic fibers enter between the superficial layer and the more central layers and send fibers to terminate in both dorsal and ventral directions. This trait joins a long list of others, neural and non-neural, where monotremes retain the reptilian state in contrast to a more derived state seen in marsupials and placentals, showing monotremes to be near the stem ancestor of the mammals (Fig. 14.6). As with the other traits of this nature, the appearance of

a superficial gray layer in the tectum appears to be the contribution of a common ancestor of marsupials and placentals. There is a disturbing flaw in the satisfying picture of the primitive reptiloid tectum of the monotremes superseded neatly by the new model, with the superficial gray layer in the marsupials and placentals. The placental edentate armadillo *Dasypus novemcinctus* turns up with the old monotreme–reptile model of tectum: optic fibers on top. Well, edentates are near the placental stem, can this be a reversion to primitive mode before the new model is firmly fixed? Some notion about the possible function of the superficial gray would assist evolutionary speculation here. What, visually, do armadillos have in common with the reptiles and monotremes that allows them to dispense with the superficial gray? What is the visual function of this superficial layer? Is it a new entity generated by the more-derived mammals? Or is this another case of translocation of homologous cell and fiber layers, due to subtle shifts in ontogenetic heterochrony? Such a shift means that the fibres and gray matter have the same functions, but due to differences in growth and migration times the cells come to lie dorsal to, rather than ventral to, the optic fibers. Two other of our variable-state traits can be ascribed to such heterochrony: the variable position

Fig. 14.6. The shift from the primitive state (solid line) of optic fibres most superficial in the tectum, to the derived state (dashed line) of a layer of gray matter superficial to the optic fibres, occurred in an ancestor of all the marsupials and placentals. A reversion to the primitive state occurs in one of the edentates: the armadillo *Dasypus novemcinctus* carries its optic fibres superficialmost in the tectum.

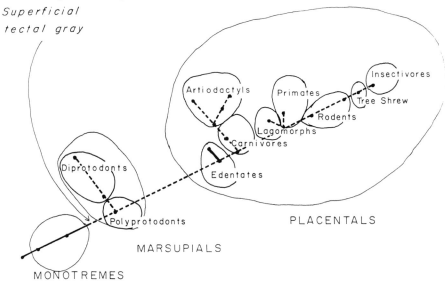

Superficial tectal gray

Artiodactyls   Primates   Insectivores   Tree Shrew   Rodents   Lagomorphs   Carnivores   Diprotodonts   Edentates   Polyprotodonts   PLACENTALS   MARSUPIALS   MONOTREMES

0   3
SCALE IN
EVOLUTIONARY STEPS

Fig. 14.7. Widely separated terminal fields of optic tract in the dorsal thalamus as seen in two monotreme species. (*a*) Tracings of parasagittal sections through the thalamus of a spiny anteater, reproduced from Campbell & Hayhow (1971) with arrows added to show the terminal fields, LGNa and LGNb. (*b*), (*c*), (*d*) Tracings of transverse sections through a platypus thalamus, reproduced with permission from Campbell & Hayhow (1972), with arrows added to locate LGNa and LGNb.

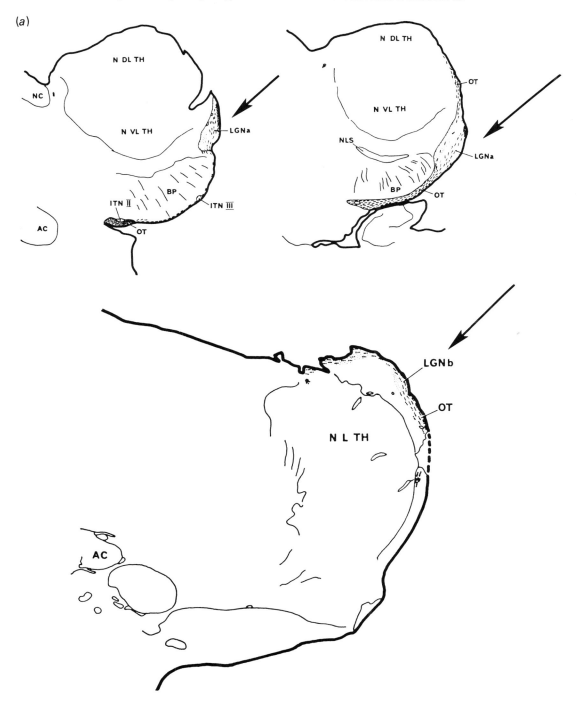

(*b*)

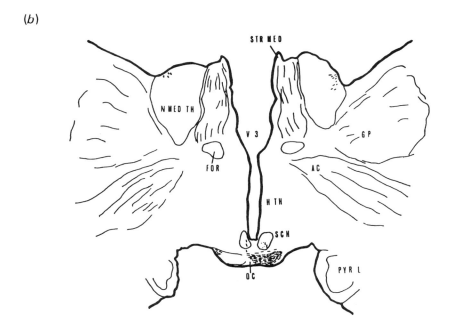

Projection drawings of representative transverse sections of the brain of platypus 2 following left ocular enucleation and processing by the Nauta-Gygax silver impregnation technique. The figures proceed in a rostrocaudal direction. Short lines and coarse dots represent degenerating axons of passage, while fine stipple represents areas of preterminal and terminal degeneration.

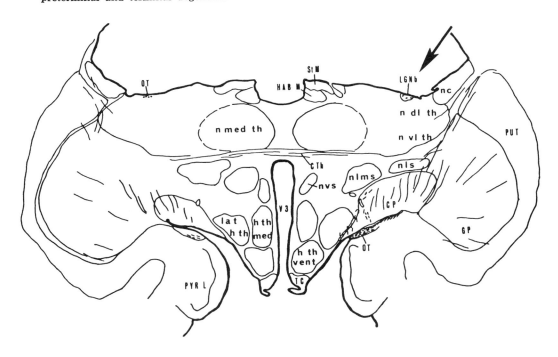

(c)

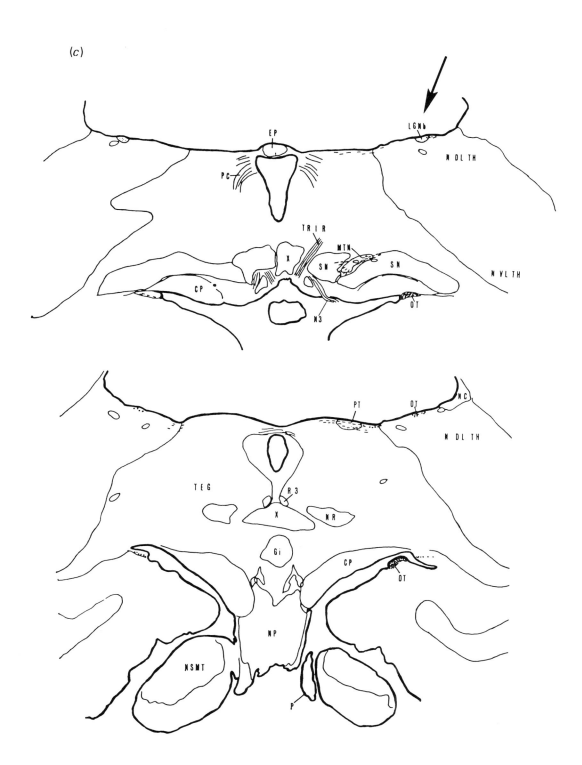

Fig. 14.7. For caption see p. 203.

(d)

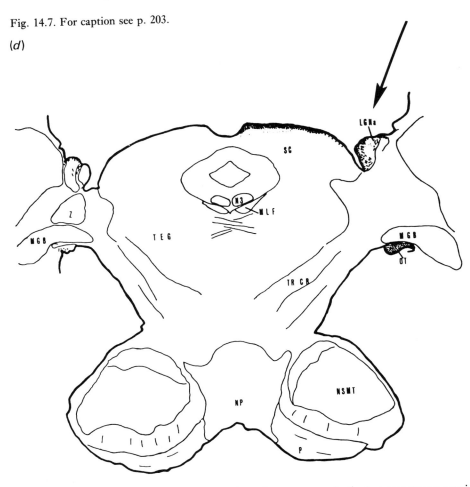

Fig. 14.8. The phylogenetic tree redrawn to show the occurrence in the two monotreme species of a shared derived trait (dashed line), widely separated terminals (LGNa and LGNb) of the optic tract in the dorsal thalamus.

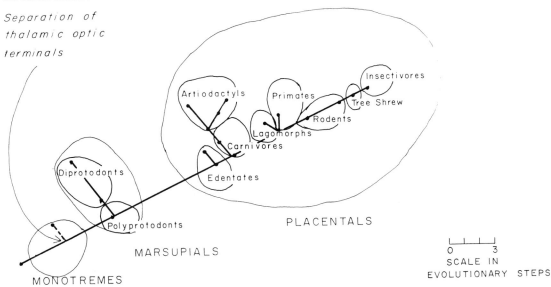

of the dorsal lateral olfactory tract and the accessory olfactory formation (Switzer *et al.*1980), and the medial versus lateral position of the ventral nucleus of the inferior olive (Watson & Herron, 1977).

Separated optic terminals in the thalamus

Campbell & Hayhow (1971, 1972) noted another striking difference between monotremes and other mammals in the course of their retinofugal fibers. Rather than termination in a juxtaposed set of thalamic nuclei, as is seen in other mammals, and in birds and reptiles (Johnson *et al.*, 1982*a*,*b*), the thalamic branch of the monotreme optic tract terminates in two widely separated cellular groups, which they designated LGNa and LGNb (Fig. 14.7) Thus we have here a case where the monotremes are different from the ancestral as well as from the advanced forms, and what appears to be one of the very rare shared derived characters that would indicate a common ancestry for these two surviving monotreme groups.

This is the only one of our neural traits showing such a shared derivation for the monotremes, and to illustrate what it means we must redraw our tree a bit (Fig. 14.8), since in the cross-correlation it was completely overwhelmed by the traits showing monotremes to be primitive stem types. We are probably not going to learn the significance of the separation of LGNs. at least until something like laboratory breeding of spiny anteaters advances to the point that large populations of experimental subjects become available.

Other potential variable-state characters

I hope in this essay to stimulate visual scientists to be alert for other variable-state characters. These will aid in the perpetual refinement and improvement of our phylogenetic schemes, and will also generate the good scientific questions so frequently drawn from the comparative approach. One ripening example is in the emerging picture of the variable lamination of the LGNs of the various mammals. This trait has been used in the past to classify the tree shrew as a primate, generating immense interest and an extraordinary amount of research using these peculiar animals. Concepts of laminations, and their function, have been in such flux in recent years we have been unable to classify them satisfactorily for use in our phylogenetic

analyses. We hope that clarifications now in progress will soon remedy the problem. Ken Sanderson (Chapter 10) and Jon Kaas (Chapter 21) have offered very promising material which may soon develop into a whole family of lateral geniculate traits useful for phylogenetic analysis.

References

Abbie, A. A. (1937) Some observations on the major subdivisions of the Marsupialia, with especial reference to the position of the Peramelidae and Caenolestidae. *J. Anat. (Lond.)*, 71, 429–36

Abbie, A. A. (1939) The origin of the corpus callosum and the fate of the structures related to it. *J. Comp. Neurol.*, 71, 9–44

Campbell, C. B. G. & Hayhow, W. R. (1971) Primary optic pathways in the echidna, *Tachyglossus aculeatus*: an experimental degeneration study. *J. Comp. Neurol.*, 143, 119–36

Campbell, C. B. G. & Hayhow, W. R. (1972) Primary optic fibers in the duckbill platypus, *Ornithorhynchus anatinus*: an experimental degeneration study. *J. Comp. Neurol.*, 145, 195–208

Johnson, J. I. (1977) Central nervous system of marsupials. In *Biology of Marsupials*, ed. D. Hunsaker, pp.157–278. New York: Academic Press

Johnson, J. I., Kirsch, J. A. W. & Switzer, R. C. (1982*a*) Phylogeny through brain traits: fifteen characters which adumbrate mammalian genealogy. *Brain Behav. Evol.*, 20, 72–83

Johnson, J. I., Switzer, R. C. & Kirsch, J. A. W. (1982*b*) Phylogeny through brain traits: the distribution of categorizing characters in contemporary mammals. *Brain Behav. Evol.*, 20, 97–117

Kirsch, J. A. W. (1983) Phylogeny through brain traits: objectives and method. *Brain Behav. Evol.*, 22, 53–9

Kirsch, J. A. W. & Johnson, J. I. (1983) Phylogeny through brain traits: trees generated by neural characters. *Brain Behav. Evol.*, 22, 60–9

Kirsch, J. A. W., Johnson, J. I. & Switzer, R. C. (1983) Phylogeny through brain traits: the mammalian family tree. *Brain Behav. Evol.*, 22, 70–4

O'Day, K. (1936) A preliminary note on the presence of double cones and oil droplets in the retina of marsupials. *J. Anat. (Lond.)*, 70, 465–8

O'Day, K.(1938) The visual cells of the platypus (*Ornithorhynchus*). *Br. J. Ophthalmol.*, 22, 321–8

Switzer, R.C., Johnson, J. I. & Kirsch, J. A. W. (1980) Phylogeny through brain traits: relation of lateral olfactory tract fibers to the accessory olfactory formation as a palimpsest of mammalian descent. *Brain Behav. Evol.*, 17, 339–63

Walls, G. L. (1939) Notes on the retinae of two opossum genera. *J. Morphol.*, 64, 67–87

Watson, C. R. R. & Herron, P. (1977) The inferior olivary complex of marsupials. *J. Comp. Neurol.*, 176, 527–38

15

The evolution of binocular vision

JOHN D. PETTIGREW

Introduction

We are all familiar with the coordinated use of both eyes to examine the same visual scene. After all, this is the habitual mode of operation of our own visual systems and, except for the 5% of the population who are stereoblind (Richards, 1970), most of us are familiar with the fact that this mode of operation, binocular vision, enables us to extract subtle details which are not available to one eye alone. This is illustrated by random-dot stereograms, where an embedded figure is invisible to monocular inspection but stands out vividly in stereoscopic depth on binocular inspection (Julesz, 1971).

Binocular vision has evolved independently at least twice amongst the vertebrates, in both mammals and birds, and in the present account I shall attempt to reconstruct these two independent evolutionary events, largely from examination of the living forms and the way in which they achieve binocular vision. Since all mammals, with the exception of the Cetacea, appear to have at least the rudiments of binocular vision, particular emphasis will be placed upon avian binocular vision. Comparison of the bird groups which appear to have achieved binocular vision with those where binocular vision does not seem to have been achieved can give some insight into the evolutionary pressures which have given rise to binocular vision.

Hallmarks of binocular vision

Binocular overlap

A first prerequisite in the achievement of binocular vision is clearly a satisfaction of the optical and oculomotor restraints which would otherwise prevent the eyes from cooperating. In other words, an animal with functional binocular vision would be expected to have at least some degree of binocular overlap in the visual fields of each eye. It must be emphasized, however, that a large degree of binocular overlap is neither necessary nor sufficient for the achievement of functional binocular vision. For example, the owl achieves a high degree of specialization in its binocular vision and yet the region of binocular overlap represents only half the total visual field of one eye and only a third of the total visual field of both eyes together (see Fig. 15.1).

Nor does the degree of binocular overlap need to be constant, since a number of species have oculomotor strategies which enable them to change from binocular visual processing to panoramic, monocular vision at will. For example, the tawny frogmouth (*Podargus strigoides*), in the camouflage posture which it adopts when threatened, completely eliminates binocular overlap between the visual fields of each eye to maximize its panoramic field of view. At other times, such as when hunting, the frogmouth can align its eyes frontally and achieve functional binocular vision of a grade equal to the owl (Wallman & Pettigrew, 1985). This kind of behaviour seems at first sight very strange to animals like ourselves in which a marked divergence of the visual axis is regarded as pathological and given the name *strabismus*. 'Voluntary strabismus' may not be confined to birds, however, since wide divergence of the visual axes occurring under some circumstances has been described for ungulates (cf. the whites-of-the-eyes posture sometimes adopted by frightened horses and the large divergent saccades shown by sheep).

That binocular overlap is not sufficient for functional binocular vision is shown by a number of bird species which have significant overlap of the visual fields of each eye but yet show no evidence of functional

Fig. 15.1. Schematic dorsal view of a bifoveate bird to show *angle alpha* (α) subtended between the binocular visual axis of the right eye (RVA), passing through the fovea located on the temporal retina, and the optical axis (ROA), passing through the fovea located on the central retina (monocular visual axis). Anatomical and physiological studies confirm that functional binocular vision exists along the binocular visual axis, despite the small proportion of the total visual field which is involved in binocular overlap and despite the high visual acuity possible along the monocular visual axis.

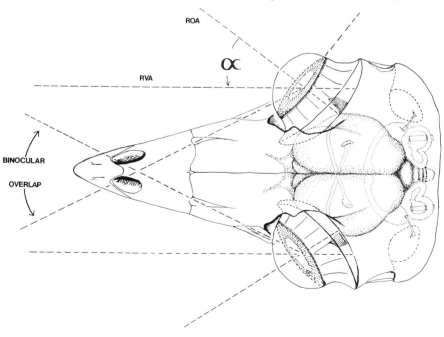

binocular vision. Such species include the swift and the oil bird (see below).

Area centralis

Perhaps more important for binocular vision than the optical presence of binocular overlap *per se* is the development of retinal specializations in the temporal retina subserving the binocular field. Since binocular vision demands a degree of visual acuity sufficient for the small differences between the two retinal images to be measured, a universal feature of animals with functional binocular vision is the presence of a retinal specialization with increased cell densities in the area of the retina which projects forwards into the binocular visual field. In other words, the presence of a temporal retinal specialization in each eye is a strong indication of the presence of funtional binocular vision. Defining the presence of such an area of retinal specialization may require extra information about the oculomotor strategies of the species concerned (as already pointed out for those species which sometimes adopt extremely divergent eye position) and is not a surefire guarantee since there are some birds which appear not to make stereoscopic use of foveas which project forward into binocular space (e.g. the swift). The degree of specialization required of the temporal *area centralis* or fovea will obviously depend upon such species-specific factors as the distance at which binocular vision is normally used and the size of the targets to be detected. For these reasons, the increase in density there may not be dramatic and escape attention without special techniques, as in the case of the rabbit. The low degree of specialization in its temporal area is perhaps related to the low demands placed upon the rabbit's need for binocular vision as it feeds on targets at very close range with a consequent larger angular target size and retinal disparity. In rabbits, the temporal area specialization is limited to one cell class which, in ordinary preparations without special labelling, tends to be overshadowed by the high density of neurons in the monocular, central retina (Provis, 1979).

Bifoveate retina: *angle alpha*

In contrast to the monofoveate condition of mammals, we may cite some of the bifoveate birds which have two foveal specializations with high cell densities, one looking into the monocular field and one into the binocular field. This dual foveal specialization in the same eye raises an important point about what

can be called *angle alpha*, *viz.* that there need not be an alignment of the optical axis of the eye and the visual axis of the eye. Since the optical axis and the visual axis are separated only by a very small amount in man (*angle alpha* = 6 degrees) there is a common tendency to treat these two axes as the same. For this reason it is commonly assumed that animals with a very marked degree of divergence of their *optical* axes will have a correspondingly high degree of divergence of their *visual* axes. That this is not the case is well illustrated by the kingfishers, which have the largest *angle alpha* so far measured in the vertebrate kingdom (around 40–50 degrees). As shown in Fig. 15.1, the binocular visual axis which passes through the nodal point of the eye and the binocular fovea is some 40 degrees away from the optical axis of the eye which is also aligned with the monocular fovea. As described below, kingfishers achieve full functional binocular vision despite this very wide separation of the binocular visual axis from the optical axis of the eye.

The question might reasonably be asked, since it appears to be possible to have simultaneous panoramic vision and binocular vision as shown by so many birds, what are the selection pressures which have operated to bring such a close alignment of the optical axis and visual axis in primates? This question has been addressed elsewhere and it appears to be a complicated one involving perhaps the optical constraints imposed by both the demands of binocular vision and the demands of a nocturnal niche, with its accompanying requirements for a large-aperture optical system (Pettigrew, 1978; Allman, 1977).

In summary, we may say then that if we see a considerable degree of specialization in the temporal retina subserving binocular vision, then there is a good chance this animal achieves functional binocular vision. Perhaps the best example of this is the owl, where there is a high degree of specialization with increased ganglion cell densities in the temporal retina. On the other hand, if there is a strong emphasis on the acquisition of high visual acuities in the monocular field, the specializations in temporal retina may be swamped and it may take special techniques to demonstrate the temporal specialization. This is perhaps well illustrated in the case of the burrowing owl which has a high-density visual streak of retinotectal ganglion cells which tends to overshadow the peak retinothalamic-ganglion cell density in temporal retina. Separate labelling of these two populations of ganglion ells can make the binocular specialization more vident (Bravo & Pettigrew 1981), as in the case of the abbit (Provis, 1979).

Convergence of the visual pathways from each eye

Sir Isaac Newton in his treatise on optics (1730) vas one of the first scientists to point out the necessity, n functional binocular vision, for information for both yes to converge on the same site in the brain. Newton easoned that a partial decussation of the fibres in the ptic nerves would be a necessary consequence in a ystem with functional binocular vision, and his natomical prediction was verified not long afterwards n gross dissections of the mammalian optic chiasm (see olyak, 1957). That a partial decussation of the optic hiasm is not the only way to achieve binocular ntegration is shown by the birds, amongst which all o far described have a total decussation. This total ecussation has led many authors (e.g. Walls, 1942) to onclude that functional binocular vision is impossible n birds, but recent work has shown that avian binocu- ar vision is accomplished by different anatomical neans. In the owls, some nightjars, kingfishers and iurnal birds of prey, the visual thalamus projects ilaterally to the visual Wulst where binocular in- gration occurs (Pettigrew & Konishi, 1976; Petti- rew, 1979: Bravo & Pettigrew, 1981)

It is rather difficult to conceive of some inter- nediate stage of evolution where there is both a partial ecussation of the optic chiasm and a bilateral rojection from the thalamus, if topographic relations re to be maintained. There seems, therefore, to be little nance that birds and mammals share a common redecessor with functional binocular vision. This may e regarded as an example of convergent evolution, oth birds and mammals having independently hieved functional binocular integration by different echanisms. Since the mammalian partial decussation d the avian double decussation by no means exhaust ie possibilities for binocular interaction, it may be xpected that further search may reveal other echanisms for convergence of binocular information. or example, although there is little in the way of other ridence to support functional binocular vision in the el and the shark, they both have rather unusual ecussations from one side of the visual pathway to the her (Ebbesson & Schroeder, 1971; Ekström 1982).

Binocular nerve cells

One of the last steps in the demonstration of functional binocular vision is a description of the properties of nerve cells which receive input from both eyes. Since the requirements of a functional binocular system capable of extracting the disparity information necessary for stereopsis have been well described, there appear to be a limited number of solutions to this problem. Evidence to support this view comes from the gradual evolution of algorithms for machine stereopsis which appear to be gradually approaching a solution rather like that achieved in the visual cortex of birds and mammals (see below). The essential two steps in the achievement of stereopsis from binocular inputs are (1) the solution of the 'matching' or 'correspondence' problem, whereby the parts of the two images belonging to the same outside object are identified, and (2) the measurement of the slight differences in these paired parts of the image (disparity detection).

The solutions to these two problems are, in a certain sense, incompatible, since the recognition of the matching elements involves at some level the detection of similarity whereas the disparity detection task involves the measurement and recognition of small differences. It therefore appears necessary for these two tasks to be carried out gradually, in parallel with one another, to avoid the difficulties which may arise, for example, when the matching process is taken too far before the disparity measurement. In physiological terms, what this means is that, despite the presence in the retina of highly specialized retinal ganglion cells which are capable of carrying out some of the first stages of the matching task by identifying common elements in the two images, these retinal ganglion cells are *not* used to construct binocular receptive fields. In preference, the concentrically organized retinal ganglion cells are used, which do not 'jump to conclusions' about the nature of the retinal stimulus except in so far as to identify the exact position of the local contrast change. Specialization for the exact form of the local part of the retinal image occurs only where there is convergence of information from the two eyes and appears to proceed gradually, hand in hand with the increasing degree of binocularity seen at higher levels in the visual pathway.

For example, at the lateral geniculate nucleus, where the dominant influence on the properties of the cells is from the concentrically organized retinal input, there are the beginnings of binocular interaction, with an inhibitory field in the non-dominant eye which has the beginnings of some selectivity for the form of the stimulus (Sanderson, Darian-Smith & Bishop, 1969). At the level of the visual cortex where fully fledged binocular neurons are seen, these binocular neurons are also highly selective for the nature of the visual stimulus, in contrast to the immediately preceding stage where they are selective only for location. These binocular neurons are also orientation-selective. This orientation selectivity is tightly matched for both receptive fields (one from each eye) within tightly controlled statistical limits (Nelson, Kato & Bishop, 1977). It is this close matching between the selectivity in the receptive fields in each eye which may go some way towards solving the matching problem. Since there are a variety of cells with the same matched receptive field properties for a variety of features, such as orientation, size and direction of motion, but with differing disparity selectivity, then we can see that at this stage in the visual pathway there are individual binocular neurons which are capable of providing the important elements for stereoscopic processing.

These highly selective binocular neurons, with tightly matched receptive field properties in each eye, are to be contrasted with the large, diffusely organized binocular receptive fields which have been described in the optic tectum of a variety of species (e.g. Gordon, 1973). There are a number of functions which these large binocular fields might subserve, such as the detection of motion-in-depth, but they seem quite unsuited to the demanding and subtle task of achieving stereopsis and therefore, in my view, do not constitute any compelling basis for the existence of binocular visual processing such as is involved in stereopsis. Of course the final proof should come with the behavioural demonstration of stereoscopic abilities. One must, however, be aware that the latency between the demonstration of neurons which are capable of stereoscopic depth discrimination and the demonstration of the same abilities in the whole, cantankerous animal may be quite long. In the case of the cat, over 10 years elapsed between the first demonstrations of disparity-selective binocular neurons (Barlow, Blakemore & Pettigrew, 1967; Pettigrew, Nikara & Bishop, 1968) and the convincing behavioural demonstrations of the same abilities using the whole cat (e.g. Mitchell, Kaye & Timney, 1979).

Binocular vision in mammals

It has been possible to demonstrate all th hallmarks of functional binocular vision in each of th mammals which have been studied with the aim investigating this aspect (e.g. cat: Bishop (1973); shee Clarke, Donaldson & Whitteridge (1976); monkey: s review in Poggio & Poggio (1984); and rabbit: Hugh & Vaney (1982). Since there is also anatomical eviden for binocular vision in all of the mammals which ha been so far studied, with the possible exception of th Cetaceae (see Chapter 13), it seems unlikely that it w be possible to reconstruct an evolutionary scenario f binocular vision in mammals. It seems more likely th binocular vision was an attribute of the earlie mammals and may in fact have played a large part the successful radiation of this group. The earliest fos primates such as *Purgatorius* and *Tetonius* (e.g. Allma 1977; Archer & Clayton, 1984) show evidence of larg frontally placed eyes and there is every reason to thi that they had functional binocular visual systems adequate as those possessed by prosimian primat living today. There are a number of special advantag of binocular vision which accrue to animals whi occupy the nocturnal niche. These include increas signal-to-noise ratio, camouflage breaking, and th absence of the need to move in order to generate moti parallax. In view of the prevailing evidence that th early mammals were nocturnal, it may further speculated that binocular vision was one of the featur which contributed to their success (as pointed out Polyak, 1957). The functional anatomy and physiolo of binocular vision in mammals is remarkably simil in all the species which have been studied, such as th cat, various primates, the sheep and the rabbit. Th is perhaps not so surprising in view of the stro likelihood that they all arose from a comm mammalian ancestor with binocular vision, but it m be valuable to summarize quickly what has been foun in terms of the organization of binocular vision in the various mammals to help emphasize some of th general features of binocular visual systems.

Binocular vision in the rabbit

The rabbit forms a convenient counterpoint other mammals since it was long considered th functional binocular vision was not a feature of its visu system. Indeed the rabbit, like many birds, ca eliminate its frontal region of functional binocul overlap by the appropriate oculomotor posture, such as the one it adopts when in the freeze position (Hughes & Vaney, 1982). Recent work has shown, however, that rabbits do converge their eyes when feeding (Zuidam & Collewijn, 1979) and that there are binocular neurons in the visual cortex which have most of the properties enumerated above to indicate that they are involved in functional binocular vision. For example, these binocular neurons have orientation selectivity which is tightly matched for both receptive fields (Hughes & Vaney, 1982). Previous work had failed to demonstrate this nice selectivity, which is matched on both retinas, and even led to the suggestion that the binocular connections of the rabbit might be maladaptive (Van Sluyters & Stewart, 1974). The conclusion that the binocular fields were not matched may well have been based upon inappropriate eye position, or failure to correct the eccentric optics of the rabbit, or both, and this serves as a good example of the need to establish optical and eye position parameters in any species for which binocular vision is being investigated (cf. the long line of baseline investigations of the cat's optics and binocular rest position of the eyes carried out by Peter Bishop during his career).

Retinal trigger features not used for binocular vision

There is another important message which can be gleaned from the work on binocular vision in the rabbit, and it concerns the way in which the visual system is organized to provide binocular neurons which have receptive fields on both retinae. It was in the rabbit retina that the highly complicated receptive field properties such as orientation selectivity, direction selectivity and local edge detection were first described for mammalian retinal ganglion cells (Barlow, Hill & Levick, 1964; Levick, 1967). Although the specialized receptive field properties were thought to be characteristic of the 'lower' vertebrates such as birds and anuran amphibians, their demonstration in the rabbit opened the way for the later experiments which demonstrated that such highly specialized receptive field properties are probably found in retinae right across the vertebrates. Differences in the proportion of specialized retinal ganglion cells turn out, then, to be quantitative rather than qualitative, and provide the following useful generalization: the proportion of concentrically organized, non-specialized retinal ganglion cells, as a function of the highly specialized retinal

ganglion cells, tends to increase with the importance of binocular vision for the animal. This was first noticed and pointed out by Levick with respect to the rabbit, in which the number of concentrically organized retinal ganglion cells is rather small, as is the proportion of the visual cortex devoted to binocular vision. More recent work has extended this generalization further, since it can be shown that the ipsilateral input to the lateral geniculate nucleus which provides the basis for binocular interaction is contributed almost exclusively by the concentrically organized retinal ganglion cells (Takahashi & Ogawa, personal communication). In other words, despite the large numbers and variety of specialized retinal ganglion cells in the rabbit retina, including the orientation-selective variety which might be expected to be used for solving the matching problem with binocular vision, none of these can be shown to contribute to the binocular neurons' receptive field properties in the visual cortex. Instead, the minority population of concentrically organized retinal ganglion cells and their lateral geniculate relay cells are used to bring about binocular receptive fields with matched orientation selectivity in the two eyes.

That this is the case in primates and carnivores was perhaps not so surprising, in view of the fact that it was long felt that the only retinal ganglion cells available for the task were of the concentric variety. The recent evidence that primate and carnivore retinae may also contain significant proportions of the specialized retinal ganglion cells (see Stone, 1983, for a review), coupled with the fact that rabbits achieve a degree of functional binocular vision, serves to underline the generalization that, even when they are available, highly specialized retinal ganglion cells do not appear to make an important contribution to the genesis of binocular receptive fields with highly specialized properties on both retinas.

A moment's consideration will show why this might be the case. Firstly, if this were otherwise there would have to be an exceedingly large number of different retinal ganglion cells to cover all the permutations and combinations of orientation, direction, size, etc., at all of the different relative retinal locations to provide disparity selectivity. In addition, as pointed out below with reference to machine stereopsis, there may be underlying theoretical constraints upon the optimal solution to the matching problem which require that it avoid 'jumping to conclusions' about the important features for stereopsis which are present in the monocular stimulus. Whatever the underlying reasons, we can say with some generality, based on work on the binocular visual pathways of cat, various primates, sheep and goats, rabbits and various birds, that specialized receptive field construction is delayed in the visual pathways subserving binocular vision until information from both eyes converges.

Cats, owls, monkeys and machines

In a facetious reference to disparity-selective binocular neurons in aardvarks, Mayhew & Frisby (1979) question the significance of physiological studies of disparity-selective binocular neurons carried out so far, because of their apparent failure to address important issues such as the particular computational strategy used for stereopsis. These demeaning remarks fall short on two counts, since (1) they fail, characteristically, to acknowledge the important stimulus to machine stereopsis which the preceding neurophysiological studies provided, and (2) they miss one of the major points of the comparative work on binocular vision, i.e. *that the computational strategies adopted for stereopsis by unrelated species are identical in the sense that all avoid feature extraction until information from both eyes has converged.* The hierarchical step from monocular, concentric organization to binocular, matched, feature-specific organization is the same in both mammals and birds, despite the considerable differences in their respective neural apparatus which might have led to other solutions, such as the synthesis of binocular receptive fields from monocular elements which were already feature-specific; an abundance of the latter exist in lower levels of the visual pathway, particularly in birds, so it is a fair question to ask why monocular concentric rather than feature-selective elements are used as inputs to the binocular neurons. The answer to such a question, while not illuminating the whole computational strategy, certainly eliminates some possibilities. The avoidance of a high degree of monocular preprocessing has also 'evolved' in the later machine algorithms for stereopsis, so it may be valuable to give some specific consideration to the machine–animal comparison.

Machine *versus* animal stereopsis

In recent years there has been a variety of machine algorithms which successfully compute depth from a pair of Julesz random-dot stereograms (reviewed in Poggio & Poggio, 1984). These models have shown a process of evolution, since they have been subject to successive refinements which have improved their versatility and efficiency. For example, one of the early models was very sensitive to small-scale differences between the two images, and could not achieve stereopsis if one image had reversed contrast (Marr & Poggio, 1976; Grimson & Marr, 1979), neither of which problems presents any difficulty for the human stereoscopic system (see Julesz, 1971). Later models have overcome the reversed-contrast problem (Marr & Poggio, 1979) but still hang up if there is size anisotropy or large vertical disparity.

It may be instructive to examine the continuing evolution of machine algorithms for stereopsis to gain further insights into the constraints operating upon the evolution of animal stereopsis. We may ask, for example, whether the striking similarities between the avian and mammalian stereoscopic systems which exist in spite of their different origins and structural components reflect a fundamental constraint upon successful algorithm. Is there, perhaps, a single most-efficient solution to the problem of stereopsis?

The designers of machine stereopsis algorithms are usually reluctant to attempt any measure of the efficiency of the successful ones, although this could be a desirable objective if they are to be measured against each other and against the performance of human subjects on a range of stereoscopic tasks. We may nevertheless, infer that the algorithms so far devised fall short of some ideal because of the ferment in this area and the fairly steady stream of new versions (e.g. Frisby & Mayhew, 1980; Mayhew & Frisby, 1980).

One of the key areas of ferment concerns the choice of strategy for the solution of the *correspondence* or *matching* problem, *viz.* the identification in each image of the paired parts which correspond to the same outside objects. The human visual system shows extraordinary versatility in solving this problem, to a degree that makes it unlikely that a search for a particular primitive is always used to match up the two images. For example, Frisby & Julesz (1975) have shown that, under some circumstances, edge orientation can be ruled out as the local primitive for matching since successful fusion and stereopsis can follow if the stereo pair is constructed of randomly oriented line segments. That this particular problem can be solved (this writer can do it only with the greatest difficulty) does not imply that the special resources brought to pear on this occasion are used as a matter of course in the operation of stereopsis under more normal circumstances. After all, a dancing poodle, however impressive its performance, does not necessarily lead to the conclusion that normal canine locomotion is bipedal. The experiment does imply, however, that if orientation is commonly used as a primitive for matching (as Barlow, *et al.* (1967) suggested for the cat visual cortex), it must be done in a facultative, rather than an obligatory, fashion so that other strategies for matching can also operate in parallel or as alternatives. It is therefore of some interest that one of the more recent machine algorithms for stereopsis has some features which avoid the rigidity in the matching strategy which hampers some earlier models and at the same time comes closer in conceptual design to the avian and mammalian blueprints (Frisby & Mayhew, 1980).

The key strategy, then, may be the avoidance of commitment to a particular feature for matching before the disparity information is extracted. In this way we can understand the avoidance, in both avian and mammalian binocular pathways, of feature extraction before information from both eyes comes together. One reason for this has already been suggested in terms of the impossibly large number of monocular feature detector neurons which would have to be located in the retina, where size and space limitations would mitigate against it. A second reason may be that too early commitment to a particular feature may preclude later solution of the disparity problem. After all retinal disparity involves any measurable difference between the images related to depth and can involve slight mismatches in size, position, velocity of movement, orientation, or even (when one considers the rivalrous inputs from lustrous surfaces) brightness and colour. Matching the images, on the other hand, inevitably requires some estimation of similarity between parts of the images which could run the risk of losing information valuable to subsequent disparity detection task. It may be for this reason that avian, mammalian and the recent machine algorithms for stereopsis all avoid feature analysis at early stages before information from both eyes is compared. Once the inputs from both have converged on binocular neurons, a variety of features can then be used for matching, such as size (the X and Y streams), edge orientation, direction of movement, velocity, and end-stopping – a few properties which are known to be matched in the receptive field pairs of binocular disparity-sensitive neurons, whether these are found in cats (Barlow *et al.*, 1967, Pettigrew *et al.*, 1968), monkey (Poggio & Talbot, 1981), sheep (Clarke *et al.*, 1976) or owls (Pettigrew & Konishi, 1976; Pettigrew, 1979).

Vertical disparity

Vertical retinal disparities are as accurately detected by binocular cortical neurons as are horizontal disparities, despite the commonly-accepted view that only the latter can contribute to stereopsis. This finding appears to be a general one, since all studies of disparity-selective neurons, from the pioneering work on the cat (Pettigrew, *et al.* 1968) to the more recent work on the monkey (Poggio & Talbot, 1981), have failed to demonstrate any prominent difference between the respective codings for disparities in the two dimensions.

The absence of the expected anisotropy in favour of horizontal disparity was felt by some to be fatal to the proposition that the disparity-selective binocular neurons could form a basis for stereopsis. This was in spite of the arguments raised to the effect that vertical disparities were an inevitable consequence of both eye movements and the optics of the situation for which provision would have to be made by the system even if there were no obvious use for them. Recent work has revived interest in the question of vertical disparities by drawing attention to old psychophysical data which clearly demonstrated a role for them, at the same time as showing that they can make a contribution to depth perception which is independent of that made by horizontal disparities.

Mayhew & Longuet-Higgins (1982) proposed a scheme whereby a gradient of vertical disparities could be used to judge distances independent of the vergence position of the eyes, if there were surfaces within reasonable viewing distance and if vertical disparity detection were as accurate as horizontal. This scheme accurately predicts a number of psychophysical phenomena in binocular vision such as the 'induced effect' first described by Ogle (1950).

Although it still remains to be shown by the neurophysiologists how the horizontal and vertical disparity information is treated separately, the new insights help to account for the accurate preservation of *both* by binocular disparity-selective neurons.

Binocular inhibition in the lateral geniculate nucleus

Another important feature of all the mammalian binocular systems which have been studied so far is the phenomenon of binocular inhibition in the lateral geniculate nucleus (LGN). Although true convergence of excitatory inputs from both eyes is first seen at the level of the visual cortex, limited binocular interaction also occurs at the LGN before the generation of binocular receptive fields with matched properties. This takes the form of an inhibitory influence from the non-dominant eye and is a subtle feature which has required special attention for its demonstration. Documentation is best in the cat where all laminae appear to be subject to binocular inhibition (Sanderson *et al.* 1969, 1971), but it has also been demonstrated in the rhesus monkey where the magnocellular laminae, but not the parvocellular laminae, show this influence (Dreher, Fukada & Rodieck, 1976). This species difference is very likely due to the fact that there is a much higher proportion of concentrically organized non-oriented receptive field properties in lamina IV of the monkey visual cortex and it seems very likely that further study of lamina IVc beta cells receiving parvocellular input will reveal significant binocular interaction there. One could regard the monkey's IVc beta with its enhanced segregation of inputs from both eyes, as representing comparable processing to that which takes place in the cat's LGN, except that it is 'encephalized' one synapse. The significance of this binocular inhibition lies in the way in which the orientation-selective properties of the cortex can subtly influence binocular processing at the very earliest stage before highly specialized binocular receptive fields are constructed. In theoretical terms this subtle influence may have its counterpart in recent machine algorithms for stereopsis, where it has been found that efficiency is increased by allowing a small degree of binocular interaction to occur at the very earliest stages, just before the matching problem is solved (Mayhew & Frisby, 1979).

Lateral geniculate lamination

A universal feature of mammalian LGNs is the segregation of inputs from the two eyes into separate laminae. The apparent paradox by which the inputs from both eyes appear to be segregated in such a precisely aligned fashion after so much trouble has been expanded to bring both eyes together, has been discussed extensively by Kaas, Guillery & Allman (1972). This paradox may be resolved partially in terms of the new findings which indicate a significant degree of binocular interaction, of an inhibitory kind, within the LGN. These have already been discussed in the previous section. In addition, there is recent evidence which suggests that geniculate lamination patterns may be an important mechanism by which disparity selectivity is generated.

The new observations derive from the work showing that different retinal ganglion cell classes have different patterns of decussation within the optic chiasm (see Levick, 1977). For example, the alpha cells which have been identified with physiological class Y-cells in the cat have a decussation pattern which is significantly to the temporal side of the zero meridian through the centre of the area centralis. In contrast, the beta cells have a decussation line which lies nasal to that of the alpha cells. Recent evidence suggests that one of the W-cell subclasses has a decussation pattern which is even more nasal than that of the beta cells. The significance of these different decussation patterns when translated into the LGN is shown in Fig. 15.3 where it can be seen that if the medial edge of the LGN represents the most medial retinal ganglion cells within the decussation, then projection lines drawn through the different layers of the LGN will result in binocular pairings which have different retinal disparities. For example, because of the more temporal position of the decussation for Y-cells, the projection line passing through adjacent laminae representing the two eyes would intersect at a retinal correspondence which would give rise to a very convergent retinal disparity. In contrast, the epsilon class of W-cells would tend to code for divergent disparities and the beta cells for disparities close to the fixation plane. The finding that different retinal ganglion cell classes connect with specific targets within the LGN and that these specific targets, in turn, relay on to specific sublaminae and even specific visual cortical areas, suggests that one important aspect of parallel processing concerns retinal disparity. The nature of the decussation patterns and the pattern of retinogeniculate lamination will lead to the generation of binocular neurons connected to different classes of retinal ganglion cells which also have different retinal disparities.

The segregation of disparity processing to different laminae could also be relevant to the question of corticofugal feedback which could thereby be used ɔ enhance a particular disparity plane at the expense f others.

Ipsilateral visual field representation

A feature of both the ungulate and the feline inocular visual pathway not evident in the primate is significant representation of the ipsilateral visual eld. This representation is much more prominent in he sheep and goat than it is in the cat (Pettigrew *et al.*, 984) and can be seen at the level of the retina, where he decussation pattern for the alpha cells involves pread into the temporal retina of the contralateral eye, ı the medial interlaminar nucleus, which is a rominent feature of both the cat and sheep visual ıalamus and which has a representation of the ɔsilateral visual field, and, finally, in the visual cortex ı the boundary zone where areas 17 and 18 adjoin. The ınctional significance of this representation of the ɔsilateral visual field can be suggested from the fact ıat the binocular neurons with receptive fields that ave a large ipsilateral component tend to have typically large and convergent disparities (Pettigrew *t al.*, 1984). This fact, in combination with the large ısjunctive eye movements which are known to occur, articularly for ungulates, suggests a role for this lack f differentiation in the medial edge of the decussation ı allowing for a relative lack of precision of binocular lignment. In contrast, the very sharp decussation of he primate would, without a large callosal mechanism or mediating midline integration, result in a complete ailure of binocular fusion when the eyes were ıisaligned following a disjunctive eye movement. This ıterpretation is supported by the finding of both very ırge convergent disparities and a large ipsilateral epresentation of the visual field within the optic ectum of the oppossum (Ramôa *et al.* 1983). Like ıgulates, marsupials have a relatively poor degree of ·inocular coordination of eye movements, and large ısparities across the vertical midline will occur as a esult of the peculiar disjunctive eye position adopted ·y marsupials under some circumstances. The ısparities created by large disjunctive eye movements ontrast markedly with the tiny retinal disparities ıvolved in fine binocular depth discriminations and ıne may speculate that the generation of binocular ·eurons coding for such large disparities would have different mechanism. Such a mechanism, involving large ipsilateral field representation and interhemi-pheric connections, will lead to the generation of very

large field disparities. It is possible that a similar mechanism, operating from ocular dominance column to ocular dominance column instead of from hemisphere to hemisphere, may explain the generation of the small receptive field disparities which characterize the majority of binocular neurons so far described.

The naso-temporal decussation

One unsuspected feature of the naso-temporal decussation in both cats and ungulates was that it did not form a right-angle at its intersection with the horizontal meridian. In these two species as well as in the opossum there is an outward tilt of naso-temporal decussation such that the superior arm lies in more temporal retina. The significance of this tilt was first realized from work on the comparison of binocular visual systems in a terrestrial owl with that of the cat and can be explained in the context of Helmholtz's vertical horopter (Cooper & Pettigrew, 1979). In the vertical plane, as predicted by Helmholtz, the horopter is tilted to pass through the fixation point and the ground below the subject's feet. The tilt can be verified by a number of independent methods, both anatomical and physiological. Psychophysical studies in man also support the nature of the tilt. The magnitude of the tilt will be greatest for terrestrial animals which have a large pupiliary separation in relation to the distance from the eyes to the ground and may help explain the fact that decussations in some of the small terrestrial mammals, such as the mouse, tend to have such an oblique inclination with respect to the horizon (e.g. Drager & Olsen, 1980).

The presence of the tilt has ramifications in a number of different areas, such as the area of binocular visual development. For example, in altricial animals born with a large pupillary separation in relation to height there will be a constant change with growth of angle θ. This constant change in θ will require some remarkable readjustments within the binocular visual system if binocular vision is to remain functional throughout the period of growth, and it is of great interest, therefore, to find that the dynamics of changes in θ correspond exactly to those for the critical period in those binocular species where data on both of these changes are available (Pettigrew *et al.*, 1984). In addition to changing during development, θ may also change markedly during the daily life of an animal. During hunting, cats may adopt a posture with the head very close to the ground in which θ would have to be markedly increased. Is it possible that during this time, when a greater degree of excyclotortion of the eyes would be necessary, the resulting change in the binocular correspondence would necessitate a switch to a form of visual processing which is different from that which occurs when the cat is in the normal posture? I suggest that one of the roles for the many different visual cortical areas may be to accommodate such a change and, further, that satellite nuclei, such as the visual claustrum, may play an important role in enabling the switch from one mode of visual processing to another to occur concomitant with changes in the total motor patterning adopted by the animal.

Evolution of avian stereopsis

Owls have evolved an elaborate neural substrate for binocular vision which is comparable with that which has already been described for cats, primates and ungulates. The parallels in physiological organization are quite striking when one considers that the anatomical organization is very much different, owls having a double decussation instead of the partial decussation of mammals (Fig. 15.2). Thus, the avian thalamic neurons which contribute to the construction of binocular receptive fields at the level of the visual cortex are all of the concentrically organized receptive field type, just as one finds in mammals, despite the fact that the avian retina has a very large number of the highly specialized receptive field types such as direction-selective, orientation-selective and edge-selective varieties. In other words, in the owl visual cortex, just as in the striate cortex of mammals, the process of binocular interaction proceeds hand in hand with the elaboration of highly specialized receptive field properties, and binocular neurons have tightly matched receptive field properties in each eye. Indeed, most of the features so far described for the mammalian binocular visual pathway have their counterparts in the binocular visual pathway of the owl. So far no functional equivalent of the visual claustrum has been described in the owl, nor does the owl have ocular dominance stripes in the normal situation. With regard to the latter, however, it must be remembered that ungulates likewise have no ocular dominance stripes, yet these appear following different forms of visual deprivation during early development. The latter is true also for the New World primates and also appears to be the case for owls which do show the phenomenon of ocular dominance banding patterns if visual input is restricted during early development.

Binocular visual organization in relatives of the owl

The finding of a neural substrate for binocular vision in the owl prompted the investigation of close relatives of the owl in an effort to sketch some kind of evolutionary history of functional binocular vision in

Fig. 15.2. Double decussation pattern which mediates binocular interaction in owls and some other predatory birds. Despite the total decussation of the optic nerve fibres, information from corresponding parts of each retina converge in the brain by means of a second decussation. Note that the fibres projecting back to the opposite side of the brain arise from the representation in the thalamic relay nucleus of retina temporal to the fovea (subserving the region of binocular overlap). This pattern achieves a similar end result, at the final cortical destination, to the more familiar partial decussation of optic nerve fibres seen in mammals. f = the binocular fovea on the temporal retina and its representation in the central visual pathway. b = the limit of binocular overlap, which corresponds roughly to the position of the blind spot at the pecten/optic nerve head.

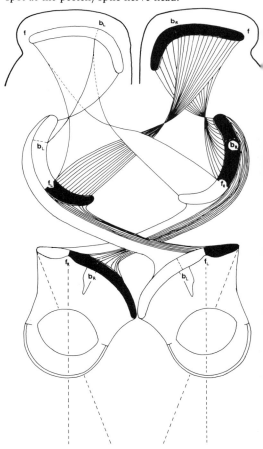

rds. All members so far studied in the families
rigidae and Tytonidae have comparable patterns of
ganization subserving binocular vision. In the diverse
der Caprimulgiformes, there is some heterogeneity
th respect to binocular vision however. For example,
e oilbird *Steatornis caripensis* is a large, Neotropical
ugivorous bird which lives in totally dark caves to
nich it returns after nocturnal forays to feed on palm
uit. It has a highly developed visual system in
dition to an echolocation system which enables it to
oid obstacles in the dark, so long as they are more
an 20 cm or so in diameter. Studies of its visual
stem have failed to reveal any evidence for a neural
bstrate for binocular vision. Instead, the visual cortex
s large numbers of orientation-selective cells which
uster around the horizontal and vertical axis and
nich can be driven only by the contralateral eye

(Pettigrew & Konishi, 1984). Likewise, a number of
members of different genera in the true nightjar
families, Caprimulginae and Chordeilinae have failed
to provide any evidence for binocular vision after
detailed electrophysiological and anatomical investigation like those carried out on the owl.

Some of the Caprimulgiformes have provided
evidence that they can achieve binocular vision like the
owls. Two such families are the Podargidae and
Aegothelidae, both Australasian groups, which have
highly developed visual systems which are used for
predation, mostly on invertebrates, at night. Like the
owls, these two families of nightjar-like birds have the
ability to take prey from the substrate. This ability sets
them apart from the other nightjar-like families which
have neither the ability to take prey from the substrate
nor any apparent mechanism for stereopsis.

Fig. 15.3. (*a*) Schematic illustration of the
postulated nasotemporal decussation for different
retinal ganglion cell classes. Note that temporal
extension is greatest for the Y-cell class.
(*b*) Consequences of different ganglion cell
decussation patterns for formation of receptive fields
of binocular neurons in the geniculostriate pathway,
if the edge of each decussation is aligned with the
medial edge of the LGN. The three different
ganglion cell classes are shown projecting to six
different laminae within the LGN. The arrow
indicates a 'projection line' along which lateral

geniculate neurons representing each eye will
converge onto the same binocular neuron. Binocular
neurons with Y input will tend to have more
convergent disparities and respond best to nearer
depth planes than binocular cortical neurons
receiving X input because of the different
decussation patterns. The limited data on the
epsilon W class suggest that they may have a
decussation pattern and central connectivity (in area
19 of the cat visual cortex) appropriate to a depth
plane at infinity, more distant than the fixation
plane represented by the X-driven system.

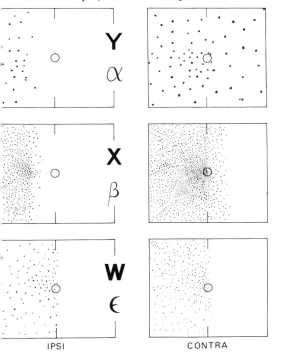

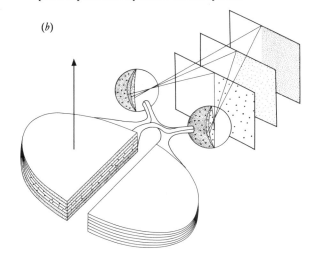

(*b*)

In this large assemblage of nocturnal birds at least, it appears that the selection pressures operating for the emergence of stereopsis appear to be greatest in the niche involving predation from a substrate such as the ground. These pressures do not appear to be particularly strong in predators which are exclusively aerial, such as the true nightjars. One can imagine a number of reasons for this. First, it is easy to see that the consequences of mislocalizing prey are much greater for an avian predator hurtling towards the ground at high speed than they would be for a nightjar taking a moth in the air. Since stereopsis is such an accurate and direct means of judging depth, it is easy to see why it might be more important in the former case. Secondly, one can see that the powerful monocular depth cue of movement parallax is much more readily available to an aerial predator as it moves in relation to its prey than it is for a 'perch-and-pounce' kind of predator which may be at a disadvantage if it moves around too much and thereby reveals itself to the prey. Finally, as pointed out by Julesz (1971), it is possible for a predator with stereoscopic ability to 'break' camouflage since a target which is invisible to monocular inspection may be visible stereoscopically if it has a slightly different depth plane from the background. Breaking camouflage is not a pervasive problem for aerial predators, for which prey will either be silhouetted or highlighted against the background. Conspicuity of prey is much more of a problem for predators taking prey on a substrate and it seems likely that the advantage conferred upon the predator which could break camouflage would be great enough to increase greatly its survival value.

The acquisition of stereopsis can be gauged roughly in birds by looking at the size of the visual Wulst which forms a prominent bulge on the dorsal surface of the brain. In the nightjars and swifts, which appear to use monocular cues exclusively for prey capture, the visual Wulst is relatively small and the vallecular groove, which marks its lateral boundary, is quite close to the midline. In contrast, in the owls, podargids and aegothelids, which have developed stereopsis, the visual Wulst is enormous and the vallecular margin extends almost out to the lateral margin of the brain.

The presence of the large Wulst, in combination with the rather small tectum, allows one to infer the presence of specialization for binocular vision in an indirect way from fossil material. The relative paucity of fossil avian material makes it difficult to go far with this approach but the existing evidence is consistent with the idea that the primitive condition amongst avian predators is stereoblindness and that the advanced condition involves the evolution of stereoscopic processing. For example, the fossil *Quipollornis* is an aegothelid nightjar which is intermediate between the existing aegothelids and the aerial insectiferous nightjars in that it has relatively longer wings (Rich & McEvey, 1977).

Stereopsis in other vertebrate groups?

There is some evidence for the existence of binocular visual processing capable of subserving stereopsis in vertebrate groups apart from the birds and mammals considered above. A number of teleost fish and anuran amphibians have anatomical evidence for binocular integration within the tectum and this evidence, coupled with behavioural evidence for binocular depth discrimination in the toad (Collet, 1977), suggests that further work may reveal fully fledged binocular visual processing in these groups (Finch & Collett, 1983; Gaillard, 1985). Certainly the diverse visual adaptations of the teleosts make it likely that at least some of them have evolved the capability for binocular vision. Whether this has an anatomical substrate similar to that of mammals, or to that of birds, or to a third system remains to be discovered.

References

Archer, M. & Clayton, G. (1984) *Vertebrate Zoogeography and Evolution in Australasia*. Carlisle, Western Australia: Hesperian Press

Allman, J. M. (1977) Evolution of the visual system in the early primates. In *Progress in Psychobiology and Physiological Psychology*, Vol. 7, ed. J. M. Sprague, pp. 1–53. New York & San Francisco: Academic Press

Barlow, H. B., Blakemore, C. & Pettigrew, J. D. (1967) The neural mechanism of binocular depth discrimination. *J. Physiol.* **193**, 327–42

Barlow, H. B., Hill, R. M. & Levick, W. R. (1964) Retinal ganglion cells responding selectively to direction and speed of image motion in the rabbit. *J. Physiol.*, **173**, 377–407

Bishop, P. O. (1973) Neurophysiology of single vision and stereopsis. In *Handbook of Sensory Physiology*, Vol. VII/2, *Central Processing of Visual Information. A: Integrative Functions and Comparative Data*, ed. R. Jung, pp. 255–305. Berlin, Heidelberg & New York: Springer, Verlag

Bravo, H. & Pettigrew, J. D. (1981) The distribution of neurons projecting from the retina and visual cortex to the thalamus and tectum opticum of the barn owl, *Tyto alba* and the burrowing owl, *Speotyto cunicularia. J. Comp. Neurol.,* 199, 419–41

Clarke, P. G. H., Donaldson, I. M. L. & Whitteridge, D. (1976) Binocular visual mechanisms in cortical areas I & II of the sheep. *J. Physiol.,* 256, 509–26

Collett, T. (1977) Stereopsis in toads. *Nature (Lond.),* 267, 349–51

Cooper, M. L., & Pettigrew, J. D. (1979) A neurophysiological determination of the vertical horopter in the cat and owl. *J. Comp. Neurol.* 184, 1–26

Drager, V. C. & Olsen, J. F. (1980) Origins of crossed and uncrossed retinal projections in pigmented and albino mice. *J. Comp. Neurol.,* 191, 383–412

Dreher, B., Fukada, Y. & Rodieck, R. W. (1976) Identification, classification and anatomical segregation of cells with X-like and Y-like properties in the lateral geniculate nucleus of old world monkeys. *J. Physiol.,* 258, 433–52

Ebbesson, S. O. E. & Schroeder, D. M. (1971) Connections of the nurse shark's telencephalon. *Science,* 173, 254–6

Ekström, P. (1982) Retinofugal projections in the eel, *Anguilla anguilla* L. (Teleostei), visualised by the cobalt-filling technique. *Cell Tiss. Res.,* 225, 507–24

Finch, D. J. & Collett, T. S. (1983) Small-field, binocular neurons in the superficial layers of the frog optic tectum. *Proc. R. Soc. Lond. B,* 217, 491–7

Frisby, J. P. & Julesz, B. (1975) The effect of orientation difference on stereopsis as a function of line length. *Perception,* 4, 179–86

Frisby, J. P. & Mayhew, J. E. W. (1980) Spatial frequency tuned channels: implications for structure and function from psychophysical and computational studies of stereopsis. *Phil. Trans. R. Soc. Lond., Ser. B,* 290, 95–116

Gaillard, F. (1985) Binocularly-driven neurons in the rostral part of the frog optic tectum. *J. Comp. Physiol. A.,* 157, 47–56

Gordon, B. (1973) Receptive fields in deep layers of cat superior colliculus. *J. Neurophysiol.,* 36 157–78

Grimson, W. E. L. & Marr, D. (1979) A computer implementation of a theory of human stereo vision. In *Image Understanding Workshop,* (April 1979, A. I. Lab.) M.I.T., Cambridge, Mass.

Hughes, A. & Vaney, D. I. (1981) Contact lenses change the projection of visual field onto rabbit peripheral retina. *Vision Res.,* 21, 955–6

Hughes, A. & Vaney, D. (1982) The organisation of binocular cortex in the primary visual area of the rabbit. *J. Comp. Neurol.,* 204, 151–64

Julesz, B. (1971) *Foundations of Cyclopean Perception.* Chicago: University of Chicago Press

Kaas, J. H., Guillery, R. W. & Allman, J. M. (1972) Some principles of organisation in the lateral geniculate nucleus. *Brain Beh. Evol.,* 6, 253–99

Kato, H., Bishop, P. O. & Orban, G. A. (1978) Hypercomplex and simple/complex cell classifications in the cat striate cortex. *J. Neurophysiol.,* 41, 1071–95

Kaye, M., Mitchell, D. E. & Cynader, M. (1981) Selective loss of binocular depth perception after ablation of cat visual cortex. *Nature (Lond.),* 293, 60–2

Levick, W. R. (1967) Receptive fields and trigger features of ganglion cells in the visual streak of the rabbit's retina. *J. Physiol.,* 188, 285–307

Levick, W. R. (1977) Participation of brisk-transient retinal ganglion cells in binocular vision – an hypothesis. *Proc. Aust. Physiol. Pharmacol. Soc.,* 8, 9–16

Luiten, P. G. M. (1981) Two visual pathways in the telencephalon of the nurse shark, *Ginglymostoma cirratum,* II. Ascending thalamo-telencephalic connections. *J. Comp. Neurol.,* 196, 539–48

Marr, D. (1982) *Vision.* San Francisco: W. H. Freeman

Marr, D. & Poggio, T. (1976) Cooperative computation of stereo disparity. *Science,* 194, 283–7

Marr, D. & Poggio, T. (1979) A computational theory of human stereo vision. *Proc. R. Soc. Lond., Ser. B,* 204, 301–28

Mayhew, J. E. W. & Frisby, J. P. (1979) Surfaces with steep variations in depth pose difficulties for orientationally tuned disparity filters. *Perception,* 8, 691–8

Mayhew, J. E. W. & Frisby, J. P. (1980) The computation of binocular edges. *Perception,* 9, 69–86

Mayhew, J. E. W. & Longuet-Higgins, H. C. (1982) A computational model of binocular depth perception. *Nature (Lond.),* 297, 376–378

Mitchell, D. E., Kaye, M. & Timney, B. (1979) A behavioural technique for measuring depth discrimination in the cat. *Perception,* 8, 389–96

Nelson, J. I., Kato, H. & Bishop, P. O. (1977) The discrimination of orientation and position disparities by binocularly-activated neurons in cat striate cortex. *J. Neurophysiol.,* 40, 260–4

Newton, I. (1730) *Opticks.* 1952 issue based on 4th ed. New York: Dover Publications

Ogle, K. N. (1950) *Researches in Binocular Vision.* New York: W. B. Saunders

Pettigrew, J. D. (1979a) Comparison of the retinotopic organisation of the visual Wulst in noctural and diurnal raptors, with a note on the evolution of frontal vision. In *Frontiers of Visual Science,* ed. S. J. Cool & E. L. Smith, pp. 328–35. New York: Springer Verlag

Pettigrew, J. D. (1979b) Binocular visual processing in the owl's telencephalon. *Proc. R. Soc. Lond., Ser. B,* 204, 435–54

Pettigrew, J. D. & Konishi, M. (1976) Neurons selective for orientation and binocular disparity in visual Wulst of the barn owl (*Tyto alba*). *Science,* 193, 675–8

Pettigrew, J. D., Nikara, T. & Bishop, P. O. (1968) Binocular interaction on single units in cat striate cortex: simultaneous stimulation by single moving slit with receptive fields in correspondence. *Exp. Brain Res.,* 6, 391–410

Pettigrew, J. D. & Konishi, M. (1984) Some observations on the visual system of the oilbird (*Steatornis caripensis*). *Natl Geogr. Res. Rep.,* 16, 439–50

Pettigrew, J. D., Ramachandran, V. S. & Bravo, H. (1984) Some neural connections subserving binocular vision in ungulates. *Brain Behav. Evol.,* 24, 65–93

Poggio, G. F. & Poggio, T. (1984) The analysis of stereopsis. *Ann. Rev. Neurosci,* 7, 379–412

Poggio, G. F. & Talbot, W. H. (1981) Mechanisms of static and dynamic stereopsis in foveal striate cortex of the rhesus monkey. *J. Physiol.,* 315, 469–92

Polyak, S. (1957) *The Vertebrate Visual System.* Chicago: University of Chicago Press

Provis, J. M. (1979) The distribution and size of ganglion cells in the retina of the pigmented rabbit: a quantitative study. *J. Comp. Neurol.*, **185**, 121–39

Ramôa, A. S., Rocha-Miranda, C. E., Méndez-Otero, R. & Josuá, K. M. (1983) Visual receptive fields in the superficial layers of the opossum's superior colliculus. *Exp. Brain Res.*, **49**, 373–81

Rich, P. V. & McEvey, A. (1977) A new owlet-nightjar from the early to mid-Miocene of eastern New South Wales. *Mem. Nat. His. Mus. Victoria*, **38**, 247–53

Richards, W. (1970) Stereopsis and stereoblindness. *Exp. Brain Res.*, **10**, 380–8

Sanderson, K. J., Darian-Smith, I. & Bishop, P. O. (1969) Binocular corresponding receptive fields of single units in the cat dorsal lateral geniculate nucleus. *Vision Res.*, **9**, 1297–303

Sanderson, K. J., Darian-Smith, I. & Bishop, P. O. (1971) The properties of binocular receptive fields of lateral geniculate neurons. *Exp. Brain Res.*, **13**, 178–207

Stone, J. (1983) *Parallel Processing in the Visual System.* New York: Plenum Press

Van Sluyters, R. C. & Stewart, D. L. (1974) Binocular neurons on the rabbit's visual cortex. Receptive field characteristics. *Exp. Brain Res.*, **19**, 166–95

Wallman, J. & Pettigrew, J. D. (1985) Conjugate and disjunctive saccades in two avian species with contrasting oculomotor strategies. *J. Neurosci.* **5**, 1418–28

Walls, G. L. (1942) *The Vertebrate Eye and its Adaptive Radiation* 1967 fascimile of 1942 edn. New York & London: Hafner Publishing Co.

Zuidam, I. & Collewijn, H. (1979) Vergence eye movements of the rabbit in visuomotor behavior. *Vision Res.*, **19**, 185–94

16

Sensory processing in the mammalian auditory system – some parallels and contrasts with the visual system

LINDSAY M. AITKIN

Introduction

It is my purpose in this review to examine the organization of the auditory pathway, particularly below the thalamocortical auditory system, and to make comparisons with some features of the visual pathway. I will concentrate on the manner in which the receptor surface (the cochlea) is represented within the primary brain auditory pathway, the emergence of parallel processing channels and the representation of sensory space. These issues have been crucial, at one time or another, in our understanding of the visual system. Space restraints dictate that this essay will only hint at such aspects as transduction mechanisms, the discharge patterns of single neurons in response to auditory and visual stimuli, and the different functions subserved by the multiple representations of the receptor surfaces on the visual and auditory cortices. This essay is, in some respects, an attempt to compare the two systems from the standpoint of my own research in auditory physiology.

It is possible at the receptor level to recognize some similarities in the way appropriate physical stimuli are handled by the two systems. First, both photoreceptors and cochlear hair cells are true receptor cells in which the level of polarization, and thus receptor current, varies with the amount of stimulus energy received, and in which ultrastructural features (synaptic rods) and physiological properties are indicative of chemical transmission between receptor cell and the distal processes of bipolar cells (and amacrine cells of the retina).

223

Comparable patterns of receptor connectivity are related, at a simplistic level, to comparable functions. Thus, it is well known that the peripheral retina is sensitive to levels of illumination lower by many orders of magnitude than the fovea. This property is imparted, in the primate, by the essentially 'private line' connections of cones to ganglion cells in the fovea, while rods in the peripheral retina have highly convergent connections to ganglion cells. Similarly, it has been suggested by a number of lines of evidence that the outer hair cell system of the cochlea may provide the greatest sensitivity to sound intensity, while elements of the inner hair cell system are sharply tuned to sound frequency. Information about the differences between afferents from outer and inner hair cells is hampered by the likelihood that outer hair cell afferents may be unmyelinated and too small to have been recorded using the standard techniques for the study of coding in cochlear nerve fibres (Kiang *et al.*, 1982).

Before examining the organization of the central auditory system, it should be noted that one outstanding feature of retinal ganglion cell receptive field organization – the presence of lateral inhibition – has no common counterpart at the auditory periphery. While it is still possible, in the face of changing terminology, to talk about retinal receptive fields with ON centres and OFF surrounds, and *vice versa* (Kuffler, 1953), no unequivocal evidence has yet been presented for neural interactions between receptor cells or their afferents in the cochlea. On the other hand, there is no large efferent pathway to the mammalian retina equivalent to the olivocochlear bundle. The function of the olivocochlear bundle remains a matter of debate, but it is possible that it may serve functions in some way comparable to those served by the intraocular muscles in visual focussing.

Topographic organization in the auditory pathway

The three major sensory systems – visual, somatosensory and auditory – all demonstrate a precise organization according to the topography of the appropriate receptor surface. While this fact was established early in this century for the visual and somaesthetic systems, on the basis of studies on human cerebral function, it remained for Woolsey & Walzl (1942) to carry out the crucial experiments to demonstrate this fact for the auditory system. They stimulated the cochlear nerve electrically at different points along the cochlear spiral of the cat and showed that the locus of potentials evoked in the auditory cortex shifted as a function of stimulus location in the cochlea.

Since that time, with the introduction of extracellular microelectrode recording techniques, most of the nuclei of the auditory pathway have been shown to contain at least one complete representation of the cochlea (e.g. Rose, 1960; Merzenich & Reid, 1974; Aitkin, 1976; Reale & Imig, 1980; Calford & Webster 1981). In all of these studies, the usual procedure has been to establish for each neuron the sound frequency at which mimimun sound pressure is needed to evoke a response (threshold best frequency or characteristic frequency), and then to relate that frequency to the presumed recording site in the nucleus investigated.

It is noteworthy that while the resultant tonotopic or cochleotopic organization is analogous to retinotopic organization in the visual system, the represented feature in the auditory system – sound frequency – is a temporal and not a spatial attribute. Furthermore, a given auditory neuron is not excited by an *unique* range of frequencies; instead, as sound pressure is increased above threshold, the neuron becomes responsive to a steadily increasing range of frequencies. For example, neurons deriving their most sensitive input from the base of the cochlea (and consequently possessing high best frequencies) may be caused to discharge to almost the entire audiofrequency range at high stimulus levels. In contrast, lateral inhibition at the retinal level ensures that the area of a visual receptive field is relatively insensitive to light intensity, so that the retinal site of origin of activity is 'labelled', even at high light intensities.

The unit receptive field in the auditory system, the tuning curve, is for most neurons a V-shaped plot of frequency against threshold intensity. Modern computer methods enabling the automated measurement of tuning curves have shown that the latter do not become appreciably sharper at increasingly higher levels of the auditory neuraxis (Calford, Webster & Semple, 1983).

There is little evidence to suggest hierarchical changes in tuning characteristics of auditory neurons of the sort proposed by Hubel & Wiesel (1962, 1965) for the visual pathway, although some data are compatible with this suggestion (Oonishi & Katsuki,1965). However, some neurons, particularly in the dorsal cochlear nucleus and inferior colliculus, do show considerable complexity in their tuning characteristics

istics, because stimuli falling outside the excitatory tuning curve may inhibit spontaneous activity (Goldberg & Brownell, 1973; Semple, 1981). Additional inhibitory regions may also be demonstrated at high intensities in the excitatory frequency region (non-monotonic units) (Greenwood & Maruyama, 1965; Semple & Aitkin, 1980). However, inhibitory interactions in the auditory system are most commonly seen as temporal, rather than spatial, features of an auditory response, particularly in the form of inhibitory 'pauses' occurring after an excitatory response.

The retrograde transport of horseradish peroxidase (HRP) has proven invaluable for demonstrating the topographic connectivity of the auditory pathway (Roth *et al.*, 1978; Andersen, 1979; Calford & Aitkin, 1983). This is illustrated in Fig. 16.1 where the topographic relationships between the major brainstem auditory nuclei and the central nucleus of the cat inferior colliculus are schematically illustrated.

The central nucleus of the inferior colliculus in most mammals is perhaps the largest auditory nucleus below the thalamocortical auditory system, and has a characteristically stratified neuronal organization (Rockel & Jones, 1973; Fitzpatrick, 1975; Aitkin & Kenyon, 1981; Morest & Oliver, 1984). It is likely that each neuronal 'lamina' has strong connectional associations with a focus on the basilar membrane of the cochlea, such that apical cochlear segments (low sound frequencies) are represented dorsally and basal segments (high frequencies) ventrally in the central nucleus. This pattern of cochlear representation has been elegantly confirmed using functional labelling with 2-deoxyglucose (Servière & Webster, 1981).

Thus, a large injection of HRP into the low-frequency part of the central nucleus (Fig. 16.1) provides retrograde label in topographically discrete regions of four brainstem nuclei ipsilaterally (the dorsal and ventral nuclei of the lateral lemniscus and the medial and lateral superior olives) and five contralateral nuclei (the dorsal nucleus of the lateral lemniscus, the lateral superior olive and the three nuclei of the cochlear nuclear complex). A similar pattern, with the focus of the label appropriately shifted, occurs following the high-frequency ventral injection (Fig. 16.1).

While a complete cochlear representation has been identified in each brainstem nucleus, the sizes of some of these nuclei vary from species to species, and the proportion of a nucleus occupied by a given cochlear segment also varies. In the cat and dog, high frequencies (> 2 kHz) are relatively compressed in the ventral most segment of the medial superior olive, while low frequencies are even more compressed into the dorsolateral limb of the S-shaped lateral superior olive (Fig. 16.1) (Tsuchitani & Boudreau, 1966; Goldberg & Brown, 1968).

It is generally accepted that the central representation of a receptor surface is related to the peripheral innervation density (e.g. Holmes, 1945; Rose & Mountcastle, 1959). Thus, the representation of the fovea in primates occupies a much greater proportion of the visual cortex than does the peripheral retina, since there are many more ganglion cells per unit area in the fovea than in the periphery. An 'acoustic fovea' has been suggested for the horseshoe bat in which a disproportionate innervation of the basal parts of the basilar membrance has been observed, with a corresponding increase in the proportion of central neurons driven by ultrasonic frequencies, used in

Fig. 16.1. Schematic frontal sections of the cat midbrain (upper) and brainstem (lower), including the principal lower auditory nuclei. Large injections of HRP have been placed in the dorsal (hatched) and ventral (stippled) thirds of the central nucleus of the inferior colliculus, where low and high best frequencies (LOW and HIGH) are, respectively, concentrated. The hatched or stippled areas in the lower auditory nuclei represent the locations of labelled cells following the two injections. (Summarized from Roth *et al.*, 1978 and Aitkin, unpublished observations.)

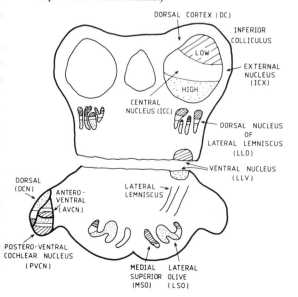

echolocation (e.g. Schuller & Pollak, 1979). An acoustic fovea appears to have been evolved independently by two groups of bats, each of which requires very precise frequency sensitivity to detect the wing beats of fluttering prey. Both these bats practise Doppler-shift compensation of their emitted constant-frequency echolocation signal so that the echo always has a frequency which is on the acoustic fovea, despite the frequency variations which might otherwise be produced by alterations in the relative motions of the bat and its prey. In this way, the small frequency modulations of the echo produced by the prey's wing flutter can be detected in the enormously expanded brain representation of the carrier frequency (Neuweiler, 1984).

The central representation of the cochlea in some mammals is not a linear function of basilar membrane innervation density. Instead, certain octaves may be more densely represented than are others. In the cat, for example, the cochlea is progressively more strongly represented in the auditory cortex as a function of apical to basal stimulus location (Merzenich, Knight & Roth, 1975). Species such as the rhesus monkey (Merzenich & Brugge, 1973) and the grey squirrel (Merzenich, Kaas & Roth, 1976) show a more uniform representation of cochlear place in the auditory cortex.

Thus, topographic organization in the auditory system has many features in common with the retinotopic organization of the visual pathway. A major difference would appear to be that information about the retinal locus of stimulation is preserved, even by higher-order receptive fields in the visual cortex over a wide continuum of light intensity, while the cochlear place of stimulation becomes increasingly more ambiguous with increasing sound pressure level.

Parallel pathways in the lower auditory system

The visual pathway consists of a number of separate streams which either travel in parallel through the geniculocortical system or supply the superior colliculus and pulvinar. Three different types of retinal ganglion cell have been recognized in the cat as starting points for these pathways on the basis of physiology and morphology – X, Y and W. It is not the purpose of this article to examine the details of these systems in the mammalian visual pathway (see e.g. Stone, Dreher & Leventhal, 1979) but it is appropriate to note some

possible functional differences between the systems in vision which may have parallels in audition. Ganglion cells classified as X cells, found in greatest numbers in and around the area centralis or fovea, have properties compatible with a role in high-resolution vision. Their central projections include a major component to Area 17 of the cat. The Y cells are relatively abundant in the peripheral retina, seem particularly suited to detection of movement, and project ultimately to the visual cortex (particularly area 18) and the superior colliculus. Finally, the W cells are concentrated in the visual streak and project mainly to the superior colliculus and, to a lesser extent, the visual cortex (particularly area 19). Their functions in vision are less clear than for X and Y cells but they may, for example, be concerned with visuomotor reflexes.

Separate anatomical channels have long been recognized as comprising the lower auditory pathway. Strictly speaking, these are not 'parallel' because they may derive their input from common spiral ganglion cells – the type I cells which form 95% of all cochlear ganglion cells (Spoendlin, 1972). These in turn are likely to relate to inner hair cells, and it is not clear what contribution the remaining ganglion cells, connected to outer hair cells, make in audition (Kiang *et al.*, 1982). However, if the number of synaptic interruptions between photoreceptors and retinal ganglion cells is equated for the peripheral auditory pathway, it might be more appropriate to consider cells in the cochlear nuclei as analogous to retinal ganglion cells as the starting points for parallel pathways in the auditory system. Thus, the locus of acoustic feature extraction could be considered as the cochlear nucleus, rather than the receptor sheet itself.

It is convenient to subdivide brainstem auditory connections leading to the midbrain into three main groups: direct crossed projections from all divisions of the cochlear nuclear complex (principally the dorsal cochlear nucleus) and indirect connections from the cochlear nucleus either via the superior olivary nuclei or the nuclei of the lateral lemniscus (Fig. 16.1). While this is probably a considerable oversimplification, certain features of the responses of neurons comprising each of these channels are suggestive of neural selection within each channel of some acoustic feature rather than others. Of these pathways, three have been examined in sufficient detail to warrant some speculations about functions in audition: those relaying hin the lateral and medial superior olives, and the ect crossed connection from the dorsal cochlear cleus to the inferior colliculus.

The cochlear nerve bifurcates upon entering the instem; one branch of each fibre proceeds socaudally to the dorsal cochlear nucleus (DCN), ile the other travels ventrally to the ventral cochlear cleus (Fig. 16.2). Within the DCN, cochlear nerve minals either contact the basal dendrites of amidal cells directly or may innervate these neurons irectly via interneurons whose local axons also ply pyramidal cells (Osen & Mugnaini, 1981). amidal cell axons, forming the output of the DCN, ject mainly to the opposite inferior colliculus. It is ther likely that the interneurons of DCN have ibitory actions upon pyramidal cells (Young & gt, 1981). The excitatory–inhibitory responses of se neurons and of their target neurons in the inferior iculus have been alluded to in the first section. It ikely that these neurons will detect the changing ctral characteristics of a complex auditory stimulus; ir non-monotonic properties with respect to ulus intensity could also give them a role in ulus range detection.

The ventral cochlear nucleus is complex in rnal organization but two principal cell types ulate its anteroventral pole: the large and small

spherical cells (Osen, 1969*a*). Cochlear nerve terminals from the apex of the cochlea (low sound frequency) penetrate most anteriorly (Rose, 1960) to supply the large spherical cells, many of which receive large synaptic endings – the end bulbs of Held (Brawer & Morest, 1975). The small spherical cells spread as a band across much of the cochlear input to the anteroventral cochlear nucleus; some neurons receive end bulbs (Osen, 1969*b*) but many may not (Brawer & Morest, 1975). There is no suggestion of lateral inhibition with large or small spherical cells, and they appear to relay cochlear input with great fidelity.

The axons of large spherical cells innervate the spindle-shaped neurons of the medial superior olive bilaterally, while those of the small spherical cells project to the ipsilateral lateral superior olive and, after relay in the nucleus of the trapezoid body, to the contralateral lateral superior olive (Fig. 16.2) (Osen, 1969*b*). As described in the previous section, the two superior olivary nuclei have disproportionate cochlear representations which reflect the origin of their connections in the cochlear nucleus; their physiological properties are distinctively different and will be described in the subsequent section on spatial localization.

Brainstem auditory pathways collect together at the central nucleus of the inferior colliculus. It is

Fig. 16.2. Simplified view of the afferent connections of the lateral (LSO) and medial (MSO) superior olives. The large spherical cells of the anteroventral cochlear nucleus (AVCN) are shown as open circles and the small spherical cells as dots. Neurons depicted in the dorsal cochlear nucleus (DCN) are pyramidal cells, while those in the MSO

and LSO are spindle-shaped neurons. B, M, A: representation of base, middle and apical parts of cochlea, respectively; PVCN: posteroventral cochlear nucleus; MNTB: medial nucleus of trapezoid body; ICC: central nucleus of inferior colliculus. (From Aitkin, 1983, with permission.)

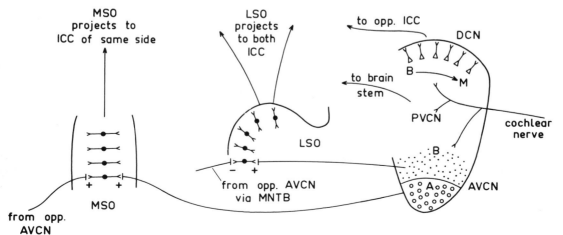

possible to consider each part of the central nucleus as receiving a 'core' projection from one or two brainstem nuclei, with different cores overlapping. Thus, some degree of segregation persists in the midbrain; the pattern of connectivity is also strongly correlated with cochlear representation in the inferior colliculus.

These patterns can be deduced by the use of small iontophoretic injections of HRP, made from micropipettes, previously used to characterize the physiological properties of inferior collicular neurons. In Fig. 16.3, the locations of six iontophoretic deposits and one small microsyringe injection of HRP (Site 1) are referenced to mid-frontal (A) and mid-sagittal (B) planes through the inferior colliculus. The threshold best frequency in kHz (BF) at each site and the

corresponding brainstem locations of labelled neurons are shown in the table (Fig. 16.3c), with the brainstem nucleus containing the largest number of labelled cells identified for each injection with a cross.

The larger microsyringe deposit (1), made in the absence of physiological characterization, is of interest because it reveals that the lateral part of the central nucleus receives projections from the dorsal nuclei of the lateral lemniscus and the superior olivary nuclei (all involved in binaural integration), as well as the contralateral cochlear nucleus (mainly the DCN). The second site (2) was associated with responses to low frequencies (0.7 kHz) but its location (dorsomedial) places it outside the 'laminated' part of the central nucleus, to a region which does not receive direct connections from the lateral lemniscus (Rockel & Jones 1973). Instead, its connections originate mainly from the low-frequency part of the opposite central nucleus.

The other injections reveal that lateral low-frequency sites preferentially label the medial superior olive (sites 3 and 4), while high-frequency loci are related to the lateral superior olive (6 and 7). All regions in the lateral half of the central nucleus receive additional projections from the cochlear nucleus and the dorsal nucleus of the lateral lemniscus. These data are compatible with the concept of 'core' zones in the inferior colliculus receiving parallel pathways from the brainstem. Further data are needed in order to determine the extent to which this apparent segregation is related to disproportionate tonotopic maps in brainstem nuclei.

Cochleotopic, diffuse and acousticomotor pathways

The preceding section has attempted to show that, within the primary auditory pathway, a number of parallel streams of information reach the inferior colliculus where they are bound together by a common cochleotopic organization. There are at least superficial similarities to the retinotopically integrated X, Y and W streams ascending to the visual cortex and superior colliculus.

The primary, cochleotopic, auditory pathway continues through the medial geniculate body to the auditory cortex, where at least four separate cochlear representations have been observed in the cat (Reale & Imig, 1980). Neurons in the primary auditory cortex have response properties to binaural stimuli which bear a close resemblance to those in the two pathways

Fig. 16.3. Locations of small iontophoretic injections of HRP in the inferior colliculus, viewed in mid-frontal (A) or mid-sagittal (B) planes. The table (C) correlates the numbered injection loci with the distribution of labelled cells in ipsilateral (I) and contralateral (C) brain stem auditory nuclei in the cat. The threshold best frequency (BF) in kHz at each injection locus is shown in the left-hand column; dots indicate nuclei containing labelled cells, while those with crosses indicate the nucleus containing the largest number of labelled cells in a given experiment. CNC: cochlear nuclear complex; other abbreviations as in Fig. 16.1. (From unpublished observations of Aitkin, Schuck & Semple.)

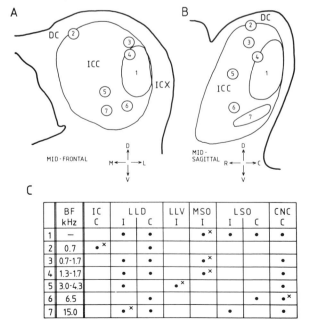

	BF kHz	IC C	LLD I	LLD C	LLV I	MSO I	LSO I	LSO C	CNC C
1	—		•	•		•×	•	•	•
2	0.7	•×		•					
3	0.7-1.7		•	•		•×			•
4	1.3-1.7		•	•		•×			•
5	3.0-4.3		•		•×				•
6	6.5			•				•	•×
7	15.0		•×	•			•		•

laying through the lateral and medial superior olives (Imig & Adrian, 1977; Middlebrooks, Dykes & Merzenich, 1980; Middlebrooks & Zook, 1983), and these neuron types appear to be segregated on the cortical surface. Another cortical region, field P, contains a high proportion of neurons whose responses to tonal stimuli as a function of intensity are similar to those of neurons comprising the DCN–inferior colliculus pathway (Orman & Phillips, unpublished). However, aside from these features, little physiological data exist to help clarify the differences between these cochleotopic auditory fields or to enable comparisons to be drawn with visual cortical fields. At present there are no hypotheses for the functions of the various auditory cortical regions analogous to those put forward for the visual cortices (for review, see e.g. van Essen, 1979).

Diffuse auditory pathways have been described running adjacent to the cochleotopic pathways (Andersen, Knight & Merzenich, 1980; Calford & Aitkin, 1983). Linkages in this stream in the cat include the dorsal cortex of the inferior colliculus, the dorsocaudal part of the medial geniculate body and area AII of the cat cerebral cortex. They are characterized by neurons with poorly defined frequency discrimination and are only weakly tonotopically organized; they are usually more readily activated by stimuli with greater spectral complexity than pure tones. The functions of these pathways are obscure; they may represent the auditory counterpart to the lateral posterior complex/pulvinar system of the visual pathway which, in the cat, receives no direct retinal input but has complex interrelations with the visual cortex and suprasylvian association areas (e.g. Updyke, 1977).

A major contrast between the visual and auditory systems lies in the position and roles of the two colliculi. Ramon y Cajal (1955) believed that the inferior colliculus received connections for reflex functions from the ascending auditory pathway, which was considered mainly to bypass the inferior colliculus on its way to the medial geniculate body. Thus, in Ramon y Cajal's view, the superior and inferior colliculi had more or less similar positions in their respective pathways.

Since the advent of the Nauta technique and retrograde tracing, it is now clear that in the cat the central nucleus of the inferior colliculus is an obligatory way for all but a handful of ascending auditory afferents (Goldberg & Moore, 1967). For example, in one recent experiment from this laboratory, in which the ascending axons to the cat medial geniculate body on one side (the brachium of the inferior colliculus) appeared to be entirely filled with HRP, 23 797 labelled neurons were counted in the inferior colliculi, compared with less than 100 in the nuclei of the brainstem (lateral lemniscus, superior olive and cochlear nucleus). In other words, in this experiment, 99.5 % of the ascending afferents terminated in the inferior colliculus and only about 0.5 % (mostly originating from the nuclei of the lateral lemniscus) continued to the medial geniculate body.

Although the inferior and superior colliculi thus differ in their relationships to the two ascending sensory pathways, the inferior colliculus, like the superior colliculus, is important for sensorimotor behaviour. Neurons in the external nucleus and dorsal cortex of the inferior colliculus (see Fig. 16.1) project to the superior colliculus and to the lateral pontine nucleus (Hashikawa & Kawamura, 1983). The latter region provides mossy fibres to the cerebellar vermis, where many neurons responding to acoustic stimuli have auditory receptive fields in the direction of frontal gaze (Aitkin & Rawson, 1983). Connections to the superior colliculus also provide the possibility of the spatial register of acoustic and visual stimuli (Gordon, 1973; Wise & Irvine, 1983) and for modulation of reflex behaviour.

Spatial localization

The spatial localization of a visual stimulus requires a knowledge of the portion of the retinal receptor array activated by that stimulus and knowledge of the positions of the eyes and head during visual tracking. Accurate auditory spatial localization necessitates binaural comparisons, since a focus of activity in one cochlea is ambiguous with respect to stimulus spatial position. However, some degree of monaural localization is possible; as for the visual system there is a dominant representation of the left spatial hemifield in the right cortex and *vice versa* (e.g. Jenkins & Masterton, 1982). There is a considerable difference between the acuities of human spatial resolution in the two systems: while the eye is able to discriminate spatial positions separated by as little as 0.5–1 min of arc (e.g. Westheimer, 1963), auditory spatial localization is probably no better than about 2 degrees of arc (e.g. Klumpp & Eady, 1956; Brown, 1982).

It has been well known since the time of Rayleigh (1909) that a sound source generates two types of interaural disparity. Wavefronts will arrive at the closer ear at an earlier time than at the further ear. For very low frequencies (e.g. for a human listener, 100 Hz), when the wavelength of the sound (330 cm) is very long compared with the interaural separation (25 cm), pressure changes at the two ears at any one time will never be very different. As frequency increases (e.g. 1000 Hz) the interaural distance becomes a significant fraction of the wavelength and *interaural phase differences* will become marked. At high frequencies (e.g. 8000 Hz) the wavelenth (4 cm) is such that a number of waves will occur between the two ears and wavelength therefore becomes an ambiguous cue. However, at these frequencies the acoustic impedance of the head becomes significantly greater than that of air, leading to the reflection of much sound energy and the consequent establishment of *interaural intensity differences*. At high frequencies, too, the pinna becomes increasingly more directional in its capacity to amplify sound and can thus enhance these intensity differences.

It would therefore be predicted that sound localization in humans is frequency-dependent. This has been demonstrated by a number of authors (e.g. Stevens & Newman, 1934). It has also been shown in both man and monkey that minimum audible angles are measured near 1000 and at frequencies in excess of about 4000 Hz (Klumpp & Eady, 1956; Brown, 1982). In the cat, lowest angular thresholds appear to occur between 1 and 2 kHz (Casseday & Neff, 1973).

That the two brainstem channels relaying through the medial (MSO) and lateral (LSO) superior olives are able, respectively, to extract interaural time and intensive cues, is suggested by a number of observations. First, lesions of the trapezoid body, which carries fibres from the cochlear nuclei to the superior olives, seriously disrupt the ability of cats to make behavioural judgements of sound location (e.g. Moore, Casseday & Neff, 1974). On the other hand, lesions of one lateral lemniscus (carrying one half of the binaurally integrated information from the olives) produce a less severe deficit, restricted to the spatial hemifield contralateral to the lesion (Jenkins & Masterton, 1982). These results show that the superior olives are essential for accurate sound localization, although the experiments do not discriminate the frequency ranges over which the constituent nuclei operate.

A second line of experimentation has shown that the relative sizes of the medial and lateral superior olives differ across phylogeny and that this difference is related to hearing range or spectrum of localization (Irving & Harrison, 1967). In animals with essentially high-frequency hearing (e.g. bats, dolphins), the LSO is large and the MSO is small or absent. In animals with low-frequency hearing (e.g. primates), the MSO is always larger than the LSO; the latter, however, is usually present. The largest MSO in the 14 species examined by Irving and Harrison was found in the cat which also has a large LSO.

Finally, a limited body of physiological data shows that neurons in the MSO are often excited by stimuli presented to each ear (Fig. 16.2) and some of these show marked changes in discharge rate as a function of interaural time or phase difference (Goldberg & Brown, 1969). Similar observations have been made by a number of authors at levels above the superior olive. For the majority of these neurons threshold best frequencies are below 2 kHz. The common shape of discharge rate versus interaural time delay functions is shown schematically in Fig. 16.4A.

More information is available for the LSO, in which most neurons are excited by ipsilateral and inhibited by contralateral stimuli (Fig. 16.2) (Boudreau & Tsuchitani, 1968) and can thus monitor the interaural intensity difference. There is already disproportionate representation of high frequencies in the LSO (Fig. 16.1) (Tsuchitani & Boudreau, 1966) and neurons with similar binaural properties, studied by many authors at higher levels of the auditory system, usually have high best frequencies. The typical shape of a discharge rate versus interaural intensity difference function is schematically illustrated in Fig. 16.4C; note that much of the output of the LSO crosses to the opposite side, so that contralateral and ipsilateral stimuli at higher levels are often, respectively, excitatory and inhibitory.

Spatial receptive fields of auditory neurons

There is thus good evidence that two channels through the brainstem encode separately the frequency-dependent cues of interaural time and intensity difference. The physiological experiments providing this evidence have used diotic stimulation where stimuli are presented to each ear through separate receivers implanted into each ear canal.

In contrast to the wealth of diotic information, there has until recently been a dearth of data obtained from experiments carried out with a single sound source moved in anechoic space about an animal's head – i.e. free-field stimulation. Experiments from this laboratory and from Moore's laboratory at Oxford have been concerned with the measurement of spatial receptive fields in the inferior colliculus of the cat. Some of the findings are summarized in Fig. 16.4B,D.

Only a proportion of neurons are directionally sensitive – many respond equally well, irrespective of stimulus location (omnidirectional units: Middlebrooks & Pettigrew, 1981; Semple *et al.*, 1983). All directionally sensitive units appear to have one common

Fig. 16.4. Idealized single-unit firing-rate functions plotted as though obtained in diotic (A, C) and free-field (B, D) experiments. The neural firing profile in A is similar to many reported in which the interaural time delay between low-frequency tones, presented diotically, has been varied, while that for C results from variation in the interaural intensity difference, for another unit, with high-frequency tones. The spatial receptive fields that might be predicted from these diotic functions are plotted in B (corresponding to A) and D (corresponding to C). The two circles depict the frontal region of space, with the animal's head located at X; C and I show the poles of the circles facing the contralateral and ipsilateral ears, respectively. Heavy lines show the firing-rate boundaries calculated from the diotic functions; vertical light lines are equi-azimuthal and horizontal light lines are equi-elevational. The latter are separated by 10-degree intervals. (From Aitkin, 1983, with permission.)

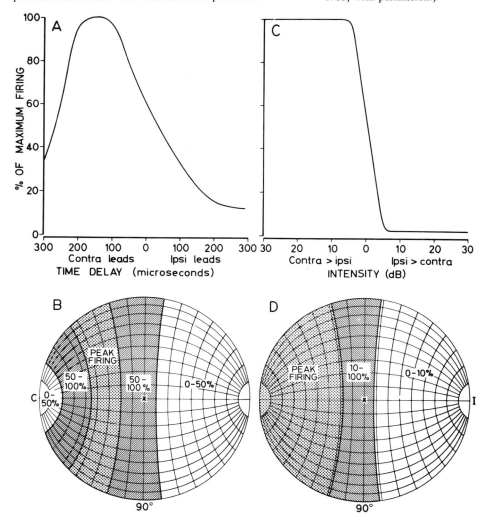

feature – a 'discharge border' where firing rate as a function of azimuth falls steeply, sometimes to zero (e.g. Fig. 16.4B,D). For the vast majority of units, maximum firing occurs at contralateral azimuths. For low-frequency directionally sensitive units, an azimuth or range of azimuths can be defined at which peak firing occurs (Fig. 16.4B; 20–45 degrees). With high-frequency neurons, peak firing is manifested over a broader range of contralateral azimuths (Fig. 16.4D); this range and the border itself may be sharply dependent on the position of the pinnae (Fig. 16.5) (Phillips *et al.*, 1982).

While there is now some evidence on the coding of stimulus position in the median plane by inferior collicular neurons, it is not clear how stimulus elevation is represented. The schematic receptive fields of Fig. 16.4 summarize the fact that, for most units

Fig. 16.5. Discharge rate plotted against speaker location in azimuth at 10 degrees above the horizontal plane, for a unit (82-15-5) recorded in the inferior colliculus of the anaesthetized brush-tailed possum. The ordinate plots the number of spikes in response to 20 tone pips (duration 200 ms, frequency 20.5 kHz, intensity 22 dB sound pressure level) presented at 1 per second. The inserts indicate, schematically, the positions of the pinnae of the possum, viewed from above. Moving the contralateral pinna from its resting position, aligned to the interaural plane (closed circles) to an angle of approximately 45 degrees to this plane (open circles) results in the discharge border shifting towards the ipsilateral hemifield to a corresponding extent. Note that, in the latter position, spatial points behind the pinna lead to reduced firing or inhibition below spontaneous (SPON) firing levels. (Unpublished observations of Aitkin, Gates & Phillips.)

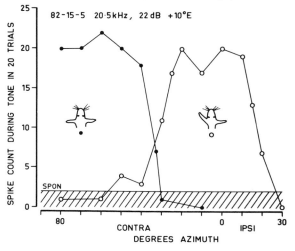

studied, discharge rates change little when the speaker is elevated or lowered at a constant azimuth. However, it is pointless to speculate how stimulus elevation might be coded given the total lack of behavioural information about the ability of cats to discriminate the elevation of a stimulus.

The organization of directionally sensitive neurons within an auditory nucleus remains to be determined for mammals. One strong possibility is that the azimuthal location of a stimulus is represented across one dimension of the isofrequency plane deduced from electrophysiological and anatomical studies. Much more information is, however, available for the barn owl, a creature whose auditory system appears specialized for the acoustic location of prey (Konishi, 1973; Knudsen, Blasdel & Konishi, 1979; Knudsen & Konishi, 1979).

A class of neuron in the inferior colliculus and optic tectum of the barn owl has acoustic receptive fields which are limited in both azimuth and elevation. The spatial locations of the centres of these fields are topographically represented in part of the inferior colliculus to from a neural map of acoustic space (Knudsen & Konishi, 1978). The spatial position of an acoustic stimulus is set into register with visual space in the optic tectum of the barn owl where most units are activated bimodally (Knudsen, 1982). Penetrations through the tectum reveal that the receptive field centres to both visual and auditory stimuli are closely aligned, that the anteroposterior dimensions of the tectum map stimulus azimuth, and that increases in penetration depth reveal systematic changes in elevation at a constant azimuth (Fig. 16.6).

It is not clear how auditory receptive field topography is mapped out in the mammalian superior colliculus, although it is known that at least a rough correspondence exists between visual and auditory space in this nucleus (Gordon, 1973; Wise & Irvine, 1983). Further work is being carried out at Monash University to examine the question of auditory receptive field topography in the superior and inferior colliculi.

Conclusion

It is usually considered that vision has primacy as a sensory system, and there is no doubt that the last several decades have seen a flowering of visual neuroscience to match the position spinal cord and peripheral nerve neurophysiology achieved at earlier times. However, the importance of audition in vocalization and speech has little counterpart in the visual system, and much of the neuroscience of audition has had some connection with temporal features of acoustic stimuli, of pitch and loudness, rather than form and spatial position.

Certain features of both systems have been compared – the precise representations of the receptor sheets, the existence of feature extraction at a peripheral level, and the transmission of different aspects of an auditory stimulus along separate, 'parallel',

pathways. However, the auditory system utilizes many synaptic mechanisms which operate in the temporal, rather than the spatial, dimension. Auditory space is not encoded at the receptor level but is deduced by interaural comparison. Resultant capacities for spatial localization are much poorer than for the visual system, where finely tuned spatial position is already manifested at the retinal level.

Fig. 16.6. Visual and auditory receptive fields of sequentially recorded, bimodal units from four separate electrode penetrations in the left optic tectum. The penetrations were made at the locations indicated by the solid circles on the dorsal view of the tectum (centre). The visual receptive fields are hatched; the auditory receptive fields are unhatched ellipses; the centres of auditory best areas are numbered. The numbers represent the order in which the units were encountered during each

dorsoventral penetration. The receptive fields from the most medial penetration (top centre) jumped from high to low. This is because the tectum curves around underneath, causing such medial penetrations to intersect both the dorsomedial (high fields) and ventromedial (low fields) edges of the tectum at two discontinuous portions of the track. a, Anterior, p, posterior; m, medial; l, lateral. (From Knudsen, 1982, with permission.)

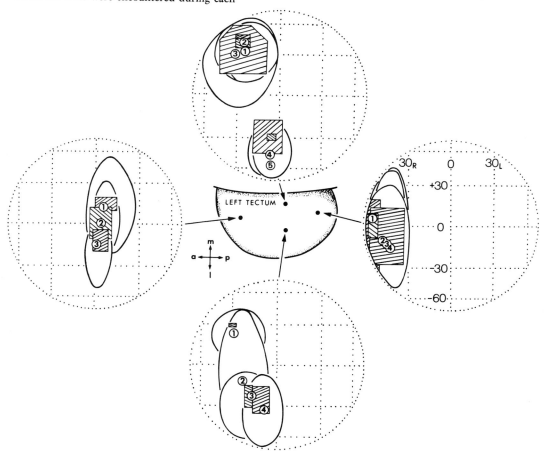

Acknowledgements

I should like to thank Lynne Hepburn for typing this manuscript; Jill Poynton, Karen Styles and Judy Sack for help with the illustrations; and my many colleagues for collaboration in this work.

References

Aitkin, L. M. (1976) Tonotopic organization at higher levels of the auditory pathway. In *International Review of Physiology. Neurophysiology II*, Vol.10, ed. R. Porter, pp. 249–79. Baltimore: University Park Press

Aitkin, L. M. (1983) Making sense of the mammalian auditory pathway. *Proc. Aust. Physiol. Pharmacol. Soc.* 14, 125–36

Aitkin, L. M. & Kenyon, C. E. (1981) The auditory brain stem of a marsupial. *Brain Behav. Evol.*, 19, 126–43

Aitkin, L. M. & Rawson, J. A. (1983) Frontal sound source location is represented in the cat cerebellum. *Brain Res.*, 265, 317–21

Andersen, R. A. (1979) *Patterns of connectivity of the auditory forebrain of the cat*. Ph.D.thesis, University of California, San Francisco

Andersen, R. A., Knight, P. L. & Merzenich, M. M. (1980) The thalamocortical and corticothalamic connections of AI, AII and the anterior auditory field (AAF) in the cat: evidence for two largely segregated systems of connections. *J. Comp. Neurol.*, 194, 663–701

Boudreau, J. C. & Tsuchitani, C. (1968) Binaural interaction in the cat superior olive S segment. *J. Neurophysiol.*, 31, 442–54

Brawer, J. R. & Morest, D. K. (1975) Relations between auditory nerve endings and cell types in the cat's anteroventral cochlear nucleus seen with the Golgi method and Nomarski optics. *J. Comp. Neurol.*, 160, 491–506

Brown, C. H. (1982) Primate auditory localization. In *Localization of sound: Theory and Applications*, ed. R. W. Gatehouse, pp. 136–54, Groton: Amphora

Calford, M. B & Aitkin, L. M. (1983) Ascending projections to the medial geniculate body of the cat: evidence for multiple, parallel auditory pathways through thalamus. *J. Neurosci.*, 3, 2365–80

Calford, M. B. & Webster, W. R. (1981) Auditory representation within principal division of cat medial geniculate body: an electrophysiological study. *J. Neurophysiol.*, 45, 1013–28

Calford, M. B., Webster, W. R. & Semple, M. N. (1983) Measurement of frequency selectivity of single neurons in the central auditory pathway. *Hearing Res.*, 11, 395–401.

Casseday, J. H. & Neff, W. D. (1973) Localization of pure tones. *J. Acoust. Soc. America*, 54, 365–72

Fitzpatrick, K. A. (1975) Cellular architecture and topographic organization of the inferior colliculus of the squirrel monkey. *J. Comp. Neurol.*, 164, 185–208

Goldberg, J. M. & Brown, P. B. (1968) Functional organization of the dog superior olivary complex: an anatomical and electrophysiological study. *J. Neurophysiol.*, 31, 639–56

Goldberg, J. M. & Brown, P. B. (1969) Responses of binaural neurons of dog superior olivary complex to dichotic tonal stimuli: some physiological mechanisms of sound localization. *J. Neurophysiol.*, 32, 613–36

Goldberg, J. M. & Brownell, W. E. (1973) Discharge characteristics of neurons in anteroventral and dorsal cochlear nuclei. *Brain Res.*, 64, 35–54

Goldberg, J. M. & Moore, R. Y. (1967) Ascending projections of the lateral lemniscus in the cat and monkey. *J. Comp. Neurol.*, 129, 143–56

Gordon, B. G. (1973) Receptive fields in deep layers of cat superior colliculus. *J. Neurophysiol.*, 36, 157–78

Greenwood, D. D. & Maruyama, N. (1965) Excitatory and inhibitory response areas of auditory neurons in the cochlear nucleus. *J. Neurophysiol*, 28, 863–92

Hashikawa, T. & Kawamura, K. (1983) Retrograde labeling of ascending and descending neurons in the inferior colliculus. A fluorescent double labeling study in the cat. *Exp. Brain Res.*, 49, 457–61

Holmes, G. (1945) The organization of the visual cortex in man. *Proc. R. Soc. Lond., Ser. B.*, 132, 348–61

Hubel, D. H. & Wiesel, T. N. (1962) Receptive fields, binocular interaction and functional architecture in the cat's visual cortex. *J. Physiol.*, 160, 106–54

Hubel, D. H. & Wiesel, T. N. (1965) Receptive fields and functional architecture in two non-striate visual areas (18 and 19) of the cat. *J. Neurophysiol.*, 28, 229–89

Imig. T. J. & Adrian, H. O. (1977) Binaural columns in the primary field (AI) of cat auditory cortex. *Brain Res.*, 138, 241–57

Irving, R. & Harrison, J. M. (1967) The superior olivary complex and audition: a comparative study. *J. Comp. Neurol.*, 130, 77–86

Jenkins, W. M. & Masterton, R. B. (1982) Sound localization: effects of unilateral lesions in central auditory system. *J. Neurophysiol.*, 47, 987–1016

Kiang, N. Y. S., Rho, J. M., Northrop, C. C., Liberman, M. C. Ryugo, D. K. (1982) Hair cell innervation by spiral ganglion cells in adult cats. *Science*, 217, 175–7

Klumpp, R. G. & Eady, H. R. (1956) Some measurements of interaural time difference thresholds. *J. Acoust. Soc. America*, 28, 859–60

Knudsen, E. I. (1982) Auditory and visual maps of space in the optic tectum of the owl. *J. Neurosci.*, 2, 1177–94

Knudsen, E. I., Blasdel, G. G. & Konishi, M. (1979) Sound localization by the barn owl (*Tyto alba*) measured with the search coil technique. *J. Comp. Physiol.(A)*, 133, 1–12

Knudsen, E. I. & Konishi, M. (1978) Space and frequency are represented separately in the auditory midbrain of the owl. *J. Neurophysiol.*, 41, 870–84

Knudsen, E. I. & Konishi, M. (1979) Mechanisms of sound localization in the barn owl (*Tyto alba*). *J. Comp. Physiol. (A)*, 133, 13–22

Konishi, M. (1973) How the owl tracks its prey. *Am. Scientist*, 61, 414–27

Kuffler, S. W. (1953) Discharge patterns and functional organization of mammalian retina. *J. Neurophysiol.*, 16, 37–68

Merzenich, M. M. & Brugge, J. F. (1973) Representation of the cochlear partition on the superior temporal plane of the macaque monkey. *Brain Res.*, 50, 275–96

Merzenich, M. M., Kaas, J. H. & Roth, G. L. (1976) Auditory cortex in the gray squirrel: tonotopic organization and architectonic fields. *J. Comp. Neurol.*, 166, 387–402

Merzenich, M. M., Knight, P. L. & Roth, G. L. (1975) Representation of cochlea within primary auditory cortex in the cat. *J. Neurophysiol.*, 38, 231–49

Merzenich, M. M. & Reid, M. D. (1974) Representation of the cochlea within the inferior colliculus of the cat. *Brain Res.*, 77, 397–415

Middlebrooks, J. C., Dykes, R. W. & Merzenich, M. M. (1980) Binaural response-specific bands within primary auditory cortex (AI) in the cat: topographical organization orthogonal to isofrequency contours. *Brain Res.*, 181, 31–48

Middlebrooks, J. C. & Pettigrew, J. D. (1981) Functional classes of neurons in primary auditory cortex of the cat distinguishable by sensitivity to sound location. *J. Neurosci.*, 1, 107–20

Middlebrooks, J. C. & Zook, J. M. (1983) Intrinsic organization of the cat's medial geniculate body identified by projections to binaural response-specific bands in the primary auditory cortex. *J. Neurosci.*, 3, 203–24

Moore, C. N. Casseday, J. H. & Neff, W. D. (1974) Sound localization: the role of the commissural pathways in the auditory system of the cat. *Brain Res.*, 83, 13–26

Morest, D. K. & Oliver, D. L. (1984). The neuronal architecture of the inferior colliculus in the cat. Defining the functional anatomy of the auditory midbrain. *J. Comp. Neurol.*, 222, 209–36

Neuweiler, G. (1984) Auditory basis of echolocation in bats. In *Comparative Physiology of Sensory Systems*, ed. L. Bolis, R. D. Keynes & S. H. P. Maddrell, pp. 115–41. Cambridge: Cambridge University Press.

Onishi, S. & Katsuki, Y. (1965) Functional organization and integrative mechanism on the auditory cortex of the cat. *Jap. J. Physiol.*, 15, 342–65

Osen, K. K. (1969a) Cytoarchitecture of the cochlear nuclei in the cat. *J. Comp. Neurol.*, 136, 453–84

Osen, K. K. (1969b) The intrinsic organization of the cochlear nucleus in the cat. *Acta Oto-Laryngologica*, 67, 352–9

Osen, K. K. & Mugnaini, E. (1981) Neuronal circuits in the dorsal cochlear nucleus. In *Neuronal Mechanisms of Hearing*, ed. J. Syka, & L. M. Aitkin, pp. 119–25. New York: Plenum

Phillips, D. P., Calford, M. B., Pettigrew, J. D., Aitkin, L. M. & Semple, M. N. (1982) Directionality of sound pressure transformation at the cat's pinna. *Hearing Res.*, 8, 13–28

Ramon y Cajal, S. (1955) *Histologie du Système Nerveux de l'Homme et des Vertébrés*. Madrid: Instituto Ramon y Cajal

Rayleigh, Third Baron. (1909) On the perception of the direction of sound. *Proc. R. Soc. Lond., Ser. A*, 83, 61–4

Reale, R. A. & Imig, T. J. (1980) Tonotopic organization in auditory cortex of the cat. *J. Comp. Neurol.*, 192, 265–91

Rockel, A. J. & Jones, E. G. (1973) The neuronal organization of the inferior colliculus of the adult cat. I. The central nucleus. *J. Comp. Neurol.*, 147, 11–60

Rose, J. E. (1960) Organization of frequency-sensitive neurons in the cochlear nuclear complex of the cat. In *Neural Mechanisms of the Auditory and Vestibular Systems*, ed. G. L. Rasmussen, & W. F. Windle, pp. 116–36. Springfield, Ill.: Charles C. Thomas

Rose, J. E. & Mountcastle, V. B. (1959) Touch and kinesthesis. In *Handbook of Physiology. Neurophysiology*, Sect. 1, Vol. 1, ed. J. S. Fields & H. W. Magoun, pp. 387–429. Washington: American Physiological Society

Roth, G. L., Aitkin, L. M., Andersen, R. A. & Merzenich, M. M. (1978) Some features of the spatial organization of the central nucleus of the inferior colliculus of the cat. *J. Comp. Neurol.*, 182, 661–80

Schuller, G. & Pollak, G. (1979) Disproportionate frequency representation in the inferior colliculus of Doppler-compensating Greater Horseshoe bats: evidence for an acoustic fovea. *J. Comp. Physiol.*, 132, 47–54

Semple, M. N. (1981) *Organization of the central nucleus of the cat inferior colliculus*, Ph.D. thesis, Monash University

Semple, M. N. & Aitkin, L. M. (1980) Physiology of pathway from dorsal cochlear nucleus to inferior colliculus revealed by electrical and auditory stimulation. *Exp. Brain Res.*, 41, 19–28

Semple, M. N., Aitkin, L. M., Calford, M. B., Pettigrew, J. D. & Phillips, D. P. (1983) Spatial receptive fields in the cat inferior colliculus. *Hearing Res.*, 10, 203–15

Servière, J. & Webster, W. R. (1981) A combined electrophysiological and [^{14}C]-2-deoxyglucose study of the frequency organization of the inferior colliculus of the cat. *Neurosci. Lett.*, 27, 113–18

Spoendlin, H. (1972) Innervation densities of the cochlea. *Acta Otolaryngol.*, 73, 235–48

Stevens, S. S. & Newman, E. D. (1934) The localization of pure tones. *Proc. Natl Acad. Sci. USA*, 20, 593–6

Stone, J., Dreher, B. & Leventhal, A. (1979) Hierarchical and parallel mechanisms in the organization of visual cortex. *Brain Res. Rev.*, 1, 345–94

Tsuchitani, C. & Boudreau, J. C. (1966) Single unit analysis of cat superior olive S segment with tonal stimuli. *J. Neurophysiol.*, 29, 684–97

Updyke, B. V. (1977) Topographic organization of the projections of cortical areas 17, 18 and 19 onto the thalamus, pretectum and superior colliculus in the cat. *J. Comp. Neurol.*, 173, 81–122

van Essen, D. C. (1979) Visual areas of the mammalian cerebral cortex. *Ann. Rev. Neurosci.*, 2, 227–63

Westheimer, G. (1963) Optical and motor factors in the formation of the retinal image. *J. Optic. Soc. America*, 53, 86–93

Wise, L. Z. & Irvine, D. R. F. (1983) Auditory response properties of neurons in deep layers of cat superior colliculus. *J. Neurophysiol.*, 49, 674–85

Woolsey, C. N. & Walzl, E. M. (1942) Topical projection of nerve fibers from local regions of the cochlea to the cerebral cortex of the cat. *Bull. Johns Hopkins Hosp.*, 71, 315–44

Young, E. D. & Voigt, H. F. (1981) The internal organization of the dorsal cochlear nucleus. In *Neuronal Mechanisms of Hearing*, ed. J. Syka & L. Aitkin, pp. 127–33. New York: Plenum

V

Visual cortex

At one of the early peaks of optimism about the
'Artificial Intelligence' approach to vision, David
Marr gave himself just three years to solve the image
analysis problem before moving on to 'more significant'
problems like object recognition. More than a decade
later, the fuzzy outlines still being generated by the
best computer algorithms for image analysis testify to
the great difficulties inherent in this essential early step
in visual processing. Two other great names in vision
were also impatient to account for the 'higher' visual
functions with their serial model of information
processing, which involved a hierarchy from 'simple
cells' through 'complex cells' to 'hypercomplex cells'.
For Hubel and Wiesel, as for Marr, the logical
imperatives of hierarchical processing acted to distract
attention away from the basic questions at their feet.

Nowhere is this more true than for Hubel and
Wiesel's revolutionary finding of orientation-selective
cortical neurons, a finding still without a basic
explanation more than a quarter of a century later. Just
as there are still no computer algorithms which can
detect contours in images as effectively as the
orientation-selective cortical neurons of live animals,
so there is no agreed-upon account of how orientation-
selective neurons achieve their properties.

There is a link between these two great gaps in
our knowledge if we accept that there are enormous
difficulties in defining the boundaries of objects in real
images, difficulties which are belied by the cartoonist's
ease in defining the essential edges of familiar objects.
Both gaps would probably be filled if we understood

the various tricks used to define edges during the operation of the visual cortex.

In Section V can be found many new hints about the principles of cortical organization which underlie these tricks used to define the boundaries of objects, even when the edges are not explicit as obvious luminance contours in the image. There are many parallel channels which can provide information about edges to orientation-selective neurons in the cortex, and it is this parallel mode of processing which is given special attention by Bishop's disciples here. Hubel and Wiesel's serial model of information processing in the visual cortex, while not completely invalidated, has not proved to have the heuristic power of this parallel approach.

The difficulties of any simple-minded serial or parallel model are immediately made apparent in Chapter 17. Here Bullier presents the striking findings that the same visual information can be delivered simultaneously to different levels in the system via collateral axonal branches. Bullier finds that bifurcating axons are ubiquitous in the visual pathway. The functional consequences of so many collaterals remain to be worked out, but one consequence is inescapable... the specific properties found at different levels imply that the same kind of information is being handled in a different and specific way by each cortical area.

Some of the intracortical wiring details are explored by Ogawa and Henry in Chapters 18 and 19. Understanding the flow of information through the visual cortex will probably be necessary before a clear picture of the basis of orientation selectivity emerges particularly when one realizes the astonishing fact (see p. 297) that the setting up of orientation selectivity can occur independently in different layers! Ogawa uses the *in vitro* approach to examine the role of inhibitory neurotransmission by GABA, while Henry combines natural visual stimulation with classical electrical stimulation in an attempt to separate the different streams of information flowing through the visual cortex.

There are so many streams and so many different visual cortical areas that one can easily despair at the prospect of ever being able to master the details particularly in the absence of an agreed nomenclature. Kaas, in the last chapter of the section, helps with the mnemonic problem by providing a clear overview of the primate visual system with parallel processing in mind. Dreher does the same with the cat's visual system in Chapter 20. Some general principles emerge from the details, such as the common occurrence of dichotomies (local/global, what/where) in the division of labour among the different cortical systems. Amongst these divisions one can detect hints of the 'Twenty Questions' game which McKay in Chapter 24, suggests must be played by the visual system.

17

Axonal bifurcation in the afferents to cortical areas of the visual system

JEAN BULLIER

Introduction

Since the end of the nineteenth century, it has generally been assumed that each sensory modality occupies a separate territory in the neocortex. With the advent of electrophysiological recording techniques in the 1940s, it became clear that the cortical territory devoted to each sensory modality contains numerous functional areas, each corresponding to a partial representation of the sensory surface (retina, cochlea, body surface). Recent mapping studies have further increased the number of known cortical areas for each sense (Van Essen, 1979; Merzenich & Kass, 1980; Tusa, Palmer & Rosenquist, 1981). The evolution of the models of cortical organization has followed a similar course. In the early days, when only a few cortical areas were thought to exist, the serial model was in favour. In this scheme, which was similar to that established by Flechsig, the sensory information relayed in the thalamus was thought to terminate in a single cortical area. Thus, Areas 17 (or V_1), S_1 and A_1 were considered as the relay stations for the visual, somatosensory and auditory signals arriving in the cortex. The sensory information was then thought to be sent from these stepping stones to other cortical areas for further processing. For example, in the visual system, the sensory messages would be cascaded through Areas 17, 18 and 19. A description of this type of model can be found in the work of Hubel & Wiesel (1965).

With the discovery, in the 1970s, of the hetero-geneity of the functional properties of neurons in the retina and the dorsal lateral geniculate nucleus (LGNd), came the view that different channels of information are kept separate from the retina to the cortex and that each cortical area is the main processing centre of a given channel. This parallel model of organization has been recently described in a review article by Stone, Dreher & Leventhal (1979).

Both serial and parallel models have generally assumed that a given neural structure sends projections to only a few cortical areas. This assumption has, however, been challenged with the development of modern neuroanatomical techniques which demonstrated a large amount of divergence in the connections between different structures devoted to a given sensory modality. Thus, the LGNd, which was traditionally thought to project to Area 17 only, has now been shown to terminate in numerous other visual cortical areas (see below). The presence of a large amount of convergence and divergence in the afferents to the various cortical areas makes it difficult to retain strictly serial or parallel models of organization unless certain subpopulations of neurons are the source of each set of connections to the different target structures. This leads to the following question: in a structure which projects to several cortical areas, are there neurons which send bifurcating axons to more than one area, or are the connections made by different populations of neurons each projecting to a given cortical area? The recent development of double tracer techniques has made it possible to answer this question and this has led to a complete revision of the general scheme of cortical organization. In this chapter, the connections to the different cortical areas of the visual system are reviewed with a special emphasis on the question of bifurcating axons terminating in more than one cortical area. An earlier review on the question of bifurcating axons in the visual system can be found in Giolli & Towns (1980).

Lateral geniculate projections to Areas 17 and 18 in the cat

The first suggestion that, in the cat, Area 18 as well as 17 receives a direct input from the LGNd was made by Talbot (1942), whose results were confirmed and expanded by Doty (1958). In a cortical region broadly corresponding to Area 18, Doty observed a strong visually evoked potential present even in animals in which Area 17 had been removed. Since the neuronal degeneration in the LGNd was very limited after

extirpation of Area 18, Doty concluded that Area 1 probably receives its thalamic input mainly v collaterals of LGNd axons terminating in Area 17. Te years later, Garey & Powell (1967) arrived at a simil conclusion after their anatomical study of tl geniculo-cortical pathway in the cat. They observe that the large cells in the LGNd degenerate only if bo Areas 17 and 18 are lesioned. They interpreted th result, later confirmed by Niimi & Sprague (1970), indicating that the large LGNd neurons se bifurcating axons to both areas.

With the advent of modern neuroanatomic techniques, such as the retrograde transport horseradish peroxidase (HRP) and the anterograd transport of radioactive amino acids, it was confirm that, in the cat, the LGNd projects onto both Areas and 18 (Rosenquist, Edwards & Palmer, 1974; Gilbe & Kelly, 1975; Maciewicz, 1975; Holländer & Vanega 1977). Furthermore, it was shown that HRP injectio into Area 17 led to the retrograde labelling of LGN neurons of all sizes, whereas the neurons labelled aft injections to Area 18 were found among the lar LGNd cells (Gilbert & Kelly, 1975; Holländer Vanegas, 1977). From this observation it follows th if some LGNd neurons send branching connections Areas 17 and 18, they have to be of large size.

This hypothesis was recently tested directly experiments using the retrograde transport of tv different markers placed in Areas 17 and 18. Aft injection of HRP and tritiated deactivated HF ([^{3}H]-apo-HRP) into Areas 17 and 18, Geisert (198 observed that a number of the large cells in the LGN were labelled by the two tracers. These double-label neurons constituted approximately 10% of the tot population of labelled neurons of laminae A and A_1 a 60% in the C laminae (terminology of Hickey Guillery (1974)). These results were confirmed another double labelling study, using fluoresce retrograde tracers (Bullier, Kennedy & Saling 1984b). There is, however, a discrepancy between t two studies regarding the proportion of LGNd neuro projecting to Area 18 alone. Geisert found that virtual every cell from layers A and A_1 which projected to Ar 18 also sent a branch to Area 17. Bullier et al., (1984 found, however, that the numbers of genicula neurons which terminate in Area 18 only are simi to the numbers of neurons whose axons bifurcate reach both Areas 17 and 18. The differences in t results are difficult to resolve, but it appears fro

udies using intra-axonal filling of HRP that indeed a
rge number of axons terminating in Area 18 do not
nd a collateral to Area 17 (Humphrey *et al.*, 1985).
his would tend to support the conclusion of Bullier
al. that Area 18 is not supplied exclusively from
:onal branches of axons terminating in Area 17.

Following the initial work of Enroth-Cugell &
obson (1966) in the retina, it has been shown by a
umber of groups that the cat LGNd contains at least
ree types of physiologically identified neurons, the X
d Y cells, which are found predominantly in laminae
and A$_1$, and the W cells, which belong mostly to the
arvocellular C laminae (C$_1$, C$_2$ and C$_3$) (for reviews,
e Rodieck, 1979; Stone *et al.* 1979; Lennie, 1980; for
1 alternative classification see Levick, 1975; Cleland
al., 1976). A number of authors have used the
fferent conduction velocities of the axons afferent to
1ese three types of LGNd neurons to identify, by
ectrical stimulation, the type of afferent LGNd drive
> the cortical areas. From these experiments it appears
1at the X cells in the LGNd only project to Area 17,
hereas both Areas 17 and 18 receive an input from
cells (Hoffmann & Stone, 1971; Stone & Dreher,
)73; Singer, Tretter & Cynader, 1975; Tretter,
ynader & Singer, 1975; Mitzdorf & Singer, 1978;
ullier & Henry, 1979; Harvey, 1980; Dreher,
eventhal & Hale, 1980). By electrically stimulating
reas 17 and 18 and recording in the LGNd, Stone &
reher (1973) found that a number of Y cells could be
:tivated at low threshold from both sites. This
1ggests that some Y cells send bifurcating axons to
reas 17 and 18, a conclusion also supported by the
ct that the double-labelled neurons in Geisert's
980) study were among the largest LGNd neurons,
hich are known to be Y cells (Friedlander *et al.*, 1981).
inally, the results of Humphrey *et al.* (1985) directly
emonstrate that some Y axons terminate in both Areas
7 and 18.

Indirect evidence also suggests that some W cells
:nd branching connections to Areas 17 and 18.
1 both Geisert's (1980) and Bullier *et al.* (1984*b*)
udies, double-labelled neurons were found in the
arvocellular C laminae (C$_1$, C$_2$, C$_3$, terminology of
[ickey & Guillery, 1974). Since these laminae are
1own to contain almost exclusively W (Wilson, Rowe
Stone, 1976) or sluggish and non-concentric units
Cleland *et al.*, (1976), it is likely that this cell type also
rovides some common input to Areas 17 and 18. Thus,
1e phenomenon of branching projections to Areas 17

and 18 is observed for neurons of all classes and all
laminae in the LGNd, with the exception of the X cells
which project to Area 17 only.

LGNd projections to other visual cortical areas in the cat

In contrast to the neurons in laminae A and A$_1$,
which project only to Areas 17 and 18 (Gilbert & Kelly,
1975; Maciewicz, 1975; LeVay & Gilbert, 1976;
Kennedy & Baleydier, 1977; Geisert, 1980; Hughes,
1980; Niimi *et al.*, 1981), the cells in the C laminae have
wide-ranging projections involving Areas 17, 18, 19, 20,
21 and the visual areas of the lateral suprasylvian sulcus
(LeVay & Gilbert, 1976; Hughes, 1980; Raczkowski &
Rosenquist, 1980; Dreher *et al.*, 1984; Fig. 17.1). Since
injection of retrograde tracers in Areas 17–18 and Area
19 led, in both cases, to a majority of neurons in these
laminae being labelled, Geisert (1980) concluded that
axons of some neurons must project to both Areas 18
and 19 or to Areas 17 and 19. This was tested directly
by injecting retrograde fluorescent tracers in retinotopically corresponding regions in pairs of all three
areas. Numerous double-labelled neurons (10–40%of
the population of labelled cells) were found in the C
laminae after injection into 17–18, 17–19 and 18–19
(Bullier *et al.*, 1984*b*), thus demonstrating that a large
number of lamina C neurons send bifurcating axons to
these areas. It remains to be determined whether
branching axons also involve other cortical areas
receiving from the C laminae of the LGNd. An attempt
at answering this question by double-label experiments
was made by Norita & Creutzfeldt (1982) who came to
the conclusion that different LGNd neurons provide
input to 17 and 19 and to various areas of the lateral
suprasylvian sulcus. However, the small number of
labelled neurons in the region of overlap of the two
populations of marked cells in this study leaves open
the possibility that double-labelled neurons may have
been overlooked. Indeed, double-labelled neurons are
usually found only in limited proportions in this
overlap zone (Bullier *et al.*, 1984*b*). Tong & Spear
(1984) recently reported that some neurons in the C
laminae are double-labelled after injections in the
posteromedial area of the lateral suprasylvian sulcus
(PMLS) and Areas 17 and 18.

The medial interlaminar nucleus (MIN) is
located close to the medial border of the LGNd. It
contains a separate representation of a large portion of
the contralateral hemifield (Sanderson, 1971; Kratz,

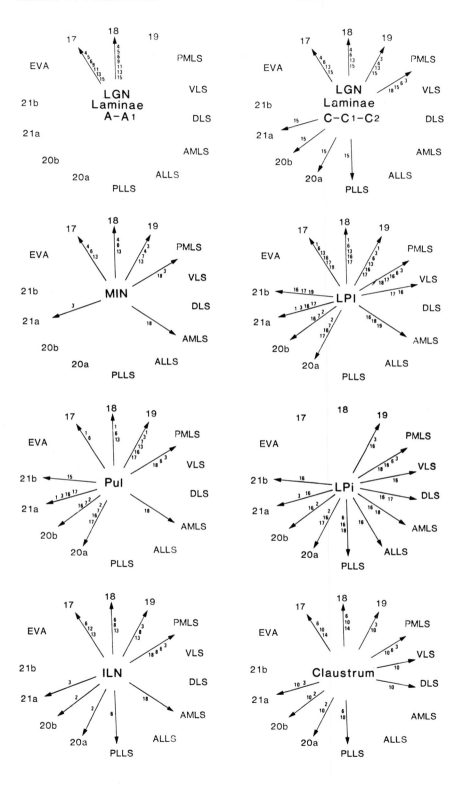

Vebb & Sherman, 1978; Guillery *et al.*, 1980; Rowe & Dreher, 1982) and is generally considered, on morphological and functional grounds, as part of the LGN. The MIN contains mostly Y cells with a small proportion of X and W cells (Mason, 1975; Kratz *et al.*, 1978; Dreher & Sefton, 1979; Rowe & Dreher, 1982). Neurons in the MIN project to Areas 17, 18 and 19 and to several visual areas in the lateral and posterior suprasylvian gyrus (Fig. 17.1; Maciewicz, 1974; Rosenquist *et al.*, 1974; Gilbert & Kelly, 1975; Maciewicz, 1975; Kennedy & Baleydier, 1977; Geisert, 1980; Hughes, 1980; Niimi *et al.*, 1981; Tong, Kalil & Spear, 1982; Dreher *et al.* 1984). Some of these projections have recently been shown to arise via axons bifurcating to Areas 17 and 18, 18 and 19 and to 17 and 19 (Bullier *et al.*, 1984*b*). Because of the prominence of the Y population in the MIN, it is likely that this cell type is implicated in the MIN, as well as in the LGNd, in sending branching axons to these cortical areas.

Cortical projections from the thalamic nuclei other than the LGNd

Besides the LGNd and the MIN, the major thalamic input to the visual cortical areas comes from the lateral posterior-pulvinar complex. In contrast to the LGN, which is almost entirely innervated by retinal afferents, the lateral posterior-pulvinar complex contains only a few small regions of direct retinal input (Berman & Jones, 1977; Berson & Graybiel, 1978; Itoh *et al.*, 1979; Guillery *et al.*, 1980; Leventhal, Keens & Tork, 1980; Itoh, Mizuno & Kudo, 1983). The cytoarchitectonic homogeneity of the lateral posterior nucleus impeded its subdivision into functional blocks until Updyke (1977) arrived at the conclusion that the lateral posterior-pulvinar complex is composed of four major divisions: the pulvinar; the lateral part of the lateral posterior nucleus (LPl); the interjacent part of

the lateral posterior nucleus (LPi), and the medial part of the lateral posterior nucleus (LPm). More recently, Berson & Graybiel (1983) showed that the subdivisions of the lateral posterior-pulvinar complex can be distinguished with acetylcholinesterase histochemistry. From the pattern of corticothalamic projections from Areas 17, 18 and 19, Updyke (1977) assumed that each of these subdivisions corresponded to a representation of the contralateral visual hemifield. This has been confirmed for LPl, LPi and the pulvinar by electrophysical mapping techniques (Mason, 1978, 1981; Raczkowski & Rosenquist, 1981) and by anatomical studies of thalamocortical projections (Berson & Graybiel, 1978, 1983; Hughes, 1980; Symonds *et al.*, 1981; Raczkowski & Rosenquist, 1983). Although each of these zones is not necessarily homogeneous with respect to its connectivity pattern and cytoarchitecture (Updyke, 1983), the four subdivisions exhibit specific patterns of connectivity with other visual structures. Thus, the pulvinar receives its input from the pretectum, the retina, the cerebellum, Area 19 and a number of other visual areas (Berman, 1977; Berman & Jones, 1977; Berson & Graybiel, 1978; Itoh *et al.* 1979). The LPl subdivision receives its input from, and projects to, most cortical visual areas (17, 18, 19, 20a, 20b 21a, 21b, AMLS*, PMLS, VLS*, PS*) Updyke, 1977, 1981; Graybiel & Berson, 1980; Hughes, 1980; Symonds *et al.*, 1981; Tong, *et al.*, 1982; Raczkowski & Rosenquist, 1983; Cavada & Reinoso-Suarez, 1983; Dreher *et al.*, 1984). LPi is the major recipient zone of tectal afferents (Updyke, 1977; Berson & Graybiel, 1978) and is interconnected with cortical areas which do not receive direct afferents from the LGN (Hughes, 1980; Mucke, *et al.*, 1982; Symonds

* AMLS = The anteromedial lateral suprasylvian area; VLS = the ventral lateral suprasylvian area (Palmer, Rosenquist & Tusa, 1977). PS = the posterior suprasylvian sulcus area (Heath & Jones, 1971).

Fig. 17.1. Summary of subcortical afferent connections to the visual cortical areas in the cat. LP1: lateral part of the lateral posterior nucleus; LPi: interjacent part of the lateral posterior nucleus; Pul: pulvinar (terminology of Updyke (1977); ILN: intralaminar nuclei; PMLS, VLS, DLS, AMLS, ALLS, PLLS: cortical areas of the lateral suprasylvian sulcus (terminology of Tusa *et al.*, 1981). EVA: ectosylvian ventral area (Mucke *et al.*, 1982). See Fig. 17.2 for location of functional areas on the cortical surface of the cat brain. The numbers indicate the following bibliographic

references. 1: Berson & Graybiel, 1978; 2: Cavada & Reinoso-Suarez, 1983; 3: Dreher *et al.*, 1984; 4: Geisert, 1980; 5: Gilbert & Kelly, 1975; 6: Hughes, 1980; 7: Itoh *et al.*, 1979; 8: Kennedy & Baleydier, 1977; 9: Le Vay & Gilbert, 1976; 10: Le Vay & Sherk, 1981*a*; 11: Maciewicz, 1975; 12: Miller & Benevento, 1979; 13: Niimi *et al.*, 1981; 14: Olson & Graybiel, 1980; 15: Raczkowski & Rosenquist, 1980; 16: Raczkowski & Rosenquist, 1983; 17: Symonds *et al.*, 1981; 18: Tong *et al.*, 1982; 19: Updyke, 1983.

et al., 1981; Updyke, 1981, 1983; Tong *et al.*, 1982; Cavada & Reinoso-Suarez, 1983; Raczkowski & Rosenquist, 1983). Finally, the retinotopic organization and connectivity of the LPm are less well known than for the more lateral parts of the lateral posterior complex (Updyke, 1977, 1983).

Each of the three major subdivisions of the lateral posterior complex (LPi, LPl and pulvinar) thus projects to a large number of cortical areas (Fig. 17.1). It is therefore to be expected that a number of neurons in these thalamic regions send branches to more than one cortical area. This is particularly the case for LPl, in which as many as 75% of the neurons have been found to project to the 17–18 border (Hughes, 1980). Since this subdivision projects to nine additional areas (see above), it is almost certain that some of these projections involve bifurcating axons. This has been tested by the retrograde transport of two different labels injected in different cortical areas. Norita & Creutzfeldt (1982) injected Areas 17, 18, 19 and different areas of the lateral suprasylvian sulcus, and concluded that the neurons feeding the different cortical areas occupy the same territory and are intermingled in the lateral posterior nucleus but do not project to more than one visual area. As mentioned earlier, however, the small extent of the overlap zone between the populations of neurons marked by the two labels leaves open the possibility that double-labelled neurons were not observed because of poor retinotopic correspondence between the two injection sites. In a comprehensive study of the thalamic projections to Areas 17, 18 and 19, Bullier *et al.*, (1984*b*) found that 10–40% of the neurons in the LPl project to both 17 and 18 and that similar proportions send branching connections to 18 and 19 and to 17 and 19. Kaufman, Rosenquist & Raczkowski (1984) also reported that neurons in the LPl send branching connections to Areas 17 and 20a and to Area 19 and the PMLS. Finally, Tong & Spear (1984) showed that some LPl neurons have bifurcating axons terminating in the PMLS and in 17–18.

The study of cortical efferents from the pulvinar also revealed the presence of numerous branching connections. A sizeable proportion of neurons in the retina-recipient zone of the pulvinar was found to send bifurcating connections to Areas 18 and 19 (Bullier *et al.*, 1984*b*). Because of the small extent of the projection from the pulvinar to Area 17 (Gilbert & Kelly, 1975; Hughes, 1980), the presence of axons bifurcating to Areas 17 and 18 or Areas 17 and 19 is not a relevant question in relation to pulvinar projections. Bifurcating axons projecting from the pulvinar to Areas 20a and 7 and to 20a and the cingulate gyrus have also been demonstrated by the retrograde transport of fluorescent dyes (Kaufman *et al.*, 1982). The study of Bullier *et al.*, (1984*b*) was limited to the projections to Areas 17, 18 and 19, which do not receive afferents from the LPi subdivision. It is not known, therefore, whether this part of the lateral posterior complex also contains neurons projecting to more than one cortical area. In view of the ubiquity of the pattern of branching in thalamocortical projections to Areas 17, 18 and 19 (Bullier *et al.*, 1984*b*), it would be surprising if a similar phenomenon did not occur in the cortical projection of the LPi.

The intralaminar nuclei (central lateral, paracentral and central medial nuclei) are more rostrally placed than the thalamic nuclei specifically related to vision. They have been considered as part of a different thalamocortical projection system since the observations of Morison & Dempsey (1942). These authors observed that, by stimulating electrically the region of the intralaminar nuclei, they could elicit an evoked cortical potential which was different from the response obtained when the thalamic nuclei associated to each sensory modality were stimulated electrically. They further reported that the potential evoked by stimulation of the intralaminar nuclei could be recorded on the major part of the cortical surface and concluded that the intralaminar nuclei give rise to diffuse thalamocortical connections innervating the whole cortical surface. This interpretation was reinforced by several anatomical studies showing that cellular degeneration could be induced in the intralaminar nuclei by lesions in various parts of the neocortex (Nauta & Whitlock, 1954; Powell & Cowan, 1964; Murray, 1966). A similar conclusion was reached by Jones & Leavitt (1974) who showed that HRP-filled cells were found in the intralaminar nuclei after injections in widely separated cortical regions. Finally, double-label retrograde studies have demonstrated that the intralaminar nuclei are quite exceptional in showing no topographical separation between neurons projecting to widely separated regions of the cortex (Bentivoglio, Macchi & Albanese, 1981). Most authors studying the interconnections of the intralaminar nuclei with sensory regions of the cortex have commented that there appears to be little, if any, topographic organization in these connections (Jones & Leavitt

1974; Galletti, Squatrito & Battaglini, 1979; Miller & Benevento, 1979).

The intralaminar nuclei have recently been shown to project to numerous cortical visual areas including Areas 17, 18, 19, 20, 21a and several areas of the lateral suprasylvian sulcus (Fig. 17.1; Kennedy & Baleydier, 1977; Miller & Benevento, 1979; Hughes, 1980; Niimi *et al.*, 1981; Tong *et al.*, 1982; Cavada & Reinoso-Suarez, 1983; Dreher *et al.*, 1984). Because of the 'diffuse' nature of the thalamocortical projections of the intralaminar nuclei, it was not surprising to find that a number of neurons in these structures send bifurcating axons to Areas 17 and 18, 18 and 19, and 17 and 19 (Bullier *et al.*, 1984*b*). Kaufman *et al.*, (1984) also reported that neurons in the intralaminar nuclei send branching connections to Areas 19, 20a, 7 and the PMLS. It remains to be demonstrated whether a similar branching pattern occurs in the projection to cortical regions related to different sensory modalities or whether there is a strict segregation of corticopetal information according to sensory modality in the intralaminar nuclei.

Cortical projections from non-thalamic structures

Three major groups of non-thalamic subcortical afferents to the cortical visual areas can be distinguished: the connections from (1) the claustrum, (2) the brainstem nuclei, and (3) the hypothalamus and basal forebrain. Of these, the structure providing by far the strongest input to the cortical visual areas is the claustrum. This structure has been shown to contain subdivisions connected with the auditory, somatosensory and visual portions of the cortex (Olson & Graybiel, 1980). The visual part of the claustrum projects exclusively to the visual cortical areas from which it receives a return projection (Riche & Lanoir, 1978; Sanides & Bucholtz, 1979; Squatrito *et al.*, 1980; LeVay & Sherk, 1981*a*; Cavada & Reinoso-Suarez, 1983). It has been shown to contain a retinotopically organized representation of the contralateral visual hemifield (LeVay & Sherk, 1981*b*) and, because of its dense connections with the visual cortical areas and the receptive field properties of its neurons (Sherk & LeVay, 1981), it has been considered as a remotely displaced satellite of the visual cortex (Olson & Graybiel, 1980; LeVay & Sherk, 1981*a*).

The cortical projections of the visual part of the claustrum involve the large majority of the neurons. In

fact, Hughes (1980) and LeVay & Sherk (1981*a*) found that 75–87% of the claustrum neurons project to Areas 17 and 18 and this percentage corresponds broadly to the proportion of neurons sending an axon out of the claustrum (LeVay & Sherk, 1981*a*). In view of this high density of claustrofugal neurons and of the plethora of cortical targets, it is no surprise that 20–30% of the neurons project to both Areas 17 and 18, and that similar percentages send branching connections to Areas 18 and 19 and to 17 and 19 (Bullier *et al.*, 1984*b*).

The hypothalamic and basal forebrain afferents to the visual cortex are relatively scarce and have received only limited attention (Albus, 1981). Afferents from the brainstem nuclei, on the other hand, have been studied quite extensively. Noradrenergic fibres which are thought to originate in the locus coeruleus have been found to travel mostly in the grey matter and to provide a fine grid pattern over most areas of the rat neocortex (Morrison *et al.*, 1978). Each of these fibres covers a large area of cortex, as shown by the experiments of Morrison, Molliver & Grzanna (1979) who made lesions in various parts of the neocortex and observed a subsequent disappearance of noradrenergic fibres over large areas of surrounding neocortex. It appears, therefore, that of all the subcortical afferents to the cortical areas, the connections from the brainstem nuclei are the least topographically organized. This is consistent with the fact that neurons in the same region of the locus coeruleus can be labelled by HRP injections placed in widely separated regions of the neocortex (Freedman, Foote & Bloom, 1975; Gatter & Powell, 1977; Bentivoglio *et al.*, 1978) and that some neurons in the locus coeruleus can be activated antidromically from the cerebellum, the frontal cortex and the occipital cortex (Nakamura, 1977). Direct demonstration of bifurcating axons innervating the cerebellum and the visual cortex was provided in the mouse by the double-labelling study of Steindler (1981). In the cat, projections from the brainstem nuclei to Areas 17, 18 and 19 were observed by Törk, Leventhal & Stone, (1979) and by Hughes (1980). The interpretation of the presence of double-labelled neurons in the brainstem nuclei after injection in 17, 18 and 19 in our study (Bullier *et al.*, 1984*b*) was complicated by the fact that one of the markers (the fast blue) tended to be more readily transported in the brainstem nuclei than the other tracer (diamidino yellow). In consequence, we found that most labelled neurons in these structures were blue. We also observed a small number of double-labelled neurons but no neuron labelled only by the diamidino yellow. This finding indicates that the projection of the brainstem nuclei to Areas 17, 18 and 19 contains some bifurcating axons. Whether all of this projection is done via branching connections remains to be determined by improving the efficiency of the retrograde transport of the diamidino yellow.

A review of the subcortical afferents to the visual areas in the cat neocortex demonstrates, therefore, that there is consistently, in *all* these connections, a certain amount of axonal bifurcation enabling single neurons to send a projection to more than one visual area. Before considering the functional implications of this organization, it seems appropriate to review the results of similar studies in other species and other sensory modalities.

Subcortical afferents to V1 and V2 in Old World monkeys

It was traditionally thought that, in the primate, the LGNd projects only to Area 17, also called the striate area or V1. It is only recently that, following improvements in the technique of HRP labelling, a projection was demonstrated from the LGNd to Area V2 and the cortical areas of the prelunate gyrus of Old World monkeys (Yukie & Iwai, 1981; Benevento & Yoshida, 1981; Fries, 1981; Bullier & Kennedy, 1983). In contrast to the cat, however, the extrastriate projections of the LGNd are very light compared to the striate connections, and they arise mostly from the interlaminar zones and S layers (terminology of Kaas *et al.*, 1978). These regions of the LGNd receive afferents from the superior colliculus and contain only a small number of neurons projecting to the striate cortex (Harting *et al.*, 1980; Benevento & Yoshida, 1981; Yukie & Iwai, 1981; Bullier & Kennedy, 1983). In keeping with this partial segregation of striate and prestriate-projecting neurons in the LGNd, only a few double-labelled neurons were found when V1 and V2 were injected with two different retrograde fluorescent tracers (Bullier & Kennedy, 1983). Preliminary results suggest that a larger number of neurons may send branching axons to V2 and cortical areas in the prelunate gyrus (Bullier & Kennedy, 1983). The results of this study demonstrate that the projection to V1 and V2 from the LGNd in the monkey is quite different from the LGNd projection to Areas 17 and 18 in the cat. In the cat LGNd, neurons projecting to Area 18

are found in all layers and there is a substantial proportion of cells sending bifurcating connections to both Areas 17 and 18. By contrast, in the monkey LGNd, cells projecting to V1 and V2 are mostly segregated and rarely send axonal branches to the two areas. The projection from the interlaminar zones and S layers to the prestriate cortex in the monkey appears to be homologous to the projections of the C layers in the cat LGNd to the extrastriate visual areas (Benevento & Yoshida, 1981; Weber *et al.*, 1983). In both cases, the LGN territory involved is recipient of collicular input (Harting *et al.*, 1980; Torrealba, Partlow & Guillery, 1981) and projects widely to striate and extrastriate cortical areas (Carey, Fitzpatrick & Diamond, 1979; Raczkowski & Rosenquist, 1980; Weber *et al.*, 1983; see also Chapter 21). It is to be expected therefore that, as in the C laminae of the cat, a large number of neurons in the interlaminar zones and S layers of the monkey LGN send bifurcating connections to more than one visual cortical area.

Besides the LGN, the major thalamic input to the monkey visual cortical areas comes from the inferior pulvinar and a part of the lateral pulvinar. Both nuclei are known to contain a retinotopic representation of the contralateral hemifield (Bender, 1981; Graham, 1982) and to be connected reciprocally with the striate and extrastriate visual areas (Benevento & Rezak, 1976; Ogren & Hendrickson, 1976; Trojanowski & Jacobson, 1976; Rezak & Benevento, 1979). The inferior pulvinar is innervated by the superior colliculus (Benevento & Fallon, 1975; Partlow, Colonnier & Szabo, 1977) and receives a small direct retinal input (Campos-Ortega, Hayhow & Cluver, 1970; Mizuno *et al.*, 1982; Itaya & Van Hoesen, 1983), whereas the lateral pulvinar is mostly connected with the visual cortical areas (Trojanowski & Jacobson, 1976; Rezak & Benevento, 1979). Following injection of different retrograde fluorescent tracers in V1 and V2, substantial amounts of double-labelled neurons (10–20%) were observed in these two structures (Kennedy & Bullier, 1985). This indicates that, despite the fact that the afferents to V1 and V2 terminate in different laminae in these two cortical areas (Ogren & Hendrickson, 1977; Rezak & Benevento, 1979), they are partially supplied by the same neurons in the pulvinar.

As in the cat, the claustrum in the monkey projects to the visual areas (Riche & Lanoir, 1979; Mizuno *et al.*, 1981; Doty, 1983). The retrograde transport of fluorescent dyes injected in V1 and V2 demonstrated that 20% of the claustral neurons send bifurcating connections to these cortical areas (Kennedy & Bullier, 1985).

Branching axons in the thalamocortical connections in other species and other sensory modalities

The presence of branching axons in the subcortical afferents to the visual areas in the monkey as well as in the cat suggests that this pattern may be a general one for all species and all sensory systems. Unfortunately, very little experimental data are available to test this possibility in the visual system of other species. Multiple visual areas have been demonstrated by anatomical or electrophysiological mapping in the rat (Montero, Bravo & Fernandez, 1973; Montero, Rojas & Torrealba, 1973; Espinoza & Thomas, 1983), the mouse (Wagor, Mangini & Pearlman, 1980) and a few other species (Van Essen, 1979; Montero, 1981). The pattern of thalamocortical connections to these areas is divergent (Hughes, 1977; Coleman & Clerici, 1980; Simmons, Lemmon & Pearlman, 1982) and it can therefore be expected that, in these animals, a number of thalamic neurons will be found to send bifurcating connections to more than one visual cortical area. It would be particularly interesting to determine whether certain species exhibit more branching in their thalamocortical connections than others, and to use these data to test whether a decrease of axon bifurcation occurs with the increase in the number of visual areas and the general segregation of systems known to occur with the development of certain sensory modalities during phylogeny (Diamond & Hall, 1969; Kaas, 1980).

Bifurcation in the thalamocortical afferents has been demonstrated in sensory modalities other than the visual system. As early as 1958, Rose & Woolsey concluded from their studies on retrograde neuronal degeneration in the auditory thalamus that 'only a small fraction of the auditory thalamo–cortical connections may be considered essential (i.e. unbranched) projections while most of them are sustaining (i.e. bifurcating)'. This was recently confirmed and extended by studies by Imig and Morel who showed that several thalamic auditory nuclei contain double-labelled neurons when A1 and AAF (anterior auditory field) or A1 and PAF (posterior auditory field) are injected with two different retrograde tracers (Imig & Morel, 1983).

The presence of bifurcation in the thalamocortical cortical projections of the somatosensory system was originally inferred from the results of physiological experiments involving the antidromic activation of thalamic neurons by electrical stimulation of Areas S1 and S2 in the cat (Anderson, Landgren & Wolsk, 1966; Rowe & Sessle, 1968; Mason, 1969). This interpretation has found support in the results of retrograde transport of HRP and [3H]-apo-HRP in the cat thalamus after injections in Areas S1 and S2 (Spreafico, Hayes & Rustioni, 1981). In the macaque monkey, Jones (1983) failed to find double-labelled cells in the thalamus after paired injection of two retrograde fluorescent tracers in Areas 3a, 3b, 1 and 2. The lack of overlap of the populations of labelled cells in the thalamus in this study raises, however, the possibility that the regions of cortex injected were not in somatotopic correspondence and that this factor may be responsible for the negative result. In fact, Cusick et al. (1985) recently provided evidence that some thalamic neurons send bifurcating axons to Areas 3a, 3b, 1 and 2 in the squirrel monkey.

In conclusion, it can be said that bifurcation in thalamocortical axons appears to be a general phenomenon. Not only are all subcortical structures found to send branching connections to more than one cortical area in the visual system, but this phenomenon is also observed in the thalamocortical connections of other sensory modalities. Furthermore, it is likely that branching subcortical afferents will be found in most, if not all, species. A few years ago, this phenomenon was hardly noticed and it was thought to be restricted to a few thalamic and cortical regions. Now it appears to be a general phenomenon, the functional consequence of which will have to be examined.

Functional implications of branching in subcortical afferents

The functional importance of bifurcating axons is stressed not only by their presence in several sensory modalities but also by their numbers and the precision of their termination patterns. Although in double tracer experiments the percentages of double-labelled neurons in subcortical structures is generally in the range of 10–30% (Bullier et al., 1984b), these figures are almost certainly an underestimate of the true proportions of neurons sending bifurcating axons. This results from a number of technical factors (Bullier et al., 1984b) and from the fact that only two areas received injections of tracers at a time. Some single-labelled neurons could therefore be sending axonal branches to other non-injected cortical regions. The proportions of double-labelled cells reported by Bullier *et al.*, (1984*b*) sometimes represent the majority of input to a given structure. For example, among the claustral afferents to Area 18, there are twice as many axons supplying an additional branch to Area 17 than axons terminating in 18 and not in 17. All these factors suggest that, in general, the actual numbers of unbranched connections from one subcortical nucleus to a given cortical area may represent only the minority of afferents.

The precision of the 'wiring' of both branches of a bifurcating axon is revealed by the fact that, in double tracer experiments, double-labelled neurons were found almost exclusively in the regions of overlap between the two populations of single-labelled cells. As the neural connections link retinotopically corresponding regions and since all the structures in the visual system, except possibly the intralaminar nuclei, are retinotopically organized, if follows that both branches of a branching axon terminate in retinotopically corresponding regions of the two injected areas. The lack of double-labelled cells in cases where the two injections were not precisely retinotopically corresponding demonstrates that the branches of bifurcating axons terminate only in cortical regions representing the same zones of the visual field (Kennedy & Bullier, 1985).

The fact that the same thalamic information is connections may be seen as a challenge to the model of parallel processing. For example, in the cat, according to this model Areas 17, 18 and 19 each process mainly one of the three channels of retinogeniculate afferents: the X, Y and W channels (Stone *et al.*, 1979: Dreher *et al.*, 1980). However, it was already recognized in the description of this model (Stone *et al.*, 1979) that information from the Y and W channels reaches all three cortical areas. The presence of bifurcating axons only points out that these connections are not made exclusively via separate subpopulations of the Y and W populations. Furthermore, the fact that the inputs to two different cortical areas may terminate in different cortical laminae suggests that the common information may be used for different purposes in the different areas. Examples of such differential laminar termination can be found in the projections of the C laminae of the cat LGN to Areas 17 and 18 (LeVay &

Gilbert, 1976) and the pulvinar input to V1 and V2 in the monkey (Ogren & Hendrickson, 1977; Rezak & Benevento, 1979). Thus, a common input could be used in one cortical area as the main drive to granular and infragranular layers, influencing the properties of subcortical feedback neurons, while in another cortical area it could influence a population of cells involved in corticocortical connections.

The different functional role of the two branches of a bifurcating axon is further substantiated by the evidence that the two branches do not necessarily carry the same message. Several investigators have reported that, at high firing frequencies, one of the axonal branches stops conducting for a given period of time (Parnas, 1972; Grossman, Spira & Parnas, 1973; Van Essen, 1973). The mechanisms involved in this conduction block range from hyperpolarization of the membrane (Van Essen, 1973) to extracellular accumulation of potassium ions (Spira, Yarom & Parnas, 1976). It has been proposed that the conduction block in branching axons leads to a filtering action of the neural messages, the most resistant branches reproducing faithfully the firing of the neurons, while other branches may drop out at different frequencies of firing (Chung, Raymond & Lettvin, 1970; Parnas, 1972; Hatt & Smith, 1975).

In conclusion, the presence of bifurcating thalamo-cortical connections to areas 17, 18 and 19 does not really constitute a challenge to the concept of parallel processing of thalamic information in different cortical areas. It does, however, suggest that each cortical area uses the neural information coming through different functional channels and different thalamic nuclei in a specific way. It is possible that the large amount of branching connections in the thalamo-cortical connections is an economical answer to the problem of distributing the same thalamic information to numerous visual areas which, in turn, use this information in a specific manner.

If this hypothesis for the functional role of axon bifurcation in thalamo-cortical connections is correct, one may wonder why there are rarely more then 30% of axons which branch to innervate two cortical areas. In the case of most thalamic nuclei, this may be due to the fact that there are other cortical projection areas and that what appears as a single-projection neuron in double-label experiments may in fact be sending branching connections to other non-injected cortical regions. This, however, cannot be the case for the neurons of laminae A and A$_1$ of the cat LGN which project only to Areas 17 and 18. It appears established that the X cells only project to Area 17 and that some of the Y cells bifurcate to innervate Areas 17 and 18 (see above). It would be interesting to know whether there are functional differences between the Y LGN neurons giving rise to a bifurcating axon and those which project exclusively to Area 17 or to Area 18. This would help in tracing the functional reasons governing the presence or absence of bifurcation in a given set of axons.

On a more general level, the presence of bifurcating axons in the connections from all thalamic visual nuclei to the visual cortical areas also bears on the distinction between specific and non-specific thalamic projections to the neocortex. This distinction was made by Morison & Dempsey (1942) who recognized 'one specific projection system with a more or less point to point arrangement' between the sensory relay nuclei and the neocortex and 'a secondary non-specific system with diffuse connections'. These two groups of thalamic afferents recognized by electrophysiological methods (Morison & Dempsey, 1942) were equated with the two anatomically identified groups of thalamic afferents, one group terminating profusely in laminae 3 and 4 ('specific afferents') and the others ('unspecific' or 'pluriareal') giving off collaterals before entering the cortex, and terminating sparsely in all layers, mainly 1 and 6 (Lorente de No, 1983). This suggested dichotomy remains largely unchallenged to the present day, despite the demonstration that the intralaminar nuclei, thought to be a major source of unspecific afferents, terminate in laminae 5 and 6 and not in lamina 1 (Herkenham, 1980) and that some nuclei (like the C laminae of the cat LGN) terminate in lamina 1 in some areas and laminae 3 and 4 in others (LeVay & Gilbert, 1976). For Lorente de No (1983) and Morison & Dempsey (1942), the specific projection system meant a one-to-one relationship between thalamic nuclei and cortical areas, whereas the 'diffuse' or 'pluriareal' nature of the unspecific projection system was evidenced by the simultaneous innervation of several cortical areas by a bifurcating axon. The results described above on the thalamo-cortical projections to Areas 17, 18 and 19 (Bullier *et al.*, 1984*b*) demonstrate that there is a similar degree of bifurcation to two cortical areas in the specific system (like the LGN projections) as in the unspecific one (intralaminar

nuclei). In view of this and of the complexity of the laminar distribution of the cortical termination of different thalamic nuclei (Herkenham, 1980), it is obvious that the terms specific and unspecific should be abandoned. An alternative classification based on the branching pattern and level of termination of the projection in the cortex (Herkenham, 1980) would include (1) the nuclei having a focussed projection to laminae 3–4 of the primary sensory areas (e.g. the A laminae of the cat LGN); (2) the nuclei having a widespread projection mainly restricted to cortical areas of one sensory modality, terminating in different laminae for different cortical areas (e.g. the C laminae of the LGN); and (3) the nuclei having a widespread cortical projection including most of the neocortex, to laminae 5 and 6 (intralaminar nuclei) or lamina 1 (ventro-medial nucleus).

Cortico-cortical connections in the primate visual system

The system of cortico-cortical connections has traditionally been considered as a different level of neural processing from the group of subcortical afferents. In view of the ubiquitous presence of bifurcating axons in the subcortical afferents, it is interesting to examine if a similar degree of branching is observed in this system of connection.

The projections from V1 to other cortical areas in the Old World monkeys have been the subject of numerous anatomical studies. It is now well established that V1 projects to V2, V3 and the middle temporal area (Kuypers *et al.*, 1965; Cragg & Ainsworth, 1969; Zeki, 1969, 1978; Rockland & Pandya, 1979; Ungerleider & Mishkin, 1979; Van Essen, Maunsell & Bixby, 1981; Weller & Kaas, 1983). In addition, the foveal part of V1 appears to project to V4, on the prelunate gyrus (Cragg & Ainsworth, 1969; Zeki, 1969), whereas the region of V1 representing the periphery of the visual field sends a projection to V3A (Zeki, 1980). These projections appear to be reciprocal, as labelled neurons can be found in cortical regions corresponding to V2, V3, V3A, V4 and MT after injections of fluorescent dyes in V1 (Kennedy & Bullier, 1985).

The connections between the other visual areas are less well known but it is apparent that V2, V3, V3A, V4 and MT are linked by reciprocal connections (Zeki, 1971; Van Essen 1979; Maunsell & Van Essen,1983; Van Essen & Maunsell, 1983). A similar situation is observed in the New World monkeys (squirrel and owl monkeys) with the difference that, in these species, V1 appears to project only to V2 and MT (Weller & Kaas, 1981).

The different visual cortical areas in the primate appear therefore to be intensely interconnected. The traditional model of a serial progression of sensory messages through 17, 18 and 19 should therefore be abandoned (Merzenich & Kaas, 1980). The present picture of the cortical connectivity includes far more distribution of the sensory messages than previously thought, although there is a certain 'hierarchical' progression. This is reflected in the fact that V1 projects only to a limited number of areas which, in turn, connect with cortical areas further removed from the main recipient zone of geniculo-cortical axons (Weller & Kaas, 1981; Van Essen, 1979; Van Essen & Maunsell, 1983).

This anisotropic organization is also evidenced by the fact that the neurons giving rise to, or receiving, cortico-cortical connections are not situated in the same laminae, depending on whether the connections are of the 'feed-forward type' (i.e. going away from V1) or of the 'feed-back' type (i.e. projecting back to an area closer to V1). Indeed, neurons giving rise to feedforward connections are mostly situated in the upper laminae and send projections to laminae 3 and 4, whereas neurons involved in the feedback connections mainly belong to the infragranular layers and project to layers outside of lamina 4 (Wong-Riley, 1978; Rockland & Pandya, 1979; Weller & Kaas, 1981; Tigges *et al.*, 1981; Lund *et al.*,1981; Lin, Weller & Kaas 1982; Van Essen & Maunsell, 1983). Using the laminar distributions of efferent neurons and of the termination of cortico-cortical afferents, one can therefore arrive at a classification of visual cortical areas in different hierarchical levels (Van Essen & Maunsell, 1983). In such a scheme, V1, V2 and V3 would occupy increasingly higher levels, while MT and V4 would be placed at the same hierarchical level above V3 (Van Essen & Maunsell, 1983). This scheme, which is a modern version of the traditional model of serial processing of information in Areas 17, 18 and 19, takes into account the recent findings in the cortico-cortical connections and provides a working model to test hypotheses on the cortical processing of sensory information (see also Chapter 21).

The question of axonal bifurcation in cortico-cortical connections in the primate visual system has only been studied for the connections to V1 and V2.

Following injections of two different fluorescent retrograde labels in these areas, it was found that cortical regions, presumably corresponding to V3, V3A, V4 and MT, contained variable amounts of double-labelled cells (Kennedy & Bullier, 1985). In cortical areas further and further removed from V2, a higher proportion of labelled neurons was found in the infragranular layers and the proportion of double-labelled neurons tended to increase. It is therefore possible that there is a higher degree of axonal bifurcation when the source of the projections is further and further removed from the region of termination. More experiments are needed to test this possibility and also to determine whether more axonal branching is found in feedback than in feedforward connections.

Cortico-cortical connections in the cat visual system

The visual areas in the cat cortex have been more clearly defined electrophysiologically and cytoarchitectonically than those in Old World monkeys. It has been known for some time that Areas 17, 18, 19 and the PMLS are interconnected densely and reciprocally (Hubel & Wiesel, 1965; Garey, Jones & Powell, 1968; Wilson, 1968; Kawamura, 1973; Gilbert & Kelly, 1975). In addition, direct and reciprocal connections from Areas 17, 18 and 19 to Areas 20a and 20b and to 21a and 21b have recently been described (Squatrito *et al.*, 1981*a*; Bullier *et al.*, 1984*c*; Dreher *et al.*, 1984). Additional connections of the visual areas of the lateral suprasylvian sulcus have also been reported (Kawamura & Naito, 1981; Squatrito *et al.*, 1981*b*: Cavada & Reinoso-Suarez, 1983). As a result, there appear to be few exceptions in the complete interconnectivity of all the known visual areas in the cat cortex (Fig. 17.2).

It is possible to class the different visual cortical areas of the cat in a similar fashion as in Old World monkeys, using the laminar distribution of the source and termination of cortico-cortical connections. Thus, Areas 17 and 18, which exchange connections from and to similar laminae (mainly supragranular), appear to belong to the same hierarchical level (Bullier *et al.*, 1984*c*: Orban, 1984). The PMLS and Area 19, which send cortical connections from a substantial proportion of lower-layer neurons, appear to belong to higher hierarchical levels, while Area 20, which projects to 17, 18 and 19 via infragranular layers, would belong to a still higher hierarchical level (Dreher *et al.*, 1984; Bullier *et al.*, 1984*c*; Symonds & Rosenquist, 1984; See also Chapter 20). More experimental data, particularly on the level of termination of cortico-cortical connections, are needed to establish more firmly such a scheme of organization in the cortical visual areas of the cat.

The issue of axonal branching in the cortical connections to Areas 17, 18 and 19 has been studied by double labelling experiments (Bullier *et al.*, 1984c).

After paired injections into two of these areas, double-labelled neurons were found in all cortical areas examined. The double-labelled neurons were confined to the regions of overlap between the two populations of labelled cells and were found in the same cortical laminae as the neurons containing only a single label. This indicates that the branches of bifurcating cortico-cortical axons terminate in the injected areas in

Fig. 17.2. Summary of the cortico–cortical connections between cortical visual areas in the cat. For abbreviations see legend of Fig. 17.1. The numbers indicate the following bibliographic references: 1: Cavada & Reinoso-Suarez, 1983; 2: Dreher *et al.*, 1984; 3: Garey *et al.*, 1968; 4: Gilbert & Kelly, 1975; 5: Heath & Jones, 1971; 6: Hubel & Weisel, 1965; 7: Kawamura, 1973; 8: Kawamura & Naito, 1981; 9: Mucke *et al.*, 1982; 10. Shoumura, 1972; 11: Shoumura & Itoh, 1972; 12: Squatrito *et al.*, 1981a; 13: Squatrito *et al.*, 1981b; 14: Wilson, 1968.

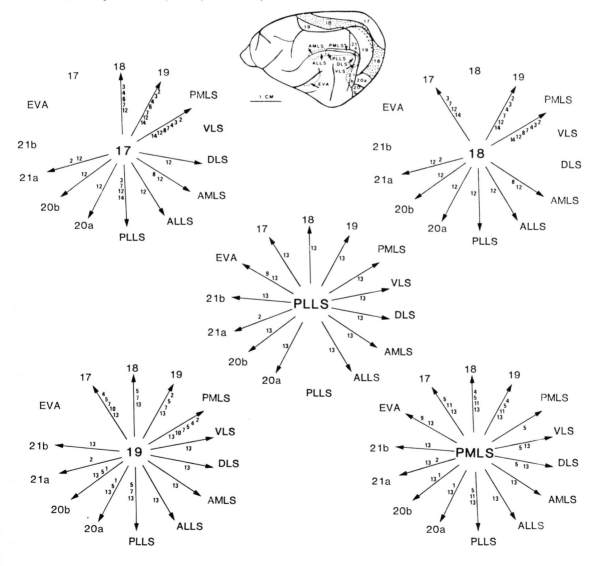

retinotopically corresponding regions, as was the case for the subcortical afferents.

The proportion of double-labelled neurons varied depending on the injected areas and the area under observation. Thus Area 17 sends few branching connections to Areas 18 and 19. Similarly, Area 18 sends few branching connections to Areas 17 and 19. On the other hand, Areas 17 and 18 receive a significant input from bifurcating axons arising in Areas 19, 20 and in the PMLS (Bullier *et al.*, 1984*c*). It appears, therefore, that there is more birfucation in the feedback connections (towards 17 and 18) than in the opposite direction. Area 20 was found to be, via its infragranular layers, the source of numerous branching axons projecting to Areas 17 and 18 and to Areas 18 and 19. This result is reminiscent of the situation in the cortical areas of the monkey where numerous double-labelled neurons were found in the deep layers of areas far removed from Area 17 after injection in V1 and V2. The small percentage of double-labelled neurons found in Area 20 after injection in Areas 17 and 19 indicates that the amount of bifurcating connections does not depend only on the direction (feedback or feedforward) of the connection, but also on the areas of projection. Experiments involving injections in the other cortical areas are needed to determine the factors contributing to the presence of bifurcating axons in the cortico-cortical connections of the visual system.

The pattern of axonal bifurcation in callosal connections was also studied with the double labelling technique (Segraves & Innocenti, 1985). In the PMLS and Areas 17 and 18, no neuron was found to send a bifurcating callosal axon to the contralateral PMLS and the 17–18 border. Neurons sending one branch to the ipsilateral cortex and another to the contralateral hemisphere were found only rarely in Areas 17, 18 and in the PMLS.

In conclusion, bifurcating axons are generally less frequently observed in cortico-cortical connections than in the subcortical afferents and the incidence of axonal branching appears to be linked with the source and target regions of the connections. These results suggest that, in contrast to the subcortical afferents, certain cortico-cortical connections exhibit a high degree of functional specificity whereas others carry the same signals to several cortical areas.

Axonal bifurcation in cortico-cortical connections during development

It is generally accepted that during the development of the mammalian nervous system there is an initial phase of exuberance during which the number of neurons and connections is much greater than occurs in the mature animal (see for review Cowan, 1973, 1979; Changeux & Danchin, 1976; Purves & Lichtman, 1980). Examples of this phenomenon have been found in the retina and the optic nerve (Ng & Stone, 1982; Rakic & Riley, 1983), the visual cortex (Cragg, 1975; O'Kusky & Colonnier, 1982) and the corpus callosum (Koppel & Innocenti, 1983). It is presumably during such stages of exuberance that one can observe a transitory innervation of a single neuron by several fibers. This has been found in the neuromuscular junction (Redfern, 1970), the connections between the climbing fibers and Purkinje cells in the cerebellum (Crepel, Mariani & Delhaye-Bouchaud, 1976; Mariani & Changeux, 1981) and in a few other systems (Purves & Lichtman, 1980). This multiple innervation phase is followed by a retraction of collaterals leading to a one-to-one relationship between afferent fiber and target neuron (Changeux & Danchin, 1976; Purves & Lichtman, 1980; Mariani & Changeux, 1981).

The presence of a transitory exuberance of axons during the postnatal development of cortical connections has been demonstrated in the callosal connections of the cat visual cortex (Innocenti, 1981; Koppel & Innocenti, 1983) and of the parietal cortex in the rat (O'Leary, Stanfield & Cowan, 1981). These experiments showed that in newborn animals many more neurons sent callosal collaterals than in the adults. Most of these neurons survived to adulthood and lost their callosal collaterals during development. In view of these findings it could be expected that in young animals more cortical neurons send collaterals to several cortical areas than in adults. Indeed, this happens in the somatosensory system of the rat. Thus, in the newborn rat, a substantial number of neurons in the somatosensory system send collaterals to the ipsilateral motor cortex. These collaterals appear to be absent in the adult animal (Ivy & Killackey, 1982). Similarly, in the newborn rat, many neurons in the neocortex send axons in the pyramidal tract and this projection is later restricted to the motor region (Stanfield, O'Leary &

ricks, 1982). In newborn cats and monkeys, the proportions of collaterals are similar to the numbers observed in the adult animals (Schwartz & Goldman-Rakic, 1982; Innocenti & Clarke, 1983; Bullier, *et al.* 1984a). These observations are not surprising since in the cat and monkey the brain is much more mature at birth than it is in the rat.

Conclusion

A number of recent studies have revealed that the axons of many neurons (which project to different functional areas of the neocortex) bifurcate. This was shown to be true for several species and sensory modalities and it appears certain that axonal branching is an important factor that will have to be included in all models of cortical organization. In contrast, until several years ago, axonal branching was thought to be restricted to the diffuse thalamo–cortical projection system and it was generally considered that the connections between nervous structures were made by axons going to only one structure. Now the situation is somewhat reversed and the lack of axon collaterals is seen as the exception rather than the rule. Such a low level of axon branching is seen among X-type LGNd cells in the cat and all LGNd cells in the monkey. There is a similar lack of axon branching among output cells in Area 17. Thus, it appears that one role of the lemniscal pathway going through the LGN and Area 17 is to keep separate the different information channels circulating in it. Indeed the results of electrophysiological experiments show that in both cats and monkeys the different streams of thalamocortical afferents to Area 17 are mostly kept separate (Bullier & Henry, 1979, 1980; Mustari, Bullier & Henry 1982).

Since the proportions of branched connections appear similar in the neonatal and the adult cat, sensory experience does not seem to be necessary for the selection of the parent cells of bifurcating axons. On the other hand, it may be that visual experience is necessary to adjust the retinotopic correspondence of the two branches of an axon. If this was shown to be the case, it would further emphasize the functional importance of axonal collaterals in the mature system.

The presence of ubiquitous axonal bifurcation to the different cortical areas suggests that both strictly hierarchical and strictly parallel models of organization represent an oversimplification of the real system. It appears that a significant amount of neural information is not processed in a serial or parallel fashion but is distributed to several cortical areas to be processed simultaneously. In this model, the different cortical areas get their specific properties, not from their position in a serial chain of information processing, nor by the specific type of afferents they receive, but by the way the different types of afferents interact in their neural networks.

Acknowledgements

This chapter was written while the author was a visiting fellow in the Department of Physiology at the John Curtin School of Medical Research in Canberra. The financial assistance of the Fondation pour la Recherche Médicale is gratefully acknowledged.

References

Albus, K. (1981) Hypothalamic and basal forebrain afferents to the cat's visual cortex: a study with horseradish peroxidase. *Neurosci. Lett.*, **24**, 117–21

Anderson, S. A., Landgren, S. & Wolsk, D. (1966) The thalamic relay and cortical projection of group 1 muscle afferents from the forelimb of the cat. *J. Physiol.*, **183**, 576–91

Bender, D. B. (1981). Retinotopic organization of macaque pulvinar. *J. Neurophysiol.*, **46**, 672–93

Benevento, L. A. & Fallon, J. H. (1975) The ascending projections of the superior colliculus in the rhesus monkey (*Macaca mulatta*). *J. Comp. Neurol.*, **160**, 339–62

Benevento, L. A. & Rezak, M. (1976). The cortical projections of the inferior pulvinar and adjacent lateral pulvinar in the rhesus monkey (*Macaca mulatta*). *Brain Res.*, **108**, 1–24

Benevento, L. A. & Yoshida, K. (1981). The afferent and efferent organization of the lateral geniculo-prestriate pathways in the macaque monkey. *J. Comp. Neurol.*, **203**, 455–74

Bentivoglio, M., Macchi, G. & Albanese, A. (1981). The cortical projections of the thalamic intralaminar nuclei, as studied in cat and rat with the multiple fluorescent retrograde tracing technique. *Neurosci. Lett.*, **26**, 5–10

Bentivoglio, M., Macchi, G., Rossini, P. & Tempesta, E. (1978). Brain stem neurons projecting to neocortex: a HRP study in the cat. *Exp. Brain Res.*, **31**, 489–98

Berman, N. (1977). Connection of the pretectum in the cat. *J. Comp. Neurol.*, **174**, 227–54

Berman, N. & Jones, E. G. (1977). A retino-pulvinar projection in the cat. *Brain Res.*, **134**, 237–48

Berson, D. M. & Graybiel, A. M. (1978). Parallel thalamic zones in the LP-pulvinar complex of the cat identified by the afferent and efferent connections. *Brain Res.*, **147**, 139–48

Berson, D. M. & Graybiel, A. M. (1983). Organization of the striate-recipient zone of the cat's lateralis posterior-pulvinar complex and its relations with the geniculostriate system. *Neurosci*, **9**, 337–72

Bullier, J., Dehay, C. & Kennedy, H. (1984a). Axonal bifurcation and corticocortical connectivity in the kitten visual cortex. *J Physiol.*, **353**, 22p

Bullier, J. & Henry, G. H. (1979). Neural path taken by afferent streams in striate cortex of the cat. *J. Neurophysiol.*, **42**, 1264–70

Bullier, J. & Henry, G. H. (1980). Ordinal position and afferent input of neurons in monkey striate cortex. *J. Comp. Neurol.*, **193**, 913–35

Bullier, J. & Kennedy, H. (1983). Projection of the lateral geniculate nucleus onto cortical area V2 in the macaque monkey. *Exp. Brain Res.*, **53**, 168–72

Bullier, J., Kennedy, H. & Salinger, W. (1984*b*). Bifurcation of subcortical afferents to visual areas 17, 18 and 19 in the cat cortex. *J. Comp. Neurol.*, **228**, 309–28.

Bullier, J., Kennedy, H. & Salinger, W. (1984*c*). Branching and laminar origin of projections between visual cortical areas in the cat. *J. Comp. Neurol.*, **228**, 329–41.

Campos-Ortega, J. A., Hayhow, W. R. & Cluver, P. F. de V. (1970). A note on the problem of retinal projections to the pulvinar nucleus of primates. *Brain Res.*, **22**, 126–30

Carey, R. G., Fitzpatrick, D. & Diamond, I. T. (1979). Layer I of striate cortex of *Tupaïa glis* and *Galago senegalensis*: Projections from thalamus and claustrum revealed by retrograde transport of horseradish peroxidase. *J. Comp. Neurol.*, **186**, 393–438

Cavada, C. & Reinoso-Suarez, F. (1983). Afferent connection of area 20 in the cat studied by means of the retrograde axonal transport of horseradish peroxidase. *Brain Res.*, **270**, 319–24

Changeux, J. P. & Danchin, A. (1976). Selective stabilisation of developing synapses as a mechanism for the specification of neural networks. *Nature*, **264**, 705–12

Chung, S., Raymond, S. A. & Lettvin, J. Y. (1970). Multiple meaning in single visual units. *Brain Behav. Evol.*, **3**, 72–101

Cleland, B. G., Levick, W. R., Morstyn, R. & Wagner, H. G. (1976). Lateral geniculate relay of slowly conducting retinal afferents to cat visual cortex. *J. Physiol.*, **255**, 299–320

Coleman, J. & Clerici, W. J. (1980). Extrastriate projections from thalamus to posterior occipital-temporal cortex in rat. *Brain Res.*, **194**, 205–9

Cowan, W. M. (1973). Neuronal death as a regulative mechanism in the control of cell number in the nervous system. In *Development and aging in the nervous system*, ed. M. Rockstein, pp. 19–41. New York: Academic Press

Cowan, W. M. (1979). Selection and control in neurogenesis. In *The Neurosciences. Fourth Study Programme*, ed. F. O. Schmitt & F. G. Worden, pp. 59–79. Cambridge, Mass.: MIT Press

Cragg, B. G. (1975). The development of synapses in the visual system of the cat. *J. Comp. Neurol.*, **160**, 147–66

Cragg, B. G. & Ainsworth, A. (1969). The topography of the afferent projections in the circumstriate visual cortex of the monkey studied by the Nauta method. *Vision Res.*, **9**, 733–47

Crepel, F., Mariani, J. & Delhaye-Bouchaud, N. (1976). Evidence for a multiple innervation of Purkinje cells by climbing fibres in the immature rat cerebellum. *J. Neurobiol.*, **7**, 567–78

Cusick, C. G., Steindler, D. A. & Kaas, J. H. (1985). Corticocortical and colateral thalamocortical connection of central somatosensory cortical areas in squirrel monkeys: a double-labeling study with radiolabeled wheatgerm agglutinin and wheatgerm agglutinin conjugated to horseradish feroxidase. *Somatosensory Res.*, **3**, 1–31.

Diamond, I. T. & Hall, W. C. (1969). Evolution of neocortex. *Science*, **164**, 251–62

Doty, R. W. (1958). Potentials evoked in cat cerebral cortex by diffuse and by punctiform photic stimuli. *J. Neurophysiol.*, **21**, 437–64

Doty, R. W. (1983). Non geniculate afferents to striate cortex in macaques. *J. Comp. Neurol.*, **218**, 159–73

Dreher, B., Ho, H. T., Lee, F. C. W. & Leventhal. A. G. (1984). Comparison of prosencephalic afferents and receptive field properties of cells in areas 19 and 21a of cat extrastriate cortex. *J. Comp. Neurol.* (In press)

Dreher, B., Leventhal, A. G. & Hale, P. T. (1980). Geniculate input to cat visual cortex: a comparison of area 19 with areas 17 and 18. *J. Neurophysiol.*, **44**, 804–26

Dreher. B. & Sefton, A. J. (1979). Properties of neurones in cat's dorsal lateral geniculate nucleus: a comparison between medial interlaminar and laminated parts of the nucleus. *J. Comp. Neurol.*, **183**, 47–64

Enroth-Cugell, C. & Robson, J. G. (1966). The contrast sensitivity of retinal ganglion cells of the cat. *J. Physiol.*, **187**, 517–52

Espinoza, S. E. & Thomas, H. C. (1983). Retinotopic organization of striate and extrastriate visual cortex in the hooded rat. *Brain Res*, **272**, 137–44

Freedman, R., Foote, S. L. & Bloom. F. (1975). Histochemical characterization of a neocortical projection of the nucleus locus coeruleus in the squirrel monkey. *J. Comp. Neurol.*, **164**, 209–32

Friedlander, M. J., Lin, C. S., Stanford, L. R. & Sherman, S. M. (1981). Morphology of functionally identified neurons in lateral geniculate nucleus of cat. *J. Neurophysiol.*, **46**, 80–129

Fries, W. (1981). The projection from the lateral geniculate nucleus to the prestriate cortex of the macaque monkey. *Proc. R. Soc. Lond.*, **213**, 73–80

Galletti, C., Squatrito, S. & Battaglini, P. P. (1979). Visual cortex projections to thalamic intralaminar nuclei in the cat. An autoradiographic study. *Arch. Ital. Biol.*, **117**, 280–5

Garey, L. J., Jones, E. G. & Powell, T. P. S. (1968). Interrelationships of striate and extrastriate cortex with th primary sites of the visual pathway. *J. Neurol. Neurosurg. Psychiatry*, **31**, 135–57

Garey, L. J. & Powell, T. P. S. (1967). The projection of the lateral geniculate nucleus upon the cortex of the cat. *Proc. R. Soc. Lond., Ser. B.*, **169**, 107–26

Gatter, K. C. & Powell, T. P. S. (1977). The projection of the locus coeruleus upon the neocortex in the macaque monkey. *Neuroscience.*, **2**, 441–5

Geisert, E. E., Jr (1980). Cortical projections of the lateral geniculate nucleus in the cat. *J. Comp. Neurol.*, **190**, 793–812 ilbert, C. D. & Kelly, J. P. (1975). The projection of cells in different layers of the cat's visual cortex. *J. Comp. Neurol.,* **163,** 81–106

iolli, R. A. & Towns, L. C. (1980). A review of axon collateralization in the mammalian visual system. *Brain Behav. Evol.,* **17,** 364–90

raham, J. (1982). Some topographical connections of the striate cortex with subcortical structures in *Macaca fascicularis. Exp. Brain Res.,* **47,** 1–14

raybiel, A. M. & Berson, D. M. (1980). Histochemical identification and afferent connections of subdivisions in the lateralis posterior pulvinar complex and related thalamic nuclei in the cat. *Neuroscience,* **5,** 1175–238

rossman, Y., Spira, M. E. & Parnas, I. (1973). Differential flow of information into branches of a single axon. *Brain Res.,* **64,** 379–86

uillery, R. W., Geisert, E. E. Jr, Polley, E. H. & Mason, C. A. (1980). An analysis of the retinal afferents to the cat's medial interlaminar nucleus and to its rostral thalamic extension, the 'geniculate wing'. *J. Comp. Neurol.,* **194,** 117–42

arting, J. K., Huerta, M. F., Frankfurter, H. J., Strominger, N. L. & Royce, G. J. (1980). Ascending pathways from the monkey superior colliculus: an autoradiographic analysis. *J. Comp. Neurol.,* **192,** 853–82

arvey, A. R. (1980). The afferent connections and laminar distribution of cells in area 18 of the cat. *J. Physiol.,* **302,** 483–505

att, H. & Smith, D. O. (1975). Axon conduction block: differential channeling of nerve impulses in the crayfish. *Brain Res.,* **87,** 85–8

eath, C. J. & Jones, E. G. (1971). The anatomical organization of the suprasylvian gyrus of the cat. *Ergebn. Anat. Entwickl. Gesch.,* **45,** 64pp

erkenham, M. (1980) Laminar organization of thalamic projections to the rat neocortex. *Science,* **207,** 532–4

ickey, T. L. & Guillery, R. W. (1974). An autoradiographic study of retinogeniculate pathways in the cat and in the fox. *J. Comp. Neurol.,* **156,** 239–54

offmann, K. P. & Stone, J. (1971). Conduction velocity of afferents to cat visual cortex: a correlation with cortical receptive field properties. *Brain Res.,* **32,** 460–6

olländer, H. & Vanegas, H. (1977). The projections from the lateral geniculate nucleus onto the visual cortex in the cat. A quantitative study with horseradish peroxidase. *J. Comp. Neurol.,* **173,** 519–36

ubel, D. H. & Wiesel, T. N. (1965). Receptive fields and functional architecture in two nonstriate visual areas (18 and 19) of the cat. *J. Neurophysiol.,* **28,** 229–89

ughes, H. C. (1977). Anatomical and neurobehavioral investigation concerning the thalamocortical organization of the rat's visual system. *J. Comp. Neurol.,* **175,** 311–36

ughes, H. C. (1980). Efferent organization of the cat's pulvinar complex, with a note on bilateral claustrocortical and reticulocortical connections. *J. Comp. Neurol.,* **193,** 937–63

umphrey, A. L., Sur, S. M., Ulrich, D. J. & Sherman, S. M. (1985). Termination patterns of individual X- and Y-cell axons in the visual cortex of the cat: Projections to Areas 18, to the 17/18 border region, and to both Areas 17 and 18. *J. Comp. Neurol.,* **223,** 190–212.

Imig, T. J. & Morel, A. (1983). Organization of the thalamocortical auditory system in the cat. *Ann. Rev. Neurosci.,* **6,** 95–120

Innocenti, G. M. (1981). Growth and reshaping of axons in the establishment of visual callosal connection. *Science,* **212,** 824–7

Innocenti, G. M. & Clarke, S. (1983). Multiple sets of visual cortical neurons projecting transitorily through the corpus callosum. *Neurosci. Lett.,* **41,** 27–32

Itaya, S. K. & Van Hoesen, G. W. (1983). Retinal projections to the inferior and medial pulvinar nuclei in the old world monkey. *Brain Res.,* **269,** 223–30

Itoh, K., Mizuno, N. & Kudo, M. (1983). Direct retinal projections to the lateroposterior and pulvinar nucleus complex (LP-Pul) in the cat, as revealed by the anterograde HRP method. *Brain Res.,* **276,** 325–8

Itoh, K., Mizuno, N., Sugimoto, T., Nomura, S., Nakamura, Y. & Konishi, A. (1979). A cerebello-pulvino-cortical and a retino-pulvino-cortical pathway in the cat as revealed by the use of the anterograde and retrograde transport of horseradish peroxidase. *J. Comp. Neurol.,* **187,** 349–58

Ivy, G. O. & Killackey, H. P. (1982). Ontogenetic changes in the projections of neocortical neurons. *J. Neurosci.,* **2,** 735–43

Jones, E. G. (1983). Lack of collateral thalamocortical projections to fields of the first somatic sensory cortex in monkeys. *Exp. Brain Res.,* **52,** 375–84

Jones, E. G. & Leavitt, R. Y. (1974). Retrograde axonal transport and the demonstration of non-specific projections to the cerebral cortex and striatum from thalamic intralaminar nuclei in the rat, cat and monkey. *J. Comp. Neurol.,* **154,** 349–77

Kaas, J. H. (1980). A comparative survey of visual cortex organization in mammals. In *Comparative neurology of the telencephalon,* Ed. S. O. E. Ebbesson, pp. 483–502. New York: Plenum Press.

Kaas, J. H., Huerta, M. F., Weber, J. T. & Harting, J. K. (1978). Patterns of retinal terminations and laminar organization of the lateral geniculate nucleus of primates. *J. Comp. Neurol.,* **182,** 517–54

Kaufman, E. F. S., Rosenquist, A. C. & Raczowski, D. (1984). The projections of single thalamic neurons onto multiple visual cortical areas in the cat. *Brain Res..,* **298,** 171–4

Kawamura, K. (1973). Cortico-cortical fibers connections of the cat cerebrum. III. The occipital region. *Brain Res.,* **51,** 41–60

Kawamura, K. & Naito, J. (1981). Cortico-cortical neurons projecting to the medial and lateral banks of the middle suprasylvian sulcus in the cat. An experimental study with the horseradish peroxidase method. *J. Comp. Neurol.,* **193,** 1009–22

Kennedy, H. & Baleydier, C. (1977). Direct projections from thalamic intralaminar nuclei to extrastriate visual cortex in the cat traced with horseradish peroxidase. *Exp. Brain Res.,* **28,** 133–9

Kennedy, H. & Bullier, J. (1985). A double-labeling investigation of the afferent connectivity to cortical areas V_1 and V_2 of the macaque monkey. *J. Neurosci.*, **5**, 2815–30.

Koppel, H. & Innocenti, G. M. (1983). Is there a genuine exuberancy of callosal projections in development? A quantitative electron microscopic study in the cat. *Neurosci. Lett.*, **41**, 33–40

Kratz, K. E., Webb, S. V. & Sherman, S. M. (1978). Studies of the cat's medial interlaminar nucleus: a subdivision of the dorsal lateral geniculate nucleus. *J. Comp. Neurol.*, **181**, 601–14

Kuypers, H. G. J. M., Szwarcbart, M. K., Mishkin, M. & Rosvold, H. E. (1965). Occipitotemporal cortico-cortical connections in the rhesus monkey. *Exp. Neurol.*, **11**, 245–62

Laties, A. M. & Sprague, J. M. (1966). The projection of optic fibers to the visual centers in the cat. *J. Comp. Neurol.*, **127**, 35–70

LeVay, S. & Gilbert, C. (1976). Laminar patterns of geniculocortical projection in the cat. *Brain Res.*, **113**, 1–19

LeVay, S. & Sherk, H. (1981a). The visual claustrum of the cat. I. Structure and connections. *J. Neurosci.*, **1**, 956–80

LeVay, S. & Sherk, H. (1981b). The visual claustrum of the cat. II. The visual field map. *J. Neurosci.*, **1**, 981–92

Lennie, P. (1980). Parallel visual pathways: a review. *Vision Res.*, **20**, 561–94

Leventhal, A. G., Keens, J. & Tork, I. (1980). The afferent ganglion cells and cortical projections of the retinal recipient zone (RRZ) of cat's 'pulvinar complex'. *J. Comp. Neurol.*, **194**, 535–54

Levick, W. R. (1975). Form and function of cat retinal ganglion cells. *Nature*, **254**, 659–62

Lin, C. S., Weller, R. E. & Kaas, J. H. (1982). Cortical connections of striate cortex in the owl monkey. *J. Comp. Neurol.*, **211**, 165–76

Lorente de No, R. (1983). The cerebral cortex: architecture, intracortical connections and motor projection. In *Physiology of the nervous system*, Ed. J. F. Fulton, pp. 274–313. Oxford: Oxford University Press

Lund, J. S., Hendrickson, A. E., Ogren, M. P. & Tobin, E. A. (1981). Anatomical organization of primate visual cortical area VII. *J. Comp. Neurol.* **202**, 19–45

Maciewicz, R. J. (1974). Afferents to the lateral suprasylvian gyrus of the cat traced with horseradish peroxidase. *Brain Res.*, **78**, 139–43

Maciewicz, R. J. (1975). Thalamic afferents to areas 17, 18 and 19 of cat cortex traced with horseradish peroxidase. *Brain Res.*, **84**, 308–12

Mariani, J. & Changeux, J. P. (1981). Ontogenesis of olivocerebellar relationships. I. Studies by intracellular recordings of the multiple innervation of Purkinje cells by climbing fibers in the developing rat cerebellum. *J. Neurosci.*, **1**, 696–702

Mason, J. (1969). The somatosensory cortical projection of single nerve cells in the thalamus of the cat. *Brain Res.*, **12**, 489–92

Mason, R. (1975). Cell properties in the medial interlaminar nucleus of the cat's lateral geniculate nucleus complex in relation to the sustained/transient classification. *Exp. Brain Res.*, **22**, 327–9

Mason, R. (1978). Functional organization in the cat's pulvinar complex. *Exp. Brain Res.*, **31**, 51–66

Mason, R. (1981). Differential responsiveness of cells in the visual zones of the cat's LP-Pulvinar complex to visual stimuli. *Exp. Brain Res.*, **43**, 25–33

Maunsell, J. H. R. & Van Essen, D. C. (1983). The connections of the middle temporal visual area (MT) and their relationship to a cortical hierarchy in the macaque monkey. *J. Neurosci.*, **3**, 2563–86

Merzenich, M. M. & Kaas, J. H. (1980). Principles of organization of sensory-perceptual systems in mammals. *Prog. Psychobiol Physiol. Psychol.*, **9**, 1–42

Miller, J. W. & Benevento, L. A. (1979). Demonstration of a direct projection from the intralaminar central lateral nucleus to the primary visual cortex. *Neurosci. Lett.*, **14**, 229–34

Mitzdorf, U. & Singer, W. (1978). Prominent excitatory pathways in the cat visual cortex (A 17 and A 18): a current source density analysis of electrically evoked potentials. *Exp. Brain Res.*, **33**. 371–94

Mizuno, N., Itoh, K., Uchida, K., Uemura-Sumi, M. & Matsushima, R. (1982). A retino-pulvinar projection in the macaque monkey as visualized by the use of anterograde transport of horseradish peroxidase. *Neurosci. Lett.*, **30**, 199–203

Mizuno, N., Uchida K., Nomura, S., Nakamura, Y., Sugimoto, T. & Uemura-Sumi, M. (1981). Extrageniculate projections to the visual cortex in the macaque monkey: an HRP study. *Brain Res.*, **212**, 454–9

Montero, V. M. (1981). Comparative studies on the visual cortex. In *Cortical Sensory Organization*, vol. 2 *Multiple Visual Areas*, ed. C. N. Woolsey, pp. 33–81. Clifton, N. J.: Humana Press.

Montero, V. M., Bravo, H. & Fernandez, V. (1973a). Striate-peristriate corticocortical connections in the albino and gray rat. *Brain Res.*, **53**, 202–7

Montero, V. M., Rojas, A. & Torrealba, F. (1973b). Retinotopic organization of striate and peristriate visual cortex in the albino rat. *Brain Res.*, **53**, 197–201

Morison, R. S. & Dempsey, E. W. (1942). A study of thalamo-cortical relations. *Am. J. Physiol.*, **135**, 281–92

Morrison, J. H., Grzanna, R., Molliver, M. E. & Coyle, J. T. (1978). The distribution and orientation of noradrenergic fibers in neocortex of the rat: an immunofluorescent study. *J. Comp. Neurol.*, **181**, 17–40

Morrison, J. H., Molliver, M. E. & Grzanna, R. (1979). Noradrenergic innervation of cerebral cortex: widespread effects of local cortical lesions. *Science*, **205**, 313–16

Mucke, L., Norita, M., Benedek, G. & Creutzfeldt, O. (1982). Physiologic and anatomic investigation of a visual cortical area situated in the ventral bank of the anterior ectosylvian sulcus of the cat. *Exp. Brain Res.*, **46**, 1–11.

Murray, M. (1966). Degeneration of some intralaminar thalamic nuclei after cortical removal in the cat. *J. Comp. Neurol.*, **127**, 344–68

Mustari, M. J., Bullier, J. & Henry, G. H. (1982) Comparison of response properties of three types of monosynaptic S cell in cat striate cortex. *J. Neurophysiol.*, **47**, 439–54 akamura, S. (1977). Some electrophysiological properties of neurones in rat locus coeruleus. *J. Physiol.*, **267**, 641–58

auta, W. J. H. & Whitlock, D. G. (1954). An anatomical analysis of the non-specific thalamic projection system. In *Brain mechanisms and consciousness*, ed. J. F. Delafresnaye, pp. 81–116. Oxford: Blackwell

g, A. Y. K. & Stone, J. (1982)The optic nerve of the cat: appearance and loss of axons during normal development. *Dev. Brain Res.*, **5**, 263–71

iimi, K. H., Matsuoka, Y., Yamazaki, Y. & Matsumoto, H. (1981). Thalamic afferents to the visual cortex in the cat studied by retrograde axonal transport of horseradish peroxidase. *Brain Behav. Evol.*, **18**, 114–39

iimi, K. & Sprague, J. M. (1970). Thalamo–cortical organization of the visual system in the cat. *J. Comp. Neurol.*, **138**, 219–50

orita, M. & Creutzfeldt, O. D. (1982). An HRP and ^{3}H-apo-HRP study of the thalamic projections to visual cortical areas in the cat. *Acta Biol. Acad. Sci. Hung.*, **33**, 269–75

gren, M. P. & Hendrickson, A. (1976). Pathways between striate cortex and subcortical regions in *Macaca mulatta* and *Saimiri sciureus*: evidence for a reciprocal pulvinar connection. *Exp. Neurol.*, **53**, 780–800

gren, M. & Hendrickson, A. (1977) The distribution of pulvinar terminals in visual areas 17 and 18 of the monkey. *Brain Res.*, **137**, 343–53

Kusky, J. & Colonnier, M. (1982) Postnatal changes in the number of neurons and synapses in the visual cortex (Area 17) of the macaque monkey: a stereological analysis in normal and monocularly deprived animals. *J. Comp. Neurol.*, **210**, 291–306

'Leary, D. D. M., Stanfield, B. B. & Cowan, W. M. (1981) Evidence that the early postnatal restriction of the cells of origin of the callosal projection is due to the elimination of axonal collaterals rather than to the death of neurons. *Dev. Brain Res.*, **1**, 607–17

son, C. R. & Graybiel, A. M. (1980). Sensory maps in the claustrum of the cat. *Nature*, **288**, 479–81

rban, G. A. (1984). *Neuronal Operations in the Visual Cortex*. Heidelberg, New York & Tokyo, Springer

lmer, L. A., Rosenquist, A. C. & Tusa, R. J. (1977) The retinotopic organization of lateral suprasylvian visual areas in the cat. *J. Comp. Neurol.*, **177**, 237–56

rnas, I. (1972) Differential block at high frequency of branches of a single axon innervating two muscles. *J. Neurophysiol.*, **35**, 903–14

rtlow, G. D., Colonnier, M. & Szabo, J. (1977) Thalamic projections of the superior colliculus in the rhesus monkey *Macaca mulatta*. A light and electron microscopic study. *J. Comp Neurol.*, **171**, 285–318

owell, T. P. S. & Cowan, W. M. (1964) The interpretation of the degenerative changes in the intralaminar nuclei of the thalamus. *J. Neurol. Neurosurg. Psychiatry*, **30**, 140–53

rves, D. & Lichtman, J. W. (1980) Elimination of synapses in the developing nervous system. *Science*, **210**, 153–7

aczkowski, D. & Rosenquist, A. C. (1980) Connections of the parvocellular C laminae of the dorsal lateral geniculate nucleus with the visual cortex of the cat. *Brain Res.*, **199** 447–51

Raczkowski, D. & Rosenquist, A. C. (1981) Retinotopic organization in the cat lateral posterior-pulvinar complex. *Brain Res.*, **221**, 185–91

Raczkowski, D. & Rosenquist, A. C. (1983) Connections of the multiple visual cortical areas with the lateral posterior pulvinar complex and adjacent thalamic nuclei in the cat. *J Neurosci.*, **3**, 1912–42

Rakic, P. & Riley, K. P. (1983) Overproduction and elimination of retinal axons in the fetal rhesus monkey. *Science*, **219**, 1441–4

Redfern, P. A. (1970) Neuromuscular transmission in new-born rats. *J. Physiol.*, **209**, 701–9

Rezak, M. & Benevento, L. A. (1979) A comparison of the organization of the projections of the dorsal lateral geniculate nucleus, the inferior pulvinar and adjacent lateral pulvinar to primary visual cortex (area 17) in the macaque monkey. *Brain Res.*, **167**, 19–40

Riche, D. & Lanoir, J. (1978) Some claustro–cortical connections in the cat and baboon as studied by retrograde horseradish peroxidase transport. *J. Comp. Neurol.*, **177**, 435–44

Rockland, K. S. & Pandya, D. N. (1979) Laminar origins and terminations of cortical connections to the occipital lobe in the rhesus monkey. *Brain Res.*, **179**, 3–20

Rodieck, R. W. (1979) Visual pathways. *Ann. Rev. Neurosci.*, **2**, 193–225

Rose, J. E. & Woolsey, C. N. (1958) Cortical connections and functional organization of the thalamic auditory system of the cat. In *Biological and Biochemical Bases of Behavior*, ed. H. F. Harlow & C. N. Woolsey, pp. 127–50. Madison: University of Wisconsin Press

Rosenquist, A. C., Edwards, S. B. & Palmer, L. A. (1974) An autoradiographic study of the projections of the dorsal lateral geniculate nucleus and the posterior nucleus in the cat. *Brain Res.*, **80**, 71–93

Rowe, M. H. & Dreher, B. (1982) Retinal W-cell projections to the medial interlaminar nucleus in the cat: implications for ganglion cell classification. *J. Comp. Neurol.*, **204**, 117–33

Rowe, M. J. & Sessle, B. J. (1968) Somatic afferent input to posterior thalamic neurons and their axon projection to the cerebral cortex in the cat. *J. Physiol.*, **196**, 19–35

Sanderson, K. J. (1971) The projection of the visual field to the lateral geniculate and medial interlaminar nuclei in the cat. *J. Comp. Neurol.*, **143**, 101–18

Sanides, D. & Bucholtz, C. S. (1979) Identification of the projection from the visual cortex to the claustrum by anterograde axonal transport in the cat. *Exp. Brain Res.*, **34**, 197–200

Schwartz, M. L. & Goldman-Rakic, P. S. (1982) Single cortical neurons have axon collaterals to ipsilateral and contralateral cortex in foetal and adult primates. *Nature*, **299**, 154–5

Segraves, M. A. & Innocenti, G. M. (1985) A comparison of the distributions of ipsilaterally and contralaterally projecting cortico-cortical neurons in cat visual cortex using two fluorescent tracers. *J. Neurosci.*, **5**, 2107–18.

Sherk, H. & LeVay, S. (1981) The visual claustrum of the cat. III. Receptive field properties. *J. Neurosci.*, **1**, 993–1002

Shoumura, K. (1972) Patterns of fiber degeneration in the lateral wall of the suprasylvian gyrus, the Clare–Bishop area of the cat. *Brain Res.*, 43, 264–7

Shoumura, K. & Itoh, K. (1972) Intercortical projections from the lateral wall of the suprasylvian gyrus, the Clare–Bishop area of the cat. *Brain Res.*, 39, 536–9

Simmons, P. A., Lemmon, V. & Pearlman, A. L. (1982) Afferent and efferent connections of the striate and extrastriate visual cortex of the normal and reeler mouse. *J. Comp. Neurol.*, 211, 295–308

Singer, W., Tretter, F. & Cynader, M. (1975) Organization of cat striate cortex. A correlation of receptive field properties with afferent and efferent connections. *J. Neurophysiol.*, 38, 1080–98

Spira, M. E., Yarom, Y. & Parnas, I. (1976) Modulation of spike frequency by regions of special axonal geometry and by synaptic inputs. *J. Neurophysiol.*, 39, 882–99

Spreafico, R., Hayes, N. L. & Rustioni, A. (1981) Thalamic projections to the primary and secondary somatosensory cortices in cat: single and double retrograde tracer studies. *J. Comp. Neurol.*, 203, 67–90

Squatrito, S., Battaglini, P. P., Galletti, C. & Riva Sanseverino, E. (1980) Autoradiographic evidence for projections from cortical visual areas 17, 18, 19 and the Clare–Bishop area to the ipsilateral claustrum in the cat. *Neurosci. Lett.*, 19, 265–9

Squatrito, S., Galletti, C., Battaglini, P. P. & Riva Sanseverino, E. (1981a) Bilateral cortical projections from cat visual areas 17 and 18. An autoradiographic study. *Arch. Ital. Biol.*, 119, 1–20

Squatrito, S., Galletti, C., Battaglini, P. P. & Riva Sanseverino, E. (1981b) An autoradiographic study of bilateral cortical projections from cat area 19 and lateral suprasylvian visual area. *Arch. Ital. biol.*, 119, 21–42

Stanfield, B. B., O'Leary, D. D. M. & Fricks, C. (1982) Selective collateral elimination in early postnatal development restricts cortical distribution of rat pyramidal tract neurones. *Nature*, 298, 371–3

Steindler, D. A. (1981) Locus coeruleus neurons have axons that branch to the forebrain and cerebellum. *Brain Res.*, 223, 367–73

Stone, J. & Dreher, B. (1973) Projection of X and Y cells of the cat's lateral geniculate nucleus to areas 17 and 18 of visual cortex. *J. Neurophysiol.*, 36, 551–67

Stone, J., Dreher, B. & Leventhal, A. (1979) Hierarchical and parallel mechanisms in the organization of visual cortex. *Brain Res. Rev.*, 1, 345–94

Symonds, L. L. & Rosenquist, A. G. (1984). Laminar origins of visual corticocortical connections in the cat. *J. Comp. Neurol.*, 229, 39–47.

Symonds, L. L., Rosenquist, A. C., Edwards, S. B. & Palmer, L. A. (1981) Projections of the pulvinar-lateral posterior complex to visual cortical areas in the cat. *Neuroscience*, 6, 1995–2020

Talbot, S. A. (1942) A lateral localization in the cat's visual cortex. *Fed. Proc.*, 1, 84

Tigges J., Tigges, M., Anschel, S., Cross, N. A., Letbetter, W. D. & McBride, R. L. (1981) Areal and laminar distribution of neurons interconnecting the central visual cortical areas 17,

18, 19 and MT in squirrel monkey (*Saimiri*). *J. Comp. Neurol.*, 202, 539–60

Tong, L., Kalil, R. E. & Spear, P. D. (1982) Thalamic projections to visual areas of the middle suprasylvian sulcus in the cat. *J. Comp. Neurol.*, 212, 103–17

Tong, L. & Spear, P D. (1984) Single thalamic neurons project to both lateral suprasylvian cortex and areas 17/18: a retrograde fluorescent double labeling study. *Invest. Ophthalmol.*, 25 (Suppl.), 211

Törk, I., Leventhal, A. G. & Stone, J. (1979) Brain stem afferents to visual cortical areas 17, 18 and 19 in the cat, demonstrated by horseradish peroxidase. *Neurosci. Lett.*, 11, 247–52

Torrealba, F., Partlow, G. D. & Guillery, R. W. (1981) Organization of the projection from the superior colliculus to the dorsal lateral geniculate nucleus of the cat. *Neuroscience*, 6, 1341–60

Tretter, F., Cynader, M. & Singer, W. (1975) Cat parastriate cortex. A primary or secondary visual area? *J. Neurophysiol.*, 38, 1099–113

Trojanowski, J. Q. & Jacobson, S. (1976) Areal and laminar distribution of some pulvinar cortical efferents in rhesus monkey. *J. Comp. Neurol.*, 169, 371–92

Tusa, R. J., Palmer, L. A. & Rosenquist, A. C. (1981) Multiple cortical visual areas. Visual field topography in the cat. In *Cortical Sensory Organization*, Vol. 2 '*Multiple Visual Areas*', Ed. C. N. Woolsey, (Ed.). pp. 1–31. Clifton, N.J.: Humana Press

Ungerleider, L. G. & Mishkin, M. (1979) The striate projection zone in the superior temporal sulcus of *Macaca mulatta*: location and topographic organization. *J. Comp. Neurol.*, 188, 347–66

Updyke, B. V. (1977) Topographic organization of the projections from cortical areas 17, 18 and 19 onto the thalamus, pretectum and superior colliculus in the cat. *J. Comp. Neurol.*, 173, 81–122

Updyke, B. V. (1981) Projections from visual areas of the middle suprasylvian sulcus onto the lateral posterior complex and adjacent thalamic nuclei in cat. *J. Comp. Neurol.*, 201, 477–506

Updyke, B. V. (1983) A reevaluation of the functional organization and cytoarchitecture of the feline lateral posterior complex with observations on adjoining cell groups. *J. Comp. Neurol.*, 219, 143–81

Van Essen, D. C. (1973) The contribution of membrane hyperpolarization to adaptation and conduction block in sensory neurons of the leech. *J. Physiol.*, 230, 509–34

Van Essen, D. C. (1979) Visual areas of the mammalian cerebral cortex. *Ann. Rev. Neurosci.*, 2, 227–63

Van Essen, D. C. & Maunsell, J. H. R. (1983) Hierarchical organization and functional stream in the visual cortex. *TINS*, Sept., 370–5

Van Essen, D. C., Maunsell, J. H. R. & Bixby, J. L. (1981) The middle temporal visual area in the macaque: myeloarchitecture, connections, functional properties and topographic organization. *J. Comp. Neurol.*, 199, 293–326

Wagor, E., Mangini, N. J. & Pearlman, A. L. (1980) Retinotopic organization of striate and extrastriate visual cortex in the mouse. *J. Comp. Neurol.*, **193**, 187–202

Weber, J. T., Huerta, M. F., Kaas, J. H. & Harting, J. K. (1983) The projections of the lateral geniculate nucleus of the squirrel monkey: studies of the intralaminzar zones and the S layers. *J. Comp. Neurol.*, **213**, 135–45

Weller, R. E. & Kaas, J. H. (1981) Cortical and subcortical connections of visual cortex in primates. In *Cortical Sensory Organization*, Vol. 2, '*Muliple Visual Areas*', ed. C. N. Woolsey, pp. 121–55. Clifton, N.J.; Humana Press

Weller, R. E. & Kaas, J. H. (1983). Retinotopic patterns of connections of area 17 with visual areas VII and MT in macaque monkeys. *J. Comp. Neurol.*, **220**, 253–79.

Wilson, M. E. (1968) Cortico-cortical connections of the cat visual areas. *J. Anat.*, **102**, 375–86

Wilson, P. D., Rowe, M. H. & Stone, J. (1976) Properties of relay cells in the cat's lateral geniculate nucleus: a comparison of W cells with X and Y cells. *J. Neurophysiol.*, **39**, 1193–209

Wong-Riley, M. (1978) Reciprocal connections between striate and prestriate cortex in squirrel monkey as demonstrated by combined peroxidase histochemistry and autoradiography. *Brain Res.*, **147**, 159–64

Yukie, M. & Iwai, E. (1981) Direct projection from the dorsal lateral geniculate nucleus to the prestriate cortex in macaque monkeys. *J. Comp. Neurol.*, **201**, 81–97

Zeki, S. M. (1969) Representation of central visual fields in prestriate cortex of monkey. *Brain Res.*, **14**, 271–91

Zeki, S. M. (1971) Cortical projections from two prestriate areas in the monkey. *Brain Res.*, **34**, 19–35

Zeki, S. M. (1978) The cortical projections of foveal striate cortex in the rhesus monkey. *J. Physiol.*, **277**, 227–44

Zeki, S. M. (1980) A direct projection from area V1 to area V3A of rhesus monkey visual cortex. *Proc. R. Soc. Lond. Ser. B*, **207**, 499–506

18

Streaming in the striate cortex

GEOFFREY H. HENRY

Streaming as a concept

The notion of neural streaming in the vis[ual] system is far from new. It may be said to have had [its] birth around 1866 when Max Schultz identified r[ods] and cones in the retina and then proposed the dup[lex] theory for night and day vision (cited by Polyak, 19[57]). By the turn of the century, Ramon y Cajal had fou[nd] that rods and cones were linked to different kinds [of] bipolar cells and proposed a separation of rod and c[one] pathways through into the brain; a crucial suggest[ion] that was accepted unquestioningly for the n[ext] half-century (cited by Polyak, 1957).

Later, around 1942, Le Gros Clark (1943, 19[ ) set out on a theoretical roundabout by proposing t[hat] streaming was also to be observed in the way in wh[ich] colour vision signals made their passage to the cort[ex]. He suggested that in animals with a six-layered lat[eral] geniculate nucleus (LGN), made up of three cros[sed] and uncrossed pairs, each pair subserved one of [the] three primary colours. The evidence for this the[ory] came from an experiment in which two of th[ese] monkeys showed degeneration in layer 1 of the L[GN] after being kept in a blue-free atmosphere for [a] month. In Le Gros Clark's terminology, layer 1 is [the] most ventral layer of the magnocellular subdivision [and] receives an input from the nasal retina of the oppo[site] side.

The suggestion that the absence of blue li[ght] would modify colour vision from the nasal half but [not] from the temporal half of the retina was unaccepta[ble] to Walls (1953) and he set out to find an alterna[tive] xplanation. He quickly hit on the idea that the red (or lue-free) light used by Le Gros Clark would fail to ctivate rods and that the nasal/temporal distinction in ne LGN degeneration came about because there was pure rod input from the nasal retina and a mixed od/cone input from the temporal retina. The nasal etina, with its more archaic crossed projection, was isplaying a remnant of the retinal design of a nocturnal ncestor. This antique trace was lost in the revamped rganization of the newer uncrossed pathway. This igenious explanation lost some of its potency when, years later, Chow (1955) was unable to duplicate Le ros Clark's results.

Not all was lost, however, as Walls, in his logical erambulations, had made some pertinent points. He w that crosstalk between streams might cause colours the percept to 'run' but that in keeping signals apart was not necessary to separate each stream spatially. he axons might intertwine and still remain functionally dependent, and spatial segregation was not essential.

Rushton (1960), in a way, took Walls to task, hen he set to rest the belief that the rods and cones ere projecting to separate streams. At a seminar onouring Henri Pieron, Rushton began by asking What assistance would you give a telegraph company aat proposed, with limited length of cable, to build parate night and day lines for sending messages?'. ithout waiting for the obvious answer, Rushton, llowing in the steps of Barlow, Fitzhugh & Kuffler 957), went on to describe experiments showing that ngle ganglion cells in the frog retina conducted signals om rods and cones. The only potentially pure athway in the rod–cone dichotomy appeared to be that ssociated with foveal cones in birds and mammals.

In his Pieron Lecture, Rushton added to the idea f receptor pooling by reporting on opponent colour rocessing in fish retina. When similar coloured pairing as later identified in cells of the retina (DeMonasterio Gouras, 1975), LGN (DeValois, 1965) and cortex Hubel & Wiesel, 1968) of the primate visual system, had to be said that each paired combination ontributed to a stream. Walls, of course, would not ave been surprised with the finding that these streams re not spatially segregated from one another but itertwine as they make their way to the cortex.

In the 25 years since Rushton's Pieron talk, there as been acceptance of a duplex line for rods and cones nd also that colour coding is carried in intertwining ables. In recent years, however, a number of

additional response parameters have been found to flow in spatially segregated streams along the visual pathway. We must now set aside Wall's scepticism about the virtues of spatial segregation and see what design advantage it may carry.

Morphological streams
In the retina

The first indication of these new segregated streams came in 1966 with the recognition of differences in the level of linearity in the response patterns of X and Y ganglion cells in the retina. Enroth-Cugell & Robson (1966) showed that X cells displayed linear summation in their responses while the firing of Y cells was non-linear. Thereafter, most of the story developed in Canberra in laboratories headed by Peter Bishop. By 1968, when we first arrived in Canberra, Levick had become devoted to unravelling the properties of X and Y ganglion cells and a year or so later he was joined in this enterprise by Cleland and Dubin (see Cleland, Dubin & Levick, 1971). A year further on, Stone came to Canberra to commence work on functional subclasses of ganglion cells. Subsequently, he and Fukuda added the W category to the classification (see Stone & Fukuda, 1974) and produced papers with associates such as Hoffmann (Hoffmann & Stone, 1971), Dreher (Stone & Dreher, 1973) and Sherman (Sherman, Hoffmann & Stone, 1972). In the meantime, Wässle came to work with Levick and Cleland (Cleland, Levick & Wässle, 1975), after he had spent a period in London looking at Boycott's Golgi slides of wholemounts of the cat retina. This started the phase when Y, X and W cells were to be linked, respectively, with the alpha, beta and gamma cells of the morphological subdivisions. In its most recent development (see Chapter 2), this relationship has led to the identification of alpha and beta ganglion cells with ON- and OFF-centre receptive fields. From the regular distribution of the cells in the ON- and OFF-centre subgroups (Wässle, Boycott & Illing, 1981; Wässle, Peichl & Boycott, 1981), it has been concluded that each contributes to an independent stream.

In the optic pathway

Although, in the cat, the spatial segregation of cells with ON- and OFF-centre receptive fields seems to find expression in the different levels of occurrence of ganglion cells, there is little evidence that the spatial displacement of the streams carries on into the LGN. Thus X and Y cells, with either ON- or OFF-centre receptive fields, appear to intermingle in the A and A_1 and upper C laminae of the LGN of the cat. Only in the laminae of the C complex are there indications that cells of the Y stream (in the upper or C segment) are separated from W cells in the lower segment, which is made up of laminae C_1 and C_2 (Wilson & Stone, 1975; Cleland *et al.*, 1976).

By contrast, the X and Y streams appear to be spatially disparate in the monkey LGN, with the prospect of the four dorsal or parvocellular layers subserving the X stream and the lower two ventral or magnocellular layers accommodating the Y stream (Dreher, Fukada & Rodieck, 1976). Homologies are yet to be established between the LGNs of cats and monkeys but there is little doubt that dorsal layers of the LGNs in the two are not analogous with regard to their incumbent functional streams. After leaving the LGN, streaming follows a parallel course in cats and monkeys; we shall concentrate on the visual cortex of the cat, as the case for closer scrutiny.

Before leaving the comparative studies, however, it is useful, as an interim step, to list some of the unanswered questions that arise in this field. The first, which has been alluded to above, asks why the X and Y streams are segregated spatially in the LGN of some species and not in others. In the absence of an obvious answer, we seem to be confronted with the unlikely possibility that segregation of the monkey LGN is a functionally benign move but there may be an advantage in preparing for the stream separation that occurs in the striate cortex. Certainly, the monkey has superior form and colour vision to other laboratory animals but there are many, more appropriate reasons than X and Y stream segregation in the LGN, to account for this superiority.

The next species variation relates to the level of interweaving that occurs in the ON- and OFF-centre streams as they pass through the LGN. Spatial segregation, present in the LGN of the mink where the ON-and OFF-centre streams pass to different layers (LeVay & McConnell, 1982), has not been observed in cats or monkeys. Again, it is difficult to appreciate how the mink's visual performance benefits from this early segregation in the LGN. Indeed, in general the functional advantages of spatial segregation in the LGN are just as mysterious now as they were to Walls in the early 1950s.

In the afferent terminals

As mentioned above, spatial segregation occurs when the X and Y streams arrive at the striate cortex in both cats and monkeys. The evidence for this separation is two-fold and comes, first, from the disposition of axon terminals arising from the optic radiations (Ferster & LeVay, 1978), and secondly, from the laminar distribution of the recipient cells in the cortex (Bullier & Henry, 1979*b*). The disposition of axon terminals is revealed through the anterograde transport of horseradish peroxidase (HRP) after injection into the optic radiations. The resulting picture shows that large diameter axons, presumed to come from Y cells in the LGN, arborize in the upper third of lamina 4, while the smaller axons, of the X stream, terminate in the lower two-thirds of the same lamina.

Fig.18.1, a collage produced from more than one preparation, shows characteristic afferent terminals anterogradely infiltrated with HRP and two spiny stellate neurons redrawn from Golgi studies (Lund *et al.*, 1979). The two axons to the left, judged to be thick and belonging to the Y stream, terminate in upper lamina 4 while the one to the right, thought to come from the X stream, arborizes in the lower half of lamina 4. An indication of the extent to which the distribution of thick and thin afferents (thick axons $> 1.7\ \mu$m; thin axons $< 1.7\ \mu$m, $> 0.9\ \mu$m) follow this pattern may be obtained from Fig.18.3A where the frequency of axon branching is plotted against laminar position for six thick and six thin axons (Bullier & Henry, 1979*c*).

Very fine axons ($< 0.9\ \mu$m), also observed following the injection of HRP, pass through the striate cortex to terminate, with restricted branching, in lamina 1. It has been suggested that these axons arise from W cells in the LGN but such a terminal arrangement would not be inconsistent for axons coming from cells in the extrastriate cortex and their origin is more uncertain than that of the thicker axons.

In the striate neurons

In lamina 4, the major site of afferent termination, there is little evidence of stream-dependent differences in the morphology of the putative receiving elements. Spiny stellate neurons inhabit this region and from the examination of HRP-labelled recorded cells they appear to have the simple cell receptive fields (Gilbert & Wiesel, 1979; Martin & Whitteridge, 1982) which have long been regarded as characteristic of first-order cortical neurons (Hubel & Wiesel, 1962;

65). At the different levels of entry of the X and Y reams, the findings from Golgi-impregnated material ggest there are only minor differences in cell design .und *et al.*,1979). The spiny stellates that lie high in mina 4 at the entry of the Y stream are larger in their rikaryon and dendritic spread than their counterparts the X stream in lower lamina 4 (for representative amples see fig. 18.1). The destination of the axons of ese two cells also differs slightly in that although they th project to lamina 3, axonal pathways running from lls in lower lamina 4 to the upper parts of the lamina are more common than pathways in the reverse rection. In other words, there is a greater potential r signal flow from the X stream to the Y than from to X.

Stream identified from rate of signal conduction in afferent pathways

In the historical course of events, evidence for eaming in the striate cortex was first sought in distinctions of the response characteristics of striate neurons. The search for sustained *versus* transient (Ikeda & Wright, 1975) or for linear as opposed to non-linear (Movshon, Thomson & Tolhurst, 1978*a*) responses has not produced the same certainty of identification that is possible for X and Y streams in the retina and LGN. Later, recourse was taken, instead, to differentiating the stream of cortical neuron from the rate of signal conduction after electrical stimulation in the afferent pathway; LGN–Y cells, with their larger diameter axons, have faster signal conducting times than their counterparts in the X stream (Bullier & Henry, 1979*c*). Over the years, attempts have also been made to assess directly the afferent stream of a striate neuron by recording concurrently from the parent cell in the LGN (Lee, Cleland & Creutzfeldt, 1977). Recently, Tanaka (1983) achieved considerable success with this cross-correlation technique and the results of his experiments are now becoming available.

Our work in Canberra, carried out in association with Jean Bullier and Mike Mustari, has concentrated

Fig. 18.1. A collage from two preparations showing the disposition of Golgi-stained cells and HRP-filled axons from the optic radiations. The cells are typical examples of spiny stellates with the larger one in the upper, and the smaller one in the lower, half of lamina 4. The two axons to the left have been classed as large (> 1.7 μm) and belonging to the Y stream (see text) while the one to the right is regarded as being of medium size (> 0.9 μm, < 1.7 μm) and of the X stream. The large axons terminate in the upper part of lamina 4 and the medium-sized axon in the lower part.

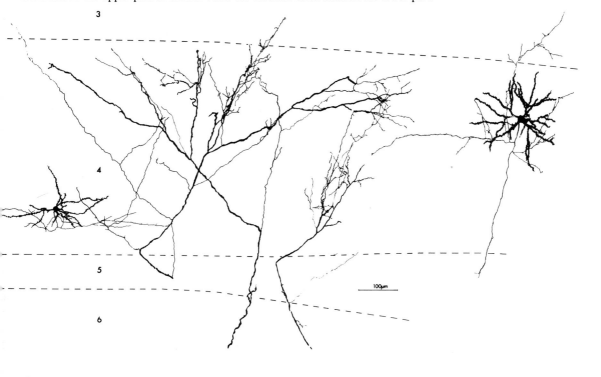

on stream identification from differences in signal conduction times in the primary afferent pathways. The experimental design for measuring conduction times is shown in Fig. 18.2. Here, the stimulating electrodes are positioned to obtain the latency difference between responses arising from stimulation at OX (optic chiasm) and OR_1 (low in optic radiations). From this measurement comes the rate of signal conduction in the afferent stream of the cortical cell. At the same time the latency from OR_2 (high in optic radiations) directly reflects the number of synaptic delays as the signal passes to the cortical neuron and, therefore, whether the cell receives a direct or indirect input from the LGN (Bullier & Henry, 1979a).

The entrance of streams to the striate cortex

First-order striate neurons of each stream, identified from their latency measurements, would be

Fig. 18.2. Arrangement of stimulating electrodes for experiments to deteretmine the afferent stream and ordinal position of a monitored cell in the striate cortex. Electrode OX is in the optic chiasm; OR_1 low in optic radiations and OR_2 high in optic radiations.

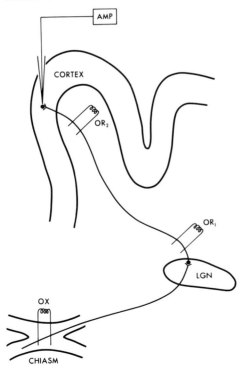

expected to distribute in lamina 4 to align with th appropriate terminal arbors as they emerge from th optic radiations. This proves to be the case, and Fi 18.3 (Bullier & Henry, 1979c) shows how well th stream alignment holds in histograms that compare th distribution of terminal branches (Fig. 18.3A) with th of first-order neurons. The thick-line bars in the low histograms represent the incidence of cells that recei only a direct input (a single spike in response to sing shock) while the added thin-line bars represent ce that receive secondary indirect inputs in addition their direct drive. This convergence of inputs, multiple spiking in response to a single shock, is common and, perhaps, inevitable characteristic complex cells, in the thin-line segment, almost all th cells came from the complex category.

Fig. 18.4 takes the interpretation of cortic streaming a stage further by showing the lamin distribution of different cell types in each strea (Bullier & Henry, 1979c; Henry et al., 1983). describe these cells in terms of their stream, each ty has been designated by its initial letter (simple by complex by C) and superscripts (X and Y) have be added to register the stream (cf. Stone, Dreher Leventhal, 1979). For example, S^X indicates an S c belonging to the X stream. As in the past, H, appli as a subscript, indicates that the cell has a preferen for foreshortened stimuli or, in other words, displa the hypercomplex property. The terms X and Y a used in their broadest sense and are treated as bei equivalent to the brisk-sustained and brisk-transie subdivisions (Cleland et al., 1971).

For the cat Fig. 18.4 shows that S and C ce occur in both the X and the Y streams and for ea afferent stream there is an appropriate distribution each cell type in lamina 4. A similar pattern, al reproduced in Fig. 18.4 is apparent for the monk striate cortex (Bullier & Henry, 1980). Here, howeve the cells of the Y stream distribute low in lamina 4 a little above the appropriate terminal arbors in lami $4C_{\alpha}$, while cells of the X stream align, as would expected, with their afferent terminals in lamina 4C For the monkey, another possible stream differen present in the sample in Fig. 18.4, is to be seen in lami 4 where S cells are found only in the Y stream wh cells in the X stream have non-oriented receptive field

Stream differences in the responses of first-order striate neurons

Receptive field size

A prior knowledge of the resident stream of a cell, s obtained from electrical stimulation, has facilitated he recognition of stream-dependent differences in the ell's response. Between streams, however, there is little o separate the responses of cells of a particular class S or C). The most noticeable distinction is that cells f the Y stream (both S and C) have larger receptive elds than their counterparts in the X stream (Mustari, ullier & Henry, 1982). The extent to which S cell eceptive fields, in the two streams, differ in size may e appreciated from Fig. 18.5, which shows four

examples of average response histograms to optimally oriented narrow bars (0.09 degrees wide) moving over the receptive field of S^Y (A and B) and S^X (C and D) cells. These are extreme patterns in that they represent the largest and smallest discharge regions of the sampled cells (Mustari *et al.* 1982).

A similar or even more marked distinction in receptive field size has been observed between C^X and C^Y cells. Fig. 18.6 shows four examples of responses to narrow moving bars (0.2 degrees wide) from C cells at different cortical depths (numbering corresponds to the subgroups in Fig. 18.14). The average response histograms in 1 and 2 show the larger discharge regions seen in C^Y cells while 3 and 4 are from C^X cells (Henry *et al.*, 1983). In contrast to the responses from the S cells

Fig. 18.3. Distribution histograms running across the laminae of the striate cortex. A shows the incidence of axonal branching in six large (diameter at white matter $> 1.7\ \mu$m) and six medium (diameter $< 1.7\ \mu$m, $> 0.9\ \mu$m) axons after infiltration with HRP. The histograms confirm that the terminal arbors of large (putative Y stream) and medium (putative X stream) axons occupy the upper and lower parts of lamina 4, respectively (adapted from Bullier & Henry, 1979c). B shows the incidences, in samples of first-order neurons with fast and slowly conducting afferents, of cells presumed to belong to the Y and X streams, respectively. The thin-lined bars represent cells receiving more than one input, as demonstrated by their multiple spiking in response to a single electrical stimulus (adapted from Bullier & Henry, 1979c and Henry *et al.*, 1983). A comparison of A and B shows how well the afferents and recipient cells of each align with each other.

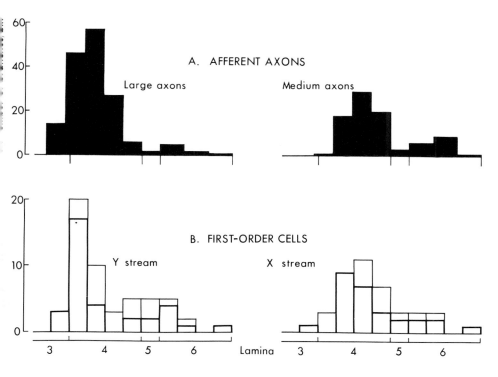

LAMINAR DISTRIBUTION

A. AFFERENT AXONS

Large axons   Medium axons

B. FIRST-ORDER CELLS

Y stream   X stream

3   4   5   6   Lamina   3   4   5   6

(in Fig. 15.5), the C cells (in fig. 18.6) respond, from all points in their receptive fields, to both the light and dark moving bars and also, with a composite ON/OFF discharge, to flashing stimuli. In the S cell receptive field, ON or OFF firing comes from spatially segregated regions. These generic distinctions hold across streams and within each of the S and C cell families there appears to be a consistency of receptive field design.

The S cell histograms of Fig. 18.5 include the responses in both directions of bar movement but no significance should be placed on any of the between-stream differences in this aspect of the responses in these examples. These are consistent with the variation seen within each stream and, to date, stream-dependent differences have not been observed in direction selectivity or in orientation specificity.

Fastest effective stimulus

Like receptive field size, the value of the fastest effective stimulus also shows some degree of stream dependence (Mustari *et al.*, 1982; Henry *et al.*, 1983) although, in S cells, this distinction may be secondary to the size difference itself. It has been proposed that retrospective inhibition from the far flank of a discharge region limits the cell's responsiveness (Goodwin & Henry, 1975). The promptness with which the stimulus enters this flank therefore determines the

Fig. 18.4. Distribution histograms from samples recorded across the striate cortex, showing the incidence of different cell types in the cat and monkey. (N = number of cells.) In the cat, the distribution runs from lamina 2 to lamina 6 and shows how, in lamina 4, both S and C cells belonging to the X and Y streams (i.e. with slow- and fast-conducting afferents, respectively) align with the point of entry of their respective stream (see Fig. 18.3). For the monkey, the distribution is restricted to lamina 4 and again the NO (cells with non-oriented receptive fields) and the S cells are distributed in a stream- dependent arrangement with cells of the Y stream being in lamina $4C_\alpha$ and lower 4B and those of the X stream in lamina $4C_\beta$. Note the absence of S cells in the X-stream. (Adapted from Bullier & Henry, 1979*c* and 1980.)

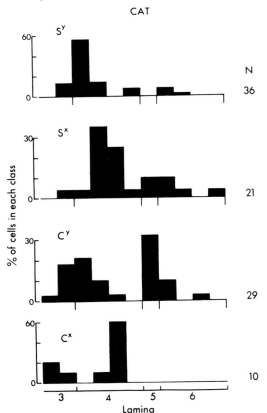

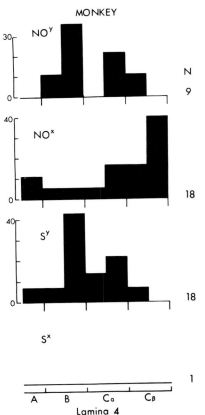

extent to which the excitatory response is aborted. The larger the discharge region, the further the stimulus must travel to reach the flank and begin the suppression of the excitatory response. If the interval for effective suppression, T, is assumed to be constant for S cells of both X and Y streams then: $T = D_x/V_x = D_y/X_y$ where, for S^X and S^Y cells, D represents the distance separating the discharge region from its inhibitory flank (taken as the width of the dominant discharge region in Fig. 18.7) and V is the cut-off velocity for the cell.

On rearrangement, the following relation should hold: $D_x/D_y = V_x/V_y$. In one experimental population (Mustari *et al.*, 1982), these ratios, derived from the sample means, took on the following values: $D_x/D_y = 2.70$ and $V_x/V_y = 2.85$. The similarity of these two values appears to justify the initial assumption that the velocity of the fastest effective stimulus depends only on the time taken by the

stimulus to reach the suppressive flank. Fig. 18.7 plots the dimension of the primary discharge region against cut-off velocity to show the scatter of individual points around the regression line, which was fitted to pass through the point of origin, and was derived from the pooled values of both streams. An indication of how closely the samples from the two streams share a common alignment and, perhaps therefore, a similar mechanism for velocity cut-off, is given by the position of their mean values (crosses in Fig. 18.7) which both fall close to the regression line.

A similar analysis of cut-off velocities has not been attempted for C cells, largely because the C cell receptive field seldom displays an inhibitory flank and the concept of retroactive inhibition lacks obvious application. Despite this and other differences in receptive field organization, there are parallels in the velocity dependence in the two families of striate

Fig. 18.5. Average response histograms to moving light and dark bars (0.09 degrees wide) recorded from first-order S cells belonging to the Y stream (A and B) and X stream (C and D). Designation of Y and X streams based on the presence of either fast or slowly conducting afferents; arrows indicate the direction of bar movement; thin and thick lines represent the responses to light and dark bars, respectively. On average, S^X cells have smaller discharge regions than S^Y cells. (Adapted from Mustari *et al.*, 1982.)

Layer 4 , S cells

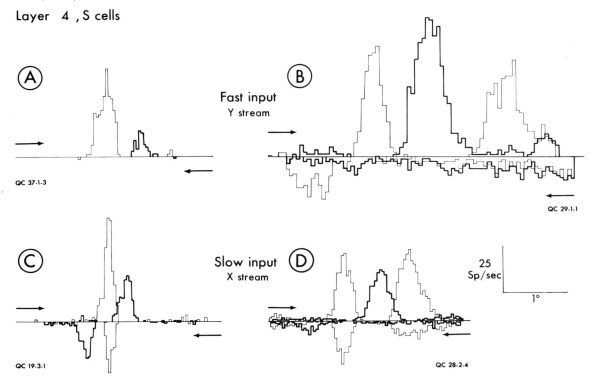

Fig. 18.6. Average response histograms to light and dark-moving bars (0.2 degrees wide), recorded from C cells belonging to Y (1 and 2) and X (3 and 4) streams. Each cell comes from the group with the same numerical representation as in Fig. 18.14. Details are the same as in Fig. 18.5 and, similarly, C^Y cells have larger receptive fields than C^X cells. (Adapted from Henry *et al.*, 1983.)

Response to moving bars

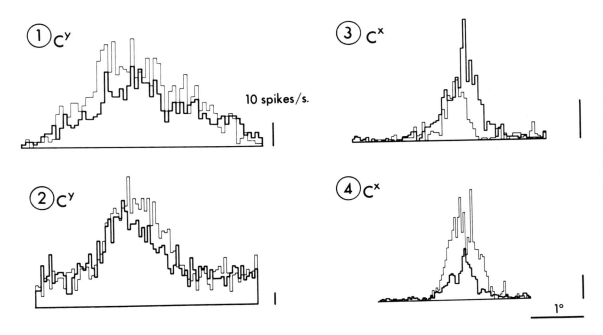

Fig. 18.7. Graph of the fastest effective stimulus against receptive field size (estimated from primary dimension, front to back, of the dominant discharge region), in lamina 4 S cells belonging to X and Y streams. Stream designation assessed from rate of signal conduction in the afferent pathway. The regression line is fitted to pass through the point of origin and crosses mark the mean values for the two samples (see text).

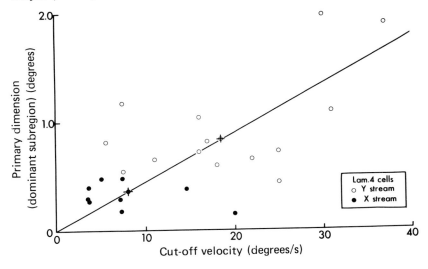

neurons. Thus, like the S family, C cells of the Y stream respond to faster stimuli than their counterparts in the X stream. To make a general comment on this point, most cells (C or S) from the X stream cease to fire to stimuli moving faster than 10 degrees, while for those in the Y stream the fastest effective stimulus moves at velocities between 10 and 30 degree/s (Mustari *et al.*, 1982; Henry *et al.*, 1983).

Receptive field design

Cells of a particular family (S or C), but of a different stream, display similar patterns in their responses to light or dark bars that are either moved across the receptive field, or flashed at different points. Although the discharge regions may vary in number and disposition in individual receptive fields of the S cell, the range of varieties is similar for the two streams. As mentioned above, the principal stream-based distinction resides in the dimensions of the discharge

region but it may be simply a carry-over of size differences already present in the responses of parent cells in the LGN. To test this prospect it is necessary first to find the way in which LGN receptive fields unite to create the responses of S cells in the striate cortex.

From a comparison of response signatures in LGN and S cells it has been possible to decide on the construction design for S cells (Bullier, Mustari & Henry, 1982). An example of how the contribution of LGN cells may be identified in the S cell firing pattern is shown in Fig. 18.8. Here, an ON-centre LGN neuron (see the response to flashing bars) presents a typical discharge region sequence to moving bars. First comes the response to a light bar, then to a dark bar and finally to a light bar again. In the symmetrical LGN receptive field, this pattern is repeated for the opposite direction of bar movement (see inverted histograms in Fig. 18.8). A similar pattern occurs in the upper histograms representing the S cell response, although the

Fig. 18.8. Response signatures recorded as average response histograms in an LGN neuron and a single-row S cell. The upper curves record the responses to moving light (thin) and dark (thick) bars (arrows indicate direction of movement) and the lower curves the ON (thin) and OFF (thick) responses to stationary flashing light bars. The response signature is similar for the LGN and the S cell suggesting that the S cell receptive field is made up of a single row of ON-centre LGN units. The measurements for assessing row length are contained in Fig. 18.10. (Adapted from Bullier *et al.*, 1982.)

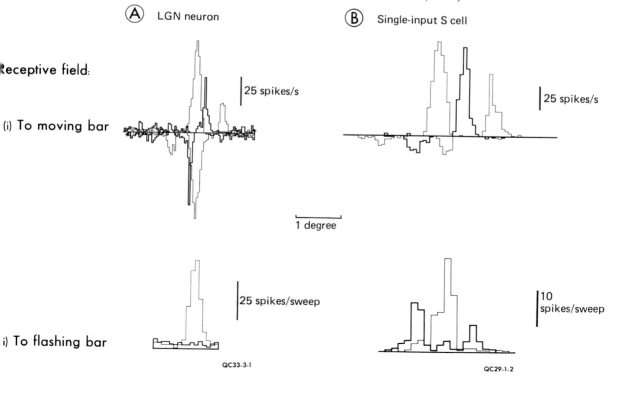

Fig. 18.9. Comparison between the receptive field centre sizes of LGN neurons and S cells from the same stream (X to the left and Y to the right). Determination of centre size from length summation curve in LGN units and from the distance separating the row centres in two row S cells. Plots indicate, at the population level, that the front-to-back dimension (perpendicular to the optimal orientation) for S cells is the same as that of LGN neurons belonging to the same stream. Points are plotted against the distance from the midpoint of the area centralis to avoid discrepancies that could arise with eccentric receptive fields. (Adapted from Bullier *et al.*, 1982.)

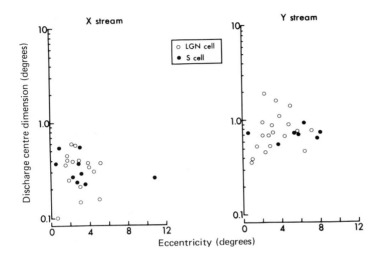

Fig. 18.10. A similar comparison to that in Fig. 15.9 but, this time, the length of a row (along the line of optimal orientation) is compared with the LGN receptive field centre. Row length and LGN receptive field centre were both determined from length summation curve. Taken as groups, the S cells have row lengths equal to twice the dimension of LGN centres of the same stream. Bracketed arrows are displaced by a factor of 2 in each graph. (Adapted from Bullier *et al.*, 1982.)

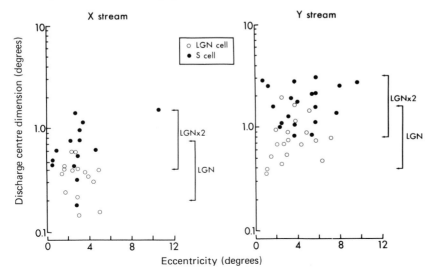

attenuated discharge peaks for the reverse direction of stimulus movement reflect the cell's direction selectivity. The discharge peaks in Fig. 18.8 are slightly wider for the S cell than the LGN cell, but they were still judged to come from a single row of ON-centre LGN receptive fields, largely because the OFF response in the S cell had the temporal characteristics of a surround rather than a centre (Bullier *et al.*, 1982).

Results from experiments, like those in Fig. 18.8 Bullier *el al.*, 1982), suggest that the S cell receptive field is constructed from either one or two rows of LGN receptive fields. Each row runs parallel to the optimal orientation and, once the row has been identified from its contribution to the response pattern, its dimensions may be compared with those of parent cells in the same stream. Thus, Fig. 18.9 shows how the width of LGN discharge regions compares with that of S cells, which have been measured perpendicular to the optimal orientation or, in other words, across what has been judged to be a single row. For each stream, the dimensions of parent and daughter cells are alike in mean value and range (Fig. 18.9); a finding supporting the decision that each measurement was taken from a single row.

The length of the row in the S cell receptive field, the measurement along or parallel to the optimal orientation, is estimated from the bar length–response curve for the cell. Fig. 18.10 shows that, although the range of row lengths in S cells is similar to that for the widths of LGN discharge regions, it is displaced upwards by a factor of two. For S cells in each stream, therefore, the row is equivalent, in its length, to the receptive fields of two parent cells. This does not mean that the input comes from only two LGN cells but that any overlapping is restricted to cells with receptive fields falling within the space occupied by two LGN discharge centres lying side by side.

In summary, then, the experimental results are consistent with the concept that S cells of each stream have a similar structure to their receptive field design. Both appear to be composed of one (an ON- or an OFF-centre) or two (an ON- and an OFF-centre) rows of LGN units and, in both, the rows are of similar length relative to their parent cells in the LGN. To this stage, no certain examples have been detected where more than two rows of LGN units contribute to the S cell receptive field. It is possible, therefore, that all S cells could be accounted for with the one- or two-row construction. If so, the Gabor model, with its

symmetrical and antisymmetrical conformations, may provide a mathematical description of the response patterns for the full gamut of S cells (Kulikowski, Marcelja & Bishop, 1982; see also Chapter 26). The curves for the Gabor function are reproduced in Fig. 18.11 where the symmetrical, or cosine, version depicts the response profile for an S cell made up of a single row of ON-centre LGN units. The antisymmetrical, or sine, version describes the responses

Fig. 18.11. Response predictions, based on the Gabor model, for a single row (ON-centre) S cell. The upper curve, which represents the response to stationary flashing bars, takes the form of the symmetrical Gabor function. From this first curve, its derivative (middle curve) is taken as representing the responses to a narrow light bar moving over the receptive field and its integral (lower curve) as the response to moving edges. These curves produce approximate representations and inaccuracies arise if the stimuli are non-optimal. (cf. Kulikowski *et al.*, 1982.) The open and shaded areas represent the responses from opposing aspects of the stimulus; in this instance the open areas are responses to light ON, a light edge and a light bar, while the shaded areas are the responses to light OFF, a dark edge and a dark bar.

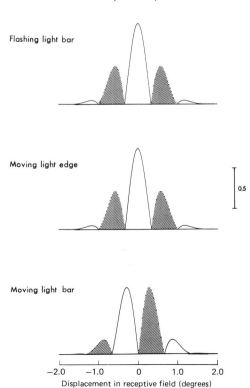

Receptive field profile

Flashing light bar

Moving light edge

0.5

Moving light bar

Displacement in receptive field (degrees)

of the S cell in which both ON- and OFF-centre rows contribute to the receptive field. As a reasonable approximation, the displayed curves represent the responses to flashing light bars and a moving light edge. The differential of each curve, achieved by interchanging the sine for the cosine representation, denotes the response to a narrow moving light bar.

Turning now to the C cell, Fig. 18.12 is included to give some idea of the relative size of the receptive fields of C^X and C^Y cells. No structural significance is attributed to the numerical relation that emerges from the width comparision between LGN and C cell receptive fields (as measured perpendicular to the optimal orientation across the C cell receptive field). As a group, receptive fields of C cells are three times as large as those of LGN cells in the same stream. The length (parallel to the optimal orientation) of C cell receptive fields is similar to that of a single LGN receptive field centre but, as yet, no stream-dependent distinctions in this dimension have been reported for the two types of C cell.

Response linearity

As mentioned above, differences in response linearity, so important in stream detection in the retina and LGN, have also been sought in the responses of cortical neurons. Although some grouping according to streams has been proposed, there is no evidence that cells with linear responses have afferents from the X stream, or that those with non-linear responses receive from the Y stream. To date, afferent streams have not been identified from impulse conduction rates in the same cells that have been tested for linearity.

Indirect evidence, however, has been applied to identify the afferent stream. It has been shown, for example, that cells responsive to fast moving stimuli and, therefore, likely to belong to the Y stream, display non-linear response summation (Kulikowski & Bishop, 1981). Likewise, C cells, which from the dimensions of their receptive fields appear to belong to the Y stream, are nearly always reported to have non-linear responses (Movshon, Thompson & Tolhurst, 1978*b*; Pollen, Andrews & Feldon, 1978). However, in one study, C^X cells (referred to as B cells), identified from their small receptive fields and lack of responsiveness to fast stimuli (faster than 10 degrees/s), were considered to have non-linear summation (Kulikowski & Bishop, 1982). This is an unexpected result for a cell residing in the X stream and it may reflect a non-linearity that develops with the convergence on numerous afferents to the C^X cell, even though each of these inputs is linear.

Linear responses appear to be limited to S cells and presumably to those members of the S family that inhabit the X stream. In a sample of 47 S cells, 62% were found to have linear responses (Movshon *et al.*, 1978*a*) and therefore could have been S^X cells. This presents an unexpected relative encounter rate in the S family since S^X cells, thought to be the smaller of the

Fig. 18.12. Comparison between the receptive field centre size of an LGN neuron and the front-to-back dimension of C cells in the same stream (X to the left and Y to the right). The LGN measurement is the same as in Fig. 18.9 and the C cell receptive field is taken from the onset and termination of firing caused by an optimally oriented moving bar.

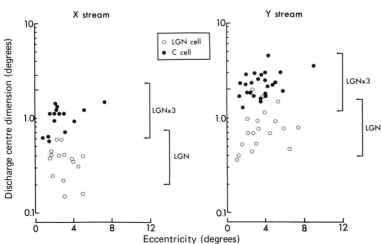

two spiny stellate types in lamina 4, are generally isolated less frequently than S^Y cells. Studies relating the afferent stream to the level of response linearity are required before linearity can be adopted as a differentiation property for cortical streaming.

The course taken by streams within the striate cortex

Little has been achieved in the identification of mophological streams running through the striate cortex. Advances have been made, in that identified C and S cells belonging to different streams have been classed morphologically from the intracellular injection of HRP (Martin & Whitteridge, 1982). The picture is compounded, however, since each of the functional types may take on more than one form. In general, however, C cells are pyramidal and most of the S cells relaying excitatory signals in lamina 4 are spiny stellates. S cells appear to take on a greater diversity of forms than C cells and may be pyramidal cells in lamina 6 or non-spiny stellates in other laminae.

Support for the prospect of C cells being pyramidal cells may also come from acute physiological experiments in which C cells have been found to respond by multiple spiking to a single electric shock in the optic radiations. Multiple spiking appears to be a potential classing feature for C cells, although it is possibly a more general manifestation of the properties of pyramidal cells. To develop this theme, multiple spiking may result from signals taking alternative afferent paths to the basal and apical dendrites of the pyramidal cell. One possible arrangement, shown for a lamina-5 pyramidal cell in Fig. 18.13, has the direct input (1) going to the apical dendrite, while the indirect drives pass through a spiny stellate (2) to project to the basal dendrite, with or without passing through a lamina-3 pyramid (3).

In early suggestions, the spiny stellate was regarded as the anatomical correlate of the S cell (Kelly & Van Essen, 1974) and was placed accordingly at the first stage of cortical processing. C cells were placed second in the hierarchical sequence (Hubel & Wiesel, 1965) and, from their association with the C cell, pyramidal cells were long regarded as second-order neurons. Now anatomical and physiological evidence suggests that the C cell (Henry *et al.*, 1983) and/or the pyramidal cell (White, 1979) receive(s) a direct input from the afferent pathway (for an alternative interpretation see Ferster & Lindstrom, 1983). Moreover, as

mentioned above, this direct drive may come from either the X or the Y stream. Thus, Fig. 18.14 (Henry *et al.*, 1983) shows the laminar distribution of C^X and C^Y cells and demonstrates that the two exist at the required levels in lamina 4 to receive a direct input from the appropriate stream (C^Y in the upper part and C^X in the lower).

The role of the C cell as a recipient from S cells may not be excluded, however, by the discovery of the direct input from the LGN to the C cell. Multiple spiking and also the nature of the composite responses in the C cell receptive field both point to a convergence of inputs and one of these secondary drives may well emanate from the S cell. Evidence is lacking on whether all the inputs to the C^X and C^Y cells come via a single stream, but the patterns in multiple spiking are often consistent with maintenance of an exclusive input from one stream. The evidence from cross–correlation of the firing of LGN and striate neurons also suggests that the input from the C^Y cell comes exclusively from the Y stream (Tanaka, 1983).

With the discovery of C^Y and C^X cells in the top and botton of lamina 4, respectively, and with the general acceptance that C cells are pyramidal in form, there is now a need to identify the concerned pyramids in lamina 4. In classical descriptions star pyramids

Fig. 18.13. A diagrammatic representation for the possible course taken by afferent signals to produce multiple spiking in a lamina-5 pyramidal cell. The numbers indicate paths by which successive signals may reach the cell.

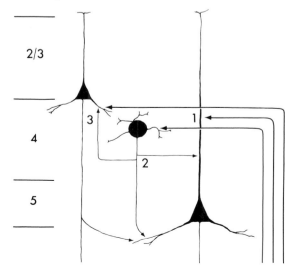

reside in upper lamina 4 (O'Leary 1941) and many of the border pyramids at the junction of laminae 3 and 4 send basal dendrites into the upper half of lamina 4. Pyramidal cells are sparse in the lower parts of lamina 4 but it is possible that the pyramids with arciform axons, said to concentrate in upper lamina 5, are more dispersed into lamina 4 than commonly believed (cf. Ramon y Cajal, 1922).

In summary, therefore, the small and large spiny stellates of lamina 4 may act as relays for excitatory signals in the X and Y streams, respectively, and thereby display the properties of S^X and S^Y cells. Pyramidal cells with arciform axons may perform as C^X cells and in so doing display a similar general design to the spiny stellate, from which they differ only in the presence of their apical dendrite. The pyramids that act as C^Y cells have axons that leave the striate cortex but they also have an abundance of axon collaterals relaying signals into laminae 3 and laminae 5. It is significant for this C^Y population that, in recent experiments using the retrograde transport of HRP, cells projecting to Area 18 from Area 17 (see below) have been found to extend out of lamina 3 into the upper part of lamina 4 (Bullier, McCourt & Henry, unpublished), the point of termination of the Y stream in the optic radiations.

It should not be concluded from the assessment above that C^Y cells are alone in sending an axon from the striate cortex. C^X cells and cells of all categories possessing the H property are found in lamina 3 where they exist, presumably as pyramidal cells with axons going to areas of the extra striate cortex. Likewise the S cells of lamina 6 are likely to be pyramidal cells with an extrinsic axon.

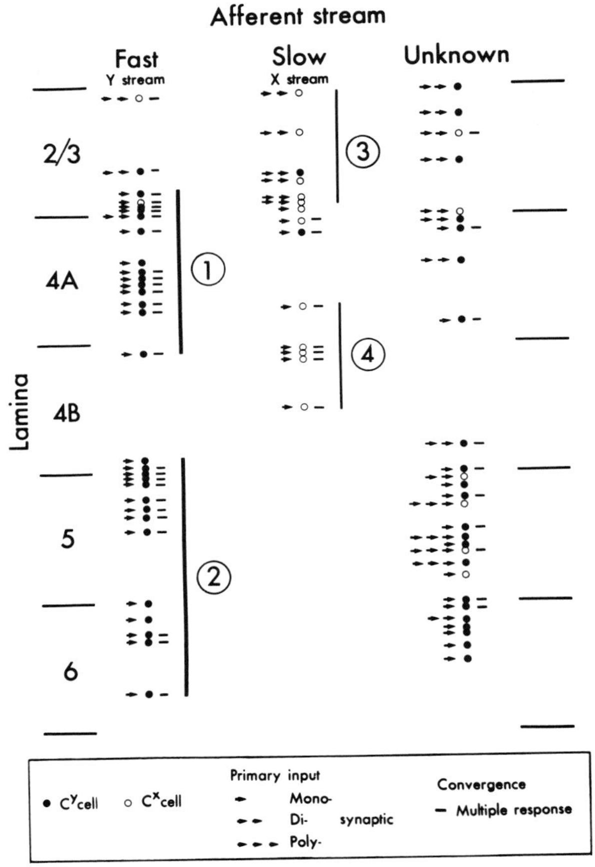

Streams leaving the striate cortex

With streams so difficult to trace within the confines of the striate cortex a good deal of attention has been given to the more accessible output paths. These may be considered in terms of their laminae of origin.

From lamina 3

The retrograde transport of HRP has shown that the pyramidal cells of lamina 3, which presumably receive their input either directly or indirectly from lamina 4, send their axons to extrastriate cortical regions (Gilbert & Kelly, 1975). Attempts to identify subpopulations of lamina-3 cells projecting to specific extrinsic sites have not proved very successful, although slight differences have been observed in distributions of lamina-3 cells sending axons to Areas 18 and 19. Fig. 18.15 makes a comparison of these two populations by showing the distribution of retrogradely labelled cells in Area 17 after injecting HRP in one instance into Area 18 and, in the other, into Area 19. The histograms at the top of Fig. 18.15 provide a graphical representation of the distribution of these two populations based on each cell's position relative to the borders of lamina 3. In this experiment it is apparent

Fig. 18.14. Laminar distribution of C cells in the striate cortex. Cells are grouped according to the nature of their afferent stream (Y for fast and X for slowly conducting). The horizontal arrows indicate that the most direct or primary input was either mono-, di- or polysynaptic. A bar tag to the right indicates that the cell responded with multiple spiking to a single electrical shock from one of the stimulating electrodes. Vertical lines, accompanied by a number, show the cortical range of proposed subgroups of C^X and C^Y cells. (Adapted from Henry *et al.*, 1983.)

that cells projecting to Area 18 tend to congregate lower in lamina 3 than those going to Area 19 (cf Gilbert & Wiesel, 1981).

If distinctive populations of lamina-3 cells send axons to particular cortical areas, then the projection to Area 18, a region that receives a pure Y input directly from the LGN (Harvey, 1980a), could maintain its exclusive character with a Y input from Area 17. The lamina-3 cells projecting to Area 19 may then form part of the X stream. In this interpretation, the striate cortex receives information from the X, Y and W streams and, through the cells in lamina 3, distributes signals to extra-striate cortical regions that act as more exclusive interpreters of activity in individual streams.

In theory, the cells leaving lamina 3 may be identified by registeing antidromic firing from stimulating electrodes sited in the recipient cortex. This has been done with some success by stimulating at the Clare–Bishop cortex (Henry, Lund & Harvey, 1978).

Only 11 lamina-3 cells were identified from their antidromic drive but, of these, nine belonged to the C^X category. It was concluded, therefore, that a branch of the X stream passes from the striate to the Clare–Bishop cortex. Otherwise, very limited success has been achieved in the search for antidromic responses in lamina-3 neurons projecting to the extrastriate cortex. Frequently, despite the fact that cells of lamina 3 are driven orthodromically from an extrinisic cortical site, none can be driven antidromically. This deficiency may well be associated with difficulties in activating the terminals of fine axons but, whatever the reason, the absence of results leaves a serious gap in tracing the outflow streams from the striate cortex.

From lamina 5

Antidromic firing is much more readily detected in cells projecting from lamina 5 to the superior colliculus. So far, the receptive fields recorded in all reports (Palmer & Rosenquist, 1974; Harvey, 1980b) indicate that these are C^Y cells. As a group, most of these C^Y cells are now thought to receive at least a direct input from Y cells in the LGN (see Fig. 18.14) (Henry *et al.*, 1983). It is possible that converging secondary inputs to these cells also pass along branches of the Y stream and that the Y stream passes through lamina 5,

Fig. 18.15. Results of an experiment showing the distribution of cells in laminae 2 and 3 of the striate cortex which project to Areas 18 and 19. Cells identified by retrograde transport of HRP injected, as indicated, into area of axonal projection. The distributions show that cells projecting to Areas 18 and 19, respectively, tend to congregate in the lower and upper reaches of laminae 2/3.

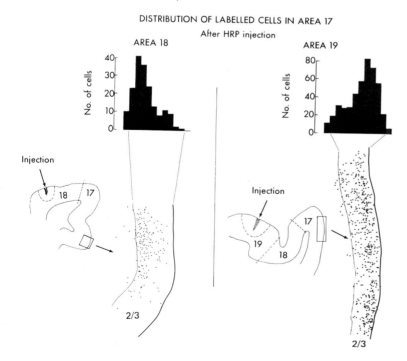

DISTRIBUTION OF LABELLED CELLS IN AREA 17

After HRP injection

and thence to the superior colliculus, without any intrusion from the X stream.

From lamina 6

The pyramidal cells of lamina 6 project to the claustrum and to the LGN. The existence of streaming in these two projections has not been reported, although Harvey (1980*b*) has shown that both S and C (almost certainly C^Y) cells from lamina 6 can be antidromically activated from a stimulating electrode in the LGN. The distribution of antidromic latencies in Fig. 18.16 shows a difference in the rate of signal conduction in the two cell types with the C cells being faster conductors than S cells. Cleland & Dubin (1977) found a similar distinction in the orthodromic drive from the striate cortex to the thalamus and further demonstrated that the fast input went to the perigeniculate nucleus (PGN) and the slow one to the laminar LGN. Harvey then made the suggestion that the C cells of lamina 6, with their faster axonal conduction, project to the PGN while the slower conducting S cells go to the laminar LGN. Recent studies with anterograde transfer of HRP, however,

have shown, as indicated in Fig. 18.17, that large and small axons distribute to all levels of the PGN and LGN (Boyapati & Henry, 1984). If large axons conduct more rapidly than small axons, then this result, at least, fails to confirm the existence of segregated streaming in the outflow from lamina 6.

Cells projecting to the claustrum, like those going from lamina 3 to the extrastriate cortex, are difficult to drive electrically (Boyapati & Henry, unpublished). Here, however, the problem may arise from a low encounter rate, since only 3.5% of lamina-6 cells

Fig. 18.17. Four drawings of centrifugal axons (anterogradely infiltrated with HRP) passing into the LGN and PGN. I, II: Thick axons (> 1.0 μm) with clusters of terminal swellings at various levels in the LGN. III, IV: Fine axons (< 1.0 μm) with terminal clusters in the LGN and PGN. The disposition of large and small axons is not consistent with the view that C cells project to the PGN and S cells to the LGN (see text). Scale bar = 50 μm. A, A$_1$ and C: laminae of LGN; CIN: central interlaminar nucleus of the LGN. (Boyapati & Henry, 1984).

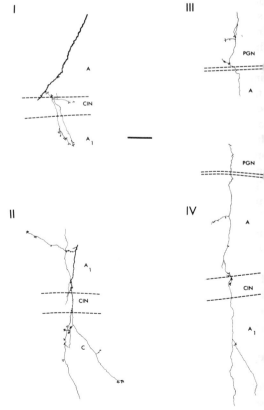

Fig. 18.16. Summary diagram comparing (A) the range of orthodromic latencies for cells in the lateral geniculate nucleus (LGN) and perigeniculate nucleus (PGN) following stimulation of the striate cortex (adapted from Cleland & Dubin, 1977) with (B) the distribution of antidromic latencies of lamina-6 striate neurons following stimulation at OR$_1$, just above the LGN (adapted from Harvey, 1980*b*). Filled bars in the histogram represent C cells; open bars represent S cells. This comparison promoted the suggestion that C cells projected to the PGN and S cells to the LGN (Harvey, 1980*b*).

A. Orthodromic (cortical stimulation)

PGN cells   LGN cells

B. Antidromic (thalamic stimulation)

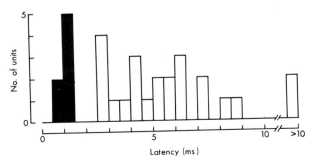

roject to the claustrum (LeVay & Sherk, 1981). In
those instances where it is difficult to elicit antidromic
responses, some insight may come from the identification of the striate neurons receiving reciprocal drive
from the region. Generally, spikes are initiated more
successfully by stimulating at or close to the cell body
and studies are now under way at Canberra to look for
possible streaming implications in these reciprocal
connections to Area 17.

Summary

In overview, the striate cortex of the cat, and
indeed all laboratory animals tested, emerges as a region
traversed by all the streams known to emerge from the
retina. It is the point where ON- and OFF-centre
streams may converge as they come together in the
creation of dual-row S cells (cf. Schiller, 1982) or
perhaps the C cell, with its composite ON and OFF
response to flashing stimuli. At the same time, ON and
OFF streaming may still be maintained in those
instances where the stream passes through single-row
S cells to terminate or merge at some, as yet, unknown
destination. Certainly segregation in the striate cortex
is a characteristic of the X and Y streams and it is here,
in the cat, that these two streams become spatially
separate for the first time.

Walls, if reincarnated, would not be surprised to
find that the functional need for spatial dispersion of
streams remains unresolved. On the structural side,
segregation may be preparing the streams for their
journey to regions that perform more exclusive roles in
the abstraction of visual information. Within these
preparatory steps, however, there still exists an
opportunity for the streams to act in a cohesive fashion.
This may be achieved in two ways – by crosstalk or by
duplicating the processing steps in both the X and the
Y stream. Crosstalk, as Walls suggested, may lead to
the smearing of information and while there are
suggestions of morphological links between the two
streams (for example through the small spiny stellates
of lower lamina 4) there appears to be a remarkable level
of stream segregation. When consideration is also given
to the high degree of replication of receptive field design
in each stream, then it appears that the most likely way
to achieve an interstream awareness is to duplicate the
steps taken in abstracting information.

If, in line with a commonly held view, it is
accepted that the X stream is involved in high-resolution
spatial vision and the Y stream in motion detection,
then the receptive field of striate neurons in either
stream does not seem to be dedicated exclusively to
meeting only one of these requirements. Cells,
whatever their stream, are simultaneously extracting
information about the shape and the motion of the
stimulus, so that the organization of the S^Y cell
receptive field, for example, ensures that the influence
of the stimulus shape is being forged into the Y stream
signal as it passes through the striate cortex. Similarly,
S^X cells are direction-selective, even though directional
dependence has been interpreted as a property of
motion detectors (see also Malpeli, Schiller & Colby,
1981).

The precision with which form and motion are
represented varies in the two streams, and S^X and C^X
cells both respond, in a modulated fashion, to higher
spatial frequencies than their counterparts in the Y
stream. On the other hand, striate neurons of the Y
stream are responsive to faster stimuli than those in the
X stream although, in their velocity dependence, cells
of both streams seldom respond to stimuli moving
faster than 25 degrees/s. It follows that movement
detected in the striate cortex may be limited to the slow
drifts that accompany high-resolution form vision. In
this achievement, the responses of cells in the X stream,
while being chiefly influenced by the form of the retinal
image are, at the same time, sensitive to the slow drifts
it makes across the eye. Similarly, information on form
and motion is extracted by cells in the Y stream
although they are more involved in processing firing
initiated by the slow drifts of the image. In the cat, these
Y-stream signals are probably passed on to Area 18
where they can merge with messages on fast eye
movements that come directly from the LGN.

If the striate cortex has developed a specialty,
therefore, it is in the initiation of high-resolution vision
after due allowance is made for slow movements of the
image across the retina. The later phases of
high-resolution vision and a more selective interpretation of movement take place outside the striate cortex.
In these achievements, the streaming in the striate
cortex may not serve any immediate purpose other than
to prepare for the oncoming and more exclusive phases
of visual processing.

References

Barlow, H. B., Fitzhugh, R. & Kuffler, S. W. (1957) Change of
    organization in the receptive fields of the cat's retina
    during dark adaptaion. *J. Physiol.*, **137**, 388–54

Boyapati, J. & Henry, G. (1984) Corticofugal axons in the lateral geniculate nucleus of the cat. *Exp. Brain Res.*, 53, 335–40

Bullier, J. & Henry, G. H. (1979*a*) Ordinal position of neurons in cat striate cortex. *J. Neurophysiol.*, 42 1251–63

Bullier, J. & Henry, G. H. (1979*b*) The neural path taken by afferent streams in striate cortex of the cat. *J. Neurophysiol.*, 42, 1264–70

Bullier, J. & Henry, G. H. (1979*c*) Laminar distribution of first-order neurons and afferent terminals in cat striate cortex. *J. Neurophysiol.*, 42, 1271–81

Bullier, J. & Henry, G. H. (1980) Ordinal position and afferent input of neurons in monkey striate cortex. *J. Comp. Neurol.*, 193, 913–35

Bullier, J., Mustari, M. J. & Henry, G. H. (1982) Receptive field transformations between LGN neurons and S cells of cat striate cortex. *J. Neurophysiol.*, 47, 417–38

Chow, K. L. (1955) Failure to demonstrate changes in the visual system of monkeys kept in darkness or in colored lights. *J. Comp. Neurol.*, 102, 597–606

Cleland, B. G. & Dubin, M. W. (1977) Organization of vision inputs to interneurons of lateral geniculate nucleus of the cat. *J. Neurophysiol.*, 40, 410–27

Cleland, B. G., Dubin, M. W. & Levick, W. R. (1971) Sustained and transient neurons in the cat's retina and lateral geniculate nucleus. *J. Physiol. (Lond.)*, 217, 473–96

Cleland, B. G., Levick, W. R., Morstyn, R. & Wagner, H. G. (1976) Lateral geniculate relay of slowly conducting retinal afferents to cat visual cortex. *J. Physiol. (Lond.)*, 255, 299–320

Cleland, B. G., Levick, W R. & Wässle, H. (1975) Physiological identification of a morphological class of cat retinal ganglion cells. *J. Physiol.(Lond).*, 248, 151–71

DeMonasterio, F. M. & Gouras, P. (1975) Functional properties of ganglion cells of the rhesus monkey retina. *Physiol. (Lond).*, 251, 167–65

DeValois, R. L. (1965) Analysis and coding of colour vision in the primate visual system. *Symp. Quant. Biol.*, 30, 567–79

Dreher, B., Fukada, Y. & Rodieck, R. W. (1976) Identification, classification and anatomical segregation of cells with X-like properties in the lateral geniculate nucleus of old-world primates. *J. Physiol.*, 258, 433–52

Enroth-Cugell, C. & Robson, J. G. (1966) The contrast sensitivity of retinal ganglion cells in the cat. *J. Physiol. (Lond.)*, 187, 517–52

Ferster, D. & LeVay, S. (1978) The axonal arborizations of lateral geniculate neurons in the striate cortex of the cat. *J.Comp. Neurol.*, 182, 923–44

Ferster, D. & Lindstrom, S. (1983) An intra-cellular analysis of geniculo–cortical connectivity in area 17 of the cat. *J. Physiol.*, 342, 181–215

Gilbert, C. D. & Kelly, J. P. (1975) The projections of cells in different layers of the cat's visual cortex. *J. Comp. Neurol.*, 163, 81–106

Gilbert, G. D. & Wiesel, T. N. (1979) Morphology and intra-cortical projections of functionally characterized neurons in the cat visual cortex. *Nature.*, 280, 120–5

Gilbert, C. D. & Wiesel, T. N. (1981) Laminar specialization and intracortical connections in cat primary visual cortex. In *The Organization of the Cerebral Cortex.*, ed. F. O. Schmitt,

F. G.Worden, G. Adelman & S. G.Dennis, pp. 163–91. Cambridge, Mass.: MIT Press.

Goodwin, A. W. & Henry, G. H. (1975) Direction selectivity of complex cells in a comparison with simple cells. *J. Neurophysiol.*, 38, 6

Harvey, A. R.(1980*a*) The afferent connexions and laminar distribution of cells in area 18 of the cat. *J. Physiol. (Lond.)*, 302, 483–505

Harvey, A. R. (1980*b*) A physiological analysis of subcortical and commissural projections in areas 17 and 18 of the cat. *J. Physiol. (Lond.)*, 302, 507–34

Henry, G. H., Lund, J. S. & Harvey, A. R. (1978) Cells of the striate cortex projecting to the Clare–Bishop area of the cat. *Brain Res.*, 151, 154–8

Henry, G. H., Mustari, M. J. & Bullier, J. (1983) Different geniculate inputs to B and C cells of cat striate cortex. *Exp. Brain Res.*, 52, 179–89

Hoffmann, K.-P. & Stone, J. (1971) Conduction velocity of afferents to cat visual cortex: a correlation with cortical receptive field properties. *Brain Res.*, 32, 460–6

Hubel, D. H. & Wiesel, T. N. (1962) Receptive fields, binocular interaction and functional architecture in the cat's visual cortex. *J. Physiol.*, 160, 105–54

Hubel, D. H. & Wiesel, T. N. (1965) receptive fields and functional architecture in two non-striate visual areas (18 and 19) of the cat. *J. Neurophysiol.*, 28, 229–89

Hubel, D. H. & Wiesel, T. N. (1968) Receptive fields and functional architecture of monkey striate cortex. *J. Physiol. (Lond.)*, 195, 215–43

Ikeda, H. & Wright, M. J. (1975) Spatial and temporal properties of 'sustained' and 'transient' neurones in area 17 of the cat's visual cortex. *Exp. Brain Res.*, 22, 363–83

Kelly, J. P. & Van Essen, D. C. (1974) Cell structure and function in the visual cortex of the cat. *J. Physiol.*, 238, 515–47

Kulikowski, J. J. & Bishop, P. O. (1981) Linear analysis of the responses of simple cells in the cat visual cortex. *Exp. Brain Res.*, 44, 386–400

Kulikowski, J. J. & Bishop, P. O. (1982) Silent periodic cells in the cat striate cortex. *Vision Res.*, 22, 191–200

Kulikowski, J. J., Marcelja, S. & Bishop, P. O. (1982) Theory of spatial position and spatial frequency relations in the receptive fields of simple cells in the visual cortex. *Biol. Cybern.*, 43, 187–98

Lee, B. B., Cleland, B. G. & Creutzfeldt, D. D. (1977) The retinal input to cells in area 17 of the cat's cortex. *Exp. Brain Res.*, 30, 527–38

Le Gros Clark, W. E. (1943) The anatomy of cortical vision. *Trans. Ophthalmol. Soc. U.K.*, 62, 229–45

Le Gros Clark, W. E. (1949) The laminar pattern of the lateral geniculate nucleus considered in relation to colour vision. *Doc. Optht.*, 3, 57–64

LeVay,S. & McConnell, S.K.(1982) ON and OFF layers in the lateral geniculate nucleus of the mink. *Nature.*, 300, 350–

LeVay, S. & Sherk, H. (1981) The visual claustrum of the cat. I. Structure and connections. *J. Neurosci.*, 1, 956–80

Lund, J. S.,Henry, G. H., MacQueen, C. L. & Harvey, A R. (1979) Anatomical organization of the primary visual cortex (area 17) of the cat. A comparison with area 17 of the macaque monkey. *J. Comp. Neurol.*, 184, 599–618

Malpeli, J. G., Schiller, P. H. & Colby, C. L. (1981) Response properties of single cells in monkey striate cortex during reversible inactivation of individual lateral geniculate laminae. *J. Neurophysiol.*, **46**, 1102–19

Martin, K. A. C. & Whitteridge, D. (1982) The morphology, function and intracortical projections of neurones in area 17 of the cat which receive monosynaptic input from the lateral geniculate nucleus (LGN). *J. Physiol.*, **328**, 37P

Movshon, J. A., Thompson, I. D. & Tolhurst, D. J. (1978a) Spatial summation in the receptive fields of simple cells in the cat's striate cortex. *J. Physiol.*, **283**, 53–77

Movshon, J. A., Thompson, I. D. & Tolhurst, D. J. (1978b) Receptive field organization of complex cells in the cat's striate cortex. *J. Physiol.*, **283**, 79–99

Mustari, M. J., Bullier, J. & Henry, G. H. (1982) Comparison of the response properties of three types of monosynaptic S cell in cat striate cortex. *J. Neurophysiol.*, **47**, 439–54

O'Leary, J. L. (1941) Structure of the area striata of the cat. *J. Comp. neurol.*, **75**, 131–61

Palmer, L. A. & Rosenquist, A. C. (1974) Visual receptive fields of single striate cortical units projecting to the superior colliculus in the cat. *Brain Res.*, **67**, 27–42

Pollen, D. A., Andrews, B. W. & Feldon, S. E. (1978) Spatial frequency selectivity of periodic complex cells in the visual cortex of the cat. *Vision Res.*, **18**, 665–82

Polyak, S. (1957) *The Vertebrate Visual System*. Chicago: University of Chicago Press

Ramon y, Cajal, S. (1922) Studien über die Sehrinde der Katze. *J. Psychol. Neurol. (Leipzig)*, **217**, 437–96

Rushton, W. A. H. (1960) At the retinal level. In *Mechanisms of Colour Discrimination*, A symposium honouring Henri Pieron, pp. 69–100. Oxford: Pergamon Press

Schiller, P. H. (1982) Central connections of the retinal ON and OFF pathway. *Nature*, **297**, 580–2

Sherman, S. M., Hoffmann, K.-P. & Stone, J. (1972) Loss of a specific cell type from dorsal lateral geniculate nucleus of visually deprived cats. *J. Neurophysiol.*, **35**, 532–41

Stone, J. & Dreher, B. (1973) Projections of X- and Y-cells of the lateral geniculate nucleus to areas 17 and 18 of the visual cortex. *J. Neurophysiol.*, **36**, 551–67

Stone, J., Dreher, B. & Leventhal, A. (1979) Heirarchical and parallel mechanisms in the organization of the visual cortex. *Brain Res. Rev.*, **1**, 345–94

Stone, J. & Fukuda, Y. (1974) Properties of cat retinal ganglion cells: a comparison of W-cells with X- and Y-cells. *J. Neurophysiol.*, **37**, 722–48

Tanaka, K. (1983) Cross correlation analysis of geniculostriate neuronal relationships in cats. *J. Neurophysiol.*, **49**, 1303–18

Walls, G. L. (1953) The lateral geniculate nucleus and visual histophysiology. *Univ. Calif. Publ. Physiol.*, no. 9, 1–100

Wässle, H., Boycott, B. B. & Illing, R.-B. (1981a). Morphology and mosaic of on- and off-beta cells in the cat retina and some functional considerations. *Proc. R. Soc. Lond. B.*, **212**, 177–95

Wässle, H., Peichl, L. & Boycott, B. B. (1981b) Morphology and topography of on- and off-alpha cells in the cat retina. *Proc. R. Soc. Lond. B.*, **212**, 157–75

White, E. L. (1979) Thalamocortical synaptic relations: a review with emphasis on the projections of specific thalamic nuclei to the primary sensory areas of the neocortex. *Brain Res.*, **1**, 275–311

Wilson, P. D. & Stone, J. (1975) Evidence of W-cell input to the cat's visual cortex via the C laminae of the lateral geniculate cortex. *Brain Res.*, **92**, 472–8

19

Studies on inhibitory neurotransmission in visual cortex *in vitro*

TETSURO OGAWA, HIROSHI
KATO AND SEISHO ITO

Introduction

Evidence is accumulating that GABA (
aminobutyric acid) acts as an inhibitory transmitter
the mammalian cerebral cortex (Krnjevic & Schwar
1966, 1967; Curtis *et al.*, 1971; Iversen, Mitchell
Srinivasan, 1971; Krnjevic, 1974; Rose & Blakemo
1974; Sillito, 1975; Emson & Lindvall, 1979; Tsumo
Eckart & Creutzfeldt, 1979). In the visual cort
Iversen *et al.* (1971) have shown that an increase
calcium-dependent release of GABA occurs
association with an intracellular inhibition. It h
subsequently been demonstrated that the GAI
antagonist bicuculline blocks inhibitory process
which are responsible for the specific responses of m
visual cortical neurons (Pettigrew & Daniels, 197
Rose & Blakemore, 1974; Sillito, 1975, 1977, 197
Tsumoto *et al.*, 1979; Sillito *et al.*, 1980*a*, *b*). Th
electrophysiological and pharmacological studies
GABA-mediated intracortical inhibition have be
supported by the demonstration of the presence
GABAergic interneurons in all layers of visual cort
of the rat, monkey and cat by immunocytochemi
methods (Ribak, 1978; Hendrickson, Hunt & W
1981; Freund *et al.*, 1983; Somogyi *et al.*, 1983) a
by autoradiographic studies of high-affinity uptake
[³H]-GABA (Somogyi *et al.*, 1981*a*, *b*; Somogyi *et a
1984*a*, *b*). Although these studies support the view th
GABA is an inhibitory transmitter in the visual cort
the evidence is somewhat indirect. To obtain mo
direct evidence the present experiments using slices
cat's visual cortex *in vitro* were undertaken. The resu

show that inhibitory postsynaptic potentials (IPSPs) evoked by stimulation of the white matter are mimicked by exogeneously applied GABA in many respects (Ogawa & Kato, 1982).

Methods

Our general methods have been described in detail elsewhere (Kato & Ogawa, 1981). The major modification adopted here was that first a small block of brain was excised with a razor blade from the posterior part of the gyrus lateralis, including the anterior part of the gyrus lateralis posterior, of a cat anaesthetized with i.p. injected pentobarbital sodium (35 mg/kg). The block, quickly rinsed with a well-oxygenated warm artificial standard solution, was placed on the top of a Ringer–agar block, with an orientation for coronal sections. The block was

sectioned at 0.3–0.7 mm normal to the cortical surface with a specially designed cutter. The slices were rapidly transferred to a small container where they were preincubated for at least 1 h in the standard solution (NaCl 124mM, KCl 5.0mM, NaH$_2$PO$_4$ 1.25mM, MgSO$_4$ 1.0mM, CaCl$_2$ 3.0mM, NaHCO$_3$ 26.0mM, glucose 10.0mM, sucrose 3.0mM, kanamycin sulphate 50 mg/l) at 30° C and pH 7.4. At a recording session the slice was transferred to an experimental chamber that was continuously perfused at a rate of 1.5 ml/min with an oxygenated bathing solution whose composition was varied depending on experimental conditions. The temperature of the bathing solution was thermostatically controlled at 35–38° C. The slice was fixed to the bottom of the chamber with a pair of stimulating electrodes (ST) made of acupuncture needles which were pressed on the white matter, as shown in Fig. 19.1.

Fig. 19.1. Schematic diagram of experimental arrangements. (A) Excision of a small block of visual cortex from which slices 0.3–0.7 mm thick were cut in the coronal plane with a specially designed knife. (B) In an experimental chamber a slice was fixed on the bottom with bipolar stimulating electrodes (ST) which were pressed into the white matter. Electrode E$_1$, which was made from an acupuncture needle, was thrust into the grey matter (layer III) to monitor field responses evoked by stimulation of the white matter. E$_2$, a glass micropipette filled with 4M potassium acetate, is for intracellular recording. GABA was locally applied through a glass micropipette (tip diameter, 50 μm) fitted to a microsyringe (MS) which was placed close to the recording micropipette. Inset records show (1) field response, (2) intracellularly recorded response to stimulation of the white matter, and (3) action potentials triggered by intracellular injection of current pulse.

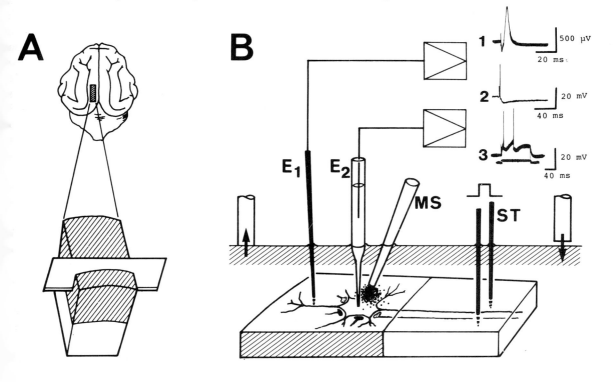

Another acupuncture needle electrode (E₁) was placed at the cortex for recording field potentials evoked by stimulation of the white matter. This recording electrode also served to fix the slice in place. For intracellular recording, glass micropipettes filled with 4M potassium acetate were inserted into cortical layer III under visual control. Data were collected only from those neurons which were maintained for sufficiently long periods of time with a stable resting membrane potential (in excess of -40 mV) and showed IPSPs either in isolation or preceded by an excitatory postsynaptic potential (EPSP) on stimulation of the white matter. Some of the neurons were stained intracellularly with either procion yellow or Lucifer yellow to verify their location in intracortical layers and cell morphology. They were all small pyramidal cells located in layer III.

Small amounts of (0.5–0.7 ml) of GABA (10^{-5}–10^{-6} M) were applied to the surface of a slice close to the recording micropipette through a glass micropipette (tip diameter 50 μm) filled with the drug solution and connected to a syringe. In some experiments drugs were ejected by pressure pulses from a glass pipette (tip diameter 5 μm) filled with the drugs and positioned close to the recording microelectrode. The ejection time was controlled from 20 ms to 1 s with

a pulse-gated magnetic valve. Bicuculline, the GABA antagonist, was bath-applied at a concentration of 10^{-5}–10^{-6} M. To block synaptic transmission, a low-Ca²⁺–high-Mg²⁺ medium (0.1mM Ca²⁺, 5.0mM Mg²⁺) was used instead of the standard solution.

Results

Extracellular recordings

In visual cortical slices it was commonly observed that cellular excitation caused by electrical shocks to the white matter was followed by a suppression of cellular activity for 20–400 ms. This suppression could be demonstrated in two ways, as shown in Fig. 19.2. First, double shocks were delivered to the white matter at various intershock intervals. The intensity of each shock was set at supramaximum for an orthodromic spike. A latency of the spike, measured from a shock artefact to the foot of the upstroke of the spike, was 4 ms. As shown in Fig. 19.2A, test shocks were unable to generate a spike discharge for 28 ms after cellular excitation caused by conditioning stimulation (1–3). Secondly, a single shock delivered to the white matter produced an orthodromic spike followed by a longlasting (about 200 ms) suppression of spontaneous discharges (Fig. 19.2B). In some cells white-matter stimulation resulted in a longlasting suppression of spontaneous discharges without eliciting orthodromic spike discharges.

Intracellular recordings

As expected from the extracellular recordings, the majority of cortical cells showed a longlasting

Fig. 19.2. Inhibition induced by stimulation of the white matter as revealed by extracellular recordings. (A) Test shocks did not trigger a unitary spike for about 28 ms after a conditioning shock triggered an action potential. (B) Spontaneous discharges were suppressed for about 200 ms after conditioning stimulation.

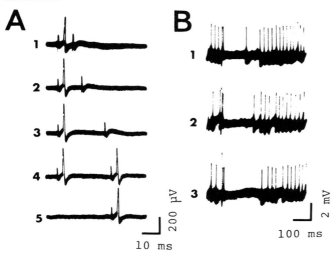

perpolarizing potential which occurred either immeately following an initial EPSP or in isolation in sponse to stimulation of the white matter. Fig. 16.3B ows that a weak shock elicited only an EPSP; the ºSP was increased in size and was followed by a nglasting hyperpolarization if there was an increase shock intensity. In Fig. 19.3C stimulation of the ite matter first elicited only a hyperpolarization; this s increased in size and duration with an increase in mulus intensity. At an intensity three times the reshold level for the IPSP there was an antidromic :ion potential preceding the hyperpolarization. In g. 19.3A, the cell responded to stimulation of the ite matter with an EPSP–IPSP sequence. A mulus twice the threshold for the EPSP triggered an :hodromic action potential. The latency of the EPSP, :asured from the shock artefact to the take-off of the ºSP, was 2.4 ms. When the intensity of shock was :reased to 2.5 times the threshold level for the EPSP, antidromic action potential preceded the EPSP with undetectable latency. The orthodromic action :ential was almost entirely blocked by the preceding :idromic action potential, even though the rising ase of the EPSP was accelerated and the amplitude s increased. At three times the threshold level, an :hodromic action potential in abortive form was :erved to be triggered on top of the EPSP.

The longlasting hyperpolarization evoked by nulation of the white matter was regarded as an SP for the following reasons. (1) The polarity

reversed when a micropipette filled with 3M potassium chloride was used instead of a micropipette filled with 4M potassium acetate. This change took place so rapidly after the impalement that no photographic records were taken. (2) When the standard bathing solution was replaced with a Cl^--free solution in which sodium propionate replaced sodium chloride in equal molarity, the hyperpolarization was converted to a depolarization on which action potentials were sometimes superimposed (Fig. 19.4B). This indicates that the hyperpolarization is Cl^--dependent. (3) The membrane resistance of impaled neurons decreased during the hyperpolarization (Fig. 19.4A), as revealed by the reduction of the amplitude of voltage response to a constant current pulse of either polarity applied through the intracellular micropipette. A depolarizing current pulse which triggered an action potential from a voltage response before the postsynaptic potential was elicited, produced only attenuated voltage responses which could trigger no action potential during the hyperpolarization. With a gradual return of the membrane potential pulse-generated voltage, responses became large enough to trigger an action potential (Fig. 19.4A3). (4) when the standard bathing solution was replaced with a low-Ca^{2+}–high-Mg^{2+} solution, the shock-induced hyperpolarization was completely abolished. (5) Bicuculline added to the standard solution at a concentration of 4×10^{-6} M abolished the hyperpolarization and unmasked the longlasting EPSP which was abolished by the low-Ca^{2+}–high-Mg^{2+} bathing

Fig. 19.3. Intracellularly recorded responses to stimulation of the white matter. (A) An EPSP–IPSP sequence. An action potential was triggered by the EPSP. An increase in stimulus intensity produced an antidromic action potential before the EPSP (2 and 3). (B) The weakest stimulus evoked only an EPSP. With an intensity of stimulus a longlasting IPSP followed the EPSP. (C) White-matter stimulation produced only a longlasting IPSP (2 and 3). An increase in stimulus intensity elicited an antidromic action potential preceding the IPSP.

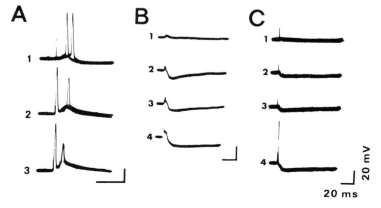

20 mV

20 ms

solution (Fig. 19.4C and D). (6) The hyperpolarization reversed its polarity when the membrane potential was hyperpolarized by passing steady current through the intracellular micropipette (Fig. 19.5). The reversal potential averaged -73 ± 8.1 mV ($n = 7$). Based on these analyses a longlasting hyperpolarization which was induced by stimulation of the white matter is simply designated an IPSP in this study.

Effects of GABA

Fig. 19.6A illustrates the most commonly observed effects of GABA. In the standard bathing solution a neuron (resting membrane potential = -40 mV) responded to stimulation of the white matter

with an EPSP–IPSP sequence (Fig. 19.6Aa). When small amount of GABA (10^{-5} M) was added to the bathing solution for about 50 s, the membrane potential was hyperpolarized immediately after the application and remained hyperpolarized for another 7 min with a gradual return to the initial level. At the peak hyperpolarization, membrane resistance was reduced as manifested by the reduction of the voltage excursion induced by an extrinsic current pulse (compare 3 and 4 with 1). The membrane resistance returned to the initial level with the recovery of the membrane potential.

The action of GABA was completely antagonized by bicuculline (4×10^{-6} M, bath application), as shown

Fig. 19.4. Some properties of the longlasting IPSP. (A) White-matter stimulation elicited a longlasting IPSP (1). Voltage responses to either hyperpolarizing (2) or depolarizing (3) constant current pulses injected through the intracellular micropipette were reduced in amplitude during the IPSP. Note that the depolarizing current pulse was strong enough to trigger an action potential before and after the IPSP. Upper trace of each record monitors current. (B) The IPSP was converted to depolarization (2 and 3) which became large enough to generate action potentials (4–6) while the slice was perfused with Cl⁻-free solution. (C) Bicuculline (4×10^{-6} M) added to the standard bathing solution eliminated the longlasting IPSP and unmasked the longlasting EPSP. (D) Orthodromic responses of the same cell as in C were blocked when the slice was perfused with low-Ca²⁺–high-Mg²⁺ solution.

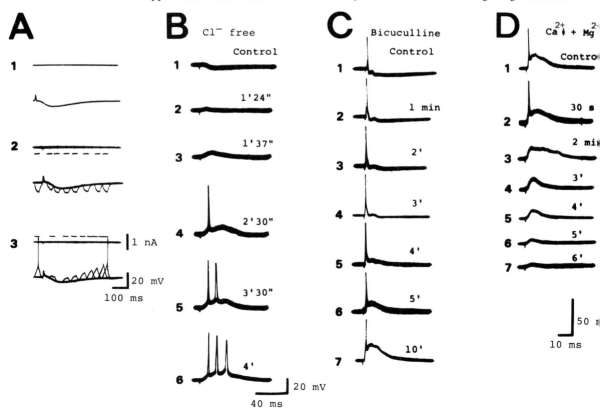

in Fig. 19.6B. In this experiment, depolarizing current pulses were used for checking changes in membrane resistance and excitability of the neuron during the manipulation. Before application of bicuculline, the depolarizing pulses elicited two spikes superimposed on the electrotonic voltage responses. These current-induced responses and the resting membrane potential were virtually unaffected by bicuculline alone. In the presence of bicuculline, GABA did not affect current-induced responses. Bicuculline evidently antagonizes GABA.

When the standard bathing solution was replaced with a low-Ca^{2+}–high-Mg^{2+} solution, orthodromic responses were completely abolished (records 1–3 in the column to the left of Fig. 19.7), but depolarizing current pulses injected intracellularly still produced action potentials (Fig. 19.7). In this condition, GABA was able to hyperpolarize the membrane potential and to decrease the membrane resistance. These findings indicate that GABA acts directly on the postsynaptic membrane.

The effects of GABA on the neuronal membrane were further studied by the local application of a tiny amount of the drug close to the target. As illustrated in Fig. 19.8, GABA (2.2 pl at a concentration of 1M)

applied by pressure pulses (20 ms, 5.3 atm) caused a transient hyperpolarization with an associated reduction of membrane resistance (resting membrane potential − 56 mV). In this cell the distance between the tip of the recording micropipette and that of the GABA-filled micropipette was estimated at about 100 μm. Comparison of the time-course of a change in membrane resistance with that of the membrane potential reveals that the maximal reduction of membrane resistance (approximately 60%) led by about 8 s the peak of a hyperpolarizing response which was at − 64 mV (Fig. 19.8 top). Increasing the number of GABA puffs (from one to four) which were delivered at an interpuff interval of 2 s resulted in prolongation of changes in membrane resistance and potential without noticeable changes in size. The lower traces in Fig. 19.8 are actual records of these changes as a function of the number of GABA puffs.

Next, we wished to determine the reversal potential of GABA responses. Nine cells were used for this purpose. Four of the nine showed the usual hyperpolarizing GABA response but five cells showed a depolarizing monophasic response. With hyperpolarizing responses changes in membrane potential toward hyperpolarization caused a reduction of the amplitude

Fig. 19.5. Measurement of a reversal potential of the IPSP. (A) Chart records of membrane potentials (a–e). The initial membrane potential is − 68 mV and membrane potential was varied by passing steady current of either polarity through the micropipette. (B) IPSPs evoked by white-matter stimulation at different membrane potentials as indicated with a–e in A. (C) Amplitude of IPSP (horizontal axis) is plotted as a function of membrane potential. The reversal potential of the IPSP is estimated at − 60 mV.

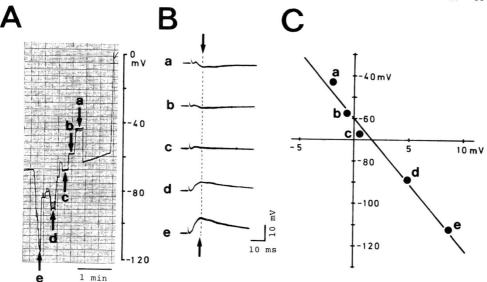

Fig. 19.6. Effects of GABA on the membrane potential and input resistance. (A) When GABA (10^{-5} M) was applied locally on the surface of the slice close to the recording electrode, the resting membrane potential was gradually hyperpolarized by about 10 mV with associated reduction of input resistance. Sample records of voltage responses to constant current pulses (0.7 nA, 50 ms) are numbered and shown below a chart recording. An orthodromically evoked intracellular response consisting of an EPSP–IPSP sequence is shown on the right. Record a was obtained before GABA application and b was during hyperpolarization induced by GABA. (B) Bicuculline did not cause any changes in membrane potential and input resistance but abolished GABA effects. Bicuculline was bath-applied (4×10^{-6} M) and GABA was applied locally.

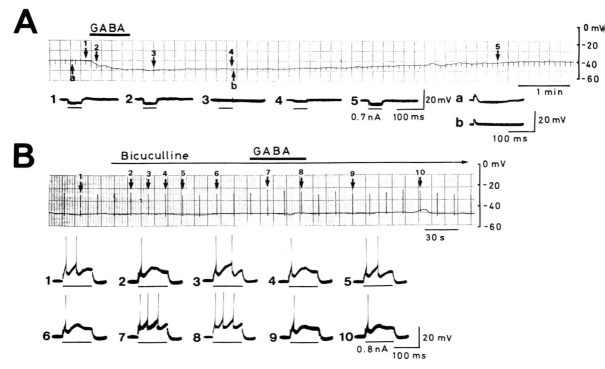

Fig. 19.7. Effects of GABA in a bathing solution, containing a reduced concentration of Ca^{2+} and an increased concentration of Mg^{2+}, which blocks synaptic transmission. Three photographic records (1–3) of intracellular responses to white-matter stimulation are shown to the left. Record 1 was obtained in the standard bathing solution and 2 and 3 were obtained at 2-min intervals after the low-Ca^{2+}–high-Mg^{2+} solution was perfused. Under these conditions, the small amount of GABA (10^{-5} M) which was continuously applied locally by pressure for 2 min elicited a hyperpolarization and reduction of membrane resistance. Sampled records of voltage responses to depolarizing current pulses (100 ms, 0.8 nA; 1–7 below the chart) indicate that they decreased in amplitude and triggered no more action spikes before the membrane potential and resistance recovered.

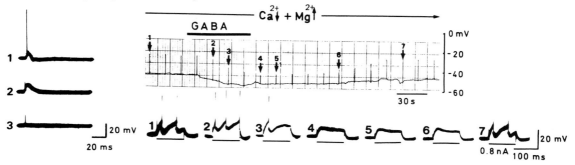

of the response until their polarity was reversed. Thus, the reversal potential for four cells averaged -76.4 ± 2.5 mV. This value is quite close to that of the IPSP. With the five cells which gave depolarizing GABA responses at the resting membrane potential, a

shift of the membrane potential from a resting level toward depolarization caused a reduction of the responses until their polarity was reversed. A representative example is shown in Fig. 19.9. By plotting GABA responses against membrane potentials

Fig. 19.8. Effects of a tiny amount of GABA (2.2 pl at a concentration of 1M) applied close to the impaled neuron. The top trace shows that GABA, which was ejected by a pressure pulse (20 ms, 5.3 atm), induced a hyperpolarization associated with reduction of membrane resistance. Vertical downward deflections represent electrotonic

responses produced by injection of constant cathodal current pulses for monitoring membrane resistance. The bottom traces show that increased numbers of GABA puffs, which were ejected by one, two, three and four pressure pulses at interpulse intervals of 2 s, produced a corresponding prolongation of GABA effects.

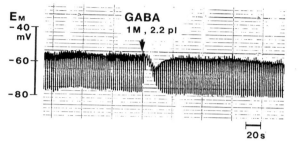

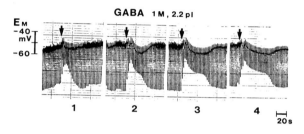

Fig. 19.9. Measurement of a reversal potential of GABA response. The top traces (chart recordings) show voltage responses produced by puffs of GABA at various membrane potentials. The membrane potential was changed by injection of steady current of various intensities. At -80 mV (resting membrane potential) a puff of GABA elicited a depolarizing response, but with an increasing depolarization, the GABA response decreased in amplitude until it reversed its polarity at -64 mV.

The relation between membrane potential and amplitude of GABA response is shown in the graph. A chart record to the left of the graph shows that membrane resistance was reduced during the depolarizing GABA response (membrane potential, -80 mV). A chart record to the right shows that the depolarizing GABA response was also obtained when synaptic transmission was blocked by a low-Ca^{2+}–high-Mg^{2+} solution.

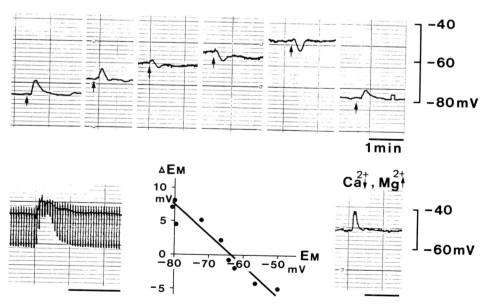

the reversal potential was estimated at -64 mV (graph in Fig. 19.9). A trace to the left of the graph shows an increase in membrane conductance associated with the depolarizing GABA response. To the right of the graph a depolarizing GABA response obtained from another cell is shown to remain unchanged when the standard bathing solution was replaced by a low-Ca^{2+}–high-Mg^{2+} solution. The mean value of the reversal potential for the five cells was -52 ± 10.2 mV.

Discussion

In these experiments, using *in vitro* slices of cat's visual cortex, we obtained the following results. (1) Most of the cells situated in layer III of Area 18 responded orthodromically to white-matter stimulation with an EPSP–IPSP sequence or sometimes with an IPSP alone. Some of them discharged an antidromic action potential preceding the EPSP. (2) The IPSP was associated with a reduction of membrane resistance and was abolished in a low-Cl^- bathing solution and in a bicuculline-containing solution. (3) The reversal potential of the IPSP was about -73 ± 8.1 mV. (4) Local application of GABA usually caused a hyperpolarization similar to the IPSP but in some cells produced a depolarization. In either case membrane potential changes were associated with an increase in membrane conductance. (5) These GABA effects were abolished by bicuculline.

Since the introduction of the *in vitro* slice preparation technique into electrophysiological research, its experimental advantages have been apparent. In particular, to *in vitro* slices we owe successful high-quality intracellular recordings for a long period of time, and the ease with which pharmacological characterization of drugs and of putative neurotransmitters can be carried out on neurons. We designed the present experiments to take advantage of these features of brain slices.

Layer III of Area 18, to which we confined our attention here, consists of pyramidal cells, sending their axons to the ipsilateral or contralateral cortex via the white matter, and various types of stellate cells whose axons terminate locally in the cortex. These cells are activated by incoming afferent fibres either monosynaptically or polysynaptically (Toyama *et al.*, 1974; Harvey, 1980). In the present experiments all of the cells studied were not driven antidromically by stimulation of the white matter. Those cells which were not antidromically driven may be regarded either as

interneurons which have no long axon descending t the white matter, or as pyramidal cells whos descending axon has been destroyed in preparation. T check these two possibilities we carried out intracellula staining of some cells with procion yellow or Lucife yellow. Five cells were stained successfully and all them were identified as pyramidal cells. This findin is favourable to the latter possibility but does n entirely exclude the former. Recent immunocytochem ical studies combined with the Golgi impregnatio technique have demonstrated that glutamic aci decarboxylase (GAD)-positive cells are relevant to class of smooth stellate cells and that pyramidal ce in layer III are invested with GAD-positive termina (Freund *et al.*, 1983; Somogyi *et al.*, 1983). The findings provide morphological evidence that pyramid cells in layer III are innervated by GABAerg interneurons.

In the present experiments we generally observe that GABA produced a hyperpolarization associate with an increase in membrane conductance in targ cells. These GABA effects remained unchanged whe synaptic transmission was blocked by a low-Ca^{2+} high-Mg^{2+} bathing solution. This led us to conclu that GABA acts directly on the membrane of pyramid cells. In some cells we found that GABA produced on a depolarization or a biphasic response consisting of depolarization followed by a hyperpolarization. typical example is shown in Fig. 19.8. This depolarizin GABA response was not affected by a blockade synaptic transmission in a low-Ca^{2+}–high-Mg solution. Hence, the depolarizing response wa regarded as being due to GABA's direct action on th membrane of pyramidal cells.

Recently a number of reports have presente evidence that GABA has a dual action on CN neurons. Specifically, this issue has been studied hippocampal pyramidal cells. GABA applied close the soma of a pyramidal cell from which intracellul recordings were being made caused a hyperpolarizatio while GABA applied to the dendrites of the cell cause a depolarization (Alger & Nicoll, 1979, 1982; Anders *et al.*, 1980; Thalmann, Peck & Ayala, 1981). W attempted similar experiments on our cells but to da no clear-cut results have been obtained. The depolar zing GABA response reversed when the membran potential was shifted to depolarization by injectin steady current. Thus, a reversal potential of th depolarizing GABA response was obtained, averagin

-52 ± 10.2 mV. This value was significantly different om the reversal potential for hyperpolarizing GABA sponses. We thought initially that the polarity of GABA responses observed at the resting membrane otential was dependent upon its displacement relative o their reversal potential. However, this proved not to e the case. In the present experiments different eversal potentials were detected for depolarizing and yperpolarizing GABA responses. It is presumed that ifferent ionic mechanisms underlie these two types of sponses.

References

ger, B. E. & Nicoll, R. A. (1979) GABA-mediated biphasic inhibitory responses in hippocampus. *Nature (Lond.)*, **281**, 315–17

ger, B. E. & Nicoll, R. A. (1982) Pharmacological evidence for two kinds of GABA receptor on rat hippocampal pyramidal cells studied *in vitro. J. Physiol.*, **328**, 125–41

ndersen, P., Dingledine, R., Gjerstad, L., Langmoen, I. A. & Laursen, A. M. (1980) Two different responses of hippocampal pyramidal cells to application of gamma-amino butyric acid. *J. Physiol.*, **305**, 279–96

urtis, D. R., Duggan, A. W., Felix, D. & Johnston, G. A. R. (1971) Bicuculline, an antagonist of GABA and synaptic inhibition in the spinal cord of the cat. *Brain Res.*, **32**, 69–96

nson, P. C. & Lindvall, O. (1979) Distribution of putative neurotransmitters in the neocortex. *Neuroscience*, **4**, 1–30

eund, T. F., Martin, K. A. C., Smith, A. D. & Somogyi, P. (1983) Glutamate decarboxylase-immunoreactive terminals of Golgi-impregnated axoaxonic cells and of presumed basket cells in synaptic contact with pyramidal neurons of the cat's visual cortex. *J. Comp. Neurol.*, **221**, 263–78

arvey, A. R. (1980) The afferent connexions and laminar distribution of cells in area 18 of the cat. *J. Physiol.*, **302**, 483–505

endrickson, A. E., Hunt, S. P. & Wu, J.-Y. (1981) Immunocytochemical localization of glutamic acid decarboxylase in monkey striate cortex. *Nature (Lond.)*, **292**, 605–9

ersen, L. L., Mitchell, J. F. & Srinivasan, V. (1971) The release of γ-aminobutyric acid during inhibition in the cat visual cortex. *J. Physiol.*, **212**, 519–34

ato, H. & Ogawa, T. (1981) A technique for preparing *in vitro* slices of cat's visual cortex for electrophysiological experiments. *J. Neurosci. Methods*, **4**, 33–8

rnjevic, K. (1974) Chemical nature of synaptic transmission in vertebrates. *Physiol. Rev.*, **54**, 418–540

rnjevic, K. & Schwartz, S. (1966) γ-aminobutyric acid, an inhibitory transmitter? *Nature (Lond.)*, **211**, 1372–4

rnjevic, K. & Schwartz, S. (1967) The action of γ-aminobutyric acid on cortical neurones. *Exp. Brain Res.*, **3**, 320–36

gawa, T. & Kato, H. (1982) GABA action on visual cortical neurons in *in vitro* slices. *Biomed. Res.*, **3** (suppl.), 101–6

ettigrew, J. D. & Daniels, J. D. (1973) Gamma-aminobutyric

acid antagonism in visual cortex: different effects on simple, complex and hypercomplex neurons. *Science (N.Y.)*, **182**, 81–3

Ribak, C. E. (1978) Aspinous and sparsely-spinous stellate neurons in the visual cortex of rats contain glutamic acid decarboxylase. *J. Neurocytol.*, **7**, 461–78

Rose, D. & Blakemore, C. (1974) Effects of bicuculline on functions of inhibition in visual cortex. *Nature (Lond.)*, **249**, 375–7

Sillito, A. M. (1975) The effectiveness of bicuculline as an antagonist of GABA and visually evoked inhibition in the cat's striate cortex. *J. Physiol.*, **250**, 287–304

Sillito, A. M. (1977) Inhibitory processes underlying the directional specificity of simple, complex and hypercomplex cells in the cat's visual cortex. *J. Physiol.*, **271**, 699–720

Sillito, A. M. (1979) Inhibitory mechanisms influencing complex cell orientation selectivity and their modification at high resting discharge levels. *J. Physiol.*, **289**, 33–53

Sillito, A. M., Kemp, J. A., Milson, J. A. & Berardi, N. (1980*a*) A re-evaluation of the mechanisms underlying simple cell orientation selectivity. *Brain Res.*, **194**, 517–20

Sillito, A. M., Kemp, J. A. & Patel, H. (1980*b*) Inhibitory interactions contributing to the ocular dominance of monocularly dominated cells in the normal cat striate cortex. *Exp. Brain Res.*, **41**, 1–10

Somogyi, P., Cowey, A., Halasz, N. & Freund, T. F. (1981*a*) Vertical organization of neurones accumulating H-GABA in visual cortex of rhesus monkey. *Nature (Lond.)*, **294**, 761–3

Somogyi, P., Freund, T. F., Halasz, N. & Kisvarday, Z. F. (1981*b*) Selectivity of neuronal H-GABA accumulation in the visual cortex as revealed by Golgi staining of the labeled neurons. *Brain Res.*, **225**, 431–6

Somogyi, P., Freund, T. F. & Kisvarday, Z. F. (1984*a*) Different types of H-GABA accumulating neurons in the visual cortex of the rat. Characterization by combined autoradiography and Golgi impregnation. *Exp. Brain Res.*, **54**, 45–56

Somogyi, P., Freund, T. F., Wu, J.-Y. & Smith, A. D. (1983) The section-Golgi impregnation procedure. 2. Immunocytochemical demonstration of glutamate decarboxylase in Golgi-impregnated neurons and in their afferent synaptic boutons in the visual cortex of the cat. *Neuroscience*, **9**, 475–90

Somogyi, P., Kisvarday, Z. F., Freund, T. F. & Cowey, A. (1984*b*) Characterization by Golgi impregnation of neurons that accumulate H-GABA in the visual cortex of monkey. *Exp. Brain Res.*, **53**, 295–303

Thalmann, R. H., Peck, E. J. & Ayala, G. F.(1981) Biphasic response of hippocampal neurons to GABA. *Neurosci. Lett.*, **21**, 319–24

Toyama, K., Matsunami, K., Ohno, T. & Tokashiki, S. (1974) An intracellular study of neuronal organization in the visual cortex. *Exp. Brain Res.*, **21**, 45–66

Tsumoto, T., Eckart, W. & Creutzfeldt, O. D. (1979) Modification of orientation sensitivity of cat visual cortex neurons by removal of GABA-mediated inhibition. *Exp. Brain Res.*, **34**, 351–63

20

Thalamocortical and corticocortical interconnections in the cat visual system: relation to the mechanisms of information processing

BOGDAN DREHER

The 'classical' concepts (largely elaborated by Hubel and Wiesel during the early 1960s, see Hubel & Wiesel 1962, 1965) concerning the mechanisms underlying information processing in the mammalian visual prosencephalon have to be substantially modified in order to accommodate the wealth of the new data gathered during the last 15 years or so (for reviews see Chapter 21 and Zeki, 1978; Stone, Dreher & Leventhal, 1979; Van Essen, 1979; Lennie, 1980; Graybiel & Berson, 1981a, b; Sherman & Spear, 1982; Stone & Dreher, 1982; Van Essen & Maunsell, 1983; Mishkin, Ungerleider & Macko, 1983).

Although many of the new insights derive from the studies conducted on a number of mammalian species (especially of the order of primates) this chapter concentrates on the visual system of the most commonly studied carnivore – the cat. Furthermore, although this chapter is firmly based on the concepts related to the existence of distinct morphological and functional classes of retinal ganglion cells and their unique patterns of projection to the retinorecipient nuclei (for recent reviews see Sherman & Spear, 1982; Rodieck & Brening, 1983; Stone, 1983), I will discuss only the thalamocortical parts of the cat visual system. Despite this narrowing of perspective it is believed that the basic principles of organization discussed in this chapter apply to other mammalian orders.

It is now well established that the cat's visual cortex (Fig. 20.1B) contains a number of retinotopically organized areas (Tusa, Palmer & Rosenquist, 198

Mucke *et al.*, 1982; Olson & Graybiel, 1983). However, only one of these areas, striate area (or Area 17), appears to contain a complete representation of the cat's entire contralateral hemifield (Tusa *et al.*, 1981) while the remaining areas contain representations of only certain portions of the visual hemifield. Furthermore, in different areas, different parts of the represented portions of visual hemifield are emphasized. Apart from Area 17, in which the central part of the visual field is emphasized, nine other cortical areas (Areas 18, 19, 20b, 21a, 21b, the posteromedial lateral suprasylvian (PMLS), posterolateral lateral suprasylvian (PLLS), ventral LS (VLS) and dorsal LS (DLS)) either seem to contain only the representation of the central part of the visual hemifield, or the amount of cortex devoted to the representation of the central portion of the visual hemifield is proportionally substantially greater than that devoted to the representation of the peripheral portion of the visual hemifield (Tusa *et al.*, 1981). By contrast, since the peripheral parts of the visual field are emphasized in cortical Area 20a, and the

anteromedial LS (AMLS) and anterolateral LS (ALLS), these areas seem to be specialized for processing information from the peripheral parts of the visual hemifield (Tusa *et al.*, 1981).

The 'output maps' from a given area do not necessarily reflect the distortions in the visual-field representation characteristic for a given area. Thus, while all retinotopically organized areas in the cat's cortex project to the rostral part of the pontine gray, and via the relay in the pons contribute to the major visual input to the cerebellum, the visual maps presented by each area to the cerebellum are very similar (Albus *et al.*, 1981). Thus in all corticopontine projections from the retinotopically organized areas, the more peripheral parts of the visual field are emphasized and the parts of a given cortical area in which the area centralis is represented hardly contribute to the corticopontine projections (Albus *et al.*, 1981).

Fig. 20.1. (A) A schematic coronal section through the 'visual thalamus' of the cat. OT = optic tract; LGNv = ventral lateral geniculate nucleus; A, A₁, C, C$_{1-3}$ = geniculate laminae; MIN = medial interlaminar nucleus; Pul = pulvinar; RRZ = retinorecipient zone of the pulvinar; PRZ = pretectorecipient zone of the pulvinar; CRZ = corticorecipient zone; LP₁ = lateral part of the lateral posterior-pulvinar complex; TRZ = tectorecipient zone; LP$_i$ = interadjacent part of the LP–pulvinar complex; LP$_m$ = medial part of the LP–pulvinar complex; Sg = suprageniculate; VM = representation of vertical meridian. Terminology for subdivisions of LP–pulvinar complex after Updyke (1977, 1981*a*, 1983), Berson & Graybiel (1978, 1983) and Graybiel

& Berson (1980). Retinotopic organization of LP–pulvinar complex outlined after Raczkowski & Rosenquist (1981). (B) Drawing of the lateral aspect of the cat brain. The parts of the cortex occupied by the 14 retinotopically organized areas are outlined (after Tusa *et al.*, 1981; Mucke *et al.*, 1982, Olson & Graybiel, 1983). AMLS, ALLS, PMLS, PLLS, DLS and VLS are, respectively, the anteromedial, anterolateral, posteromedial, posterolateral, dorsal and ventral lateral suprasylvian areas. The PMLS is largely equivalent to the Clare–Bishop area in the older literature. EVA = ectosylvian visual area (Olson & Graybiel, 1983). It is the equivalent of the anterior ectosylvian visual area (AEV) of Mucke *et al.* (1982).

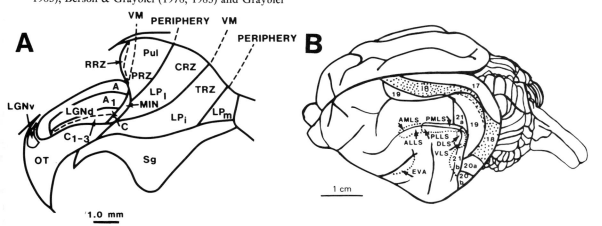

Thalamocortical connections

The retinotopic organization of at least some of the cortical areas seems to be largely determined by the retinotopic organization of their principal afferent thalamic nucleus. Thus, the retinotopic organization of striate cortex (Area 17) seems largely to reflect the retinotopic organization of the A laminae of the dorsal lateral geniculate nucleus (LGNd; see Sanderson, 1971a, b; Tusa et al., 1981). Similarly, the retinotopic organization of Area 18 seems largely to reflect the retinotopic organization of the medial interlaminar nucleus (MIN) (Sanderson, 1971a; Guillery et al., 1980; Tusa et al., 1981; Rowe & Dreher, 1982; Lee et al., 1984).

Numerous studies (Rosenquist, Edwards & Palmer, 1974; Gilbert & Kelly, 1975; Maciewicz, 1975; LeVay & Gilbert, 1976; Hollander & Vanegas, 1977; Geisert, 1980; Raczkowski & Rosenquist, 1983; Bullier, Kennedy & Salinger, 1984a) indicate that only two of the retinotopically organized cortical areas (corresponding to cytoarchitectonic Areas 17 and 18 distinguished by Otsuka & Hassler (1962) should be considered parts of the primary visual cortex since only these two areas appear to receive direct input from the main laminae (laminae A and A_1) of the ipsilateral LGNd (Figs 20.1A, 20.2). However, while in the case of Area 17 the great majority (over 90%) of all thalamic afferents originate from the A laminae (Hollander & Vanegas, 1977; D. I. M. Robinson, unpublished), in the case of Area 18 less than 50% of thalamic afferents originate from these laminae (Hollander & Vanegas, 1977).

Both Areas 17 and 18 also receive direct input from other parts of the ipsilateral LGNd complex: magnocellular lamina C, parvocellular laminae C, C_1, C_2, and the MIN. In addition Area 18, but not Area 17, receives direct input from the ipsilateral retino-recipient zone of the pulvinar (RRZ) (Raczkowski & Rosenquist, 1983; Bullier et al., 1984a; see, however, Leventhal, Keens & Törk, 1980) which some investigators consider a distinct part of the LGNd complex ('geniculate wing') (Guillery et al., 1980; Raczkowski & Rosenquist, 1983).

Area 18, and to a lesser extent Area 17, also receive direct input from the corticorecipient part (CRZ) of the lateral posterior nucleus–pulvinar (LP–pulvinar) complex (Berson & Graybiel, 1978, 1983; Graybiel & Berson, 1980, 1981a, b; Hughes, 1980; Symonds et al., 1981; Raczkowski & Rosenquist, 1983; Bullier et al., 1984a). In addition, both Areas 17 and 18 receive direct input from the intralaminar thalamic nuclei related to vision (central lateral paracentral and central medial nuclei) (Kennedy & Baleydier, 1977; Miller & Benevento, 1979; Hughes, 1980; Niimi et al., 1981; Bullier et al., 1984a).

Fig. 20.2. Schematic illustration of interconnections between the retinotopically organized cat cortical areas and the retinotopically organized parts of the cat thalamus. (Based on the data of Kawamura *et al.*, 1974; Maciewicz, 1974, 1975; Rosenquist *et al.*, 1974; Gilbert & Kelly, 1975; Updyke, 1975, 1977, 1981, 1983; Hollander & Vanegas, 1977; Berson & Graybiel, 1978, 1983; Geisert, 1980; Graybiel & Berson, 1980; Hughes, 1980; Leventhal *et al.*, 1980; Miller *et al.*, 1980; Raczkowski & Rosenquist, 1980, 1983; Niimi *et al.*, 1981; Symonds *et al.*, 1981; Lee *et al.*, 1982; Mucke *et al.*, 1982; Naito & Kawamura, 1982; Tong *et al.*, 1982; Cavada & Reinoso–Suarez, 1983; Olson & Graybiel, 1983; Bullier *et al.*, 1984a; Kaufman *et al.*, 1984; Kuchiiwa *et al.*, 1984; Dreher *et al.*, unpublished; C. W. F. Lee, unpublished; D. I. M. Robinson, unpublished.)

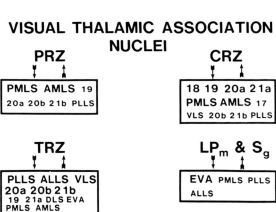

The connections between Areas 17 and 18 and the retinorecipient thalamic nuclei, as well as their connections with the CRZ of the LP–pulvinar complex (also called the lateral division of the lateral posterior complex (LPL) are reciprocal and only retinotopically corresponding regions appear to be interconnected. Indeed, the rule of reciprocity seems to apply to all connections between the thalamus and the retinotopic-ally organized cortical areas which will be described below (Updyke, 1975, 1977, 1981a, 1983).

Since Area 19 does not seem to receive direct input from the A laminae, it should not be considered a part of the primary visual cortex. However, Area 19 receives a substantial proportion of its thalamic input from the other retinorecipient components of the LGNd complex (lamina C, laminae C_{1-2}, MIN) as well

Fig. 20.3. Percentage distribution of thalamic and cortical neurons projecting to cortical Areas 19, 21 and the PMLS. The neurons were retrogradely labelled after the injections of the enzyme horseradish peroxidase (HRP) into those parts of Areas 19, 21a or the PMLS (indicated by different symbols) in which the central 10 degrees of the contralateral visual field are represented. (A)

Distribution of neurons in the ipsilateral thalamus. VA/VL = ventral anterior and ventral lateral nuclei. Vertical bars indicate standard deviations (three animals in the cases of Area 19 and 21a injection sites). (B) Distribution of neurons in the ipsilateral cortex. (Based on the data of Lee *et al.*, 1982, and B. Dreher, H. T. Ho, C. W. F. Lee & A. G. Leventhal, unpublished.)

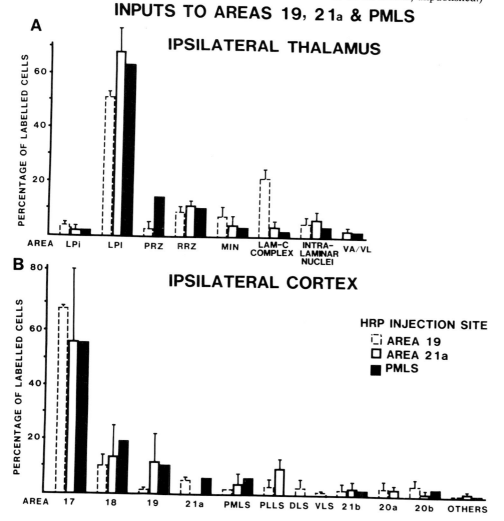

as from the RRZ of the pulvinar (Figs 20.2, 20.3A) (Holländer & Vanegas, 1977; Geisert, 1980; Leventhal *et al.*, 1980). Of all retinotopically organized cortical areas outside the primary visual cortex, Area 19 receives the highest proportion of its thalamic afferents from the retinorecipient nuclei. Thus, as indicated in Fig. 20.3A, almost 40% of the thalamic input to Area 19 originates in the retinorecipient nuclei of the ipsilateral thalamus. However, unlike in the case of Areas 17 and 18, the principal direct thalamic input to Area 19 originates from the LPL part of the LP–pulvinar complex (CRZ).

In addition, a small proportion of thalamic afferents to Area 19 originate from the interadjacent (LP$_i$) (Updyke, 1977, 1981*a*, 1983) or the tectorecipient zone of the LP–pulvinar complex (TRZ) (Berson & Graybiel, 1978, 1983; Graybiel & Berson 1980, 1981*a*, *b*). A small proportion of thalamic afferents to Area 19 also originates from the pretectorecipient part of the pulvinar (PRZ) (Figs 20.2, 20.3). Finally, Area 19 receives its direct thalamic input from the 'visual intralaminar nuclei' (Fig. 20.3) (Kennedy & Baleydier, 1977; Niimi *et al.*, 1981; Bullier *et al.*, 1984*a*).

Area 21a and the PMLS seem to be even further removed from the primary visual cortex since they receive less than 20% of their thalamic afferents from the retino-recipient parts of the thalamus (laminae C$_{1-2}$, MIN, RRZ) (Figs 20.2, 20.3A). The bulk of the visual thalamic afferents to Area 21a and the PMLS originate from the lateral, that is, the CRZ of the LP–pulvinar complex (Figs 20.2, 20.3A), which in turn receives its principal input from the primary (especially the striate) visual cortex (Graybiel & Berson, 1981*a*, *b*; Berson & Graybiel, 1983). Both areas also receive some direct thalamic input from the TRZ of the LP–pulvinar complex (Figs 20.2, 20.3A), which in turn receives its principal input from the superficial laminae of the superior colliculus. Finally, both Area 21a and the PMLS receive direct input from the 'visual intralaminar nuclei' (Fig. 20.3A) (Kennedy & Baleydier, 1977; Hughes, 1980; Tong, Kalil & Spear, 1982; Raczkowski & Rosenquist, 1983) and the 'motor thalamus' (ventral anterior and ventral lateral nuclei) (Fig. 20.3A).

Although the majority of the remaining retino-topically organized areas (Areas VLS, AMLS, PLLS, 20a, 20b) receive some direct input from the retinorecipient parts of the LGNd complex (laminae C$_{1-2}$, MIN), the proportion of the thalamic afferents contributed by the LGNd complex seems to be

extremely small (Raczkowski & Rosenquist, 1980, 1983). Finally, Areas ALLS, 21b and the ectosylvian visual area (EVA) do not seem to receive direct input from the retinorecipient parts of the thalamus (Raczkowski & Rosenquist 1980, 1983; Mucke *et al.*, 1982; Olson & Graybiel, 1983.) In addition, all these 'non-primary' areas receive a certain proportion of their thalamic afferents from the 'visual' intralaminar nuclei as well as from several non-visual thalamic nuclei (Naito & Kawamura, 1982; Tong *et al.*, 1982; Cavada & Reinoso-Suarez, 1983; Raczkowski & Rosenquist, 1983; Kuchiiwa *et al.*, 1984; D. I. M. Robinson, unpublished).

As indicated in Fig. 20.2, of the areas receiving a very small proportion (if any) of their input from the retinorecipient nuclei, only the AMLS receives its principal thalamic input from the CRZ of the LP–pulvinar complex (Tong *et al.*, 1982; Raczkowski & Rosenquist, 1983). By contrast, almost all cortical areas located laterally to the lateral suprasylvian sulcus (LSS) (with the exception of the EVA) receive their principal thalamic input from the tectorecipient or LP$_i$ region of the LP–pulvinar complex (Fig. 20.2) (Mucke *et al.*, 1982; Tong *et al.*, 1982; Raczkowski & Rosenquist, 1983; Olson & Graybiel, 1983).

Thus, it appears that in the cat visual cortex there is mediolateral 'hierarchy' of the retinotopically organized areas – the more medially located the area, the greater proportion of its thalamic afferents originates from the RRZ of the ipsilateral thalamus. Furthermore, areas located medial to the LSS receive a substantial (often principal) input from the CRZ of the LP–pulvinar complex, while almost all areas located lateral to the LSS receive their principal thalamic input from the TRZ of the LP–pulvinar complex.

However, recent data indicate a substantial degree of overlap between cortical (Area 17) and tectal inputs to the LP–pulvinar complex (Benedek, Norita & Creutzfeldt, 1983). Indeed, both the CRZ and the TRZ of the LP–pulvinar complex appear to receive converging input from the striate cortex and superior colliculus (Benedek *et al.*, 1983).

Despite the fact that a substantial proportion of thalamic neurons send branches to the retinotopically corresponding regions of a number of cortical areas (10–30% according to Bullier *et al.*, 1984*a*; see also Garey & Powell, 1967; Stone & Dreher, 1973; Geisert, 1980; Kaufman, Rosenquist & Raczkowski, 1984), the visual information conveyed by different axonal ranches of the same neuron seems to be used by each rea in a specific way. Apart from the uniqueness of the omplement of the thalamic afferents which each area eceives, the specific way in which each area appears o process information conveyed by its afferents must lso be related to the fact that in different cortical areas erminals of the neurons from the same thalamic nuclei end to terminate in different cortical laminae Rosenquist *et al.*, 1974; LeVay & Gilbert, 1976; Leventhal, 1979; Miller, Buschmann & Benevento, 980). Thus, for example, afferents from the MIN in reas 18 and 19 tend to terminate in lamina 4, while n Area 17 they terminate in lamina 1 and the border etween laminae 3 and 4.

Associational connections

The primary visual areas (Areas 17 and 18) have ssociational connections with each other as well as ssociative reciprocal interconnections with retinotopi-ally corresponding parts of other cortical areas eceiving a substantial thalamic input from the etinorecipient nuclei (Fig. 20.4 – Areas 19, 21a and the MLS).

In the case of Areas 19, 21a and PMLS, again, eir principal associational interconnections are with ther cortical areas receiving either principal or a ubstantial thalamic input from the retinorecipient uclei (Figs 20.3B, 20.4). However, unlike areas onstituting the primary visual cortex, Areas 19, 21a nd the PMLS, also have some associational intercon-ections with the retinotopically organized cortical reas which receive very small (if any) thalamic input om the retinorecipient thalamic nuclei (Figs 20.3B,).4). With the exception of the DLS (which is terconnected with Area 19) and the EVA (interconn-cted with the PMLS), all those areas receive thalamic put from the corticorecipient part of the LP–pulvinar omplex (Fig. 20.2). Thus, the cortical areas receiving substantial input from the RRZ of the thalamus and ose cortical areas receiving a substantial input from e CRZ of the LP–pulvinar complex form related amily clusters' (Graybiel & Berson, 1981a). Further-ore, corticocortical associational connections are rgely limited to the members of the same thalamic amily' (Graybiel & Berson, 1981a, b).

It is also worthwhile to note that the areas ceiving a substantial proportion of their thalamic ferents from the retinorecipient nuclei (the 'retino-ipient family') receive the overwhelming majority of

their associational afferents from the retinotopically organized areas (Fig. 20.3B). By contrast, retinotopically organized areas which receive only a very small proportion of their thalamic afferents from the retinorecipient nuclei have strong associational inter-connections with cortical areas which are not retino-topically organized (associational areas, areas related to other sensory modalities, premotor and limbic areas; see Kawamura & Naito, 1980; Cavada & Reinoso-Suarez, 1983; Kuchiiwa *et al.*, 1984; D. I. M. Robinson, unpublished).

Within a given 'family cluster' of cortical areas there is a characteristic trend in the laminar distribution of associative neurons. Thus, in the case of the 'retinorecipient family cluster', associative neurons projecting from Area 17 to Area 18 are

Fig. 20.4. Schematic illustration of associational interconnections between retinotopically organized areas in the cat's visual cortex. (Based on the data of Maciewicz, 1974; Gilbert & Kelly, 1975; Sugiyama, 1979; Kawamura & Naito, 1980; Graybiel & Berson, 1981a, b; Meyer & Albus, 1981; Montero, 1981; Lee *et al.*, 1982; Mucke *et al.*, 1982; Cavada & Reinoso-Suarez, 1983; Olson & Graybiel, 1983; Bullier *et al.*, 1984b; Kuchiiwa *et al.*, 1984; Dreher, Ho, Lee & Leventhal, unpublished; C. W. F. Lee, unpublished.)

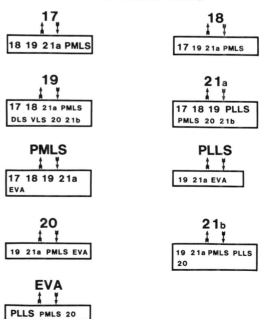

ASSOCIATIONAL CORTICOCORTICAL INTERCONNECTIONS

17
18 19 21a PMLS

18
17 19 21a PMLS

19
17 18 21a PMLS
DLS VLS 20 21b

21a
17 18 19 PLLS
PMLS 20 21b

PMLS
17 18 19 21a
EVA

PLLS
19 21a EVA

20
19 21a PMLS EVA

21b
19 21a PMLS PLLS
20

EVA
PLLS PMLS 20

predominantly pyramidal neurons located in supra-granular laminae 2 and 3 (Gilbert & Kelly, 1975; Meyer & Albus, 1981; Bullier *et al.*, 1984*b*). Only a very small proportion of Area 17 neurons projecting to Area 18 are spiny stellate cells located in lamina 4ab or pyramidal cells located in the infragranular laminae 5 and 6 (Meyer & Albus, 1981; Bullier *et al.*, 1984*b*). Similarly, as summarized in Fig. 20.5, the great majority of associational neurons projecting from the primary visual areas to areas 19, 21a and the PMLS (that is, to the areas receiving a smaller proportion of their thalamic afferents from the retinorecipient nuclei) are pyramidal neurons located in supragranular cortical laminae (Gilbert & Kelly, 1975; Bullier *et al.*, 1984*b*). While the bulk of the associative neurons projecting from Area 19 to Area 21a and the PMLS are located in the supragranular laminae, associative neurons projecting from Area 21a to Area 19 are almost evenly distributed between the supragranular and infragranular laminae (Fig. 20.5). On the other hand, the majority of associational neurons projecting from Area 21a to the PMLS are located in the supragranular laminae.

Only a small majority of PMLS neurons

Fig. 20.5. Laminar distribution of HRP-labelled neurons after injection into parts of cortical Areas 19, 21a and the PMLS in which the central 10 degrees of the contralateral visual field are represented. (Based on the unpublished data of Dreher, Ho, Lee & Leventhal.)

projecting to ipsilateral Areas 19 and 21a are located in the supragranular laminae. Similarly, a high proportion (although still a minority) of associative PMLS neurons projecting to the other part of the PMLS, in which the same part of the retina is represented (Tusa *et al.*, 1981), is located in the infragranular laminae.

Thus, it appears that the cortical areas which are located more medially and receive the higher proportion of their thalamic afferents from the retinorecipient nuclei, tend to send the information to the more laterally located retinotopically organized areas (receiving a smaller proportion of their thalamic afferents from the retinorecipient areas) mainly via supragranular laminae. By contrast, less 'primary' or 'higher' cortical areas (that is, areas receiving fewer afferents from the retinorecipient thalamic nuclei) tend to have a high proportion of the associative neurons projecting to 'more primary or lower areas' located in the infragranular laminae. This trend is even more striking in the case of areas receiving a very small (if any) proportion of their thalamic input from the retinorecipient nuclei (Fig. 20.5). Thus, of the associative cells projecting to Areas 19, 21a and the PMLS, the proportion of associative neurons located in infragranular laminae is greater in more laterally located Areas 20a, 20b and 21b than in the more medially located areas DLS, VLS and PMLS (Fig. 20.5). Furthermore, the proportion of associational cells located in the infragranular layers is greater

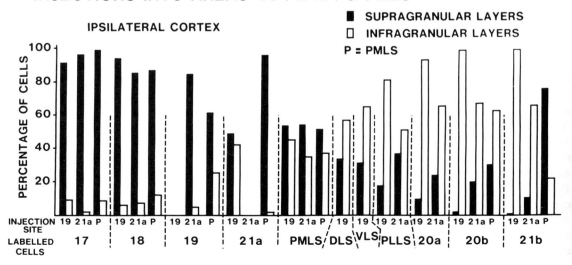

LAMINAR DISTRIBUTION OF LABELLED CELLS AFTER HRP INJECTIONS INTO AREAS 19 , 21a , & PMLS

among the cells projecting to Area 19 than among those projecting to Area 21a and the PMLS (Fig. 20.5).

This gradual mediolateral shift of the associative neurons in the cat visual cortex from their 'classical' location in the supragranular laminae (Jones, 1981) to the infragranular laminae is very similar to the dorsorostral shift of associative neurons from supra- to infragranular laminae apparent in the visual cortices of both Old World (Rockland & Pandya, 1979, 1981; Van Essen & Maunsell, 1983) and New World Primates (Tigges *et al.*, 1981). Since in the primates associative projections which originate from the supragranular laminae, like the thalamocortical (for example, geniculostriate) projections, terminate predominantly in lamina 4, they have been designated as 'feedforward' or ascending projections (Rockland & Pandya, 1979, 1981; Tigges *et al.*, 1981; Van Essen & Maunsell, 1983). On the other hand, since associational projections which originate from the infragranular laminae tend to terminate outside lamina 4, they have been designated descending or 'feedback' projections (Rockland & Pandya, 1979, 1981; Tigges *et al.*, 1981; Van Essen & Maunsell, 1983). The idea of a 'feedback' nature of the infragranular associational projections is further reinforced by the fact that in the cat (as in primates), the infragranular laminae of the visual cortex contain 'feedback' corticothalamic, corticotectal, corticopretectal and corticopontine projections (Hollander, 1974; Kawamura, Sprague & Niimi, 1974; Gilbert & Kelly, 1975; Magalhaes-Castro, Saraiva & Magalhaes-Castro, 1975; Updyke, 1977; Kawamura & Konno, 1979; Lund *et al.*, 1979; Albus *et al.*, 1981; Swadlow, 1983).

Interestingly, a similar mediolateral trend is also apparent in the level of termination of the corticotectal (presumed 'feedback') fibres. Thus, while corticocollicular fibres originating from the medially located, more primary areas (17, 18 and 19) terminate almost exclusively on the superficial (above the stratum opticum) retinotopically organized laminae of the ipsilateral superior colliculus (Kawamura *et al.*, 1974; McIlwain, 1977; Updyke, 1977; Kawamura & Konno, 1979; Behan, 1984), those originating in the lateral ('less primary') areas, around the LSS, terminate in both superficial and deep collicular laminae (Segal & Beckstead, 1984). Furthermore, while the majority of corticotectal neurons located in laminae 5 of the AMLS, PMLS, DLS and VLS project to the superficial layers of the ipsilateral superior colliculus, the majority of those located in lamina 5 of the ALLS

and PLLS project to the deep layers of the ipsilateral colliculus (Segal & Beckstead, 1984).

Finally, virtually all corticotectal cells located in the most lateral retinotopically organized area – the anterior ectosylvian visual area (EVA) – project to the deep layers of the ipsilateral colliculus (Segal & Beckstead, 1984).

In view of substantial differences in the receptive-field properties of cells located in different laminae of the same cortical area (Gilbert, 1977; Leventhal & Hirsch, 1978; Bullier & Henry, 1979*c*; Henry, Harvey & Lund, 1979; Harvey, 1980; Gilbert & Wiesel, 1981; Ferster & Lindström, 1983; Gilbert, 1983; Martin & Whitteridge, 1984; Mullikin, Jones & Palmer, 1984), the specific laminar distributions of associational and commissural neurons projecting from a given area to other cortical areas suggest great specificity of visual information conveyed by associational and commissural neurons. Indeed, a high degree of independence of different cortical laminae in at least one of the retinotopically organized cortical areas (Area 17) has been demonstrated by Malpeli (1983) in a set of very elegant experiments. Thus it appears that a number of receptive-field properties, including orientation and direction selectivities of cortical cells located in the supragranular layers 2 and 3 and those of at least some cells located in the infragranular layer 5, are hardly affected by the abolition (through selective inactivation of the lamina A of LGNd) of all visual activity in cortical layers 4 and 6. The supragranular laminae (and to a lesser extent infragranular lamina 5) seem to establish a number of receptive-field properties independently from the information conveyed to them from laminae 4 and 5.

Taking into account the fact that the proportion of 'feedforward' (or ascending) associational projections in a given cortical area is positively correlated with the proportion of thalamic afferents to this area originating from the retinorecipient nuclei, one can construct a sort of 'hierarchy' of cat visual cortical areas by arranging the cortical areas in a sequence in which the 'lower' or primary areas (e.g. Areas 17 and 18) send 'feedforward' information to the 'higher' areas, while gradually 'higher' areas send a gradually greater proportion of their associational efferents via 'feedback' infragranular laminae (Fig. 20.5).

Commissural connections

Each of the retinotopically organized areas studied receives its principal callosal input from its homotopic area. In addition, all areas receive some callosal input from heterotopic areas of the opposite hemisphere (Figs 20.6, 20.7; see also Segraves & Rosenquist, 1982*a*, *b*). Each of the areas receiving principal or substantial input from the retinorecipient nuclei of the ipsilateral thalamus (Areas 17, 18, 19, 21a and PMLS) receives also the bulk of its callosal afferents from the areas in the opposite hemisphere belonging to the 'retinorecipient family cluster' (Fig. 20.7). Furthermore, the callosal connections tend to involve a greater number of retinotopically organized areas in the opposite hemisphere than the number of such areas with which a given cortical area has associative connections (Figs 20.5, 20.6).

Fig. 20.6. Schematic-illustration of afferent callosal projections to different retinotopically organized areas of the cat visual cortex. (Based on the data of Shatz, 1977; Sanides, 1978; Innocenti, 1980; Keller & Innocenti, 1981; Meyer & Albus, 1981; Segraves & Rosenquist, 1982*a*, *b*; Cavada & Reinoso-Suarez, 1983; Dreher, Ho, Lee & Leventhal, unpublished; H. T. Ho, unpublished; D. I. M. Robinson, unpublished.)

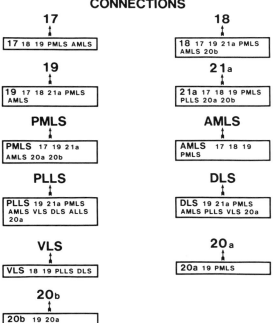

COMMISSURAL CORTICOCORTICAL CONNECTIONS

There are two mediolateral trends in the distribution of callosal neurons which correlate well with mediolateral decrease in the proportion of thalamic afferents originating from the retinorecipient nuclei. First, there is a progressive mediolateral increase in the area of the visual field representation containing callosal neurons (Ebner & Myers, 1965; Sanides, 1978; Segraves & Rosenquist, 1982*a*). Thus, while in the case of Area 17 callosal neurons are present only in the region where the part of the visual field 0–5 degrees from the vertical meridian is represented, in the case of callosal neurons in Area 18, they are present in the region where the part of the visual field located 0–10 degrees from the vertical meridian is represented (Segraves & Rosenquist, 1982*a*). Furthermore, in Areas 19 and 21a, callosal neurons are present in the regions in which the part of the visual field visual areas the callosal neurons are located in the regions where the part of the visual field located 0–40 degrees from the vertical meridian is represented. In Areas 20a, 20b and 21b the callosal neurons seem to be present not only in the regions representing part of the visual field close to the vertical meridian but also in the part in which the far periphery of the visual field is represented (Segraves & Rosenquist, 1982*a*). However, even in the cortical areas in which callosal neurons are widely distributed, most of these neurons originate in the region of the given area in which the vertical meridian of the visual field is represented (Sanides, 1978; Innocenti, 1980; Keller & Innocenti, 1981; Segraves & Rosenquist, 1982*a*).

Secondly, although callosal neurons projecting to the homotopic areas in the opposite hemisphere are predominantly located in supragranular laminae (Fig. 20.7B), there is a clear-cut mediolateral trend in laminar distribution of heterotopic callosal neurons. Thus, while heterotopic callosal projections from Area 17 to Areas 18, 19, 21a and the PMLS in the opposite hemisphere originate mainly from lamina 3 pyramidal neurons (supragranular and lamina 4), in Areas 18, 19, 21a and the PMLS increasing percentages of heterotopic callosal neurons originate in the infragranular laminae (mainly lamina 6) (Fig. 20.7B) (Glickstein & Whitteridge, 1976; Shatz, 1977; Keller & Innocenti, 1981; Segraves & Rosenquist, 1982*a*). About half of the callosal neurons projecting from Area 20a to Area 21a in the opposite hemisphere and the majority (virtually all) of the callosal neurons projecting from the PLLS and Area 20b to Area 21a in the opposite hemisphere

Fig. 20.7. (A) Percentage distribution of telencephalic cells projecting to the contralateral Areas 19 and 21s and the PMLS. The cells were labelled retrogradely by HRP injections into the parts of the respective areas in which the central 10 degrees of the contralateral visual field are represented. (B) The laminar distribution of cortical cells projecting to the contralateral Areas 19 and 21a and the PMLS. Again the HRP injections into

Areas 19 and 21a and the PMLS were restricted to those parts of the areas in which the central 10 degrees of the contralateral visual field are represented. (Based on unpublished data of H. T. Ho and Dreher, Ho, Lee & Leventhal.) (C) Schematic diagram of the lateral aspect of the cat cerebral cortex. The retinotopically organized areas are outlined after Tusa *et al.* (1981); Mucke *et al.* (1982) and Olson & Graybiel (1983).

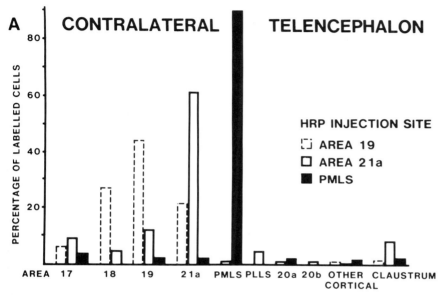

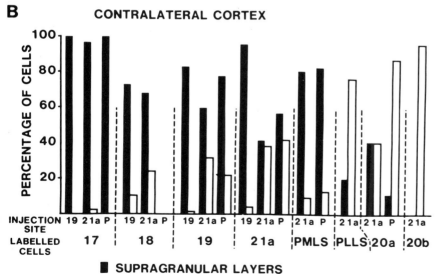

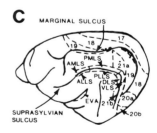

Fig. 20.8. (A) Schematic illustration of the laminar distribution of geniculate terminals and associational neurons in Area 17 of the cat visual cortex. The distribution of LGNd terminals is from Ferster & LeVay (1978); Leventhal (1979); Bullier & Henry (1979c); Gilbert & Wiesel (1981); Martin & Whitteridge (1984). The distribution of associational neurons projecting to Areas 19, 21a and PMLS is based on unpublished data of Lee and Dreher, Ho, Lee & Leventhal. (B) Upper diagram: percentages of cells in Areas 17, 18, 19, 21a and the PMLS of the normal cat visual cortex which exhibit either low cut-off velocity (respond only to slowly moving stimuli, $\leqslant 10$ degrees/s or high cut-off velocity (> 100 degrees/s). The values for Areas 17, 18, 19 and 21s are based on the data of Dreher *et al.* (1980) and unpublished data of Dreher, Ho, Lee & Leventhal. The values for the PMLS are based on the data of Spear & Baumann (1975). Lower diagram: percentages of cells in the PMLS after chronic lesions of retinotopically corresponding region of ipsi- and contralateral Areas 17 and 18. (Based on the data of Spear & Baumann, 1979.) (C) The mean width of receptive fields (discharge regions) of neurons from Areas 17, 18, 19 and 21a subserving different parts of the visual field. (Based on data of Dreher *et al.*, 1980, and unpublished data of Dreher, Ho, Lee & Leventhal.)

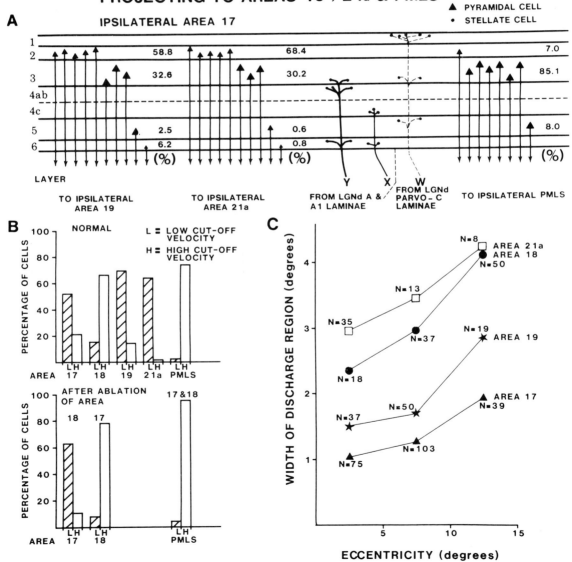

LAMINAR DISTRIBUTION OF CORTICAL NEURONES
PROJECTING TO AREAS 19 , 21a & PMLS

re located in infragranular (mainly lamina 6) laminae Fig. 20.7B).

Receptive-field correlates

Despite a substantial heterogeneity in the eceptive-field properties of single neurons located in given retinotopically organized cortical area, the eurons located in such an area exhibit a substantial egree of similarity in their receptive-field properties hich sets them apart from the neurons in the other eas. Indeed, it has been postulated on that basis that ach area specializes in the analysis of specific aspects f the visual environment (Zeki, 1978; Stone *et al.*, 979; Van Essen, 1979; Baker *et al.*, 1981; Stone & reher, 1982; Stone, 1983; Van Essen & Maunsell, 983; Orban, 1984).

As illustrated in Fig. 20.8C, cells in each of the eas comprising the 'retinorecipient family cluster' ffer substantially in their receptive-field sizes. Thus rea 17 cells tend to have smaller receptive fields than ose of cells located in the retinotopically corresponding gions in Areas 19, 18 and 21a. The receptive fields Area 19 cells tend to be smaller than those of cells cated in the retinotopically corresponding regions in reas 18 and 21a (Hubel & Wiesel, 1965; Duysens *al.*, 1982*a*, *b*; Rapaport, Dreher & Rowe, 1982; rban, 1984). Finally, the receptive fields of cells cated in the PMLS tend to be even larger than those Area 21a cells (Hubel & Wiesel, 1969; Wright, 1969; ear & Baumann, 1975; Turlejski, 1975; Camarda & zzolatti, 1976; Guedes, Watanabe & Creutzfeldt, 83; Spear, Miller & Ohman, 1983.) Similarly, there e substantial differences between the areas belonging the 'retinorecipient family' in the proportions of lls exhibiting low (≤ 10 degrees/s) and high cut-off > 100 degrees/s) stimulus velocities. Thus, as ustrated in Fig. 20.8B, in Areas 19 and 21a substantial ajorities of cells respond only to slowly moving imuli (low cut-off velocity) while in Area 18 and the MLS, substantial majorities of cells respond to st-moving stimuli (high cut-off velocity). The oportions of Area 17 cells exhibiting low and high t-off velocities falls somewhat between these tremes. Thus, while slightly over 50% of Area 17 lls respond only to slowly moving stimuli, about 20% spond to fast-moving stimuli (for review see Orban, 84).

Correlating the receptive-field sizes and cut-off locities of cortical cells and their thalamic afferents

is most reliable in the case of cortical cells with small receptive fields and those exhibiting high cut-off velocities. Thus, cortical neurons with large receptive fields might receive convergent input from a number of thalamic cells with small receptive fields. Similarly, cortical neurons receiving thalamic input exhibiting high cut-off velocities might, due to some intracortical inhibitory mechanisms, respond only to slowly moving stimuli (Goodwin & Henry, 1978; Mustari, Bullier & Henry, 1982; Duysens, Orban & Cremieux, 1984).

Nevertheless, both receptive-field sizes and velocity selectivity of cells located in the cortical areas belonging to the 'retinorecipient family' appear to correlate well with the receptive-field sizes and velocity selectivity of thalamic cells which provide the principal input to a given cortical area. In particular, only Area 17 seems to receive direct input from X-type geniculate cells (restricted to laminae A of LGNd) and X-type cells have smaller receptive fields than either Y- or W-type geniculate cells which provide, respectively, principal geniculate inputs to Areas 18 and 19 (Stone & Dreher, 1973, 1982; Dreher, Leventhal & Hale, 1980; Sherman & Spear, 1982; Stone, 1983; Orban, 1984). Furthermore, within Area 17, which in addition to X-type geniculate input receives also Y-type and some W-type geniculate input (Stone & Dreher, 1973; Singer, Tretter & Cynader 1975; Leventhal, 1979; Dreher *et al.*, 1980), cells receiving X-type input tend to have smaller receptive fields than those receiving Y-type input (Bullier & Henry, 1979*b*; Dreher *et al.*, 1980; Mustari *et al.*, 1982; Ferster & Lindström, 1983; Henry, Mustari & Bullier, 1983; Mullikin *et al.*, 1984).

By contrast, cells in Area 21a and the PMLS, which receive their principal thalamic inputs from the LPl (or CRZ) part of the LP–pulvinar complex, tend to have the largest receptive fields. In turn, cells in the CRZ or LP–pulvinar complex also have large receptive fields (Mason, 1978, 1981; Benedek *et al.*, 1983). Similarly, the cut-off velocity of cells in different cortical areas correlates reasonably well with the cut-off velocity of cells providing their principal thalamic input. Thus, the majority of cells in Area 18, which receive its main geniculate input from Y-type cells, that is, from the cells exhibiting high cut-off velocities (Stone & Dreher, 1973, 1982; Tretter, Cynader & Singer, 1975; Dreher *et al.*, 1980; Harvey, 1980; Sherman & Spear, 1982; Stone, 1983; Orban, 1984), also exhibit high cut-off velocities. By contrast, Area 19, which receives its principal geniculate input from

W-type cells with their characteristically low cut-off velocities, contains very few cells exhibiting high cut-off velocity (Dreher *et al.*, 1980). It is more difficult, however, to correlate a high proportion of cells exhibiting high cut-off velocity in the PMLS with properties of cells in its main thalamic afferent nucleus. The CRZ of the LP–pulvinar complex, in which the bulk of thalamic neurons projecting to the PMLS is located, also provides principal thalamic input to Areas 19 and 21a. However, Areas 19 and 21a contain only a very small proportion of neurons exhibiting high cut-off velocity.

Several factors might be responsible for this apparent discrepancy. First, as mentioned earlier, a substantial proportion of cells located in the CRZ of the LP–pulvinar complex also receive strong input from the tectum (Benedek *et al.*, 1983). Secondly, LP cells receiving strong tectal input exhibit high cut-off velocities (Chalupa, Williams & Hughes, 1983). Thirdly, the PMLS, unlike Areas 19 and 21a, might receive its principal input from cells in the cortico-recipient zone which also receive a strong tectal input. Fourthly, the PMLS seems to receive some Y-type input (high cut-off velocity) from the MIN (Leventhal *et al.*, 1980). Furthermore, although all three areas have input from the supragranular laminae of Area 17 (Figs 20.3B, 20.5), the associational 'feedforward' projections are from Area 17 to lamina 3, close to the terminals of Y-type geniculate afferents (Fig. 20.8A). Indeed, associational neurons projecting from Area 17 to Area 18, which like the PMLS contains a high proportion of cells exhibiting high cut-off velocities, are also located mainly in lamina 3, that is, in the vicinity of Y-type geniculate terminals (Gilbert & Kelly, 1975; Bullier *et al.*, 1984*b*).

Two lines of evidence challenge the idea that associational projections from Area 17 to Area 18 and the PMLS provide those areas with afferents exhibiting high cut-off velocity. First, the majority of cells in the striate cortex which can be activated antidromically by the electrical stimulation of the PMLS tend to prefer slowly moving stimuli (Henry, Lund & Harvey, 1978). Secondly, removal, or cooling, of Area 17 does not reduce the proportions of cells exhibiting high cut-off velocity in the ipsilateral Area 18 (Fig. 20.8B lower diagram) (Dreher & Cottee, 1975; Sherk, 1978) or the ipsilateral PMLS (Fig. 20.8B) (Spear & Baumann, 1979). Similarly, cooling of Areas 17 and 18 does not seem to affect velocity selectivity of cells in the ipsilateral Area 19 (Kimura *et al.*, 1980). On the other hand, removal of 'feedback' associational projection from Area 18 (which contains a high proportion of cell exhibiting a high cut-off velocity) to Area 17 seems t reduce the proportion of Area 17 cells exhibiting hig

Fig. 20.9. (A) Percentage of S-type and C-type cells in Areas 17, 18, 19 and 21a of normal cats (upper diagram) and in Areas 17 and 18 of cats in which retinotopically corresponding parts of, respectively, Areas 18 and 17 have been removed (lower diagram). The identification of S and C cells was based on the criteria of Henry (1977).
(B) Percentages of cells with strong inhibitory end-zones in Areas 17, 18, 19 and 21a of normal cats (upper diagram) and in Areas 17 and 18 of cats in which retinotopically corresponding parts of, respectively, Area 18 and 17 have been removed (lower diagram). For both A and B the values for normal animals are based on data of Hale *et al.* (1978) and unpublished data of Dreher, Ho, Lee & Leventhal. The values for animals in which either Area 17 or 18 has been ablated are based on the data of Dreher & Winterkorn (1974) and Dreher & Cottee (1975). All cells had receptive field centres within 15 degrees radius of the area centralis. The numbers of cells in each group are indicated in Fig. 20.11.

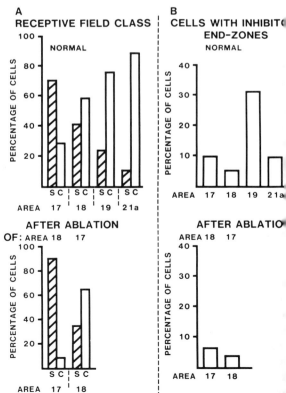

ut-off velocity (Fig. 20.8B, lower diagram) (Dreher & Winterkorn, 1974).

As illustrated in Fig. 20.9A, cortical areas belonging to the 'retinorecipient family' differ substantially from each other in the proportions of S and C-type cells they contain (for identifying criteria of S and C cells see Henry, 1977; see also simple and complex cells of Hubel & Wiesel, 1962, 1965). Although within the primary visual cortex both S (simple) and C (complex) cells can be monosynaptically activated from the LGNd (Hoffmann & Stone, 1971; Stone & Dreher, 1973; Singer et al., 1975; Tretter et al., 1975; Toyama, Maekawa & Takeda, 1977; Bullier & Henry, 1979a; Harvey, 1980; Ferster & Lindström, 1983; Henry et al., 1983; Martin & Whitteridge, 1984; cf however Gilbert, 1983), it is apparent from Fig. 20.9A that cortical areas which receive a higher proportion of their thalamic afferents from the retinorecipient nuclei also have a higher proportion of C cells. In view of the fact that thalamic cells with concentric, antagonistic centre-surround (ON-centre, OFF-surround or OFF-centre, ON-surround) receptive fields are largely restricted to the retinorecipient nuclei (Mason, 1981; Chalupa et al., 1983), this trend is consistent with the fact that while only about 20–40% of C cells in the primary visual cortex can be driven monosynaptically from the LGNd, the majority (60–80%) of S-type cells can be driven monosynaptically from the LGNd.

The removal of 'feedforward' associational projections from Area 17 to Area 18 does not seem to change significantly the proportions of S and C cells in Area 18 (Fig. 20.9A lower diagram) (Dreher & Cottee, 1975). However, the removal of 'feedback' associational projections from Area 18 to Area 17 reduces dramatically the percentage of C-type cells and, correspondingly, increases the percentage of S cells in Area 17 (Fig. 20.9A lower diagram) (Dreher & Winterkorn, 1974).

There are also apparent differences between cortical areas belonging to the 'retinorecipient family' in the proportion of cells with strong inhibitory end-zones in their receptive fields (hypercomplex cells of Hubel & Wiesel, 1965; see also Dreher, 1972; Kato et al., 1978; Camarda, 1979; Rose, 1979; Duysens et al., 1982a, b; Orban, 1984). While the proportion of cells with strong end-zone inhibition is small (10% or less) in Areas 17, 18 and 21a, in Area 19 they constitute over 30% of cells. Although a number of studies

indicate that, at least within Area 17, inhibitory end-zones are elaborated by intracortical inhibitory inputs (Sillito & Versiani, 1977; Orban, 1984), there is also some evidence indicating that end-zone inhibition is a reflection of the LGNd surround antagonism (Rose, 1979; Cleland, Lee & Vidyasagar, 1983). Thus, the higher proportion of cells with strong end-zone inhibition in Area 19 might be, at least partially, a reflection of the input to this area from thalamic cells with a strong antagonistic surround.

In the primary visual cortex the percentages of cells exhibiting strong end-zone inhibition are not significantly affected by removal of either 'feed forward' associational projections from Area 17 to Area 18 or 'feedback' associational projection from Area 18 to Area 17 (Fig. 20.9B lower diagram) (Dreher & Winterkorn, 1974; Dreher & Cottee, 1975).

The majority of cortical cells in all the areas belonging to the 'retinorecipient family', with exception of cells in the PMLS, are orientation-selective (Fig. 20.10A). Similarly, the majority of cells in Areas 17, 18, 21a and the PMLS are either completely direction-selective or exhibit clear-cut directional preferences (Fig. 20.10B). Only in Area 19 do direction-selective cells constitute a minority (Fig. 20.10B).

Both orientation and direction selectivity seem to be largely elaborated by the intracortical mechanisms restricted to a given cortical area (Sillito, 1975, 1977; Tsumoto, Eckart & Creutzfeldt, 1979; Sillito et al., 1980a). Furthermore, it appears that orientation selectivity in the cat striate cortex is to a large extent independently elaborated in supragranular laminae, in lamina 4 (Malpeli, 1983) and infragranular laminae (Bauer, 1982). Other data suggest, however, that orientation selectivity of cortical cells in Areas 17, 18 and 19 is partially determined by the orientation preferences of their excitatory LGNd afferents (Daniels, Norman & Pettigrew, 1977; Vidyasagar & Urbas, 1982; Leventhal 1983; Leventhal, Schall & Wallace, 1984; Vidyasagar & Heide, 1984). Corticogeniculate 'feedback' projection might in turn, affect the orientation biases apparent in the receptive fields of many LGNd neurons (Vidyasagar & Urbas, 1982).

Orientation selectivity of cells in Areas 17, 18 and 19 does not seem to be affected by removal of associational 'feedforward' and 'feedback' projections between those areas (Fig. 20.10B) (Dreher & Winterkorn, 1974; Dreher & Cottee, 1975; Sherk, 1978; Kimura et al., 1980).

Removal of 'feedforward' associational projections from Areas 17 and 18 reduces dramatically the proportion of direction-selective cells in the PMLS (Fig. 20.10D) (Spear & Baumann, 1979). There is also some reduction in the proportion of direction-selective cells in Area 17 after the removal of 'feedback' projections from Area 18 (Fig. 20.10D) (Dreher & Winterkorn, 1974).

Direction selectivity of cortical cells located in

Areas 19 and 21a and in the PMLS, as well as the direction selectivity of cells in other cortical areas receiving a substantial thalamic input from the LP–pulvinar complex, might be at least partially due to the direction-selective input from the LP–pulvinar complex (Mason, 1978, 1981; Benedek et al., 1983; Chalupa et al., 1983). In turn, direction selectivity of cells located in the CRZ and TRZ of the LP–pulvinar complex is probably imposed by the direction-selectivity

Fig. 20.10. Percentages of orientation-selective and direction-selective cells in Areas 17, 18, 19, 21a and the PMLS of normal cats (A and C) and Areas 17, 18 and the PMLS of cats in which retinotopically corresponding parts of, respectively, Areas 18, 17 and both Areas 17 and 18 have been removed (B and D). The values for Areas 17, 18, 19 and 21a of normal cats are based on the data of Hale et al. (1978) and unpublished data of Dreher, Ho, Lee & Leventhal. Only cells with receptive field centres within 15 degrees radius of the area centralis were considered. The values for the PMLS are based on the data of Spear & Baumann (1975). The values for Areas 17 and 18 in animals with ablated parts of the visual cortex are based on the data of Dreher & Winterkorn (1974) and Dreher & Cottee (1975). All cells had their receptive field centres within 15 degrees radius of the area centralis. The values for the PMLS after ablation of Areas 17 and 18 are based on the data of Spear & Baumann (1979). The numbers of cells in each group are indicated in Fig. 20.11.

Fig. 20.11. Percentages of monocular weakly and strongly binocular cells in Areas 17, 18, 19, 21a and the PMLS in normal cats (A) and in cats in which the retinotopically corresponding parts of ipsi- and contralateral Areas 18, 17 or 17 plus 18 have been ablated. Monocular group comprises classes 1 and 7 of Hubel & Wiesel (1962, 1965). Weakly binocular group comprises classes 2 and 6 of Hubel & Wiesel (1962, 1965). Strongly binocular groups comprises classes 3, 4 and 5 of Hubel & Wiesel (1962, 1965). The values for Areas 17, 18, 19 and 21a of normal cats are based on the data of Hale et al. (1978) and unpublished data of Dreher, Ho, Lee & Leventhal. All cells had receptive field centres within 15 degrees radius of the area centralis. The values for the PMLS are replotted from the data of Spear & Baumann (1975). The values for Areas 17 and 18 after ablation, respectively, of Areas 18 and 17 are based on the data of Dreher & Winterkorn (1974) and Dreher & Cottee (1975).

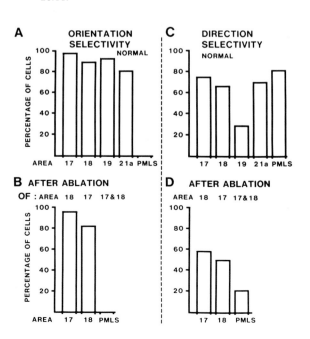

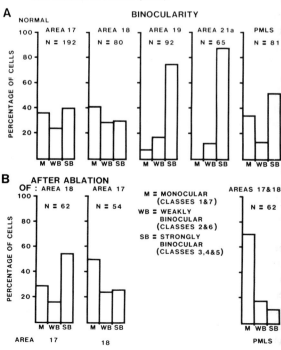

rtical and tectal inputs to these areas (Chalupa *et al.*, 083).

There are also substantial differences in the proortions of monocular and binocular cells between corcal areas belonging to the 'retinorecipient family' ig. 20.11A). In the primary visual cortices (Areas 7 and 18), high proportions of cells (about 40%) :long to classes 1 and 7 of Hubel & Wiesel (1962), at is, they can be activated only by stimulation ' one eye (contra-class 1, ipsi-class 7). By contrast, in reas 19 and 21a and, to a lesser extent, in the PMLS, e clear majority of cells receives strong excitatory put from both eyes (classes 3, 4 and 5 of Hubel & 'iesel, 1962, 1965).

Although the great majority of the cells in the t LGNd have binocular receptive fields (Sanderson *al.*, 1969, 1971; Singer, 1970; Rodieck & Dreher, 179; Kato *et al.*, 1981), the input from the non-ominant eye is almost always inhibitory. Furthermore, e axon terminals of the geniculate cells located in the minae A conveying information from different eyes e segregated into distinct patches in lamina 4 of the imary visual cortex (Ito, Sanides & Creutzfeldt, '77; Shatz, Lindström & Wiesel, 1977; LeVay, ryker & Shatz, 1978; Shatz & Stryker, 1978; Tieman Tumosa, 1983). The cortical patches dominated by e input from one eye are interconnected by intrinsic tracortical connections (Gilbert & Wiesel, 1983; lartin & Whitteridge, 1984). The binocular conver-nce on single neurons of primary visual cortex is not cessarily excitatory and virtually all neurons in Area ' have binocular receptive fields (Kato *et al.*, 1981). here is also some indication that the intracortical hibitory mechanism within Area 17 contributes to the :ular dominance of cells in Area 17 (Sillito *et al.*, 80*b*).

It has been proposed that binocular neurons in e cat primary visual cortex, via corticogeniculate ʻojections, influence the binocularity of the geniculate lls (Schmielau & Singer, 1977; cf, however, Sander-n *et al.*, 1971; Kato *et al.*, 1981).

Axon terminals of the geniculate cells located in e C laminae of the LGNd are poorly segregated in rea 19 (Tieman & Tumosa, 1983). Furthermore, cells Area 19, like those in Area 21a and the PMLS, ceive substantial excitatory binocular input from the alamus – a high proportion of cells located in the P–pulvinar complex can be driven through either eye *Aason*, 1978; Chalupa *et al.*, 1983). In turn,

binocularity in the LP–pulvinar complex is probably imposed by the cortical and tectal afferents. There is little evidence suggesting that either 'feedforward' associational projections from Area 17 to Area 18 or 'feedback' associational projections from Area 18 to Area 17 affect the binocularity of cells in the respective areas (Fig. 20.11B) (Dreher & Winterkorn, 1974; Dreher & Cottee, 1975).

Removal of Areas 17 and 18 and, thus, removal of 'feedforward' associational projections from those areas affects dramatically the binocularity of cells in the ipsilateral PMLS (Fig.20.11B) (Spear & Baumann, 1979; cf, however, Guedes *et al.*, 1983). There is also some, albeit indirect, evidence suggesting that associational projections from Areas 17 and 18 to Area 19 affect the binocularity of cells in Area 19 (Rapaport *et al.*, 1982; Leventhal & Hirsch, 1983; cf, however, Kimura *et al.*, 1980).

The excitatory binocular convergence in Areas 17 and 18, as well as that in the PMLS, is strongly reduced when callosal commissural connections are destroyed (Dreher & Cottee, 1975; Payne *et al.*, 1980; Berlucchi, 1981; Antonini, Berlucchi & Lepore, 1983; Blakemore *et al.*, 1983; Payne, Pearson & Berson, 1984). The effect is largely limited to those parts of the respective areas in which the part of the visual field located 0–12 degrees from the vertical meridian is represented. The long-term effects of callosal section seems to be limited to the neurons with receptive fields within 4 degrees of the vertical meridian. Furthermore, the effect is largely restricted to cells located outside lamina 4 (Payne *et al.*, 1984). The relative contribution of homotopic and heterotopic interconnections to the binocularity of cells in a given cortical area remains largely unknown.

Functional specializations

Area 17 is the only cortical area which receives X-type geniculate input and contains many cells with small receptive fields and low cut-off velocity. Consistently, Area 17 seems to play a crucial role in high-acuity tasks such as vernier acuity, fine-orientation acuity and fine stereopsis in the fixation plane (Sprague, Berkley & Hughes, 1979; Kaye, Mitchell & Cynader, 1981; Sprague, Hughes & Berlucchi, 1981; Ferster, 1981; Pettigrew & Dreher, 1982). In addition, Area 17 seems to play a crucial role in determining direction selectivity, velocity selectivity, spatial organization of receptive fields and the degree of excitatory binocular convergence of the cells located in the superficial (retinorecipient) laminae of the ipsilateral superior colliculi (Wickelgren & Sterling, 1969; Rosenquist & Palmer, 1971; Ogasawara, McHaffie & Stein, 1984).

There is, however, no indication that specific 'feature detectors' present in Area 17 (Hubel & Wiesel 1962, 1965) are crucially involved in global aspects of pattern vision (Sprague *et al.*, 1979, 1981). Thus, Area 17 of carnivores does not seem to constitute the base of 'form or pattern vision hierarchy' (cf, however, Area 17 of the primates – Mishkin *et al.*, 1983; Van Essen & Maunsell, 1983).

Area 18, which receives its principal thalamic input from Y-type geniculate cells, seems to be mainly concerned with processing information about motion and low-acuity pattern vision (Tretter *et al.*, 1975; Movshon, Thompson & Tolhurst, 1978; Harvey, 1980) as well as processing convergent binocular disparities (stereoscopic vision of the region between the fixation plane and the animal – Levick, 1977; Ferster, 1981; Pettigrew & Dreher, 1982). There is also substantial evidence indicating that the cat Area 18 is strongly involved in dynamic depth perception (Cynader & Regan, 1978, 1982).

Area 19, with its strong W-type thalamic input, as well as many cells exhibiting low cut-off velocity and divergent binocular retinal disparities, is likely to be concerned with stereoscopic vision beyond the fixation plane (Pettigrew & Dreher, 1982). There is also some evidence indicating that Area 19 participates in form and depth perception and discrimination (Sprague *et al.*, 1977).

Area 21a, which contains mainly cells with relatively large receptive fields and low cut-off velocity, and sends strong projections to the medial, lateral and dorsal terminal accessory optic nuclei (Marcotte & Updyke, 1982), is likely to play a significant role in modulating activities of the accessory optic system which, in turn, is implicated in the control of optokinetic nystagmus and visual-vestibular inter-actions (Grasse, Cynader & Douglas, 1984; for review see Simpson, 1984). However, the projection from Area 21a to the superficial collicular layers is very weak (Kawamura & Konno, 1979; Segal & Beckstead, 1984). Furthermore, Area 21a does not project to the deep collicular layers (Segal & Beckstead, 1984).

In the PMLS, as well as in Areas 20a, 20b, 21b, not only central (along the zero vertical meridian) but also peripheral representations of the visual field are callosally interconnected. Consistent with these wide-spread callosal interconnections, these areas have been implicated in the interhemispheric transfer of visual information in normal cats as well as interocular transfer in cats with split chiasm (Berlucchi, 1981; Berlucchi & Sprague, 1981; cf, however, Ptito & Lepore, 1983).

The AMLS and PMLS, which contain mainly direction-selective cells with large receptive fields and high cut-off velocity, are likely to be involved in the analysis of movement. Many cells in those areas 'respond selectively to the relative motion of stimuli in the center and surround of the receptive field and thus provide a possible mechanism for figure–ground motion abstraction' (Spear *et al.*, 1983; Von Grünau & Frost, 1983). The PMLS (together with the PLLS, the DLS and the VLS) seems to play a crucial role in determining direction and velocity selectivities as well as the binocularity and spatial organization of receptive fields of the collicular neurons located in the deep laminae of the ipsilateral colliculus (Ogasawara *et al.*, 1984). The deep collicular laminae are in turn directly implicated in visually guided behavior (Sprague, 1975). The PMLS has also been implicated in coordinating lens accommodation during fixation of visual objects following saccadic eye movements (Bando *et al.*, 1981).

The AMLS and PMLS are the only retinotopically organized cortical areas which contain a high proportion of cells exhibiting high cut-off velocity and project massively to all three terminal accessory nuclei (Marcotte & Updyke, 1982). Cells located in the lateral and dorsal terminal nuclei of the cat accessory optic system lose their responsiveness to fast-moving stimuli after ipsilateral cortical ablations involving both the AMLS and PMLS (Grasse *et al.*, 1984). Thus, it appears that the AMLS and PMLS play a highly significant role in determining the responses of cells in the accessory optic nuclei to fast-moving stimuli. Furthermore, the PMLS seems to relay, presumably via 'feedback' infragranular projection, 'efference copy signals' of eye movements from the oculomotor system to the striate and possibly other cortical areas (Toyama, Komatsu & Shibuki, 1984; see also Kennedy & Magnin, 1977; Vanni-Mercier & Magnin, 1982; Komatsu, Shibuki & Toyama, 1983).

There is rather a limited amount of information concerning the receptive-field properties and the functional role played by Areas 20a and 20b. Both areas have been implicated in regulating the pupillary constriction accompanying lens accommodation and convergence in the near reflex (Shoumura, chiiwa & Sukekawa, 1982; Kuchiiwa *et al.*, 1984). rthermore, Area 20 has been implicated in learning t not in retention of the pattern and form discrimina- n (Sprague *et al.*, 1977, 1981). Interestingly, Area 20 ms to be the only retinotopically organized cortical a which does not project to the ipsilateral superior liculus (Kawamura *et al.*, 1974; Segal & Beckstead, 34). It is therefore unlikely that Area 20 is nificantly involved in the visual–oculomotor egrations.

The EVA (Olson & Graybiel, 1983) or anterior osylvian visual area (AEV) (Mucke *et al.*, 1982) tains mainly cells with very large receptive fields ich are highly sensitive to small visual stimuli ving rapidly in a particular (preferred) direction. e receptive-field properties of cells in the EVA, ong reciprocal connections between the EVA and the LS (Fig. 20.4), strong projection from the EVA to deep layers of the ipsilateral superior colliculus gal & Beckstead, 1984) in which sensory information pparently converted into a signal appropriate for tor control (Berson & McIlwain, 1983), as well as rojection from the EVA to periaqueductal grey tter (Bandler, McCulloch & Dreher, 1985) and the ximity of the somatotopically organized fourth atosensory cortical area (Clemo & Stein, 1983), licate the EVA in some sort of visuomotor gration. Indeed, Mucke *et al.* (1982) suggested that is area may play a role in recording movements to visual environment relative to the body and in the ustment of motor behaviour to such movements'. wever, in addition to the properties of cells and nections which implicate the EVA in visuomotor gration, the EVA is interconnected with Area 20 g. 20.4), which in turn seems to be implicated in al pattern discrimination.

The multiplicity of functions in which the ority (if not all) of the retinotopically organized as in the cat visual cortex seem to be involved, makes construction of a rigid sequential (from the point iew of the visual information processing) hierarchy the cat visual cortex a not very worthwhile position (at least for the time being). On the other d, there is a clear indication that while some areas principally involved in the analysis of form (e.g. as 17, 19, 21a, 20a and 20b), the others (e.g. AMLS, LS, ALLS, AMLS and EVA) are principally olved in the analysis of motion and visuomotor

integration. Such specializations are somewhat similar to those observed in primates (see chapter 21; Baker *et al.*, 1981; Mishkin *et al.*, 1983, Van Essen & Maunsell, 1983).

Summary and conclusions

Each of the 14 or so retinotopically organized areas in the cat cortex receives a unique complement of afferents from the ipsilateral thalamus as well as from the ipsi- and contralateral cortices.

The areas which receive their principal thalamic inputs from the retinorecipient thalamic nuclei tend to have their associational and commissural neurons concentrated in the supragranular 'feedforward' cortical laminae. Furthermore, those areas receive very small (if any) input from the 'non-visual' thalamic nuclei and from the cortical areas which are not retinotopically organized.

The opposite pattern is apparent in the cortical areas which receive very small (if any) proportions of their thalamic afferents from the retinorecipient nuclei. These areas tend to receive a substantial proportion of their thalamic afferents from the 'non-visual' thalamic nuclei and a substantial proportion of their cortical afferents from the cortical areas which are not retinotopically organized (association, other sensory areas, limbic and motor). These 'higher' areas tend to send their associational and commissural projections via infragranular ('feedback') cortical laminae.

Retinotopically organized cortical areas which share a significant proportion of their thalamic input form 'family clusters' (Graybiel & Berson, 1981*a*, *b*). Only the members of the same 'family cluster' have strong associational interconnections, while the inter-connections between cortical areas belonging to different family clusters are much less numerous.

It is possible to build up a 'hierarchy' of cat cortical visual areas on the basis of the proportion of the thalamic afferents which each area receives from the retinorecipient nuclei and from the proportion of their associational and commissural neurons located in the supragranular ('feedforward') cortical laminae.

However, from the point of view of processing of visual information, it is probably most useful to think about the interrelationships between different retinotopically organized areas in terms of the regions of the neocortex specializing in the analysis of different aspects of visual environment and relating this analysis to the analysis taking place in other areas.

Such an approach does not preclude some flexible 'hierarchy' between different areas, with a given area being 'higher' in respect to another area from the point of view of processing further certain aspects of visual information, and at the same time 'lower' to that area from the point of view of processing other aspects of visual information. Indeed, a certain, albeit small, proportion of associational neurons projecting from the 'highest' area in the putative 'hierarchy' to the 'lowest' one is located in 'feedforward' supragranular laminae.

Arranging cortical areas into the rigid 'hierarchy' of elements in sequential information processing (Van Essen & Maunsell, 1983) does not seem warranted, in view of the available data concerning the functional roles played by different areas in the analysis of different aspects of the visual environment. Furthermore, it appears that the primary visual cortex (Area 17) is the only retinotopically organized cortical area which is homologous in all mammalian orders studied so far, with the possible exception of monotremes (Kaas, 1980). The homologies between any of the other visual areas in any of the mammalian orders studied are, at the moment at least, uncertain or even unlikely (Kaas, 1980; Baker *et al.*, 1981). Thus it appears very likely that Area 17 is the oldest visual area basic to all mammalia while the other areas developed largely independently in different evolutionary lines. The apparent 'hierarchy' of visual cortical areas present in both primates and carnivores is most likely related to the evolutionary interrelationships between different retinotopically organized areas, with 'newer' areas communicating to 'older' areas mainly via infragranular ('feedback') laminae, and the 'older' areas communicating to 'newer' areas mainly via 'feedforward' supragranular laminae.

Acknowledgements

First of all I would like to express my gratitude to Professor Peter O. Bishop for bringing me to Australia; for allowing me to use the first rate facilities in his laboratories in Canberra; for inspiring me with his complete devotion to science, his hard work, attention to the precise measurements and the appreciation of other people's efforts and talents. Lastly, but not least, I thank Professor Bishop for gathering around him an 'Australian visual mafia', since many of the 'mafiosi' became my close and cherished friends.

Thanks are due to my BSc(Med) students, H. T. (Richard) Ho, C. W. (Francis) Lee and David I. M. Robinson for allowing me to present some of the unpublished da which we collected together. Murray J. McCall help greatly in collecting the references and preparation illustrations, while Jennifer Allen-Narker 'word processe the manuscript. The research efforts of my laborato have been supported by grants from the Australian Natior Health and Medical Research Council.

References

Albus, K., Donate-Oliver, F., Sanides, D. & Fries, W. (1981) T distribution of pontine projection cells in visual and association cortex of the cat: an experimental study with horseradish peroxidase. *J. Comp. Neurol.*, **201**, 175–89

Antonini, A., Berlucchi, G. & Lepore, F. (1983) Physiological organization of callosal connections of a visual lateral suprasylvian cortical area in the cat. *J. Neurophysiol.*, **49**, 902–21

Baker, J. F., Petersen, S. E., Newsome, W. T. & Allman, J. A. (1981) Visual response properties of neurons in four extrastriate visual areas of the owl monkey (*Aotus trivirgatus*): a quantitative comparison of medial, dorsomedial, dorsolateral, and middle temporal areas. *J. Neurophysiol.*, **45**, 397–416

Bandler, R., McCulloch, T. & Dreher, B. (1985) Afferents to a midbrain periaqueductal grey region involved in the 'Defense Reaction' in the cat as revealed by horseradish peroxidase: I. The telencephalon. *Brain Res.* 330, 109–1

Bando, T., Tsukuda, K., Yamamot, N., Maeda, J. & Tsukahara, N. (1981) Cortical neurons in and around the Clare–Bishop area related with lens accommodation in th cat. *Brain Res.*, **225**; 195–9

Bauer, R. (1982) A high probability of an orientation shift between layers 4 and 5 in central parts of the cat striate cortex. *Brain Res.*, **48**, 245–55

Behan, M. (1984) An EM-autoradiographic analysis of the projection from cortical areas 17, 18, and 19 to the superior colliculus in the cat. *J. Comp. Neurol.*, **225**, 591–604

Benedek, G., Norita, M. & Creutzfeldt, O. D. (1983) Electrophysiological and anatomical demonstration of an overlapping striate and tectal projection to the lateral posterior-pulvinar complex of the cat. *Exp. Brain Res.*, 5 157–69

Berlucchi, G. (1981) Recent advances in the analysis of the neu substrates of interhemispheric communication. In *Brain Mechanisms and Perceptual Awareness*, ed. O. Pomeiano & C. Ajmone Marsan, pp. 133–52. New York: Raven Press

Berlucchi, G. & Sprague, J. M. (1981) The cerebral cortex in visual learning and memory, and in interhemispheric transfer in the cat. In *The Organisation of the Cerebral Cortex*, ed. F. O. Schmitt, F. G. Worden, G. Adelman & S. G. Dennis, pp. 415–40. Cambridge, Mass.: MIT Pre

Berson, D. M. & Graybiel, A. M. (1978) Parallel thalamic zones in the LP-pulvinar complex of the cat identified by their afferent and efferent connections. *Brain Res.*, **147**, 139–4

Berson, D. M. & Graybiel, A. M. (1983) Organization of the striate-recipient zone of the cat's lateralis posterior-pulvinar complex and its relations with the geniculostriate system. *Neuroscience*, 9, 337–72

Berson, D. M. & McIlwain, J. T. (1983) Visual cortical inputs to deep layers of cat's superior colliculus. *J. Neurophysiol.*, 50, 1143–55

Blakemore, C., Diao, Y.-C., Pu, M.-L., Wang, Y.-K. & Xiao, Y.-M. (1983) Possible functions of the interhemispheric connections between cortical areas in the cat. *J. Physiol.*, 337, 331–50

Bullier, J. & Henry, G. H. (1979a) Ordinal position of neurons in cat striate cortex. *J. Neurophysiol.*, 42, 1251–63

Bullier, J. & Henry, G. H. (1979b) Neural path taken by afferent streams in striate cortex of the cat. *J. Neurophysiol.*, 42, 1264–70

Bullier, J. & Henry, G. H. (1979c) Laminar distribution of first-order neurons and afferent terminals in cat striate cortex. *J. Neurophysiol.*, 42, 1271–81

Bullier, J., Kennedy, H. & Salinger, W. (1984a) Bifurcation of subcortical afferents to visual areas 17, 18, and 19 in the cat cortex. *J. Comp. Neurol.*, 228, 309–28

Bullier, J., Kennedy, H. & Salinger, W. (1984b) Branching and laminar origin of projections between visual cortical areas in the cat. *J. Comp. Neurol.*, 228, 329–41

Bullier, J., Mustari, M. J. & Henry, G. H. (1982) Receptive-field transformations between LGN neurons and S-cells of cat striate cortex. *J. Neurophysiol.*, 47, 417–37

Camarda, R. M. (1979) Hypercomplex cell types in area 18 of the cat. *Exp. Brain Res.*, 36, 191–4

Camarda, R. M. & Rizzolatti, G. (1976) Visual receptive fields in the lateral suprasylvian area (Clare–Bishop area) of the cat. *Brain Res.*, 101, 427–43

Cavada, C. & Reinoso-Suarez, F. (1983) Afferent connections of area 20 in the cat studied by means of the retrograde axonal transport of horseradish peroxidase. *Brain Res.*, 270, 319–24

Chalupa, L. M., Williams, R. W. & Hughes, M. J. (1983) Visual response properties in the tectorecipient zone of the cat's lateral posterior-pulvinar complex: a comparison with the superior colliculus. *J. Neurosci.*, 3, 2587–96

Cleland, B. G., Lee, B. B. & Vidyasagar, T. R. (1983) Response of neurons in the cat's lateral geniculate nucleus to moving bars of different length. *J. Neurosci.*, 3, 108–16

Clemo, H. R. & Stein, B. E. (1983) Organization of a fourth somatosensory area of cortex in cat. *J. Neurophysiol.*, 50, 910–25

Cynader, M. & Regan, D. (1978) Neurones in cat parastriate cortex sensitive to the direction of motion in three-dimensional space. *J. Physiol.*, 274, 549–69

Cynader, M. & Regan, D. (1982) Neurons in cat visual cortex tuned to the direction of motion in depth: effect of positional disparity. *Vision Res.*, 22, 967–82

Daniels, J. D., Norman, J. L. & Pettigrew, J. D. (1977) Biases for orientated moving bars in lateral geniculate nucleus neurones of normal and stripe-reared cats. *Exp. Brain Res.*, 29, 155–72

Dreher, B. (1972) Hypercomplex cells in the cat's striate cortex. *Invest. Ophthalmol. Vis. Sci.*, 11, 355–6

Dreher, B. & Cottee, L. (1975) Visual receptive-field properties of

cells in area 18 of cat's cerebral cortex before and after acute lesions in area 17. *J. Neurophysiol.*, 38, 735–50

Dreher, B., Leventhal, A. G. & Hale, P. T. (1980) Geniculate input to cat visual cortex: a comparison of area 19 with areas 17 and 18. *J. Neurophysiol.*, 44, 804–26

Dreher, B. & Winterkorn, J. M. S. (1974) Receptive field properties of neurones in cat's cortical area 17 before and after acute lesions in area 18. *Proc. Aust. Physiol. Pharmacol. Soc.*, 5, 63P

Duysens, J., Orban, G. A. & Cremieux, J. (1984) Functional basis for the preference for slow movement in area 17 of the cat. *Vision Res.*, 24, 17–24

Duysens, J., Orban, G. A. van der Glas, H. W. & de Zeagher, F. E. (1982a) Functional properties of area 19 as compared to area 17 of the cat. *Brain Res.*, 231, 279–91

Duysens, J., Orban, G. A., van der Glas, H. W. & Maes, H. (1982b) Receptive field structure of area 19 as compared to area 17 of the cat. *Brain Res.*, 231, 293–308

Ebner, F. B. & Myers, R. E. (1965) Distribution of corpus callosum and anterior commissure in cat and raccoon. *J. Comp. Neurol.*, 124, 353–65

Ferster, D. (1981) A comparison of binocular depth mechanisms in areas 17 and 18 of the cat visual cortex. *J. Physiol.*, 311, 623–55

Ferster, D. & LeVay, S. (1978) The axonal arborizations of lateral geniculate neurons in the striate cortex of the cat. *J. Comp. Neurol.*, 182, 923–44

Ferster, D. & Lindström, S. (1983) An intracellular analysis of geniculo-cortical connectivity in area 17 of the cat. *J. Physiol.*, 342, 181–215

Galletti, C., Squatrito, S. & Battaglini, P. P. (1979) Visual cortex projections to thalamic intralaminar nuclei in the cat. An autoradiographic study. *Arch. Ital. Biol.*, 117, 280–5

Garey, L. J. & Powell, T. P. S. (1967) The projection of the lateral geniculate nucleus upon the cortex in the cat. *Proc. R. Soc. B.*, 169, 107–26

Geisert, E. E. Jr. (1980) Cortical projections of the lateral geniculate nucleus in the cat. *J. Comp. Neurol.*, 190, 793–812

Gilbert, C. D. (1977) Laminar differences in receptive field properties of cells in cat primary visual cortex. *J. Physiol.*, 268, 391–421

Gilbert, C. D. (1983) Microcircuitry of the visual cortex. *Ann. Rev. Neurosci.*, 6, 217–47

Gilbert, C. D. & Kelly, J. P. (1975) The projections of cells in different layers of the cat's visual cortex. *J. Comp. Neurol.*, 163, 81–106

Gilbert, C. D. & Wiesel, T. N. (1981) Laminar specialization and intracortical connections in cat primary visual cortex. In *The organization of the Cerebral Cortex*, ed. F. O. Schmitt, F. G. Worden, G. Adelman & S. G. Dennis, pp. 163–91. Cambridge, Mass.: MIT Press

Gilbert, C. D. & Wiesel, T. N. (1983) Clustered intrinsic connections in cat visual cortex. *J. Neurosci.*, 3, 1116–33

Glickstein, M. & Whitteridge, D. (1976) Degeneration of layer III pyramidal cells in area 18 following destruction of callosal input. *Brain Res.*, 104, 148–51

Goodwin, A. W. & Henry, G. H. (1978) The influence of stimulus velocity on the responses of single neurones in the striate cortex. *J. Physiol.*, **277**, 467–82

Grasse, K. L., Cynader, M. S. & Douglas, R. M. (1984) Alterations in response properties in the lateral and dorsal terminal nuclei of the cat accessory optic system following visual cortex lesions. *Exp. Brain Res.*, **55**, 69–80

Graybiel, A. M. & Berson, D. M. (1980) Histochemical identification and afferent connections of subdivisions in the lateralis posterior-pulvinar complex and related thalamic nuclei in the cat. *Neuroscience*, **5**, 1175–238

Graybiel, A. M. & Berson, D. M. (1981*a*) Families of related cortical areas in the extrastriate visual system. In *Cortical Sensory Organization. Vol. 2. Multiple Visual Areas*, ed. C. N. Woolsey, pp. 103–120. Clifton, NJ: Humana Press

Graybiel, A. M. & Berson, D. M. (1981*b*) On the relation between transthalamic and transcortical pathways in the visual system. In *The organization of the Cerebral Cortex*, ed. F. O. Schmitt, F. G. Worden, G. Adelman & S. G. Dennis, pp. 285–319. Cambridge, Mass.: MIT Press

Guedes, R., Watanabe, S. & Creutzfeldt, O. D. (1983) Functional role of association fibres for a visual association area: the posterior suprasylvian sulcus of the cat. *Exp. Brain Res.*, **49**, 13–27

Guillery, R. W., Geisert, E. E. Jr, Polley, E. H. & Mason, C. A. (1980) An analysis of the retinal afferents to the cat's medial interlaminar nucleus and to its rostral thalamic extension, the 'geniculate wing'. *J. Comp. Neurol.*, **194**, 117–42

Hale, P. T., Dreher, B. & Leventhal, A. G. (1978) A comparison of the receptive field properties of cells in areas 17, 18 and 19 of the cat. *Proc. Aust. Physiol. Pharmacol. Soc.*, **9**, 193P

Harvey, A. R. (1980) The afferent connections and laminar distribution of cells in area 18 of the cat. *J. Physiol.*, **302**, 483–505

Henry, G. H. (1977) Receptive field classes of cells in the striate cortex of the cat. *Brain Res.*, **133**, 1–28

Henry, G. H., Harvey, A. R. & Lund, J. S. (1979) The afferent connections and laminar distribution of cells in the cat striate cortex. *J. Comp. Neurol.*, **187**, 725–44

Henry, G. H., Lund, J. S. & Harvey, A. R. (1978) Cells of the striate cortex projecting to the Clare–Bishop area of the cat. *Brain Res.*, **151**, 154–8

Henry, G. H., Mustari, M. J. & Bullier, J. (1983) Different geniculate inputs to B and C cells of cat striate cortex. *Exp. Brain Res.*, **52**, 179–89

Ho, H. T., Lee, C. W. F. & Dreher, B. (1982) Receptive field properties of neurones in area 21a of cat visual cortex. *Proc. Aust. Physiol. Pharmacol. Soc.*, **13**, 196P

Hoffmann, K.-P. & Stone, J. (1971) Conduction velocity of afferents to cat visual cortex: a correlation with cortical receptive field properties. *Brain Res.*, **32**, 460–6

Holländer, H. (1974) On the origin of the corticotectal projections in the cat. *Exp. Brain Res.*, **21**, 433–9

Holländer, H. & Vanegas, H. (1977) The projection from the lateral geniculate nucleus onto the visual cortex in the cat. A quantitative study with horseradish peroxidase. *J. Comp. Neurol.*, **173**, 519–36

Hubel, D. H. & Wiesel, T. N. (1962) Receptive fields, binocular interaction, and functional architecture in the cat's visual cortex. *J. Physiol.*, **160**, 106–54

Hubel, D. H. & Wiesel, T. N. (1965) Receptive fields and functional architecture in two nonstriate visual areas (18 and 19) of the cat. *J. Neurophysiol.*, **28**, 229–89

Hubel, D. H. & Wiesel, T. N. (1969) Visual area of the lateral suprasylvian gyrus (Clare–Bishop area) of the cat. *J. Physiol.*, **202**, 251–60

Hughes, H. C. (1980) Efferent organization of the cat pulvinar complex, with a note on bilateral claustrocortical and reticulocortical connections. *J. Comp. Neurol.*, **193**, 937–63

Innocenti, G. M. (1980) The primary visual pathway through the corpus callosum: morphological and functional aspects in the cat. *Arch. Ital. Biol.*, **118**, 124–88

Ito, M., Sanides, D. & Creutzfeldt, O. D. (1977) A study of binocular convergence in cat visual cortex neurons. *Exp. Brain Res.*, **28**, 21–35

Jones, E. G. (1981) Anatomy of cerebral cortex: columnar input-output organization. In *The Organization of the Cerebral Cortex*, ed. F. O. Schmitt, F. G. Worden, G. Adelman & S. G. Dennis, pp. 199–235. Cambridge, Mass. MIT Press

Kaas, J. H. (1980) A comparative survey of visual cortex organization in mammals. In *Comparative Neurology of the Telencephalon*, ed. S. O. E. Ebbesson, pp. 483–502. New York: Plenum.

Kato, H., Bishop, P. O. & Orban, G. A. (1978) Hypercomplex and simple/complex classifications in cat striate cortex. *J. Neurophysiol.*, **41**, 1071–95

Kato, H., Bishop, P. O. & Orban, G. A. (1981) Binocular interaction on monocularly discharged lateral geniculate and striate neurons in the cat. *J. Neurophysiol.*, **46**, 932–5

Kaufman, E. F. S., Rosenquist, A. C. & Raczkowski, D. (1984) The projections of single thalamic neurons onto multiple visual areas in the cat. *Brain Res.*, **298**, 171–4

Kawamura, K. & Konno, T. (1979) Various types of corticotectal neurons of cats as demonstrated by means of retrograde axonal transport of horseradish peroxidase. *Exp. Brain Res.* **35**, 161–75

Kawamura, K. & Naito, J. (1980) Corticocortical neurons projecting to the medial and lateral banks of the middle suprasylvian sulcus in the cat: an experimental study with the horseradish peroxidase method. *J. Comp. Neurol.*, **193**, 1009–22

Kawamura, S., Sprague, J. M. & Niimi, K. (1974) Corticofugal projections from the visual cortices to the thalamus, pretectum and superior colliculus in the cat. *J. Comp. Neurol.*, **158**, 339–62

Kaye, M., Mitchell, D. E. & Cynader, M. (1981) Selective loss of binocular depth perception after ablation of cat visual cortex. *Nature*, **293**, 60–2

Keller, G. & Innocenti, G. M. (1981) Callosal connections of the suprasylvian visual areas of the cat. *Neurosci.*, **6**, 703–12

Kennedy, H. & Baleydier, C. (1977) Direct projections from thalamic intralaminar nuclei to extrastriate visual cortex in the cat traced with horseradish peroxidase. *Exp. Brain Res.*, **28**, 133–9

Kennedy, H. & Magnin, M. (1977) Saccadic influences on single neuron activity in the medial bank of the cat's suprasylvian sulcus (Clare–Bishop area). *Exp. Brain Res.*, **27**, 315–17

mura, M., Shida, T., Tanaka, K. & Toyama, K. (1980) Three classes of area 19 cortical cells of the cat classified by their neuronal connectivity and photic responsiveness. *Vision Res.*, **20**, 69–77

matsu, Y., Shibuki, K. & Toyama, K. (1983) Eye movement-related activities in cells of the lateral suprasylvian cortex of the cat. *Neurosci. Lett.*, **41**, 271–6

chiiwa, S., Shoumura, K., Kuchiiwa, T. & Imai, H. (1984) Afferents to the cortical pupillo–constrictor areas of the cat, traced with HRP. *Exp. Brain Res.*, **54**, 377–81

e, C., Malpeli, J. G., Schwark, H. D., and Weyand, T. G. (1984) Cat medial interlaminar nucleus: retinotopy, relation to tapetum and implications for scotopic vision. *J. Neurophysiol.*, **52**, 848–69

e, C. W. F., Ho, H. T. & Dreher, B. (1982) Area 21a and posteromedial lateral suprasylvian area (PMLS) of the cat visual cortex: one or two areas? A horseradish peroxidase study. *Proc. Aust. Physiol. Pharmacol. Soc.*, **13**, 195P

nnie, P. (1980) Parallel visual pathways: a review. *Vision Res.*, **20**, 561–94

Vay, S. & Gilbert, C. D. (1976) Laminar patterns of geniculocortical projection in the cat. *Brain Res.*, **113**, 1–19

Vay, S., Stryker, M. P. & Shatz, C. J. (1978) Ocular dominance columns and their development in layer IV of the cat's visual cortex: a quantitative study. *J. Comp. Neurol.*, **179**, 223–44

venthal, A. G. (1979) Evidence that the different classes of relay cells of the cat's lateral geniculate nucleus terminate in different layers of the striate cortex. *Exp. Brain Res.*, **37**, 349–72

venthal, A. G. (1983) Systematic relationship between preferred orientation and receptive field position of neurons in cat striate cortex. *J. Comp. Neurol.*, **220**, 476–83

venthal, A. G. & Hirsch, H. V. B. (1978) Receptive field properties of neurons in different laminae of visual cortex of the cat. *J. Neurophysiol.*, **41**, 948–62

venthal, A. G. & Hirsch, H. V. B. (1983) Effects of visual deprivation upon the geniculocortical W–cell pathway in the cat: area 19 and its afferent input. *J. Comp. Neurol.*, **214**, 59–71

venthal, A. G., Keens, J. & Törk, I. (1980) The afferent ganglion cells and cortical projections of the retinal recipient zone (RRZ) of the cat's 'pulvinar complex'. *J. Comp. Neurol.*, **194**, 535–54

venthal, A. G., Schall, J. D. & Wallace, W. (1984) Relationship between preferred orientation and receptive field position of neurons in extrastriate cortex (area 19) in the cat. *J. Comp. Neurol.*, **222**, 445–51

vick, W. R. (1977) Participation of brisk-transient retinal ganglion cells in binocular vision – an hypothesis. *Proc. Physiol. Pharmacol. Soc.*, **8**, 9–16

nd, J. S., Henry, G. H., MacQueen, C. L. & Harvey, A. R. (1979) Anatomical organization of the primary visual cortex (area 17) of the cat. A comparison with area 17 of the macaque monkey. *J. Comp. Neurol.*, **184**, 599–618

ciewicz, R. J. (1974) Afferents to the lateral suprasylvian gyrus

of the cat traced with horseradish peroxidase. *Brain Res.*, **78**, 139–43

Maciewicz, R. J. (1975) Thalamic afferents to areas 17, 18 and 19 of cat cortex traced with horseradish peroxidase. *Brain Res.*, **84**, 308–12

McIlwain, J. T. (1977) Topographic organization and convergence in corticotectal projections from areas 17, 18, and 19 in the cat. *J. Neurophysiol.*, **40**, 189–98

Magalhaes-Castro, H. H., Saraiva, P. E. S. & Magalhaes-Castro, D. (1975) Identification of corticotectal cells of the visual cortex of cats by means of horseradish peroxidase. *Brain Res.*, **83**, 474–9

Malpeli, J. F. (1983) Activity of cells in area 17 of the cat in absence of input from layer A of lateral geniculate nucleus. *J. Neurophysiol.*, **49**, 595–610

Marcotte, R. R. & Updyke, B. V. (1982) Cortical visual areas of the cat project differentially onto the nuclei of the accessory optic system. *Brain Res.*, **242**, 205–17

Martin, K. A. C. & Whitteridge, D. (1984) Form, function and intracortical projections of spiny neurones in the striate visual cortex of the cat. *J. Physiol.*, **353**, 463–504

Mason, R. (1978) Functional organization in the cat's pulvinar complex. *Exp. Brain Res.*, **31**, 51–66

Mason, R. (1981) Differential responsiveness of cells in the visual zones of the cat's LP–pulvinar complex to visual stimuli. *Exp. Brain Res.*, **43**, 25–33

Meyer, G. & Albus, K. (1981) Spiny stellates as cells of origin of association fibres from area 17 to area 18 in the cat's neocortex. *Brain Res.*, **210**, 335–41

Miller, J. W. & Benevento, L. A. (1979) Demonstration of a direct projection from the intralaminar central lateral nucleus to the primary visual cortex. *Neurosci. Lett.*, **14**, 229–34

Miller, J. W., Buschmann, M. B. T. & Benevento, L. A. (1980) Extrageniculate thalamic projections to the primary visual cortex. *Brain Res.*, **189**, 221–7

Mishkin, M., Ungerleider, L. G. & Macko, K. A. (1983) Object vision and spatial vision: two cortical pathways. *Trends Neurosci.*, **6**, 414–17

Montero, V. M. (1981) Topography of the cortico–cortical connections from the striate cortex in the cat. *Brain Behav. Evol.*, **18**, 194–218

Movshon, J. A., Thompson, I. D. & Tolhurst, D. J. (1978) Spatial and temporal contrast sensitivity of neurones in areas 17 and 18 of the cat's visual cortex. *J. Physiol.*, **283**, 101–20

Mucke, L., Norita, M., Benedek, G. & Creutzfeldt, O. D. (1982) Physiologic and anatomic investigation of a visual cortical area situated in the ventral bank of the anterior ectosylvian sulcus of the cat. *Exp. Brain Res.*, **46**, 1–11

Mullikin, W. H., Jones, J. P. & Palmer, L. A. (1984) Receptive-field properties and laminar distribution of X-like and Y-like simple cells in cat area 17. *J. Neurophysiol.*, **52**, 350–71

Mustari, M. J., Bullier, J. & Henry, G. H. (1982) Comparison of the response properties of three types of monosynaptic S-cells in cat striate cortex. *J. Neurophysiol.*, **47**, 439–54

Naito, J. & Kawamura, K. (1982) Thalamocortical neurons projecting to the areas surrounding the anterior and middle suprasylvian sulci in the cat. *Exp. Brain Res.*, **45**, 59–70

Niimi, K. H., Matsuoka, H., Yamazaki, Y. & Matsumoto, H. (1981) Thalamic afferents to the visual cortex in the cat studied by retrograde axonal transport of horseradish peroxidase. *Brain Behav. Evol.*, **18**, 114–39

Ogasawara, K., McHaffie, J. G. & Stein, B. E. (1984) Two visual corticotectal systems in cat. *J. Neurophysiol.*, **52**, 1226–45

Olson, C. R. & Graybiel, A. M. (1983) An outlying visual area in the cerebral cortex of the cat. *Progr. Brain Res.*, **58**, 239–45

Orban, G. A. (1984) *Neuronal Operations in the Visual Cortex.* Berlin: Springer

Otsuka, R. & Hassler, R. (1962) Uber Aufbau und Gliederung der corticalen Sehsphare bei der Katze. *Arch. Psychiat. Nervenkrh.*, **203**, 212–34

Payne, B. R., Elberger, A. J., Berman, N. & Murphy, E. H. (1980) Binocularity in the cat visual cortex is reduced by sectioning the corpus callosum. *Science*, **207**, 1097–9

Payne, B. R., Pearson, H. E. & Berman, N. (1984) Role of corpus callosum in functional organization of cat striate cortex. *J. Neurophysiol.*, **52**, 570–94

Pettigrew, J. D. & Dreher, B. (1982) Parallel processing of binocular disparity in the cat's geniculo–cortical pathway. *Neurosci. Abs.* **8**, 810P

Ptito, M. & Lepore, F. (1983) Effects of unilateral and bilateral lesions of the lateral suprasylvian area on learning and interhemispheric transfer of pattern discrimination in the cat. *Behav. Brain Res.*, **7**, 211–27

Raczkowski, D. & Rosenquist, A. C. (1980) Connections of the parvocellular C laminae of the dorsal lateral geniculate nucleus with the visual cortex in the cat. *Brain Res.*, **199**, 447–51

Raczkowski, D. & Rosenquist, A. C. (1981) Retinotopic organization in the lateral posterior-pulvinar complex. *Brain Res.*, **221**, 185–91

Raczkowski, D. & Rosenquist, A. C. (1983) Connections of the multiple visual cortical areas with the lateral posterior-pulvinar complex and adjacent thalamic nuclei in the cat. *J. Neurosci.*, **3**, 1912–42

Rapaport, D. H., Dreher, B. & Rowe, M. H. (1982) Lack of binocularity in cells of area 19 of cat visual cortex following monocular deprivation. *Brain Res.*, **246**, 319–24

Rockland, K. S. & Pandya, D. N. (1979) Laminar origins and terminations of cortical connections of the occipital lobe in the rhesus monkey. *Brain Res.*, **179**, 3–20

Rockland, K. S. & Pandya, D. N. (1981) Cortical connections of the occipital lobe in the rhesus monkey: interconnections between areas 17, 18, 19 and the superior temporal sulcus. *Brain Res.*, **212**, 249–70

Rodieck, R. W. & Brening, R. K. (1983) Retinal ganglion cells: properties, types, genera, pathways and trans-species comparisons. *Brain Behav. Evol.*, **23**, 121–64

Rodieck, R. W. & Dreher, B. (1979) Visual supression from nondominant eye in the lateral geniculate nucleus: a comparison of cat and monkey. *Exp. Brain Res.*, **35**, 465–77

Rose, D. (1979) Mechanisms underlying the receptive field

properties of neurons in cat visual cortex. *Vision Res.*, **19**, 533–44

Rosenquist, A. C., Edwards, S. B. & Palmer, L. A. (1974) An autoradiographic study of the projections of the dorsal lateral geniculate nucleus and the posterior nucleus in the cat. *Brain Res.*, **80**, 71–93

Rosenquist, A. C. & Palmer, L. A. (1971) Visual receptive field properties of cells of the superior colliculus after cortical lesions in the cat. *Exp. Neurol.*, **33**, 629–52

Rowe, M. H. & Dreher, B. (1982) Retinal W-cell projections to the medial interlaminar nucleus in the cat: implications for ganglion cell classification. *J. Comp. Neurol.*, **204**, 117–33

Sanderson, K. J. (1971*a*) The projection of the visual field to the lateral geniculate and medial intralaminar nuclei in the cat. *J. Comp. Neurol.*, **143**, 101–18

Sanderson, K. J. (1971*b*) Visual field projection columns and magnification factors in the lateral geniculate nucleus. *Exp. Brain Res.*, **13**, 159–77

Sanderson, K. J., Bishop, P. O. & Darian-Smith, I. (1971) The properties of the binocular receptive fields of lateral geniculate neurons. *Exp. Brain Res.*, **13**, 178–207

Sanderson, K. J., Darian-Smith, I. & Bishop, P. O. (1969) Binocular corresponding receptive fields of simple units in the cat dorsal lateral geniculate nucleus. *Vision Res.*, **9**, 1297–1303

Sanides, D. (1978) The retinotopic distribution of visual callosal projections in the suprasylvian visual areas compared to the classical visual areas (17, 18, 19) in the cat. *Exp. Brain Res.*, **33**, 435–43

Schmielau, F. & Singer, W. (1977) The role of visual cortex for binocular interactions in the cat lateral geniculate nucleus. *Brain Res.*, **120**, 354–61

Segal, R. L. & Beckstead, R. M. (1984) The lateral suprasylvian corticotectal projection in cats. *J. Comp. Neurol.*, **225**, 259–75

Segraves, M. A. & Rosenquist, A. C. (1982*a*) The distribution of the cells of origin of callosal projections in cat visual cortex. *J. Neurosci.*, **2**, 1079–89

Segraves, M. A. & Rosenquist, A. C. (1982*b*) The afferent and efferent callosal connections of retinotopically defined areas in cat cortex. *J. Neurosci.*, **2**, 1090–107

Shatz, C. J. (1977) Anatomy of interhemispheric connections in the visual system of Boston Siamese and ordinary cats. *J. Comp. Neurol.*, **173**, 497–518

Shatz, C. J., Lindström, S. & Wiesel, T. N. (1977) The distribution of afferents representing the right and left eye in the cat's visual cortex. *Brain Res.*, **131**, 103–16

Shatz, C. J. & Stryker, M. P. (1978) Ocular dominance in layer IV of the cat's visual cortex and the effects of monocular deprivation. *J. Physiol.*, **281**, 267–83

Sherk, H. (1978) Area 18 cell responses in the cat during reversible inactivation of area 17. *J. Neurophysiol.*, **41**, 204–15

Sherman, S. M. & Spear, P. D. (1982) Organization of visual pathways in normal and visually deprived cats. *Physiol. Rev.*, **62**, 740–855

Shoumura, K., Kuchiiwa, S. & Sukekawa, K. (1982) Two pupillo-constrictor areas in the occipital cortex of the cat. *Brain Res.*, **247**, 134–7

Sillito, A. M. (1975) The contribution of inhibitory mechanisms to the receptive field properties of neurones in the striate cortex of the cat. *J. Physiol.*, 250, 305–29

Sillito, A. M. (1977) Inhibitory processes underlying the directional specificity of simple, complex and hypercomplex cells in the cat's visual cortex. *J. Physiol.*, 271, 699–720

Sillito, A. M., Kemp, J. A., Milson, J. A. & Beradi, N. (1980a) A reevaluation of the mechanisms underlying simple cells orientation selectivity. *Brain Res.*, 194, 517–20

Sillito, A. M., Kemp, J. A. & Patel, H. (1980b) Inhibitory interactions contributing to the ocular dominance of monocularly dominated cells in the normal cat striate cortex. *Exp. Brain Res.*, 41, 1–10

Sillito, A. M. & Versiani, V. (1977) The contribution of excitatory and inhibitory input to the length preference of hypercomplex cells in layer II and III of cat's striate cortex. *J. Physiol.*, 273, 775–90

Simpson, J. I. (1984) The accessory optic system. *Ann. Rev. Neurosci.*, 7, 13–41

Singer, W. (1970) Inhibitory binocular interaction in the lateral geniculate body of the cat. *Brain Res.*, 18, 165–70

Singer, W., Tretter, F. & Cynader, M. (1975) Organization of cat striate cortex: a correlation of receptive field properties with afferent and efferent connections. *J. Neurophysiol.*, 38, 1080–98

Spear, P. D. & Baumann, T. P. (1975) Receptive-field characteristics of single neurons in lateral suprasylvian visual area of the cat. *J. Neurophysiol.*, 38, 1403–20

Spear, P. D. & Baumann, T. P. (1979) Effects of visual cortex removal on receptive-field properties of neurons in lateral suprasylvian visual area of the cat. *J. Neurophysiol.*, 42, 31–56

Spear, P. D., Miller, S. & Ohman, L. (1983) Effects of lateral suprasylvian visual cortex lesions on visual localization, discrimination, and attention in cats. *Behav. Brain Res.*, 10, 339–59

Sprague, J. M. (1975) Mammalian tectum: intrinsic organization, afferent inputs and integrative mechanisms. *Neurosci. Res. Prog. Bull.*, 13, 204–13

Sprague, J. M., Berkley, M. A. & Hughes, H. C. (1979) Visual acuity functions and pattern discrimination in the destriate cat. *Acta Neurobiol. Exp.*, 39, 643–82

Sprague, J. M., Hughes, H. C. & Berlucchi, G. (1981) Cortical mechanisms in pattern and form perception. In *Brain Mechanisms and Perceptual Awareness*, ed. O. Pompeiano & C. Ajmone Marson, pp. 107–32. New York: Raven Press

Sprague, J. M., Levy, J., DiBerardino, A. & Berlucchi, G. (1977) Visual cortical areas mediating form discrimination in the cat. *J. Comp. Neurol.* 172, 441–88

Squatrito, S., Galletti, C., Battaglini, P. P. & Riva Sanseverino, E. (1981a) Bilateral cortical projections from cat visual areas 17 and 18. An autoradiographic study. *Arch. Ital. Biol.*, 119, 1–20

Squatrito S., Galletti, C., Battaglini, P. P. & Riva Sanseverino, E. (1981b) An autoradiographic study of bilateral cortical projections from area 19 and lateral suprasylvian visual area. *Arch. Ital. Biol.*, 119, 21–42

Stone, J. (1983) *Parallel Processing in the Visual System. The Classification of Retinal Ganglion cells and its Impact on the Neurobiology of Vision.* New York: Plenum Press

Stone, J. & Dreher, B. (1973) Projection of X- and Y-cells of the cat's lateral geniculate nucleus to areas 17 and 18 of visual cortex. *J. Neurophysiol.*, 36, 551–67

Stone, J. & Dreher, B. (1982) Parallel processing of information in the visual pathways. A general principle of sensory coding? *Trends Neurosci.*, 5, 441–6

Stone, J., Dreher, B. & Leventhal, A. G. (1979) Hierarchical and parallel mechanisms in the organization of visual cortex. *Brain Res. Rev.*, 1, 345–94

Sugiyama, M. (1979) The projection of the visual cortex on the Clare–Bishop area in the cat. A degeneration study with the electron microscope. *Exp. Brain Res.*, 36, 433–43

Swadlow, H. A. (1983) Efferent systems of primary visual cortex: a review of structure and function. *Brain Res. Rev.*, 6, 1–24

Symonds, L. L., Rosenquist, A. C., Edwards, S. B. & Palmer, L. A. (1981) Projections of the pulvinar-lateral posterior complex to visual cortical areas in the cat. *Neuroscience*, 6, 1995–2020

Tieman, S. B. & Tumosa, N. (1983) [^{14}C]2-Deoxyglucose demonstration of the organization of ocular dominance in areas 17 and 18 of the normal cat. *Brain Res.*, 267, 35–46

Tigges, J., Tigges, M., Anschel, S., Cross, N. A., Letbetter, W. D. & McBride, R. L. (1981) Areal and laminar distribution of neurons interconnecting the central visual cortical areas 17, 18, 19, and MT in squirrel monkey (*Saimiri*). *J. Comp. Neurol.*, 202, 539–60

Tong, L., Kalil, R. E. & Spear, P. D. (1982) Thalamic projections to visual areas of the middle suprasylvian sulcus in the cat. *J. Comp. Neurol.*, 212, 103–17

Toyama, K., Komatsu, Y. & Shibuki, K. (1984) Integration of retinal and motor signals of eye movements in the striate cortex cells of the alert cat. *J. Neurophysiol.*, 51, 649–65

Toyama, K., Maekawa, K. & Takeda, T. (1977) Convergence of retinal inputs onto visual cells: I. A study of the cells monosynaptically excited from the lateral geniculate body. *Brain Res.*, 137, 207–20

Tretter, F., Cynader, M. & Singer, W. (1975) Cat parastriate cortex: a primary or secondary visual area? *J. Neurophysiol.*, 38, 1099–113

Tsumoto, T., Eckart, W. & Creutzfeldt, O. D. (1979) Modification of orientation sensitivity of cat visual cortex neurons by removal of GABA-mediated inhibition. *Exp. Brain Res.*, 34, 351–63

Turlejski, K. (1975) Visual responses of neurons in the Clare–Bishop area of the cat. *Acta Neurobiol. Exp.*, 35, 189–208

Tusa, R. J., Palmer, L. A. & Rosenquist, A. C. (1981) Multiple cortical visual areas. In *Cortical Sensory Organization.* Vol. 2. *Multiple Visual Areas*, ed. C. N. Woolsey, pp. 1–31. Clifton, NJ: Humana Press

Updyke, B. V. (1975) The patterns of projection of cortical areas 17, 18, and 19 onto the laminae of the dorsal lateral geniculate nucleus in the cat. *J. Comp. Neurol.*, 163, 377–96

Updyke, B. V. (1977) Topographic organization of the projections from cortical areas 17, 18, and 19 onto the thalamus, pretectum and superior colliculus in the cat. *J. Comp. Neurol.*, **173**, 81–111

Updyke, B. V. (1981*a*) Projections from visual areas of the middle suprasylvian sulcus onto the lateral posterior complex and adjacent thalamic nuclei in cat. *J. Comp. Neurol.*, **201**, 477–506

Updyke, B. V. (1981*b*) Multiple representations of the visual field. In *Cortical Sensory Organization.* Vol. 2. *Multiple Visual Areas*, ed. C. N. Woolsey, pp. 83–101. Clifton, NJ: Humana Press

Updyke, B. V. (1983) A reevaluation of the functional organization and cytoarchitecture of the feline lateral posterior complex, with observations on adjoining cell groups. *J. Comp. Neurol.*, **219**, 143–81

Van Essen, D. C. (1979) Visual areas of the mammalian cerebral cortex. *Ann. Rev. Neurosci.*, **2**, 227–63

Van Essen, D. C. & Maunsell, J. H. R. (1983) Hierarchical organization and functional streams in the visual cortex. *Trends Neurosci.*, **6**, 370–5

Vanni-Mercier, G. & Magnin, M. (1982) Retinotopic organization of extra-retinal saccade-related input to the visual cortex in the cat. *Exp. Brain Res.*, **46**, 368–76

Vidyasagar, T. R. & Heide, W. (1984) Geniculate orientation biases seen with moving sine wave gratings: implications for a model of simple cell afferent connectivity. *Exp. Brain Res.*, **57**, 196–200

Vidyasagar, T. R. & Urbas, J. V. (1982) Orientation sensitivity of cat LGN neurones with and without inputs from visual cortical areas 17 and 18. *Exp. Brain Res.*, **46**, 157–69

Von Grünau, M. & Frost, B. J. (1983) Double-opponent process mechanism underlying RF-structure of directionally specific cells of cat lateral suprasylvian visual area. *Exp. Brain Res.*, **49**, 84–92

Wicklegren, B. G. & Sterling, P. (1969) Influence of visual cortex on receptive fields in the superior colliculus of the cat. *J. Neurophysiol.*, **32**, 16–23

Wright, M. J. (1969) Visual receptive fields of cells in cortical area remote from the striate cortex in the cat. *Nature*, **223**, 973–5

Zeki, S. M. (1978) Uniformity and diversity of structure and function in rhesus monkey prestriate cortex. *J. Physiol.*, **277**, 273–90

21

The structural basis for information processing in the primate visual system

JON H. KAAS

Introduction

The traditional view of the processing sequence for the visual system of primates has been that retinal input is relayed by neurons in the lateral geniculate nucleus (LGN) to the primary sensory cortex, where elementary sensory functions occur. This information is then projected to two band-like secondary areas of 'psychic' cortex, Areas 18 and 19 of Brodmann (1909), with more complex sensory and perceptual functions, and finally relayed to large multimodal 'association areas' for all higher-order abilities. Recent advances have made it obvious that this traditional view is too simple. First, in important studies initiated in cats and more recently extended to monkeys, there is compelling evidence for the parallel processing of three channels of information from the retina to at least striate cortex in the 'X', 'Y' and 'W' subsystems (see Stone, Dreher & Leventhal, 1979, for review). In monkeys and other primates, these three subsystems may be even more segregated than in cats, although the evidence in many ways is less complete. Another advance has been in gaining an understanding of the internal organization of striate cortex. Thus, we have ocular dominance 'columns', orientation 'columns', cytochrome oxidase 'blobs' and periodic patterns of intrinsic connections in striate cortex to incorporate in schemes of information processing, and there are clear suggestions that similar types of internal organization exist in other subdivisions of visual cortex. In addition, the significance of cortical lamination in the striate cortex has become more obvious since laminar connection patterns and the response properties of neurons in the layers have been better understood. Finally, much more is known about the organization of extrastriate cortex, its subdivision into a number of visual areas, the interconnections of many of these areas, the properties of single neurons in some of these areas, and the functional consequences of lesions in some visual regions.

The present review attempts to integrate some of this information into a comprehensive scheme of visual system organization, that for reasons of space and incomplete information is necessarily oversimplified. Regrettably, the important subcortical visual centers, the pulvinar complex and the superior colliculus, are hardly considered, and the many cortical outputs to subcortical centers are neglected, but the roles of these structures and connections in perception are uncertain. This review is concerned with the structural basis of visual perception in primates. The proposal borrows extensively from those of other researchers, especially from concepts of parallel and hierarchical processing as reviewed by Rodieck (1979), Stone et al. (1979), Van Essen (1979), Norton & Casagrande (1982), Sherman &

Spear (1982), Ungerleider & Mishkin (1982), Maunsell & Van Essen (1983b).

Major classes of retinal ganglion cells and their projections

One of the more significant advances in understanding the mammalian visual system was the division of retinal ganglion cells of cats into 'X', 'Y' and 'W' classes. These classes were first characterized physiologically, and then associated with anatomical differences so that X, Y and W classes eventually became identified with a wide range of distinct features, well beyond those revealed by the limited tests that originally defined these classes (see Rowe & Stone, 1977; Rodieck, 1979; Stone et al., 1979; Sherman & Spear, 1982; Rodieck & Brening, 1983, for reviews). This progress inspired research on other mammals; there is now good evidence for comparable classes in the retina of primates, and the similarites are so great that a number of researchers have concluded that the primate classes are indeed homologous to the cat classes. However, the issue of homology is not critical for the present review and, in recognition of existing

Fig. 21.1. Projections of cell classes in the retina to the LGN and the superior colliculus of monkeys. The pattern is similar in prosimian primates except that W cells also project to the koniocellular layers which are inserted between the parvocellular layers (see Fig. 17.2). In monkeys, the W cells project to interlaminar regions and to the superficial S layers. W–cell information also seems to be relayed from the superior colliculus to the interlaminar zones and the koniocellular layers. External and internal parvocellular (PE and PI) and magnocellular (ME and MI) layers are named after the labelling system of Kaas et al. (1978). Based on evidence reviewed and presented in Weller, Kaas & Wetzel (1979), Leventhal et al. (1981), Itoh et al. (1982) and Weber et al. (1983).

MAJOR CONNECTIONS OF THE PRIMATE RETINA

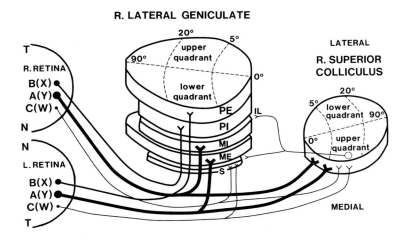

differences in opinion, the terms X-like, Y-like and W-like are used for the three classes (Fig. 21.1, Tables 21.1, 21.2). While some of the Y-like cells would be classified as X cells by the single criterion of having linear spatial summation (see Shapley, Kaplan & Soodak, 1981; Kaplan & Shapley, 1982), such an extension of 'cell type' nomenclature across species on the basis of a common property does not imply homology and is of limited use (see Rodieck & Brening, 1983). The argument for tentatively homologizing the X-like, Y-like and W-like pathways in primates with the X, Y and W pathways in cats is valid and has important theoretical consequences (see Stone *et al.*, 1979; Norton & Casagrande, 1982; Stone & Dreher, 1982; Rodieck & Brening, 1983).

The major features of X-like, Y-like and W-like ganglion cells are summarized in Table 21.1, and their projections are shown in Fig. 21.1. The vast majority of ganglion cells are X-like cells. They are highly concentrated in and near the fovea, project to the LGN and not to the superior colliculus, and typically have opponent color receptive field organization and sustained responses to standing contrast. These and other features, such as small receptive fields, indicate that X-like cells are likely to be the source of information essential for form or object vision. Similar proposals have been made for X cells of cats (e.g. Stone *et al.*, 1979).

The Y-like cells project to both the LGN and the superior colliculus. They are much fewer in number and only moderately concentrated in the central retina. In cats, the Y cells, which have branched projections to both the LGN and the superior colliculus, have receptive fields that cover the retina without significant overlap or gaps, once for on–center Y cells and once for off–center Y cells (Wässle, Peichl & Boycott, 1981). A similar distribution may exist for primate Y-like cells. The sparse distribution, lack of color sensitivity, large receptive fields, sensitivity to rapidly moving stimuli and stimuli of low contrast, transient nature of the excitatory response, rapid conduction of information, and dual projection to the LGN and superior colliculus, are all features of Y-like cells that are consistent with a role in the detection and localization

Table 21.1. *Properties of retinal ganglion cells in primates*

	Anatomical[a]	Physiological[a]
X–like or B Majority	Small soma (1, 2, 3, 4) Small dendritic field (1, 2, 3) Medium axon (1, 5) Lost after 17 lesions (6) Strong foveal concentration (8) Projects to parvocellular LGN (1, 3, 4)	Typically color opponent (7, 8) Moderate conduction velocity (7) Linear spatial summation (8) Typically sustained (7)
Y–like or A Minority	Large soma (1, 2, 3, 4) Medium dendritic field (1, 2, 3) Coarse axon (1, 5) Sustained after 17 lesions (6) Projects to magnocellular LGN and SC (1, 3)	Unselective for color (7, 8) Rapid conduction velocity (7, 9) Linear[b] and non-linear summation (8) Typically transient (7)
W–like or C[c] Minority	Small soma (1, 3, 4) Large dendritic field (1, 4) Fine axon (1, 5) Sustained after 17 lesions (6) Projects to interlaminar and koniocellular LGN or SC (1, 3)	Appear to be heterogeneous, with good or poor color sensitivity, slow conduction, and sometimes low response rates (7, 8)

[a] Soma, dendritic field, and receptive field sizes increase with eccentricity.
[b] Classed as X-cells on tests of linearity alone.
[c] Cells projecting to the pretectum are not included.
Conclusions are based on: 1. Leventhal, Rodieck & Dreher (1981); 2. Perry & Cowey (1981); 3. Itoh, Conley & Diamond (1982); 4. Casagrande & DeBruyn (1982); 5. Fitzpatrick, Itoh & Diamond (1983); 6. Weller, Kaas & Wetzel (1979); 7. Schiller & Malpeli (1977a); 8. deMonasterio (1978a, b, c); 9. Marrocco (1978).

of stimulus motion and change, as proposed for cat Y cells (e.g. Stone *et al.*, 1979).

In contrast to cats, where about half of the ganglion cells are W cells (see Stone *et al.*, 1979), as few as 10% of the ganglion cells in primates may be W-like cells (Schiller & Malpeli, 1977*a*; deMonasterio, 1978*c*). They are rather heterogeneous in response properties, and undoubtedly functionally distinct subtypes or types will be recognized (see Rodieck & Brening, 1983). They project to the LGN and the superior colliculus, and some may project to only one target rather than both. Because they are so few in number, project so sparsely to the LGN of higher primates, conduct information slowly, and often respond poorly to visual stimuli, they would appear to have no important role in either object vision, or the detection and localization of stimulus change. Yet, there is some evidence, discussed below, that they contribute by activating or modulating the striate cortex as a whole, and that they also contribute, perhaps in other ways, to the subsystem for object vision.

The structure, connections and cell types of the LGN

One of the conspicuous features of the LGN of primates is the presence of distinct layers of cells. Although the laminar pattern can be complex and variable across and even within species, much of the variability is in sublamination patterns of the parvocellular or small-celled layers (see Kaas, Guillery & Allman, 1972; Kaas *et al.*, 1978). The division of the parvocellular layers into interdigitated sublayers of differing ocular inputs has no obvious functional significance (however, the sublamination may partly segregate cell types; see Schiller & Malpeli, 1978). If this sublamination is ignored for the present, there are only two basic patterns of lamination in primates. The pattern found in monkeys and hominoids consists of a large dorsal parvocellular mass, constituting most of the nucleus, a pair of smaller magnocellular layers, one for each eye, and scattered small cells between layers and ventral to the magnocellular layers, where they form the S or superficial layers (Figs 21.1 and 21.2). In all primates, the parvocellular mass is divided into a more dorsal zone or layer with input from the contralateral

Fig. 21.2. Laminar patterns of retinal inputs and cortical projections for the LGN. Geniculate relays appear to be X-like (from B cells), Y-like (from A cells) or W-like (from C cells), and each laminar class terminates in its own characteristic laminar pattern in Area 17. Cortical layers are after Hassler (1966), and geniculate layers are from Kaas *et al.* (1978). The geniculate projections to the cortex are based on evidence presented and reviewed in Hubel

& Wiesel (1972); Glendenning, Kofron & Diamond (1976); Hendrickson, Wilson & Ogren (1978); Winfield, Rivera-Dominguez & Powell (1981); Fitzpatrick *et al.* (1983); Weber *et al.* (1983). See Lin & Kaas (1977) and Symonds & Kaas (1978) for descriptions of cortical projections to the LGN. See Fitzpatrick, Carey & Diamond (1980) and Harting, Casagrande & Weber (1978) for superior colliculus projections to the LGN.

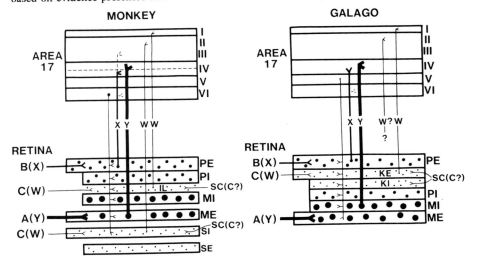

Table 21.2. *Properties of lateral geniculate relay cells*

	Anatomical	Physiological
Parvo B (X-like)	Medium soma (1, 2, 3, 4, 5) Tendency for dendrites in plane of isorepresentation in galago (4), human (6) Medium axons, and restricted arbors in striate cortex (7, 8)	Most spectrally dependent and color opponent[a] (9, 11, 12, 15, 16, 17) Sustained response to standing contrast (10, 11, 12, 19) Smaller receptive fields[b] (10) Medium latency to optic chiasm shock (10, 11, 12, 16, 17, 19) Poor activation by fast moving bars (10, 11, 19) Medium velocity conducting axons to cortex (10, 12, 19) Monophasic excitation *or* suppression to moving bar (13) No suppression from nondominant eye (14) Linear spatial summation (16, 17) Poor contrast sensitivity (17, 18)
Magno A (Y-like)	Large soma (1, 2, 3, 4, 5) Radially symmetric dendrites in galago (4) More variety in dendritic arbors and many translaminar dendrites (6) Large axons and terminal arbors in striate cortex (7, 8)	Spectrally broadband (9, 11, 12, 15, 16, 17) Transient response to standing contrast (10, 11, 12, 19) Large receptive fields[b] (10) Short latency to optic chiasm shock (10, 11, 12, 16, 17, 19) Vigorous response to fast-moving bars (10, 11, 19) Fast-conducting axons to cortex (10, 12, 19) Biphasic (excitation *and* suppression) to moving bar (13) with suppression from non-dominant eye (14) Both linear[c] and non-linear spatial summation (16, 17) High-contrast sensitivity (17, 18)
Konio C (W-like)	Small soma (3, 4) Thin dendrites and few branches (4)	Long to medium optic chiasm shock latencies Large receptive fields,[b] heterogeneous response properties, brisk to sluggish responsiveness, many nonconcentric receptive fields (19)
Interlaminar + S	Small soma (1)	

[a] Probably less so in nocturnal primates.
[b] Increases with eccentricity.
[c] Many are X-like based on spatial linearity (16, 17).

Based on: 1. Norden & Kaas (1978); 2. Headon *et al.*, (1981); 3. Casagrande & DeBruyn (1982); 4. Birecree & Casagrande (1981); 5. Wilson & Hendrickson (1981); 6. Hickey & Guillery (1981); 7. Florence, Sesma & Casagrande (1983); 8. Blasdel & Lund (1983); 9. Wiesel & Hubel (1966); 10. Sherman *et al.* (1976); 11. Dreher, Fukada & Rodieck (1976); 12. Schiller & Malpeli (1978); 13. Lee, Creutzfeldt & Elepfandt (1979); 14. Rodieck & Dreher (1979); 15. Creutzfeldt, Lee & Elepfandt (1979); 16. Kaplan & Shapley (1982); 17. Marrocco, McClurkin & Young (1982b); 18. Lennie & Derrington (1981); 19. Norton & Casagrande (1982).

eye, and a more ventral layer with input from the ipsilateral eye. However, these two basic layers subdivide and interdigitate to form more complex laminar patterns in many higher primates.

In the prosimian galagos, lorises and lemurs, the basic pattern is modified by the presence of two layers of very small cells, the koniocellular layers, inserted between the two parvocellular layers (Fig. 21.2). The dorsal or external koniocellular layer receives input from the contralateral eye, while the internal koniocellular layer receives input from the ipsilateral eye.

The laminar organization of the LGN relates directly to the three classes of retinal ganglion cells. Parvocellular layers receive inputs from the X-like cells and magnocellular layers are activated by Y-like cells. Judging from the fine caliber of axons terminating in the interlaminar zones and S layers of monkeys (Fitzpatrick *et al.*, 1983) and the small sizes of the cells projecting to the koniocellular layers of prosimians (Itoh *et al.*, 1982), these layers and zones appear to receive inputs from the heterogeneous W-like cell class.

Of course, lateral geniculate cells in particular layers, even if activated by a single class of ganglion cell, need not be homogeneous in form or function. For example, individual layers contain both relay cells projecting to striate cortex, and a few intrinsic neurons (Norden & Kaas, 1978) that probably modulate the activity of relay cells through inhibition (Ogren *et al.*, 1982). Another cautionary concern is that species probably differ in the proportion of neuron subtypes in layers. Thus, it is likely that nocturnal prosimians and nocturnal owl monkeys have fewer color-selective cells in the LGN than diurnal squirrel and macaque monkeys. However, little is known about differences in proportions of cell subtypes for different primates (see Jones, 1966).

As one might expect, cells in the parvocellular layers tend to be X-like (see Table 21.2). Most are sustained in their response to standing contrast, almost every cell shows linear summation, a large majority have color-opponent responses, and they have smaller receptive fields. Similarly, cells in the magnocellular layers tend to be Y-like. They are exclusively broadband in spectral sensitivity, predominantly phasic in response to standing contrast, and sensitive to rapidly moving stimuli and stimuli of low contrast. Because there are few cells in the interlaminar zones and in the S layers, the responses of these cells have not been extensively studied. However, cells in the

koniocellular layers of galagos are generally W-like in that they have large receptive fields, long response latencies, and heterogeneous receptive field properties and sometimes they are sluggish in responsiveness (Norton & Casagrande, 1982). The S-layer cells, interlaminar cells, and koniocellular layer cells appear to be directly activated by W-like ganglion cell inputs and W cell inputs to the superficial gray layer of the superior colliculus apparently are relayed to these same geniculate cells (see Harting *et al.*, 1978; Fitzpatrick *et al.*, 1980; Weber *et al.*, 1983). In addition, a brief report on interlaminar cells in galagos indicates that they are W-like (Irvin *et al.*, 1984).

The observation that parvocellular layers, magnocellular layers, and the interlaminar zones plus the S layers have different relative sizes in primates (LeGros Clark, 1941; Hassler, 1966) supports the contention that these subdivisions have different functional roles. In all primates the majority of neurons are in the parvocellular layers (Hassler, 1966). Old World monkeys, for example, have seven to eight times as many neurons in the parvocellular layers as in the magnocellular layers, and they have very few interlaminar and S-layer cells. Advanced New World monkeys, such as *Cebus*, have similar proportions. However, nocturnal owl monkeys have only three times as many parvocellular as magnocellular neurons but have more interlaminar neurons. Similarly, in nocturnal prosimians, the magnocellular and koniocellular layers are proportionately larger than in diurnal prosimians. These values are consistent with the suggestion that the parvocellular layers have a major role in the detailed object vision that would be more important in diurnal and, perhaps, advanced primates.

Besides direct inputs from the retina, the major afferents to the LGN are from the superior colliculus, visual cortex, and the reticular nucleus of the thalamus. As mentioned, afferents from the superior colliculus consist of fine axons that terminate in the interlaminar zones, koniocellular layers and S layers, and therefore appear to be a second source of W-like information. Almost all of the direct cortical input is from layer VI cells of the striate cortex (Fig. 21.3), although there is evidence of slight inputs from Area 18 and the Middle Temporal Visual Area, MT (Lin & Kaas, 1977). These visual areas and others (Graham, Lin & Kaas, 1979*a*) project to GABAergic neurons in the reticular nucleus (Hendrickson *et al.*, 1983) which relay to the LGN to provide, in all likelihood, an inhibitory modulatory

Fig. 21.3. The laminar organization of Area 17 at the border with Area 18 in primates. The distinct layer IIIb of macaque monkeys is less developed in owl monkeys and hardly distinguishable in galagos. Layers IIIb, IIIc and IV sometimes give the false appearance of merging into layer IV of Area 18 in some primates, and hence these sublayers of layer III have been considered as sublayers of layer IV. Brodmann's (1909) traditional nomenclature for layer IV is shown (4a, b and c) along with the format adopted from Hassler (1966). See Weller & Kaas (1982b) for further details.

GALAGO

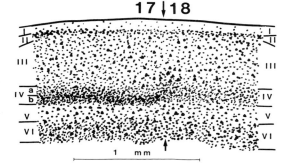

OWL MONKEY

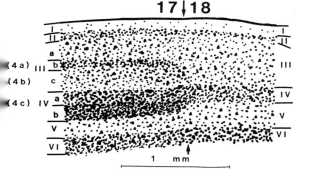

MACAQUE MONKEY

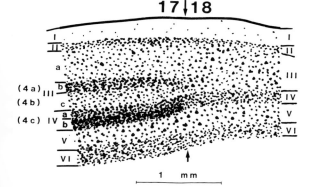

influence (see Marrocco *et al.*, 1982*a*). Except for a few scattered larger cells projecting to the prestriate cortex (Benevento & Yoshida, 1981; Yukie & Iwai, 1981), and collaterals to the reticular nucleus (Jones, 1975), all projections of the LGN are to layers of the striate cortex (Figs 21.2 and 21.3).

The striate cortex: lamination, modular organization, intrinsic and extrinsic connections, and neuron response characteristics

The striate cortex, as the name denotes, is a conspicuously laminated structure. Since cells in different layers are distinguished by different physiological properties, cellular structure and connections, it is important for further discussion to label the layers. Two basic schemes of labeling layers and sublayers are currently in use, one stemming from the early and monumental descriptions of Brodmann (1909) and the other coming from the more recent reports of Hassler (1966). Hassler's scheme is used here because comparative studies of lamination in Area 17 of primates and other mammals and studies of laminar patterns of connections seem clearly to indicate that Brodmann incorrectly identified two sublayers of layer III as sublayers of layer IV (see Fig. 21.3).

Fig. 21.4 shows the relation of cortical layers to patterns of terminal inputs, distributions of cells with known efferent connections, and some intrinsic connections. As an example of the significance of using one system of labeling layers over another, note that layer IIIc of Hassler's system projects strongly to the extrastriate cortex (MT). A similar projection to the extrastriate cortex characterizes the bottom of layer III in cats and other non-primates, and we conclude that this is a consistent feature across species. In contrast, using Brodmann's system, the cells projecting to MT in primates are in layer IVb and one would be forced to conclude, incorrectly we believe, that primates are distinctly different than other mammals in the course of extrastriate projections.

A number of recent studies have described the laminar terminations of geniculate inputs in the striate cortex (e.g. Hubel & Wiesel, 1972; Livingstone & Hubel, 1982; Fitzpatrick *et al.*, 1983; Weber *et al.*, 1983) and, to a limited extent, the termination patterns of single geniculate axons (Blasdel & Lund, 1983; Florence *et al.*, 1983). The magnocellular, parvocellular and interlaminar-S layer systems project to different cortical layers, supporting the argument that these layers have different functional roles. Magnocellular or Y-like axons branch widely in upper layer IV. Each axon is apparently capable of activating a broadly distributed group of cells that relay to lower layer III neurons (Lund *et al.*, 1979; Lund, 1980). These layer IIIc neurons are labeled by the same monoclonal antibody that selectively labels magnocellular over parvocellular geniculate neurons (Hendry *et al.*, 1984). The layer IIIc cells, in turn, project to the middle temporal visual area, MT, of the extrastriate cortex (see Tigges *et al.*, 1981). A collateral branch of layer IIIc cells may terminate on lower layer VI cells of striate cortex, which in turn provide feedback to the magnocellular layers (Lund *et al.*, 1979). The parvocellular input terminates in relatively restricted arborizations in sublayer IVb, thus preserving retino-topic fidelity. The stellate cells in sublayer IVb relay to

upper layer III cells (Lund *et al.*, 1979; Fitzpatrick *et al.*, 1983), which in turn project to Area 18 or V-II (e.g. Tigges *et al.*, 1981). The axons of these stellate cells also branch to upper layer VI to terminate on cells feeding back to the parvocellular layers, and to layer V, and layer III. Axons of the W-like cells in the S layers and interlaminar zones terminate in layer I, where they apparently have a modulatory and, perhaps, potentiating influence on the ends of apical dendrites of pyramidal cells in other layers, and in spaced clusters of cells in layer III which project to V-II. Other connections of striate cortex (Fig. 17.4) include feedback inputs from V-II and MT to layers I, V, and IV, modulatory layer I inputs to striate cortex from the pulvinar complex, connections with the claustrum and basal ganglia, and outputs to the superior colliculus.

Hubel & Wiesel have prompted considerable interest in the intrinsic organization of striate cortex

Fig. 21.4. Laminar connections of Area 17 of primates. Major features include the relatively broad distribution of the terminal arbors of thick axons from the magnocellular geniculate layers in IVa, the more restricted terminations of medium diameter axons from the parvocellular layers in IVb and IIIb, and the distributions of thin axon inputs from S layers and interlaminar geniculate cells to layer I and to puff-like regions of layer IIIa with high cytochrome oxidase activity (shading). The intrinsic relay of information by layer-IV stellate cells favours the influence of cells projecting to MT by magnocellular geniculate inputs, to striate cortex and the influence of cells projecting to Area 18 by parvocellular geniculate inputs to striate cortex. CLAUST., claustrum; DM, dorsomedial visual area; LGN, lateral geniculate nucleus; MT, middle temporal visual area; M, magnocellular input; P, parvocellular input; S, superficial layer input; PUL., pulvinar complex; S.C., superior colliculus; RET., reticular nucleus; V-II, second visual area or Area 18. Cortical layers after Hassler (1966). For recent evidence and reviews on connections, see Graham, Wall & Kaas (1979*b*); Carey, Bear & Diamond (1980); Weller & Kaas (1981, 1982*b*); Tigges *et al.* (1982); Fitzpatrick *et al.* (1983); Weber *et al.* (1983).

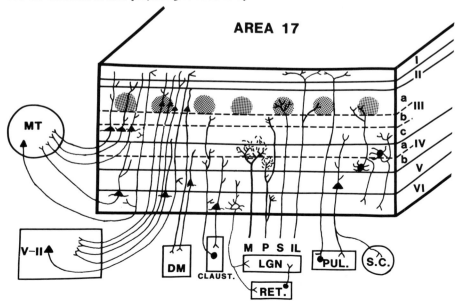

with their elegant proposal of a repeating processing module called a 'hypercolumn' (see Hubel & Wiesel, 1977, for review). In macaque monkeys, geniculate inputs relaying activity from the right eye are segregated from those relaying from the left eye in alternating 400–500 μm-wide bands in layer IV. In addition, most cells in the striate cortex are best activated by narrow stimuli of particular orientations, and cells preferring a given orientation are aligned vertically and grouped in long bands that repeat every 1 mm or so for macaque monkeys (Hubel, Wiesel & Stryker, 1976). Hubel & Wiesel define a hypercolumn as a block of cortical tissue containing a pair of ocular dominance 'columns' (1 mm length of two blocks) or one set of orientation 'columns' (1 mm length of the full 180-degree range of orientation preference bands).

While the band-like segregation of ocular inputs in macaque monkeys suggests modular processing units of about 1 mm dimensions, the segregation itself may not be functionally important, since its presence is quite variable across species. A strong segregation of ocular inputs characterizes layer IV of several Old World monkeys, at least some hominoids and one New World

monkey (*Ateles*) (Hubel & Wiesel, 1972, 1977, Wiesel, Hubel & Lam, 1974; Florence & Casagrande, 1978; Hendrickson & Wilson, 1979; Tigges & Tigges, 1979; Hitchcock & Hickey, 1980). However, ocular inputs are coextensive in New World squirrel, owl and cebus monkeys (Kaas, Lin & Casagrande, 1976; Hendrickson *et al.*, 1978; Rowe, Benevento & Rezak, 1978; Tigges, Hendrickson & Tigges, 1984), and ocular inputs are only weakly segregated in the striate cortex of the prosimian galagos (Glendenning *et al.*, 1976; Casagrande & Skeen, 1980; Casagrande & DeBruyn, 1982). The variability in the presence or absence of ocular dominance bands, and the observation that they can be induced by directing afferents from two eyes to a frog's tectum (Constantine-Paton, 1982), argue against ocular dominance bands having functional significance *per se*.

Other evidence that the striate cortex regularly repeats modular processing units is the even spacing of puffs of layer III cells (Fig. 21.4) with high levels of metabolic activity (e.g. Wong-Riley & Carroll, 1984*a*, *b*) that are associated with geniculate axon terminations and are revealed by staining for cytochrome oxidase (Livingstone & Hubel, 1982; Fitzpatrick *et al.*,

Fig. 21.5. Some intrinsic connections of Area 17 of monkeys. A large injection of HRP reveals a lattice-like pattern of connections around the injection in layers II and III. More uniform but still periodic connections extend in layer IIIc, and lateral connections are also found in layers V and VI. Based on Rockland & Lund (1983). A small superficial injection of HRP labels cells in layers IIIc and V, and results in some lateral spread of axons and dendrites in those layers (see Fitzpatrick *et al.*, 1983).

AREA 17
INTRINSIC CONNECTIONS

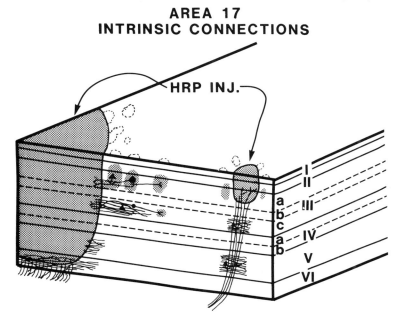

1983; Weber *et al.*, 1983). These so-called cytochrome oxidase blobs have been found in a range of primate species (macaque monkey, baboon, and galago: Horton & Hubel, 1981; squirrel monkey: Humphrey & Hendrickson, 1983; Livingstone & Hubel, 1982; Fitzpatrick *et al.*, 1983; owl monkey and marmosets: Cusick & Kaas, unpublished). Such blobs are spaced in parallel rows that are centered on ocular dominance bands in macaque monkeys, but do not appear to form rows in squirrel monkeys or marmosets, which do not have ocular dominance bands. The spacing varies in large and small primates so that large macaque monkeys, with about 1000 mm² of striate cortex (see Weller & Kaas, 1983), have about 5.3 blobs per mm², while small marmosets, with approximately 60 mm² of striate cortex, have about 8.0 blobs per mm². If cytochrome oxidase blobs reflect some unit of modular organization these observations suggest that macaque monkeys have more processing modules than marmosets, and that marmosets partially compensate by having smaller modules. Surprisingly, macaque monkeys and squirrel monkeys appear to have about the same spacing of blobs (Humphrey & Hendrickson, 1983).

Table 21.3. *Properties of Area* 17 *neurons*

	Few neurons	
	W-like	
I	Inputs modulate dendrites of cells in other layers	
II	Probably W-like properties	
IIIa	Cytochrome patches: non-oriented, color-selective and non-selective (2)	W-like domination
	Outside patches: orientation selective (2) color-coded clusters (7)	
	B-response: more sustained, small RF (3), complex	X-like domination
b	Non-oriented (1)	
	Many color-selective (1, 5)	
	Mostly monocular (1)	X-like domination
	Low contrast sensitivity (1)	
c	Color non-selective (1, 3)	
	Most direction-selective (3)	
	High contrast sensitivity (1)	
	Large binocular RFs (1)	Y-like domination
	Short latency to optic radiation shock (3)	
	Many C cells with larger receptive fields (RFs), transient ON, and preference for rapid movement (3)	
IVa	Y-like: larger RFs (1)	
	Short latency to optic radiation shock (3)	
	High contrast sensitivity (1)	
	Spectrally non-selective (1, 3)	Small, non-oriented
	Magnocellular monosynaptic driving (3)	monocular receptive fields (6)
b	X-like: smaller RFs (1)	LGN-like
	Parvocellular monosynaptic driving (3)	
	Low contrast sensitivity (1)	
	Both color-selective and non-selective (1, 3)	
V	B-cells: more sustained and small RF (3) complex	
	Corticotectal cells: larger RFs	
	Strongly binocular, more directional (4)	Y-like domination
	Broader orientation tuning	
VIa		
b	Non-oriented – broad band (3)	

Based on: 1. Blasdel & Fitzpatrick (1984); 2. Livingstone & Hubel (1982); 3. Bullier & Henry (1980); 4. Schiller, Malpeli & Schein (1979); 5. Dow (1974); 6. Hubel & Wiesel (1968); 7. Michael (1981).

The regular patterns of intrinsic connections that have been recently revealed in monkeys (Rockland & Lund 1983), and more clearly in tree shrews (Rockland, Lund & Humphrey, 1982; Sesma, Casagrande & Kaas, 1984), further support the concept of modular processing units in the striate cortex (Fig. 21.5). A restricted injection of horseradish peroxidase (HRP) labels cells and terminals in clumps and sometimes short bands around the injection site. The spacing of such bands raises the possibility that some of the intrinsic connections join orientation columns of similar (see Mitchison & Crick, 1982; Rockland *et al.*, 1982) or opposite preferences (Matsubara & Cynader, 1983), but it is not yet apparent whether the injections are revealing periodic subsets of a continuously distributed pattern, or a fixed periodic pattern.

Some of the response properties of Area 17 neurons are summarized in Table 21.3. The main conclusion to be derived from the table is that neurons in the laminar zones of geniculate inputs related to the Y-like, X-like, and W-like ganglion cell subsystems closely reflect those properties, and that output zones can be characterized as dominated by one or another of the three retinal subsystems. The relation of this functional segregation of outputs to further processing in the extrastriate cortex is considered in a subsequent section of this review.

An overview of extrastriate visual cortex

The visual areas

A number of subdivisions of the extrastriate visual cortex have been proposed, and somewhat different concepts of organization have been developed for New World (Fig. 21.6) and Old World (Fig. 21.7) monkeys (for reviews see Allman & Kaas, 1976; Kaas, 1978; Van Essen & Zeki, 1978; Zeki, 1978; Van Essen, 1979; Weller & Kaas, 1981, 1982*b*; see Weller & Kaas, 1982*b* for prosimians). Neither scheme is completely satisfactory in that the evidence for some of the visual areas is equivocal and open to other interpretations. Some of the proposed subdivisions have been established by multiple criteria and others are supported by more limited evidence. In the New World owl monkey, Area MT (the middle temporal visual area) is characterized by a systematic representation of the visual hemifield, a densely myelinated appearance in brain sections stained for fibers, a particular pattern of reciprocal connections with other visual areas, a

unique pattern of dense interconnections with a specific subdivision of the inferior pulvinar, other distinctive connections, and a preponderance of directionally selective neurons (see Felleman & Kaas, 1984; Weller, Wall & Kaas, 1984, for reviews). Thus, the evidence is so extensive that it seems certain that the MT is a valid subdivision of the visual cortex. Likewise, the location, extent and retinotopic organization of V-II or Area 18 has been clearly established from mapping studies, patterns of connections and architectonic features. Area DM (the dorsomedial visual area) has almost equal status, while areas DL (the dorsolateral visual area) and M (the medial visual area) are known

Fig. 21.6. Proposed subdivisions of cortex in New World owl monkeys. The primary and secondary fields, V-I or Area 17 and V-II or Area 18, as well as the dorsolateral (DL), dorsointermediate (DI), dorsomedial (DM), medial (M), posterior parietal (PP), and middle temporal (MT) visual areas are from Allman & Kaas (1976). The ventral area, V, corresponds to the two ventral fields of Newsome & Allman (1980). The dorsal (IT_D), ventral (IT_V), rostral (IT_R), medial (IT_M), and polar (IT_P) divisions of inferotemporal cortex are from Weller & Kaas (1982*a*). Superior temporal (ST) and temporal parietal (TP) visual regions are based on Weller *et al.* (1984). Somatosensory representations in Areas 3a, 3b, 1 and 2 have been described by Merzenich *et al.* (1978). Rostral (4_R) and caudal (4_C) motor fields in Area 4, the supplementary motor field, (S. MOTOR), and the frontal eye field (FEF) are based on Gould *et al.* (1983). The frontal visual region (FV) is a region of visual connections (unpublished studies). Auditory fields, primary auditory (AI), rostral (R), anterior lateral (AL) and posterior lateral (PL) are from Imig *et al.* (1977).

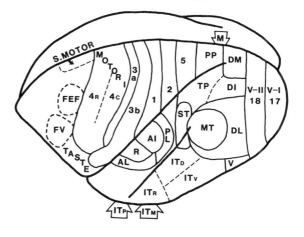

OWL MONKEY

from retinotopic maps, patterns of connections, and some defined architectonic borders where they adjoin known subdivisions, but all borders have not been identified with certainty. Area DI (the dorsointermediate visual area) is known from very limited mapping data, and definable borders with DL, DM, D and V-II. The superior temporal visual area, ST, (Weller *et al.*, 1984) and the dorsal, ventral and rostral divisions of the inferotemporal cortex, IT_D, IT_V and IT_R, (Weller & Kaas, 1982*a*) have been deduced largely from patterns of connections. The ventral region, V, was postulated on the basis of limited mapping data (see Newsome & Allman, 1980), but patterns of cortical connections are consistent with the concept of the area (Weller *et al.*, 1984). However, it may consist of more than one area,

and its borders, other than that with V-II, are uncertain. The posterior and temporal parietal areas, PP and TP, are regions that respond to visual stimuli and visual connections have been identified, but each region probably contains more than one visual area.

In macaque monkeys, MT has been identified by all the features used to identify MT in owl monkeys. The organization and extent of V-II, including only part of the traditional 'Area 18', is well known especially from recent mapping (Gattass *et al.*, 1981) and 'architectonic' studies (2-[^{14}C]-deoxy-D-glucose, cytochrome oxidase, and myelin stain procedures) (Tootell, Silverman & De Valois, 1938; Livingstone & Hubel, 1984; Wong-Riley & Carroll, 1984*b*). V3 is a narrow band of cortex along a limited portion of V-II (apparently much more restricted than originally proposed); its existence as a separate visual area is best supported by the evidence for a separate projection pattern from V-I. It seems to represent only the lower visual quadrant (for evidence for a more traditional V3 in macaque monkeys, see Ungerleider *et al.*, 1983). As of yet, there is no evidence for a V3 in other primates. V4 is a large region of uncertain organization and boundaries. It appears to contain more than one visual area; hence V4 has been called a 'visual complex' (see Zeki, 1978, 1983; Baizer & Maguire, 1983). Because of comparable location, and similar connections with V-II and IT, much of V4 is probably the homologue of DL. V3A is known from limited electrophysiological mapping and studies of connections, but its borders have not been architectonically defined. V3A resembles DM in several respects (see Lin, Weller & Kaas, 1982), including its dorsomedial location, relative size, visual field map with the upper quadrant lateral and the lower quadrant medial, and sparse projections from the striate cortex. The parieto-occipital area, PO, is a small representation of the visual hemifield with distinguishing architectonic features. Its location and organization suggest that it is homologous to Area M of owl monkeys (Covey *et al.*, 1982). The MST (medial superior temporal area), like ST in owl monkeys (Weller & Kaas, 1983), has been defined as a region with input from the middle temporal visual area, MT, (Maunsell & Van Essen, 1983*b*). It may contain one or more visual areas in both New and Old World monkeys. VIP (the ventral intraparietal area) is a posterior parietal projection zone of MT in macaque monkeys (Maunsell & Van Essen, 1983*b*) that may correspond to a MT projection zone in part of PP of owl monkeys. VP appears to represent

Fig. 21.7. Proposed subdivisions of cortex in macaque monkeys. The brain has been 'expanded' to open the superior temporal sulcus and the lunate sulcus. V-I = the first visual area or Area 17. The location and extent of the second visual area, V-II, is largely based on Gattass, Cross & Sandell (1981). V3, V3A and V4 are from the reports of Zeki (see Van Essen & Zeki, 1978; Zeki, 1978). The middle temporal visual area (MT) has been described by a number of investigators (see Weller & Kaas, 1982*b*). The parieto-occipital area (PO) is from Covey, Gattass & Gross (1982). The ventral posterior area (VP) is from Newsome, Maunsell & Van Essen (1980). The ventral intraparietal area (VIP) and the medial superior temporal area (MST) are from Maunsell & Van Essen (1983*b*). The superior temporal polysensory area (STP) is from Bruce, Desimone & Gross (1981). Cytoarchitectonic divisions of inferior temporal cortex are those of Bonin and Baily (see Mishkin, 1982). Auditory fields are the same as for Fig. 17.6. Areas 3a, 3b, 1, 2 and 5 are subdivisions of the somatosensory cortex; Area 7 is visual and somatic (see Nelson *et al.*, 1980; Hyvärinen, 1982). Motor and premotor fields as in Fig. 17.6.

MACAQUE MONKEY

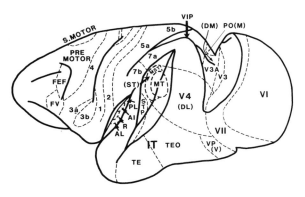

he upper visual quadrant (see Van Essen, Newsome
Bixby, 1982); its boundaries are uncertain. In
cation and known organization, it resembles VP of
wl monkeys (V in Fig. 21.6 includes 'VA' and 'VP'
f Newsome & Allman, 1980).

Patterns of connections and cortical architecture
stify dividing the IT cortex in macaque monkeys into
posterior TE zone with input from V4 and an anterior
E zone with input from posterior TE (see Mishkin,
982). Similar connections suggest that IT_D and IT_V
gether correspond to TE in macaque monkeys
Weller & Kaas, 1983). The polysensory area, STP,
orsal to IT, is from Desimone & Gross (1979). No
mparable area has yet been defined in owl monkeys.

Patterns of connections

The ipsilateral connection pattern of the
xtrastriate visual cortex is astonishingly complex (see
aas, 1978; Graham *et al.*, 1979*a*; Weller & Kaas, 1981,
982*b*). Each visual area projects to a number of other
sual areas and to a number of subcortical centers. In
ddition, each visual area may receive inputs from
veral subdivisions of the pulvinar complex, the
austrum and, sparsely, from a number of other
rainstem centers (see Tigges *et al.*, 1982). MT, for
ample, projects cortically to V-I, V-II, DL, DM, PP,
I, V, ST and possibly M (Weller *et al.*, 1984).
ubcortically, MT projects to the caudate, putamen,
austrum, reticular nucleus, dorsal LGN, pregeniculate
ucleus, two nuclei of the inferior pulvinar, two
bdivisions of the superior pulvinar, pretectum,
perior colliculus, and the pons (Graham *et al.*,
979*a*). MT receives cortical inputs from V-I, DL,
M, M, PP and ST (see Weller & Kaas, 1982*b*), and
bcortical inputs from the inferior pulvinar complex
in & Kaas, 1979), the claustrum, lateral hypothalamus,
sal nucleus of Meynert, pons, raphe nucleus, and
cus coeruleus (Tigges *et al.*, 1982). Obviously,
urons in any visual area are potentially influenced by
puts from many sources. However, connections differ
laminar arrangement and density, suggesting
ffering functional roles. Presumably, dense inputs
ve the greater effects, and terminations in the
anular layers are more effective than terminations in
yer I. Afferents from the striate cortex to the
trastriate cortex are obviously the most important,
nce most of the extrastriate cortex is visually
responsive after striate cortex removal (Rocha-
iranda *et al.*, 1975; Desimone *et al.*, 1979).

The functional significance of the different
laminar patterns of cortical terminations and outputs
has been a matter of speculation (e.g. Allman & Kaas,
1974; Kaas, Lin & Wagor, 1977; Tigges *et al.*, 1977,
1981; Wong-Riley, 1978; Rockland & Pandya, 1979,
1981; Maunsell & Van Essen, 1983*b*), but a general
consensus has emerged. Because the principal activating
influences of primary sensory areas are from thalamocortical sensory relays terminating largely in layer IV,
and because the majority of layer IV neurons are
intrinsic neurons relaying to supra- and infragranular
layers, layer IV terminations have been characterized
as feedforward inputs that provide the basic information
for further processing. Feedback of information from
primary sensory areas to sensory relay nuclei in the
thalamus originates in layer VI, and therefore layer VI
can be thought of as a major feedback layer. Such
feedback from the striate cortex to the LGN appears
to provide modulation rather than activation, so that a
response to direct activation from the retina is
enhanced or inhibited (see Marrocco *et al.*, 1982*a*).

Area 17 provides dense feedforward terminations
to layer IV of both Area 18 and MT. Since Area 18
neurons (Schiller & Malpeli, 1977*b*) and, most
probably, MT neurons are visually unresponsive after
Area 17 is deactivated by cooling, the feedforward
projections of Area 17 are clearly the major activating
source. The feedback connections from layers III, V
and VI in Area 18 terminate largely in layers I, V and
VI of Area 17. Cooling Area 18 only modifies the
excitability of most Area 17 neurons (Sandell &
Schiller, 1982), so again the feedback is modulatory.
Interestingly, much of the feedback from MT is from
layer VI cells. Thus, in some cases corticocortical as
well as corticothalamic feedback comes from layer VI
neurons. MT terminations are concentrated in layers
I, IIIc and VI of Area 17. These patterns of connection
suggest that modulatory feedback terminations concentrate either in layer I, where they may synapse on the
ends of apical dendrites and perhaps enhance activity,
or in infragranular and supragranular layers where they
may synapse on, and enhance the activity of, efferent
cells, or synapse on intrinsic neurons that inhibit the
activity of neighboring cells. Feedforward connections
relate to early stages of intrinsic processing, while
feedback types relate to late or final stages.

There are several anatomical variations of
feedforward and feedback corticocortical projections
(Weller & Kaas, 1981; Maunsell & Van Essen, 1983*b*).

Feedforward projections may be concentrated in layer IV, or they may strongly involve other layers as well. Feedback connections appear to always include layer I, and avoid layer IV, but other terminations may concentrate in different patterns near the output cells of layers III, V and VI.

Callosal connections

The interhemispheric connections of extrastriate visual areas are extensive (Newsome & Allman, 1980; Van Essen *et al.*, 1982; Cusick, Gould & Kaas, 1984). Callosal visual connections have been traditionally portrayed as connecting the representations of the two visual hemifields to form bilateral representations of the region of the vertical meridian. However, the total distribution of callosal connections indicates that they must also play a broader functional role.

Fig. 21.9 shows the distribution of callosally labeled cells relative to cortical visual areas in an owl monkey. Several features of this distribution are worth stressing. First, a scattering of callosal cells is found in Area 17 (V-I) within 2 mm of the border. Thus, Area 17 does have some callosal projections, even from cortex slightly away from the border zone representing the zero vertical meridian. The callosally projecting cells displaced from the border are so few, however, that their functional significance can be questioned, and they may represent no more than errors in development. Callosal terminations, which in general have a more restricted pattern in general than callosal cells, extend

only slightly into Area 17. Secondly, there is concentration of callosally projecting cells at the 17/1 (V-I/V-II) border, as traditionally described. Suc cells could contribute to receptive fields spanning th zero vertical meridian. Other locations representing th zero vertical meridian, such as the outer border of M and the rostral border of DM, also have callosal cell Thirdly, extrastriate visual areas typically have callos connections for the representation of much or all of th visual field. The caudal half of the belt-like Area 18 densely populated with callosal cells, and this distanc can correspond to considerable displacements from th representation of the zero vertical meridian at the Are 17/18 border (Allman & Kaas, 1974). Furthermore, the dorsolateral part of Area 18, which is devoted to th central 8 degrees of vision near the horizontal meridia callosal cells densely populate all of Area 18. Oth fields such as MT, DL and DI have dense, althoug not uniform, callosal connections. These connectior clearly include parts of the visual field representatior away from the zero vertical meridian. Strangely, the le dense regions in DL are near the representation of th zero vertical meridian. Areas ST and IT hav extremely dense and more uniform connection Fourthly, some regions besides Area 17 have rathe sparse callosal connections. DM has connectior largely limited to the border region corresponding the zero vertical meridian. Much of PP has few callos connections, and a zone of cortex at the DI/T junction is consistently almost devoid of callosal cel

Fig. 21.8. Some of the connections of the visual cortex in owl monkeys. Visual areas as in Fig. 17.6. See Weller & Kaas (1981) for review.

SOME OF THE VISUAL CONNECTIONS OF OWL MONKEY CORTEX

Fig. 21.9. The distribution of callosally projecting neurons in the visual cortex of an owl monkey. See Cusick *et al.* (1984) for further details. Visual areas as in Fig. 17.6.

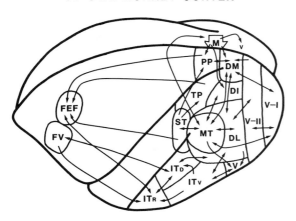

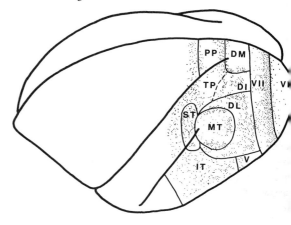

d terminations. Fifthly, there are slight hints of a riodic pattern for callosal connections in Area 18, ere bands are suggested by termination patterns, and MT, where dense terminations form large clumps ng the outer border. A sixth point is not evident from e total callosal pattern. When individual areas are ected with anatomical tracers, it is apparent that a ven visual area typically projects to more than one ual area in the other hemisphere. MT, for example, ojects densely to MT of the other hemisphere, and s densely to ST and DL (Weller *et al.*, 1984). Such nnections can be of the feedforward or feedback type. nally, such experiments also show that projections m one visual area to its counterpart in the other misphere are typically concentrated in positions that e at least roughly mirror symmetric in regard to the ual hemifield representations. Thus, upper MT oject to upper MT.

While direct evidence for the significance of llosal connections is limited, several conclusions seem lid. First, in regions representing the zero vertical eridian, callosal connections may contribute to the citatory receptive field centers as traditionally pposed (e.g. Choudhury, Whitteridge & Wilson, 65). Secondly, for neurons in the inferior temporal rtex, callosal connections have been shown to ntribute the ipsilateral hemifield component of ceptive fields with extensive crossings of the zero rtical meridian (Rocha-Miranda *et al.*, 1975). Finally, ost visual areas have neurons with excitatory ceptive fields that are only in the contralateral mifield, and thus most of the callosal connections uld not be directly expressed in the excitatory ceptive fields. However, in MT (Miezen, McGuinness Allman, 1982) and V4 (Moran *et al.*, 1983; Schein, esimone & deMonasterio, 1983), neurons have tensive antagonistic surrounds that include the silateral hemifield. Callosal connections undoubtedly ntribute to these surrounds, perhaps even by minating on non-spiking interneurons.

Processing channels in the extrastriate cortex

There is considerable evidence for particular ocessing sequences in the cortex. Feedforward ojections have long been used to formulate pothetical processing chains. In a recent effort, aunsell & Van Essen (1983*b*) used the extensive formation available on connections in macaque

monkeys to position cortical visual areas in a processing hierarchy with several visual areas at comparable levels for all but the initial cortical stage (Area 17). A similar, but simpler, hierarchical system is shown for the owl monkey in Fig. 21.10. The arrangement is limited to two major feedforward subsystems, each with a different postulated functional role. The proposal is basically an elaboration of the pathways of the 'two cortical visual systems' as described by Mishkin (1972, 1982) and Ungerleider & Mishkin (1982) for macaque monkeys. One cortical system involves projections that ultimately reach the inferior temporal cortex, and this system is concerned with visual recognition of objects. The other system consists of stations and pathways finally reaching posterior parietal cortex, and this system mediates functions related to spatial localization. The anatomical evidence supports the concept of these two major processing sequences (Fig. 21.10), but they are complicated by interconnections between parts of the subsystems at several levels, and by interrelations

Fig. 21.10. Hypothetical mainline processing sequences for object vision and recognition (left) and orientation and location in visual space (right). The concept of 'two cortical visual systems' stems from Mishkin and co-corkers and is influenced by other discussions of hierarchical and parallel processing (see text for references). Visual areas as in Fig. 6.

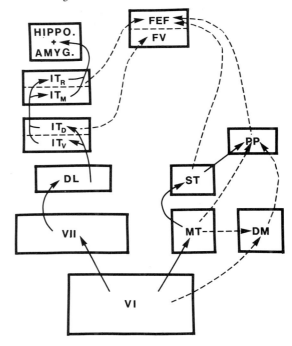

with visual areas (see Fig. 21.6) not shown in the sequences.

The two processing sequences start with V-I. Two visual areas, V-II and MT, are the major cortical projection targets of V-I. The inputs are dense and of the feedforward type. Both areas are presumably dependent on this direct striate cortex input for activation. Thus, V-II and MT are at equal levels in the cortical processing hierarchy. Processing sequences from V-II and from MT are directed toward the temporal and parietal lobes, respectively. Since lesions of temporal and parietal lobes lead to quite different visual impairments, it is useful to consider these two sequences in detail.

The MT–PP processing sequence

MT is the first separate station in the cortical processing hierarchy directed toward the posterior parietal cortex. Posterior parietal cortex lesions in monkeys and man result in disorders of visual guidance of movements and in visual neglect (e.g. Lynch, 1980; Hyvärinen, 1982). Deficits in pursuit eye movement follow chemical lesions of the MT in monkeys (Newsome *et al.*, 1983).

The functions of the MT and ultimately of the posterior parietal cortex appear to depend on predominance of Y-like ganglion cell information th is relayed from Area 17. The cells projecting from Are 17 to MT have been identified by a number investigators (see Weller & Kaas, 1982*b*). The maj projection is from pyramidal cells and some stellate ce in the superficial half of layer IIIc (Fig. 21.11), whe the majority of neurons project to MT (e.g. Tigg *et al.*, 1981). These layer IIIc neurons are in a positi to be strongly influenced by layer IVa stellate cells th relay to layer IIIc and have magnocellular (Y-lik geniculate input. A few large pyramidal cells near t junction of layer V with layer VI also project to M (Figs 21.4 and 21.11).

The known response properties of layer IIIc cel in Area 17 are completely consistent with Y-li domination (Table 21.3). In particular, these neuro (Y-like ganglion cells, magnocellular geniculate cell and IVa cortical cells) lack color selectivity and are ve sensitive to low levels of luminance contrast (Blasdel Fitzpatrick, 1984). In addition, several new properti are added that are consistent with the properties MT neurons. The layer IIIc cells have large bin cular receptive fields, and they have high selectivi for orientation and often direction of moveme

Fig. 21.11. Hypothetical projection patterns of Area 17 to MT. Bands of similar orientation, axis, or direction-selective neurons in layers IIIc and VI of Area 17 project to bands of neurons in MT of matched axis and direction preferences. Each neuron in MT diverges to several bands of matched axis selectivity, and each band in MT receives inputs from several orientation bands in Area 17. Thus, larger receptive fields in MT are created without a loss of axis selectivity. For simplicity, systematic arrays of four different orientation- or axis-selective bands are shown for each area, but actual arrangements are more complex. Orientation 'bands' may be imposed on a continuum of change rather than representing step-wise change. SC, superior colliculus. See text for supporting references.

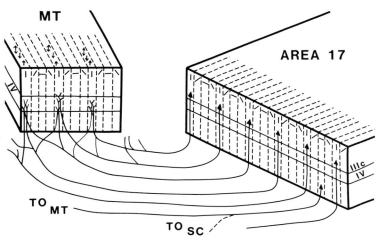

'ig. 21.11). Perhaps some of these new properties ²sult from the widespread and somewhat periodic ¹trinsic connections of layer IIIc (Fig. 21.5).

In addition to layer IIIc cells, a smaller number ⁺layer VI cells project to MT. These neurons are ²ated along the junction of layer V with layer VI and ᵉy include most or all of the largest pyramidal cells, ᵉ solitary Meynert cells. Apparently these same cells ᵒject to the superior colliculus (Fries & Distel, 1982). ⁻periments with reversible inactivation of geniculate ᵉʳs indicate that neurons in Area 17 projecting to the ᵖerior colliculus are dependent on magnocellular, not ʳvocellular layer activation (Schiller *et al.*, 1979), and ²refore we can surmise that the Meynert cells ᵒjecting to both the superior colliculus and MT are, ᵉ the layer IIIc cells, relaying magnocellular (Y-like) ᶦormation to MT. Since corticotectal cells are also

orientation-selective (Finlay, Schiller & Volman, 1976), these properties can be assumed for Meynert cell inputs to MT.

Neurons projecting from a given location in Area 17 diverge to several periodically spaced locations in MT, and several locations in Area 17 converge upon a given location in MT (Montero, 1980; Lin *et al.*, 1982). A reasonable interpretation of this anatomical pattern is shown in Fig. 21.11. In Area 17, orientation- and direction-selective neurons are arranged in bands of similar selectivities (e.g. Hubel & Wiesel, 1977). Cells in MT are predominantly direction-selective (Table 21.4), and there is evidence that they are also arranged in bands (Albright, Desimone & Gross, 1984). In addition, the neurons in MT have receptive fields that are 4–10 times larger than those for neurons in Area 17. The larger receptive fields of neurons in MT

ᵇle 21.4. *Properties of neurons in the extrastriate visual cortex*

	References
Γ 70–85% high directionally selective (1, 2, 3, 4) 5–10% bidirectional (1) Selective for narrow bars (1)	1. Felleman & Kaas (1984) 2. Baker *et al.* (1981) 3. Maunsell & Van Essen (1983*a*)
Velocity tuning higher than Area 17 (1, 2) Transient ON and OFF responses (1) Large RFs (2–25 degrees width) (1, 2, 3, 4) Very large bilateral antagonistic surrounds (5)	4. Albright *et al.* (1984) 5. Miezen *et al.* (1982)
ST) Direction-selective (1) Eye movement related (1) Enhanced responsiveness during visual fixation and tracking Very large, often bilateral visual RF responses before and during visually evoked saccades, directional selectivity to visual stimuli	1. Newsome & Wurtz (1982) See Lynch (1980); Motter & Mountcastle (1981); Hyvärinen (1982); Sakata *et al.* (1983)
II) 50% color-sensitive (1) 90% orientation-selective (1) 15–20% direction-selective (1) Broadened spectral selectivity (1)	1. Burkhalter & Van Essen (1982)
⁴) Large, silent, spectrally matched suppressive surrounds (1) Similar or higher color selectivity as V-I (2, 3, 4) selective for stimulus length and width (5, 6)	1. Schein *et al.* (1983) 2. Kruger & Gouras (1980) 3. deMonasterio (1983) 4. Zeki (1983) 5. Baker *et al.* (1981) 6. Petersen, Baker & Allman (1980)
Larger, bilateral RFs (1, 2) Selective for shape, texture or color (1, 2) Responds better to complex objects than slits or edges (1, 2, 3, 4)	1. Desimone & Gross (1979) 2. Gross, Rocha-Miranda & Bender, (1972) 3. Leonard *et al.* (1983) 4. Schwartz *et al.* (1984)

could result directly from a convergence of Area 17 inputs. In order to maintain axis and direction selectivity, cells in orientation-matched sets of bands in Area 17 would project to cells in similar bands in MT (Fig. 21.11). These properties of direction selectivity and large receptive fields may be reinforced by the periodic banding of intrinsic connections that occurs in MT (Weller *et al.*, 1984).

MT is an area where 90% of the neurons show some direction selectivity, and 80% are highly selective (Table 21.4). MT neurons respond to a flashed bar of the preferred orientation with transient excitation at both stimulus onset and offset for all positions in the receptive field. However, stimuli moving in the preferred direction are much more effective than flashed stimuli, and neurons have preferred rates of movement. The new features added in MT are the wide receptive field (5–20 degrees), an apparently enhanced preference for moving stimuli, and the creation of a large surround that is antagonistic for the same direction of movement. Since this surround extends into the ispilateral visual hemifield, it is undoubtedly mediated in part by the callosal connections of MT (Fig. 21.9). Despite the large excitatory receptive fields, the neurons prefer very narrow stimuli (Felleman & Kaas 1984).

The major cortical output of MT is to the adjoining ST (or MST) cortex (Fig. 10; see also Van Essen *et al.*, 1981; Weller *et al.*, 1984). Little is known about the response properties of neurons in the MT projection zone but, not surprisingly, they appear to be predominantly direction-selective, like MT neurons. In addition, the responses of many neurons are affected by eye movements (Newsome & Wurtz, 1982).

The ST region sends major projections to the frontal eye fields, presumably related to the control of eye movements, and to part of the posterior parietal cortex. While the response properties of neurons in the specific parietal lobe target zone of ST are uncertain, neurons in the posterior parietal cortex have been extensively studied (for review, see Hyvärinen, 1982; Sakata, Shibutani & Kawano, 1983). Many of these neurons (about 30%) respond to visual stimuli, are typically direction-selective, have large, sometimes bilateral receptive fields, usually excluding the fovea, have little velocity selectivity, and exhibit enhanced visual responses during visual fixation. Other neurons respond during visual saccades (> 10%), visual tracking (> 10%), and visual fixation (about 30%).

Overall, these properties are consistent with th proposed role of the posterior parietal cortex in visu orientation, visual guidance and attention, and with major input from ST. In general, visual neurons appe to be related to the contralateral visual hemifield, an this may reflect the restricted callosal input (Fig. 21.9

Of course, the V-I to MT to ST to PP pathw does not function in isolation, and other visual inpu reach the posterior parietal cortex. For example, son input comes from DM. DM also has directional selective neurons (Baker *et al.*, 1981) and some dire inputs from Area 17 (Lin *et al.*, 1982), as well as inp from MT (Weller *et al.*, 1984). A major cortical outpu of PP is to the frontal eye fields (Kaas *et al.*, 1977).

The V-II–IT processing sequence

The projections of Area 17 to V-II originate fro quite different cells than do those to MT. The majori of output cells to Area 18 or V-II are pyramidal ce in layer IIIa (Tigges *et al.*, 1981; Rockland & Lun 1983). Small injections of a neuroanatomical tracer V-II will label several distinct clusters of neurons in V indicating that a given location in V-II receives convergence of input from spaced locations in V (Livingstone & Hubel, 1984). In addition, a sm injection of tracer in Area 17 will label patches terminals in V-II (Weller & Kaas, 1983), so a giv location in V-I projects to several locations in V-II. locations in Area 17 project to V-II, since lar injections in V-II label a continuous band of layer II cells in V-I. A few cells along the lower margin of lay IIIb and in lower layer V also project to V-II.

The response properties of cells in the layer II projection zone of the striate cortex are not homoge eous. Some of the projecting cells undoubte originate in the cytochrome oxidase stained blobs layer IIIa (Fig. 21.10). Cells in the blobs have little no orientation selectivity (Livingstone & Hubel, 198 1984), and some are rather broadband and unselecti for color, while others are color-coded (Dow, 1974). part, these properties may reflect heterogeneous inpu relayed from W-like ganglion cells (Fig. 21.1) via t interlaminar and S layer geniculate cells (Figs 21.2 a 21.4). In contrast, cells located between the cytochro oxidase blobs in layer IIIa are orientation-selecti (Livingstone & Hubel, 1982, 1984), and clusters bands of these cells are strongly color-coded, wh other clusters or bands are activated by white lig (Michael, 1981). The color-coded cells with orientati lectivity, at least, can be expected to be dominated y the parvocellular geniculate inputs, though indirectly, d even the non-selective or broadband cells may be rongly related to parvocellular geniculate inputs as ell (in part through the convergence of different olor-selective types; see Kruger & Gouras, 1980), though mixed parvocellular and magnocellular fluences certainly occur (Malpeli, Schiller & Colby, 981). Thus, it appears that the output of V-I to V-II rgely consists of processed information relayed from –like and, to some extent, W-like ganglion cells of the tina.

The major cortical output of V-II is to DL (V4) f the same cerebral hemisphere (Zeki, 1971, 1978; aas *et al.*, 1977; Ungerleider *et al.*, 1983; Weller & aas, 1985). The properties of identified output cells

Fig. 21.12. Hypothetical projection pattern of Area 17 to Area 18 (V-II). Projecting neurons are located in layers IIIa and V. The layer IIIa projecting neurons include those in cytochrome puffs (shaded) and those between the puffs, apparently related to the C (W) and B (X) projection systems, respectively (see Fig. 4). Each projecting neuron projects to several distinct locations, and each location in V-II receives neurons from several separated locations in Area 17. Thus, there is convergence in the projections, to produce larger receptive fields, and divergence, so that several separated modules of neurons receive the same information. The regions of dense cytochrome oxidase reaction in Area 17 may have neurons projecting to the dense reaction regions in Area 18. The cytochrome oxidase dense regions or puffs in Area 18 form bands (shading) which appear to alternate as thick and thin (Tootell *et al.*, 1983; Livingstone & Hubel, 1984; Wong-Riley & Carroll, 1984*b*). Inputs from the inferior pulvinar are discontinuous (Curcio & Harting, 1978) and they may relate specifically to the cytochrome bands (Livingstone & Hubel, 1982).

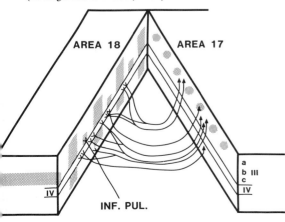

of V-II are unknown, but perhaps half of the neurons in V-II are color-sensitive (Burkhalter & Van Essen, 1982), and Kruger & Gouras (1980) suggest that most V-II cells reflect the earlier convergence of connections of different types of color-opponent cells in Area 17. Most V-II cells are orientation-selective, and some (10–20%) are direction-selective (Table 21.4). DL or V4 has a high proportion of color-sensitive cells, apparently with broadened spatial selectivity. Although various proportions of color-sensitive cells have been described, there is general agreement that the area is at least as involved in color processing as V-II (Table 21.4). DL neurons appear to be more highly selective for lesser stimulus lengths and widths than neurons in MT and DM (Petersen *et al.*, 1980; Baker *et al.*, 1981); however, see Felleman & Kaas (1984) on the selectivity of MT neurons for narrow stimulus widths. As in V-I and V-II, neurons in DL have receptive fields in the contralateral visual hemifield, but they also have large, silent, spectrally matched suppressive surrounds that sometimes extend into the ipsilateral hemifield (Schein *et al.*, 1983), apparently as a result of callosal connections, since most of DL (V4) receives moderate to dense callosal connections from DL of the opposite hemisphere (Weller & Kaas, 1982*a*).

The major output of DL or V4 (see Desimone *et al.*, 1980; Weller & Kaas, 1982*a*, 1985; Felleman & Van Essen, 1983) is to the caudal portion of the inferior temporal cortex (IT_D and IT_V, Fig. 21.6; caudal IT, Fig. 21.7), and this output goes to both cerebral hemispheres, thereby creating the commonly observed extensions of excitatory receptive fields into the ipsilateral hemifield (Gross *et al.*, 1972; Desimone & Gross, 1979). Receptive fields are typically 20–30 degrees in diameter, and neurons respond better to complex objects or shapes than to slits or edges (Table 21.4). The caudal IT projects to the rostral IT (Mishkin, 1972; Weller & Kaas, 1982*a*) and neurons in the rostral IT have even larger receptive fields (Desimone & Gross, 1979). The rostral IT cortex in turn projects to the amygdala and hippocampus (Turner, Mishkin & Knapp, 1980). The important roles of these structures in object vision and visual memory, judging from ablation studies, have been outlined by Mishkin (1982), Ungerleider & Mishkin (1982), and Zola-Morgan, Squire & Mishkin (1982). In macaque monkeys, lesions in Area TEO (Fig. 21.7) produce severe impairments in shape discriminations. On the other hand, animals with TE lesions can discriminate shapes, but they exhibit greatly impaired object recognition, apparently by depriving the amygdala and hippocampus of critical information for memory storage. Animals with amygdalohippocampal ablations are severely deficient in recognition memory.

Summary and conclusions

Among the several significant recent advances in an understanding of the functional organization of the visual system of primates are: (1) the evidence for parallel processing pathways from the retina to the cortex that are highly similar to the X, Y, and W subsystems in cats; (2) the discovery that approximately the posterior half of the neocortex is visual, and that this vast expanse of visual cortex contains a number of sharply defined visual areas, most of which are systematic representations of the visual field; (3) the acquisition in remarkable detail of a comprehensive picture of the complex connection pattern of the visual system; (4) a concept of several important features of the modular and laminar organization of visual structures, in particular the LGN and Area 17, with intriguing suggestions for Area 18; and (5) considerable information on response properties of neurons at several processing levels in the visual system, including those in several critical extrastriate visual areas.

Some of this newer information has been related to the longstanding observations that lesions in the temporal and parietal lobes of higher primates result in quite different behavioral impairments. Following the proposal of Mishkin and co-workers (see Ungerleider & Mishkin, 1982) for 'two cortical visual systems', we describe the perceptual visual system of primates as two somewhat dissociated subsystems, one mediating form or object vision, and relating importantly to visual recognition and memory, and the other concerned with aspects of vision roughly characterized as detection, localization and orientation in space. The object vision pathway depends on the inputs from the X-like ganglion cells that are concentrated near the fovea, relayed in the parvocellular geniculate layers to particular sublaminae of Area 17, and form a major basis for input into Area 18. The second visual area or Area 18 relays in turn to more rostral extrastriate cortex, previously described as DL or V4, and then information is sent over several stations in inferior temporal cortex to relate ultimately to memory storage, retrieval, and reinforcement functions of the hippocampus and amygdala. Part of the heterogenous W-like

ganglion cell pathway appears to relay to the striate cortex information that is important in the object recognition subsystem, and this information remains relatively segregated, at least into Area 18. Other functions of the W-like input to Area 17 are more obscure, but a general activating mode is consistent with widespread terminations in cortical layer I. The visual detection, localization and orientation pathway starts with the Y-like ganglion cells of the retina, which are characterized by relatively large receptive field, sensitivity to low luminance contrast and rapid stimulus movements, transient responsiveness, and lack of color selectivity. The Y-like pathway is relayed through the magnocellular geniculate layers to specific sublayers of Area 17, where higher-order, direction-selective cells, apparently with little influence from the X-like pathway, project to MT which in turn relays to ST or MST, providing a major input to part of the posterior parietal cortex.

The concept of two cortical subsystems is obviously a simplification. First, both X-like and Y-like inputs via the parvocellular and magnocellular geniculate layers jointly affect many, perhaps the majority, of neurons in Area 17 (Malpeli *et al.*, 1981). There are many opportunities for interactions at subsequent processing stages through the multitude of interconnections of visual areas (Fig. 21.8). Yet, the two major streams of flow seem, from single neuron recordings, to be remarkably pure in the sources of their dominant driving. The proposed model would be greatly enhanced by a better understanding of how the X-like, Y-like and W-like channels interact.

Secondly, the roles of the visual areas and subcortical visual centers outside of the two cortical subsystems are not well understood. DM, as already suggested, appears from connections to be most strongly related to the posterior parietal cortex subsystem. DM depends to some extent on both MT and direct V-I inputs (Fig. 21.10). The limited single neuron recordings to date have not yet suggested a distinctive role for DM (Baker *et al.*, 1981). Because the ventral visual region (VP and VA) is strongly implicated in color processing in macaque monkey (Burkhalter & Van Essen, 1982), it appears to be most directly related to the temporal lobe system, with which it connects (Weller & Kaas, 1982*a*). Area M provides the feedback type of connections that presumably modulate visual areas largely in the posterior parietal cortex subsystem, but Area M also projects to Area 18

Graham *et al.*, 1979*a*). The possible roles of other subdivisions of visual cortex are less apparent, and little is known of the functional significance of the pulvinar complex or of the superior colliculus to pulvinar relay of visual information to portions of extrastriate cortex (Desimone *et al.*, 1979).

Thirdly, it is not clear why subdivisions of the visual system are so widely interconnected. Even connections that terminate in the middle layers of cortex, when sparse, may not be capable of driving neurons without other sources of activation, and thus many of the widespread connections of the visual cortex may be modulatory. In addition, direct influences on subcortical motor programming centers are undoubtedly important, since they characterize all subdivisions of visual cortex. Most or all higher-order cortical visual areas (IT, ST, PP) also project to visually responsive and apparently motor-related regions of the frontal lobes. Although these are some of the obvious limitations of the simplification of X-like and Y-like channels and two major cortical streams, such a concept provides a useful framework for the development of a more comprehensive theory of the functional organization of the primate visual system.

Acknowledgments

Research by the author was supported by NEI Grant YO2686. Helpful comments on the manuscript were made by B. Dreher, P. E. Garraghty, M. F. Huerta, M. A. Sesma, L. G. Ungerleider and R. E. Weller.

References

Albright, T. D., Desimone, R. & Gross, C. G. (1984) Columnar organization of directionally selective cells in visual area MT of the macaque. *J. Neurophysiol.*, 51, 16–31

Allman, J. M. & Kaas, J. H. (1974) The organization of the second visual area (V-II) in the owl monkey: a second order transformation of the visual hemifield. *Brain Res.*, 76, 247–65

Allman, J. M. & Kaas, J. H. (1976) Representation of the visual field on the medial wall of occipital-parietal cortex in the owl monkey. *Science*, 191, 572–5

Baizer, J. S., & Maguire, V. M. (1983) Double representation of lower visual quadrant in prelunate gyrus of rhesus monkey. *Invest. Ophthalmol. Vis. Sci.*, 24, 1436–9

Baker, J. F., Petersen, S. E., Newsome, W. T. & Allman, J. M. (1981) Visual response properties of neurons in four extrastriate visual areas of the owl monkey (*Aotus trivirgatus*): a quantitative comparison of medial, dorsomedial, and middle temporal areas. *J. Neurophysiol.*, 45, 397–416

Benevento, L. A. & Yoshida, K. (1981) The afferent and efferent organization of the lateral geniculoprestriate pathways in the Macaque monkey. *J. Comp. Neurol.*, 203, 455–74

Birecree, E. & Casagrande, V. A. (1981) Laminar differences in the morphology of lateral geniculate nucleus cells in galago. *Soc. Neurosci. Abstr.*, 7, 421

Blasdel, G. G. & Fitzpatrick, D. (1984) Physiological organization of layer 4 in macaque striate cortex. *J. Neurosci.*, 4, 880–95

Blasdel, G. G. & Lund, J. S. (1983) Terminations of afferent axons in macaque striate cortex. *J. Neurosci.*, 3, 1389–413

Brodmann, K. (1909) *Vergleichende Lokalisationslehre der Grosshirnrinde.* Leipzig: Barth

Bruce, C., Desimone, R. & Gross, C. G. (1981) Visual properties of neurons in a polysensory area in superior temporal sulcus of the macaque. *J. Neurophysiol.*, 46, 369–84

Bullier, J. & Henry, G. H. (1980) Ordinal position of afferent input of neurons in monkey striate cortex. *J. Comp. Neurol.*, 193, 913–36

Burkhalter, A. & Van Essen, D. C. (1982) Processing of color, form and disparity in visual areas V2 and VP of ventral extrastriate cortex in the macaque. *Soc. Neurosci. Abstr.*, 8, 811

Carey, R. G., Bear, M. F. & Diamond, I. T. (1980) The laminar organization of the reciprocal projections between the claustrum and striate cortex in the tree shrew, *Tupaia glis*. *Brain Res.*, 184, 193–8

Casagrande, V. A. & DeBruyn, E. J. (1982) The galago visual system: aspects of normal organization and developmental plasticity. In *The Lesser Bushbaby (Galago) as an Animal Model: Selected Topics*, ed. D. E. Haines, pp. 138–68. Boca Raton, Florida: CRC Press

Casagrande, V. A. & Skeen, L. C. (1980) Organization of ocular dominance columns in galago demonstrated by autoradiographic and deoxyglucose methods. *Soc. Neurosci. Abstr.*, 6, 315

Choudhury, B. P., Whitteridge, D. & Wilson, M. E. (1965) The function of the callosal connections of the visual cortex. *Quart. J. Exp. Physiol.*, 50, 2140–9

Constantine-Paton, M. (1982) The retinotectal hookup: the process of neural mapping. In *Development Order: Its Origin and Regulation*, ed. S. Subtelny & P. Green, pp. 317–49. New York: Alan R. Liss, Inc.

Covey, E., Gattass, R. & Gross, C. G. (1982) A new visual area in the parieto–occipital sulcus of the macaque. *Soc. Neurosci. Abstr.*, 8, 681

Creutzfeldt, O. D., Lee, B. B. & Elepfandt, A. (1979) A quantitative study of chromatic organization and receptive fields of cells in the lateral geniculate body of the rhesus monkey. *Exp. Brain Res.*, 35, 527–45

Curcio, C. A. & Harting, J. K. (1978) Organization of pulvinar afferents to Area 18 in the squirrel monkey: evidence for stripes. *Brain Res.*, 143, 155–61

Cusick, C. G., Gould, H. J. & Kaas, J. H. (1984) Interhemispheric connections of visual cortex of owl monkeys (*Aotus trivirgatus*), marmosets (*Callithrix jacchus*), and galagos (*Galago crassicaudatus*). *J. Comp. Neurol.*, 230, 311–36

deMonasterio, F. M. (1978*a*) Properties of concentrically

organized X and Y ganglion cells of macaque retina. *J. Neurophysiol.*, 41, 1394–417

deMonasterio, F. M. (1978b) Center and surround mechanisms of opponent-color X and Y ganglion cells of retina of macaques. *J. Neurophysiol.*, 41, 1418–34

deMonasterio, F. M. (1978c) Properties of ganglion cells with atypical receptive-field organization in retina of macaques. *J. Neurophysiol.*, 41, 1435–49

deMonasterio, F. M. (1983) Color selectivity of V4 cells and induced spectral properties. *Soc. Neurosci. Abstr.*, 9, 153

Desimone, R., Bruce, C. J. & Gross, C. G. (1979) Neurons in the superior temporal sulcus of the macaque still respond to visual stimuli after removal of striate cortex. *Soc. Neurosci. Abstr.*, 5, 781

Desimone, R., Fleming, J. & Gross, C. G. (1980) Prestriate afferents to inferior temporal cortex: an HRP study. *Brain Res.*, 184, 41–55

Desimone, R. & Gross, C. G. (1979) Visual areas in the temporal cortex of the macaque. *Brain Res.*, 178, 363–80

Dow, B. M. (1974) Functional classes of cells and their laminar distribution in monkey visual cortex. *J. Neurophysiol.*, 37, 927–46

Dreher, B., Fukada, Y. & Rodieck, R. W. (1976) Identification, classification and anatomical segregation of cells with X-like and Y-like properties in the lateral geniculate nucleus of Old World primates. *J. Physiol. (Lond.)*, 258, 433–52

Felleman, D. J. & Kaas, J. H. (1984) Receptive field properties of neurons in the middle temporal visual area (MT) of owl monkeys. *J. Neurophysiol.*, 52, 488–513

Felleman, D. J. & Van Essen, D. C. (1983) The connections of Area V4 of macaque monkey extrastriate cortex. *Soc. Neurosci. Abstr.*, 9, 153

Finlay, B. L., Schiller, P. H. & Volman, S. F. (1976) Quantitative studies of single-cell properties in monkey striate cortex. IV. Corticotectal cells. *J. Neurophysiol.*, 39, 1352–61

Fitzpatrick, D., Carey, R. G. & Diamond, I. T. (1980) The projection of the superior colliculus upon the lateral geniculate body in *Tupaia glis* and *Galago senegalensis*. *Brain Res.*, 194, 494–9

Fitzpatrick, D., Itoh, K. & Diamond, I. T. (1983) The laminar organization of the lateral geniculate body and the striate cortex in the squirrel monkey (*Saimiri sciureus*). *J. Neurosci.*, 3, 673–702

Florence, S. L. & Casagrande, V. A. (1978) A note on the evolution of ocular dominance columns in primates. *Invest. Ophthalmol. Vis. Sci.*, 17, 291–2

Florence, S. L., Sesma, M. A. & Casagrande, V. A. (1983) Morphology of geniculo-striate afferents in a prosimian primate. *Brain Res.*, 270, 127–30

Fries, W. & Distel, H. (1982) Large layer VI cells in the macaque striate cortex labeled after HRP injection into the superior colliculus. *ARVO Abstracts. Invest. Ophthal. Vis. Sci.*, 22, (suppl.), 243

Gattass, R., Gross, C. G. & Sandell, J. H. (1981) Visual topography of V2 in the macaque. *J. Comp. Neurol.*, 201, 519–39

Glendenning, K. K., Kofron, E. A. & Diamond, I. T. (1976) Laminar organization of projections of the lateral geniculate nucleus to the striate cortex in *Galago. Brain Res.*, 105, 538–46

Gould, H. J., III, Cusick, C. G., Pons, T. P. & Kaas, J. H. (198 The relation of callosal connections to microstimulation maps of precentral motor cortex in owl monkeys. *Soc. Neurosci. Abstr.*, 9, 309

Graham, J., Lin, C.-S. & Kaas, J. H. (1979a) Sub-cortical projections of six visual cortical areas in the owl monkey, *Aotus trivirgatus. J. Comp. Neurol.*, 187, 557–80

Graham, J., Wall, J. & Kaas, J. H. (1979b) Cortical projections of the medial visual area in the owl monkey, *Aotus trivirgatu Neurosci. Lett.*, 15, 109–14

Gross, C. G., Rocha-Miranda, C. E. & Bender, D. B. (1972) Visual properties of neurons in inferotemporal cortex of the macaque. *J. Neurophysiol.*, 35, 96–111

Harting, J. K., Casagrande, V. A. & Weber, J. T. (1978) The projection of the primate superior colliculus upon the dorsal lateral geniculate nucleus: autoradiographic demonstration of interlaminar distribution of tectogeniculate axons. *Brain Res.*, 150, 593–9

Hassler, R. (1966) Comparative anatomy of the central visual systems in day- and night-active primates. In *Evolution the Forebrain*, ed. R. Hassler & H. Stephen, pp. 419–34. Stuttgart: Thieme

Headon, M. P., Sloper, J. J., Hiorns, R. W. & Powell, T. P. S. (1981) Cell sizes in the lateral geniculate nucleus of norm infant and adult rhesus monkeys. *Brain Res.*, 229, 183–96

Hendrickson, A. E., Ogren, M. P., Vaughn, J. E., Barber, R. P. Wu, J.-Y. (1983) Light and electron microscopic immunocytochemical localisation of glutamic acid decarboxylase in monkey geniculate complex: evidence for GABAergic neurons and synapses. *J. Neurosci.*, 3, 245–62

Hendrickson, A. E. & Wilson, J. R. (1979) A difference in [^{14}C] deoxyglucose autoradiographic patterns in striate cortex between *Macaca* and *Saimiri* monkeys following monocular stimulation. *Brain Res.*, 170, 353–8

Hendrickson, A. E., Wilson, J. R. & Ogren, M. P. (1978) The neuroanatomical organization of pathways between the dorsal lateral geniculate nucleus and visual cortex in old world and new world primates. *J. Comp. Neurol.*, 182, 123–36

Hendry, S. H. C., Hockfield, S., Jones, E. G. & McKay, R. (1984) Monoclonal antibody that identifies subsets of neurons in the central visual system of monkey and cat. *Nature*, 307, 267–9

Hickey, T. L. & Guillery, R. W. (1981) A study of Golgi preparations from the human lateral geniculate nucleus. *J Comp. Neurol.*, 200, 545–77

Hitchcock, P. F. & Hickey, T. L. (1980) Ocular dominance columns: evidence for their presence in humans. *Brain Res.*, 182, 176–9

Horton, J. C. & Hubel, D. H. (1981) Regular patchy distribution of cytochrome oxidase staining in primary visual cortex of macaque monkey. *Nature*, 292, 762

Hubel, D. H. & Wiesel, T. N. (1968) Receptive fields and functional architecture of monkey striate cortex. *J. Physio (Lond.)*, 195, 215–43

Hubel, D. H. & Wiesel, T. N. (1972) Laminar and columnar

distribution of geniculo-cortical fibers in the macaque monkey. *J. Comp. Neurol.*, **146**, 421–50

Hubel, D. H. & Wiesel, T. N. (1977) Functional architecture of macaque monkey visual cortex. *Proc. R. Soc. Lond. Ser. B*, **198**, 1–59

Hubel, D. H., Wiesel, T. N. & Stryker, M. P. (1976) Anatomical demonstration of orientation columns in macaque monkey. *J. Comp. Neurol.*, **172**, 563–84

Humphrey, A. L. & Hendrickson, A. E. (1983) Background and stimulus-induced patterns of high metabolic activity in the visual cortex (Area 17) of the squirrel and macaque monkey. *J. Neurosci.*, **3**, 345–58

Hyvärinen, J. (1982) Posterior parietal lobe of the primate brain. *Physiol. Reviews*, **62**, 1060–129

Imig, T. J., Ruggero, M. H., Kitzes, L. M., Javel, E. & Brugge, J. F. (1977) Organization of auditory cortex in the owl monkey (*Aotus trivirgatus*). *J. Comp. Neurol.*, **171**, 111–28

Irvin, G. E., Norton, T. T., Sesma, M. A. & Casagrande, V. A. (1984) W-like receptive-field properties of interlaminar cells in primate lateral geniculate nucleus. *Soc. Neurosci. Abstr.*, **10**, 297

Itoh, K., Conley, M. & Diamond, I. T. (1982) Retinal ganglion cell projections to individual layers of the lateral geniculate body in *Galago crassicaudatus*. *J. Comp. Neurol.*, **205**, 282–90

Jones, A. E. (1966) Wavelength and intensity effects on the response of single lateral geniculate nucleus units in the owl monkey. *J. Neurophysiol.*, **29**, 125–38

Jones, E. G. (1975) Some aspects of the organization of the thalamic reticular complex. *J. Comp. Neurol.*, **162**, 284–308

Kaas, J. H. (1978) The organization of the visual cortex of primates. In *Sensory Systems of Primates*, ed. C. R. Noback, pp. 151–70. New York: Plenum Press

Kaas, J. H., Guillery, R. W. & Allman, J. M. (1972) Some principles of organization in the dorsal lateral geniculate nucleus. *Brain Behav. Evol.*, **6**, 253–99

Kaas, J. H., Huerta, M. F., Weber, J. T. & Harting, J. K. (1978) Patterns of retinal terminations and laminar organization of the lateral geniculate nucleus of primates. *J. Comp. Neurol.*, **182**, 517–54

Kaas, J. H. & Lin, C.-S. (1977) Cortical projections of area 18 in owl monkeys. *Vis. Res.*, **17**, 739–41

Kaas, J. H., Lin, C.-S. & Casagrande, V. A. (1976) The relay of ipsilateral and contralateral retinal input from the lateral geniculate nucleus to striate cortex in the owl monkey: a transneuronal transport study. *Brain Res.*, **106**, 371–8

Kaas, J. H., Lin, C.-S. & Wagor, E. (1977) Cortical projections of posterior parietal cortex in owl monkeys. *J. Comp. Neurol.*, **171**, 387–408

Kaplan, E. & Shapley, R. M. (1982) X and Y cells in the lateral geniculate nucleus of macaque monkeys. *J. Physiol.*, **330**, 125–43

Kruger, J. & Gouras, P. (1980) Spectral selectivity of cells and its dependence on slit length in monkey visual cortex. *J. Neurophysiol.*, **43**, 1055–70

Lee, B. B., Creutzfeldt, O. D. & Elepfandt, A. (1979) The responses of magno- and parvocellular cells of the monkey's lateral geniculate body to moving stimuli. *Exp. Brain Res.*, **35**, 457–557

LeGros Clark, W. E. (1941) The laminar organization and cell content of the lateral geniculate body in the monkey. *J. Nat.*, **75**, 419–32

Lennie, P. & Derrington, A. (1981) Spatial contrast sensitivity in the macaque's LGN. *Invest. Ophthalmol. Vis. Sci.*, **20**, 14

Leonard, C. M., Rolls, E. T., Baylis, G. C., Wilson, F. A. W., Williams, G. V., Griffiths, C. & Murzi, E. (1983) Response properties and distribution of neurons which respond to faces in the monkey. *Soc. Neurosci. Abstr.*, **9**, 958

Leventhal, A. G., Rodieck, R. W. & Dreher, B. (1981) Retinal ganglion cell classes in the old world monkey: morphology and central projections. *Science*, **213**, 1139–42

Lin, C.-S. & Kaas, J. H. (1977) Projections from cortical visual areas 17, 18, and MT onto the dorsal lateral geniculate nucleus in owl monkeys. *J. Comp. Neurol.*, **173**, 457–74

Lin, C.-S. & Kaas, J. H. (1979) The inferior pulvinar complex in owl monkeys: architectonic subdivisions and patterns of input from the superior colliculus and subdivisions of visual cortex. *J. Comp. Neurol.*, **187**, 655–78

Lin, C.-S., Weller, R. E. & Kaas, J. H. (1982) Cortical connections of striate cortex in the owl monkey. *J. Comp. Neurol.*, **211**, 165–76

Livingstone, M. S. & Hubel, D. H. (1982) Thalamic inputs to cytochrome oxidase-rich regions in monkey visual cortex. *Proc. Natl Acad. Sci. USA*, **79**, 6098–101

Livingstone, M. S. & Hubel, D. H. (1984) Anatomy and physiology of a color system in the primate visual cortex. *J. Neuroscience*, **4**, 309–56

Lund, J. S. (1980) Intrinsic organization of the primate visual cortex, Area 17, as seen in Golgi preparations. In *The Organization of the Cerebral Cortex*, ed. F. O. Schmitt, F. G. Worden, G. Adelman & S. G. Dennis, pp. 105–24. Proceedings of a Neuroscience Research program colloquium. Cambridge, Mass.: MIT Press

Lund, J. S., Henry, G. H., MacQueen, C. L. & Harvey, A. R. (1979) Anatomical organization of the primary visual cortex (Area 17) of the cat. A comparison with Area 17 of the macaque monkey. *J. Comp. Neurol.*, **184**, 599–618

Lynch, J. C. (1980) The functional organization of posterior parietal associations cortex. *Behav. Brain Sci.*, **3**, 485–534

Malpeli, J. G., Schiller, P. H. & Colby, C. L. (1981) Response properties of single cells in monkey striate cortex during reversible inactivation of individual lateral geniculate laminae. *J. Neurophysiol.*, **46**, 1102–19

Marrocco, R. T. (1978) Conduction velocities of afferent input to superior colliculus in normal and decorticate monkeys. *Brain Res.*, **140**, 155–8

Marrocco, R. T., McClurkin, J. W. & Young, R. A. (1982a) Modulation of lateral geniculate nucleus cell responsiveness by visual activation of the corticogeniculate pathway. *J. Neurosci.*, **2**, 256–63

Marrocco, R. T., McClurkin, J. W. & Young, R. A. (1982b) Spatial summation and conduction latency classification cells of the lateral geniculate nucleus of macaques. *J. Neurosci.*, **2**, 1275–91

Matsubara, J. & Cynader, M. S. (1983) The role of orientation tuning on the specificity of local intrinsic connections in cat visual cortex: an antomical and physiological study. *Soc. Neurosci. Abstr.*, **9**, 475

Maunsell, J. H. R. & Van Essen, D. C. (1983*a*) Single unit responses in the Middle Temporal Area of the macaque: I. Direction, speed, and orientation. *J. Neurophysiol.*, **49**, 1148–67

Maunsell, J. H. R. & Van Essen, D. C. (1983*b*) The connections of the middle temporal visual area (MT) and their relationship to a cortical hierarchy in the macaque monkey. *J. Neurosci.*, **3**, 2563–86

Merzenich, M. M., Kaas, J. H., Sur, M. & Lin, C.-S. (1978) Double representation of the body surface within cytoarchitectonic Areas 3b and 1 in 'SI' in the owl monkey (*Aotus trivirgatus*). *J. Comp. Neurol.*, **181**, 41–74

Michael, C. R. (1981) Columnar organization of color cells in monkey striate cortex. *J. Neurophysiol.*, **46**, 587–604

Miezen, F., McGuinness, E. & Allman, J. (1982) Antagonistic direction-specific mechanisms in Area MT in the owl monkey. *Soc. Neurosci. Abstr.*, **8**, 681

Mishkin, M. (1972) Cortical visual areas and their interaction. In *The Brain and Human Behavior*, ed. A. G. Karczmar & J. C. Eccles, pp. 187–208. Springer-Verlag: New York.

Mishkin, M. (1982) A memory system in the monkey. *Phil. Trans. R. Soc. Lond., Ser. B*, **298**, 85–95

Mitchison, G. & Crick, F. (1982) Long axons within the striate cortex: their distribution orientation, and patterns of connection. *Proc. Natl Acad. Sci. USA*, **79**, 3661–5

Montero, V. M. (1980) Patterns of connections from striate cortex to cortical visual areas in superior temporal sulcus of macaque and middle temporal gyrus of owl monkey. *J. Comp. Neurol.*, **189**, 45–59

Moran, V., Desimone, R., Schein, S. J. & Mishkin, M. (1983) Suppression from ipsilateral visual field in Area V4 of the macaque. *Soc. Neurosci. Abstr.*, **9**, 957

Motter, B. C. & Mountcastle, V. B. (1981) The functional properties of the light-sensitive neurons of the posterior parietal cortex studied in waking monkeys: foveal sparing and opponent vector organization. *J. Neurosci.*, **1**, 3–26

Nelson, R. J., Sur, M., Felleman, D. J. & Kaas, J. H. (1980) Representation of the body surface in postcentral parietal cortex of *Macaca fascicularis*. *J. Comp. Neurol.*, **192**, 611–43

Newsome, W. T. & Allman, J. M. (1980) Interhemispheric connections of visual cortex in the owl monkey, *Aotus trivirgatus*, and the bushbaby, *Galago senegalensis*. *J. Comp. Neurol.*, **194**, 209–33

Newsome, W. T., Maunsell, J. H. R. & Van Essen, D. C. (1980) Areal boundaries and topographic organization of the ventral posterior area (VP) of the macaque monkey. *Soc. Neurosci. Abstr.*, **6**, 579

Newsome, W. T. & Wurtz, R. H. (1982) Identification of architectonic zones containing visual tracking cells in the superior temporal sulcus (STS) of macaque monkeys. *Invest. Ophthalmol. Vis. Sci.*, **22**, 238

Newsome, W. T., Wurtz, R. H., Dursteler, M. R. & Mikami, A. (1983) Deficits in pursuit eye movements after chemical lesions of motion-related visual areas in the superior temporal sulcus of the macaque monkey. *Soc. Neurosci. Abstr.*, **9**, 154

Norden, J. J. & Kaas, J. H. (1978) The identification of relay neurons in the dorsal lateral geniculate nucleus of monkeys using horseradish peroxidase. *J. Comp. Neurol.*, **182**, 707–26

Norton, T. T. & Casagrande, V. A. (1982) Laminar organization of receptive-field properties in the lateral geniculate nucleus of bushbaby (*Galago crassicaudatus*). *J. Neurophysiol.*, **47**, 714–41

Ogren, M. P., Hendrickson, A. E., Vaughn, J., Barber, R. P. & Wu, J.-Y. (1982) GABAergic neurons and synapses in monkey dorsal lateral geniculate nucleus: a light- and electron-microscopic immunohistochemical analysis. *Soc. Neurosci. Abstr.*, **8**, 262

Perrett, D. I., Rolls, E. T. & Caan, W. (1982) Visual neurones responsive to faces in the monkey temporal cortex. *Exp. Brain Res.*, **47**, 324–42

Perry, V. H. & Cowey, A. (1981) The morphological correlates of X- and Y-like retinal ganglion cells in the retina of monkeys. *Exp. Brain Res.*, **43**, 226–8

Petersen, S. E., Baker, J. F. & Allman, J. M. (1980) Dimensional selectivity of neurons in the dorsolateral visual area of the owl monkey. *Brain Res.*, **197**, 507–11

Rocha-Miranda, C. E., Bender, D. B., Gross, C. G. & Mishkin, M. (1975) Visual activation of neurons in inferotemporal cortex depends on striate cortex and forebrain commissures. *J. Neurophysiol.*, **38**, 474–91

Rockland, K. S. & Lund, S. J. (1983) Intrinsic laminar lattice connections in primate visual cortex. *J. Comp. Neurol.*, **216**, 303–18

Rockland, K. S., Lund, J. S. & Humphrey, A. L. (1982) Anatomical banding of intrinsic connections in striate cortex of tree shrews (*Tupaia glis*). *J. Comp. Neurol.*, **209**, 41–58

Rockland, K. S. & Pandya, D. N. (1979) Laminar origins and terminations of cortical connections of the occipital lobe in the rhesus monkey. *Brain Res.*, **179**, 3–20

Rockland, K. S. & Pandya, D. N. (1981) Cortical connections of the occipital cortex in rhesus monkey: interconnections between areas 17, 18, 19 and the superior temporal sulcus. *Brain Res.*, **212**, 249–70

Rodieck, R. W. (1979) Visual pathways. *Ann. Rev. Neurosci.*, **2**, 193–225

Rodieck, R. W. & Brening, R. K. (1983) Retinal ganglion cells: properties, types, genera, pathways, and trans-species comparisons. *Brain Behav. Evol.*, **23**, 121–64

Rodieck, R. W. & Dreher, B. (1979) Visual suppression from nondominant eye in lateral geniculate nucleus: a comparison of cat and monkey. *Exp. Brain Res.*, **35**, 465–77

Rowe, M. H., Benevento, L. A. & Rezak, M. (1978) Some observations on the patterns of segregated geniculate inputs to the visual cortex in New World primates: an autoradiographic study. *Brain Res.*, **159**, 371–8

Rowe, M. H. & Stone, J. (1977) Naming of neurons: classification and naming of cat retinal ganglion cells. *Brain Behav. Evol.*, **14**, 185–216

Sakata, H., Shibutani, H. & Kawano, K. (1983) Functional properties of visual tracking neurons in posterior parietal association cortex of the monkey. *J. Neurophysiol.*, **49**, 1364–80

Sandell, J. H. & Schiller, P. H. (1982) The effect of cooling area

18 on striate cortex cells in the squirrel monkey. *J. Neurophysiol.*, **48**, 38–48

chein, S. J., Desimone, R. & deMonasterio, F. M. (1983) Spectral properties of area V4 cells of macaque monkey. *ARVO Abstracts. Invest. Ophthalmol. Vis. Sci.*, **24**, 107

chiller, P. H. & Malpeli, J. G. (1977a) Properties and tectal projections of monkey retinal ganglion cells. *J. Neurophysiol.*, **40**, 428–45

chiller, P. H. & Malpeli, J. G. (1977b) The effect of striate cortex cooling on area 18 cells in the monkey. *Brain Res.*, **126**, 366–9

chiller, P. H. & Malpeli, J. G. (1978) Functional specificity of lateral geniculate nucleus laminae of the rhesus money. *J. Neurophysiol.*, **41**, 788–97

chiller, P. H., Malpeli, J. G. & Schein, S. J. (1979) Composition of geniculostriate input through the superior colliculus of the rhesus monkey. *J. Neurophysiol.*, **42**, 1124–33

chwartz, E. L., De Simone, R., Albright, T. P. & Gross, C. G. (1984) Shape recognition and inferior temporal neurons. *Proc. Natl. Acad. Sci. USA*, **80**, 5776–8

esma, M. A., Casagrande, V. A. & Kaas, J. H. (1984) Cortical connections of striate cortex in tree shrews. *J. Comp. Neurol.*, **230**, 337–51

hapley, R., Kaplan, E. & Soodak, R. (1981) Spatial summation and contrast sensitivity of X and Y cells in the lateral geniculate nucleus of the macaque. *Nature*, **292**, 543–5

herman, S. M. & Spear, P. D. (1982) Organization of visual pathways in normal and visually deprived cats. *Physiol. Reviews*, **62**, 783–855

herman, S. M., Wilson, J. R., Kaas, J. H. & Webb, S. V. (1976) X- and Y-cells in the dorsal lateral geniculate nucleus of the owl monkey (*Aotus trivirgatus*). *Science*, **192**, 475–7

tone, J. & Dreher, B. (1982) Parallel processing of information in the visual pathways. *Trends Neurosci.*, **5**, 441–6

tone, J., Dreher, B. & Leventhal, A. (1979) Hierarchical and parallel mechanisms in the organization of visual cortex. *Brain Res. Rev.*, **1**, 345–94

ymonds, L. L. & Kaas, J. H. (1978) Connections of striate cortex in the prosimian, *Galago senegalensis. J. Comp. Neurol.*, **181**, 477–512

igges, M., Hendrickson, A. E. & Tigges, J. (1984) Anatomical consequences of long-term monocular eyelid closure on lateral geniculate nucleus and striate cortex in squirrel monkey. *J. Comp. Neurol.*, **227**, 1–13

igges, J. & Tigges, M. (1979) Ocular dominance columns in the striate cortex of chimpanzee (*Pan troglodytes*). *Brain Res.*, **166**, 386–90

igges, J., Tigges, M., Anschel, S., Cross, N. A., Ledbetter, W. D. & McBride, R. L. (1981) Areal and laminar distribution of neurons interconnecting the central visual cortical areas 17, 18, 19 and MT in squirrel monkey (*Saimiri*). *J. Comp. Neurol.*, **202**, 539–60

igges, J., Tigges, M., Cross, N. A., McBride, R. L., Ledbetter, W. D. & Anschel, S. (1982) subcortical structures projecting to visual cortical areas in squirrel monkey. *J. Comp. Neurol.*, **209**, 29–40

igges, J., Tigges, M. & Perachio, A. A. (1977) Complementary laminar terminations of afferents to area 17 originating in area 18 and the LGN in squirrel monkey. *J. Comp. Neurol.*, **176**, 87–100

Tootell, R. B. H., Silverman, M. S. & DeValois, R. L. (1983) Functional organization of the second cortical visual area in primates. *Science*, **220**, 737–9

Turner, B. H., Mishkin, M. & Knapp, M. (1980) Organization of the amygdalopetal projections from modality-specific cortical association areas in the monkey. *J. Comp. Neurol.*, **191**, 515–43

Ungerleider, L. G., Gattass, R., Sousa, A. P. B. & Mishkin, M. (1983) Projections of Area V2 in the macaque. *Soc. Neurosci. Abstr.*, **9**, 152

Ungerleider, L. G. & Mishkin, M. (1982) Two cortical visual systems. In *Analysis of Visual Behavior*, ed. D. J. Ingle, M. A. Goodale & R. J. W. Mansfield, pp. 549–86. Cambridge, Mass.: MIT Press

Van Essen, D. C. (1979) Visual areas of the mammalian cerebral cortex. *Ann. Rev. Neurosci.*, **2**, 227–63

Van Essen, D. C., Maunsell, J. H. R. & Bixby, J. L. (1981) The middle temporal visual area in the macaque; myeloarchitecture, connections, functional properties and topographic organization. *J. Comp. Neurol.*, **199**, 293–326

Van Essen, D. C., Newsome, W. T. & Bixby, J. L. (1982) The pattern of interhemispheric connections and its relationship to extrastriate visual areas in the macaque monkey. *J. Neurosci.*, **2**, 265–83

Van Essen, D. C. & Zeki, S. M. (1978) The topographic organization of rhesus monkey prestriate cortex. *J. Physiol. (Lond.)*, **277**, 193–226

Wässle, H., Peichl, L. & Boycott, B. B. (1981) Morphology and topography of on- and off-alpha cells in the cat retina. *Proc. R. Soc. Lond., Ser. B*, **200**, 441–61

Weber, J. T., Huerta, M. F., Kaas, J. H. & Hartin, J. K. (1983) The projections of the lateral geniculate nucleus of the squirrel monkey: studies of the interlaminar zones and the S layers. *J. Comp. Neurol.*, **213**, 135–45

Weller, R. E. & Kaas, J. H. (1981) Cortical and subcortical connections of visual cortex in primates. In *Multiple Cortical Somatic Sensory-Motor, Visual, and Auditory Areas and Their Connectivities*, ed. C. N. Woolsey, pp. 121–55. Clifton, New Jersey: Human Press

Weller, R. E. & Kaas, J. H. (1982a) Subdivisions and connections of inferior temporal cortex in owl monkeys. *Soc. Neurosci. Abstr.*, **8**, 812

Weller, R. E. & Kaas, J. H. (1982b) The organization of the visual system in Galago: comparisons with monkeys. In *The Lesser Bushbaby (Galago) as an Animal Model: Selected Topics*, ed. D. E. Haines, pp. 107–35. Boca Raton, Florida: CRC Press, Inc.

Weller, R. E. & Kaas, J. H. (1983) Retinotopic patterns of connections of Area 17 with visual areas V-II and MT in macaque monkeys. *J. Comp. Neurol.*, **220**, 253–79

Weller, R. E. & Kaas, J. H. (1985) Cortical projections of the dorsolateral visual area in owl monkeys: The prestriate relay to inferior temporal cortex. *J. Comp. Neurol.*, **233**, (in press)

Weller, R. E., Kaas, J. H. & Wetzel, A. B. (1979) Evidence for the loss of X-cells of the retina after long-term ablation of visual cortex in monkeys. *Brain Res.*, **160**, 134–9

Weller, R. E., Wall, J. T. & Kaas, J. H. (1984) Cortical connections of the middle temporal visual area (MT) and the superior temporal cortex in owl monkeys. *J. comp. Neurol.*, **228**, 81–104

Wiesel, T. N. & Hubel, D. H. (1966) Spatial and chromatic interactions in the lateral geniculate body of the rhesus monkey. *J. Neurophysiol.*, **29**, 1115–56

Wiesel, T. N., Hubel, D. H. & Lam, D. M. K. (1974) Autoradiographic demonstration of ocular-dominance columns in the monkey striate cortex by means of transneuronal transport. *Brain Res.*, **79**, 273–9

Wilson, J. R. & Hendrickson, A. E. (1981) A Golgi and quantitative electron microscopic analysis of the normal Macaca dorsal lateral geniculate nucleus with a comparative EM study of the monocularly deprived nucleus. *J. Comp. Neurol.*, **197**, 517–39

Winfield, D. A., Rivera-Dominguez, M. & Powell, T. P. S. (1981) The termination of geniculocortical fibres in area 17 of the visual cortex in the macaque monkey. *Brain Res.*, **231**, 19–32

Wong-Riley, M. (1978) Reciprocal connections between striate and prestriate cortex in squirrel monkey as demonstrated by combined peroxidase histochemistry and autoradiography. *Brain Res.*, **147**, 159–64

Wong-Riley, M. & Carroll, E. (1984*a*) Effect of impulse blockage on cytochrome oxidase activity in monkey visual system. *Nature*, **307**, 262–4

Wong-Riley, M. T. T. & Carroll, E. W. (1984*b*) Quantitative light and electromicroscopic analysis of cytochrome oxidase-rich zones in V-II prestriate cortex of the squirrel monkey. *J. Comp. Neurol.*, **222**, 18–37

Yukie, M. Y. & Iwai, E. (1981) Direct projections from the dorsal lateral geniculate nucleus to the prestriate cortex in Macaque monkeys. *J. Comp. Neurol.*, **201**, 81–97

Zeki, S. M. (1971) Cortical projections from two prestriate areas in the monkey. *Brain Res.*, **34**, 19–35

Zeki, S. M. (1978) Functional specialization in the visual cortex of the rhesus monkey. *Nature (Lond.)*, **274**, 423–8

Zeki, S. M. (1980) The response properties of cells in the middle temporal area (area MT) of owl monkey visual cortex. *Proc. R. Soc. Lond., Ser. B.*, **207**, 239–48

Zeki, S. M. (1983) The distribution of wavelength and orientation selective cells in different areas of monkey visual cortex. *Proc. R. Soc. Lond., Ser. B.*, **217**, 449–70

Zola–Morgan, S., Squire, L. R. & Mishkin, M. (1982) The neuroanatomy of amnesia: amygdala-hippocampus versus temporal stem. *Science*, **218**, 1337–9

VI

Integrative aspects

The great power of reductionist neuroscience is evident in the increasingly rapid appearance of new molecules, such as 'neuropeptides', the various growth factors and monoclonal antibodies, all of which have unprecedented specificity for different functional and morphological subsets of neuronal elements. Peter Bishop's successor at the Australian National University is a molecular neuroscientist and nearly everyone these days seems to have their own pretty micrograph obtained with a monoclonal antibody directed against some sharply distributed molecule in the nervous system. But neuroscience also needs pioneers with their noses pointed firmly in the opposite direction; those holists with the hermeneutic skill to recognize the same message in many languages without being distracted by the fact that the English version might be composed of gypsum particles on a blackboard whilst the Japanese version might be some horribly complex organic emanation from a felt pen.

This last section attempts to represent some viewpoints of neuroscientists working at the holistic, integrative end of the neuroscience spectrum. It will be no surprise to find that most of these workers have direct experience with psychophysics, a field of endeavour which has always guided the neurophysiologist in his choice of the interesting questions to answer (*vide* the work of Bishop's laboratory on stereopsis). As with stereopsis, where complex stereoscopic processing was found very early in the visual pathway by a Bishop team, here one of Bishop's disciples (Peterhans: Chapter 22) presents evidence for the processing of illusory contours by single neurons at early stages in the visual pathway. This account is set alongside Day's (Chapter 23) psychophysical studies of illusory contours. Day's conclusion that physical attributes of the stimulus, such as the local brightness distributions, cannot account for the phenomenon, is nicely complemented by the behaviour of individual neurons in the monkey visual cortex which were studied by Peterhans. The response of these neurons to a contour which was not physically present on the retina is nevertheless explicable in terms of the properties of other cortical neurons sensitive to terminations (end-stopped neurons) if convergence occurred from end-stopped neurons whose fields were precisely aligned along the illusory contour. The reductionists might claim that their building blocks can explain it all, but it seems likely that the perceptual phenomenon worthy of explanation would have been missed altogether without the holists.

In addition to psychophysics, another most important guiding light for the neuroscientist comes from clinical experience. Lance (Chapter 25) recounts his clinical experience with visual hallucinations, phenomena which may ultimately reveal much about the mapping of visual functions when emission-scanning studies of a living hallucinating brain are linked to knowledge of the increasingly well-defined extrastriate visual cortical areas.

McKay (Chapter 24) presents his viewpoint on the very important question of what is the task of vision. What are visual systems designed to do? The answer to such questions are vital at this stage of the enquiry, when there are so many puzzling facts known about visual systems but so little in the way of coherent schemata to link them. Like David Marr's book, *Vision*, MacKay offers some provocative thoughts to guide vision research beyond the postage-stamp collection phase into a search for coherent themes of visual organization. The venue for this search seems likely to be at the interface between visual psychophysics, artificial intelligence and neuroscience.

22

Neuronal responses to illusory contour stimuli reveal stages of visual cortical processing

E. PETERHANS,
R. VON DER
HEYDT and
G. BAUMGARTNER

Introduction

Since Schumann (1900) described the phenomenon of illusory contours, stimulus configurations in which contours are perceived at sites where the stimulus is actually homogeneous (see examples in Fig. 22.1) have been studied extensively (see Chapter 23 for a review). Illusory contours seem to reveal principles of perception and are valuable tools also in the investigation of visual cortical processing at the neuronal level. We have already reported that many cells in monkey prestriate cortex respond to illusory contour stimuli as if real lines or edges were present at the site of the contour, whereas cells in the striate cortex apparently lack this ability (von der Heydt, Peterhans & Baumgartner, 1984). In this chapter, we examine the properties of stimuli that evoke responses in the prestriate cortex when contours are presented at the cells' preferred orientations, but fail to do so in the striate cortex. We were particularly interested in the question of how far the responses are determined by the spatial

Fig. 22.1. Examples of illusory contours, reproduced from Kanizsa (1979).

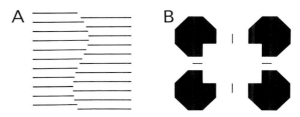

filtering, or length-summation, properties of receptive fields.

Methods

Rhesus monkeys were trained to fixate a small target consisting of two vertical lines presented on an oscilloscope display. At the beginning of fixation the monkey pulled a lever. After a variable delay of 0.3–5.0 s, the lines of the target turned horizontal and the animal had to release the lever within 0.4 s. Correct trials were rewarded with a small amount of liquid. Besides the fixation target other stimuli were presented, usually oscillating back and forth over the receptive field of the cell under study. Responses during fixation were recorded and depicted immediately in the form of a dot display. Groups of different stimulus patterns were arranged in blocks. Similarly, the groups of stimulus parameters such as different dimensions,

contrasts or orientations were sorted. For each stimulus parameter, responses to eight stimulus cycles (16 sweeps) were recorded during fixation. The classes of one group were presented once in a pseudorandom order and the sequence was then repeated one to three times. After completion of a block, the dot display was recorded on instant film and the mean numbers of spikes per stimulus presentation with standard deviations were displayed by computer. To confirm the relevant observations, several blocks with the same experiment were recorded from each unit, thus accounting for response variability. Glass-insulated platinum–iridium microelectrodes were prepared according to Wolbarsht, MacNichol & Wagner (1960) except that we did not coat the tips with platinum black. The electrodes were inserted through the intact dura posterior to the lunate sulcus and the 17/18 border into the region of the visual cortex representing the central

Fig. 22.2. Responses of four neurons to moving bars (rows 1 and 3) and to stimuli with moving illusory contours (rows 2 and 4), as shown on the left. Units A and B were recorded in the striate cortex, C and D in the prestriate cortex. Units A and C responded to dark bars, but only C gave a response to the border between gratings. Units B and D responded to light bars, but only D was activated by the illusory bar stimulus in which a central strip covering the response field was blanked out. The ellipses represent schematically the response field of unit D. Each frame of the dot displays represents responses to 24 cycles of stimulus motion. Responses to forward sweeps are represented in the left half of the frames with time axis from left to

right, responses to backward sweeps in the right half with time axis reversed. The mean numbers of spikes per cycle (both sweeps) are given on the right. The mean spontaneous activity during the time of a stimulus cycle was 0.0 for units A, B and D, and 0.16 for unit C. The dimensions of the bars were 0.17 by 1.3 degrees (A), 0.17 by 2.7 degrees (B), 0.10 by 2.8 degrees (C), and 0.27 by 5.4 degrees (D). The lines of the gratings were 1/60 degrees wide and spaced 0.25 degrees (A) or 0.38 degrees (C) apart. In the illusory bar stimuli, the blanked strips were 1 degree (B) or 2 degrees (D) wide. To simplify the figure, all stimuli are shown with vertical orientation of references, and in reversed contrast.

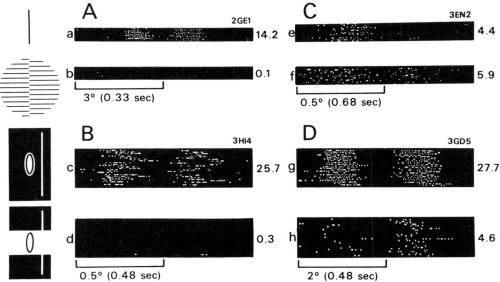

degrees of the visual field. Subsequent histological reconstructions confirmed that recordings done immediately after piercing the dura were within the striate cortex (area 17) and, after having crossed a strip of white matter, were in the prestriate cortex (area 18) within the posterior bank of the lunate sulcus.

Results

The illusory contour stimuli that we have used in this study are illustrated in Fig. 22.2. One consisted of two abutting gratings of thin lines offset by half a cycle (second row). The outer ends were covered by a circular or rectangular field stop while the border between the gratings was being moved back and forth over the receptive field under study. The other stimulus (row 4) was derived from a light bar in a black rectangle by blanking out a central strip, thus reducing the bar to a pair of notches. It gives the illusion of a light bar laying over two black rectangles. (To simplify the figures, the stimuli were reproduced with reversed contrast and the text has been made consistent with the figures.) Neuron activity was recorded in response to the notches moving back and forth in synchrony.

Edges, bars and illusory contours

Fig. 22.2 shows dot displays of responses to illusory and real contour stimuli, illustrating the typical results obtained in striate (A and B) and prestriate (C and D) cells. For each example, the responses to a moving bar of the cell's preferred orientation are shown in the upper row, and the responses obtained with an illusory contour of the same orientation are depicted below. The strengths of the responses are indicated to the right of the traces in spikes per stimulus cycle. For the illusory bar stimulus, we have chosen the width of the blanked strip such that the stimulus straddled the response field. In a number of cells we have determined the length of the response field quantitatively by scanning the field with a bar just long enough to elicit a stable response. The distance between the bars in the two positions where the cell just failed to respond was taken as the length of the response field. The width was measured from the length of the response trains on the dot display. The ellipses in Fig. 22.2 represent the response field as determined by this method for the unit in D.

Cells located in the striate cortex responded to bars but not to the illusory contour stimuli. The unit in Fig. 22.2A gave a strong response to the dark bar

(a) but was not activated by the border between gratings (b). Similarly, the unit in Fig. 22.2B responded to the light bar (c) but not to the illusory bar stimulus (d). By contrast, the cells recorded in the prestriate cortex (Fig. 22.2C and D) were activated also by illusory contour stimuli. The unit in Fig. 22.2C preferred a dark bar among real contour stimuli (e), and responded even better to the border between gratings (f). An example with the second type of stimulus is given in D. Here, the responses to the illusory bar stimulus (h) were weaker than the responses to the light bar (g). In 11 of the 28 cells studied quantitatively, responses to illusory contour stimuli were about equal to or better than those to bars or edges. Illusory contour responses were weaker by a mean factor of 2.3 ($N = 28$). Of the 96 cells studied in the prestriate cortex, 31 cells (32%) gave responses related to illusory contours, but none of the 74 cells that were recorded in the striate cortex did so.

Fig. 22.3 shows orientation tuning curves, again for cells in the striate (A and B) and prestriate cortices (C and D). Filled symbols represent the bar responses, open symbols the responses to illusory contour stimuli. The border between gratings was used in A, B and C, the illusory bar stimulus in D. Two typical examples of cells located in the striate cortex are illustrated in A and B. Both were sharply tuned to the orientation of the bar. Neither of them responded when the border between gratings was presented at the preferred orientation. The cell in A responded when the stimulus was rotated 90 degrees, i.e. the cell signalled the orientation of the physical lines. This was the most common result obtained in cells studied in the striate cortex. Many cells failed completely to respond to this stimulus as shown by the example of Fig. 22.3B. Cells in this cortical area did not respond to the illusory bar stimulus either.

Units C and D of Fig. 22.3 were recorded in the prestriate cortex. Unit C showed nearly identical orientation tuning for the border between gratings and for the dark bar. It is another example of a cell that gave stronger responses to such a border than to the bar. D demonstrates orientation tuning to the illusory bar stimulus. The responses were weaker than to the light bar, but the optimal orientation was similar for both kinds of stimulus.

Illusory contour stimuli and linear filtering

Many cells in the visual cortex that are sensitive to bars and edges show summation for stimulus length. Their receptive fields are often conceived as anisotropic two-dimensional filters with a low-pass characteristic along the optimal stimulus orientation and a band-pass characteristic perpendicular to it. This filtering scheme implies that units should respond not only to continuous edges or bars, but also when only parts of their summation zone were stimulated, and would thus provide a parsimonious explanation for illusory contours of the type seen at the Kanizsa triangle (Kanizsa, 1979).

With the abutting gratings, there is no change in mean luminance at the border between the gratings if one integrates along lines parallel to the border. Therefore, it is not surprising that cells in the striate cortex showing length summation did not respond to this stimulus. In the following we report results obtained with stimuli made up of line gratings in which

'edges' and 'bars' could be detected by a length-summation mechanism.

Fig. 22.4A shows the responses of a cell in the striate cortex to an optimally oriented edge of relatively high contrast (a; for the actual luminances of stimuli and background see figure legend). The cell did not respond to the border between gratings (b). We then compared the responses to edges of reduced luminance (c,e) with the responses to borders of single line gratings (d,f). Although the gratings approximately matched the edges in mean luminance, they evoked almost no activity. Responses in the striate cortex consistently fell short of the length-summation prediction, as demonstrated in Fig. 22.4A. In only one of six units tested with grating borders and edges were nearly equal strengths of response obtained. The other units gave much weaker responses to 'edges' produced by gratings than to conventional edges with continuous borders.

An example of the results obtained in the prestriate cortex is given in Fig. 22.4B. This cell

Fig. 22.3. Comparison of the orientation tuning curves obtained with bars (filled symbols) and illusory contour stimuli (open symbols). A and B, cells in the striate cortex; C and D, cells in the prestriate cortex. In A, B and C, dark bars and the border between two abutting gratings were used; in D, a light bar and the corresponding illusory bar with a blank strip of 2 degrees placed over the response field of the unit. For each curve, 16 orientations were presented in pseudorandom fashion and responses during eight stimulus cycles at each orientation were averaged.

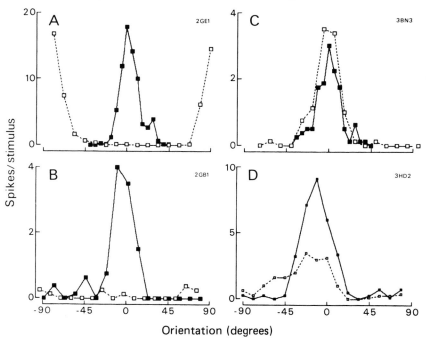

Spikes/stimulus

Orientation (degrees)

responded to bars and edges about equally well; the dark bar response is shown in g. The traces h, i and k show that the border between gratings and the borders of either grating alone were more effective than bars and edges. Length summation would predict responses in i and k, but not in h, where the mean luminance is equal on both sides of the contour. With this stimulus the strongest response was obtained.

In another experiment, we varied the alignment of two gratings making them overlap or be separated by a small amount. The results are illustrated in Fig. 22.5 for three cells, one in the striate (A), and two in the prestriate cortex (B and C). The result of Fig. 22.5A is typical for striate cells. They responded when the gratings were either overlapping or gapping, or in both situations, as shown here, but not when the

Fig. 22.4. Comparison of edges and lines with edges of gratings or borders between gratings. Conventions of data presentation are as in Fig. 22.2. A, unit in striate cortex: (a) preferred edge (luminance 17 cd/m²); (b) border between gratings (mean luminance 5.9 cd/m²); (c) and (e), edges of reduced luminance (6.5 cd/m²); (d) and (f), borders of single gratings (mean luminance 5.9 cd/m²). Background 4.3 cd/m². Neither the border between gratings (b) nor the borders of either grating (d, f) produced a response, although edges of similar mean luminance did (c, e). Spontaneous activity was zero. B, unit in prestriate cortex: (g) dark bar; (h) border between gratings; (i) and (k) borders of single gratings. Each frame represents 16 stimulus cycles. The unit responded nearly equally well to each of the borders formed by gratings (h, i, k). These stimuli were more effective than conventional stimuli (g). Spontaneous activity was zero.

Fig. 22.5. Responses to line gratings with various alignments. The stimuli shown in the insets correspond to −40 min arc (overlap), zero, and +40 min arc (gap). The grating lines were perpendicular to the receptive field orientations. A, unit in striate cortex. The unit responded best when the gratings were separated by 15 min arc and gave no response when the gratings were abutting one another. These responses were weak compared to the response to a light bar (not shown). The alignment was changed in steps of 2.5 min arc. B and C, units in prestriate cortex. The responsiveness of unit B was nearly unaffected by changes of the alignment. Unit C gave the strongest response when the gratings were exactly abutting. The alignment was changed in steps of 5 min arc.

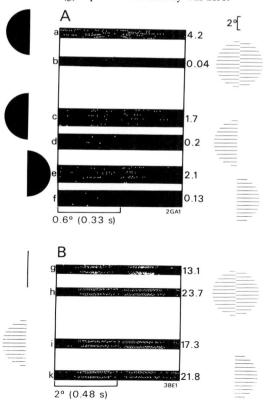

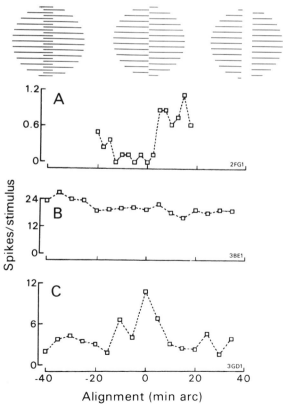

gratings were abutting. The unit represented here responded better when the gratings were slightly separated, in agreement with its preference for light bars. (In general, cells preferring dark bars responded to overlapping gratings, and units preferring light bars were activated by gapping gratings.) Note that the responses recorded with gaps between the gratings were weak (0.6–1.1 spikes/stimulus); stronger responses (3.6 spikes/stimulus) were obtained with a 10 min arc-wide light bar.

By contrast, the majority of cells in the prestriate cortex that responded to the border between gratings often did not show significant changes in strength of response (total spike count per stimulus cycle) when the alignment of the gratings was varied (Fig. 22.5B). However, the shortest responses appeared on the dot display when the gratings were abutting. As the overlap or gap became larger, double responses were sometimes observed. The cell of Fig. 22.5C gave the strongest response with abutting gratings.

Discussion

The analysis of neuronal responses as demonstrated by the examples in Figs 22.2 and 22.3 revealed a difference in the coding of visual information between the striate and prestriate cortex. Regarding position and orientation, luminance gradients appear to be similarly encoded in both striate and prestriate cortex, whereas illusory contours are represented only in prestriate cortex. In this area, cells signal position and orientation of such contours as accurately as of luminance gradients. This was shown in Fig. 22.2C and D, where the neuronal responses signal the passage of a border (illusory contour or luminance step) over the receptive field, and in Fig. 22.3C and D, where similar orientation tunings were obtained for either kind of border. About one-third of the cells (31/96) in the prestriate cortex gave responses related to such illusory contours, but none of the 74 cells studied in the striate cortex did so.

Responses in the striate cortex

Some types of illusory contour, such as the border between abutting gratings, cannot be brought out by linear filtering. The strong response obtained with this stimulus in some prestriate cells thus indicates non-linear processing, whereas the 'unresponsiveness' of striate cells seems to agree with the concept of summation over the elongated subregions of receptive

fields (Hubel & Wiesel, 1962). However, we have seen in Figs 22.4 and 22.5 that even when stimuli were used in which luminance gradients (bars or edges) would be detectable by the mean of length summation, such as the overlapping or gapping pairs of line gratings, or the border of a single grating, these stimuli often did not produce the expected responses in the striate cortex, but were less effective than the corresponding continuous borders. These results may be explained by the cross-orientation inhibition as demonstrated by Morrone, Burr & Maffei (1982) in the cat striate cortex. It was shown that cells are inhibited by the presence of Fourier components in the stimulus at orientations outside the cells' tuning range. Correspondingly, with edges of gratings (Fig. 22.4), and gapping or overlapping pairs of gratings (Fig. 22.5), the lines of the gratings may have produced inhibition that counteracted the excitation by the 'bars' or 'edges' formed by the grating borders. The same kind of inhibition may have been involved also in some of the experiments with illusory bars (see inset of Fig. 22.2B). These stimuli sometimes failed to produce a response even when the blanked strip was made so narrow that the moving ends of the bar passed over the response field of the cell (Baumgartner, von der Heydt & Peterhans, 1984). Here, the blanking of a strip produced edges perpendicular to the preferred orientation, which probably had an inhibitory influence. Note that in the illusory bar experiments, as well as in those with line gratings, the presumed inhibitory edges or lines were not moved broadside over the response field, but were kept stationary except for the moving notches or tips of lines. However, in experiments with awake and actively fixating animals, stationary stimulus features can drive cortical cells (Poggio, Doty & Talbot, 1977).

Responses in the prestriate cortex

Given the 'unresponsiveness' of striate neurons we have to look for an explanation for the responses in the prestriate cortex. A model that is compatible with the present results, as well as those of our previous study (von der Heydt et al., 1984), is outlined in Fig. 22.6. This figure also includes some features not required for the explanation of these exact data, but which become necessary if further results, mentioned briefly below, are taken into account.

We assume that in order to signal luminance gradients as well as illusory contours, a prestriate neuron needs two types of input from the striate cortex

the case of a cell with vertical preferred orientation,
e first comes from a unit or units that respond to
rtical light or dark bars or edges. Their response
lds are represented schematically by ellipse (1) in
g. 22.6. A second input (2) comes from a vertical row
receptive fields whose orientations are horizontal,
at is, perpendicular to input (1). These units are
sumed to be hypercomplex with asymmetrical end-
hibition. Their excitatory centres are represented by
rizontal ellipses and the major inhibitory end-zones
vertical ellipses. Such a unit can be activated by the
d of a horizontal line or by a corner, but not by a line
edge extending over the end-zone. There are two
oups of such units with receptive fields that are
irror images of each other, one responding to a line
d or corner extending to the left, the other respond-
g to a similar feature extending to the right. Always,
o fields, one of each group, are thought to be over-
pping. They are drawn with a vertical offset for clarity
ly. Within input channel (2), pairs of units with equal
ymmetry of the end-inhibition but with receptive

Fig. 22.6. Hypothetical circuit diagram of the
reorganization of receptive fields at the level of the
prestriate cortex. (1) and (2) represent two sets of
receptive fields of striate neurons which are
coextensive on the retina: (1) receptive field with
vertical orientation; (2) receptive fields with
horizontal orientation and asymmetric
end-inhibition. See text for further explanations.

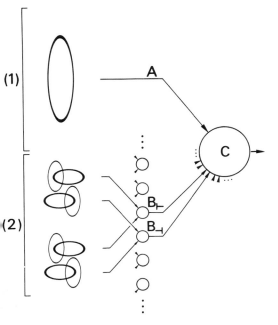

field centres separated vertically by some distance are
connected by multiplications in B ⊢ and B ⊣. The
signals B together with signal A are summed in C.

According to this model, the responses to bars
and edges depend on the type of unit that represents
input (1). These stimuli do not activate input cells (2)
at any orientation because of the orientation tuning and
end-inhibition of these cells. Input (2) is activated by
ends of lines or corners perpendicular to the optimal
stimulus orientation of input (1). As the recordings in
the striate cortex have shown, such features tend to
exclude activation of input (1).

Due to the multiplicative connection, at least two
units have to be stimulated to give a signal at B. This
corresponds to the following observations: (i) an
illusory edge stimulus could evoke a response although
either half of the stimulus gave none (von der Heydt
et al., 1984). The illusory edge was produced by two
aligned corners on opposite sides of the response field.
Only both together would give a signal at B in the
model. (ii) When the number of lines of the abutting
gratings was reduced, the response dropped and
reached zero with two lines (von der Heydt *et al.*, 1984).
Two lines on opposite sides of the contour would not
activate the two inputs of any of the units B in the
model; at least two line ends on the same side (or three
lines of the abutting gratings) are necessary. An exact
multiplication may not be necessary; other non-linear
connections can possibly do the same job. There are
observations, however, speaking against a logical AND
connection (with an all-or-none output). When the
abutting gratings were distorted so that the lines were
no longer perpendicular to the contour but tilted 45
degrees, usually weaker responses were obtained. In
our model, this would be explained by the orientation
tuning of the input cells (2). Clearly then, stage B must
be able to transmit graduated signals which an
AND-gate could not do.

The effect of tilting the lines is also the reason
why orientation-selective units must be postulated in
input channel (2). Otherwise, a number of concentric
fields with strong centre-surround antagonism would
suffice.

Finally, there are reasons for assuming summation
at C. Again, variation of the number of lines of the
abutting gratings is the crucial experiment. We have
invariably found a gradual increase of responses,
sometimes up to more than seven lines (von der Heydt
et al., 1984). The model can explain this by the

increasing number of B units that are stimulated, and whose signals add up in unit C.

The assumption that the two inputs (1) and (2) are independent and additive provides a simple explanation for the observation that the relative strengths of the responses to edges or bars and to illusory contour stimuli vary considerably between units in the prestriate cortex. This would indicate that the inputs can have different weights at the summation point C.

The puzzling results obtained with edges of gratings (Fig. 22.4) and with overlapping or gapping pairs of gratings (Fig. 22.5) can be understood at least qualitatively. In the unit of Fig. 22.4B, the border between abutting gratings with exactly equal mean luminance on both sides produced the strongest response, and the two edges of gratings were slightly less effective. Neither of these stimuli would drive input (1) of the model, as the recordings from the striate cortex indicate. In channel (2), the complete figure can excite theoretically twice as many of the B units, provided the lines are sufficiently spaced apart, and should therefore produce a greater sum at C. The relatively small differences actually observed may indicate response saturation at C. The introduction of gaps and overlaps between the gratings caused little change in the number of spikes per response of most units, as demonstrated by the example of Fig. 22.5B. Only a transition from a single volley of action potentials to a double volley was sometimes noted. This is consistent with summation of the signals $B \vdash$ and $B \dashv$ at C. When the abutting gratings are drifting over the receptive field, $B \vdash$ and $B \dashv$ are activated simultaneously; with gaps or overlaps, they are activated in sequence, but the net result is the same. The cell represented in Fig. 22.5C, on the other hand, seems to indicate some interaction between the two signals.

Theories of illusory contours

Among the various hypotheses that have been proposed for the phenomenon of illusory contours the most parsimonious is perhaps the idea that these contours are byproducts of the spatial filtering scheme employed by the visual system (Ginsburg, 1975). Although the isotropic filtering originally proposed by Ginsburg may not reveal the contours in the case of the Kanizsa triangle (cf. Tyler, 1977), more successful approaches using anisotropic (slit-shaped) spatial filters have been reported (Becker & Knopp, 1978). It may

be thought also that non-linearities in peripheral processing might bring out structures that cannot be revealed by strictly linear filtering. For example, tips of lines might be enhanced at the level of concentric receptive fields so that linear filtering at cortical level may detect a black or white line at the border between gratings, etc. However, this is apparently not the case. Our recordings have shown that the two kinds of illusory contours we employed had no correlate in the activity of cells in the striate cortex. The detection of illusory contours seems to require a convergence of excitatory input from columns of different orientation selectivities that does not exist below the prestriate cortex.

The concentric centre-surround organization of receptive fields explains a number of brightness illusions, e.g. the phenomena of simultaneous border contrast, Mach bands, and the Hermann grid illusion (Hartline, 1949; Baumgartner, 1960). Simultaneous contrast in turn has been thought to be the source of illusory contours (Brigner & Gallagher, 1974; Frisby & Clatworthy, 1975). Some brightness illusions obviously contradict the laws of simultaneous contrast, for example the Ehrenstein illusion (Ehrenstein, 1941). For these, antagonistic mechanisms in non-concentric receptive fields have been invoked (Jung & Spillmann, 1970; Frisby & Clatworthy, 1975). It is difficult to assess the relevance of our results to the brightness explanation of contours. The fact that many cells in the prestriate cortex respond to the border between gratings where brightness effects are minimal or absent as well as to bars and edges, or a gap between gratings, indicates that the basic qualities of contours, location and orientation, are treated largely independently of the brightness aspect.

References

Baumgartner, G. (1960) Indirekte Grössenbestimmung der rezeptiven Felder der Retina beim Menschen mittels der Hermannschen Gittertäuschung. *Pflügers Arch. Ges. Physiol.*, **272**, 21–2

Baumgartner, G., von der Heydt, R. & Peterhans, E. (1984) Anomalous contours: a tool in studying the neurophysiology of vision. *Exp. Brain Res.*, **9** (Suppl.), 413–19

Becker, M. F. & Knopp, J. (1978) Processing of visual illusions in the frequency and spatial domains. *Perception Psychophys.*, **23**, 521–6

Brigner, W. L. & Gallagher, M. B. (1974) Subjective contour: apparent depth or simultaneous brightness contrast? *Perceptual Motor Skills*, **38**, 1047–53

Ehrenstein, W. (1941) Ueber Abwandlungen der L. Hermannschen Helligkeitserscheinung. *Zeitschr. Psychol.*, **150**, 83–91

Frisby, J. P. & Clatworthy, J. L. (1975) Illusory contours: curious cases of simultaneous brightness contrast? *Perception*, **4**, 349–57

Ginsburg, A. P. (1975) Is the illusory triangle physical or imaginary? *Nature (Lond).*, **257**, 219–20

Hartline, H. K. (1949) Inhibition of activity of visual receptors by illuminating nearby retinal areas in the *Limulus* eye. *Fed. Proc.*, **8**, 69

Hubel, D. H. & Wiesel, T. N. (1962) Receptive fields, binocular interaction and functional architecture in the cat's visual cortex. *J. Physiol. (Lond).*, **160**, 106–54

Jung, R. & Spillmann, L. (1970) Receptive-field estimation and perceptual integration in human vision. In *Early Experience and Visual Information Processing in Perceptual and Reading Disorders*, ed. F. A. Young & D. B. Lindsley, pp. 181–97. Washington, DC: National Academy of Sciences

Kanizsa, G. (1979) *Organization in Vision. Essays on Gestalt Perception.* New York: Praeger

Morrone, M. C., Burr, D. C. & Maffei, L. (1982) Functional implications of cross-orientation inhibition of cortical visual cells. I. Neurophysiological evidence. *Proc. R. Soc. Lond., Ser. B*, **216**, 335–54

Poggio, G. F., Doty, R. W., Jr & Talbot, W. H. (1977) Foveal striate cortex of behaving monkey: single neuron responses to square-wave gratings during fixation of gaze. *J. Neurophysiol.*, **40**, 1369–91

Schumann, F. (1900) Beiträge zur Analyse der Gesichtswahrnehmungen. Erste Abhandlung. Einige Beobachtungen über die Zusammenfassung von Gesichtseindrücken zu Einheiten. *Zeitschr. Psychol.*, **23**, 1–32

Tyler, C. W. (1977) Is the illusory triangle physical or imaginary? *Perception*, **6**, 603–4

von der Heydt, R., Peterhans, E. & Baumgartner, G. (1984) Illusory contours and cortical neuron responses. *Science*, **224**, 1260–2

Wolbarsht, M. L., MacNichol, E. F., Jr & Wagner, H. G. (1960) Glass insulated platinum microelectrode. *Science*, **132**, 1309–10

23

Enhancement of edges by contrast, depth and figure: the origins of illusory contours

R. H. DAY

This chapter is about edges or contours, perhaps the most basic of all stimulus features in that they mark the boundaries between figures and their grounds and between objects and the spaces around them. In particular, this chapter is about illusory contours, curious appearances of contours in regions that are entirely uniform in physical terms. The main point to be argued on the basis of a now considerable body of data is that illusory contours derive from processes in perception that in the normal course of events serve to enhance the visibility of real edges. When, as will be shown, these processes are evoked by artifice, i.e. by means removed from the circumstances in which the processes normally operate, the observer experiences an impression of 'edgeness'. The most common condition used to evoke these enhancement effects where they are, so to speak, inappropriate, is incomplete or partially delineated borders of objects and figures. It is along these incomplete boundaries that illusory contour form.

Conceived of in this way, illusory contours together with other perceptual illusions, represent the outcome of processes that normally play a significant role in veridical perception (Day, 1984). When these processes are activated artificially they give rise to illusions that are often compelling and invariably surprising. Illusory contours are treated here as an outcome of one such process, edge enhancement; they are what remain – the residue – after the physical edge which they normally enhance is removed.

In what follows, the roles of induced contrast

apparent depth, and figural formations in edge enhancement, and, therefore, in the generation of illusory contours, are discussed in that order. In the last section, two further instances of enhancement by figural and depth features of the stimulus – contrast enhancement (the Benary effect) and the object-superiority effect – are considered as closely related phenomena.

Illusory contours

Since they were first described by Schumann (1900), illusory contours have been noted from time to time in a variety of configurations, in particular by Prantl (1927) and Ehrenstein (1942). More recently they have been subjected to close experimental analysis by Kanizsa (1955, 1974, 1976, 1979), Coren (1972) and Gregory (1972). Some typical instances are shown in Fig. 23.1. In A and B, the partially delineated white square appears brighter than the background and continuously contoured. In C and D, the square appears darker and surrounded by a continuous contour. It is to be noted that the effect is not restricted to achromatic patterns such as those in Fig. 23.1.

Changes in the appearance of the partially delineated regions and the formation of illusory contours occur also when coloured or grey elements are included as shown in Fig. 23.2. In A and B, the partially delineated square appears transparent grey and clearly visible illusory contours have formed between the grey square and white ground in each figure.

Halpern (1981) has pointed out that at least eight explanations, variously stressing sensory-neural and cognitive processes, have been proposed for illusory contours. Half of these give greatest weight to physiological processes in the visual system and half to higher-order cognitive processes. Since Halpern (1981) has carefully summarized the main features of these explanations and Pritchard & Warm (1983) have more recently compared the two levels of theory, it is unnecessary to review them again here. It is sufficient to point out that the various explanations are by no means mutually exclusive. Indeed, the point of this chapter is partly to show that perceptual processes at the levels of both sensory and cognitive functioning serve to enhance the saliency of visual edges. It is argued that the activation of these processes, in the

Fig. 23.1. Illusory contours in four figures. In A and B, the white square appears to be brighter than the background and to be surrounded by a continuous contour. In C and D, the black square appears darker than the background and similarly surrounded by a continuous contour. In all four squares the borders are only partially delineated by short lengths of black–white edge or the ends of lines.

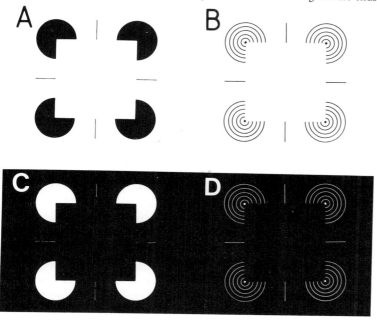

absence of physical edges with which they are normally associated, gives rise to the strong impressions of continuous edges in Figs 23.1 and 23.2.

Illusory contours and brightness contrast

Following a series of experiments in our laboratory (Day & Jory, 1978; Jory & Day, 1979), we set out an explanation of illusory contours couched in terms of the spread of brightness contrast to partially delineated contours (Day & Jory, 1980). The three types of locally generated brightness contrast and their spreading over extended regions, it was pointed out, can be indexed independently of the illusory contours that they are invoked to explain. Here, that explanation will be extended and modified. While it is clear that contrast and contrast spreading does serve to enhance edges, and that the enhancement can occur in the absence of continuous physical edges, an enhancement effect occurs also as a result of both two- and three-dimensional object characteristics of the stimulus. That is to say, the formation of flat and apparently solid objects can by itself generate illusory contours in the absence of differential contrast. These are described below. In this section illusory contours due to contrast spreading will be discussed.

Depending on the difference between the levels of luminance, their proximity and their relative areas, the brightness of one area can be perceptually increased or reduced by another. Simultaneously brightness contrast (see Ratliff (1965) for a description and historical review) and line-end contrast (Frisby & Clatworthy, 1975; Day & Jory, 1978) increase the difference in brightness between two adjacent regions. Assimilation of brightness, first noted by von Bezold (1874) and Rood (1879) and investigated in detail by Helson and his colleagues (Helson, 1943; Helson & Rohles, 1959; Helson & Joy, 1962), is a reduction in the difference in brightness between spatially contiguous regions.

Our explanation of illusory contours in terms of brightness contrast effects is that if contrast serves to increase or reduce brightness in one region and spread evenly up to borders that are partly delineated by features such as the ends of lines, short extents of real edge, dots or sharp bends in lines, then an illusory contour will form along that border. If, by means of suitable patterns, increased brightness from simultaneous or line-end brightness occurs on one side of a border and decreased brightness due to assimilation on the other, then the illusory contour will be stronger than if an increase or a reduction occurs alone. The contours in Fig. 23.1A and C probably involve only an increase, due to a combination of simultaneous and line-end contrast. Fig. 23.1B and D involve these increases in brightness but include also reduction due to assimilation of brightness between the concentric circles. The apparent darkening is plainly visible

Fig. 23.2. Illusory contours with 'neon-spreading' of grey elements. In A and B, the grey elements lend a transparent grey appearance to the squares. The illusory contours apparently surround the squares as in Fig. 23.1.

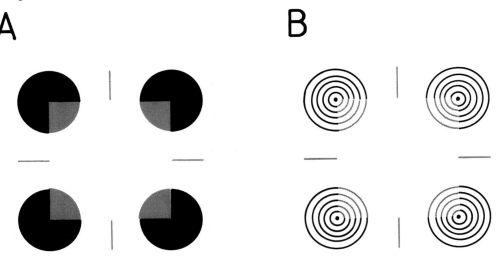

ween the circles. It is probably for this reason that illusory contours in the figures with concentric les appear more salient.

That both increased and reduced contrast ads to borders that are only partly indicated was rly demonstrated using a cross-like figure originally ised by Koffka (1935) and to which Landauer (1978) e recently drew attention. The effects of line-end htened contrast can be observed at the centre of figure (Fig. 23.3A). By arranging eight dots in the ns of a square, a diamond and a circle, as shown in 23.3B, C and D, the region of greater brightness ltered in shape. Illusory contours can be seen nd the borders of each figure, indicating the spread ontrast to the incomplete borders.

A curious colour-spreading effect has been cribed by van Tuijl (1975), mainly in lattice-like erns. It has the appearance of a region of 'misty' ur embedded in a black–white pattern. The effect been investigated in detail by van Tuijl & de Weert 79), van Tuijl & Leeuwenberg (1979), and Redies pillman (1981). The latter have pointed out that this

Fig. 23.3. Spread of line-end contrast and assimilation of brightness to borders partially delineated by line and dot elements. The region inside the figure appears lighter due to line-end contrast and those within the arms of the cross darker due to assimilation of brightness. An illusory contour coincident with the partly delineated border is visible.

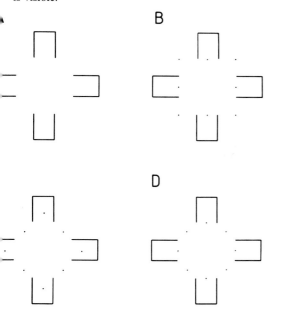

phenomenon, called by van Tuijl (1975) 'neon-color-spreading', occurs also with achromatic (light grey) elements. Recently, we confirmed the role of contrast spreading to incomplete borders by exploiting this effect (Day, 1983). In much the same manner as that shown in Fig. 23.3, we delineated the inside and outside borders of a diamond with regularly spaced dots. As can be seen in Fig. 23.4B, illusory contours form in consequence around the inside and outside edges of the diamond along the line of dots.

Whether the van Tuijl effect is simply a special case of assimilation is yet to be established. von Bezold (1874) originally described assimilation in coloured patterns. We have considered the question of whether or not the neon colour effect is the same as assimilation elsewhere (Day, 1983). The main point to note here is that van Tuijl's neon colour spreads in the manner of other contrast effects and gives rise to illusory contours along partially delineated borders.

Our explanation of illusory contours in terms of the spread of induced brightness can be summarized as follows. When increased or decreased contrast is generated in the form of simultaneous or line-end contrast on the one hand, or assimilation of brightness on the other, it spreads throughout a partially delineated region and renders it more prominent. Illusory contours derive from these increases or decreases in apparent contrast along the partially delineated borders of the figure. The strongest illusory contours occur when contrast is increased on one side of a border and decreased on the other. It can be reasonably assumed that in figures and objects with unbroken borders the heightening and reduction of

Fig. 23.4. The 'neon-colour-spreading' effect reported by van Tuijl (1975). In A, the effect spreads into the squares on either side of the grey elements and in B stops more perceptibly at the borders delineated by the dots.

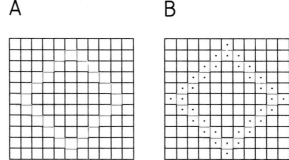

contrast serves to enhance the visibility of their borders. However, when the latter are incomplete, as in Figs 23.1 and 23.2, the enhancing component appears as an illusory contour in the breaks in the physical contour. In brief, what normally enhances an edge can by itself give rise to an appearance of real edge.

Illusory contours and apparent depth

Recently, we reported a series of experiments (Day & Kasperczyk, 1983*a*) designed to test an alternative explanation of illusory contours proposed by Kanizsa (1974, 1976, 1979). Kanizsa has argued that the elements in figures with illusory contours appear incomplete. A process of perceptual completion accordingly occurs and this requires the formation of an overlying, seemingly opaque form. Such an emergent form in turn requires perceptible edges. The squares in Fig. 23.1 are cases in point; the circular elements with 90-degree sectors forming the corners of the square have the appearance of incompleteness and the four lines impingeing on the sides of the square form an incomplete cross. The incomplete elements result in the formation of an apparently opaque square that appears to lie over complete but partly occluded elements. Perception of the square results in the emergence of apparent edges, i.e. of illusory contours. Changes in perceived brightness are regarded as a consequence of apparent depth and surface density of overlying figures.

We tested this explanation using magnitude estimation of the strength of illusory contours in figures made up of various combinations of apparently complete and incomplete elements devised by Kanizsa (1974) himself. The figures are shown in Fig. 23.5. The areas of the octagon shapes and the crosses were the same. Magnitude estimation (Stevens, 1956) relative to the standard (also shown in Fig. 23.5) has consistently proved the most satisfactory technique for quantifying the strength of illusory contours (Day & Kasperczyk, 1983*a*). As can be seen, one group of elements (octagons with 90-degree sectors removed and lines forming part of a cross) were apparently incomplete and another group (black crosses and diamond forms) apparently complete. Throughout five experiments of which the figures shown were used in the first, the illusory contours in the Kanizsa (1955) triangle figure served as a standard (of strength 10) for estimating the strength of the illusory contours in other figures.

The results from the first and subsequent experiments were clear and consistent; although the estimated strength of illusory contours was consistently greater with incomplete elements forming the corner

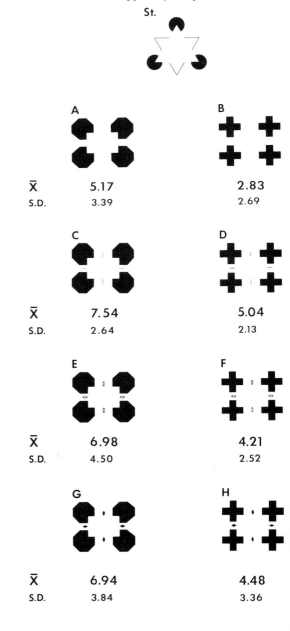

Fig. 23.5. Figures used by Day & Kaspercyzk (1983*a*) to test Kanizsa's element completion explanation of illusory contours. The octagon elements with missing sectors and the four short lines are supposedly incomplete and the crosses and the diamonds supposedly complete.

St.

A		**B**	
$\bar{X}$	5.17		2.83
S.D.	3.39		2.69
C		**D**	
$\bar{X}$	7.54		5.04
S.D.	2.64		2.13
E		**F**	
$\bar{X}$	6.98		4.21
S.D.	4.50		2.52
G		**H**	
$\bar{X}$	6.94		4.48
S.D.	3.84		3.36

of a partially delineated rectangle than with complete elements, illusory contours *always occurred* with complete elements. Furthermore, although both the line elements forming an incomplete cross and the diamond elements, complete in themselves, added to the strength of the contours, they were not different in the degree to which they caused illusory contours. The mean estimates of contour strength are shown in Fig. 23.5 for all the combinations of complete and incomplete elements.

These outcomes clearly do not support an interpretation of illusory contours in terms of perceptual completion of incomplete elements. The results of the later experiments in the series were consistent with this conclusion. However, the data do suggest an alternative explanation for the greater strength of the contours around the rectangles whose corners are delineated by the missing sectors of octagons. Inspection of Fig. 23.5 suggests that in these

figures the appearance of a rectangle in front of the elements is notably stronger than for the figures in which crosses form the corners. The addition of either four lines or four diamonds with their ends or points impingeing on the boundary of the rectangle strengthens the impression of a white opaque rectangle in front of black elements.

This informal observation led us to the view that the appearance of depth due to overlay in some way, perhaps by making the appearance of a figure more compelling, does not result necessarily in the formation of illusory contours but rather strengthens them. If this strengthening is attributable to apparent stratification and not to some quirky quality of the figures with incomplete octagons at the corners, then it would be expected also in figures apparently in depth due to 'cues' other than overlay. We tested this possibility using the figures shown in Fig. 23.6, in which relative depth derived from simulated overlay (A) or simulated perspective (B and C) (Day & Kasperczyk, 1983*b*). The perspective in the latter two figures produced a strong impression of 'outside' and 'inside' corners in depth. In each figure there was a vertical border that was partially delineated by the ends of lines with the same

Fig. 23.6. Figures used by Day & Kaspercyzk (1983*b*) to compare the relative effects of apparent depth due to overlay (A) and perspective (B and C) on the strength of illusory contours. Illusory contours are visible along the vertical borders in A, B and C.

ST.

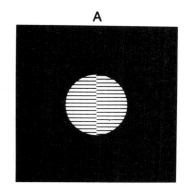
A

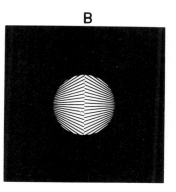
B

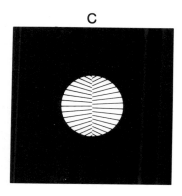
C

spaces between them in all figures. Again using magnitude estimation we found that relative to the standard (also shown in Fig. 23.6), whose contours were assigned a value of 10, the strength of the illusory contours in A, B and C were, respectively, 9.25, 9.19 and 10.42. An analysis of variance revealed that there is a significant difference between these means. More detailed analysis (Newman–Keuls) indicated that the mean for C was significantly greater than those for A and B. The difference between A and B was not statistically different.

The outcomes of this experiment indicate that apparent overlay of stimulus elements by a partly delineated figure is not necessary for the formation of illusory contours as stated by Kanizsa (1974, 1976, 1979). They also show that apparent depth due to cues other than overlay is at least as effective as overlay in increasing the strength of illusory contours that presumably derive in the first instance from the spread of line-end contrast.

Taken together the data from our recent experiments (Day & Kasperczyk, 1983*a*,*b*) show convincingly that while illusory contours involving the spread of contrast occur with only a very weak impression of depth from simulated overlay (Fig. 23.5), they are markedly stronger when there is a strong impression of depth. The impression can derive from

either simulated overlay or simulated perspective (Fig. 23.5 and 23.6).

At this point two questions can be posed. First, do illusory contours occur in the absence of *any* appearance of depth in the stimulus figure? Secondly, can depth alone, i.e in the absence of induced contrast, give rise to illusory contours?

A clear answer to the first question emerges from the figure devised by Kennedy (1979) and shown in Fig. 23.7. A curved illusory contour appears to follow the border delineated by the sharply pointed ends of the black intrusions from the background. The contour extends continuously across the figure. It can be presumed that line-end contrast generated by pointed elements spreads to occupy the whole of the anterior (right) region of the figure. At the same time assimilation of brightness occurs in the white posterior strips and is also limited by the border formed by the points. Moreover, the whole figure appears distinctly planar with no apparent depth, an impression confirmed informally by 20 observers all of whom reported the illusory contour. Thus Kennedy's figure provides a clear instance of an illusory contour coincident with a partly delineated border in the absence of even a weak impression of stratification or any other depth effect.

Can apparent depth by itself, i.e. without contrast spread, generate illusory contours? This question cannot yet be answered unequivocally. There are no quantitative data available and what evidence exists is anecdotal. It is presented in Fig. 23.8. In A, the lower white rectangle appears to overlay the upper one; the illusory contours of the lower appear to extend across the upper. At the same time there is no reason to suppose that the levels of line-end contrast generated by the ends of the incomplete concentric circles or the assimilation of contrast between them is any different between the upper and lower white rectangles.

Another pattern suggestive (but not confirmatory) of the formation of illusory contours by apparent depth alone is shown in Fig. 23.8B. An illusory contour is visible across the border delineated by the angular bends. It is more prominent towards the two sides than it is in the central region. The arrangement, which is taken from Gibson (1950), represents a change in gradient in depth. However, in this case it is conceivable that assimilation of brightness leading to darkening between the lines may contribute to the appearance of an edge. Because of the closer spacing

Fig. 23.7. An illusory contour in the absence of apparent depth. The contour is visible along the border formed by the sharply pointed ends of the black intrusions from the background and is presumably due to line-end contrast at the ends of the intrusions and assimilation of brightness between them.

the lines in the upper part of the pattern and wider acing below, the greater inter-line darkening above ay have spread and averaged and so contributed to e formation of the illusory contour along the border lineated by the obtuse-angle bends. The possibility an averaging as well as a spreading process is itself some interest.

Nevertheless, the illusory contours in the overlay rangement in Fig. 23.8A and in the gradient in depth Fig. 23.8B are strongly suggestive of the enhancement edges by depth effects alone. That is to say, apparent pth, like apparent brightness, may well enhance ntours so that when the enhancement factor is parated and occurs alone it may well give rise to an pression of 'edgeness'.

Illusory contours and figures

Illusory contours are clearly visible along the igonal borders of Fig. 23.9, delineated by the right-angle corners of the concentric black and white squares. There is a compelling impression of a continuous edge running along the line of the corners. What is particularly interesting about this figure is that while apparent darkening due to assimilation can be expected to occur (and in all likelihood does so) in the white interspaces, it can not be expected to differ on either side of the border formed by the corners. That is to say, there is no reason to suppose that there is a brightness differential between two regions which would account for the apparent edge, as is the case in Figs 23.1, 23.3, 23.5, 23.6 and 23.7. Even if, as is likely, the degree of darkening were greater in the immediate vicinity of the angles (Brigner & Gallagher, 1974) and were to spread outward from them, there is no expectation that it would do so to a greater degree on one side of the incomplete border than on the other.

There is another conceivable basis for a brightness difference. It is known that blurring of

Fig. 23.8. Illusory contours possibly attributable to apparent depth alone. In A, the level of contrast is the same throughout but a contour is visible along the border of the apparently near square overlaying the far square. In B, a contour is visible along the horizontal border formed by the obtuse-angle bends. It is conceivable that different levels of averaged assimilation of brightness contribute to the formation of this contour.

A   B

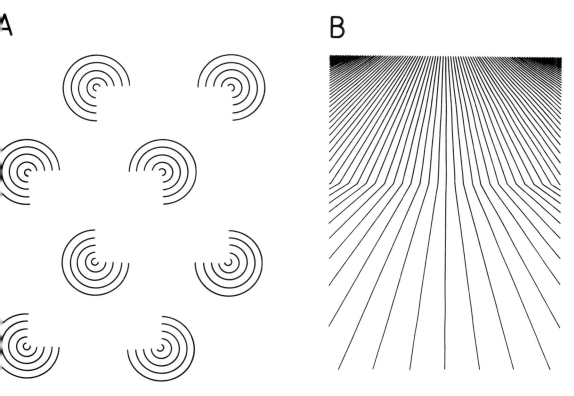

orthogonal contours can result from regular astigmatism of the eye. If this is so, then, as Wade & Day (1978) have pointed out, focussing on the lines in one orientation would render the others blurred. In consequence, black and white bars would appear greyer so that consequent chromatic aberration would yield colours. In other words, a difference in brightness and colour between the vertical and horizontal bars in Fig. 23.9 could derive from regular astigmatism. However, the illusory contours appear equally strong when the square is rotated through 45 degrees so that the lines are oblique and the borders and illusory contours are vertical and horizontal. Blurring due to astigmatism is relatively uncommon for oblique orientatations yet all observers shown the oblique pattern reported that the illusory contours were as strong when arranged in the cardinal axes as they were in the oblique axes.

These observations lead tentatively to the view that illusory contours occur in two-dimensional arrays in the absence of different levels of induced contrast on either side of a particularly delineated border. It is yet to be established exactly which characteristics of partially delineated borders are associated with illusory

Fig. 23.9. Illusory contours in the absence of contrast. The contours are visible along the diagonal borders delineated by the corners of the squares. It is unlikely that these derive from blurring in one direction due to irregular astigmatism since the illusory contours are equally prominent when the partially delineated borders are turned to the vertical and horizontal.

contours in the absence of different contrast levels apparent depth. It is conceivable that the right angl in Fig. 23.9 are implicated, rather than mere regulari of spacing. This issue is yet to be investigated.

Moving visual phantoms

Tynan & Sekuler (1975) reported a curiou phenomenon involving the formation of illusor contours with moving stimulus patterns. If a horizont strip of opaque material is arranged so that it covers t central region of a vertical grating pattern movir laterally, the contours of light and dark bars a perceived as continuous across the occluded regio Phantom-like bars across the mask are clearly visibl The effect occurs also with rotary motion (Sekuler Levinson, 1977) and, like real motion, gives rise to motion aftereffect (Weisstein, Mathews & Berbaur 1977). We have frequently observed the movi 'phantoms' (as Tynan & Sekuler have dubbed ther with vertical gratings of low spatial frequency movir laterally and with rotating black and white sectors demonstrated by Sekuler & Levinson (1977). T arrangement of laterally moving low-frequency ba and opaque mask is represented in Fig. 23.10.

Weisstein & Maguire (1982) showed that t depth relations between the moving elements and t

Fig. 23.10. Tynan & Sekuler's (1975) moving phantoms. When the black and white bars move laterally they appear to be continued across the horizontal opaque mask. The phantoms have the same contrast relations as the bars but appear markedly less bright and dark. It is possible that these effects are due to apparent depth from simulated motion parallax.

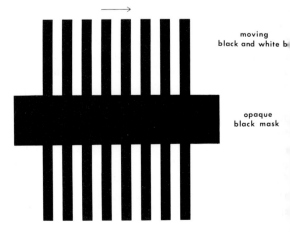

moving
black and white b

opaque
black mask

opaque occluding mask is critical for the generation of the moving phantoms. If the opaque mask is seen in front of the grating, i.e. overlying the moving bars, the effect does not occur. When the mask appears to be in the same plane as the bars or behind it, the illusion is visible. Weisstein *et al.* (1977) have shown that when the mask appears in front of the bars, the illusory effect of moving phantoms is markedly reduced with binocular compared to monocular viewing.

One interpretation of these illusory contours that, as far as is known, has not been considered involves simulated motion parallax as the basis of apparent depth. Although speculative, such an interpretation accords well with the data, identifies the moving phantoms with stationary illusory contours, and further implicates perceived depth as a basis for edge enhancement. This view deserves some consideration.

Motion (or monocular) parallax is the smaller extent and lower velocity of movement of the retinal representations of far stimulus features relative to the representations of near features during head or body movement. It is one of a number of stimulus correlates ('cues') associated with variation in the distance of objects relative to the observer. Like other such correlates, motion parallax can be simulated. If some of a number of coplanar elements are made to oscillate through greater extents at a higher velocity than others, they appear nearer when viewed with one eye. That is to say, like overlay and perspective, motion parallax can be simulated in a two-dimensional array.

The essential condition for the occurrence of visual phantoms is the occlusion of a region of the moving bars by an opaque mask (Tynan & Sekuler, 1975). It can be supposed that, as is the case with apparent depth from simulated overlay and perspective, moving edges are thereby perceptually enhanced. It can also be supposed, again as in the overlay and perspective arrangements, that when borders are incomplete, the enhancement component associated with perceived depth persists and is observable in the spaces between the moving elements.

In this regard the moving bars, which are apparently nearer due to motion parallax in the arrangement represented in Fig. 23.10, are equivalent to the squares in Fig. 23.1, which are apparently nearer due to overlay. It can be assumed that in the absence of actual edges, enhancement effects form illusory contours collinear with the real edges. Thus it can be supposed that edge enhancement, the function of which

is to render the edges of objects more prominent, is virtually separated from the edges that it normally renders more perceptually salient. The process appears to be triggered by simulated overlay (Fig. 23.1), perspective (Fig. 6), and motion parallax (Fig. 23.10). There remains the interesting question of whether other stimulus features correlated with depth might also give rise to the residues of edge enhancement.

This proposed explanation of moving visual phantoms, speculative as it is, offers an interpretation of why the phantoms do not occur in stationary patterns and why they are absent when the occluding mask is apparently nearer than the bars. In the first of these two circumstances motion parallax would not be simulated so that apparent depth would be lacking. In the second, the bars would presumably be perceived as occluded by a nearer object, a condition that would not be expected to generate the enhancement process. The motion parallax explanation also suggests why the effect is reduced or absent with binocular viewing; stereoscopic information with binocular viewing could be expected to eliminate the appearance of depth due to simulated motion parallax.

General discussion

The essence of the argument set out here is that perceived contours are enhanced relative to their physical counterparts. The enhancement occurs in respect of contrast, depth and figure. The function of enhancement can be presumed to be that of rendering edges and boundaries more perceptually salient. When the borders of objects and figures are broken, i.e. only partially delineated, these enhancement processes are nevertheless activated. In consequence, contours due to the enhancement component alone are visible in the spaces between the border elements. In these terms illusory contours are conceived of as the outcomes of normal perceptual processes evoked by artifice. The artificial stimulus conditions give rise to an illusion of edges where none exist. From this standpoint, illusory contours are similar to a diversity of non-veridical effects in perception – phenomena that under normal perceptual conditions serve a function in veridical perception but that when artificially, i.e. inappropriately, evoked give rise to illusions of reality rather than veridical representations of it (Day, 1984).

Two points can be made in conclusion. The first concerns the bearing of the explanation of illusory contours on the nature of perception and the second, edge enhancement as one instance of perceptual enhancement.

Perception cannot be regarded simply as the *exact* representation of the external world. It is well established that perception is highly selective and frequently goes well beyond the physical properties of the stimulus. The perceptual process is selective of events, objects and features that are relevant to the business of adapting to and surviving in a world of physical and social events. In these terms it should not be surprising that features such as edges are highlighted. They are critical not only for distinguishing objects and events from their ground (figure–ground discrimination) but for identifying and recognizing them. Edges, after all, are correlated with sizes, shapes, textures and patterns. All of these are common

properties of the environment. It is thus to be expected that they are enhanced the better to perceive the nature of the world.

Edge enhancement is only one instance of the perceptual highlighting of features by global structures. Another is the Benary effect (Benary, 1924), the greater degree of simultaneous brightness contrast when a small region of different luminance is superimposed on a figure compared with the degree of contrast when the region is located outside the figure. The effect is shown in Fig. 23.11. The grey triangle on the cross appears brighter than a similar grey triangle located in a corner. This difference in contrast, first reported nearly 60 years ago by Benary (1924), has been amply confirmed since (Jenkins, 1930; Mikesell & Bentley, 1930; Coren, 1969). Coren (1969) has pointed out that, in terms of

Fig. 23.11. The Benary effect. The grey triangular region on the cross appears lighter than that outside it. In terms of simultaneous brightness contrast the latter could be expected to be the lighter of the two since it is immediately surrounded by a greater area of black. It is unclear whether the difference is due to simulated overlay or to 'belongingness' as Benary suggested.

the area of blackness that surrounds the grey triangles, the one lying outside the triangle might reasonably be expected to exhibit a higher level of brightness contrast. In point of fact, the triangle forming part of the cross is perceived as the brighter of the two. Coren (1969) has concluded that: 'Cognitive variables such as figure–ground relationships appear to play a large part in the magnitude of brightness contrast. Figure–ground differences alone may cause a variation of up to a factor of 2 in the total amount of brightness contrast obtained.'

At this stage it is not clear whether the enhancement of brightness shown in Fig. 23.11 is due to the upper grey triangle being perceived as part of the cross (as Benary (1924) concluded), or to the triangle *overlaying* part of the cross as our preliminary observations suggest. If this second interpretation is valid, it would seem that apparent depth might affect brightness levels as well as edge. It will be of considerable interest to look into the Benary effect more closely.

Finally, and still on the second concluding point, perceptual enhancement associated with apparently three-dimensional displays is also not without precedent. A further instance is the object-superiority effect reported first by Weisstein & Harris (1974) and since frequently confirmed (Weisstein, Williams & Harris, 1979, 1982; Weisstein & Harris, 1980). When a barely visible line is briefly exposed, it can be identified more accurately when it forms part of a pattern that appears coherent, unified and three-dimensional than when it forms part of one that appears non-coherent and flat. In the original experiment, subjects were required to identify which of four target lines were present in figures varying in structural coherence and apparent depth. The main outcome was that successful identification was highest when the lines formed part of an apparently three-dimensional object and declined as the object appeared more incoherent and flat.

Recently, Weisstein *et al.* (1982) have investigated more closely which aspects of figural context are most intimately involved in the facilitation of target identification. They showed that accuracy of identification was highly correlated with ratings of mean apparent depth (accounting for 95% of the variance) and with mean ratings of 'structural relevance', i.e. the relevance of the target line to the figure of which it was part.

Clearly, then, simple features – such as lines representing edges – of objects are seen more readily when they form part of an object that is represented in depth. It is not going too far beyond the data of experiments on illusory contours and the object-superiority phenomenon to suggest that the significance of edges in three-dimensional objects leads to their enhancement in perception. The enhancement processes result in the shadowy but nevertheless perceptually compelling phenomenon of illusory contours – a residue of enhancement when the real contour is removed – and in the greater detectability of edges in the form of lines when the lines are presented under perceptually reduced conditions.

A final point is worth making. We live in an environment of stationary and moving objects of varying degrees of complexity and significance. The business of responding to and coping with these constitutes a considerable part of behaviour. That our perceptual systems have evolved in a way that highlights critical features of objects to which we respond is hardly surprising. The enhancement of these features only comes as a surprise when by ingenuity, artifice, or sheer chance we separate the enhancement component from the features it is intended perceptually to highlight. However, it is by separating the shadows from the substance, so to speak, that they can be exposed for systematic examination. What does seem clear is that while local-to-global processing of stimulus information by way of feature analyzers is part of the total process of perception, global-to-local processing is also part. The first can reasonably be regarded as the business of constructing objects from their basic features and the second that of rendering more salient certain of these features that have greater significance for the observer.

Acknowledgement

The assistance of Mr R. T. Kasperczyk in experimental work and in the preparation of this paper is gratefully acknowledged.

References

Benary, W. (1924) Beobachtungen zu einem Experiment ueber Helligkeitskontrast. *Psychol. Forsch.*, **5**, 131–42

Brigner, W. & Gallagher, M. (1974) Subjective contours: apparent depth or simultaneous brightness contrast? *Perceptual Motor Skills*, **38**, 1047–53

Coren, S. (1969) Brightness contrast as a function of figure-ground relations. *J. Exp. Psychol.*, **80**, 517–24

Coren, S. (1972) Subjective contour and apparent depth. *Psychol. Rev.*, **79**, 359–67

Day, R. H. (1984) The nature of perceptual illusions. *Interdiscipl. Sci. Rev.*, 9, 47–58

Day, R. H. (1983) Neon color spreading, partially delineated borders, and the formation of illusory contours. *Perception Psychophys.*, 34, 488–90

Day, R. H. & Jory, M. K. (1978) Subjective contours, visual acuity and line contrast. In *Visual Psychophysics: Its Physiological Basis*, ed. J. C. Armington, J. E. Krauskopf & B. R. Wooten, pp. 331–40. New York: Academic Press

Day, R. H. & Jory, M. K. (1980) A note on a second stage in the formation of illusory contours. *Perception Psychophys.* 27, 89–91

Day, R. H. & Kasperczyk, R. T. (1983*a*) Amodal completion as a basis for illusory contour. *Perception Psychophys.*, 33, 355–64

Day, R. H. & Kasperczyk, R. T. (1983*b*) Illusory contours in line patterns with apparent depth due to either perspective or overlay. *Perception*, 12, 485–90

Ehrenstein, W. (1942) *Probleme der ganzheitpsychologischen Wahrnehmungslehre.* Leipzig: Barth

Frisby, J. P. & Clatworthy, J. C. (1975) Illusory contours: curious cases of simultaneous brightness contrast? *Perception*, 4, 349–57

Gibson, J. J. (1950) *Perception of the Visual World.* Boston: Houghton Mifflin

Gregory, R. L. (1972) Cognitive contours. *Nature*, 238, 51–2

Halpern, D. F. (1981) The determinants of illusory-contour perception. *Perception*, 10, 199–213

Helson, H. (1943) Some factors and implications of color constancy. *J. Optic. Soc. America*, 33, 555–67

Helson, H. & Joy, V. L. (1962) Domains of lightness assimilation and contrast. *Psychol Beitr.*, 6, 405–15

Helson, H. & Rohles, F. H. (1959) A quantitative study of the reversal of classical lightness contrast. *Am. J. Psychol.*, 72, 530–8

Jenkins, J. G. (1930) Perceptual determinants in plane designs. *J. Exp. Psychol.*, 13, 24–36

Jory, M. K. & Day, R. H. (1979) The relationship between brightness contrast and illusory contours. *Perception*, 8, 3–9

Kanizsa, G. (1955) Margini quasi-percettivi in campi con stimolazione omogenea. *Riv. Psicol.*, 49, 7–30

Kanizsa, G. (1974) Contours without gradients or cognitive contours. *It. J. Psychol.*, 1, 93–112

Kanizsa, G. (1976) Subjective contours. *Scientific American*, 234, 48–52

Kanizsa, G. (1979) *Organization in Vision: Essays on Gestalt Perception.* New York: Praeger

Kennedy, J. M. (1979) Subjective contours, contrast, and assimilation. In *Perception and Pictorial Representation*, ed. C. F. Nodine & D. F. Fisher, pp. 167–95. New York: Praeger

Koffka, K. (1935) *Principles of Gestalt Psychology.* New York: Harcourt Brace

Landauer, A. A. (1978) Subjective states and the perception of subjective contours. In *Conceptual Analysis of Method in Psychology: Essays in Honour of W. M. O'Neil*, ed.

J. P. Sutcliffe, pp. 142–6. Sydney: Sydney University Press.

Mikesell, W. T. & Bentley, M. (1930) Configuration and brightness contrast. *J. Exp. Psychol.*, 13, 1–23

Prantl, A. (1927) Über gleichsinnige Induktion und die Lichtverteilung in gitterartigen Mustern. *Zeitschr. Sinnphysiol.*, 58, 263–307

Pritchard, W. S. & Warm, J. S. (1983) Attentional processing and subjective contour illusion. *J. Exp. Psychol.: General*, 112, 145–75

Ratliff, F. (1965) *Mach Bands: Quantitative Studies on Neural Networks on the Retina.* San Francisco: Holden-Day

Redies, C. & Spillman, C. (1981) The neon color effect in the Ehrenstein illusion. *Perception*, 10, 667–81

Rood, O. N. (1879) *Modern Chromatics.* London: Kegan Paul

Schumann, F. (1900) Beiträge zur Analyze der Gesichts-wahrnehmungen: Einige Beobachtungen über die Zusammenfassung von Gesichtseindrücken zu Einheiten. *Zeitsch. Psychol.*, 23. 1–32

Sekuler, R. & Levinson, E. (1977) The perception of moving targets. *Scientific American*, 236, 60–73

Stevens, S. S. (1956) The direct estimation of sensory magnitudes – loudness. *Am. J. Psychol.*, 69, 1–25

Tynan, P. & Sekuler, P. (1975) Moving visual phantoms: a new contour completion effect. *Science*, 188, 951–2

van Tuijl, H. F. J. M. (1975) A new visual illusion: neonlike color spreading and complementary color induction between subjective contours. *Acta Psychol.*, 39, 441–5

van Tuijl, H. F. J. M. & de Weert, C. M. M. (1979) Sensory conditions for the occurrence of the neon spreading illusion. *Perception*, 8, 211–15

van Tuijl, H. F. J. M. & Leeuwenberg, E. L. J. (1979) Neon color spreading and structural information measures. *Perception Psychophys.*, 25, 269–84

von Bezold, W. (1874) *Die Farbenlehre.* [*The Theory of Colour.*] (trans. S. R. Koehler, 1876) Boston: L. Prang

Wade, N. J. & Day, R. H. (1978) On the colors seen in achromatic patterns. *Perception Psychophys.*, 23, 231–4

Weisstein, N. & Harris, C. S. (1974) Visual detection of line segments: an object superiority effect. *Science*, 186, 752–5

Weisstein, N. & Harris, C. S. (1980) Masking and unmasking of distributed representations in the visual system. In *Visual Coding and Adaptability*, ed. C. S. Harris, pp. 317–64 Hillsdale, N.J.: Erlbaum

Weisstein, N. & Maguire, W. (1982) The effect of perceived depth on phantoms and the phantom motion aftereffect. In *Organization and Representation in Perception*, ed. J. Beck, Hillsdale, N.J.: Erlbaum

Weisstein, N., Mathews, M. & Berbaum, K. (1977) A phantom motion aftereffect. *Science*, 189, 955–8

Weisstein, N., Williams, M. C. & Harris, C. S. (1979) Line segments are harder to see in flatter patterns: the role of three-dimensionality, in object-line and object-superiority effects. *Bull. Psychonomic Soc.*, 14, 229–30

Weisstein, N., Williams, M. C. & Harris, C. S. (1982) Depth, connectedness, and structural relevance in the object-superiority effect: line segments are harder to see in flatter patterns. *Perception*, 11, 5–18

24

Vision – the capture of optical covariation

D. M. MACKAY

The need for pretheoretical analysis

The mammalian visual system, despite its multimillionfold complexity, offers itself to the neuroscientist in a deceptive simplicity. Here are two retinae, each with its photosensitive array wired with topographic precision, by way of the lateral geniculate nucleus (LGN), to the primary cortex. Other branches of the input signal array (e.g. to the superior colliculus) are obviously 'minor', at least in an anatomical sense. The main function of the system seems (at first sight) to be to reproduce in the cortex a faithful topographic neural image of the visual scene.

We have only to take a few tentative steps along this line of thought, however, to be deluged with awkward questions. Successive stages of the visual pathway seem to be 'processing the image' in curious ways. First, contrast is enhanced by lateral inhibition; so far, so good. But next, we find at various levels cells whose response depends on factors such as the geometry, texture, motion and colour of the retinal image in some restricted 'receptive field'. Admittedly, receptive fields are topographically arranged, at least in the primary cortex; but what does all this do to the notion of a 'neural image'? Do we have to think of it perhaps as something like a printed map, whose elements have name tags or descriptions attached to them? In that case, recognition of a particular pattern might be achieved by a logical synthesis of the descriptions of its elementary components afforded by the firing of the relevant 'feature detectors'. Theorists

365

who took this line presumed such a synthesis to be a hierarchical process, ending up with the excitation of some single neuron or group of neurons whose firing was supposed to signify the presence of the visual pattern in question.

An alternative line of thought started from the observation that many primary cortical cells, if stimulated by sinusoidal gratings, showed a preference for a limited range (perhaps one octave wide) of spatial frequencies. The theory was advanced that the recognition of specific patterns depends on a kind of Fourier synthesis performed on the spatial frequency information contained in the firing pattern of so-called cortical 'frequency analysers' (see Chapter 26). This mechanism, it was sometimes suggested, could allow patterns to be recognized regardless of their location on the retina, since, mathematically, Fourier analysis is agnostic with respect to the geometrical origin of the spatial coordinates. Unfortunately for this line of argument, the responses of units with a marked preference for a given spatial frequency band have turned out also to be strongly affected by the *phase* of the spatial modulating sinusoid. Indeed their properties are by and large just what would be predicted on the assumption that each primary unit had a receptive field profile of the 'Mexican hat' type such as had encouraged others to classify them as 'edge-' or 'bar-detectors'. Attempts to decide experimentally between the two interpretations have been (not surprisingly) inconclusive, and several writers (e.g. Marcelja, 1980; MacKay, 1981a) have pointed out that the issue is a bogus one, since any unit with a receptive field divided into two or more linear zones of opposite polarity of response is mathematically bound to have preference for a grating of a corresponding spatial frequency, provided that the phase has been adjusted to match the crossovers of the grating to the crossovers of the receptive field. Fourier analysis, useful though it is as a way of specifying performance mathematically, here serves only to obscure these elementary considerations.

The fact that units with such 'Mexican hat' receptive fields do not respond to checkerboard edges aligned with the 'preferred' axis has been cited (e.g. by De Valois, De Valois & Yund, 1979) in support of the 'Fourier' interpretation, on the ground that a checkerboard has no Fourier components in the direction of its edges. The lack of response is equally explicable, however, without recourse to Fourier theory, on the assumption that edges of opposite polarity of luminance gradient, lying within the same receptive field region, make antagonistic contributions to the unit's response: in other words, that the effective stimulus for a so-called 'edge-detector' is the *net* spatial gradient of luminance. Experiments in collaboration with my colleague Peter Hammond (Hammond & MacKay, 1981a, b, 1983a, b) have confirmed that aligned edges of opposite luminance gradient do indeed exert antagonistic effects, though the opponent process appears to be non-linear: i.e. adding a small edge – or bar – segment of opposite polarity has a disproportionately large suppressive effect on the response to the main stimulus.

For both types of theory, a further problem is raised by the phenomenon of perceptual stability during exploratory eye movement. Saccadic eye-jumps must result in a continual disruption of the signal pattern emitted by cells at all levels, and in continual shifts of the neural image (if there is one) around the cortical population in the receiving area. How is the visual system supposed to cope with this so as to provide a stable internal representation of the objects being scanned? Theories have been advanced (Von Holst & Mittelstaedt, 1950) according to which corollary information from the oculomotor system would be used automatically to 'subtract out', 'compensate for' or 'cancel' the changes in visual input resulting from each saccade; but extensive experiments have failed so far to reveal any signs of this operation, at least up to the primary cortex.

Finally, the 'neural image' theory has to reckon with the growing multiplicity of visual projection areas, described elsewhere in this volume. Are these like the pages of a printed atlas, with repeat maps of the same area devoted respectively to rainfall, population, elevation and so forth? If so, people ask, how are these numerous 'maps', some of them only crudely topographical, eventually recombined to form the unitary image of the visual world we perceive? How are they all kept 'in register'? Is there some more central brain area in which we should expect to find once again a needle-sharp neural picture of the visual world, free of all signs of eye movement and the like? If not, then what?

The presupposition behind such questions is that the task of the visual system is to generate eventually a single *explicit* internal neural image of the world. From an information-engineering standpoint, however, this presupposition is far from obvious. I have long argued (e.g. MacKay, 1954, 1956a, b, 1962a) that the task of the visual system, as of other perceptual systems, should be seen first and foremost as the setting up and keeping up to date of the organism's *conditional readiness for action* in its world, in matching response to sensory input, and that on this basis the perceived world could find itself adequately represented *implicitly*, by the configuration of conditional constraints on action and the planning of action which are set up to match its structure. Such ideas have, however, had relatively little sale among neurophysiologists until the last few years, when the work of Marr (1976, 1982) and others engaged in 'computational vision' has drawn fresh attention to the inadequacies of the 'neural image' concept.

The plan in this chapter will therefore be to start at ground level, asking how we can best characterize the job that the visual system has to do for the active organism. We shall then see on what principles the nervous system might best function in order to extract information relevant to this task in a useful form. Although this 'pretheoretical' exercise will involve some repetition of suggestions now 20 or 30 years old, it will I think highlight a few theoretical options that have not as yet been fully exploited.

Covariation

It is a truism that in a relatively stable and structured environment the sensory input to an organism shows massive redundancy in the information-theoretic sense. Action in that environment, however elaborate and varied, can therefore be planned and steered largely on the basis of information already received and stored. From the standpoint of information engineering, then, perception is an adaptive exploitation of the redundancy in the pattern of demand imposed by the sensory input (MacKay, 1954, 1956a, 1967, 1970). The main form taken by this redundancy can be termed *covariation* (MacKay, 1962b, 1978; Koenderink, 1984a, b; Phillips, Zeki & Barlow, 1984).

Movement of an array of receptors relative to an object in the environment gives rise to coincident changes in signals from different channels. Some receptors will signal OFF simultaneously. Some will go OFF as others go ON. Some will repeatedly go OFF or ON at a definite time interval after others; and so on. These are examples of *inter-receptor* covariations. Covariations between receptors within the same visual or tactile modality may betoken such things as the presence of a fixed or a moving edge, or of an object of a certain width. They could arise whether the relative motion producing the changes were caused by movement of the organism or movement of the object in question. Covariations between signals in different modalities (e.g. tactile and visual, or auditory and visual) offer the informational basis for the perception of location in space relative to the body as a whole; and so on.

The power of covariational analysis – asking 'what else happened when this happened?' – may be illuminated by its use in the rather different context of military intelligence-gathering. It becomes effective and economical, despite its apparent crudity, when the range of possible states of affairs to be identified is relatively small, and when the categories in terms of which covariations are sought have been selected or adjusted according to the information already gathered. It is particularly efficacious where many coincidences or covariations can be detected cheaply in parallel, each eliminating a different fraction of the set of possible states of affairs. To take an idealized example, if each observation were so crude that it eliminated only half of the range of possibilities, but the categories used were suitably orthogonalized (as in the game of 'Twenty questions'), only 100 parallel analysers would be needed in principle to identify one out of 2^{100}, or say 10^{30}, states of affairs.

Exploratory probing of the environment, as by a moving fingertip or a moving retina (projected optically on to its environment by the lens), gives rise to covariation between sensory signals and the exploratory motor actions. It is important to recognize that even with a single unextended receptor such covariation can suffice both in principle and in practice for the perception of spatial form. A blind man, for example, can recognize the location and shape of a manhole cover by probing it with the tip of his cane. A coat-button can be perceived in considerable detail by exploring it with the finger nail. Thus, it is not in general necessary (though of course it helps!) to have the geometrical form which is to be recognized projected on to an extended array of receptors such as the retina. It is enough, in principle, to have adequate analysers of the ways in which sensory signals (even from a single exploring receptor) covary with the motor programme of exploration.

Locomotion of the whole organism through the

environment gives rise to a further class of covariations. Proprioceptive signals from muscles will show strong correlations with certain sensory inputs – so much so that some of the latter can function proprioceptively, as emphasized especially by Gibson (1950, 1966). A good example is the subsystem sensitive to image-motion in the far visual periphery, which is powerfully coupled to the vestibular system for the maintenance of posture (Dichgans & Brandt, 1972).

It is already clear from these examples that the covariations of importance for the organization of action must form a hierarchical family with many levels. One further level may be mentioned – namely covariation between motor and/or sensory signals on the one hand, and signals from the various *evaluative* processes on the other, whose feedback converts mere movements into goal-directed or purposeful actions. Here covariation indicates the value (positive or negative) attached statistically to each ingredient of the ongoing programme and its sensory outcome. A simple case would be that of optical search, where a quite diverse selection of covariational clues might be expected to combine to steer the process to a successful outcome.

'Top down' *versus* 'bottom up'

It would be surprising if the need to optimize the categories of covariational analysis were not met in part by genetic preprogramming; even so it is well known that failure to expose the visual system early enough to a normal perceptual diet leads to deterioration in feature selectivity. In general, however, it would be informationally efficient to allow higher analysing levels to adjust the categories used at lower levels according to current results, much as a player of 'Twenty questions' selects his next question on the basis of earlier answers.

Such 'top down' procedures have found much application in 'computer vision' (see for example Ballard, Hinton & Sejnowski, 1983). Most two-dimensional images of the three-dimensional word are highly ambiguous. Instead of trying to compute the 3-D structure purely by 'bottom up' analysis, it can be economical to use what engineers call a 'relaxation' process. This in effect discovers, by interactive self-adjustment, which of a restricted range of real-world possibilities can best match the sensory input. For this 'perception by matching response' (MacKay, 1954, 1956a) to converge rapidly, the

higher-level system must use many simultaneous parallel processes to evaluate (and eliminate useless) combinations of parameters. To this end, the input may find itself analysed (again in parallel) in quite different categories as the iteration proceeds. Associative nets in which relative probabilities can be represented by continuously variable physical quantities interacting in 'analog' fashion, as physiologically excitatory and inhibitory inputs do, are particularly economical for such parallel computations. Cortical architecture seems well suited to perform them. Whether the rich downward projections from the cortex to the LGN and superior colliculus mediate 'top down' control in the foregoing sense remains to be seen.

Why so many visual areas?

From this new starting point we can see a good reason for the multiplicity of cortical sensory areas in each modality. Detection of coincidences or covariations requires that all signals whose covariation may be significant for the organization of action should converge upon the appropriate 'listening posts' or coincidence detectors. Obviously, a detector that signals coincidences all the time supplies little information in the statistical sense (i.e. Shannon's well-known measure $\Sigma p_i . \log (1/p_i)$ tends to zero if the probability of one outcome tends to 1). On the other hand if coincidences are too rarely signalled the information rate again tends to zero. Our coincidence detectors need to be shielded from trivial or noise-generated coincidences, yet exposed to a data stream that keeps them optimally busy signalling functionally significant covariations.

For the multiplicity of visual areas, then, a simple explanation (MacKay, 1981b; Phillips *et al.*, 1984) is that the different areas are needed in order to segregate signals in different categories (such as form, motion, texture or colour) so as to allow covariations between them to be 'listened for' and analysed by different populations of detectors, each thus being free of the background chatter of traffic in the other populations. In the 'motion' area, for example, the question asked by local analysers of covariation would be: 'What else was moving in the same direction as the stimulus in this receptive field?', or more generally, 'Are there any groupings of units that signal the same direction of motion? Covariations of velocity with respect to spatial coordinates could provide indications of figure/ground boundaries, the angle of relative approach to a textured background and the like (see Reichardt, Poggio & Hausen, 1983). The 'colour' area (Zeki, 1980) could similarly look out for coincidences between visual events linked by a common colour and so having greater than average probability of being related to a common object.

For this purpose the incoming signals in each category should be as rich in selective information content (as low in redundancy) as possible – not so much, as Barlow (1969) suggested, in order to economize on impulses as such, but in order to minimize the density of spurious or functionally meaningless coincidences. This fits with the fact that most visual cortical cells have low background firing rates. It would explain why at so many levels lateral inhibition is used to derive the Laplacean (effectively the source-density distribution) of the incoming signal intensity, since this highlights likely locations of significant covariation at the expense of non-significant covariation. The ubiquitous phenomenon of neural adaptation, whereby changes in relevant stimulus parameters are emphasized in preference to steady levels, also fits well with the hypothesis that the basic principle of CNS function is the detection and analysis of *changes* that occur together – asking 'what covaries with what?'.

The significance of 'feature sensitivity'

The hypothesis that cortical neurons are primarily detectors of covariation offers a new way of looking at the function of simple and complex cells (MacKay, 1978, 1985). Instead of asking what elementary geometrical features of optical patterns (whether edges or spatial periodicities) a given cell 'detects', we should perhaps ask *what kinds of covariation* in the signals from the retinal array it is equipped to sense. Sensitivity to particular oriented edges and/or bars and/or gratings could well be important clues to the answer; but we might well miss the main functional point if we rest content with classifying and analysing cells solely in these terms, as if their firing were meant to signify to the higher CNS the presence of those features.

The case of Area 17 'complex' cells is perhaps a salutary illustration. All of these cells respond readily to appropriately oriented bars or gratings; so it was reasonable for most of us to speculate initially that their function might be to 'generalize for location' the

'edge-detecting' function attributed to simple cells. Unfortunately for this notion, as we (Hammond & MacKay, 1975) discovered to our surprise, they can also be excited (some more vigorously than others) by appropriately moving textures such as static visual noise (Fig 24.1) that fail entirely to excite simple cells in the same cortical column. Their polar diagrams of motion sensitivity to textured and conventional bar

Fig. 24.1. Responses to bar and visual noise stimuli of simple and complex cells in the same radial penetration of Area 17 of the cat. Each row compares responses to two different stimuli presented alternately. Average response histograms illustrated (left) are for 16 consecutive stimulus pairs; dot displays for each stimulus pair are shown on the right. Motion of a bar of static visual noise (A) is compared with (B) motion of a dark bar, against the same stationary background of static visual noise, (C) motion of a whole field of static visual noise, and (D) simultaneous motion of a dark bar together with its background of static visual noise. (From Hammond & MacKay, 1976.)

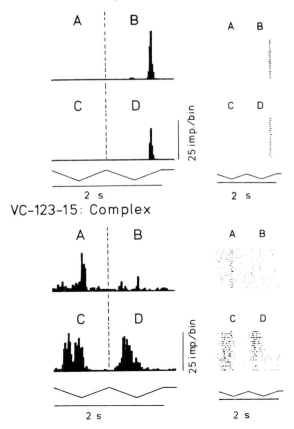

Fig. 24.2. Polar diagrams showing responses of three complex cells to moving fields of static visual noise (solid lines) and to conventional moving bars (dashed lines). (From Hammond, 1979.)

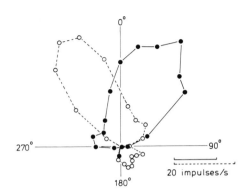

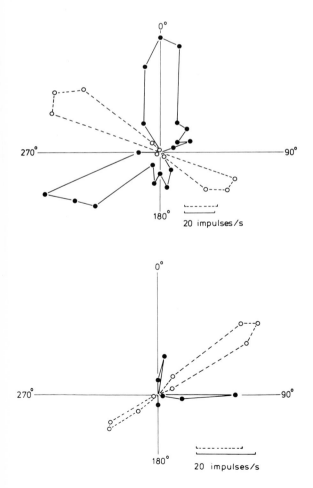

stimuli, moreover, are quite different (Fig. 24.2). Thus, the fact that such a complex cell is excited would not logically signify, on its own, the presence of any one geometric feature, or even of motion in any one direction, on the retina. This does not mean that such geometrical information cannot be represented in the firing of the complex cell population, but only that (as we suggested in 1975) it would have to be represented by the covariation of the activity in many parallel channels. It also does not deny the hypothesis of Hubel & Wiesel (1962) that complex cells receive excitatory inputs from simple cells. All it means is that the complex cells responsive to moving noise patterns must have additional inputs that bypass simple cells, and that the types of covariation they can detect must be more complex than those caused by either moving bars or moving noise.

Fig. 24.3. Spatial distribution of influence of a small 'window' of moving texture in an otherwise stationary textured background, upon the response of an Area 17 complex cell to an optimally oriented 2-degree black bar moving synchronously across the field centre (at 0 degrees). The dotted line indicates the level to which the response is reduced if the whole background moves with the bar. Note facilitatory effects of moving texture at extremes of range – i.e. outside cell's excitatory receptive field. (From Hammond & MacKay, 1981*b*.)

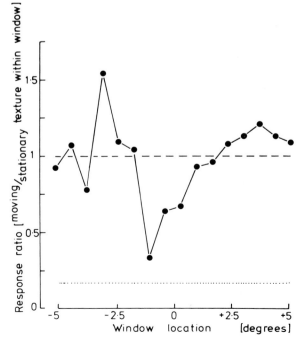

The picture was still further complicated when we also found evidence (Hammond & MacKay, 1978, 1981b) suggesting that some complex cells, sensitive to moving texture, may have inhibitory or excitatory connections to simple cells that are not directly excitable by the same textured stimuli. The response of most simple cells to conventional bar stimuli, it turns out, can be modulated by the motion of patches of texture inside or outside their receptive fields (Fig. 24.3). This of course suggests that even simple cells cannot be thought to perform a purely geometrical analysis of the portion of an image falling on their receptive fields.

In view of all this, it is hard to know what to make of the intriguing report by Peterhans (Chapter 22) that some V2 neurons in the macaque monkey respond to 'illusory contours' of the kind described by Day (Chapter 23). What one needs to know is to which other stimuli such units are (or are not) responsive. In Area 7 of the cat, Hammond and I have found no cells that responded only to the *edges* of textured fields revealed by relative motion. All complex cells responded to motion of a textured area without regard to the orientation of its edge. If there are single units in the cat which are selective for edge defined by 'texture shearing' they presumably occupy higher levels than Area 17. In the light of the arguments above, however, it would in any case be unsafe to conclude from their responsiveness that the function of such cells is simply to describe corresponding isolated elements of the visual scene.

Perceptual stability

The significance of covariational redundancy occurred to me first in relation to the problem of perceptual stability during voluntary eye movements (MacKay, 1957, 1962b, 1970, 1972, 1973). As indicated above, if the optically projected retina is thought of as exploring and sampling the visual world in a manner analogous to tactile probing, then the covariation of retinal signals with the exploratory programme contains *invariant* information as to the structure of the explored environment. Provided that the changes produced by the exploratory saccades are evaluated in the light of the changing criteria implicit in the saccadic programme, there is no need whatever to suppress, subtract or otherwise cancel those changes, since it is from their covariation with the saccades that the main positive evidence for the stability of the world can be extracted.

The analogy between tactile and optical probing is of course limited. As a probe the retina takes an enormous sample at each fixation, and the covariation of 'onset' responses to a single sample will in itself contain much information as to local geometrical features, whereas (at the other extreme) the blind man's cane can give no geometrical information apart from the covariations of its tactile input with his exploratory movements. But the principle on which I suggest that perceptual stability is maintained during exploration is the same in the two cases: the sensory changes produced are in an obvious sense a *confirmation* rather than a disconfirmation of the stability of the perceptual world being explored, as long as the relevant covariations are stable.

Vision with a stabilized image

At the other extreme from ocular exploration is the case of vision with a stabilized retinal image. As is well known, in this case the perceived image rapidly fades but can be revived either by slightly displacing or by modulating the brightness of the stabilized stimulus. Where a partially stabilized image of high contrast refuses to fade, it can be 'wiped out' as a whole by moving over it an unstabilized image, which seems to give rise to a form of 'monocular rivalry' (MacKay, 1960).

It is unlikely that simple physiological satiation can account for the fading of stabilized images (Evans, 1965). Although it undoubtedly plays a significant part, there is no evidence that such adaptation could be sufficient to abolish completely the primary cortical response to a stabilized image. Indeed anyone who has worked with conventional animal preparations must be familiar with cells in Area 17 that respond tonically for many seconds to suitably patterned stimulation, even when (by virtue of curarization) the retinal image is fully stabilized.

Why then do most patterns become invisible to human observers under complete image stabilization? The answer on our present hypothesis is clear. Stabilization, even if it does not abolish all retinal signals, eliminates all *covariation*. If no correlated changes take place, there is nothing for analysers of covariation to analyse. If, then, seeing depends on the results of covariational analysis, there will be no seeing. Furthermore, in the case where a poorly stabilized image generates a sufficiently fluctuating retinal input for analysers of covariation to work upon, the same hypothesis would predict that adding a strong background of coherently changing signals from an unstabilized image could readily saturate the population of covariance detectors, so as to swamp the feebler covariations in the poorly stabilized input, causing the original image to fade from view.

What is specific about seeing?

What then distiguishes the neural activity that mediates vision from that subserving audition, touch and the like? The classical answer has been in terms of 'local sign': the signals that give rise to seeing are those originating in the visual pathway. But why? The cytoarchitecture offers no obvious correlate of the huge qualitative differences between, say, seeing and hearing.

Our present standpoint would suggest another possible answer. What qualitatively distinguishes the visual from other modalities is the specific *repertoire of action* for which it sets up conditional readinesses, and the kinds of covariation involved in doing so. Specifically, vision depends on covariation between retinal inputs and what we might call ocular motion – which includes the use of both oculomotor and locomotor systems to displace the optical projections of the retinae over the environment. The visual world, in functional terms, is the domain within which the roving foveae have to navigate. Its structure imposes constraints primarily on the organism's conditional readiness for 'ocular navigation' (MacKay, 1981*b*). Secondarily, of course, these constraints have implications for other forms of actions – locomotor, descriptive and so forth; but what is distinctive about the neural correlate of seeing, I suggest, is that it updates our readiness for *further looking*, in matching response to the demands of the current retinal input.

It follows that if we want to find the neural representation of the world as currently seen, the place to look, at least for its coarse-grained form, should be at the interface between the incoming visual information and the higher-level planning of ocular navigation. At this level, we should not be surprised to find objects represented only implicitly, by the conditional constraints they impose on the planning of further ocular motion – rather as the stepping stones in a river might find themselves represented implicitly in the conditional motor programme for jumping from one to the next, or as a curved driveway might find itself represented

in the matching programme of rotations of the steering wheel of a car.

This of course can be thought of as an amplification rather than a denial of the classical doctrine of local sign; but it emphasizes that what matters for perceptual experience is not where signals originate, but the kinds of conditional readiness they set up. In particular it makes it obvious that even physiological events that leave the visual pathway undisturbed can give rise to specifically visual experiences (such as illusions of visual movement caused by disturbances to the evaluative criteria or the oculomotor programme).

Conclusion

My argument has been that for neurophysiological purposes it is useful to think of vision first and foremost not as a process of image-making, but as the capture of optical covariation: capture, in the sense of an analysis that sets up a matching state of conditional readiness to take advantage of the informational redundancy that the covariation represents. Visually perceptible objects are primarily potential targets for foveation, entities giving rise to retinal inputs that covary with the programme of ocular exploration and bodily locomotion. For this purpose they can find themselves adequately represented implicitly rather than explicitly. Thus, the final neural representation of the visual world in the brain may be as different from a map or a topographical image as a slide rule is different from a printed multiplication table, or as the steering angle of a car's wheels is different from the shape of the road it follows. In interpreting the fast-emerging data on 'feature sensitivity' in cortical cells, and on the multiplicity of cortical visual areas, we may miss crucial points unless we are alert to the range of possibilities thus opened up.

In short, our need is to see the visual cortex in relation to the functioning of the cerebral information system as a whole. It is the internal posture of the system as a whole that symbolizes the contents of the world-as-perceived. The visual system adequately serves its purpose if it can supply the clues needed to keep that internal posture (with all its conditional implications, especially for ocular navigation) matched to the structure of the visual world.

References

llard, D. H., Hinton, G. E. & Sejnowski, T. J. (1983) Parallel visual computation. *Nature*, **306**, 21–6

rlow, H. B. (1969) Trigger features, adaptation, and economy of impulses. In *Information Processing in the Nervous System*, ed. K. N. Leibovic, pp. 209–26. New York: Springer

: Valois, K. K., De Valois, R. L. & Yund, E. W. (1979) Response of striate cortex cells to grating and checkerboard patterns. *J. Physiol.*, **291**, 483–505

chgans, J. & Brandt, T. (1972) Visual-vestibular interaction and motion perception. In *Cerebral Control of Eye Movements and Motion Perception*, (ed. J. Dichgans & E. Bizzi, pp. 327–38. Basel: Karger

ans, C. R. (1965) Some studies of pattern perception using a stabilized retinal image. *Br. J. Psychol.*, **56**, 121–133

ibson, J. J. (1950) *The Perception of the Visual World*. Boston: Houghton, Mifflin

ibson, J. J. (1966) *The Senses considered as Perceptual Systems*. Boston: Houghton, Mifflin

ammond, P. (1979) Lability of directional tuning and ocular dominance of complex cells in the cat's striate cortex. *Developmental Neurobiology of Vision*, ed. R. D. Freeman, pp. 163–74. New York: Plenum

ammond, P. & MacKay, D. M. (1975) Differential responses of cat visual cortical cells to textured stimuli. *Exp. Brain Res.*, **22**, 427–30

ammond, P. & MacKay, D. M. (1976) Functional differences between cat visual cortical cells revealed by use of textured stimuli. In *Afferent and Intrinsic Organization of Laminated Structures in the Brain*, ed. O. Creutzfeldt, pp. 397–402. Berlin & Heidelberg: Springer-Verlag

ammond, P. & MacKay, D M. (1978) Modulation of simple cell activity in cat by moving textured backgrounds. *J. Physiol.*, **284**, 117P

ammond, P. & MacKay, D. M. (1981*a*) Suppressive effects of luminance gradient reversal on simple cells in cat striate cortex. *J. Physiol.*, **315**, 30P

ammond, P. & MacKay, D. M. (1981*b*) Modulatory influences of moving textured backgrounds on responsiveness of simple cells in feline striate cortex. *J. Physiol.*, **319**, 431–42

ammond, P. & MacKay, D. M. (1983*a*) Influence of luminance gradient reversal on simple cells in feline striate cortex. *J. Physiol.*, **337**, 69–87

ammond, P. & MacKay, D. M. (1983*b*) Effects of luminance gradient reversal on complex cells in cat striate cortex. *Exp. Brain Res.*, **49**, 453–6

Iolst, E. Von & Mittelstaedt, H. (1950) Das Reafferenzprinzip (Wechselwirkungen zwischen Zentralnervensystem und Peripherie). *Naturwissenschaften*, **37**, 464–476

Iubel, D. H. & Wiesel, T. N. (1962) Receptive fields, binocular interaction and functional architecture in the cat's visual cortex. *J. Physiol.*, **160**, 106–54

Koenderink, J. J. (1984*a*) Simultaneous order in nervous nets from a functional standpoint. *Biol. Cybern.*, **50**, 35–41

Koenderink, J. J. (1984*b*) Geometrical structures determined by the functional order in nervous nets. *Biol. Cybern.*, **50**, 43–50

MacKay, D. M. (1954) Operational aspects of some fundamental concepts of human communication. *Synthese*, **9**, 182–98 (Reprinted in D. M. MacKay (1969) *Information, Mechanism and Meaning*, Cambridge, Mass.: M.I.T.)

MacKay, D. M. (1956*a*) Towards an information-flow model of human behaviour. *Br. J. Psychol.*, **47**, 30–43

MacKay, D. M. (1956*b*) Cerebral activity III. *Advancement Sci.*, **42**, 392–5

MacKay, D. M. (1957) The stabilization of perception during voluntary activity, In *Proceedings of the 15th International Congress of Psychology*, pp. 284–5. Amsterdam: North Holland

MacKay, D. M. (1960) Monocular 'rivalry' between stabilized and unstabilized retinal images. *Nature*, **185**, 834

MacKay, D. M. (1962*a*) Theoretical models of space perception. In *Aspects of the Theory of Artificial Intelligence*, ed. C. A. Muses, pp. 83–104. New York: Plenum Press

MacKay, D. M. (1962*b*) Self-organization in the time domain. In *Self-Organizing Systems*, ed. M. C. Yovits, G. T. Jacobi & G. D. Goldstein, pp. 37–48. Washington, D.C.: Spartan Books

MacKay, D. M. (1967) Ways of looking at perception. In *Models for the Perception of Speech and Visual Form*, ed. W. Wathen-Dunn, pp. 25–43. Boston: M.I.T. Press

MacKay, D. M. (1970) Perception and brain function. In *The Neurosciences: Second Study Program*, ed. F. O. Schmitt *et al.*, pp. 303–16. New York: Rockefeller University Press

MacKay, D. M. (1972) Visual stability. *Invest. Ophthalmol.*, **11**, 518–24

MacKay, D. M. (1973) Visual stability and voluntary eye movement. In *Handbook of Sensory Physiology, Vol. VII/3A*, ed. R. Jung, pp. 307–31. Heidelberg & New York: Springer

MacKay, D. M. (1978) The dynamics of perception. In *Cerebral Correlates of Conscious Experience*, ed. P. A Buser & A. Rougeul-Buser, pp. 53–68. Amsterdam: Elsevier

MacKay, D. M. (1981*a*) Strife over visual cortical function. *Nature*, **289**, 117–18

MacKay, D. M. (1981*b*) What kind of neural image? *Freiburger Universitaetsblatter*, **74**, 67–72

MacKay, D. M. (1985) The significance of 'feature sensitivity'. In *Models of the Visual Cortex*, ed. D. Rose & V. Dobson. Chichester & New York: Wiley (In press)

Marcelja, S. (1980) Mathematical description of the responses of simple cortical cells. *J. Optic. Soc. America*, **70**, 1297–300

Marr, D. (1976) Early processing of visual information. *Phil. Trans. R. Soc. B.*, **275**, 483–524

Marr, D. (1982) *Vision*. San Francisco: Freeman

Phillips, C. G., Zeki, S. & Barlow, H. B. (1984) Localization of function in the cerebral cortex. *Brain*, **107**, 327–61

Reichardt, W. E., Poggio, T. & Hausen, K. (1983) Figure–ground discrimination by relative movement in the visual system of the fly. *Biol. Cybern.*, **46** (Suppl.), 1–30

Zeki, S. M. (1980) Representation of colours in the cerebral cortex. *Nature*, **284**, 412–18

25

Visual hallucinations and their possible pathophysiology

JAMES W. LANCE

The neurologist must turn to the neurophysiologist for assistance in understanding many of the
symptoms and signs encountered in clinical practice.
This appeal should be regarded as a stimulus rather
than a chore, for until we can explain the mechanism
of such aberrations we cannot be said to understand
fully the normal functioning of the machine.

Diplopia, impairment of the visual fields and
simple unformed hallucinations such as flashes or
zigzags of light are common clinical problems, but
metamorphopsia, palinopsia and formed visual hallucinations are as rare as they are intriguing. The purpose
of this communication is to describe some of these
phenomena with whatever anatomical and physiological
observations are available, in the hope that visual
scientists may be able to make some deduction or
speculation that eludes the mind of the clinician. For
as Walsh & Hoyt (1969) stated; 'regarding this
complex subject our present knowledge does not carry
us beyond generalizations which may ultimately, in
large part, prove erroneous'.

Non-localized visual hallucinations may be
experienced by patients who are completely blind from
any cause as part of an afferent deprivation syndrome
or by patients with toxic-confusional states as a
symptom of drug intoxication or systemic illness.
Hallucinations arising from ischaemia or irritation of
the visual pathways may occupy a quadrant, half-field
or the entire field of vision and may be formed or
unformed. Their nature and localization may throw
some light on the organization of the visual system of

least pose questions which should eventually be answerable in physiological terms.

Foerster (1931) elicited unformed hallucinations of flashes of light or colour by electrical stimulation of Brodmann Areas 17 and 18, while organized images of figures, animals or people were obtained from Area 19. Penfield & Perot (1963) observed that stimulation of the cerebral cortex, particularly the junction of the temporal and occipital lobes, replicated formed visual hallucinations in epileptic patients. In one patient, a boy aged 12 years, 'occipital cortex stimulation produced coloured flashes of light and, immediately anterior to this, stimulation produced the figures of robbers with guns'. In practice, unformed hallucinations are described most often as a prodrome of classical migraine, while formed hallucinations may be an epileptic disturbance or a release phenomenon, usually associated with a visual field defect resulting from cerebral infarction, angioma or tumour.

Visual disturbance in migraine

Visual symptoms other than simple blurring of vision occurred in 41% of 448 migrainous patients reported by Selby & Lance (1960) and 32.8% of 500 migrainous patients analysed by Lance & Anthony (1966). In the latter series, fortification spectra had been experienced by 51 patients (10.2%) of whom 17 also described photopsia on occasions. The prodrome consisted of photopsia alone in 113 patients. The majority of patients had also observed other neurological disturbances at the time of the migraine attack. Such symptoms were referable to the vertebrobasilar circulation (diplopia, vertigo, dysarthria, incoordination and ataxia) in 41%, to the internal carotid circulation alone (aphasia, unilateral paraesthesiae) in 4%, and to both circulations in 13%. Visual symptoms constituted the only neurological disturbance in the remaining 42%.

In a recent analysis of 600 headache patients attending our neurology clinic, 135 were found to suffer from classical migraine (Drummond & Lance, 1983). Fortification spectra (zigzag scintillations, white or coloured) were seen by 12 patients; a pattern of wavy, wiggly lines or shimmering of vision was noted by eight; scotomas, mottled patches, black spots or hemianopia occurred in 33; coloured blobs or colour changes in the whole visual field were described by seven patients; 'split vision' was experienced by one patient, and photopsia (white or coloured flashes of light, described as 'silvery tinsel, crystals, exploding lights, rockets or stars') were reported by the remaining 74. Such symptoms usually lasted from 10 min to 2 h and disappeared before the headache started, although they persisted into the headache phase in seven instances. Colour changes in the whole field were described as 'black to red to white' by one patient and 'black to red to black' by another. The patient who complained of 'split vision' said that one half of an observed object was seen on a lower level than the other for 30 min preceding the headache.

In the three series of patients with which the author has been associated, summarized above and including 483 patients with visual disturbances out of a total of 1548 headache patients, no patient complained of metamorphopsia or other perceptual illusions. Such illusions have been described by others (Lippman, 1952; Klee & Willanger, 1966; Sacks, 1970; Golden 1979) and have been called by some 'The Alice in Wonderland Syndrome'. Charles Dodgson (Lewis Carroll) did indeed suffer from classical migraine preceded by fortification spectra but, as far as is known, this malady came on after he had written 'Alice in Wonderland' and he had no opportunity or need to draw on his migrainous experiences.

Apart from the rare phenomenon of 'retinal migraine', the phosphenes comprising photopsia clearly arise from the visual cortex since they may appear in one half-field only and, furthermore, have been reported in a woman whose eyes were removed for bilateral retinoblastoma when she was 2 years old (Peatfield & Rose, 1981). Phosphenes are seen as moving during voluntary eye movements and appear to shift in the direction of the slow phase of induced vestibular nystagmus (Jung, 1979).

Lashley (1941) plotted the spread of his own fortification spectra across the visual field and concluded that they were formed by a wave of excitation propagating across the visual cortex at a rate of about 3 mm/min, followed by suppression of activity with a resulting scotoma. The size of each fortification spectrum did not increase as the scotoma grew but additional figures were added to one side as some vanished on the opposite side, giving the impression of five or more dazzling lines arranged in parallel or at angles to each other. The fortification figures appear to be stronger and coarser in their lower quarter (Lashley, 1941; Miles, 1958). Miles (1958) reported that the scintillations had an ON–OFF frequency of 8–12/s and faded out about 30 degrees from the fixation point, although Richards (1971) mapped the fortification spectra outwards to 50 degrees from the fixation point. The time taken for the spectrum to progress from the centre to the periphery of the visual field varied from 20 to 25 min in different observations but was constant for the one subject. Hare (1966) found that the inhalation of low doses of the vasodilator amyl nitrite reduced the vertical dimension of the spectrum by about 15% but that high doses had no effect, suggesting that the march of symptoms, once begun, was largely independent of the state of the cerebral vessels.

The causes of photopsia and fortification spectra

Since these positive visual phenomena are usually followed by negative symptoms (blurred vision, scotomas or hemianopia), it is probable that they are manifestations, respectively, of excitatory and inhibitory (or exhaustion) phenomena in the visual cortex. Electrical stimulation of the visual cortex (Areas 17 and 18) gives rise to phosphenes 'like stars in the sky' that are usually simple in shape, fixed in the visual field and move only when the patient's eyes move, in which case all phosphenes move together (Brindley & Lewin, 1968; Brindley *et al.*, 1972). Since phosphenes are distributed evenly over homonymous fields in migraine they must be caused by some diffuse influence exciting individual cells or groups of cells simultaneously. On the other hand, fortification spectra progress gradually over the visual field as a 'slow march'. Since the demonstration that the visual cortex is arranged in vertical columns of cells, some ocular dominant and some responding to a stimulus with a specific orientation (Hubel & Wiesel, 1968, 1977) it has been assumed that the angular configuration of the fortification spectra results from activation of columns of cells by some laterally spreading process.

Studies of regional cerebral blood flow with [133]xenon, inhaled or injected into the carotid artery, have shown that flow diminishes during the prodromal phase of migraine by a mean of some 20% then increases during the headache phase (earlier studies summarized by Lance, 1982). Olesen, Larsen & Lauritzen (1981) reported sequential studies on eight patients with regional cerebral blood flow recorded during an attack of classical migraine by 254 sensors after the intracarotid injection of [133]xenon. The most characteristic feature was an initial focal hyperaemia in three patients followed by occipitoparietal oligaemia starting posteriorly and spreading forwards over 15–45 min, the focal reduction in flow varying from 2 to 60% (mean 36%). In other patients oligaemia progressed anteriorly without initial focal hyperaemia and was noted to persist into the headache phase of some patients. Further studies (Lauritzen, Gjedde & Hansen, 1982) showed that the rate of spread of oligaemia was about 2 mm/min, similar to the rate of spread of scotomas calculated by Lashley (1941) and by Richards (1971) at 3 mm/min and to 'spreading depression' described by Leão (1944), which traverses the cortical mantle of the experimental animal at a rate of 2–3 mm/min (Milner, 1958). There is thus circumstantial evidence that a process akin to spreading depression may underlie the slow march of migrainous scintillations and scotomas. Whether the process is initiated by constriction of the cortical microcirculation or whether regional cerebral flow diminishes secondarily in response to reduced metabolic demand remains unknown.

It has now been clearly established that intrinsic serotonergic projections from the raphe nuclei innervate layer IVc of the visual cortex, which receives the geniculocalcarine afferent pathway, and that layers III, V and VI, which contain the cells giving rise to fibres leaving the visual cortex, receive noradrenergic efferents from the locus coeruleus (Morrison *et al.*, 1982; Morrison & Magistretti, 1983). The effect of the serotonergic and noradrenergic systems has not yet been completely investigated but there is some evidence that the coeruleocortical system may enhance evoked activity at the expense of spontaneous activity (Morrison & Magistretti, 1983). Stimulation of the locus coeruleus in the monkey diminishes spontaneous cortical activity (Katayama *et al.*, 1981) and cerebral blood flow (Raichle *et al.*, 1975; Goadsby, Lambert & Lance, 1982). It is possible that serotonergic pathways modulate input to the visual cortex, that noradrenergic pathways regulate output, and that discharge of these pathways may play a part in the initiation of the visual symptoms characteristic of classical migraine.

Formed visual hallucinations experienced in an homonymous field defect

Horrax (1923) summarized the literature on formed hallucinations up to the time of publication tarting with a report by Westphal in 1879, and escribed 12 patients from Cushing's clinic in whom uch hallucinations were a symptom of cerebral umour. This and other reports (Symonds & Mackenzie, 957; Mooney *et al.*, 1965) emphasized that the allucinations were most commonly projected to the ea of a defect in the opposite half-fields. On the other and, Weinberger & Grant (1940) concluded that visual allucinations had no localizing value and could be rovoked by lesions at any level of the visual pathways. his conclusion is open to criticism since of the six atients they reported with complex imagery, one had ymptoms suggesting hypothalamic and temporal lobe isturbance, one had a fit before the visions, one had an homonymous hemianopia, and two were blind before hallucinations began.

In 1976, I described the symptoms of 13 patients in whom objects, animals or people were 'seen' in the affected quadrant or half of the contralateral visual field (Lance, 1976) (Fig. 25.1). The cause was parioeto-occipital infarction in nine patients, a postencephalitic state in one, and epileptic discharge in three. Two of the epileptic patients were of particular interest. One was a schoolboy aged 13 years who had experienced transient blurring of the left visual field, lasting 30–60 s at a time, since he was 8 years old. During this time he could see himself as a prisoner behind bars looking downwards and to the left where he could see little people with horns who looked like 'men from outer space'. While the attack was in progress his left pupil was constricted, presumably by activation of pupillo-constrictor fibres from the opposite occipital lobe. Removal of the pupilloconstrictor zone on one side in the cat results in anisocoria, with the larger pupil on

Fig. 25.1. An artist's impression of the formed hallucinations described by 13 patients (Lance, 1976), the images being drawn in the quadrant or half-field in which they were 'seen' by each patient. (Drawing by Mrs F. Rubiu, University of New South Wales.)

the opposite side (Cogan, 1948). His electroencephalogram showed a right parietal sharp-wave focus but repeated computed tomography of the brain has not demonstrated the causative lesion. At the age of 20 he is still subject to transient episodes of left hemianopia but without any hallucinations. The second epileptic patient was of particular interest because the nature of her visual hallucinations changed after removal of a right parieto-occipital angioma. She had suffered from epilepsy since the age of 25 years, with her seizures being preceded by a sensation of a flash 'like lightning' in the left visual field, followed by tingling in the left arm and leg. On examination she was found to have a left homonymous hemianopia, sparing the macular area. After the vascular malformation was removed surgically, her hemianopia became complete and the nature of her seizures changed. She then experienced visions of people in her left half-field preceding her fits on three occasions. The apparition differed each time. Once it was her father-in-law, once her neurosurgeon and, on the final occasion, Her Majesty the Queen. The patient was able to describe in detail the clothing worn by each of the participants in her visions: the type of belt worn by her neurosurgeon, the Queen's hat and handbag, and so on. It is evident that, with the infarction of the macular area of the visual cortex as the result of operation, her hallucinations changed from unformed to formed, suggesting that the site of origin had moved anteriorly to the visual association cortex.

In this series of patients, the causative lesions all lay posterior to the temporal lobe, involving the visual association cortex. Hallucinations were repeated briefly and frequently in six patients, lasting from a few seconds to a minute and recurring from one to twenty times each day. In the remainder, they were persistent at first but became episodic later on. In seven instances there was a latent interval between the cerebral insult and the onset of hallucinations, varying from 12 h to 4 months. The apparitions, all in colour, were mostly smaller than life-size in six patients but of normal size in the remainder. In some cases there was a connection between the object or person seen and past visual experience; for example, an elderly lady saw a girl in a sailor suit like her sister wore in an old photograph. In most cases, the vision bore no obvious relation to past or present interests, such as a man who saw Roman soldiers marching past pyramids (see Fig. 25.1). The lesion was in the right hemisphere in nine patients, in the left hemisphere in three patients and in both hemispheres in one patient. The preponderance of lesions in the posterior right hemisphere may be accounted for by the fact that the non-dominant parietal lobe plays a specific role in the intellectual processing of visual data (Basso *et al.*, 1973). It may be concluded that formed hallucinations may be an irritative phenomenon, which at times is frankly epileptic in nature, or may be a release phenomenon of the visual association cortex when it is deprived of its normal afferent inflow from the calcarine cortex. The emphasis on objects, people and animals suggests that these are the common building blocks of visual memory rather than the recall of any specific series of visual events. Spontaneous discharge of the visual association cortex may evoke a formed hallucination which is representative of groups or categories of the most familiar images. More complex visual patterns, linked with memory and often associated with auditory hallucinations may arise from the temporal lobe.

Cogan (1973) distinguished between irritative and release phenomena as causes of formed hallucinations, the former being transient and the latter more continuous. In many instances the two mechanisms may coexist, since paroxysmal or continuous excitation of the association cortex is more likely to arise when its connections with the primary visual cortex are severed. Hallucinations may be diminished by anticonvulsant therapy (Lance, 1976; Cummings *et al.*, 1982).

Palinopsia

Palinopsia (paliopsia) is the recurrence of visual perceptions after the stimulus-object has been removed (Critchley, 1951). Two of the patients subject to formed hallucinations described above (Lance, 1976) had also noted palinopsia. A man aged 53 years with a right occipital embolic infarct causing a left homonymous hemianopia, noticed that a person who had just walked past his bed would appear to make the journey again a moment afterwards. Objects also appeared distorted (metamorphopsia), such as fingers becoming thicker and thinner, shorter then longer. Another patient, a woman aged 62 years with a right occipital infarct causing a complete left homonymous hemianopia experienced the same phenomenon. She described a lady in a blue dress walking past, seen 'with her right eye', followed by repetition a moment later seen 'with the left eye'. She had a similar experience with her own actions, 'seeing' herself performing the same action again or, on one occasion, encountering five or six apparitions of herself doing things she had done a short time beforehand. Bender, Feldman & Sobin (1968) observed that palinopsia usually occurred with, and may be localized to, a visual field defect resulting from a parieto-occipital lesion, usually on the right side of the brain. They considered as possible mechanisms visual after-sensations like retinal after-images organized at a cerebral level or sensory seizures, and discounted the possibilty of psychogenic disturbance being responsible. Recent studies have had the advantages of computerized tomography to aid in cerebral localization (Meadows & Munro, 1977; Michel & Troost, 1980; Cummings *et al.*, 1982). Michel & Troost considered that palinopsia resulted from disordered temporal synthesis of observed visual information but did not suggest a mechanism. Cummings *et al.* (1982) were able to study a patient with a right parieto-occipital cyst who had recurrent palinopsia over a period of 6 years. His electroencephalogram showed right parietotemporal slow and sharp wave activity which did not change during the palinoptic episodes, although the visual evoked response was altered at this time. They concluded that palinopsia was a form of release hallucinations in which the exciting stimulus occurred in the recent past. The observation of one of the patients I reported, cited above, that the image was first seen 'with the right eye' and then 'with the left eye', suggests that the visual image may have initially been registered in the intact left visual cortex, then transmitted to the visual association cortex of the damaged right occipital lobe where it was 'seen' again. This mechanism could not account for the less common instances when the image recurs after minutes or hours or recurs repetitively, unless a reverberating circuit is set up between the hemispheres.

Visual hallucinations and palinopsia raise fascinating questions about the sequence of events in the processing of sensory information and its committal to memory which cannot be answered satisfactorily at present.

Conclusions and summary

Unformed visual hallucinations may arise from the retina or primary visual cortex. In the prodromal phase of migraine, regional cerebral blood flow is diminished but whether this is the direct result of constriction of the cortical microcirculation or is a response to diminished metabolic demand remains uncertain. Diffuse ischaemia could account for the repetitive spontaneous discharge of superficial layers of cortical cells, giving rise to photopsia. The slow march of fortification spectra (teichopsia) could be explained by 'spreading depression' slowly traversing columns of cells in the visual cortex, to cause a wave of excitation, followed by inhibition or neuronal exhaustion, the scintillating angles of the spectrum being formed by the sequential involvement of columns of cells each of which normally responds to a visual stimulus of edges or bars oriented at various angles.

Formed visual hallucinations may appear in association with blindness from any cause as a deafferentation release phenomenon, or in some toxic-confusional states. They are encountered most often with occipital lesions, right more than left, which produce a contralateral field defect but spare the visual association cortex. Some transient hallucinations appear to be epileptic in origin but most are probably release phenomena caused by spontaneous activity in the visual association cortex deprived of its normal afferent inflow. The stereotyped nature of such hallucinations, confined to the area of a field defect, suggests a limited role for the visual association cortex in classifying visual percepts into broad categories such as people, animals or objects.

The phenomenon of palinopsia may be accounted for by delay in transmission from the visual cortex of the intact hemisphere to the visual association cortex of the side which has sustained damage to the occipital pole. The patient thus sees two images, one after the other, derived, respectively, from the intact and damaged hemispheres. Longer delays in repetitive imagery may be caused by reverberating interhemispheric circuits.

References

Basso, A., De Renzi, E., Faglioni, P., Scotti, G. & Spinnler, H. (1973) Neuropsychological evidence for the existence of cerebral areas critical to the performance of intelligence tasks. *Brain* **96**, 715–28

Bender, M. B., Feldman, M. & Sobin, A. J. (1968) Palinopsia. *Brain*, **91**, 321–38

Brindley, G. S., Donaldson, P. E. K., Falconer, M. A. & Rushton, D. N. (1972) The extent of the region of occipital cortex that when stimulated gives phosphenes fixed in the visual field. *J. Physiol.* (*Lond.*), **225**, 57–8P

Brindley, G. S. & Lewin, W. S. (1968) The sensations produced by electrical stimulation of the visual cortex. *J Physiol.* (*Lond.*), **196**, 479–93

Cogan, D. G. (1948) *Neurology of the Ocular Muscles*. Springfield, Ill.: Thomas

Cogan, D. G. (1973) Visual hallucinations as release phenomena. *Albrecht v. Graefes Arch. Klin. Exp. Ophthalmol.*, **188**, 139–50

Critchley, M. (1951) Types of visual perseveration: 'paliopsia' and 'illusory visual spread'. *Brain*, **74**, 267–99

Cummings, J. L., Syndulko, K., Goldberg, Z. & Treiman, D. M. (1982) Palinopsia reconsidered. *Neurology (N.Y.)*, **32**, 444–7

Drummond, P. D. & Lance, J. W. (1984) Clinical diagnosis and computer analysis of headache symptoms. *J. Neurol. Neurosurg. Psychiat.*, **47**, 128–33

Foerster, O. (1931) The cerebral cortex in man. *Lancet*, ii, 309–12

Goadsby, P. J., Lambert, G. A. & Lance, J. W. (1982) Differential effects on the internal and external carotid circulation of the monkey evoked by locus coeruleus stimulation. *Brain Res.*, **249**, 247–54

Golden, G. S. (1979) The Alice in Wonderland syndrome in juvenile migraine. *Pediatrics*, **63**, 517–19

Hare, E. J. (1966) Personal observations on the spectral march of migraine. *J. Neurol. Sci.*, **3**, 259–64

Horrax, G. (1923) Visual hallucinations as a cerebral localizing phenomenon with special reference to their occurrence in tumors of the temporal lobes. *Arch. Neurol. Psychiat., Chicago*, **10**, 532–47

Hubel, D. H. & Wiessel, T. N. (1968) Receptive fields and functional architecture of monkey striate cortex. *J. Physiol. (Lond.)*, **195**, 215–43

Hubel, D. H. & Wiesel, T. N. (1977) Functional architecture of macaque monkey visual cortex. *Proc. R. Soc. Lond., Ser. B*, **198**, 1–59

Jung, R. (1979) Translokation corticaler Migrainephosphene bei Augenbewegungen und vestibularen Reizen. *Neuropsychologia*, **17**, 173–85

Katayama, Y., Ueno, Y., Tsukiyama, T. & Tsubokawa, T. (1981) Long-lasting suppression of firing of cortical neurons and decrease in cortical blood flow following train pulse stimulation of the locus coeruleus in the cat. *Brain Res.*, **216**, 173–9

Klee, A. & Willanger, R. (1966) Disturbances of visual perception in migraine. *Acta Neurol. Scand.*, **42**, 400–14

Lance, J. W. (1976) Simple formed hallucinations confined to the area of a specific visual field defect. *Brain*, **99**, 719–34

Lance, J. W. (1982) *The Mechanism and Management of Headache*, 3rd edn. London: Butterworths

Lance, J. W. & Anthony, M. (1966) Some clinical aspects of migraine. *Arch Neurol.*, **15**, 356–61

Lashley, K. S. (1941) Patterns of cerebral integration indicated by the scotomas of migraine. *Arch. Neurol. Psychiat., Chicago*, **46**, 331–9

Lauritzen, M., Gjedde, A. & Hansen, J. H. (1982) Associations between spreading depression and migraine. In *Advances in Migraine Research and Therapy*, ed. F. C. Rose, pp. 121–6. New York: Raven Press

Leão, A. A. P. (1944) Spreading depression of activity in the cerebral cortex. *J. Neurophysiol.*, **7**, 359–90

Lippman, C. W. (1952) Certain hallucinations peculiar to migraine. *J. Nerv. Ment. Dis.*, **116**, 346–51

Meadows, J. C. & Munro, S. S. F. (1977) Palinopsia. *J. Neurol. Neurosurg. Psychiat.*, **40**, 5–8

Michel, E. M. & Troost, B. T. (1980) Palinopsia: cerebral localization with computed tomography. *Neurology*, **30**, 887–9

Miles, P. W. (1958) Scintillating scotoma. Clinical and anatomic significance of pattern, size and movement. *J. Am. Med. Assoc.*, **167**, 1810–13

Milner, P. M. (1958) Notes on a possible correspondence between the scotomas of migraine and spreading depression of Leão. *Electroenceph. clin. Neurophysiol.*, **10**, 705

Mooney, A. J., Carey, P., Ryan, M. & Bofin, P. (1965) Parasagittal parieto-occipital meningioma with visual hallucinations. *Am. J. Ophthalmol.*, **59**, 197–205

Morrison, J. H., Foote, S. L., Molliver, M. E., Bloom, F. E. & Lidov, H. G. W. (1982) Noradrenergic and serotonergic fibers innervate complementary layers in monkey primary visual cortex: an immunohistochemical study. *Proc. Natl Acad. Sci. USA*, **79**, 2401–5

Morrison, J. H. & Magistretti, P. J. (1983) Monoamines and peptides in cerebral cortex. Contrasting principles of cortical organization. *TINS*, April, 146–50

Olesen, J., Larsen, B. & Lauritzen, M. (1981) Focal hyperemia followed by spreading oligemia and impaired activation of rCBF in classic migraine. *Ann. Neurol.*, **9**, 344–52

Peatfield, R. C. & Rose, F. C. (1981) Migrainous visual symptoms in a woman without eyes. *Arch. Neurol.*, **38**, 466

Penfield, W. & Perot, P. (1963) The brain's record of auditory and visual experience. *Brain*, **86**, 595–696

Raichle, M. E., Hartman, B. K., Eichling, J. O. & Sharpe, L. G. (1975) Central noradrenergic regulation of cerebral blood flow and vascular permeability. *Proc. Natl Acad. Sci. USA*, **72**, 3726–30

Richards, W. (1971) The fortification illusions of migraine. *Scientific American*, **224**, (5), 88–96

Sacks, O. W. (1970) *Migraine. Evolution of a Common Disorder*. London: Faber & Faber

Selby, G. & Lance, J. W. (1960) Observations on 500 cases of migraine and allied vascular headache. *J. Neurol. Neurosurg. Psychiat.*, **23**, 23–32

Symonds, C. & Mackenzie, I. (1957) Bilateral loss of vision from cerebral infarction. *Brain*, **80**, 415–55

Walsh, F. B. & Hoyt, W. E. (1969) *Clinical Neuro-ophthalmology*, 3rd edn, Vol. 1. Baltimore: Williams and Wilkins

Weinberger, L. M. & Grant, F. C. (1940) Visual hallucinations and their neuro-optical correlates. *Arch. Ophthalmol., Chicago*, **23**, 166–99

26

Image analysis performed by the visual system: feature *versus* Fourier analysis and adaptable filtering

J. J. KULIKOWSKI and
K. KRANDA

Introduction

In the last two decades, two hypotheses of visual information processing have dominated the thinking of neurophysiologists. One group of protagonists has argued that the system operates as a feature analyser, whereas the other group believes that the system performs Fourier analysis. However, although much effort has gone into trying to establish the exclusive validity of one or the other hypothesis, the evidence presented is in both cases rather unconvincing.

Recent experiments have indicated that an intermediate mechanism of neural processing may operate in the geniculostriate pathway, namely adaptable spatial filtering of moderate selectivity to orientation and spatial frequency. Here we propose that the analysis of a visual image into symmetrical and antisymmetrical components (e.g. bars and edges) is based on filtering by visual cortical cells of a simple type whose receptive fields (RFs) are organized in symmetrical or antisymmetrical arrangements, approximating Gabor-type operators.

Models for encoding visual information

Retinal encoding of the visual input is essentially a transformation of the image rather than a point-by-point representation. The main function of the vertebrate retina is to extract contrast from luminance distribution, thus making the percept largely independent of the ambient illumination level. The retina and central structures of the visual system extract contrast information, defined according to Michelson as $C = (L_{max} - L_{min})/(L_{max} + L_{min})$, where L_{max} and L_{min} are luminate maxima and minima of the adjacent areas. Contrast extraction is achieved by differentiating outputs of cells. This type of spatial differentiation operates over a RF with at least two adjacent antagonistic subregions which are selectively activated by contours, respectively lighter or darker than the background. Contrast extraction by a RF with just two subregions can be expressed as a differential operator, d/dx (differentiation takes place along the x-axis only), whose digital approximations are weighting functions of either $+1$ and -1, or -1 and $+1$. An operator of double spatial differentiation, d^2/dx^2 whose weighting functions are either -1, $+2$, -1, or $+1$, -2, $+1$ describes the operation of a symmetrical RF which has at least three subregions. Fig. 26.1 shows a transformation of a luminance distribution along the x-axis (*a*) by single- (*b*, *c*) and by double-differentiating operators (*d*,*e*). Differentiation along the y-axis would be represented by a corresponding vertical distribution of weighting coefficients. Differentiation along both axes would be represented by a two-dimensional array of weighting coefficients (Fig. 26.1*f*,*g*,*h*,*i*). When th widths of the antagonistic subregions along both axe are equal and cover one receptor, then the double differentiation function can be expressed in a digita approximation as the simplest 'concentric operator (Fig. 26.1*f*,*g*).

The antagonistic subregions may receive input from several receptors, i.e. spatial summation ma occur within RFs (Fig. 26.1*h*), as has been known sinc Adrian & Matthews' paper of 1928.

Spatial differentiation *versus* contour outlining

In general, spatial summation observed in th visual system can be modelled satisfactorily along on dimension by digital differential operators tha correspond to units with several positive and negativ subregions. Here the width of the subregions woul provide a basis for size tuning. Several paralle operators with different tuning characteristics woul clearly offer an advantage over simple operators (se Kulikowski, 1966). These operators can be followed b another set of operators which are non-digital, bu

Fig. 26.1. Digital description of one-dimensional luminance distribution of an image (a) and its transformations by four operators (b, c, d, e). Two-dimensional digital operators are shown in (f) and (g), and their analog equivalents in (h) and (i). Note that all these operators introduce negative signals (dashed lines). If the negative signals are impossible to handle, the number of operators has to be doubled, e.g. (b) copes with positive signals and (c) with negative signals.

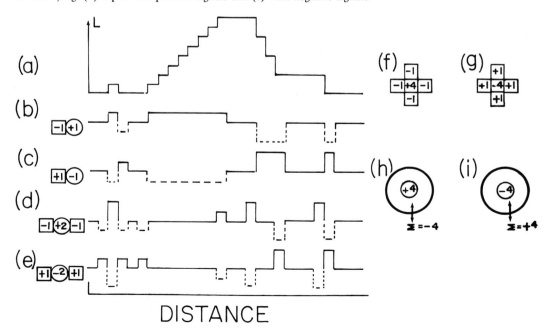

ose outputs can be digitally quantized. For instance, roducing an additional non-linear threshold device »duces a quantized output of either 0 or 1 depending whether the threshold value has been exceeded. This reshold' algorithm performs contour outlining, »vided that only the 'concentric' operators are lized (Kovasznay & Joseph, 1955; Kulikowski, 56).

We have outlined some possible types of image »cessing and shall now briefly demonstrate how a niliar picture of the *Mona Lisa* (Fig. 26.2a) is nsformed by double spatial differentiation using a »ncentric' operator. The transformed image in ;. 26.2b is roughly equivalent to a poor-quality otocopy. Figs 26.2c, d and e show, respectively, nsformed images after contour outlining with a non-ear threshold device, double spatial differentiation the opposite polarity (negative of b) and the nega-e of the original image. If one compared transforma-ns of all these images, then the transformed image in s easier to recognize than is the negative (e). The ferentiated (b) and outlined sketches (c) are short in tail but easy to recognize, irrespective of contrast larity (d).

Contour outlining

This model is favoured by theoreticians, for mple Marr (1982). Marr assumed that double tial differentiation of the image is performed by ncentric' operators coupled to a non-linear threshold vice, detecting zero crossings and outlining the con-rs. Marr also assumed that this outline constitutes ; 'primal sketch' of the visual scene which is sub-

Fig. 26.2. (a) Photograph of Mona Lisa; (b) its double spatial differentiation; (c) outlined and smoother contours; (d) negative of b; and (e) negative of a.

sequently used for further processing. Note also that contour outlining is usually assumed to be performed by a number of parallel processors, each of which functions as a 'concentric' operator tuned to a given stimulus size (Kulikowski, 1966).

Although contour outlining models seem feasible on theoretical grounds, there is no evidence that any of the processes compatible with such models occur anywhere in the primate visual system. Most cells in the retina and lateral geniculate nucleus (LGN), and some in the primate striate cortex can be considered as 'concentric' operators, yet information conveyed by such units can only be utilized for contour outlining after non-linear processing by a threshold device.

A further development of this model would require a higher degree of specialization in the detec-tion of trigger features. Yet neither simple nor complex cells in the striate cortex show any such properties. Although some feature extraction of a 'higher order' could operate in the extrastriatal visual areas (see Zeki, 1978, 1980, 1983 for a description of cell response properties), one also needs to assume the existence of higher brain centres, with hypothetical 'gnostic units' (see Konorski, 1969) which could perform pattern and shape recognition. However, the evidence for the existence of such units is not convincing, (see e.g. Gross, Rocha-Miranda & Bender, 1972), revealing in-stead assemblies of units capable of preserving per-ceptual constancies (Perrett, Rolls & Caan, 1982).

Description of images by linear and analogue functions

Any continuous signal which may, for instance, convey luminance distribution of a given image can be described by a set of orthogonal (independent) functions. The best known set of functions is the Fourier series which describe an image in terms of its

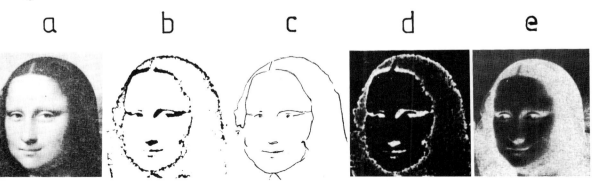

a   b   c   d   e

harmonic (space-periodic) components. Here the image is described in terms of spatial frequencies f_x, f_y rather than as a distribution along the two geometrical axes x and y. This type of transformation is often used in image processing such as holography, where the photographic plate stores information about the amplitude and phase of spatial harmonics.

Let us consider the implication of this type of image processing for the visual system. If the striate cortex indeed encodes images in terms of spatial frequency (see Pollen, Lee & Taylor, 1971, for this line of argument) then lesioning of this area in the monkey should produce an overall (global) decrease in the perceptible image quality, rather than the appearance of localized scotomas reported by Cowey (1967) and Weiskrantz & Cowey (1967). The topographic rather than global nature of the impairment is inconsistent with the spatial-frequency (holographic) image representation in the striate cortex, indicating the importance of spatial localization.

Gabor functions – a compromise between spatial and spatial-frequency approach

Gabor (1946) introduced a theory of signal processing which is applicable to information transfer in general and, as has recently been realized, to visual processing in particular (see also Helstrom, 1966; Cowan, 1977; Marcelja, 1980; Daugman, 1984). According to this theory, a complex signal, such as a visual image, can be broken into its elements, i.e. into spatial-frequency harmonics of cosine and sine waves limited in space by a Gaussian function. The Gabor functions thus combine spatial localization with spatial-frequency selectivity of image transformation.

Gabor-type analysis in the visual system would thus require linear filter units whose spatial response profiles (i.e. RFs) should be sine and cosine functions, antisymmetrical and symmetrical, multiplied by a Gaussian function. The response sensitivity of cells with symmetrical (Rs) and antisymmetrical (Ra) profiles along the x-axis is given by:

$$Rs(x) = \exp. \, (-x^2/2s^2) \cos 2\pi \, fo \, x \qquad (1a)$$

$$Ra(x) = \exp. \, (-x^2/2s^2) \sin 2\pi \, fo \, x \qquad (1b)$$

where s is the standard deviation of the Gaussian sensitivity envelope and fo is the optimal spatial frequency (cycles/degree). Here the standard deviation (s) determines the width (W) of the RF, which can be

conveniently defined as the subtense within which the response (R) exceeds 10% of the maximum ($W10\%$). In such a case $W10\% = 4.3s$ (see Appendix).

Fig. 26.3(a) shows theoretical RF profiles of Gabor-type units (which closely resemble sensitivity profiles of simple cells: see Hubel & Wiesel, 1959, 1962, and Kulikowski & Bishop, 1981, for descriptions of these units). The simple-type RFs have antagonistic subregions, denoted here as positive and negative, (light and dark areas in Fig. 26.3a), which respond respectively, to light and dark portions of the image. The width of each subregion is a half of the spatial cycle, i.e. $\frac{1}{2} fo$.

The spatial profiles (Fig. 26.3a) determine uniquely the spatial-frequency tuning curve which can be plotted on linear (d) or logarithmic coordinates. The absolute bandwidth of this curve is most directly determined as a difference of corner frequencies at half maximum ($f_2 - f_1$), i.e. the width of the tuning curve plotted on a linear scale.

The relative spatial-frequency bandwidth (defined as $Bdf = (f_2 - f_1)/f_0$ or, in octaves, as $Boct = \log_2 (f_2/f_1)$) can best be depicted as the width (at half-maximum sensitivity) of a spatial-frequency tuning curve on a logarithmic scale. A comparison of Figs 26.3G, H and I shows that the relative bandwidth is inversely proportional to the number of antagonistic subregions. The more subregions there are, the greater is the uncertainty regarding the location of a stimulus (e.g. a single bar) which activated the cell. Thus, accuracy of spatial localization competes against selectivity of spatial-frequency tuning. The most important feature of Gabor functions is that they provide image transformation with the minimum product of uncertainty of localization in space and uncertainty in spatial frequency. Either of these uncertainties can be minimized separately, depending on which factor is more important.

Fig. 26.3 illustrates some implications of these two possibilities. Either system must start with units tuned to the fundamental spatial frequency, f_0 the RF of these units must have a small number of subregions if they are not to be too large (a). Subsequently, however, for higher spatial harmonics (e.g. $3f_0$ in b and c), the number of subregions may either (b) increase filling the same space as (a), or (c) remain constant. Consequently, the system (a–b) develops higher spatial-frequency selectivity with increasing spatial quency (relative bandwidth decreases), whereas the stem (a–c) develops greater accuracy of spatial calization (smaller RFs).

The following section examines spatial *versus* atial-frequency characteristics of units, especially in e visual cortex. Among the cells examined, only mple cells can be treated as linear filters which proximate a Gabor system as in Fig. 26.3a–c. The maining cells, not showing a sufficient degree of earity, are discussed separately.

Receptive fields and spatial-frequency tuning of single units

Properties and classification of cells in the retina and LGN

Studies of the RFs of retinal ganglion cells (especially in 'lower' vertebrates) revealed not only their linear properties of spatial summation (Barlow 1953), but also suggested their (non-linear) selectivity to some stimuli, as though these cells operated as feature detectors (Lettvin *et al.*, 1959, Barlow, Hill & Levick, 1964).

Fig. 26.3. One-dimensional Gabor profiles and outlines of their RFs in two dimensions. The basic system (a) may have its equivalent third-harmonic system (tuned to three times its fundamental spatial frequency) with either the overall RF size being constant (b), or with the profile shape being constant (c). Compare equivalent tuning curves plotted on linear (d, e, f) and logarithmic (g, h, i) coordinates. Relative ($\Delta f/f_0$) and absolute (Δf) bandwidth are illustrated schematically as functions of optimal spatial frequency for system (b) in (j) and for system (c) in (k).

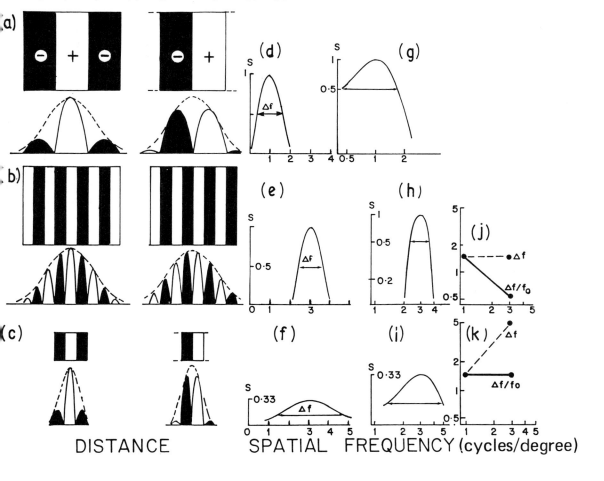

DISTANCE   SPATIAL FREQUENCY (cycles/degree)

Later studies of the visual systems of cat and monkey, however, have suggested a much smaller numerical representation of 'specialized' units (Cleland & Levick, 1974 – see below) and also that many such cells projected outside the geniculostriate pathway (Schiller & Malpeli, 1977).

The responses of most cells in the cat retina seem regular and amenable to description by a linear model (Rodieck, 1965). A particular aspect of linearity is linear spatial summation at threshold (Enroth-Cugell & Robson, 1966). A stationary flashing grating stimulus can be positioned on the cell's RF at a 'null position' such that no response is generated (i.e. the excitatory action of one part of the stimulus is compensated by the antagonistic action of the other part). The retinal ganglion cells tested in this way can be divided into X (linear) and Y (non-linear) classes. The presence of the 'null' position is the strictest test of linear spatial summation. However, the cells not passing this test need not necessarily act in a very non-linear way.

Enroth-Cugell & Robson (1966) and other authors have summarized characteristics of X retinal ganglion cells in the cat as having null position, relatively sustained responses to stationary stimuli (Cleland, Dubin & Levick, 1971) and medium axon conduction velocity (Fukada, 1971). Conversely, Y-cells have no null position, are more transient and have the fastest conducting axons. The remainder are W-cells which do not form a homogeneous group. A small fraction of W-cells are specialized cells (Cleland & Levick 1974; Stone & Fukada, 1974) but their possible role as trigger feature detectors has not been substantiated.

The proportion of ganglion cells in each group (Y,X,W) can only be roughly estimated and only in morphological studies of their neuroanatomical substrates (alpha, beta, gamma) since electrophysiological estimates are prone to gross errors of sampling. In view of a substantial overlap in some sizes the current estimates of cells, other than substrates of Y–cells – alpha type (Boycott & Wässle, 1974), are likely to lead to significant errors (Rowe & Dreher, 1982; Leventhal, Keens & Tork, 1980).

The primate retina has a greater degree of physiological complexity (Leventhal, Rodieck & Dreher 1981; Perry & Cowey 1981) made by the presence of cone-opponent ganglion cells, i.e. cells whose RFs have different cone inputs to the centre and surround and consequently different chromatic characteristics (for review see Zrenner, 1983). Gouras (1968) noticed that these cells show sustained (tonic) responses, whereas de Monasterio, Gouras & Tolhurst (1976) found that most of them exhibit linear spatial summation. W-cells form probably only about 10% of all ganglion cells.

In the monkey LGN a large majority of cells in all layers exhibit linear spatial summation (Shapley, Kaplan & Soodak 1981; Hicks, Lee & Vidyasagar 1983), even though these cells in parvocellular and magnocellular layers belong to distinctly different classes. Dreher, Fukada & Rodieck (1976) found that cells in parvocellular layers, most of which have chromatically opponent RFs and a null position (de Valois, Abramov & Mead, 1967; Wiesel & Hubel, 1966; de Monasterio et al., 1976; for review see Zrenner 1983), have slower conduction velocities as compared with the cells in the magnocellular layers. The latter cells are practically achromatic and tend to have more transient responses irrespective of whether they do or do not have a null position (even if they do not, they are still not very non-linear). It seems that linearity becomes a rather common characteristic of many types of cells, thereby losing its importance as a classificatory criterion.

Spatial-frequency tuning in the retina and LGN

Spatial-frequency tuning of retinal and LGN cells with concentric RFs is far from sharp. Although five subregions have been identified with the cat retina (Ikeda & Wright, 1972), the outer disinhibitory subregions are very weak and thus cannot sharpen the tuning characteristics significantly. The LGN could be considered as an additional stage in the lateral inhibition process (see Ratliff, 1965), and consequently should have still other (dis-disinhibitory) subregions leading to sharper tuning characteristics. However, this is apparently not the case in either the cat (e.g. Maffei & Fiorentini, 1972; Hammond, 1973) or the monkey (e.g. Hicks et al., 1983).

Classification of visual cortical cells

The test of the 'null position' is not sufficient for complete classification of visual cortical cells since linearity is not an all-or-none property, but rather a matter of degree.

There are various indications of the degree to which neuronal responses can be described as linear

Extensive investigations of visual cortical cells (since their original description by Hubel & Wiesel, 1962, 1968) have led to the following linearity criteria for simple cells (summarized after Movshon, Thompson & Tolhurst, 1978a; Kulikowski & Bishop, 1981, 1982; Kulikowski, Bishop & Kato, 1981; Kulikowski, Rao & Vidyasagar, 1982b) illustrated in Fig. 26.4.

1. Responses to moving bars and edges (optimally oriented) should have similar spatial profiles for movement in both directions and these profiles, in turn, should resemble the corresponding static profiles (Kulikowski, 1979).

2. Responses to moving edges must be integrals of the responses to bars, since an edge may be considered as a sum of bars (linear response to a sum of signals equals the sum of responses to these constituent signals).

3. Drifting sinusoidal gratings produce responses which are half-wave rectified sinusoids when spontaneous activity is absent (Movshon *et al.*, 1978); if such responses are rounded off or show signs of full-wave rectification (Kulikowski & Bishop, 1981), this indicates a contribution of a non-linear component (note that squaring produces full-wave rectification, i.e. a doubled frequency component as compared to the fundamental frequency of a discharge produced by a drifting grating).

4. Spatial-frequency tuning curves measured by using either contrast sensitivity or response characteristics should be simple Fourier transformations of responses to bars (this criterion excludes many narrow-band cells whose receptive fields do not have many subregions commensurate with their narrow bandwidth – see Fig. 26.3).

5. There should be a null position, or a spatial phase of a grating stimulus reversed in contrast at which no response or a minimum response is evoked. For convenience the phase of this minimum response may be called 0 degree phase and the minimum response might be expressed as a fraction of the maximum response obtained at $+90$ degrees phase (Kulikowski *et al.*, 1982b). This ratio can then be regarded as a measure of non-linearity: it is zero for linear simple cells and about one for C (complex) cells which respond to every contrast reversal, irrespective of the spatial phase of a grating.

Identifiable cell types and their functional significance

It seems that the responses of cortical cells of the main identifiable types (which exhibit different degrees of linearity), reveal also some functional significance of systems containing such units. Fig. 26.4 summarizes the most important properties of various cortical cells (see Henry, 1977; Kulikowski *et al.*, 1981, 1982b).

Two simple cells shown in Fig. 26.4 represent extremes of their range. The linear simple cell (SL) belongs to the group of 'slow' cells and is most common in the striate cortex. Fast simple cells (SF) are so named because they respond well to movement velocity above 50 degrees/s (Dreher, Leventhal & Hale, 1980), are transient and show partial non-linearity. There are, however, 'slow' simple cells which only differ from linear simple cells by not having a null position. These non-linear cells should be distinguished from the end-stopped simple cells (Dreher, 1972) which may exhibit linear properties when short stimuli are used, and non-linear properties when the inhibitory end-zones are also stimulated (Kulikowski & Bishop, 1981). SF-cells were encountered on both sides of the Area 17/18 border, not only in Area 18 (c.f. Ikeda & Wright, 1975; Movshon *et al.*, 1978c). Since both the 'slow' and fast simple cells have RFs of symmetrical and antisymmetrical types it seems that they both participate in contour analysis but at different velocities of moving patterns.

A-cells form a distinctive group of cells with large RFs and just-discernible subregions, responding to low spatial frequencies and having broad bandwidth. However, these are not simple cells, and in our studies they were not encountered in Area 17 of either cat or monkey. Their properties resemble those of neurons in Area 18, which were reported to be affected by the cooling of Area 17 (Sherk, 1978; cf. Dreher & Cottee, 1975). Their large RFs, often located near the vertical meridian, make them suitable for integrating vision between the two hemispheres and possibly for coarse stereopsis. Their broad spatial-frequency tuning makes their contribution to Fourier analysis unlikely.

B-cells form another group of intermediate non-linear cells with a certain neuroanatomical identity (Henry, Lund & Harvey, 1978). It seems that two types of neurons may be classified under this category: periodic cells with very narrow band tuning (Kulikowski & Bishop, 1982) and periodic complex cells (Pollen & Feldon, 1979). Both have rather small RFs and may be confused with simple cells. They respond to fine patterns (their resolution limit of about 6 cycles/degree is comparable to that of the finest simple cells) and are therefore sensitive to defocussing the retinal image. Their response properties and neural connections with the Clare–Bishop area (Henry *et al.*,

Fig. 26.4. Schematic responses of various types of visual cortical cells to flashing and moving stimuli. (a) Static RFs outlined by responses to light bar ON (+) and OFF (−). Four RF types are shown to form a quadruplet. (b) Averaged response histograms to bars moving at optimal velocities; the open areas denote responses to light bars and filled areas to dark bars, whereas the dots indicate the overlap between these two responses. (c) Response histograms to moving edges (other details as in b). (d) Idealized responses to moving gratings of an optimal spatial frequency for each cell (responses to 1.5 grating cycles are shown). (e) Tuning curves: amplitude of a modulated response (as in d) as a function of grating spatial frequency. Note that for various cell types maxima occur at different spatial frequencies and that the tuning curve for complex cells (dotted line) is obtained by contrast reversal presentation, since drifting gratings give poorly modulated responses. (f) Responses of various cell types to contrast reversal of a grating at a spatial phase of 0 degrees (minimum response) and +90 degrees (maximum response). Only linear simple cells (SL) do not respond at a zero degree phase (null position).

1978) make them suitable for control of accommodation (Bando *et al.*, 1981).

Complex (C) cells are nonlinear and seem to respond to all properly oriented stimuli, hence they could be regarded as non-selective detectors of moving texture. This may be an oversimplification, since Hammond & Smith (1983) have found that C-cells respond preferentially to moving contours, with suppression of response components to background visual noise.

Spatial-frequency tuning in the visual cortex

Cortical cells have been found to be more narrowly tuned than their LGN counterparts (see Cooper & Robson, 1968; Campbell, Cooper & Enroth-Cugell, 1969; Tolhurst & Thompson, 1981), yet the significance of these reports cannot be evaluated since no distinction was made between the responses of simple and complex cells (see Fig. 26.4 for their response characteristics). The analysis of simple- and complex-cell responses requires, however, different approaches. For instance, Maffei & Fiorentini (1973) have proposed that simple cells function as spatial-frequency analysers. Alternatively, one could look for cells which could transform spatial coordinates into spatial frequency (Pollen *et al.*, 1971). Complex cells have been considered as potential candidates for this role (see also Pollen & Ronner, 1975, 1981; Glezer, Ivanov & Tscherback, 1973; Glezer *et al.*, 1976), but their non-linear properties exclude them from linear Fourier analysis. Thus the only group of cells approximately fulfilling the criteria of linearity are simple cells and the most direct way of describing them is in terms of their spatial and spatial-frequency properties.

Spatial and spatial-frequency characteristics of simple cells

The RF sizes of simple cells and their spatial-frequency bandwidths as a function of optimal spatial frequency (f_o) are shown in Fig. 26.5 (see Kulikowski & Vidyasagar, 1982, 1984). Note that RF width (W) or the bandwidth (Δf) may vary by a factor of two for cells with similar f_o and there is a continuum of f_o values. The tuning curves of such cells may be computed from their response to gratings and also from responses to bars and edges moving at optimal velocities (Kulikowski & Vidyasagar, 1982, 1984). Response profiles to bars of the optimal velocity show

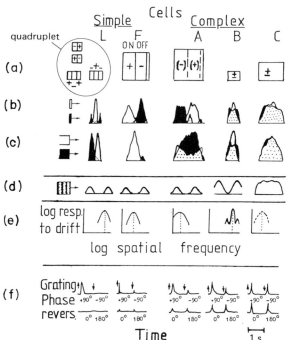

remarkable regularity of symmetrical or antisymmetrical arrangements (see Kulikowski, 1979), the ratio of which is about 1:1 (Kulikowski & Bishop, 1981; Kulikowski & Vidyasagar, 1984).

The two regression lines fitted to the data points in Fig. 26.5, RF width *versus* optimal spatial frequency (fo), and (*b*) bandwidth *versus* fo, have mean slopes of -0.92 and 0.83, respectively. The width of the RFs and the bandwidth thus show, respectively, a reciprocal and an almost directly proportional relation to fo, rather than being constant as postulated by the patch-by-patch theory proposed by Robson (1975).

The plot of the absolute bandwidth *versus* fo can be recomputed in terms of the relative bandwidth (Bdf). In such a case the slope is about -0.17, i.e. Bdf decreases only slightly with increasing spatial frequency. A similar result was obtained independently by de Valois, Albrecht & Thorell (1982): the mean relative bandwidth expressed in octaves decreased from about 1.8 to 1.3 octaves (i.e. by 0.5 octaves) when the optimal spatial frequency increased from 1 to 8 cycles/degree (i.e. by 3 octaves). Both these results are consistent with psychophysical evaluation of the relative bandwidth (King-Smith & Kulikowski, 1975; see also below).

Fig. 26.5. RF width and spatial frequency bandwidth as a function of optimal spatial frequency; macaque striate cortex, parafoveal representation. Compare with Fig. 26.3 (j, k). (After Kulikowski & Vidyasagar, 1982.)

The most common simple cells are those with RFs having three to four subregions (mean $Bdf = 1.07$ cycles/degree or 1.7 octaves), although some RFs have five to six subregions (with the narrowest Bd equal to 0.6 cycles/degree or 0.93 octaves (see Kulikowski & Vidyasagar, 1982)). This variation in properties of cells which are tuned to the same fo may, in some cases, have an advantage over processors of the same type, by providing units with overlapping properties and a built-in redundancy (for further discussion see Kulikowski, Marcelja & Bishop, 1982*a*). As was shown above, the symmetrical profiles must have a minimum of three subregions; consequently five subregions are needed to introduce some variability. In such a system with redundancy the antisymmetrical profiles may have two, four or six subregions and this has been observed (Kulikowski & Bishop, 1981; Kulikowski & Vidyasagar, 1982, 1984).

Five antagonistic subregions across the RF are present, even in the retina (Ikeda & Wright, 1972). Thus, if the visual system were to extract spatial-frequency information then this number could be expected to increase by two at each stage of consecutive lateral inhibition (Ratliff, 1965), i.e. seven in the LGN and nine at the first stage of the striate cortex. Moreover, since many simple cells are interconnected (Bullier & Henry 1979), the number of subregions of simple RFs should thus exceed nine for symmetrical and ten for antisymmetrical profiles. Since, on the contrary, the number of antagonistic subregions remains almost

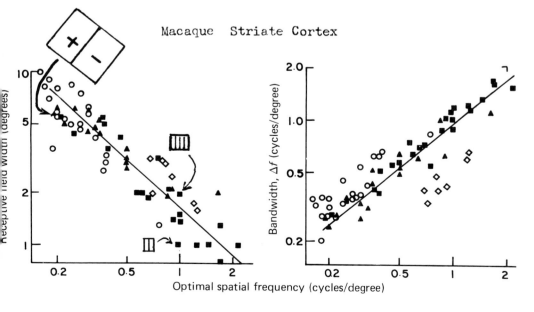

Macaque Striate Cortex

constant from the retina to the cortex, it seems that the visual system is not primarily designed for extracting harmonic components from the image, as this would require a high degree of spatial-frequency selectivity. Note that high selectivity would be most economical only for the analysis of an image composed of many spatially periodic patterns; since such images are encountered infrequently in nature, their accurate analysis has presumably a low survival value (see Kulikowski *et al.*, 1982*a*).

Although simple cell operations can be expressed in terms of Gabor-type linear analysis of the visual image, such a description is only partial and is thus considered as the first approximation of the cell function. This description is incomplete, because simple cell responses to complex or non-optimal stimuli sometimes deviate from linearity (Bishop, Coombs & Henry, 1973). Such a non-linearity may reflect non-linear interactions of suppressive or facilitatory nature (Creutzfeldt, Kuhnt & Benevento, 1974; Hammond & Mackay, 1983; Kulikowski & Carden, unpublished). We also face here a conceptual problem in that we have to consider whether simple cells may subserve one or a multitude of functions. As some simple cells may signal direction of movement or perhaps disparity, linear spatial analysis is presumably not their only function.

In conclusion, it depends largely on conditions of visual stimulation whether a Gabor-type function is an adequate description of simple-cell operation. For instance, if one considers the linear response characteristics of simple cells alone, then the simple cells are virtually Gabor-type operators of moderate spatial-frequency selectivity. Yet once their non-linearity and multifunctionality are also considered, simple cells appear more like adaptable spatial processors than Gabor-type filters and, besides, there are more simple cells than required for a Gabor system (Fig. 26.3a, c). It is possible that this adaptability is more economic than the linear system by suppressing responses of non-optimally stimulated units.

Psychophysical and perceptual correlates of visual processing

The response of single units and perceptual phenomena

In the previous section, we pointed out that nowhere in the geniculostriate pathway of cat and monkey do we encounter cells with specialized response properties similar to those found in non-mammalian vertebrates (see Lettvin *et al.*, 1959). For instance, cortical cells seem to operate as filters, extracting several features of the image such as orientation, size, movement and colour (see Barlow, 1981), rather than as selective feature detectors. Simple cells exhibit dual properties: for simple stimuli presented under optimal conditions they behave like linear processors of the Gabor type, whereas for other conditions their responses are subject to non-linear inhibition or facilitation.

Here we describe how to investigate the nature of spatial image analysis by resorting to psychophysics. This approach should, at least in theory, tap response properties of given central nervous system structures, since some factors, such as the cortical architecture and response characteristics of neurons, inevitably place constraints on perception and thus express themselves in perceptual phenomena (see Barlow, 1981). The analysis of orientation is one example. Orientation selectivity was originally reported by Hubel & Wiesel (1959) and has since been the most investigated feature of cortical cells. Subsequent psychophysical measurements of masking effects (see Campbell & Kulikowski, 1966) also established orientation selectivity and thus emphasized the compatibility of results produced by the two ways of investigation.

The analysis of frequency information

Psychophysical measurements and analysis of flicker sensitivity (de Lange, 1958) constituted the first systematic investigation into the frequency characteristic of the visual system. This type of analysis showed that the flicker sensitivity to complex waveforms can, at high frequencies, be predicted from the sensitivity to the fundamental frequency.

Robson & Campbell (1964), who introduced this concept to the linear analysis of spatial information, made a similar observation in that the sensitivity to a square-wave grating was determined by the sensitivity to the fundamental harmonic, provided that the spatial freqency of the grating exceeded 3 cycles/degree. Equally detectable square- and sine-wave gratings indeed appear identical and can only be discriminated when the third harmonic of the square-wave reaches threshold (Robson & Campbell, 1964). Yet the detectability of the third harmonic is not, as originally assumed (Robson & Campbell, 1964), equal to the detectability of an equivalent sine-wave grating olhurst, 1972*b*). Moreover, the assumption of ¬eral detection linearity, according to which 'the ¬rement in the contrast of a grating required to ⸱duce a just noticeable effect is independent of initial ⁻ntrast' (Robson & Campbell 1964), holds only for ⸱rse gratings. Strong masking rather than linear ⸱tection dominates at the middle- and high-frequency ⸱ges, making the just noticeable difference in contrast ⸱oportional to initial contrast (Campbell & Kulikowski, ⸱66; Kulikowski, 1969; Kulikowski & Gorea, 1978).

Spatial frequency masking and adaptation

Campbell & Robson (1968) argued that the ⸱ndwidth of hypothetical spatial-frequency channels ⸱s less than 3 octaves. Subsequent measurements of ⸱ttern masking effects consistently yielded bandwidth ⸱timates of about 2.3 octaves (Kulikowski, 1969) so ⸱at *B*df equals 1.6 octaves (see Appendix). Stromeyer ⸱ Julesz (1972), who used broad-band masking ⸱tterns in a simplified paradigm, reported a bandwidth ⸱about 2 octaves.

Fig. 26.6 illustrates the effect of pattern masking ⸱ contrast sensitivity. Contrast sensitivities of test ⸱tterns are strongly reduced in cases where the mask ⸱d test have similar spatial frequencies. However, ⸱nerally masking is stronger when the spatial ⸱equency of the test grating is lower than the frequency ⸱ the mask. Pattern masking is evidently less effective ⸱hen test and mask spatial frequencies are non-⸱rmonically related. For fine test gratings, masking ⸱ay even be 'negative', i.e. produce higher sensitivity ⸱ig. 26.6.*a*,*c*), as though the presence of a low ⸱atial-frequency mask made it easier to notice the high ⸱atial-frequency grating. In fact an addition of two ⸱rmonically unrelated gratings produces a new ⸱ercept of a 'beat' which can be clearly demonstrated ⸱ threshold. When two such gratings (e.g. 10 and 15 ⸱cles/degree), both at subthreshold contrasts (invisible ⸱hen presented separately), are superimposed, the ⸱bject can detect a 'beat' pattern (Kulikowski, 1969). ⸱tripes of this 'beat' are spaced at the difference in ⸱atial frequency, i.e. 5 cycles/degree. Such 'beats' ⸱ee Fig. 26.7) are clearly visible also at suprathreshold ⸱vels (note that a standard frequency analyser would ⸱ot indicate a 'difference' frequency for two additive ⸱aveforms). Moreover, this 'beat' pattern (consisting ⸱f two added gratings) when used as a mask produces ⸱oth masking and adaptation effects (Fig. 26.7*b*), not ⸱nly around its constituent spatial frequencies (10 and

Fig. 26.6. Spatial-frequency masking. Contrast sensitivity as a function of spatial frequency before (circles) and after (squares) adding masking patterns: (a) steady grating 10 cycles/degree; (b) filtered noise with a maximum at 5 cycles/degree; (c) filtered noise peaking at 15 cycles/degree. (After Kulikowski, 1969.)

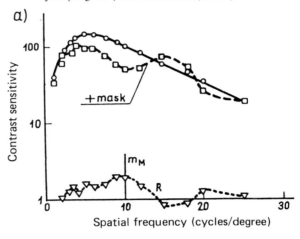

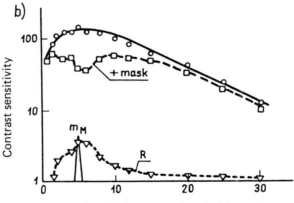

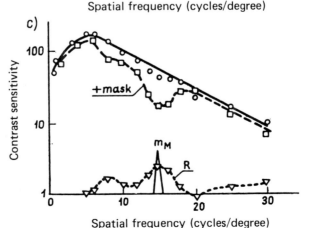

15 cycles/degree), but also at a 'difference' frequency (5 cycles/degree). Closely related phenomena have recently been reported by Henning, Hertz & Broadbent (1975). The appearance of beats and the presence of masking side-bands effectively rule out the existence of narrow-band channels. These effects would not occur if the visual system contained mainly narrow-band channels, such as the grating detectors in Fig. 26.9.

Blakemore & Campbell (1969) investigated spatial-frequency tuning by adapting the visual system to one grating and testing its sensitivity to another grating presented subsequently (see Fig. 26.7*a*). Their evaluation of bandwidth was, however, based on a linear subtraction of contrast sensitivity before and after adaptation, which narrowed the apparent tuning down to about 1.2 octaves, even though the sensitivity reduction attributable to adaptation is very similar to that of masking (compare Figs 26.6 and 26.7). Blakemore & Campbell (1969) felt, nevertheless, so encouraged by the apparent narrowness of the bandwidth that they proposed the existence of channels which sample spatial-frequency components of an image and compute their ratios. This operation was supposed to underlie the property of size invariance, which is important in pattern recognition. Although such channels could fulfil such a function, their sharp-tuning characteristic is not in this case indispensible, since invariance does not require very sharply tuned channels *per se*.

One curious aspect of the pattern adaptation story was the apparent inability to identify any spatial frequency channels below 3 cycles/degree (see Blakemore & Campbell, 1969). Yet low-frequency adaptation, down to about 0.2 cycles/degree, is easily obtained on a larger screen (Kranda & Kulikowski, 1976; see also Stromeyer *et al.*, 1982, who reported similar results).

Adaptation and masking: inhibitory non-linear interactions

Tuning curves determined with adaptation or masking techniques were, because of the spatial-frequency selectivity of such effects, often erroneously believed to represent the channel bandwidth. Consequently, channels whose frequency maxima were separated by more than their supposed bandwidth were assumed not to interact. Yet, adaptation experiments produce channels which sometimes inhibit each other, even though their maxima are 1.6 octaves apart (see Tolhurst, 1972*b*). If one assumes that adaptation partially reflects inhibition between channels with adjacent response envelopes, then the asymmetrical curve in Fig. 26.7*b* may have to be reinterpreted. This curve reflects the asymmetrical nature of the adaptation effect, i.e. adaptation to a given grating reduces the detectability of coarser gratings to a much greater extent than detectability of finer gratings. This implies that fine patterns are much more effective in suppressing the visibility of coarse patterns than *vice versa*. Since adaptation is asymmetrical, the curve in Fig. 26.8(*b*) could be presented in a more informative way if the spatial-frequency axis was reversed (see Maudarbocus

Fig. 26.7. (a) Beat luminance (*L*) profile, resulting from linear addition of non-harmonically related gratings (10 and 15 cycles/degree). (b) Contrast sensitivity (*S*) before (circles) and after (squares) adaptation to the 'beat' pattern shown in (a): sensitivity is depressed, not only for two constituent spatial frequencies 10 and 15 cycles/degree, but also for 5 cycles/degree (see ▼ symbols).

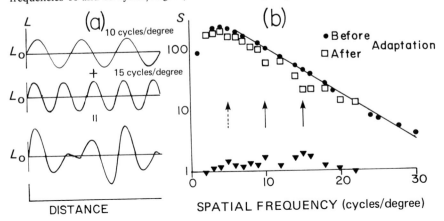

& Ruddock, 1973, who measured the adaptation effects of several different gratings on one test grating).

At present there is little doubt that both adaptation and masking reflect inhibition between channels, since under certain conditions it is possible to produce inhibition of inhibition i.e. disinhibition. For instance, Kulikowski & King-Smith (1973) reported a demasking effect, where the presence of a second mask reduces the effectiveness of the first mask. In their experiment, the ratios of the spatial frequencies of the test, first and second masks were 1 : 2 : 4. The test contrast sensitivity was reduced by half a log unit in the presence of the first mask (twice the spatial frequency), but remained unaffected by the second mask (four times the frequency). Yet when the second mask was presented simultaneously with the first, the masking effect was reduced.

A disinhibition is also observed when the first and second masks have identical spatial frequency but different orientations (Rao & Kulikowski, 1980).

In conclusion, the available evidence tends to disprove the hypothesis which assumes that the spatial-frequency range of adaptation and masking effects can specify the bandwidth of the detection channel. This hypothesis needs revision because adaptation and masking techniques produce results which reflect complex and non-linear interactions. This rules them out as suitable tools for investigating the response characteristics of channels that are assumed to participate in linear Fourier analysis.

Elementary (line/edge) detectors revealed by linear subthreshold interactions

Many neurons in the mammalian visual system exhibit linear spatial summation especially at the stimulation threshold (Barlow, 1953; Hubel & Wiesel, 1962; Enroth-Cugell & Robson, 1966) and it should be possible to demonstrate these linear responses using psychophysical methods (Westheimer, 1967; Fiorentini, 1968; Kulikowski, 1969; Sachs, Nachmias & Robson, 1971).

A psychophysical method of linear spatial summation was, therefore, used to evaluate spatial and spatial-frequency properties of the detection mechanisms for elementary stimuli such as lines or edges on the supposition that such mechanisms are able to perform linear analysis of patterns. The following findings have lead to the interpretation of these mechanisms as independent line and edge detectors, each consisting of a linear spatial filter followed by a threshold device.

1. Lines and edges are recognizable near their detection thresholds (Kulikowski & King-Smith, 1973; Tolhurst & Dealy, 1975).
2. Subthreshold background stimuli, superimposed within the RF of the detector, alter in a linear way the corresponding contrast threshold for detection

Fig. 26.8. The effect of adaptation at 7.1 cycles/degree. Note a similar reduction in contrast sensitivity to that in Fig. 26.6. (After Blakemore & Campbell, 1969.)

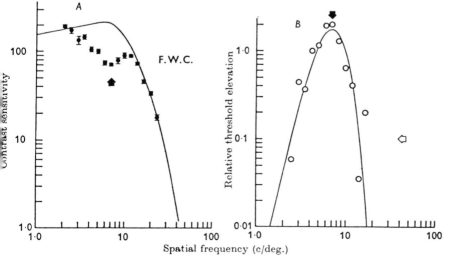
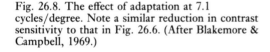

Spatial frequency (c/deg.)

of test lines and edges (without altering detector properties or test stimulus appearance near threshold).

3. The amount of alteration of threshold by the subthreshold stimuli determines the sensitivity of the detectors to these stimuli; hence, by varying the position of subthreshold lines we can evaluate spatial sensitivity profiles of a detector, and by varying subthreshold spatial frequency we can evaluate the tuning curve.

Fig. 26.9. Contrast sensitivities of line, edge and grating detectors determined by subthreshold summation. (After King-Smith & Kulikowski, 1973a.)

4. The spatial/spatial-frequency properties of a given detector are mutually predictable by linear analysis (Fig. 26.9), similar to those of a linear filter (see Appendix).

5. The properties of line and edge detectors are independent, i.e. not predictable from each other (unlike the properties of complex detectors, see Fig. 26.9 bottom row and Fig. 26.10).

Since the spatial arrangement of the line and edge detectors closely resembles Gabor's elementary operators, these two types of detector can be considered as elementary units in analysis of complex patterns (Fig. 26.10).

Moreover, separate detectors for light and dark bars (as well as light and dark edges), analogous to the

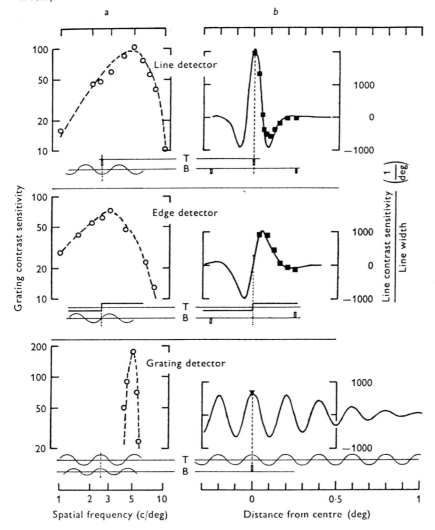

corresponding ON-centre and OFF-centre simple RFs, can be evaluated by the same method. These four elementary detectors (quadruple), evaluated using lines and edges at threshold, are the most sensitive detectors. Other less-sensitive detectors (tuned to other ranges of spatial frequencies than the most sensitive detectors) can be evaluated with either blurred (coarse) or multiline test stimuli whose profiles make them undetectable by the most sensitive detectors at threshold (see discussion of Fig. 26.11). In this way properties of detectors with optimal spatial frequencies from about 1 to 10 cycles/degree were specified (Kulikowski & King-Smith 1973; King-Smith & Kulikowski, 1975).

Relative bandwidth *versus* optimal spatial frequency

The relative spatial frequency bandwidth of tuning curves of the psychophysical detectors varies little with the optimal spatial frequency. The mean slope on the log–log scale is about −0.2, which is comparable with a slope obtained for simple cells (−0.17 for macaque – see above; −0.3 for cat – see Kulikowski & Bishop, 1981).

Thus the system of psychophysically evaluated detectors is close to the Gabor system with constant relative bandwidth (Fig. 26.3a,c), i.e. the system is biased towards accuracy of spatial localization.

What do the elementary detectors signal?

Another problem is to find out what kind of information is signalled by a given detector. We may ask whether the information contained in a detector's signal corresponds to the spatial profile of the detector's sensitivity. Fig. 26.9 shows the spatial profile of a line detector; on activation by a line, however, it signals the percept of the line rather than a pattern equivalent to its spatial profile (a bright line with two dark flanks). Similarly, activation by an edge detector leads to a simple percept and not a complex spatial profile (Fig. 26.9b). Therefore, such detectors do not necessarily signal information equivalent to their spatial profiles.

Glezer *et al.* (1977) also examined this problem by comparing detection and recognition thresholds of two lines. They noted that the actual recognition of two lines is substantially better than might be predicted by assuming the involvement of a number of line detectors. Instead they proposed that units with many antagonistic subregions (similar to grating detectors) are involved. However, a careful examination of probability of detection and recognition (King-Smith & Kulikowski, 1973b, 1975, 1981) indicates that the recognition of two lines is hardly better than might be expected from probability summation.

Not only is there no evidence for periodicity detection, but the perception of periodicity also shows a poor frequency selectivity, even at suprathreshold contrasts (MacKay, 1973).

Adaptation to gratings with variable ratios of light and dark phases also shows that the width of the stripes is a more important parameter than the spatial periodicity (de Valois, 1977).

Fig. 26.10. Contrast sensitivity profiles of detectors involved in the detection of (a) wide bars and (b) two cycles of a grating. (After King-Smith & Kulikowski, 1973b.)

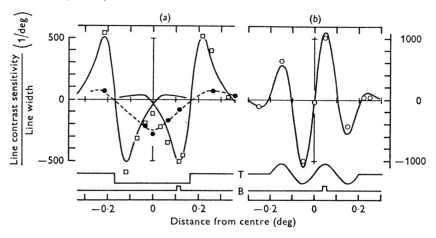

Detection of complex patterns by elementary detectors

At this stage we should stress that line and edge detectors are not selectively tuned to detect lines and edges but, rather, they seem to be elementary building blocks for analysis of more complex patterns in terms of symmetrical and antisymmetrical components. The properties of complex pattern detectors should, therefore, be explained in terms of the contribution of elementary detectors.

Many other detectors were evaluated by using the method of subthreshold summation, for example detectors of a grating (Fig. 26.9), of two spatial cycles, or of a broad bar (Fig. 26.10). It was, therefore, important to find that the grating detector is mainly determined by the contribution of many line detectors (King-Smith & Kulikowski, 1975), whereas the profile of a broad-bar detector consists of profiles of two edge detectors (Fig. 26.10a).

Moreover, the contribution of line and edge detectors to the threshold detection of complex patterns is mainly probabilistic: lateral interactions, such as non-linear disinhibition or facilitation contribute very little at threshold, although their contribution is significant under suprathreshold conditions (King-Smith & Kulikowski, 1975, 1981).

The same argument, however, may be directed against the accuracy of evaluation of line and edge detectors, since their properties result from the probabilistic contributions of certain subunits. Even at threshold there may be a range of units of similar sensitivity which are activated by, for example, a fine line, and these units may differ slightly in their optimal spatial frequency or in the number of subregions (bandwidth) see Fig. 26.12(A:a,b,c).

Thus the properties of the line and edge detectors should be corrected by accounting for probability summation not only in space but also in the spatial frequency domain (for further discussion see King-Smith & Kulikowski, 1975; Robson, 1975; Graham, 1977; Bergen, Wilson & Cowan, 1979).

The mean relative bandwidth of a subunit was evaluated as about 1.5 octaves (Bdf = 1 cycle/degree). However, the precise contributions of the probability summation in space *versus* spatial frequency are unknown, because the solution of such a non-linear equation would require *a priori* knowledge of the spatial profiles of units contributing to the detection and their role in perception.

However, it is encouraging that the mean bandwidth evaluated above is similar to the mean for simple cells in the visual cortex of the cat (about 1.5 octaves, see Movshon *et al.*, 1978c; Kulikowski & Bishop, 1981) and monkey (between 1.4 and 1.7 octaves, de Valois *et al.*, 1982; Kulikowski & Vidyasagar, 1982).

Possible contribution to perception by line and edge detectors

A psychophysical detector is a hypothetical system which can recognize a pattern at threshold. Although the spatial profiles of such detectors closely resemble those of cortical cells, there are several essential differences. First, psychophysical detectors are not equivalent to the action of a single cortical unit but rather to the activity of many (see Fig. 26.12). Secondly, the term detector implies some detection process, yet cortical cells are probably merely processors, in that the activation of these alone probably cannot constitute a percept.

Fig. 26.11 shows several stimulus profiles and their subjective percepts. All simple patterns (a–f, j) can be recognized near their detection threshold as bars or edges with the appropriate contrast polarity (light or dark): a forced-choice method yields only about 10% difference between the detection and recognition thresholds (Tolhurst & Dealy, 1975; King-Smith & Kulikowski, 1981).

However, it is impossible to judge the width of just-detectable bars which are narrower than 6 min. Such bars are equally detectable if the product of their width and contrast is constant, as shown in Fig. 26.11(c,d,e,f). When their contrast is raised slightly above threshold they all appear like fine lines, evidently because they are detected by one mechanism, namely the most sensitive line detector whose optimal spatial frequency is about 5 cycles/degree and whose central subregion is 6 min wide (Kulikowski, 1967, 1969; Kulikowski & King-Smith, 1973). At higher suprathreshold contrasts, however, finer bars (e.g. Fig. 26.11e,f) activate additional finer detection mechanisms (e.g. those with fo = 10 cycles/degree or higher) and these fine detectors presumably enable us to discriminate betwen 4- and 2-min bars, even though both bars must activate the most sensitive detector as well. Kulikowski & King-Smith (1973) suggested, as a possible explanation, the mechanism of inhibition by which finer detectors override the responses of coarse tectors (see Fig. 26.6 for masking by fine gratings as illustration of this point).

In order to reveal mechanisms other than the ost sensitive detector, the threshold stimuli cannot be nple bars or edges but have to have special profiles. arse detectors are activated by stimuli with cosine ninance profiles (Fig. 26.11a,b), whereas a detector er than the most sensitive one can only be activated a stimulus composed of a central bar flanked by two rs of opposite polarity (see Fig. 26.11g). Such stimuli g. Fig. 26.11a,g) fail to activate the most sensitive line tector. As may be expected, cosine bars appear as oad bars near threshold, whereas the composite muli (Fig. 26.11g,h) look like very fine lines. It is teresting that when one of the latter stimuli (Fig. .11g,h) is presented near its threshold simultaneously th a 6-min line, its percept (a single line) appears still er than the percept of a single bar narrower than nin. Thus, the visual system apparently assumes that rs with a width of 6 min and less (i.e. those which tivate detectors with an optional spatial frequency at cycles/degree), are always fine at threshold since no tput of still finer detector is available to compare. onsequently, this perception of fineness is no longer striking when such single lines are compared with tterns (see Fig. 26.11g,h) that activate even finer

detection mechanisms. Needless to say, the single bars in Fig. 26.11(c–f) would appear different if viewed above threshold. This indicates that the main role of detectors is to signal localized patterns and the polarity of their contrast, rather than their exact size. Probably the size of bars is evaluated as a result of inhibitory interactions between detectors at suprathreshold contrasts (Kulikowski & King-Smith, 1973; see also Blakemore & Sutton, 1969).

It is interesting that space periodic stimuli with three or even five components (see Fig. 26.11g,h)\ appear, against all expectations, as a single, localized line when viewed at threshold. The visual system thus seems to ignore the periodicity of multiple-bar patterns, rather than to facilitate their recognition, as claimed by Glezer *et al.* (1977). Since the multibar pattern (e.g. Fig. 26.11h) is detected by a mechanism with a medium bandwidth (King-Smith & Kulikowski, 1975), one may question the role of hypothetical channels with a narrow-band tuning, if such are demonstrated.

Interaction between line and edge detectors at suprathreshold contrasts

Percepts of patterns are normally assumed to correspond to that which we believe to be physical reality and this can be described in terms of the stimulus luminance distribution. Any percept which departs or deviates from this assumed correspondence

Fig. 26.11. Various luminance profiles (above) and their respective subjective appearances (below). For further details see text.

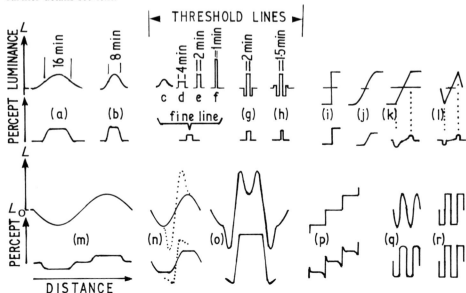

is considered to be an illusion. Yet percepts of many simple and complex patterns are, even at threshold, illusory (see Fig. 26.11). The visual system is thus not primarily organized to produce a 'high-fidelity representation' of the outer world but, as we have seen in previous sections, to transform it. This partly non-linear transformation may in fact be a necessary step in image processing and recognition. The investigation of illusory percepts could thus prove useful in revealing the nature of image transformation.

The well-known 'Mach bands' illusion (see Ratliff, 1965) generated by trapezoid ramp stimuli (see Fig. 26.11k) can serve here as an example. When viewed at threshold, this luminance distribution looks like a blurred edge (Fig. 26.11d). However, at suprathreshold contrast this ramp gives an appearance of dark and bright stripes (Mach bands) centred at borders of sharp luminance transitions. The continuous luminance gradient between these two transitions cannot be recognized (see Tolhurst, 1972a), although a similar gradient with smoother transitions (Fig. 29.11j) does give an appearance of an edge. von Bekesy (1960) observed a related phenomenon when viewing a triangular luminance distribution (see Fig. 29.11,l), namely that this pattern appears as alternating light and dark bars positioned at the peaks and troughs of the triangular wave. What we may see here is the suppression of edge detectors by two line detectors of opposite polarity, which are strongly activated by the sharp luminance transitions.

Single cycles of sine-wave luminance distribution (see Fig. 26.11m,n) appear as light and dark bars. The broad sinusoidal cycle in Fig. 26.11m (below 1 cycle/degree) gives the appearance of two bars separated by a gap, whereas the bars produced by the luminance distribution in Fig. 26.11n appear continuous but more like a defocussed square-wave grating. This perceptual distortion is consistent with the hypothesis which assumes that each half-cycle of a periodic luminance distribution activates bar detectors and not edge detectors (see Kulikowski & King-Smith, 1973). Occasionally one bar in a pair (for example, light) may fade away but reappears later, while the other (dark) bar fades away in turn, as though there was a mutual inhibition between detectors of light and dark bars. It is interesting, however, that the sinusoidal spatial cycle never appears like an edge.

The addition of a second and third harmonic to the distribution in Fig. 26.11n (dotted lines) produces

the appearance of a single edge rather than of two bars and this is the well-known Craick–Cornsweet illusion (Ratliff, 1965). The percept of an apparent edge can be augmented further by juxtaposing this pattern to its mirror image (Fig. 26.11o). This addition generates the appearance of a complementary edge of an opposite polarity and the two edges are perceived as a wide bar. (Juxtaposition of two sharp edges of an opposite polarity (Fig. 26.11i) likewise produces the perception of a bar.) Evidently, edge detectors of opposite polarity facilitate each other's operation and consequently help the detection and recognition of wide bars (see King-Smith & Kulikowski, 1973b). Conversely, edge detectors of the same polarity tend to inhibit each other (Fig. 26.11p). This interaction is then manifested by the appearance of illusory bands at the borders (see Ratliff 1965).

Gratings consisting of many spatial cycles, such as those illustrated in Fig. 26.11m,n,q,r, can, like single bars or edges, also be recognized near their detection threshold. However, the detection and recognition of gratings near threshold is to a large extent (85%) predictable from the probability summation of detection thresholds of its single-line components (see King-Smith & Kulikowski, 1975). It is questionable whether a hypothetical grating detector could account for the remaining 15% (or more at suprathreshold contrasts), since sinusoidal gratings appear distorted in many ways. For instance, a sinusoidal grating whose spatial frequency is below 0.x cycles/degree appears as a set of completely independent light and dark bars. Finer sine-wave gratings tend to appear more like defocussed rectangular-wave patterns. Even more distorted are laser-generated sinusoidal fringes (Maudarbocus & Ruddock, 1973), in which light stripes appear wider than dark stripes. It is paradoxical that these sinusoidal patterns do not seem to generate the percept of sinusoidal periodicity. Instead, these patterns give rise to the appearance of two independent sets of light and dark bars (see also de Valois, 1977).

Units activated by bars and edges

Stimuli, such as bars and edges, are known to activate cells with symmetrical and asymmetrical profiles (in fact their RFs are plotted with these stimuli, see Fig. 26.4). Yet the percept produced by these stimuli still corresponds to the original symmetrical or asymmetrical luminance distribution. It is thus

onceivable that the ultimate percept of such stimuli is
 function of many activated cells or processor units
ontributing to a detector unit. Here we attempt to
emonstrate that the operation of many units, having
Gabor-type symmetrical and antisymmetrical profiles,
an account for the discrimination and perhaps the
ltimate percept of simple stimuli, such as lines and
dges. In the model presented here, which is based on
he electrophysiological and psychophysical data, only
inear interactions of the processor units were
onsidered.

In Fig. 26.12 we show receptive field profiles of
ells which are activated by fine lines, bars, and sharp
nd blurred edges. First, the mean elementary profiles
ave to be derived by using threshold stimuli which
ctivate a minimal number of units. Fig. 26.12A shows
he sensitivity profiles of units which are activated by
wo narrow lines (2 and 6 min) at threshold, i.e. in this
ase at 50% probability of seeing. Only the
ymmetrical, most sensitive units respond to these fine
nes, because units with antisymmetrical sensitivity
rofiles have lower sensitivity to bars. Although the
nits whose central subregion subtends 6 min will be
ctivated most efficiently, we can envisage that other
nits can also be activated, namely units with central

subregions of the same size, but with five subregions,
(a, dashed line) and units whose central subregions are
either somewhat wider or narrower than 6 min with
three or five subregions (b, c). We computed the mean
activation profile (Ad) of these units by assuming that
they all contribute in a linear fashion to the detection
of a fine line. This mean activation profile corresponds
to the threshold line detector that produces a percept
of a fine line for both 6- and 2-min lines (cf.
Kulikowski, 1967; Kulikowski & King-Smith, 1973).
Analogous computations can be carried out to derive
the most sensitive (threshold) edge detector.

Once the contrast of these lines exceeds
threshold, the antisymmetrical units also become active
and thus alter the activation pattern. For reasons of
expediency, Fig. 26.12B and C show only mean sensitivity profiles (like that in Ad) of the activated units
which are assumed to be tuned to harmonically related
optimal spatial frequencies. Note that the 6-min bar,
sharp or blurred (dotted outline), activates units with
6-min subregions, but a sharp bar activates also units
with finer subregions, whose spatial arrangement is
mainly antisymmetrical. The number of activated
units, which can be of various types, naturally increases
once the stimulus exceeds threshold. The multiple

Fig. 26.12. Schematic patterns of activation of
various units by bars and edges. (A) Bars, 6 and
2 min wide, equally detectable at threshold, activate
similar units with symmetrical profiles (a–c), all
having nearly maximum sensitivity. Continuous
lines show profiles with three subregions, and
dashed lines with five subregions. The computed
mean profile is shown in (d). (B) Suprathreshold
6-min bar. Both symmetrical and antisymmetrical

units respond to this bar. Furthermore, fine
antisymmetrical units respond to the sharp edges
but not to blurred edges (dotted outline). (C)
Suprathreshold 2-min bar activates all units in (B),
as well as fine units (only symmetrical units are
shown). (D) Suprathreshold blurred edge activates
only coarse units. (E) Suprathreshold sharp edge
activates all units in (D) as well as fine units.

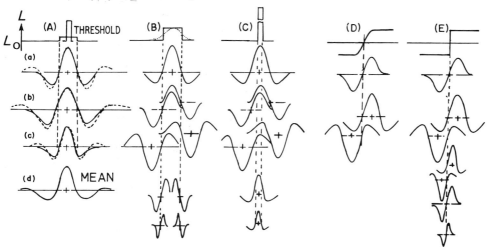

activation of various units may subserve the percept of a bar with sharp borders and allow its discrimination from a blurred (cosinusoidal) stimulus.

Narrow bars or lines of suprathreshold contrast probably also activate units with broader sensitivity profiles. Thus recognition of fine lines can only be achieved by the activation of units with very fine subregions.

Antisymmetrical stimuli, such as edges (Fig. 26.12D,E), likewise activate a wide range of units. The perception of a sharp edge, however, requires additional activation of units, with narrower subregions. In both cases (D,E), an antisymmetry leads to a percept of an edge.

It should be stressed that signalling of large areas limited by edges is difficult without an antisymmetrical mechanism, since it would require symmetrical (or concentric) units responding to low spatial frequencies. The edge detector can signal locally a step change of luminance and the system remains in this state until another spatial discontinuity is encountered. Thus no active units are needed in the middle of a wide bar (Shapley & Tolhurst, 1973; King-Smith & Kulikowski, 1973). The Craick–Cornsweet illusion (see Fig. 26.11n,o) provides a strong evidence in favour of this model by demonstrating that the local change in contrast is sufficient to produce an apparent difference in brightness in two areas.

Conclusions

There is no evidence for the hypotheses that the visual cortex performs feature or Fourier analysis. Some cells found in the geniculostriate pathway of higher mammals operate, under optimal conditions, as if they were linear filters. However, these cells also respond in a non-linear manner, perhaps in order to facilitate the recognition of certain patterns. Yet this non-linearity does not make the cells' response very selective to specific patterns so that they could function as feature detectors. Simple cells in the cortex probably constitute an oversampled multiprocessor system which performs a linear analysis of the image in such a way that there is a trade-off between the optimal signalling of accuracy of spatial localization and the signalling of spatial-frequency selectivity. This trade-off, however, seems to be biased towards better spatial localization, primarily because the system lacks narrowly tuned units. The image analysis performed in this way can at the very best be considered a rudimentary Fourier analysis (the Q-factor of the filters is below 2). The function of the cortical simple cells can be more appropriately described in terms of Gabor type units which operate in a linear fashion under optimal conditions and can signal both symmetrical and antisymmetrical image components. The parameters of these units can probably be modulated, so that their filtering characteristics are flexible and controllable. This adaptable filtering allows the extraction of information pertaining to localization, orientation and spatial-frequency content of the image. This type of filter can probably also contribute to signalling of other aspects of visual stimuli, e.g. movement. The spatial filters form detectors whose activation may explain simple perceptual phenomena.

References

Adrian, E. D. & Mathews, R. (1928) The action of light on the eye (Part III): the interaction of retinal neurones. *J. Physiol.*, **65**, 273–98

Andrews, B. W. & Pollen, D. A. (1979) Relationship between spatial frequency selectivity and response field profile of receptive fields. *J. Physiol.*, **287**, 163–76

Bando, T., Tsukuda, K., Yamamoto, N., Maeda, J. & Tsukahara, N. (1981) Cortical neurons in and around the Clare–Bishop area related with lens accommodation in the cat. *Brain Res.*, **225**, 195–9

Barlow, H. B. (1953) Summation and inhibition in the frog's retina. *J. Physiol.*, **119**, 69–88

Barlow, H. B. (1981) Critical limiting factors in the design of the eye and visual cortex. *Proc. R. Soc.*, (*B*), **212**, 1–34

Barlow, H. B., Hill, R. M. & Levick, W. R. (1964) Retinal ganglion cells responding selectively to direction and speed of image motion in the rabbit. *J. Physiol.*, **173**, 377–407

Bekesy, G. von (1960) Neural inhibitory units of the eye and skin Quantitative description of contrast phenomena. *J. Optic Soc. Am.*, **50**, 1060–70

Bergen, J. R., Wilson, H. R. & Cowan, J. D. (1979) Further evidence for four mechanisms mediating vision at threshold: sensitivities to complex gratings and aperiodic stimuli. *J. Optic Soc. Am.*, **69**, 1580–7

Bishop, P. O., Coombs, J. S. & Henry, G. H. (1973) Receptive fields of simple cells in the cat striate cortex. *J. Physiol.* (*Lond*), **231**, 31–60

Blakemore, C. & Campbell, F. W. (1969) On the existence of neurones in the human visual system selectively sensitive to the orientation and size of retinal images. *J. Physiol.*, **203**, 237–60

Blakemore, C. & Sutton, P. (1969) Size adaptation: a new after-effect. *Science*, **166**, 245–7

Boycott, B. B. & Wässle, H. (1974) The morphological types of ganglion cells of the domestic cat's retina. *J. Physiol.*, **240**, 397–419

Bullier, J. & Henry, G. H. (1979) Ordinal position of neurones in cat striate cortex. *J. Neurophysiol.*, **42**, 1251–63

Campbell, F. W. & Cooper, G. F. & Enroth-Cugell, C. (1969) The spatial selectivity of the visual cells of the cat. *J. Physiol. Lond.*, **181**, 576–93

Campbell, F. W. & Kulikowski, J. K. (1966) Orientational selectivity of the human visual system. *J. Physiol.*, **187**, 437–45

Campbell, F. W. & Robson, J. G. (1968) Application of Fourier analysis to the visibility of gratings. *J. Physiol. (Lond.)*, **203**, 223–35

Cleland, B. G., Dubin, M. W. & Levick, W. R. (1971) Sustained and transient neurone in the cat's retina and LGN. *J. Physiol.*, **217**, 473–96

Cleland, B. G. & Levick, W. R. (1974) Properties of rarely encountered types of ganglion cells in the cat's retina and an overall classification. *J. Physiol.*, **240**, 457–92

Cooper, G. F. & Robson, J. G. (1968) Successive transformations of spatial information in the visual system. *I.E.E. N.P.L. Conf. Proc.*, **42**, 134–43

Cowan, J. D. (1977) Some remarks on channel bandwidth for visual contrast detection. *MIT Neurosci. Res. Prog. Bull.*, **15**, 492–517

Cowey, A. (1967) Perimetric study of field defects in monkeys after retinal and cortical ablations. *Q. J. Physiol.*, **19**, 232–45

Creutzfeldt, O. T., Kuhnt, U. & Benevento, L. A. (1974) An intracellular analysis of visual cortical neurons to moving stimuli: responses in a co-operative neuronal network. *Exp. Brain Res.*, **21**, 251–74

Daugman, J. G. (1984) Spatial visual channels in the Fourier plane. *Vision Res.*, **24**, 891–910

de Lange, H. (1958) Research into the dynamic nature of the human fovea-cortex systems with intermittent modulated light. I. Attenuation characteristics with white and coloured light. *J. Optic Soc. Am.*, **48**, 777–84

de Monasterio, F. M., Gouras, P. & Tolhurst, D. J. (1976) Spatial summation, response pattern and conduction velocity of ganglion cells of the rhesus monkey retina. *Vision Res.*, **16**, 674–7

De Valois, K. K. (1977) Independence of black and white: phase-specific adaptation. *Vision Res.*, **22**, 209–16

De Valois, R. L., Albrecht, D. G. & Thorell, L. G. (1982) Spatial frequency selectivity of cells in macaque visual cortex. *Vision Res.*, **22**, 545–59

De Valois, R. L., Abramov, I. & Mead, M. R. (1967) Single-cell analysis of wavelength discrimination at the LGN in the macaque. *J. Neurophys*, **30**, 415–33

Dreher, B. (1972) Hypercomplex cells in the cat's striate cortex. *Invest. Ophthalmol.*, **11**, 355–6

Dreher, B. & Cottee, L. J. (1975) Visual receptive field properties of cells in area 18 of cat's cerebral cortex before and after acute lesions in area 17. *J. Neurophysiol.*, **38**, 735–50

Dreher, B., Fukada, Y. & Rodieck, R. W. (1976) Identification, classification and anatomical segregation of cells with X-like and Y-like properties in the lateral geniculate nucleus of old-world primates. *J. Physiol., (Lond.)*, **258**, 433–52

Dreher, B., Henry, G. H. & Bishop, P. O. (1972) Sustained and transient responses in the striate neurones of the cat. *Proc. Aust. Physiol. Pharmacol. Soc.*, **3**, 41–2

Dreher, B., Leventhal, A. G. & Hale, P. T. (1980) Geniculate input to cat visual cortex: a comparison of area 19 with areas 17 and 18. *J. Neurophysiol.*, **44**, 804–26

Enroth-Cugell, C. & Robson, J. G. (1966) Contrast sensitivity of retinal ganglion cells of the cat. *J. Physiol.*, **187**, 517–52

Fiorentini, A. (1968) Excitatory and inhibitory interactions in the human eye. In *Visual Science*, ed. J. Pierce & J. Levene, pp. 269–83. Indiana University Press

Fukada, Y. (1971) Receptive field analysis of cat optic nerve fibres with special reference to conduction velocity. *Vision Res.*, **11**, 209–26

Gabor, D. (1946) Theory of communication. *J. I.E.E. (Lond.)*, **13**, 429–41

Ginsburg, A. P., Moshon, J. A. & Tolhurst, D. J. (1976) Periodicity in complex cell responses. *J. Physiol.*, **254**, 69P

Glezer, V. D., Cooperman, A. M., Ivanov, V. A. & Tsherbach, T. A. (1976) An investigation of spatial frequency characteristics of the complex receptive fields in the visual cortex of the cat. *Vision Res.*, **16**, 789–97

Glezer, V. D., Ivanov, V. A. & Tsherbach, T. A. (1973) Investigation of complex and hypercomplex receptive fields of visual cortex of the cat as spatial frequency filters. *Vision Res.*, **13**, 1875–904

Glezer, V. D., Kostelyanets, N. D. & Cooperman, A. M. (1977) Composite stimuli are detected by grating detectors rather than by line detectors. *Vision Res.*, **17**, 1067–70

Gouras, P. (1968) Identification of cone mechanisms in monkey ganglion cells. *J. Physiol.*, **199**, 533–47

Graham, N. (1977) Visual detection of aperiodic stimuli by probability summation among narrow band channels. *Vision Res.*, **17**, 637–52

Gross, C. G., Rocha-Miranda, C. E. & Bender, D. B. (1972) Visual properties of neurons in interotemporal cortex of the macaque. *J. Neurophysiol.*, **35**, 96–111

Hammond, P. (1973) Contrasts in spatial organization of receptive fields at geniculate and retinal levels: centre surround and outer surround. *J. Physiol.*, **228**, 115–37

Hammond, P. & Mackay, D. M. (1977) Differential responsiveness of simple and complex cells in cat striate cortex to visual texture. *Exp. Brain Res.*, **30**, 275–96

Hammond, P. & Mackay, D. M. (1983) Influence of luminance gradient reversal on simple cells in feline striate cortex. *J. Physiol.*, **337**, 69–89

Hammond, P. & Smith, A. T. (1983) Directional tuning interactions between moving oriented and textured stimuli in complex cells of feline striate cortex. *J. Physiol.*, **342**, 35–49

Helstrom, C. W. (1966) An expansion of a signal in Gaussian elementary signals. *I.E.E.E. Trans. Inf. Theory*, **IT-12**, 81–2

Henning, G. B., Hertz, B. G. & Broadbent, D. E. (1975) Some experiments bearing on the hypothesis that the visual system analyses spatial patterns in independent bands of spatial frequency. *Vision Res.*, **15**, 887–98

Henry, G. H. (1977) Receptive field classes of cells in the striate cortex of the cat. *Brain Res.*, **133**, 1–28

Henry, G. H., Lund, J. S. & Harvey, A. R. (1978) Cells of the striate cortex projecting to the Clare–Bishop area of the cat. *Brain Res.*, **151**, 154–8

Hicks, T. P., Lee, B. B. & Vidyasagar, T. R. (1983) The responses of cells in macaque lateral geniculate nucleus to sinusoidal graphings. *J. Physiol.*, 337, 183–200

Hubel, D. H. & Wiesel, T. N. (1959) Receptive fields of single neurones in the cat's striate cortex. *J. Physiol. (Lond.)*, 148, 574–91

Hubel, D. H. & Wiesel, T. N. (1962) Receptive fields, binocular interaction and functional architecture in the cat's visual cortex. *J. Physiol.*, 160, 106–54

Hubel, D. H. & Wiesel, T. N. (1968) Receptive fields and functional architecture of monkey striate cortex. *J. Physiol. (Lond.)*, 195, 215–43

Ikeda, H. & Wright, M. J. (1972) The outer disinhibitory surround of the retinal ganglion cell receptive field. *J. Physiol. (Lond.)*, 226, 511–44

Ikeda, H. & Wright, M. J. (1975) Spatial and temporal properties of sustained and transient neurones in area 17 of the cat's visual cortex. *Exp. Brain Res.*, 22, 363–83

Kaplan, E. & Shapley, R. M. (1982) X and Y cells in the lateral geniculate nucleus of macaque monkeys. *J. Physiol.*, 330, 125–43

King-Smith, P. E. (1978) Analysis of a detection of a moving line. *Perception*, 7, 449–58

King-Smith, P. E. & Kulikowski, J. J. (1973a) Line edge and grating detectors in human vision. *J. Physiol. (Lond.)*, 230, 23–5P

King-Smith, P. E. & Kulikowski, J. J. (1973b) Lateral interaction in the detection of composite spatial patterns. *J. Physiol. (Lond.)*, 234, 5–6P

King-Smith, P. E. & Kulikowski, J. J. (1975) The detection of gratings by independent activation of line detectors. *J. Physiol. (Lond.)*, 247, 237–71

King-Smith, P. E. & Kulikowski, J. J. (1981) The detection and recognition of two lines. *Vision Res.*, 21, 235–50

Konorski, J. (1969) *The Integrative Action of the Brain*. Chicago: Chicago University Press

Kovasznay, L. & Joseph, H. (1955) Image processing. *Proc. IRE.*, 43, 560–70

Kranda, K. & Kulikowski, J. J. (1976) Adaptation of coarse gratings in the scotopic and photopic conditions. *J. Physiol.*, 257, 35–6P

Kulikowski, J. J. (1966) Adaptive visual signal preprocessor with a finite number of states. *I.E.E.E. Trans.*, SS-C.2, 96–101

Kulikowski, J. J. (1967) Model of the detection of simple patterns by the visual system. *Avtometria*, 6, 113–20

Kulikowski, J. J. (1969) Limiting conditions of visual perception. *Prace Inst. Automat. PAN, Warsaw*, 77, 1–133

Kulikowski, J. J. (1976) Effective contrast constancy and linearity of contrast sensation. *Vision Res.*, 16, 1419–431

Kulikowski, J. J. (1979) Neural stages of visual signal processing. In *Search and the Human Observer*, ed. J. N. Clare & M. A. Sinclair, pp. 74–87. London: Taylor & Francis

Kulikowski, J. J. & Bishop, P. O. (1981) Linear analysis of the responses of simple cells in the cat visual cortex. *Exp. Brain Res.*, 44, 386–400

Kulikowski, J. J. & Bishop, P. O. (1982) Silent periodic cells in the cat striate cortex. *Vision Res.*, 22, 191–200

Kulikowski, J. J., Bishop, P. O. & Kato, H. (1981) Spatial arrangement of responses by cells in the cat visual cortex

to light and dark bars and edges. *Exp. Brain Res.*, 44, 371–85

Kulikowski, J. J. & Gorea, A. (1978) Complete adaptation to patterned stimuli: a necessary and sufficient condition for Weber's law for contrast. *Vision Res.*, 18, 1223–7

Kulikowski, J. J. & King-Smith, P. E. (1973) Spatial arrangement of line, edge and grating detectors revealed by subthreshold summation. *Vision Res.*, 13, 1455–78

Kulikowski, J. J., Marcelja, S. & Bishop, P. O. (1982a) Theory of spatial position and spatial frequency relations in the receptive fields of simple cells in the visual cortex. *Biol. Cybern.*, 43, 187–98

Kulikowski, J. J., Rao, V. M. & Vidyasagar, T. R. (1982b) Spatial phase sensitivity in the cat visual cortex. *J. Physiol.*, 324, 70P

Kulikowski, J. J. & Vidyasagar, T. R. (1982) Representation of space and spatial frequency in the macaque striate cortex. *J. Physio.*, 332, 10–11P

Kulikowski, J. J. & Vidyasagar, T. R. (1984) Macaque striate cortex: pattern, movement and colour processing. *Ophthalmic. Physiol. Optics*, 4, 77–81

Lamar, E. S., Hecht, S., Shlaer, S. & Hendley, C. D. (1948) Size, shape and contrast in detection of targets by daylight vision. *J. Optic Soc. Am.*, 38, 741–55

Le Vay, S., Hubel, D. H. & Wiesel, T. N. (1975) The pattern of ocular dominance columns in the macaque visual cortex revealed by a reduced silver stain. *J. Comp. Neurol.*, 159, 559–76

Lettvin, J. Y., Maturana, H. R., McCulloch, W. & Pitts, W. H. (1959) What the frog's eye tells the frog's brain. *Proc. I.R.E.*, 47, 1940–51

Leventhal, A. G., Keens, J. & Tork, I. (1980) The afferent ganglion cells and cortical projections of the retinal recipient zone (RRZ) of the cat's 'pulvinar complex'. *J. Comp., Neurol.*, 194, 535–54

Leventhal, A. G., Rodieck, R. W. & Dreher, B. (1981) Retinal ganglion cell classes in the old world monkey: morphology and central projections. *Science*, 213, 1140–2

MacKay, D. M. (1973) Lateral interaction between neural channels sensitivity to texture density. *Nature*, 245, 159–61

MacKay, D. M. (1981) Strife over visual cortical function. *Nature*, 289, 117–18

Maffei, L. & Fiorentini, A. (1971) Retinogeniculate convergence and analysis of contrast. *J. Neurophysiol.*, 35, 65–72

Maffei, L. & Fiorentini, A. (1973) The visual cortex as a spatial frequency analyser. *Vision Res.*, 13, 1255–67

Maffei, L. & Fiorentini, A. (1977) Spatial frequency rows in the striate visual cortex. *Vision Res.*, 17, 257–64

Marcelja, S. (1980) Mathematical description of the responses of simple cortical cells. *J. Optic Soc. Am.*, 70, 1297–1300

Marr, D. (1982) *Vision*. San Francisco: W. H. Freeman

Maudarbocus, A. Y. & Ruddock, K. H. (1973) Non-linearity of visual signals in relation to shape-sensitive adaptation responses. *Vision Res.*, 13, 1713–37

Movshon, J. A., Thompson, I. D. & Tolhurst, D. J. (1978a) Spatial summation in the receptive fields of simple cells in the cat's striate cortex. *J. Physiol.*, 283, 53–77

Movshon, J. A., Thompson, I. D. & Tolhurst, D. J. (1978b)

Receptive field organisation of complex cells in the cat's striate cortex. *J. Physiol.*, **283**, 79–99

Movshon, J. A., Thompson, I. D. & Tolhurst, D. J. (1978c) Spatial and temporal contrast sensitivity of neurones in areas 17 and 18 of the cat's visual cortex. *J. Physiol.*, **283**, 101–20

Orban, G. (1984) *Neuronal Operations in the Visual Cortex*. Berlin: Springer-Verlag

Peichl, L. & Wässle, H. (1979) Size, scatter and coverage of ganglion cell receptive field centres in the cat retina. *J. Physiol.*, **291**, 117–41

Perrett, D. I., Rolls, E. T. & Caan, W. (1982) Visual neurones responsive to faces in the monkey temporal cortex. *Exp. Brain Res.*, **47**, 329–42

Perry, V. H. & Cowey, A. (1981) The morphological correlates of X- and Y-like retinal ganglion cells in the retina of monkeys. *Exp. Brain Res.*, **43**, 226–8

Pettigrew, J. D., Nikara, T. & Bishop, P. O. (1968) Responses to moving slits by single units in cat striate cortex. *Exp. Brain Res.*, **6**, 373–90

Pollen, D. A. & Feldon, S. E. (1979) Spatial periodicities of periodic complex cells. *Invest Ophthalmol. Vis. Sci.*, **18**, 429–33

Pollen, D. A., Lee, J. R. & Taylor, J. H. (1971) How does the striate cortex begin the reconstruction of the visual world? *Science (N.Y.)*, **173**, 74–7

Pollen, D. A. & Ronner, S. F. (1975) Periodic excitability changes across the receptive fields of complex cells in the striate and parastriate cortex of the cat. *J. Physiol.*, **245**, 667–97

Pollen, D. A. & Ronner, S. F. (1981) Phase relationship between adjacent simple cells in the visual cortex. *Science*, **212**, 1409–11

Rao, V. M. & Kulikowski, J. J. (1980) Disinhibition between orientation-specific channels. *Neurosci. Lett.*, Suppl. **5**, 548

Ratliff, F. (1965) *Mach Bands*. San Francisco: Holden-Day Inc

Robson, J. G. (1975) Receptive fields: spatial and intensive representations of the visual image. In *Handbook of Perception*, vol. v, ed. B. Carteret & W. Friedman, pp. 81–112. New York: Academic Press

Robson, J. G. & Campbell, F. W. (1964) A threshold contrast function for the visual system. In *The physiological basis for form discrimination*, pp. 44–8. Providence, Rhode Island: Brown University

Rodieck, R. W. (1965) Quantitative analysis of cat retinal ganglion cell response to visual stimulus. *Vision Res.*, **5**, 583–601

Rowe, M. H. & Dreher, B. (1982) Retinal W-cell projections to the medial interlaminar nucleus in the cat: implications for ganglion cell classification. *J. Comp. Neurol.*, **204**, 117–33

Sachs, M., Nachmias, J. & Robson, J. G. (1971) Spatial-frequency channels in human vision. *J. Optic Soc. Am.*, **61**, 1176–86

Schiller, P. H. & Malpeli, J. G. (1977) Properties of tectal projections of monkey retinal ganglion cells. *J. Neurophysiol.*, **40**, 428–44

Shapley, R., Kaplan, E. & Soodak, R. (1981) Spatial summation and contrast sensitivity of X and Y cells in the lateral geniculate nucleus of the macaque monkey. *Nature*, **292**, 543–5

Shapley, R. & Tolhurst, D. J. (1973) Edge detectors in human vision. *J. Physiol.*, **229**, 165–83

Sherk, H. (1978) Area 18 cell responses in cat during reversible inactivation of area 17. *J. Neurophysiol.*, **41**, 204–15

Stone, J. & Fukada, Y. (1974) Properties of cat retinal ganglion cells: a comparison of W-cells with X- and Y-cells. *J. Neurophysiol.*, **37**, 722–48

Stromeyer, C. F. & Julesz, B. (1972) Spatial frequency masking in vision: critical bands and spread of masking. *J. Optic Soc. Am.*, **62**, 1221–32

Stromeyer, C. F., Klein, S., Dawson, B. M. & Spillmann, L. (1982) Low spatial frequency channels in human vision: adaptation and masking. *Vision Res.*, **22**, 225–33

Tolhurst, D. J. (1972a) On the possible existence of edge detector neurones in the human visual system. *Vision Res.*, **12**, 797–804

Tolhurst, D. J. (1972b) Adaptation to square-wave gratings: inhibition between spatial frequency channels in human visual system, *J. Physiol.*, **226**, 231–48

Tolhurst, D. J. & Dealy, R. S. (1975) The detection and identification of lines and edges. *Vision Res.*, **15**, 1367–72

Tolhurst, D. J. & Thompson, I. D. (1981) On the variety of spatial frequency selectivities shown by neurons in area 17 of the cat. *Proc. R. Soc.*, *B*, **213**, 183–99

Weiskrantz, L. & Cowey, A. (1967) Comparison of the effects of striate cortex and retinal lesions on visual acuity in monkeys. *Science*, **155**, 104–6

Westheimer, G. (1967) Spatial interaction in human cone vision. *J. Physiol.*, **190**, 139–54

Wiesel, T. N. & Hubel, D. H. (1966) Spatial and chromatic interactions in the lateral geniculate body of the rhesus monkey. *J. Neurophysiol.*, **29**, 1115–56

Wilson, H. R. & Bergen, J. R. (1979) A four mechanism model for threshold spatial vision. *Vision Res.*, **19**, 19–32

Wilson, H. R., Phillips, G., Rentschler, I. & Hiltz, R. (1979) Spatial probability summation and disinhibition in psychophysically measured line spread functions. *Vision Res.*, **19**, 593–8

Zeki, S. (1978) Functional specialization in the visual cortex of the rhesus monkey. *Nature*, **274**, 423–8

Zeki, S. (1980) The representation of colours in the cerebral cortex. *Nature*, **284**, 412–18

Zeki, S. (1983) The distribution of wavelength and orientation selective cells in different areas of monkey striate cortex. *Proc. R. Soc. Lond. (B)*, **217**, 449–70

Zrenner, E. (1983) *Neurophysiological aspects of colour vision in primates*. Berlin: Springer-Verlag

Appendix
Basic equations

The spatial-frequency tuning curve can be represented by the Fourier transform of the RF profile (eqns 1a, 1b (p. 384)) and if the spatial phase is discounted, the curve in Fig. 26.3(D) is a Gaussian function of spatial frequency (f). The response, $R(f)$, is then given by

$$R(f) = \exp.\{-2[\Pi s(f-fo)]^2\} \qquad (A1)$$

where fo is the optimal spatial frequency at which maximal response occurs and s is the standard deviation. Spatial frequency selectivity of a tuning curve can be defined by a single parameter such as the Q-factor; $Q = fo/\Delta f$. Δf is the width of the tuning curve or the difference between high and low spatial frequencies ($f_2 - f_1$) at which the cell response becomes half of its maximum at fo. This difference is called the absolute bandwidth (Kulikowski *et al.*, 1982a) and is determined by the standard deviation:

$$\Delta f = \sqrt{(2 \ln 2)}/\Pi s \qquad (A2)$$

The RF width (W), specified conveniently at a 10% level of the maximum response, determines the bandwidth:

$$W10\% \cong 4.3\,s \qquad (A3)$$

$$\Delta f \cong 0.62/W10\% \qquad (A4)$$

The absolute bandwidth is inversely proportional to the RF width. Δf on its own, however, does not indicate the sharpness of tuning, which in turn is given by the Q-factor or its reciprocal, the relative bandwidth, Bdf:

$$Bdf = \Delta f/fo \qquad (A5)$$

or, expressed in octaves,

$$Boct = \log_2 (f_2/f_1) \qquad (A6)$$

Both measures of the relative bandwidth are related by the following equation:

$$Bdf = \log_2 [1 + Bdf/2)/(1 - Bdf/2)] \qquad (A7)$$

Two models of the Gabor system

Gabor's theory does not specify how standard deviation of the RF profile (s), or bandwidth (Δf) should vary as a function of the optimal spatial frequency (fo).

Fig. 26.3(B) and (C) illustrates two extremes of RF organization where the optimal spatial frequency equals the third harmonic of the frequency in Fig. 26.3(A). The Gaussian envelope and the RF size in Fig. 26.3(B) are the same as in Fig. 26.3(A). The standard deviation in Fig. 26.3(C), however, is one-third that in (A), so that the RF width is three times smaller as well (the length is reduced accordingly). The number of RF subregions, however, and thus the shape of the RF profile remain constant here. Fig. 26.3(E) and (F) shows tuning curves of the systems in (B) and (C) plotted on linear and logarithmic scales.

Note that for the system with the constant width the absolute bandwidth is also constant, but then the relative bandwidth narrows with increasing fo. When, however, the number of subregions remains constant, Bdf remains constant, but Δf increases with fo.

27

Unsolved problems in the cellular basis of stereopsis

JEREMIAH I. NELSON

Detection problems

Types of disparity detectors

The list of disparity detector types has been growing, with new detectors supplementing rather than superceding old ones. These binocular neurons are the substrate for all binocular mechanisms to be discussed, and are briefly reviewed.

Position disparity: depth and near/far detectors

The classic disparity detector (Pettigrew, Nikara & Bishop, 1968) is a tuned excitatory unit which responds to a *specific depth* (e.g. Joshua & Bishop, 1970, Fig. 13). What are today termed tuned *inhibitory* specific-depth cells (Poggio & Talbot, 1981) were also reported by Pettigrew *et al.* (1968). Where an excitatory specific-depth cell shows excitation to appropriate binocular stimulation, a tuned inhibitory cell shows 'occlusion' (reduction of response). Whether by excitation or inhibition, specific-depth cells tie together specific retinal points. Therefore they are the most appropriate kind of detector for specifying binocular correspondence.

Ambiguity of correspondence is an unappreciated problem caused by specific-depth detectors. With these detectors, the two requirements for disparity detection are (1) specificity of response, and (2) scatter in optimal disparity from cell to cell. Only with scatter can a range of disparities be detected. The wider the scatter, the greater the range of discriminable depth. Yet the same scatter defines a range of disparities within which it is difficult to define one set of points as more corresponding than another. We have won an understanding of depth detection at the cost of ever being able to specify correspondence uniquely. The immediate response to this problem was to define correspondence as the *mean disparity tuning* of specific depth cells (Bishop, 1970); this fixed definition has several shortcomings in dealing with dynamic properties of binocular vision (below). The problem is deepened by other detectors, which encode depth over a broad range of disparities without linking specific retinal points in the two eyes (considered next) and by the possibility of different mean values in W-, X- and Y-dominated cortical areas (Pettigrew & Dreher, 1982).

Fine color discrimination arises from a three-way system of broadly tuned detectors, the red, green, and blue cones of the retina. Evidence that depth might be coded in the three broad bands of 'near', 'in the fixation plane', and 'far' first came from confusion errors in psychophysical experiments (Richards, 1970, 1971). Some observers confused either large 'near' or large 'far' disparities with zero-depth stimuli, suggesting that a hypothetical 'near' or 'far' class of detector was absent or defective. In neurophysiological research, disparity tuning curves appropriate for near and far detectors have been found in monkeys (Maunsell & van Essen, 1983; Poggio & Fischer, 1977; general discussion in Poggio, 1979) and sheep (Clarke, Donaldson & Whitteridge, 1976). These findings are not in disagreement with earlier work in cat, where a disparity tuning curve for what today would be called a near unit was described (Bishop, Henry & Smith, 1971). The same authors present a useful and subsequently confirmed analysis of why near and far units can never be binocularly excitable. Silencing the response for a given disparity range requires a contribution from one eye which is purely inhibitory. A comparison between near, far, and especially tuned inhibitory neurons and various classic receptive field types of the Canberra group is drawn by Kato, Bishop and Orban (1981).

The detector names 'near' and 'far' are appropriate, because the tuning curve transitions are well aligned at zero disparity in the awake, fixating monkey (Poggio, 1984). However, the tuning curves are flat over large disparity ranges, unlike the sloping discriminator functions of photopigment spectral absorption curves, and are therefore unsuitable for discriminating gradations in depth. Except at the narrow near/far transition point, only crude classifica-

tions are possible. Richard's psychophysical subjects in fact saw only very large disparities and classified them only as 'nearer' or 'further' than the fixation plane. The near/far system as currently understood neurophysiologically cannot be a general depth-discrimination system. Recent psychophysical research has shown that near/far stereoanomalies occur only under special, short exposure-time conditions (Patterson & Fox, 1984), as might be expected for a special-purpose system. That purpose may be evoking vergence eye movements, or making rapid but crude depth judgments, especially of moving objects. Questions remain about the division of labor between broad-band and specific depth systems, their possible origins in Y and X visual pathways, respectively, and where and how their outputs are combined.

Orientation disparity

A straight edge tilted in depth gives rise to retinal images differing in orientation. Tilt may be specified by either this single orientation disparity or *cyclodisparity* (O'Shea, 1983) value, or by an infinite progression of increasing *position* disparity values. Blakemore, Fiorentini & Maffei (1972) suggested that the more economical orientation disparity cue is directly detected. Orientation disparity detection in the cat was critically examined by Nelson, Kato & Bishop (1977) with regard to two questions: (1) does a *range* of orientation disparities occur among binocularly excitable striate neurons; and (2) for any particular neuron, is the discrimination of stimulus orientation differences between the eyes *sharper* than the neuron's monocular orientation selectivity? The studies were in good agreement on the range of optimal stimulus orientation differences (standard deviation 6.7 degrees, 90% of the cells within a range of ± 11 degrees: Nelson *et al.*, 1977) but Nelson *et al.*, failed to find any binocular mechanism for orientation disparity discrimination. In particular, no interocular inhibition arose to make binocular orientation tuning curves sharper than conventional monocularly curves. Such sharpening inhibition, and sometimes facilitation as well, does come into play for position disparities.

Von der Heydt *et al.* (1982) have recently found a small number of cells which are in fact orientation disparity selective. In the rhesus monkey, 8 of 170 cells (all but one of which were in area V2) were particularly sensitive to tilt using conventional line stimuli. Half-width at half-height for orientation disparity tuning was as small as 2 degrees; total disparity range was 5 degrees. This range is far less than conventional monocular orientation tuning, where even in V1, a median half-width of 20 degrees at half-height (De Valois, Yund & Hepler, 1982) and a modal half-width of 20 degrees at 29% height (Schiller, Finlay & Volman, 1976) have been reported. Nelson *et al.* (1977) laboriously tested orientation disparity selectivity across a wide range of position disparities to be certain that optimal response conditions had been found. Von der Heydt *et al.* (1982) used a clever random grating stimulus comprised of a mixture of spatial frequencies and phases. This pattern always stimulates a unique orientation value but a randomized set of position disparities. Cyclodisparity selectivity with such a stimulus can not be based on position disparity mechanisms. All tilt-selective cells retained their unusual selectivity when tested with the random grating.

Cyclodisparity detectors in monkey are too scarce to be a major depth system. Yet information from these tilt detectors must somehow guide and accelerate detection in the principal disparity systems. Perhaps few cells are involved because the cue is efficient as well as specialized.

Motion in depth

Cells preferring opposite directions of motion in the two eyes were oberved in the cat by Bishop, Henry & Smith (1971), and by Pettigrew (1973; Area 18), who recognized their importance for detecting a projectile aimed at the head. In the general case, trajectories towards or away from the observer produce image motions with velocity disparities. For the special case of a trajectory along the midline between the eyes, the velocities are equal but their directions are opposite. Velocity disparity selectivity has been measured in a variety of areas (monkey: superior temporal sulcus (Zeki, 1974); striate cortex (Poggio & Talbot, 1981); middle temporal visual area (Maunsell & van Essen, 1983); cat: peristriate cortex (Cynader & Regan, 1978); 17/18 border (Cynader & Regan, 1982); Clare–Bishop area (Toyama & Kozasa, 1982)).

As with cyclodisparity, one may ask whether motion in depth produces particularly selective responses. To test this, one must first optimize position disparity on the frontoparallel plane and then introduce motion in depth (velocity differences). When this is done, the response increases are not dramatic in either

cats (Cynader & Regan, 1982) or monkeys (Maunsell & van Essen, 1983). Low-grade motion-in-depth information is coded and available at the cellular level, but neurophysiology does not yet tell us if motion-in-depth perception uses these primitives, or is synthesized at a higher visual-system level.

Summary

The variety of available disparity detectors is richer than it used to be. For static disparities, the disparity detection function can be related to receptive field topography. Sharply defined inhibitory zones, termed *sidebands*, often occur on both sides of a simple cell's discharge center, and determine in large part the orientation and disparity selectivity of the neuron. With better links between receptive field topography and neural function, newer disparity detector types could be related to the classic simple cell descriptions of Bishop and colleagues (Kato *et al.* 1981).

Constancy problems

Depth–distance constancy

Depth, or the relative separation of objects in three-dimensional space, and the absolute distance of an object from the observer, are phenomenologically different percepts based upon different cue systems. For example, if we look down the aisle of an airplane with one eye closed, many cues enable us to judge the seats' distances from us (e.g. atmospheric perspective in the smoking section), but the scene is nonetheless flat. Only retinal disparity cues and the kinetic depth effect (a laboratory curiosity based on monocular cues from object deformation under rotation; Braunstein, 1976) generate the experience of stereoscopic depth. In this section, the extraction of distance information from vertical disparity is considered, but I would no more elevate vertical disparity to a stereo-depth cue than I would atmospheric perspective.

There *is* a link between distance and stereoscopic depth: depth judgments from disparity input require constancy scaling according to absolute distance. The disparity generated by a fixed depth interval decreases with viewing distance; first, halving retinal image dimensions (by doubling viewing distance) halves disparity values. In addition, the parallax differences generated by our fixed interocular separation decrease with viewing distance. Therefore the amount of perceived depth output assigned to a fixed position disparity must be scaled upwards approximately as the square of observation distances when observing natural scenes (Wallach & Zuckerman, 1963; reviewed by Ono & Comerford, 1977).

Disparity scaling with distance is a neglected problem in stereopsis. It is generally assumed (Ogle, 1964) that classic distance cues such as accommodation, convergence, observer motion parallax (Gogel & Tietz, 1979) and atmospheric perspective provide the distance information, and that the scaling operation is a higher cognitive process.

Reliance on separate cue systems and higher cognitive functions has information processing and biological (dynamic range) disadvantages. Until the point of cue combination (constancy scaling) is reached, the activity in binocular visual pathways is uncalibrated and of limited usefulness. Until constancy scaling is applied, the feeble depth signal for distant objects could get lost in neural noise, while large depth signals from objects close at hand might become saturated.

Vertical disparity

Mayhew & Longuet-Higgins (1982; see also Mayhew, 1982) have proposed a more elegant scheme to generate the needed distance information within the disparity detection apparatus itself. The cue is vertical disparity. It must be emphasized that vertical disparities are uncorrelated with depth and do not trigger stereopsis. What is asserted is that vertical disparities contain information which can modify the depth interpretation assigned to horizontal disparity cues. These vertical disparities arise in central vision when we look to the side in asymmetric convergence: the object of regard is then closer to one eye than the other, and that eye's image of the object is on the whole larger. The interocular magnification (image size) difference shrinks as viewing distance grows large with respect to the separation between the eyes. But shrinkage of the magnification difference – and the vertical disparities along with it – also occurs as fixation becomes more 'straight ahead'. The problem is to apportion the observed vertical disparities between asymmetry of convergence and absolute distance.

Solving the two unknowns of fixation asymmetry and distance can only be done by sampling two points in the visual field. Consider points of some fixed horizontal angular separation. Increasing *difference* in vertical disparity between them indicates increasingly asymmetric gaze, while increasing vertical disparity of

both points indicates decreasing fixation distance. Three scaling operations are needed to make the conclusions quantitative: (1) the angular separation of the points; (2) the vertical positions of the points; and (3) the horizontal positions of the points. Greater angular separation obviously affords opportunity for greater differences in distance ratios, and hence greater vertical disparities for the two points. Horizontal position can carry the sample points into the peripheral visual field, increasing vertical disparity as if fixation had been more eccentric. Lastly, vertical position scales vertical disparity because of the geometry of image magnification. When we 'zoom' an image, a central image point remains stationary as the image expands centrifugally around it. The displacement between points in the original and magnified images increases towards the periphery. The three required items of information are surely available to the brain, although certain sample points are of no use (Longuet-Higgins, 1982).

In principle, the computational scheme is workable and generates information badly needed for constancy scaling of the disparity–depth relation, and for horopter constancies (below). The model does not address itself to sensory fusion; how is single vision achieved despite disparities large enough for detection and important enough for this kind of processing? Nor does the model dispose of horizontal disparities which contain no depth information, and arise from a variety of sources in addition to asymmetric convergence (Nelson, 1977).

Opportunities for vertical disparity detection

Binocular neurons have ample capability for vertical disparity detection. As with any other disparity cue, the criteria for an adequate substrate are (1) a scatter of preferred values, and (2) selectivity (sharp tunings). Vertical scatter in receptive field alignment for binocularly excitable neurons is well established (Nikara, Bishop & Pettigrew, 1968; Joshua & Bishop, 1970; each receptive field was plotted monocularly). When tested with binocular interaction methods, the vertical disparity selectivity of neurons whose optimal bar orientation is horizontal is no worse than the disparity selectivity of vertically tuned neurons (Nelson et al., 1977), although such cells may be less numerous (Von der Heydt et al., 1978), in accord with the low sampling requirements of the computational model. We think of neurons with vertical orientation references as only horizontally disparity selective, yet many of them will be simultaneously selective for vertical disparity, if they are hypercomplex. The extent to which end-stopped inhibition imposes disparity selectivity remains unexplored.

Vertical disparity detection is needed in binocular vision. Early data suggesting that binocular neurons perform such detection are not an embarassment to the cellular basis of steropsis (Rodieck, 1971). Rather, we have failed to appreciate the perceptual processing needed after initial cue detection; vertical disparity plays a role in this processing. The processing includes depth–distance constancy (above), sensory fusion, the induced effect and horopter constancies. Before considering the latter two in some detail, it is worth recalling that vertical disparity in visual direction may also be shifted by sensory fusion in the interests of binocular single vision (e.g. Perlmutter & Kertesz, 1978; Ogle, 1964) and that small vertical misalignments are also deleterious to performance in a variety of visual tasks (e.g. cyclofusional response (Wright & Kertesz, 1975). We cannot dismiss sensory fusion of vertical disparities as a failure of acuity for difference in visual direction ('vertical disparities are ignored'), and yet accept that visual system performance is degraded by the slightest vertical misalignment.

Vertical disparity and the induced effect

In the induced effect, vertical disparities created optically in one eye and absorbed (fused) by the visual system affect the correspondence relation between the eyes for all disparities, horizontal or vertical. For example, vertically disparate horizontal-bar gratings do not themselves appear tilted (Blakemore, 1970), but after viewing and fusing contours with vertical disparity, a non-disparate horizontal probe stimulus will be seen in depth. The horizontal probe has acquired an effective disparity and reveals a distortion in horizontal correspondence (Riguidiere, 1975; Nelson, 1977; van der Meer, 1978). Sensory fusional mechanisms are central in this approach to the induced effect. Because one eye received larger extents, larger extents in that eye now correspond to shorter extents in the other. The correspondence shift is not orientation-specific and therefore a horizontal probe bar which is physically equal in both eyes will be effectively disparate. Like any other, this horizontal disparity is linked to a perceived depth response.

In Nelson's (1977) disparity shift model of the induced effect, a few vertical disparities are sufficient to cause a recalibration of retinal correspondence which is (1) global across the visual field, and (2) general for disparities in horizontal and vertical extent. The recalibration is produced by sensory fusional mechanisms. In the model of Mayhew & Longuet-Higgins (1982) the induced effect is an illusion of perceived fixation angle. At least two vertical disparities are sampled to calculate the distance and asymmetry of fixation. Vertical magnification of one eye's image distorts these calculations. There is evidence against this approach to the induced effect (Epstein & Daviess, 1972) and there are disadvantages: (1) an illusion of the perceived angle of asymmetric fixation is created when accurate distance and asymmetry of convergence information is needed; (2) The illusion derived from vertical disparity information will conflict with oculomotor information about eye position. Resolving the conflict requires further mechanisms and risks further distortion of the information.

An attempt has been made to explain the induced effect away as a product of other cues (Arditi, Kaufman & Movshon, 1981). In a scene with oblique contours, vertical magnification will produce disparities with both horizontal and vertical components. The horizontal components are processed by the visual system in the usual manner to produce the induced effect. This argument has several shortcomings (O'Shea & Crassini, 1983; Mayhew & Frisby, 1982): the perceived slant of an oblique grating runs from top to bottom, whereas the induced effect produces a left-to-right slant; left and right oblique gratings will produce opposite directions of slant (the induced effect does not change), or, if both oblique components are present in a scene, their disparities will cancel one another (the induced effect does not disappear); the magnification will produce spatial frequency disparities causing tilts opposite to that produced by the induced effect.

Difficulties in measuring the induced effect contribute to the controversies surrounding it. Two inherent difficulties are subject variability and cue conflicts. van der Meer (1978) comments, 'The induced effect . . . is for many people difficult to obtain. In our experience there are some subjects who never notice it, other subjects only after long inspection times, and there are few subjects who perceive it clearly.'

A cue conflict arises between the vertically disparate contours used to induce the effect, and horizontal extents not made disparate (or given an opposite, 'nulling' disparity) and used to probe the effect. For these horizontal contours, binocular induced effect mechanisms signal a disparity, while monocular mechanisms are available to signal that the bars are of equal length. Which cue will prevail? To be perceptually important, the vertical differences should be consistent, and widely presented throughout the visual field, while the probe contors (horizontal intervals) should be small or few in number. When reduced stimulus configurations are used in the laboratory, vertical cues may be as sparse as horizontal ones; no shift in spatial values occurs.

A computational model such as that of Mayhew & Longuet-Higgins (1982) does not attempt to specify mechanisms. All points sampled for vertical disparity information are mathematically equivalent. For Nelson (1975, 1977), the induced effect is a consequence of a shift in retinal correspondence brought about by network interactions. The network or domain interactions are mutual facilitation among like-tuned disparity detectors, spreading broadly across visuotopic space, and mutual inhibition among unlike-tuned detectors, visuotopically confined to one locality. Because of these specific, non-uniform rules of connectivity, the induced effect must be sensitive to stimulus configuration, as indeed it is (Pastore, 1964). Vertical disparities which enclose or surround a single, central horizontal probe disparity are more likely to enter the computation, or, in my terms, are more likely to reset correspondence. Lastly, in a network model, the response is slow (see below). The failure of Westheimer (1978) to demonstrate an induced effect may be attributable to the factors mentioned: individual variability, cue conflicts, configurational effects (the vertical disparities were not more numerous than the non-disparate horizontal intervals and did not enclose them), and temporal factors. What did enclose Westheimer's stimuli was a zero-disparity frame presented between trials. The frame restored correspondence to normal, while the trials were too brief (50 ms) to reset it.

It is sometimes said that vertical disparity plays no role in stereopsis. It would be more accurate to say vertical disparities are 'uncorrelated with object depth', or 'convey no stereoscopic depth information', but are abundant in everyday retinal images and may be important in stereopsis. As long as the role of vertical disparity is unresolved, electrophysiological data for the disparity sensitivity of neurons with *all* orientation preferences deserves to be obtained and reported.

Horopter constancies

The horopter is the surface in three-dimensional space containing all corresponding points. Here, 'horopter' will refer to a horizontal cross-section through this surface, the *longitudinal horopter*. Because this is the surface we perceive as having the same depth as the fixation point, it is an important frame of reference for localizing objects in the world, and our relations to them. Perceptual stability demands stability of this frame of reference. Unfortunately, tilts and shape distortions arise from the geometry of corresponding points. Constancy mechanisms partially neutralize these horopter changes by imposing changes on correspondence. These constancy mechanisms are an embarrassment to the old concept of fixed retinal correspondence, and have been a neglected problem in binocular vision (further references and discussion in Nelson, 1977).

Distance and asymmetry of fixation

A flat horopter would be a useful shape, and can be created with the pattern of binocular correspondence shown in Fig. 27.1. The visual axes of corresponding points are projected from the two retinas at the bottom of the figure to their point of intersection at the top. The locus of all intersections is a longitudinal horopter. The same retinas bearing the identical set of corresponding points may be rotated to simulate a change in fixation distance. Remarkably, if correspondence is stable, the horopter will not be. Change in horopter curvature with change in fixation distance is illustrated in Fig. 27.2. Classic horopter data show less change in curvature than expected on geometrical grounds, implying the operation of a constancy mechanism.

In asymmetric convergence, a flat horopter perpendicular to the line of sight would be most useful for spatial localization. Such a surface is termed the *objective normal plane* (*ONP*). However, a flat horopter in symmetric convergence remains tangent to the Vieth–Müller circle in asymmetric convergence, and this tangent is tilted more than the ONP. (The Vieth–Müller circle is the circle defined by the two eyes' nodal points and the fixation point.) This tangency is illustrated in Fig. 27.3, where eyes bearing the same set of corresponding points as in Fig. 27.1 have been rotated clockwise to simulate asymmetric convergence. The new horopter is demonstrably curved as well as tangent. Again, modern (Reading, 1984) as well as classic horopter data shows less rotation of the horopter

Fig. 27.1. Ray diagram showing the horopter generated by a particular set of corresponding points. The two partial circles at the bottom represent the two eyes. A pair of corresponding points in these eyes may be thought of as the retinal locations of a binocular cortical neuron's two receptive fields. As these are corresponding points, the binocular neuron has a disparity tuning of zero. Just as receptive field pairs define corresponding points on the retina, lines from them through the eyes' nodal points define corresponding points in space. Such a line is termed a receptive axis. By definition, the locus of all receptive axis intersections is a horopter. In order for the horopter to be flat, points with a given displacement into temporal retina must be paired with points of much greater displacement into nasal retina ('nasal stretch').

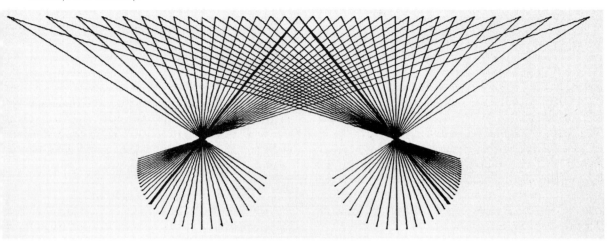

n asymmetric convergence than expected on geometrical grounds, implying a constancy mechanism for holding the horopter close to the ONP. Can simple cellular mechanisms be proposed for this complex aspect of binocular vision?

Cellular basis

The shifts in correspondence implied by the constancy of the horopter's flatness and perpendicularity to the line of fixation are not the embarrassment they once were. With the realization that a range of disparity tunings makes possible a range of possible correspondence states, all that are required are selection mechanisms for expressing one particular correspondence state or another. In sensory fusion, disparity input in the visual stimulus drives the selection mechanism. For horopter constancy, information about fixation distance and angle are needed. These, too, could come from the disparity content of the visual stimulus if the brain is able to perform the computational operations of Longuet-Higgins (1982) and Mayhew (1982). Nelson (1977) has pointed out that once we have the required information, simple modulations of responsiveness imposed upon the disparity detectors themselves will produce the subtle

shifts in correspondence required. A gentle inhibition of neurons with crossed horizontal disparity tunings, linked to ocular convergence information and stronger in the peripheral visual field, provides constancy with fixation distance (Nelson, 1977). A similar inhibition of neurons with increasing vertical disparity, also stronger in the periphery and linked to ocular position, could stabilize the tilt of the horopter in asymmetric convergence (Nelson, 1977) by amplifying vertical magnification signals from the two eyes and so heightening the compensatory induced effect.

This 'new look' in vertical disparity holds that it is concerned with magnification detection. This leads to three differences in the way horizontal and vertical disparities must be ordered. (1) The brain must order vertical disparity according to direction away from the fovea. Whereas horizontal disparity increases with increasing shifts in the same direction wherever they occur on the retina, a shift upward for a contour above the fovea and a shift downward below the fovea are both increases in vertical disparity. (2) Whereas a given horizontal shift represents the same horizontal disparity everywhere across the retina, a small vertical misalignment near the fovea and a large one at a vertical displacement away from the fovea represent the same

Fig. 27.2. Instability of the horopter curvature when fixation distance is changed. Three horopters are generated by counter-rotating the eyes to simulate three fixation distances. A set of corresponding points has been chosen which generates a flat horopter at the middle fixation distance. This set of points (i.e. retinocortical wiring) is not changed. Due to geometrical considerations alone, the horopter changes from concave to convex with change in fixation distance. If permitted to occur, this shape change would detract from the horopter's usefulness as a stereoscopic frame of reference.

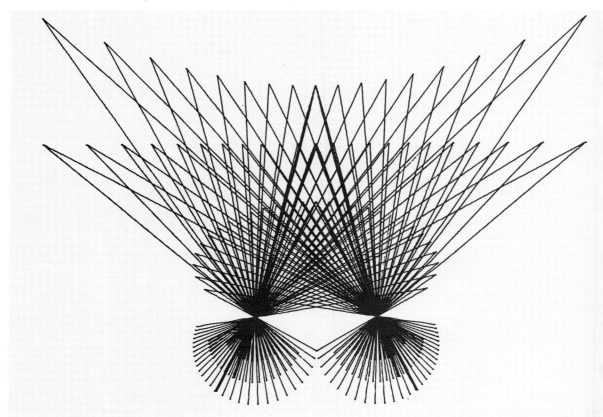

vertical disparity value. The larger vertical scatter of binocular receptive fields in the vertical periphery, particularly beyond 20 degrees (Joshua & Bishop, 1970) may reflect this requirement. (3) Whereas horizontal disparities are ordered on a near/far continuum, vertical disparities must be ordered on a left eye/right eye continuum. I define a big left vertical disparity as a receptive field misalignment appropriate to a magnified image in the left eye. The suggested horopter *tilt* constancy mechanism is an inhibition of right vertical disparity tunings when the eyes turn left. The cortical connectivity rules are not complex. If the scaling factor in (2) above is built into the visual system, the vertical disparity inhibitory fibers can ramify uniformly across the visual field. The inhibition required for *distance* constancy must be gently weighted across both the near/far horizontal disparity

continuum and the foveal/peripheral visuotopic continuum.

In summary, vergence information is needed for certain stereoscopic constancy mechanisms. The vergence information could come from the oculomotor system or from vertical disparities via computational mechanisms. The information could be applied to horizontal disparities with direct effect, or to vertical disparities which in turn generate the required shifts in correspondence. Above, compensation was applied to vertical disparities for horopter tilt, as tilt involves magnification, and there is growing suspicion that the brain has a magnification detection system based on vertical disparities. With all proposals, activity in ordered populations of neurons is modulated to produce shifts in sensory coding.

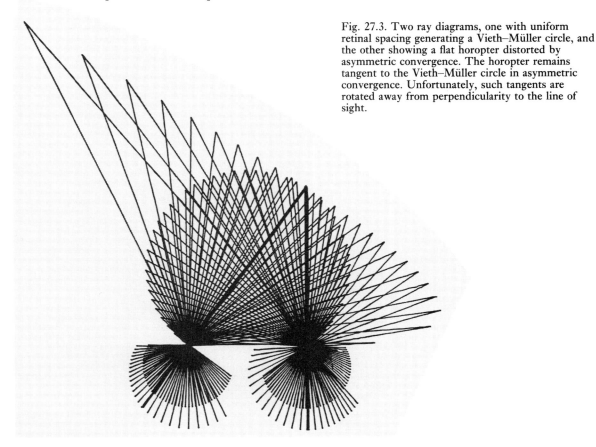

Fig. 27.3. Two ray diagrams, one with uniform retinal spacing generating a Vieth–Müller circle, and the other showing a flat horopter distorted by asymmetric convergence. The horopter remains tangent to the Vieth–Müller circle in asymmetric convergence. Unfortunately, such tangents are rotated away from perpendicularity to the line of sight.

Network problems

Population complexities are everywhere in stereopsis. The first hint of them came from the slowness of stereopsis itself.

The slowness problem

Detectors respond with brief retinocortical latencies (cat striate cortex, 60 ms: Bishop, Coombs & Henry, 1971), yet stereopsis is resplendent with lags, latch-ups and even long-term learning effects. The human visual system is unable to discriminate alternations between near and far depth occurring above 3.2 Hz (square wave modulation), even though the left and right displacements used to generate the disparity may be discriminated monocularly at much higher rates (Richards, 1951). The detection of lateral *motion* has a clear optimum at 2 Hz and extends at least to 20 Hz, while motion in *depth* detection continues to improve with *decreases* in rate of oscillation below 1 Hz (Regan & Beverley, 1973*a,b*). Stereopsis is not only slow, it is persistent. Engel (1970) has shown that effective binocular stimulation for less than 4 ms produces up to 100 ms of stereoscopic depth perception. Other work has demonstrated temporal hysteresis: once the temporal overlap conditions for perceived depth are met, the depth response will persist as binocular overlap is decreased *below* the former threshold value (Efron, 1957). Learning effects arise in sensory fusion (Hyson, Julesz & Fender, 1983) and global stereopsis, and must occur early in the visual pathways because they show orientation and retinal locus specificity (Ramachandran & Braddick, 1973; Ramachandran, 1976).

At the cellular level, analogous threshold non-linearities followed by response perseveration have yet to be observed. This implies that many layers of elaboration (processing) intervene between the detector systems currently studied electrophysiologically and the percepts they engender. Slowness is appropriate to a system with computational complexity and network

Fig. 27.4. Model for global disparity detection. Detectors of similar disparity tuning are aggregated into radial columns. Progressions of columns exist in which detectors display a sequence of disparity tunings. Mutual inhibition within these *sequence slabs* suppresses matching noise. Mutual facilitation among columns matched for disparity tuning (*isoslabs*) strengthens response to the correct disparity. Mutual facilitation is stronger if neurons are also matched in orientation preference and vernier alignment.

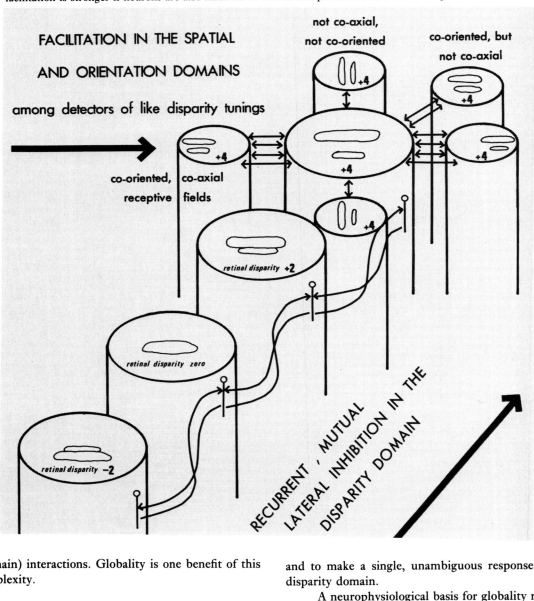

(domain) interactions. Globality is one benefit of this complexity.

The globality problem

The mechanical nature of random-dot stereo-grams would seem to make local detection of disparity easy, but the precisely repeating patterns also create ambiguities termed *matching noise*. Globality enables the binocular visual system to overcome ambiguities

and to make a single, unambiguous response in the disparity domain.

A neurophysiological basis for globality remains to be found. Nelson (1975) suggested a domain interaction model. If specific-depth detectors with *closely matched* disparity tunings but widely spread spatial positions were mutually facilitatory, response to the globally correct disparity would be strengthened. Additionally, if specific-depth detectors with a wide ange of *differing disparity tunings* but closely matched spatial (visuotopic) positions were mutually inhibitory, matching noise would be suppressed. Columnar architecture simplifies the required connectivity in the spatial and disparity domains. The neurophysiological model is illustrated in Fig. 27.4. Several computational models embodying network interactions of this sort have been proposed (reviewed in Poggio & Poggio, 1984; Nelson, in press).

The most exciting findings in this area is Poggio's demonstration of neurons capable of responding to a moving bar presented as a dynamic visual noise, random-dot stereogram; i.e. without monocular brightness cues (Poggio, 1980; Poggio & Poggio, 1984). These neurons respond to conventional contours as well; significantly, latency increases two- to three-fold when the bar is changed to a pure cyclopean stimulus. Latency goes down and response goes up if the correct disparity can be pooled across a larger cyclopean stimulus. So far, all global disparity detectors have been complex cells. This finding supports a very 'local' view of global stereopsis: complex cells are very texture-sensitive (Hammond & MacKay, 1975). Binocularly matched subunits within the complex-cell receptive field could respond to local texture elements conveying the disparity information. Simple cells are believed to be the source of these subunits, and a small patch of them (the size of the complex cell's receptive field) could converge on the cyclopean complex cell to give it limited globality. The domain interaction model, on the other hand, calls for interactions running within cortical slabs over much larger visuotopic regions (Fig. 27.4); this model also postulates inhibition as well as excitatory convergence.

A domain interaction model (Nelson, 1975) makes it possible for peristriate and even striate areas to achieve global stereopsis, as Poggio's electrophysiological findings confirm (Poggio, 1980; Poggio & Poggio, 1984). It is implausible for the visual system to wait for higher centers or to need hemispheric lateralization for global disparity detection. Yet there is evidence of right hemisphere lateralization of binocular visual function (reviewed by Ross, 1983). If global disparity detection is not lateralized, then other visual processes must be involved in these findings. These lateralized processes could involve: (1) the formation of contours (edges); (2) the synthesis of figures from a collection of edges; (3) figure-ground segmentation; and (4) non-visual language and cognitive deficits.

Random-dot stereograms spurred the search for single-unit disparity detectors by showing stereopsis could get underway without figural processes. Yet random-dot stereograms also brought the problem of globality. The network mechanisms postulated for globality have now become useful for driving sensory fusion and selecting the state of correspondence. Thus random-dot stereograms continue to occupy an ironic position, showing how much may be accomplished at the cellular level, while emphasizing the need for undiscovered interactions in a missing functional architecture.

Correspondence and sensory fusion

Disparities are differences, and we know something about their detection. But what is not different? This is the problem of correspondence.

Detector model

Correspondence may be defined by some measure of central tendency in the distribution of disparity detector tunings. Most neurophysiological studies have followed Bishop's (1970) lead by defining the mean tuning as the zero disparity value (mode and mean are essentially identical in these distributions). The modal value for all points along the horizontal meridian defines a cross-section of the horopter. Panum's area, or the disparity limits of sensory fusion, may be arbitrarily set as the disparity values one standard deviation to either side of the mean (Joshua & Bishop, 1970; Bishop, 1973, 1975). According to this model, fewer binocular neurons of the appropriate tuning may be found as increasing disparities are stimulated. At the same time, two *additional* populations of neurons, each capable of being excited monocularly, will begin to respond to the very disparate left and right half-images. The activity of these two monocularly stimulated populations signal stimulation from separate visual directions. Diplopia is perceived. When depth is perceived in the presence of diplopia, presumably both binocularly and monocularly excitable populations are active. Many refinements will be needed to fit this simple fixed-threshold model of diplopia and sensory fusion to the facts of psychophysics. The boundary between single vision and diplopia is not fixed, but varies (1) with the contour content of the stimulus, (2) as a function of past stimulation history (hysteresis effects), and (3) with the context of surrounding disparities. Thus, a given disparity stimulus and detector response may or may not be accompanied by sensory fusion. It has never been explained in detail why any detector response should cause shifts in perceived visual direction. Attempts to dismiss visual direction shifts (e.g. rivalry theories of sensory fusion) have been discredited (Nelson, 1975).

The large size of the disparities we can fuse requires an active mechanism for shifting perceived visual direction. Fender & Julesz (1967) have reported sensory fusion of disparities up to 2 degrees using stabilized image apparatus to control eye movements. Duwaer (1983) has recently found a fusional limit of 40 minutes of arc in most subjects (80 arc min in one) instead of the 120 arc min reported by Fender & Julesz. He attributes Fender & Julesz's higher values to contact lens slip and more lax subjective criteria for what constitutes a fused, global percept. Duwaer used an after-image technique: he pulled random-dot stereograms temporalward to their fusional limit and marked the retinas with an after-image. To get the disparity, Duwaer subtracted after-image location upon refixation. His maximum fusible disparity may have been erroneously low, as the subjects may not have been able to relax their eyes fully after strenuous, prolonged divergence (sessions lasted about 4 h). Alpern & Hofstetter (1948) have shown that after less than 3 h wearing 18 PD of prism power, the eyes required over 1.5 h to return to their original angle. After-image contours are also subject to sensory fusion (Wheatstone, 1838; Beloff, 1961) so that as probes of eye position they are neither completely neutral nor objective. In a new study answering earlier criticisms, Hyson *et al.* (1983) have reported fusional limits of 3 degrees average (5 degrees maximum) when a larger stimulus and free fixation are used. Similar values were obtained by Patterson (personal communication; 2 degrees) and by Crone & Hardjowijoto (1979; 4 degrees), the latter using monocular nonius lines instead of an elaborate contact lens eye-tracking system. In all studies, the binocular visual system absorbed large, field-wide, uniform horizontal disparities without disturbing the depth percept (a random-dot stereogram, conveying, for example, a three-dimensional spiral). I call this depthless fusion: the visual system received disparity stimulation, produced a fusional change in perceived contour location, and generated no depth percept. The occurrence of sensory fusion independent of depth sensation shows the need for an independent fusional mechanism. It is time to move on from the hope that if a disparity detector fires, fusion and depth will somehow take care of themselves.

Depthless fusion

There are many examples of depthless fusion (Nelson, 1977): binocularly uncoordinated microsaccades during fixation; magnification differences between the eyes; and laboratory reports such as those above. Once disparate linear extents have undergone fusion, any additional horizontal linear extents added to the visual field which do *not* have the same mismatch will have *effective disparity* and be seen in depth. Depthless fusion embodies a shift in retinal correspondence. A shift triggered by disparate spatial extents in one meridian seems able to reset disparity values for all orientations. Since vertical disparities convey no depth information and generate no depth percepts, their fusion, if it occurs, will always be depthless. Vertical disparities are therefore a good way to make distortions in the disparity–depth relation for all disparities, and the induced effect is one example of this. Retinal image displacements and magnification differences which are uniform across visuotopic space are most subject to depthless fusion.

As with the induced effect above, cellular mechanisms for sensory fusion are hardly more than conjecture, while psychophysical research has failed to quantify what must be found. It must be recognized that for any stimulus, Panum's limit is really two values: one for *spontaneous* fusion and another for the hysterically *extended* fusional limit. Both limits depend upon the stimulus. Measurements are needed from both basic research and clinical practice of this stimulus-dependency using neurophysiologically interpretable variations in stimulation (e.g. total amount of contour, orientation disparity, hue and contrast difference).

This is the case for correspondence shifts in normal binocular vision. Such shifts are far more evident in anomalous correspondence, one of the sensory adaptations to strabismus.

Plasticity and anomalous correspondence

When oculomotor abnormalities disrupt accurate fixation, then amblyopia, binocularity loss or anomalous correspondence ensue. Amblyopia occurs when an eye disadvantaged in any way during the early critical period of cortical plasticity (for example, by anisometropia) loses the competition for cortical access. If the two eyes' images are on average of equal quality but unrelated content (e.g. alternating strabismus), each will retain cortical access but will be served by separate populations of neurons. Vision survives, but the binocular substrate for stereopsis is lost. In anomalous correspondence (*AC*; Bagolini, 1967), the binocular substrate survives, but may display wildly abnormal properties. Small, constant, unilateral angles of tropia (eyeball misalignment) favor the development of AC. By definition, AC is a shift in the pairings of retinal points; doubtless there is a shift in binocular receptive field separation; i.e. in the points chosen for convergent input upon binocular cortical neurons. AC, however, is more than a simple compensatory shift in receptive field disparity. AC is accompanied by changes in cortical processing; in particular, sensory fusional range is exaggerated. Yet despite the ability to detect and process an abnormally large range of disparities, depth discrimination is deficient.

Good sensory fusion

In anomalous correspondence, correspondence may change continuously or alternate abruptly between two states; in either case, the correspondence change appears to be an immediate response to stimulus input. Rønne & Rindziunski (1953) have given a provocative description of the amount of image processing the binocular visual system is prepared to perform in the interests of single vision. In the cases below, the eye is stationary as prisms are used to displace the image of the surrounding room. The displaced images provide a disparity stimulus which the binocular visual system must fuse. Unable to rotate the eye in its socket (there is no vergence, no motor fusion), the brain resorts to exaggerated sensory fusion and shifts its image of the world. The shifts in perceived visual direction are revealed by the movement of an after-image fixed upon an unmoving retina.

The patient observes these 'negative after-images' projected on the tangent scale or a wall, and a prism-rack (with horizontal prisms) is moved quickly up and down before one of the eyes. Upon questioning, several patients report that the distance between the after-images changes while the prism-rack is moving . . . after-images originally showing normal correspondence may be separated, moving to and from each other; or after-images originally abnormal come partly or completely together during the movement of the prism rack. Often these movements are practically continuous but sometimes they proceed in a more jumping fashion. Sometimes they change rhythmically with the movement of the rack, sometimes it is observed that originally normal afterimages become anomalous and continue to be perceived abnormally, at least for some time. These findings can, of course, only be explained on the assumption of a change in correspondence . . . in several squint cases it is the only test which shows that normal correspondence is beginning to break down . . .

Such smooth correspondence shifts appear to be an exaggeration of sensory fusion shifts seen in normal observers, which, as noted above, can reach 5 degrees and can displace after-images. In AC, abrupt alternations in correspondence also occur, typically between the normal and an anomalous value. It may be postulated that a broadened distribution of disparity tunings is the cellular foundation of broad, smooth shifts in correspondence, while a bimodal distribution leads to alternation. Both behaviours can be dealt with by defining correspondence as the disparity tuning with the most activity, and then analyzing the probable responses of disparity detectors to the stimulation found in clinical tests of binocular vision. With a bimodal distribution of tunings and alternations of correspondence, the tendency for either state of correspondence to be dominant at the expense of the other may reflect mutual inhibition among disparity detectors of different tunings. This is the same disparity domain inhibition previously found necessary for globality in normal vision. Should this inhibition be weak, two states of correspondence could be expressed at once. This would be a cellular basis for monocular diplopia: two equally active tunings produce two valid sets of correspondence simultaneously. One monocular stimulus point will then have two localizations and be seen double.

Poor depth

When correspondence is variable, the loss of the capacity for depth discrimination is severe. The depth loss is out of proportion to the great amount of useful disparity processing accomplished in the service of single vision. Tschermak (1899) had a variable angle of anomaly similar to a V-syndrome, due in part to paresis. He described his inability to get any phenomenological depth from haploscopically presented disparity. With small strabismic angles and matching angles of correspondence anomaly, some stereoacuity may be observed. But the slope of the psychophysical discrimination function is only a third that of normal observers (Pasino, Maraini & Santori, 1963). The position that one can develop exaggerated fusional capability only at the expense of depth perception has been clearly stated by Pasino & Maraini (1966; see also Mateucci, Pasino & Maraini, 1967). Is this really so novel? I suggest it is only an extension of the depthless fusion seen in normal binocular vision. The range and ease of depthless fusion is extended because the scatter of receptive field disparities is greater, and, in the domain interaction model, because the range of disparity domain interactions is broader (Nelson, 1981).

Conclusion

Stereopsis remains a model system for pursuing the cellular basis of higher perceptual processes. Particularly in globality, the stimulus input, the output, and the processing which must intervene are now all well defined. In those aspects of stereopsis where neurophysiology has failed to provide cellular mechanisms, psychophysics has also failed clearly to specify and measure what behavior needs to be accounted for.

Stereopsis also remains a very demanding area of neurophysiological research. Stimulus position must be controlled to within a few minutes of arc, and some means must be found of measuring eye position with comparable accuracy. The use of free fixation in the awake animal is preferable, but brings other problems of operant control and objective fixation monitoring. Of course, the number of programmable stimulus channels will be double that required for most other experiments. The cellular basis for stereopsis was first discovered and most completely developed in Australia, in part because stable funding of the Australian National University permitted the long-term development of specialized instrumentation. Many years and discoveries later, there are still more unsolved problems in the cellular basis of stereopsis than solved ones. It was a fruitful endeavour and a good investment.

Acknowledgement

I thank Robert O'Shea, Robert Patterson, Jack Pettigrew and especially Gian Poggio for critical discussions

References

Alpern, M. & Hofstetter, H. W. (1948) The effect of prism on esotropia – a case report. *Am. J. Optom. Arch. Am. Acad. Optom.*, **25**, 80–91

Arditi, A., Kaufman, L. & Movshon, J. A. (1981) A simple explanation of the induced size effect. *Vision Res.*, **21**, 755–64

Bagolini, B. (1967) Anomalous correspondence: definition and diagnostic methods. *Doc. Ophthal.*, **23**, 346–98

Barlow, H. B., Blakemore, C. & Pettigrew, J. D. (1967) The neural mechanism of binocular depth discrimination. *J. Physiol.*, **193**, 327–42

Beloff, J. (1961) The stripe paradox. *Br. J. Psychol.*, **52**, 323–31

Bishop, P. O. (1970) Beginning for form vision and binocular depth discrimination in cortex. *The Neurosciences: Second Study Program*, ed. F. O. Schmitt, pp. 471–85. New York: Rockefeller University Press

Bishop, P. O. (1973) Neurophysiology of binocular single vision and stereopsis. In *Handbook of Sensory Physiology*, vol. VII/3, *Central Processing of Visual Information*, ed. R. Jung, pp. 255–305. New York: Springer

Bishop, P. O. (1975) Binocular vision. In *Adler's Physiology of the Eye: Clinical Application*, 6th edn, ed. R. A. Moses, pp. 558–614. St Louis: Mosby

Bishop, P. O., Coombs, J. S. & Henry, G. H. (1971) Responses to visual contours. Spatiotemporal aspects of excitation in receptive fields of simple striate neurons. *J. Physiol.*, **219**, 625–57

Bishop, P. O., Henry, G. H. & Smith, C. J. (1971) Binocular interaction fields of single units in the cat striate complex. *J. Physiol.*, **216**, 39–68

Blakemore, C. (1970) A new kind of stereoscopic vision. *Vision Res.*, **10**, 1181–99

Blakemore, C., Fiorentini, A. & Maffei, L. (1972) A second neural mechanism of binocular depth discrimination. *J. Physiol.*, **226**, 725–49

Braunstein, M. L. (1976) *Depth Perception through Motion.* New York: Academic Press

Clarke, P. G. H., Donaldson, I. M. L. & Whitteridge, D. (1976) Binocular visual mechanisms in cortical areas I and II of the sheep. *J. Physiol. (Lond.)*, **256**, 509–26

Crone, R. A. & Hardjowijoto, S. (1979) What is normal binocular vision? *Doc. Ophthalmol.*, **47**, 163–99

Cynader, M. & Regan, D. (1978) Neurones in cat parastriate cortex sensitive to the direction of motion in three-dimensional space. *J. Physiol.*, **274**, 549–69

Cynader, M. & Regan, D. (1982) Neurons in cat visual cortex tuned to the direction of motion in depth: effect of positional disparity. *Vision Res.*, **22**, 967–82

De Valois, R. L., Yund, E. W. & Hepler, N. (1982) The orientation and direction selectivity of cells in macaque visual cortex. *Vision Res.*, **22**, 531–44

Duwaer, A. L. (1983) Patent stereopsis with diplopia in random-dot stereograms. *Percept. Psychophys.*, **33**, 443–54

Efron, R. (1957) Stereoscopic vision. I. Effect of binocular temporal summation. *Br. J. Ophthalmol.* **41**, 709–30

Engel, G. R. (1970) An investigation of visual responses to brief stereoscopic stimuli. *Q. J. Exp. Psychol.*, **22**, 148–66

Epstein, W. & Daviess, N. (1972) Modification of depth judgment following exposures to magnification of uniocular image: are changes in perceived absolute distance and registered direction of gaze involved? *Percept. Psychophys.* **12**, 315–17

Fender, D. & Julesz, B. (1967) Extension of Panum's fusional area in binocularly stabilized vision. *J. Opt. Soc. Am.*, **57**, 819–30

ogel, W. C. & Tietz, J. D. (1979) A comparison of oculomotor and motion parallax cues of egocentric distance. *Vision Res.*, **19**, 1161–70

ammond, P. & MacKay, D. M. (1975) Differential responses of cat visual cortical cells to textured stimuli. *Exp. Brain Res.*, **22**, 427–30

vson, M. T., Julesz, B. & Fender, D. H. (1983) Eye movements and neural remapping during fusion of misaligned random-dot stereograms. *J. Opt. Soc. Am.*, **73**, 1665–73

hua, D. E. & Bishop, P. O. (1970) Binocular single vision and depth discrimination. Receptive field disparities for central and peripheral vision and binocular interaction on peripheral single units in cat striate cortex. *Exp. Brain Res.*, **10**, 389–416

to, H., Bishop, P. O. & Orban, G. A. (1981) Binocular interaction on monocularly discharged lateral geniculate and striate neurons in the cat. *J. Neurophysiol.*, **46**, 932–51

nguet-Higgins, H. C. (1982) The role of the vertical dimension in stereoscopic vision. *Perception*, **11**, 377–86

ateucci, P., Pasino, L. & Maraini, G. (1967) Déséquilibre moteur et vision binoculaire anormale dans l'espace. *Doc. Ophthal.*, **23**, 399–424

aunsell, J. H. R. & van Essen, D. C. (1983) Functional properties of neurons in middle temporal visual area of the macaque monkey. II. Binocular interactions and sensitivity to binocular disparity. *J. Neurophysiol.*, **49**, 1148–67

ayhew, J. E. W. (1982) The interpretation of stereo-disparity information: the computation of surface orientation and depth. *Perception*, **11**, 387–407

ayhew, J. E. W. & Frisby, J. P. (1982) The induced effect: arguments against the theory of Arditi, Kaufman & Movshon (1981). *Vision Res.*, **22**, 1225–8

ayhew, J. E. W. & Longuet-Higgins, H. C. (1982) A computational model of binocular depth perception. *Nature*, **297**, 376–9

lson, J. I. (1975) Globality and stereoscopic fusion in binocular vision. *J. Theoret. Biol.*, **49**, 1–88

lson, J. I. (1977) The plasticity of correspondence: after-effects, illusions and horopter shifts in depth perception. *J. Theoret. Biol.*, **66**, 203–66

lson, J. I. (1981) A neurophysiological model for anomalous correspondence based on mechanisms of sensory fusion. *Doc. Ophthal.*, **51**, 3–100

lson, J. I. (1985) The cellular basis of perception. In *Models of the Visual Cortex*, ed. D. Rose & V. Dobson. New York: Wiley, pp. 108–122

lson, J. I., Kato, H. & Bishop, P. O. (1977) Discrimination of orientation and position disparities by binocularly activated neurons in cat striate cortex. *J. Neurophysiol.*, **40**, 260–83

kara, T., Bishop, P. O. & Pettigrew, J. D. (1968) Analysis of retinal correspondence by studying receptive fields of binocular single units in cat striate cortex. *Exp. Brain Res.*, **6**, 353–72

gle, K. N. (1964) *Binocular Vision*. New York: Hafner

gle, K. N., Martens, T. & Dyer, J. (1967) *Oculomotor Imbalance in Binocular Vision and Fixation Disparity*. Philadelphia: Lea & Febiger

Ono, H. & Comerford, J. (1977) Stereoscopic depth constancy. In *Stability and Constancy in Visual Perception: Mechanisms and Processes*, ed. W. Epstein, pp. 91–128. New York: Wiley

O'Shea, R. P. (1983) *Spatial and Temporal Determinants of Binocular Contour Rivalry*. Dissertation, University of Queensland

O'Shea, R. P. & Crassini, B. (1983) Vertical disparities lead to the 'induced effect'. *Vision Res.*, **23**, 113–14

Pasino, L. & Maraini, G. (1966) Area of binocular vision in anomalous retinal correspondence. *Br. J. Ophthalmol.*, **50**, 646–50

Pasino, L., Maraini, G. & Santori, M. (1963) Caratteristiche della visione binoculare anomala nellámbiente. III. Acutezza visiva stereoscopica. *Ann. Ottal.*, **89**, 1005–11

Pastore, N. (1964) Induction of a stereoscopic depth effect. *Science, N.Y.*, **144**, 888

Patterson, R. & Fox, R. (1984) The effect of testing method on stereoanomaly. *Vision Res.*, **24**, 403–8

Perlmutter, A. L. & Kertesz, A. E. (1978) Measurement of human vertical fusional response. *Vision Res.*, **18**, 219–23

Pettigrew, J. D. (1965) *Binocular Interaction on Single Units of Striate Cortex of the Cat*. B. Sc.(Med). Thesis, University of Sydney

Pettigrew, J. D. (1973) Binocular neurones which signal change of disparity in area 18 of cat visual cortex. *Nature New Biology*, **241**, 123–4

Pettigrew, J. D. & Dreher, B. (1982) Parallel processing of binocular disparity in the cat's geniculo-cortical pathway. *Soc. Neurosci. Abstracts*, **8**, 810

Pettigrew, J. D., Nikara, T. & Bishop, P. O. (1968) Binocular interaction on single units in cat striate cortex: simultaneous stimulation by single moving slit with receptive fields in correspondence. *Exp. Brain Res.*, **6**, 391–410

Poggio, G. F. (1979) Mechanisms of stereopsis in monkey visual cortex. *TINS*, **2**, 199–201

Poggio, G. F. (1980) Neurons sensitive to dynamic random-dot stereograms in Areas 17 and 18 of rhesus monkey cortex. *Neurosci. Abstr.*, **6**, 672

Poggio, G. F. (1984) Processing of stereoscopic information in primate visual cortex. In *Dynamic Aspects of Neocortical Function*, ed. G. M. Edelman, W. E. Gall & W. M. Cowan, pp. 613–35. New York: Wiley

Poggio, G. F. & Fischer, B. (1977) Binocular interaction and depth sensitivity in striate and prestriate cortex of behaving rhesus monkey. *J. Neurophysiol.*, **40**, 1392–405

Poggio, G. F. & Poggio, T. (1984) The analysis of stereopsis. *Ann. Rev. Neurosci.*, **7**, 379–412

Poggio, G. F. & Talbot, W. H. (1981) Mechanisms of static and dynamic stereopsis in foveal cortex of the rhesus monkey. *J. Physiol.*, **315**, 469–92

Ramachandran, V. S. (1976) Learning-like phenomena in stereopsis. *Nature*, **262**, 382–4

Ramachandran, V. S. & Braddick, O. (1973) Orientation-specific learning in stereopsis. *Perception*, **2**, 371–6

Reading, R. W. (1984) Horopter shifts due to a magnification change. *Am. J. Optom. Physiol. Optics*, **61**, 310–17

Regan, D. & Beverley, K. I. (1973a) Some dynamic features of depth perception. *Vision Res.*, **13**, 2369–79

Regan, D. & Beverley, K. I. (1973*b*) The dissociation of sideways movements from movements in depth: Psychophysics. *Vision Res.*, **13**, 2403–15

Richards, W. J. (1951) The effect of alternating views of the test object on vernier and stereoscopic acuities. *J. Exp. Psychol.*, **42**, 376–83

Richards, W. (1970) Stereopsis and stereoblindness. *Ex. Brain Res.*, **10**, 380–8

Richards, W. (1971) Anomalous stereoscopic depth perception. *J. Opt. Soc. Am.*, **61**, 410–14

Riguidiere, F. (1975) Fusion binoculaire et localisation spatiale de mires verticales et horizontales de fréquences spatiales differentes. *Vision Res.*, **15**, 931–8

Rodieck, R. W. (1971) Central nervous system: afferent mechanisms. *Ann. Rev. Physiol.*, **33**, 203–40

Rønne, G. & Rindziunski, E. (1953) The diagnosis and clinical classification of anomalous correspondence. *Acta Ophthal mol.*, **31**, 321–45

Ross, J. E. (1983) Disturbance of stereoscopic vision in patients with unilateral stroke. *Behav Brain Res.*, **7**, 99–112

Schiller, P. H., Finlay, B. L. & Volman, S. F. (1976) Quantitative studies of single-cell properties in monkey striate cortex. II. Orientation specificity and ocular dominance. *J. Neurophysol.*, **39**, 1320–33

Toyama, K. & Kozasa, T. (1982) Responses of Clare–Bishop neurones to three dimensional movement of a light stimulus. *Vision Res.*, **22**, 571–4

Tschermak, A. (1899). Ueber anomale Sehrichtungsgemeinschaft der Netzhäute bei einem Schielenden. *von Graefes Arch. Ophthal.* **47**, 508–550 and Plates XXVI, XXVII

van der Meer, H. C. (1978) Linear combinations of stereoscopic depth effects in dichoptic perception of gratings. *Vision Res.*, **18**, 707–14

Von der Heydt, R., Adorjani, C., Hänny, P. & Baumgartner, G. (1978) Disparity sensitivity and receptive field incongruity of units in the cat striate cortex. *Exp. Brain Res.*, **31**, 523–45

Von der Heydt, R., Hänny, P., Dürsteler, M. R. & Poggio, G. F. (1982) Neuronal responses to stereoscopic tilt in the visual cortex of the behaving monkey. *Invest. Ophthal. Vis. Sci.*, **22**(3), 12 (abstract)

Wallach, H. & Zuckerman, C. (1963) The constancy of stereoscopic depth. *Am. J. Psychol.*, **76**, 404–12

Westheimer, G. (1978) Vertical disparity detection: is there an induced size effect? *Invest. Ophthal. Vis. Sci.*, **17**, 545–51

Wheatstone, C. (1838) Contributions to the physiology of vision. I: On some remarkable, and hitherto unobserved, phenomena of binocular vision. *Phil. Trans.*, 371–94 and Plates X, XI

Wright, J. C. & Kertesz, A. E. (1975) The role of positional and orientational disparity cues in human fusional response. *Vison Res.*, **15**, 427–30

Zeki, S. M. (1974) Cells responding to changing image size and disparity in the cortex of the rhesus monkey. *J. Physiol.*, **242**, 827–41

CONTRIBUTORS

Lindsay M. Aitkin

My approach to neurophysiology has been particularly influenced by two teachers – Peter Bishop, then at the University of Sydney, and Jerzy Rose of the University of Wisconsin. I have always been impressed by the enduring interest of both these scientists in experimental neurophysiology and their belief in finding solutions to 'how?' questions before approaching 'why?' problems. I owe a particular debt to Peter Bishop who, in my postgraduate days, provided valuable advice about measurement in neurophysiology and generally encouraged my research in auditory physiology.

The intellectual atmosphere in the Department of Physiology at the University of Sydney in the mid-1960s was very stimulating for a postgraduate student. Peter Bishop and Bill Levick had for some years been applying quantitative methods to visual neurophysiology, and spike average techniques were being used with considerable imagination in retinal ganglion cell physiology by Bob Rodieck and Jonathan Stone. Jon and Sandra Hansen (now Sandra Rees) were also developing neuroanatomical methods for studying the retina and its connections with the lateral geniculate body, while Liam Burke and Ann Sefton were examining inhibitory circuitry in the rat lateral geniculate. Towards the end of my period at Sydney, Jack Pettigrew, Peter Bishop and Tos Nikara were starting to unravel the mechanism of binocular vision. These influences, in one way or another, were felt in the collaboration that was taking place between Bill Webster, Colin Dunlop and myself in our studies of the medial geniculate body of the cat. I certainly feel that one of the best things that ever happened to me was to go from Sydney to Wisconsin to work with Rose – and this was largely due to Peter's advice.

Horace Barlow

I was persuaded by William Rushton that physiology is the most fruitful scientific approach to the brain, and started research on vision under E. D. Adrian in 1947. Though I did, in those early days, succeed in recording from single retinal ganglion cells of the frog, it was not until I collaborated with Bill Levick in the 1960s that I learned how the professionals – trained by Peter Bishop – did it. A few years later Jack Pettigrew, from the same stable, showed Colin Blakemore and me how to record from cortical neurons. The finishing touches to my training came when I visited Peter's lab in Canberra for a few months in 1970, but I am afraid I have done very little single-unit recording since then, partly because it is so difficult to maintain the very high standards of evidence and technique that his laboratory has set!

Jean Bullier

Like a number of other neuroscientists, my original training was in electrical engineering which I studied in Paris at the Ecole Supérieure d'Electricité. I then spent 4 years in the USA at the Duke University where I obtained a Ph.D. in Biomedical Engineering. In my research project, a quantitative study of receptive-field properties of LGN neurons, I was already influenced by the elegant studies of Bishop and Henry on cortical neurons. After my

Ph.D., I went to Canberra to do some research on the visual cortex. I worked with Geoff Henry in Lab. J, which was situated on the ground floor whereas the rest of the department mostly occupied the fourth floor. During the 3 years I spent in Canberra, I viewed the Physiology department as a great battleship, with the machine room in Lab. J and the top deck on the fourth floor. At the end of the corridor on the top deck was situated the captain's cabin where the

officers were invited for lunch every day. It is partly during these lunches that I learnt to appreciate Professor Bishop's remarkable personality. What impressed me most was the great enthusiasm that he showed for everything which touched research. It was extremely encouraging for me, a young researcher, to see somebody of his age and reputation who was so energetic and ready to back you up if he was convinced that you were doing good work.

After three very pleasant and fruitful years spent in Australia, I decided to return to France but to stay in the visual cortex field. I have now been working in Lyon for 3 years and memories of Canberra are slowly fading. Some things, however, never disappear. For example, when I feel depressed after an experiment which did not work, I can still hear Professor Bishop saying, 'There is always another cat'.

W. Burke

Liam Burke initially trained as a pharmacist. After National Service in the Royal Air Force he took a B.Sc. in special physiology at University College, London. He obtained his Ph.D. under Sir Bernard Katz in the Biophysics Department of UCL and was then appointed to the staff of the Department of Physiology at the University of Sydney in 1956. He collaborated with Peter Bishop extensively in those early years on the physiology of binocular interaction and the lateral geniculate nucleus. He was remained in the Department ever since, succeeding Peter Bishop in one of the two Chairs of Physiology in 1967.

B G. Cleland

Though I received all my education up to graduate level in the Sydney environment, it was at Northwestern University in Chicago that I first met Peter Bishop, shortly after I had arrived in the USA in 1963. At the time, Peter was making a cross-country tour during a period of study leave at MIT. That I had not met

him previously, or even heard of him, is hardly surprising, as I was only just developing my interests in physiology and was quite unaware of the many-faceted field of vision.

My first real association with Peter Bishop was in late 1968 when I went to Canberra to work with Bill Levick. By this time the Physiology Department was rapidly becoming established in its new direction of visual neurophysiology. Three of what were eventually to become seven vision laboratories were just nearing completion. Central to each was a stereotaxic holder mounted on a gun turret, reflecting Peter's career in the navy. However these were not second-hand turrets from a disposal store, but were designed by Peter and constructed in the John Curtin School workshops so that they could lay a cat for any desired trajectory with pinpoint accuracy.

Due to our different areas of immediate interest I never had the opportunity of working with Peter. However it was a rare day indeed that the academic members of the Department did not join him in his office for lunch. In 14 years of lunches one gets to know a colleague very well, even a Head of Department, and there are many anecdotes that could be recounted. What was always clear was Peter's enthusiasm for science, and many were the discussions on experimental methods and techniques. Visual neurophysiology was advancing rapidly during these years, in the mammal, probably nowhere more so than in Peter's department. [Editor's note: Cleland's debt to the superb engineering in the Bishop labs is evident in his technical achievement, with Levick and Dubin, of the first simultaneous recording from a lateral geniculate neuron and the retinal ganglion cell driving it.]

H. Day

y association with Peter
shop goes back to the
id-1950s when he had
cently been appointed to the
hair of Physiology in the
niversity of Sydney and I,
ss impressively, to a
ctureship in psychology. I
d just completed a Ph.D.
d returned from England to
y first tenured job.

The 1950s were not
ntage years for research in
ustralia. The universities
ere overcrowded and
derstaffed, teaching duties
ere onerous and took
ecedence over scholarship,
d there was a dearth of
uipment, materials and
sistance for research.
oreover, there was little or
encouragement to undertake
search; funds were difficult
impossible to come by from
her internal or external
urces.

It was in this bleak
imate that Peter Bishop
tablished one of the few rich
d vigorous research groups,
the University of Sydney.
he Brain Research Unit – I
ell remember that brave
gend emblazoned on a door
the Department of
ysiology – became an
ample to me of how research
uld and should be
nducted. Indeed, Peter
shop's group and that of
rrin Riggs in which, a few
ars later, I worked at Brown
niversity served as models
r the research units I later
lped to establish at Monash.

Although Peter Bishop
d I have not worked
gether in research – he is an
ashamed neurophysiologist

and I an unabashed
experimental psychologist – we
do share an interest in how the
world is perceived visually. He
has been concerned mostly
with the neural mechanisms
and I with perceptual
phenomena and the
higher-order processes
associated with them. He gets
his data through the sharp end
of a microelectrode and I from
the mouthparts of human
subjects. Nevertheless, because
of our shared interest in vision,
our paths have frequently
crossed. We were, as I have
mentioned, in the University
of Sydney at the same time, we
have met on numerous
occasions at assorted symposia,
colloquia and conferences, and
we were at the same time
members of that most convivial
of clubs, the Australian
Research Grants Committee.

I should also say that
Peter Bishop was
extraordinarily generous some
years ago in allowing one of
my protégés, Bill Webster, to
work in his department in
Sydney when I was unable to
provide the proper facilities for
what he wanted to do. That
generosity has long since payed
handsome dividends in the
form of the highly productive
Neuropsychology Unit that
Bill Webster has built up at
Monash.

It is my view that a
sense of commitment,
readiness to encourage, and the
maintenance of high standards
by senior academics is as
important for the sustenance of
good research as modern
laboratories, good equipment
and liberal funding. These less
tangible catalysts for
high-grade research have been
provided in full measure by
Peter Bishop for a long time
and served as an inspiration to
us all.

Bogdan Dreher

The first person from whom I
heard about Peter O. Bishop
was Vlod Kozak who had
worked during the early 1960s
in Bishop's laboratory in

Sydney. According to Vlod,
the early 1960s in Bishop's
laboratory were times full of
complex mechanical
contraptions, very solid
stereotaxic apparatus, cat
schematic eyes, posterior nodal
distances, retinotopic maps of
the cat's lateral geniculate
nucleus, retinal wholemounts
and all sorts of very precise
measurements. There were also
dynamic, sometimes rather
mischievous, interactions
between George Vakkur, Bob
Rodieck, Bill Levick and Vlod
and, later on, Jonathan Stone.
Peter Bishop stood above the
battle and inspired everyone by
his hard work and his
enthusiasm towards the work
and ideas of his students and
collaborators. He was seen by
the people working in his
laboratory as a demanding but
benevolent father-figure.

Then, in September
1966, Peter Bishop appeared in
person at the Nencki Institute
of Experimental Biology in
Warsaw. He turned out to be
very unassuming but at the
same time full of enthusiasm
for visual neurophysiology and
prepared to listen to fluent
presentations (in broken
English) of a number of
half-baked ideas. On the
second day of his visit, Peter
Bishop delivered a lecture
describing his recent work
done in collaboration with his
B.Sc. (Med.) student Jack
Pettigrew, and a visitor from
Japan, T. Nikara. The terms
'retinal correspondence',
'receptive field disparities',
'binocular interactions of

single units in the cat's striate
cortex' sounded to me like
charming musical notes
inviting me into the world of
Panum's fusional area. I knew
that, invited or not, I would
end up in Bishop's laboratory.
To my great delight,
admittedly after some discrete
prodding from Vlod, Peter
Bishop invited me to work as a
postdoctoral fellow in his lab.

It was, however, easier
said than done – especially when
you happen to live in one of the
countries which permanently
have some 'temporary
difficulties'. At the very time I
was supposed to go abroad these
chronic temporary difficulties
were somehow exacerbated. But
temporary or not, exacerbated or
not, in Peter Bishop's approach
to life the difficulties have to be
overcome. And so they were.

In September 1969, to
the great relief of many people
in central Europe, I landed in
Canberra and imposed myself
on an unsuspecting, peaceful,
hard-working Australian visual
community.

Peter Bishop and his
wife Hilare prevented my
complete breakdown by
inviting me regularly to their
beautiful home and politely
listening to my stories relating
elements of European
geography and history to the
properties of simple and
complex cells in the visual
cortex of the midpontine
pretrigeminal preparation.

During our regular
lunchtime discussions in Peter
Bishop's office all the problems
that absorbed each research
group were dragged out by
Peter Bishop into the broad
daylight. One characteristic of
Peter Bishop constantly
emerged throughout all his
prodding and questioning.
Behind Peter's rather
conservative demeanour there
is a man who loves unorthodox
approaches to new problems
and all those who are
successful in getting answers
are loved and respected by
him.

Geoffrey H. Henry

Unlike Gaul my career is clearly divisible into two rather than three. The first half began after I graduated in science from Melbourne University on my twentieth birthday. I then began a degree in optometry which was to lead me into professional practice for 15 years. During this time, 2 years after graduating in optometry, I became Chairman of the Victorian College of Optometry, something that could not happen to such a raw recruit today with the current plethora of academic optometrists in Victoria. During the next 13 years of my chairmanship the present college building was constructed, the optometry course entered Melbourne University, and I took out a higher degree and won a Churchill Fellowship that was to take me to America and to England to work with Russel DeValois and William Rushton. In 1960 I had started into the psychophysics of colour vision but the neurophysiology carried out during the visit to the DeValois laboratories brought the change in direction that was to usher in the second phase of my career.

In January 1968 I came to Canberra to begin a working association with Peter Bishop that was to continue for the next 8 years. Over this period we worked together on all projects carried out in the world-renowned Lab. A, famous for the intricacy of its electronic and mechanical design and also, I hope, for the quality of the published work that emanated from its doors. It might be thought that I came under Peter Bishop's influence well after the critical period and that I would be immune to his moulding influence. It is not possible, however, to work day and night with such a vital character and not receive some imprinting. While I'm too close to judge the worth of our scientific association I am acutely aware of the aspects of the Bishop character that most contributed to our collaboration. These were his intellectual honesty, the intensity with which he addresses scientific problems and, perhaps above all, his capacity to debate at an impersonal level and to pursue every argument, no matter how controversial, in his quest for the scientific truth.

As I look back, I certainly cannot regret the change that brought me to visual neuroscience and for the part he played in this transition I must thank Peter Bishop. For his contribution, first to our scientific association, and then to our friendship, I also express my deepest appreciation.

Austin Hughes

Austin Hughes is Director of the National Vision Research Institute of Australia (N.V.R.I.) and Professor in the University of Melbourne. Educated at The Liverpool Institute High School for Boys, with two of The Beatles, he read Physiology at Pembroke College, Oxford, and did postgraduate work in 'Engineering in Medicine' at the Electrical Engineering Department, Imperial College, London, where his interest in information theory began. He was introduced to vision research at Edinburgh where he pursued his doctoral studies as a member of David Whitteridge's 'M.R.C. Group for Research into the Central Mechanisms of Vision' and as an Assistant Lecturer in the Department of Physiology at the Medical School. In 1968 he returned to Oxford as Demonstrator in Physiology and developed a marked ecological emphasis in his work. In 1972 he met Peter Bishop and took up an invitation to visit Canberra. He began with a field trip to examine marsupial eyes in Papua–New Guinea but, in response to the leanings of his colleagues, his work became progressively more laboratory oriented and obsessive. In 1982 he resigned his post as Senior Research Fellow at the John Curtin School of Medical Research and moved to Melbourne to take over N.V.R.I. He is establishing a multidisciplinary group working on optics and retinal organisation, learning to read a balance sheet, fostering the growth of the Institute, and working to ensure the success of the newly founded National Vision Research Foundation. His prime interests are mammalian optical and retinal organisation and their relation to lifestyle.

John Irwin Johnson

I had done my Ph.D. thesis in psychology on comparative studies of visual discrimination in mammals and followed my interest in brain evolution by learning neurobiology at the University of Wisconsin Neurophysiology Laboratory doing postdoctoral work with Wally Welker and Clinton Woolsey. They were assembling their great collection of mammalian brains, and our experiments were mapping somatic sensory systems. Since the collection lacked marsupials, and since we had done no work in visual systems, I wrote to Peter Bishop about the possibility of his sponsoring me as a Fullbright scholar to come to his laboratory to collect

marsupial brains and to acquaint myself with current developments in visual neurobiology. He welcomed me with his customary enthusiasm and generosity, and was of great assistance in the achievement of both my goals. After 14 months I departed with the world's most extensive set of marsupial and monotreme brains and thoroughly conversant with visual neuroscience, ready to establish my own laboratory at Michigan State University. My experience in Sydney in Prof. Bishop's department also left me with continuing productive relationships across the years with several of the remarkable array of people working there at the time – Lindsay Aitkin, Liam Burke, Jack Pettigrew, Bob Rodieck, Ann Sefton and Jonathan Stone; I have had several additional return trips to Australia and two of my students, John Haight and Le Weller, have taken up Australian careers.

Jon Kaas

Being a modest fellow, Jon Kaas did not provide any biographical details, so these have been provided by one of the editors (KJS) who worked with Jon at the University of

isconsin in 1971–73. Jon
tained a Ph.D. at Duke
niversity in North Carolina,
nder the guidance of Irving
iamond, with work on the
ditory cortex. At Duke
niversity, and subsequently
the University of Wisconsin
Madison, Jon developed his
terest in the processing of
nsory information by the
rebral cortex and carried out,

ith John Allman, a detailed
apping study of the visual
rtex of the own monkey.
lman and Kaas showed that
e visual cortex in the owl
onkey contains quite a
mber of representations of
e visual field; pioneering
dies by Clinton Woolsey
d previously described
imary and secondary
nsorimotor, visual and
ditory areas. At Wisconsin
n collaborated also with Ray
uillery in describing the
sual system of 'Midwestern'
amese cats. In 1973 Jon
oved to Vanderbilt
niversity in Nashville,
ennessee, where he has
aintained an interest in
ocessing of information by
e cerebral cortex. In recent
ars a successful collaboration
tween Jon's laboratory in
ashville and that of Michael
erzenich in San Francisco
s shown that the primary
matosensory cortex in
imates has three or four
parate maps of the body
rface, each receiving input
om a different class of
ceptor. Jon's acquaintance
th Peter Bishop commenced

on Lord Howe Island at the
Festschrift where they met for
the first time; Jon has also
attained a sort of second-hand
acquaintance with Peter Bishop
through his collaborations with
a number of people from
Bishop's laboratory – Ken
Sanderson, Murray Sherman
and, most recently, in 1984,
Jack Pettigrew.

J. Kulikowski

I first met Peter Bishop in
1966. On that occasion he was
delivering a lecture on his
team's new discovery that
visual cortical cells display
interocular disparity. This
story (published later in 1968
with Pettigrew and Nikara)
made a lasting impression on
me, and no wonder – it now
forms the core of our present
knowledge of the neural basis
of binocular vision.
    Peter Bishop visited
Manchester in 1973 and there
we had the chance of planning
our joint research. My first
visit to Canberra was a great
experience and fruitful study,
since Peter Bishop has a
unique talent for creating
good, congenial conditions for
everybody working with him,
by introducing any newcomer
to the intricacies of his

laboratory without destroying
the individual approach of a
visitor. Peter Bishop, through
his own example, influenced
his associates to maintain
certain standards of accuracy
of measurements, by insisting
that, within reason, greater
than the minimum does not do
any harm; on the contrary, it
provides a reference framework

to discover some regularities in
the data which with poorer
accuracy could pass unnoticed.
The full advantage of this
approach came out later when
we were able to prepare six
papers. Writing these papers
was also an unforgettable
experience of a struggle for
clarity of expression in which
Peter Bishop showed his
insight and analytical power.
Ultimately this is why so many
people kept returning to work
with him and have been
pleased to welcome him to
their laboratory since he
retired.

James W. Lance

Jim Lance was introduced to
neurophysiological research by
Peter Bishop in 1951 and
completed an M.D. thesis on
the pyramidal tract in the cat
under his guidance. He later

continued postgraduate studies
in medicine and neurology at
the Postgraduate Hospital,
Hammersmith, The National
Hospital for Nervous Diseases,
Queen Square, London, and
the Massachusetts General
Hospital, Boston. He was
appointed Chairman of the
Department of Neurology, the
Prince Henry Hospital, Sydney
in 1961 and Professor of
Neurology at the University of
New South Wales in 1975. His
research interests have
included the control of
movement, disorders of
movement (such as spasticity,
Parkinson's disease, myoclonus
and choreoathetosis), and the
mechanism and management of
headache.

W. R. Levick

I was heading into a surgical
career but for some timely
words of wisdom from the
'master of subtlety'. 'Why be

a little fish in the big pond of
medicine', said he, 'when you
could be a big fish in the little
pond of research?'. How could
he have imagined an appeal to
vanity might work whereas an
appeal to logic might not? Of
course, there was a good deal
more substance to the appeal
in terms of the opportunities
for research and it was also a
happier time for research in
Australia generally. The
freedom of the training
environment Peter Bishop
created at Sydney was
breathtaking: you could really
learn the art of making things
work on your own. It may not
have been the fastest way to
learn but the experience is
coded indelibly. He, now the
'master of strategy', also
steered me imperceptibly into
another kind of experience at
Cambridge. Here, the magic of
physiological investigation was
woven with Meccano sets,
mirrors, aircraft glue and
imagination. The seeming
flimsiness was more apparent
than real, since this style of
work subsequently survived a
transatlantic transplantation to
that uniquely heady
environment on the eastern
side of San Francisco Bay
where research was conducted
as a business as well as a
pleasure. Later, with the
bravado of youth, I flew in the
face of the Australian
custodians of my Fellowship in
order to stay on at Berkeley,
but again it was the shrewd

parting words of Peter Bishop that started a slow fuse culminating in my pilgrimage home only 2 years later. I joined him once more in a totally unexpected venture at that Mecca of Australian neurophysiology in Canberra. What an experience it has been! Only now after his retirement do I really appreciate the true qualities behind his success: he had that happy knack of spotting where help would be most effective and he would give it unstintingly. It was not so much the amount of help but the notion of giving a 'fair go' that generated the inspiration that always surrounded him.

D. M. MacKay

My interest in the visual system goes back to the late 1940s. After graduating in Natural Philosophy (Physics) at St Andrew's University in 1943, I spent 3 years on radar research. When I moved in 1946 to lecture in physics at King's College, London, my wartime experience led me to work on electronic computing and on the foundations of what was later known as Information Theory. Partly in reaction to Norbet Wiener's (1948) comparison of the brain to a digital computer, I began in 1949 to explore the potentialities of mechanisms utilising non-digital principles for the control of transition-probabilities in stochastic processes, such as neural activity seemed to be. This led to some novel ideas about the way in which the visual world could be

represented internally, not by an explicit 'neural image' but implicity by the matrix of transition-probabilities or 'state of conditional readiness' set up in the organism to match the demands of the visual input.

After spending 1951 in the USA as a Rockefeller Fellow I decided to make sensory communication my main field and set up a small research group within the physics department at King's. In 1960 Keele University gave me the opportunity to build an interdisciplinary Research Department of Communication and Neurosciences, combining the approaches of neurophysiology, psychophysics and communications engineering in the study of sensory systems, particularly vision and hearing.

Peter Bishop and I have never worked together apart from a few lively and enjoyable days in 1972 in Canberra when he allowed me to add some dynamic visual noise stimuli to the more orthodox visual diet of his experimental cats. From our first meeting in the early 1960s, however, Peter has been one of the most stimulating of my neurophysiological friends, always game for an argument in a good scientific spirit and on the basis of good solid data. When I found that the ceiling of his Canberra laboratory was decorated with metre-marks to facilitate placement of visual stimuli at exact distances from the eye of the experimental animal, this seemed exactly in keeping with the image I had of him as a master craftsman for whom only the best scientific standards were good enough. It is a pleasure to join in congratulating Peter on his remarkable achievement, not only as a pioneering individual, but also as the inspiring leader of one of the brightest teams of visual neurophysiologists to adorn the scientific firmament in our time.

Tetsuro Ogawa

Tetsuro Ogawa of Akita University and Hiroshi Kato of Yamagata University represent two different phases in the influence of Bishop and his laboratory on overseas students of the visual system. Both were trained initially at the Tohoku Medical School, renowned for its neurophysiology of the visual system.

Ogawa worked with Bishop in the early years of the Brain Research Unit at Sydney when Levick, Kozak and Rodieck were carrying out the first recordings from single neurons in the lateral geniculate nucleus (LGN) as natural visual stimuli were moved about in front of the cat. At that time, Ogawa made the first demonstration at the single-neuron level of the potency of extravisual inputs on synaptic transmission through the LGN. The disinhibitory input from the mesencephalic reticular formation which he discovered is an important feature of the 'circuit diagram' of the LGN.

Kato worked with Bishop during the Canberra phase, when the focus of attention was the detailed operation of visual cortical neurons. In two separate bursts of activity, first with Bishop and J. I. Nelson and later with Bishop and Orban, Kato produced two monuments in the detailed description of visual cortical receptive fields. The first concerned a close comparison between positional and orientational disparities of binocular receptive fields. The

second provided the most comprehensive picture so far of the end-stopped neurons (previously hypercomplex cells and, in the process, completely changed our thinking about the significance of this class of neurons. Rather than being a minor subclass, end-stopped neurons were shown by Kato's work to be a large fraction of all cells present. Previous work had missed many of them because the power of the end-zone inhibition and the tiny size of the fields ensured that they would be silenced by the usual exploratory stimuli. The important role played by such neurons is covered in more detail in Section VI.

Guy A. Orban

Guy A. Orban was reborn scientifically in Canberra, Australia. After a Ph.D. obtained at the Katholieke Universiteit te Leuven, I came to Australia in autumn 1975 to do a short postdoc with Peter Bishop, then the Professor of Physiology at the John Curtin School of Medical Research. Coming from a country where visual science and neuroscience were hardly developed to such a sophisticated laboratory as Lab. A, my stay at ANU was bound to be extremely fruitful.

Not only did we finish a large amount of experimental work, but I learned from Peter Bishop several important human and scientific lessons.

Working with Peter Bishop and Hiroshi Kato, I learned to give precise – i.e. quantitative – descriptions of visual cortical properties and

esign and carry out
europhysiological tests to
nravel the receptive field
nechanisms responsible for
hese properties. At a more
eneral level Peter Bishop
howed me how important was
he link between
europhysiology and visual
erception as well as how
uch more efficient could be a
esearch effort focussing on a
ngle theme. After my return
o Leuven, where I am
urrently Professor at the
Medical School, I have tried to
pply these lessons to the
tudy of visual cortical
nechanisms involved in
notion perception.

The many discussions
round the lunch table, with
'eter Bishop, Bill Levick,
\ustin Hughes, Hiroshi Kato,
erry Nelson, Barrie Frost and,
t the end, Brian Cleland,
nade me realize the
mportance of teamwork to
olve the intricate problems
ddressed in neuroscience.
oining such a prestigious
am, I was very proud to be
dopted so quickly by my
\ustralian 'family' and it was
great occasion for me to be
eunited with them at Lord
Howe Island.

. Peterhans

had the opportunity to spend
years, from 1978 to 1980, in
eter Bishop's laboratory in
he Department of Physiology
Canberra. During that time
e studied, together with
. M. Camarda, the correlation
etween the responses in cat

striate cortex to a stationary
flashing bar and to moving
stimuli. We examined
quantitatively the spatial
relation between the static-field
plots and the response profiles
to moving bars and edges.
These results were further
analysed with regard to the
property of direction selectivity
of cortical cells.

My present work builds
on the results of the study with
bars and edges carried out in
Canberra. I thank Peter Bishop
for giving me the opportunity
to work in his laboratory and
for his continuous
encouragement and support.

John D. Pettigrew

I was not inspired by Peter
Bishop's lectures in
neurophysiology, much
preferring Jack Eccles'
stimulating overgeneralisations
at the time. But in that first
year in his laboratory (1965),
Peter enabled me to achieve a
sense of fulfilment which had
been all but lacking in the
previous 4 years of swallowing
medical minutiae and
regurgitating them for examiners
at each year's end. It would be
wrong to say that we got on
well that year, since there were
differences of opinion and my
mountaineering activities
meant that I was in the
laboratory rather fitfully in the
early months. Indeed, I was
passed over for the more
glamorous project in the
offering, which involved
spinning the stereotaxic
instrument, cat, microdrive
and all, upside down inside a

giant ball-race while recording
from a single orientation-
selective visual cortical neuron.
Nevertheless, Peter gave me
every encouragement in a
superbly equipped laboratory
staffed with helpful and
competent assistants such as
Margaret Ackary (now Stone)
who taught me numerous
laboratory skills with which I
can surprise students and
postdocs even today, Cyril Mears
who helped me build my own
low-power Risley prism, Don
Larnach and Bob Tupper who
designed beautiful equipment
such as the mirror-galvo visual
stimulator.

My fellow B.Sc. (Med.)
student Gedis Grudzinskas,
golden-haired boy at the time,
was given the stereotaxic inside
the ball-race and I was given a
Risley bi-prism to superimpose
binocular receptive fields. As
luck would have it, the
proposed vestibulovisual
interaction behind the ball-race
experiment proved to be rather
elusive, although Gedis did
introduce nitrous oxide
anaesthesia to visual physiology
that year in his search for a
drug regimen more conducive
to vestibular input to the
cortex. Unexpectedly, my
project with the Risley
bi-prism turned up
disparity-selective binocular
neurons and I soon found
myself being invited to the
USA on the strength of my
B.Sc. (Med.) thesis in which I
described them. That I was
given Peter's blessing to accept
the invitation and that Peter
had set the stage for me to
make the finding in the first
place, both reflect one of his
key qualities in the nurture of
a young scientist: his respect
for, and confidence in, their
personal autonomy.

The trip overseas in
1966 changed and complicated
my life, not to mention that of
others, for three reasons. I
returned to complete my final
2 years of medical school
without any conviction that I
would ever practice medicine.
Peter was left with the thorny
problem of sorting out priority
for the work on disparity which

I had done with Nikara and
himself in Sydney and also with
Blakemore and Barlow in
Berkeley. Finally, I had both
legs in plaster as a result of a
mountaineering accident in the
Dolomites on the way home.

I returned to research
after completing the medical
course and a residency,
spending 4 years at Berkeley
(1970–73) with Horace Barlow,
then 7 years at Caltech where I
did my first work on owl vision
and audition. I took up my
present position in Queensland
in 1983 after a period at the
National Vision Research
Institute of Australia.

R. W. Rodieck

I came to The University of
Sydney in 1962 to do graduate
work with Peter Bishop, who
was then head of the
Department of Physiology.
Nelson Kiang had taught me
single-unit recording, at The
Eaton Peabody Laboratory of
The Massachusetts Eye and
Ear Infirmary. But, as a
graduate in electrical
engineering, I knew little else
of neurophysiology and
nothing of vision or biology.
My profound ignorance failed
to impress Bishop. Anyone
generally able, and prepared to
work hard, could do research.
If one needed to know
something about some area,
one sat down and learned it. If
one needed a piece of
equipment such as a
stereotaxic apparatus or
oscilloscope, and nothing
suitable was available, one sat
down and built it. Focus one's

efforts and stick to the problem. His view was simple and direct, and it, together with his enthusiasm, seemed to permeate the thinking of everyone in the lab.

This practical approach had a strong and positive effect on me, and I remain grateful, both for what I learned from him, and for the strong support and encouragement he gave me.

K. J. Sanderson

I began as a Ph.D. student with P. O. Bishop in the Department of Physiology at Canberra, in February 1968, soon after his move from Sydney. For about a month I sat in on neurophysiological experiments with P. O. Bishop, Geoff Henry and Jim Smith and then one day, P. O. Bishop said, 'You're going to work on the lateral geniculate nucleus', which I have done ever since, with excursions into other parts of the visual system and the motor cortex.

During my first year as a Ph.D. student I worked with P. O. Bishop and Ian Darian-Smith. We looked carefully at the properties of lateral geniculate neurons and failed to find the x, y and w classes. However, we did find binocular receptive fields for lateral geniculate neurons, which P. O. Bishop had predicted from his knowledge of the anatomy and physiology. The story of this discovery is as follows. I was initially sceptical and so did not pursue the matter. One night, after a long day in which everything had gone wrong, Ian

Darian-Smith and I were labouring not too fruitfully on an experiment. P. O. Bishop came into the lab from his office next door some time after 9 pm and suggested that the time had come to map some binocular receptive fields for lateral geniculate neurons. He set up two stimuli on the tangent screen in front of the cat, one for each eye, and proceeded to analyse the response from the non-dominant eye. The first neuron was not binocular but the second one we tested had an inhibitory field in the non-dominant eye which was revealed by stimulus-response averaging. Following the initial excitement of discovery we examined many more lateral geniculate neurons and were able to show that most of them had binocular receptive fields.

My $3\frac{1}{2}$-year stay in Canberra was an enjoyable time with fruitful collaborations also with Murray Sherman and Bogdan Dreher. In addition to his direct scientific help, P. O. Bishop provided a lunch table at which many topics were argued, and a scientific atmosphere which attracted many visitors and set a high standard for scientific excellence and persistence in obtaining the desired objectives. From Canberra I went on to a 2-year position with Ray Guillery in Madison, Wisconsin, where I learned some neuroanatomy, and then to a position in the School of Biological Sciences at Flinders University in South Australia, where for the past 10 years I have indulged an interest in comparative neurology.

Ann Jervie Sefton

I first encountered P. O. Bishop when, as a 17 year old, I was one of his lecture class of some 400 second-year medical students at the University of Sydney. Enthusiastic to study the brain, after completing my third year I enrolled in the degree of B.Sc. (Med.) to study with

Bishop who had just become the Professor of Physiology. The degree, spent in research, has been the means of deflecting a number of medical students from the clinical path, and Bishop was responsible for supervising at least 15 during his time at the University of Sydney; several others are also contributors to this volume.

Despite my ignorance of electronics and workshop practice, I was set the task of recording from the visual system of the rat, so as to provide some comparative data. After modifying and designing appropriate equipment, experiments started and, for much of the year, it seemed that my thesis would be devoted to an analysis of keeping rats alive under anaesthesia (a problem, as I learned later, encountered by many others both before and since). However, some electrophysiological results were ultimately achieved. At the same time, Bill Hayhow, then in the Department of Physiology at Sydney, stimulated my interest in neuroanatomy, and we undertook some pathway-tracing experiments.

After completing Medicine, I rejoined the Department, of which Bishop was still Head, and completed a Ph.D. under the supervision of Liam Burke, who later succeeded Bishop in the Sydney Chair. I have been there ever since, continuing to explore the rat's visual system both in adults and, more recently, during development.

The experiences of the

B.Sc. (Med.) year under Bishop's supervision (a 'critical period') played a major role in determining my choice of an academic career as well as the direction of my research interests. I owe him and his family a great deal for numerous acts of kindness and for their friendship, support and assistance over many years.

Jonathan Stone

Jonathan Stone is an Associate Professor in the School of Anatomy at the University of New South Wales. He began his scientific work as an undergraduate and then postgraduate student in Peter Bishop's department at the University of Sydney. After gaining his Ph.D. there, and postdoctoral training in Israel, the USA and Germany, he returned to Australia in 1970, to take up a position in

Professor Bishop's department at that time in John Curtin School of Medical Research at the Australian National University. He moved to his present position in 1976. His interest in retinal topography stems from his earliest research work, undertaken for a B.Sc. (Med.) degree and supervised by Peter Bishop.

David I. Vaney

David Vaney was born in Christchurch, New Zealand, in 1954 and educated at Christchurch Boys' High School and the University of Canterbury, where he majored in animal physiology. He was

sequently awarded an
.N.U. postgraduate
holarship to study at the
n Curtin School of Medical
search and joined Austin
ghes' laboratory in the
ysiology Department in

75. David's research on the
ual system of the rabbit
mbined studies in
uroanatomy,
ctrophysiology and
ysiological optics, but
rhaps his most distinctive
ntribution to Peter Bishop's
partment was his nocturnal
dy on the psychophysics of
ld rabbits. In 1979 David
ney joined Heinz Wässle's
oratory at the Max Planck
stitute in Tübingen where

he was initiated into what one
American Observer has called
'the British school of retinal
anatomy'. After 2 years in
Germany, David moved to
Cambridge where he
established an independent lab.
within Horace Barlow's group
at the Physiological
Laboratory. He was awarded a
Beit Memorial Fellowship for
Medical Research and held the
Charles & Katharine Darwin
Research Fellowship, Darwin
College. David Vaney returned
to Australia in 1984 and is now
a Research Fellow of the
National Institute for Vision
Research and an Academic
Associate in the Optometry
Department, University of
Melbourne.

Heinz Wässle

I first met Peter Bishop when,
after finishing my studies in
physics, I did my Ph.D. on the
optical quality of the cat eye in
O. D. Creutzfeldt's laboratory
in Munich. One day
P. O. Bishop visited our lab.;
when Creutzfeldt mentioned
my optical measurements,
Bishop got excited, and spent a
good fraction of his time in

Munich, discussing with me
the spatial resolution of the cat
eye. From that date I was no
longer regarded as a physicist
in Creutzfeldt's lab., but as a
physiologist. After finishing my
Ph.D. I spent a few months in
B. B. Boycott's laboratory in
London, where we did a
morphological classification of
cat retinal ganglion cells. At
about the same time two
groups in Canberra were
working independently on a
physiological classification of
cat retinal ganglion cells. Thus
it was natural for me to go and
work in Canberra in order to
learn more about the function
of cat retinal ganglion cells.
The next story I know only
from rumour. Both retinal

groups in Canberra were
interested in the morphological
data which I brought over
from Europe and wanted to
meet me at the airport. To
avoid conflict Peter Bishop
himself drove to the airport.
The 14 months I spent in
Canberra working together
with Brian Cleland, Bill Levick
and Austin Hughes were
extremely exciting and
scientifically fruitful. The only
problem was that Peter Bishop
never learned to pronounce my
strange family name with the
umlaut, but for obvious
reasons had no problems with
my first name Heinz.

After returning to
Germany I first worked at the
University of Konstanz, where
I did my 'habilitation'. In
1977 I when to Tübingen and
in 1981 I moved to the
Max-Planck-Institute for Brain
Research in Frankfurt. Within
the 12 years since meeting
Peter Bishop for the first time,
I made a metamorphosis from
a physicist into a physiologist
and finally into a
neuroanatomist. During those
years I also had the pleasure of
a fruitful cooperation with
Brian Boycott in London.

PETER BISHOP

The first 65 years

JAMES W. LANCE

Peter Orlebar Bishop was born in Tamworth on June 14, 1917. Orlebar was a family name carried to Australia by his paternal grandfather Herbert Orlebar Bishop who migrated from England in the early 1870s. Peter's father, Ernest John Hunter Bishop, was born in Queensland in 1877. He was one of the first pupils at the newly established Toowoomba Grammar School, subsequently became a Government Surveyor, and married Mildred Alice Havelock Vidal in 1914. Peter was the second of their five children.

In 1925 the family moved to Armidale where Peter attended the Primary and High Schools before transferring to Barker College, Hornsby, for the last three years of his secondary education. A near neighbour and friend in Armidale was John Cornforth who subsequently won a Nobel Prize for his work in organic chemistry. Peter's best friend at Barker was Leslie Sulman, grandson of Sir John Sulman, the noted architect and town planner who founded the Sulman art prize. Leslie was killed in Malaya but his father, Arthur Sulman, continued to take a keen interest in Peter's career, helping him to undertake research after the war ended, providing financial support for his work at Sydney University and, finally, leaving a bequest to the Australian National University for the continuation of his vision research programme.

Peter played in the first XV and first XI at Barker, became Dux of the School and was awarded an Exhibition to the University of Sydney, a high honour in those days. Although his best subjects were mathematics and physics and he had intended to be an engineer, his mother influenced him to take up medicine. It is interesting to see how his subsequent life reflected his early interests and how his mathematical talents were turned to the advancement of medical knowledge. He was a student at St Paul's College during his undergraduate days and represented his College nd University in football as well as serving on the editorial taff of University, Faculty and College publications.

The neurological spark was kindled in the third year of the medical course, when Peter first dissected a brain. 'I will never forget the fascination of actually holding a human brain in my hands and realizing that the brain I was holding was once a person like myself with the same thoughts, feelings and wishes. That made a tremendous impact on me and there was never any question from then on that I would try to make a career in brain research'. The Professor of Anatomy at the time, A. N. Burkitt, had a large personal library which was at Peter's disposal and, as a result of his dissection and reading, he wrote an article, 'The nature of consciousness', that was published in the *Sydney University Medical Journal* in June 1939. He examined the function of the nervous system in three categories, as neuromuscular, neurosensory and neurointellectual systems, and stated that 'the neurones of intellect probably work according to laws which we have long since learned from neuro-muscular behaviour'. These views were apparently taken literally by the later well-known neuroanatomist A. A. Abbie, then senior lecturer, who wrote a reply in the November issue of the *Journal* stating rather grimly that the proposed distinction of three neural systems as separate entities was quite incompatible with present-day knowledge. He criticized the emphasis Peter Bishop had placed on the importance on emotion (and the place of the gyrus cinguli) in consciousness. 'Moreover, I have stimulated the gyrus cinguli on several occasions in bandicoots and possums without eliciting the slightest response. It is true that these animals were under light anaesthesia....' Well! With the hindsight of more than 40 years it does not seem to the present-day reader that the young Bishop's theoretical approach to one of mankind's greatest enigmas was demolished by his distinguished critic. In any event Peter's brave attempt attracted the attention of the Great and helped to plot his neurological pathway to the future. Happily it was also the start of a friendship with both Abbie and Burkitt.

With the outbreak of war in 1939, the medical course was shortened and students in their final year acted as Junior Resident Medical Officers to replace those that had joined the Armed Services. At his departure from undergraduate life, the College Magazine, *Pauline*, recorded that 'Peter was the most literary of our literary-minded medicos and the most philosophic of our philosophers' although there was one snide reference to him as a poet who 'by precept and example condems unnecessary clarity'. His write-up concludes: 'Later on, he will not have much patience with the routine of the general practitioner or the hypocrisy of the specialist. He will probably be happiest in a research job, for which his intellectual ability, eagerness and independence should pre-eminently fit him'. The writer was a future Prime Minister, Gough Whitlam.

After graduating in December 1940, Peter was offered a residency in neurosurgery (usually taken by a Senior Resident Medical Officer) at the Royal Prince Alfred Hospital. This gave him the opportunity to work with Gilbert Phillips, an outstanding young neurosurgeon with an understanding of the neurological sciences, and Professor (later Sir) Harold Dew, both of whom had an important influence on his future career. He also met another outstanding member of the neurosurgical nursing staff who had an even more important influence on his life, Hilare Holmes. Hilare and Peter were married in February 1942.

Peter had joined the Royal Australian Navy after graduating and was called up for service early in 1942. He served as a Surgeon Lieutenant on the cruiser *H.M.A.S. Adelaide* and subsequently on the destroyer *H.M.A.S. Quiberon*. He sailed the seven seas with pots of animal and human brains secreted under his bunk to provide material for dissection when times were quiet. The war with Japan ended in August 1945 and Peter returned to Sydney in 1946 as Clinical Assistant in Neurosurgery. A. K. (Archie) McIntyre returned from service in the Air Force and was working in the Department of Surgery at Sydney University, so they decided to join together in a research programme. The first step was to obtain equipment. Jack (later Sir John) Eccles had left a well-equipped laboratory in the Kanematsu Institute at Sydney Hospital when he left to take the Otago Chair of Physiology in Dunedin, New Zealand. No sooner had some of the equipment been acquired from the Kanematsu Institute than Eccles arrived to claim it all and transfer it to New Zealand. Thus McIntyre and Bishop were left again with empty laboratories. Just at the right time, McIntyre was awarded a Rockefeller Fellowship to work with David Lloyd at the Rockefeller Institute and Bishop obtained a Fellowship of the Sydney Postgraduate Committee to train in Oxford under Sir Hugh Cairns with whom Gilbert Phillips had worked in the North African campaign during the war.

Peter and Hilare, now blessed with 2-year-old Phillippa and Clare, aged 13 months, arrived in Oxford in August 1946 and found an old house, aptly named Bishop's Barn, in Wiltshire. Cairns had arranged for Peter to be Clinical Clerk to Sir Charles Symonds at The National Hospital for Nervous Disease, Queen Square, for a period of neurological training before he undertook, as Cairns supposed, the study of neurosurgery. Although Peter found intellectual stimulation in ward work he still had a keen nose for the scent of research. One day he found George Dawson in a basement laboratory tinkering with electronic equipment. Dawson was a pioneer in the use of averaging techniques to record cerebral potentials evoked by peripheral nerve stimulation. Before long, Peter Bishop was firmly strapped in position with electrodes at elbow and wrist as a pioneer normal subject. So great was the amplitude of Peter's evoked potentials that averaging was scarcely necessary: they are preserved for posterity in the figures illustrating one of

Dawson's papers published in 1947. His research interest being reawakened, Peter saw Professor E. A. Carmichael, the Director of the Neurological Research Unit, about a research appointment. On being told that Peter was 30 years old and had never done any research, Carmichael (with a lack of insight not characteristic of the man) remarked 'Well, I doubt if you ever will'. He could not have said anything more calculated to strengthen Peter's resolve for a research career.

Lovatt Evans, then Professor of Physiology at University College, put him in touch with Professor J. Z. Young – 'that young man who has just come down from Oxford' – because neurophysiology was 'all electronics and that sort of thing'. J. Z. Young took him on immediately and gave him a big empty room opposite a laboratory where Bernard Katz (later Sir Bernard) was then studying the local response which precedes and initiates the nerve action potentials. J. Z. Young suggested that Peter try to replicate some work done in the USSR on the alteration of the electroencephalogram by learning procedures in the rabbit. So there he was, with a research project, an empty room, no research training and no knowledge of electronics. He enrolled in a course on electronics and at the same time set out to build a high-gain DC amplifier. Three years (and seven published electronics papers) later, he and Dr E. J. Harris described their DC amplifier in the *Review of Scientific Instruments*. The performance specifications of this instrument were the best that had been achieved in the world at that time. Subsequently the amplifier was used to make recordings from the mammalian visual system but a condenser–coupled amplifier would have done as well. A high-gain DC amplifier was simply not needed,. In the process of building this equipment, as well as stimulators, power supplies and so on, Peter gained a good grounding in electronics and the use of the mechanical workshop. He put this to use at first by studying resting potentials in the optic tectum of the frog but ran into problems of injury potentials and other artefacts.

This was an era of enormous advance in neurophysiology with the advent of intracellular recordings: Graham and Gerard in 1946 first recorded the resting potential of the muscle cell membrane: Fatt and Katz reported their work on the muscle end–plate potential in 1951; while Hodgkin and Huxley described the ionic basis of nerve conduction in 1952. Eccles and his colleagues in New Zealand first recorded from within mammalian spinal motor neurons in 1950. The work that particularly appealed to Peter Bishop at this stage was the systematic survey of the visual system by another Bishop (G.H.) and O'Leary in St Louis. On returning to Sydney in May 1950, Peter joined the Department of Surgery where Professor Dew gave him four large bare rooms in which to establish his laboratory, henceforth known as the Brain Research Unit.

In that first year, 1950–51, Peter acquired four B.Sc. (Med.) students, including Richard Gye, now Dean of the

Faculty of Medicine at Sydney University, and David Be now honorary psychiatrist to St Vincent's Hospital. Fo more B.Sc. (Med.) candidates joined the following year: B Levick, now F.R.S. with a personal chair in Physiology at t John Curtin School; Jim McLeod, now Bushell Professor Neurology at the University of Sydney; David Jeremy, no medical director of Merck, Sharp and Dohme, Australia; ar Annette Walsh, who subsequently joined the Department Health. At the same time, the later Brian Turner and I sign on as M.D. candidates. After graduating in medicine I kne that I wanted a career in medical research but was not su how to go about it. Professor Hugh Ward arranged for n and Henry Harris to meet Sir Howard Florey when he visite Sydney. His advice was to spend a year or two in t department of Physiology in Melbourne and then undertake the Honours course in Physiology at Oxfor Henry Harris did this and in time was appointed to Florey Chain in Oxford on his retirement and later as Regi Professor of Medicine. I visited the Department Physiology in Melbourne but was not inspired. I then hear of the recently formed Brain Research Unit in Sydney an made an appointment to see Peter Bishop. Peter's enthusias was contagious and it took only a few minutes for me to mal up my mind to start my research life and neurophysiolog Although I began by assisting him with his current proje on the optic nerve, he encouraged me to pursue my intere in the motor system, specifically the pyramidal tract, a interest which has continued until the present day.

In 1951 Peter became a senior lecturer in physiolog and in 1954, when Professor Cotton retired from the Chai was appointed Professor of Physiology. He faced a enormous administrative task. The Department of Physiolog was responsible for teaching large numbers of students fror the Faculties of Science, Dentistry and Veterinary Scienc and Medicine, as well as students of physiotherapy, speec therapy, occupational therapy and various postgradua diplomas. In 1961 there were about 1500 studying physiolog in the various courses. The assistance of Bill Lawrence a senior lecturer in the early days was invaluable. In 1956 Pau Korner and Liam Burke joined the department as senic lecturers, to be followed later by Bill Hayhow and Ia Darian-Smith. Paul and Ian left in 1960 to found th Department of Physiology at the new Medical School of th University of New South Wales. The battle of excessiv student numbers and inadequate research funds was finall won in the 1960s. In the meantime, research work ha continued on single-unit activity in the optic tract an radiation and binocular interactions in the lateral geniculat ganglion and visual cortex. The Australian Physiologica Society (later the Australian Physiological and Pharmacolog ical Society) held its first meeting in Sydney in May 1960 R. W. Rodieck became a Ph.D. student of Peter Bishop's i 1962 and subsequently became lecturer and reader, leavin

1978 to become Bishop Professor of Ophthalmology in the
University of Washington, Seattle. From his return to Sydney
until his departure from the Chair of Physiology in Sydney,
Peter Bishop was a strong supporter of the B.Sc. (Med.)
programme which introduced a whole generation of medical
students to research and has had far-reaching effects on the
development of medical research in Australia. He was
delighted when, in the course of time, his son Roderick
became a medical student and obtained the B.Sc. (Med.)
degree. Roderick is presently a Senior Resident Medical
Officer at the Royal Prince Alfred Hospital, Sydney.

Peter was elected a Fellow of the Australian Academy
Science in 1967 and, in the same year, was appointed
Professor of Physiology at the Australian National University,
Canberra, to succeed Sir John Eccles. In the past 15 years,
he has concentrated his research on the function of the striate
cortex in general and the mechanism of binocular vision in
particular. He has established his department as one of the
best vision research laboratories in the world and has
attracted research students and colleagues from all parts of
the world to work with him. Of the people who have worked
with Peter, or under his direction, some 25 now hold full
chairs in Anatomy, Physiology, Neurology, Ophthalmology
or (remarkably) Obstetrics in the USA, UK, Europe, Japan
and Australia. Peter Bishop became a Fellow of the Royal
Society in 1977 and received an honorary doctorate in
medicine from the University of Sydney this year. The esteem
in which he is held by his clinical colleagues may be gauged
by his election to honorary membership of the Australian
Association of Neurologists, the Neurosurgical Society of
Australia and the Ophthalmological Society of New Zealand.
His devotion to his profession has been matched only by his
devotion to Hilare who has constantly given him her
wholehearted support and has tolerated his exacting work
schedule with patience and good humour. The many visitors
to the Department over the years remember Hilare with warm
affection. He concern for their welfare and that of their
families, particularly during the process of settling in to a new
environment, and the hospitality she extended to them
throughout their stay gave the Department a family
atmosphere.

This adds up to a remarkable achievement. For one
who did not have the chance to start research work until the
age of 30 to become a world leader in his field and to inspire
so many others along the way, there must be intrinsic magical
ingredients. Not only physiology but ophthalmology and
neurology are much the richer for his life's work. Those of
use whose own lives have been influenced and enriched by
Peter's can only express our thanks and try to repay our debt
of gratitude by the encouragement of others.

PETER BISHOP

The Canberra years

GEOFFREY H. HENRY

An appreciation of the contribution made by Peter Bishop during his time in the JCSMR, could simply concentrate on the man and his work. This approach would satisfy propriety but I doubt if it would please Peter, who saw himself as the leader, and very much an integral part, of a team and rightly identified himself with the achievements of the team members. In reviewing his years at Canberra, therefore, it is appropriate to follow two lines of achievement; one where he pursued his own research programme for which he was rewarded with election to the Royal Society in 1977, and the other in which he gathered together the team that contributed so significantly to visual science. Not only was the work that flowed from the JCSMR to make an implant on international thought but a number of outstanding young scientists, from around the world, were to come to benefit from the Bishop concept of science.

The Canberra years started in June 1967 when work began on the construction of laboratories in the JCSMR. Peter Bishop derived great pleasure from designing equipment and building laboratories. His compulsive attention to detail and his capacity to work a 16-hour day suited him perfectly for this role. I always felt there was a streak of Leonardo da Vinci in the pleasure he got from preparing detailed mechanical drawings. Certainly everyone who worked in Canberra benefited from this talent and all the laboratories carried the Bishop stamp.

When Peter Bishop arrived in Canberra in late 1967 to succeed Sir John Eccles as head of the Physiology Department, the Department already consisted of David Curtis, who had been promoted to Professor in the previous year, Graham Johnston, Russell Close and Jack Coombs. The Curtis and Johnston combination set out on a series of experiments that led to the discovery of bicuculline as a selective inhibitor of GABA and Russell Close went deeper

the mechanical changes accompanying the contraction of
etal muscle. Jack Coombs, who with his electronic skills
made a significant contribution to Eccles' Nobel
e-winning work, with the change to visual physiology
ed in the new experiments.

By the end of 1967, when I first arrived in Canberra,
Pettigrew was recording the first cells from the cat's
ate cortex. Jack had come to finish off the papers on work
e in Sydney with Nikara and to help set up. Bill Levick
just arrived from Berkeley after spending a couple of
nths in Sydney in transit. Clyde Oyster and Ellen
ahashi had also come from Berkeley and with Bill they
e intent on exploring the rabbit retina. At the same time,
ever, there was a good deal of talk about the X and Y
inction that John Robson and Christina Enroth-Cugell
recently drawn for retinal ganglion cells. In a sense, the
re was beginning to unfold even in the earliest days.

At the start of the academic year in 1968, Jack
tigrew returned to Sydney to complete the final year of
medical degree and, shortly after, Ken Sanderson arrived
tart work on his Ph.D. which was to produce the first
nitive maps of the cat lateral geniculate nucleus. Later in
same year, Brian Cleland joined Bill Levick and, early in
9, in association with Ken Sanderson, they began to
avel the X/Y story. Later that year Mark Dubin came
n the USA and Ken returned to the preparation of
ctional maps of the lateral geniculate nucleus. The team
Levick, Cleland and Dubin then embarked on, and
ieved, the simultaneous recording of a ganglion cell and
recipient cell in the lateral geniculate nucleus.

After Jack Pettigrew's departure, Jim Smith, who was
sabbatical leave from the State University of New York,
ffalo, Peter Bishop and I started to examine the internal
ucture of receptive fields of striate neurons. Later, with
k Coombs replacing Jim Smith, these investigations were
develop new techniques for recording inhibitory and
pressive regions in monocular and binocular receptive
ds and so mark the start of a comprehensive analysis of
response characteristics of cortical neurons. About the
ne time, Ian Darian-Smith, who was waiting to return to
ns Hopkins University, started work with Ken Sanderson
binocular interactions in the lateral geniculate nucleus. By
end of the first year, work had begun at all levels in the
ual pathways and the Bishop charabanc was underway.

In 1969 Bogdan Dreher arrived from Poland. He had
velled across the world largely on the Bishop name.
rtainly he had no money and in the early days was not a
le homesick. Some months later he was joined by his family
d together they began to live out one of the success stories
the Australian migration policy. Bogdan initially joined the
shop/Henry group in the striate cortex but later was
combine in turn with Sanderson, Hoffmann and Stone,
velling in so doing from the lateral geniculate nucleus to

the superior colliculus and thence back to the cortex to
become involved in studies into the primary afferents going
to areas 17 and 18.

In 1970, Jonathan Stone, who came back from a
postdoctoral sojourn overseas, had also become enthralled by
the subclasses of retinal ganglion cells. About the same time,
Peter Hoffmann from Germany and Murray Sherman from
the USA commenced stays that were to extend over a period
of two years. Soon, the analysis of the responses of cells in
the X and Y streams in the retina and lateral geniculate
nucleus was in full swing. Levick and Cleland were into the
project while two doors along the corridor Stone, in
association with Hoffmann and Sherman, was also deeply
committed. This was a period of intense competition and over
the next 6 years there was to be a high level of productivity
from both groups. Throughout Peter Bishop worked to
maintain accord without dampening enthusiasm and visual
science was the beneficiary.

The end of the first stanza of the Canberra story
occurred in February 1972 when 20 American visual
scientists came to Canberra to attend the US–Australian
Symposium on Vision. This interchange arose from President
Johnson's visit to Australia and Peter Bishop was responsible
for the organization at the Australian end. He grasped the
opportunity to move his team onto the world's stage and there
can be little doubt that the Symposium set the seal of approval
on the X and Y story and made the concept of parallel
processing a talking point in vision laboratories around the
world. At the same time, the steps taken in the serial
sequencing of visual information were also thrown into
question when the Canberra group of Dreher, Henry and
Bishop pointed out that hypercomplex cells need not
necessarily reside at the pinacle of a hierarchical process.
Instead 'endstopping' or 'hypercomplexity' was advanced as
a property that could occur in cells belonging either to the
simple or the complex family.

After the Symposium, Hoffmann and Sherman
departed for home and Dreher moved to Sydney. Their stay
in Canberra had been most productive; to briefly outline some
of their achievements it may be said that Murray Sherman
found that the Y cells in the lateral geniculate nucleus were
more affected by visual degeneration than cells in the X
system, Peter Hoffmann reported that W cells provided the
major retinal input to the superior colliculus, and Bogdan
Dreher, in his work with Jonathan Stone, showed that the Y
system projected to cortical areas 18 and 17 and that X cells
went only to area 17. Now the three left to develop their own
laboratories and carry their contributions to new horizons but
in the process, there was to be a distillation of the Bishop
philosophy around the world.

If 1972 marked the end of the beginning at Canberra
then it also saw the start of the next phase. The themes of
the Symposium were expanded as Austin Hughes from

Britain, Yutaka Fukuda from Japan, and Heinz Wässle, who came early in 1973 from Germany, arrived to contribute to the studies of retinal physiology and morphology. Tony Goodwin from South Africa and Jerry Nelson from the USA also moved into the team studying the response properties of striate neurons and the heady days of the Symposium gave way to the nights and days of labour that are the lot of the visual physiologist.

Over the next two years, Jon Stone and Yutaka Fukuda completed a study of the rate of signal conduction in the axons of ganglion cells and further described the properties of W cells, which Jon Stone and Peter Hoffmann had first described at the 1972 Symposium. Since the conduction of axons in X cells was slower than that in Y cells and in the new subclass it tended to be slower still, the name of W cells was proposed to align the alphabetic sequence with axonal conduction rates. The development of an acceptable scheme of terminology raised some dissention and as an alternative the Levick/Cleland group advanced a classification based on visual-response properties. Each scheme had its advocates and today each remains largely unchallenged in defining different aspects of the physiology of ganglion cells.

Throughout this period, short but significant visits were made to the Levick/Cleland group by Horace Barlow and William Rushton and experiments were carried out on the threshold requirement of ganglion cells. The link between structure and function of ganglion cells began to be forged when the Levick/Cleland team was joined by Heinz Wässle, who had observed alpha, beta and gamma cells in Golgi stained whole mounts prepared by Brian Boycott in London. The group then moved to correlate these morphological types with the functional counterparts and along with the Stone and Fukuda group made significant contributions to this aspect of the ganglion cell story. In a way, the die was cast in this year for Heinz Wässle's future career.

As we shall see, 1975 was to be a time of transition and in many ways marked the end of the second phase of our saga. In looking back over the years in Canberra there now seems a remarkable order that allowed projects to be viewed over 4-year cycles. I doubt if even the Bishop organizational ability was capable of such systematic guidance and 4 years may be the optimal time for maintaining a coherent team effort in the JCSMR.

In the run up to 1975, the cortex group, which now included Tony Goodwin, were deeply involved in interpreting the organization steps that lead to direction selectivity and orientation specificity in striate neurons. At the same time, Jerry Nelson was propounding a comprehensive explanation for binocular vision and later, with Barrie Frost from Canada, was to explore further the properties of binocular inhibition. In this period also Jonathan Stone was joined for a year by Paul Wilson and Mike Rowe while Henry Wagner and Des Kirk, who was doing his Ph.D., collaborated with the

Levick/Cleland team. Parallel processing was consolidate and distinguishing properties were established for the f range of ganglion and lateral geniculate neurons and the tw groups almost simultaneously demonstrated that the W (sluggish) class of ganglion cell projects from retina to the ce of the C complex in the lateral geniculate nucleus. The sce was to change, however, with the departure of Jonatha Stone, early in 1976, to the School of Anatomy at t University of New South Wales. Jon's contribution had bee a major one and his departure emphasized the end of t second phase; the movement into the third phase becam complete when Brian Cleland and I each established ne research groups.

In 1975, I was away from Canberra for a year with R and Jenny Lund at the University of Washington, Seattle, an in 1976 this visit was reciprocated when the Lunds spent t next year in the JCSMR. This collaboration brought t Golgi technique to Canberra and led to the creation of active neurohistological laboratory; a particularly helpf adjunct that permitted the comparison of histological an physiological interpretations of neural pathways. At this tim Allan Harvey had come from Cambridge to pursue his Ph.I and was employing electrical stimulation to study t connectivity of neurons in areas 17 and 18. Much ne information came from his study but, in particular, it reveale the types of cell projecting from cortex to the later geniculate nucleus. Later the technique of using sign conduction-times to identify the nature of the neural strea providing the input to striate neurons was carried a ste further when Jean Bullier came from Duke University collaborate with me for a period of 3 years. Mike Musta from Seattle, was later to join the team and a start was mad in tracing the X and Y streams as they ran through the stria cortex.

In this period, Brian Cleland working in associatio with Tom Harding and Ulker Tulunay-Kessey, both fro the USA, examined the relation between receptive field si and velocity dependence in the different classes of gangli cell. In the same laboratory, studies were carried out association with Sheila and David Crewther and D Mitchell to show that retinal ganglion cells in the cat retaine a high resolving capacity despite the presence of amblyop resulting from surgically induced strabismus.

Beginning in 1976 also, Larry Thibos from the US and Bill Levick joined forces to describe a systemat asymmetry they had observed in the responses of retin ganglion cells that otherwise displayed concentrical organized receptive fields. Another line of activity for Bill a Larry came through a chance misfortune that struck not on their animals but a number of strabismic cats that I h prepared with Ray Lund and Don Mitchell. These anima developed a central retinal degeneration that appeared come from a dietary deficiency of the vital (for cat) amino-aci

urine. The recording procedures developed over the years
in the laboratory were then adapted to follow the course of
this condition in the hope that these results might shed some
light on the heriditary macular dystrophies that occur in man.

The arrival of Austin Hughes, late in 1972, had added
a new dimension to the skein of research in the physiology
department. Austin had developed an intense interest in
visual optics and was also fascinated by the literature on the
history of optics – two topics that had long interested Peter
Bishop. Many lunchtime discussions revolved around these
subjects and Peter got a lot of pleasure in following the
progress of Austin and Melanie Campbell, who was
completing her Ph.D., as they traced the paths taken by rays
through the lens of the rat eye. There was a touch of *déjà vu*
in these experiments as memories came back of the early days
in Sydney when Peter, anxious to do visual physiology in the
cat, acted to fill a critical gap in the knowledge of the day,
by ray tracing and preparing a schematic eye. Late in the
time at Canberra also, Bill Jagger came from Germany to
collaborate with Austin in measuring the eye's line spread
function with an optical microprobe.

Another arm of the Hughes group which included
David Vaney from New Zealand and Elzbieta Wieniawa-
Narkiewicz from Poland, both of whom were completing their
Ph.D., was concerned also with the ganglion cell layer of the
retina. After a time, however, it was realized that this layer
contained a large number of displaced amacrine cells and that
a new approach was required to interpret the organizational
links in the inner plexiform layer. Again, a breakthrough had
emerged from those whom the Bishop net had attracted to
the wilds of Australia.

With all this activity going on around him, Peter still
maintained a vital research effort that was to carry through
until the day of his retirement. After I set out on a separate
path in 1976, he was associated with a number of overseas
co-workers in the quantitative analysis of the receptive fields
of all types of striate neuron. Then, moving according to plan,
with the aid of another group of colleagues, he set out to draw
up a simple theoretical explanation of what appeared to be
a complicated series of response patterns.

From late in 1974 when Hiroshi Kato arrived from
Japan, Peter collaborated in the characterization of the
receptive fields of striate neurons with Jerry Nelson from

USA, Guy Orban from Belgium, Marcello Camarda from
Italy, Esther Peterhans from Switzerland, Shigeru Yamane
from Japan, Richard Maske from South Africa and Janusz
Kulikowski from England. Janusz was also of great help in
developing the theoretical explanation and Peter, he and
Stjepan Marcelja from the Australian National University
developed a model based on a mathematical function
advanced in the 1940s by Nobel laureate Dennis Gabor. This
model neatly defines the properties of striate neurons by
providing a synoptic description of their visual response
patterns.

In concentrating on the scientific output of Peter
Bishop and his team in Canberra we have left unheralded the
contribution made by Hilare and the family both to Peter's
equanimity and to the welfare of the department as a whole.
All those who passed through Canberra benefited from their
friendship, kindness and graciousness, and there was a strong
sense that in the Bishop ideology the 'family' included all
those concerned with visual physiology.

As we look back, the achievements of the Canberra
years are in no way dulled by the passage of time and, indeed,
the full worth of some may yet to be realized. Beyond this
appreciation, however, comes another than emerges clearly
from the story set out above. It is that in his time at Canberra,
Peter Bishop, despite his strong Australian bonds, became
truly international. This is the phase of his life when the world
sat up and took notice and with this recognition came the
influx of visitors from every corner of the world.

There remains the need to express the feelings of his
co-workers and colleagues. This is no simple task but,
inadequate as they are, I would like to repeat the words I
wrote on the occasion of Peter's retirement from the JCSMR.
When referring to those left in the department I said, 'those
of us who have worked in close association will miss the day
to day contact, the never-abating enthusiasm and the
unswerving support in our research ventures. We shall also
miss the Bishop preparedness to start each day with renewed
vitality unmarred by any of the disappointments of the
previous day. We shall miss the intellectual honesty and
the total commitment to the scientific method for the
advancement of knowledge.' I believe it is these characteristics
that has set the man apart and made him so greatly admired
in the world of science.

PETER BISHOP

a tribute

AUSTIN HUGHES

My first encounter with the work of Peter Bishop occurred
that fateful day in 1965 on which I was initiated into the
mysteries of schematic eyes. Wandering into Mike Block's
laboratory in the Edinburgh University Medical School I
found him making measurements on rat eyes (Block, 1969).
My suggestion that it was an odd way to spend a nice
afternoon elicited a terse explanation. I departed with the
comment that I could not see much point in that sort of thing.

But the Siren had sung. The idea of modelling the
entire optical performance of the eye caught my imagination.
David Whitteridge, whose doctoral student I was at that time,
had followed the tradition of his predecessors in cossetting
the snug departmental library and I was soon enjoying the
two model studies by Vakkur, Bishop & Kozak (1963) and
Vakkur & Bishop (1963) and finding myself held at bay by
equations with Gothic script in von Helmoholtz's *Handbuch
der Physiologischen Optik* (1856). However, the development
of a subsequently published rabbit schematic eye (Hughes,
1975) soon brought me down to earth amongst the severe
limitations of the conventional models.

I met Peter Bishop 7 years later at Oxford and then,
with an invitation to visit Canberra, began my personal
indebtedness to his enthusiasm and generous support. That
visit has lasted 12 years. During my first years in Australia
a variety of utilitarian and basic work reawakened my interest
in comparative optics. Studies on the cat visual field (Hughes,
1976), the error of refraction (Glickstein & Millodot, 1970;
Hughes, 1977*a*; Hughes, 1979*a*; Hughes & Vaney, 1979) and
a model for the small eye of the rat (Hughes, 1979*b*) provided
grist for the mill in many profitable discussions at the
Professorial Lunch Table. The reviewing of this field
(Hughes, 1977*b*) gave impetus to the aim of developing a
powerful analytic eye model.

In passing, I note that there must be a mysterious

ingredient which made the famous Lunch Table function. How otherwise could certain of his colleagues have stood the strain of watching Bill Levick fold his sandwich wrapper around the remains of a blackened banana on some 3700 working days? I suspect it was Peter Bishop's magic phrase 'changing the subject' or, when I was speaking, 'seriously now' which maintained a semblance of law and order.

The prior incumbent of Peter Bishop's Chair of Physiology at the John Curtin School of Medical Research (JCSMR), Sir John Carew Eccles, has commented (Eccles, 1971) on the idyllic environment and facility offered by the Institute of Advanced Studies (IAS) at the Australian National University (ANU) in Canberra. In the 1970's and early 1980's these assets were compounded by the remarkable number of workers interested in physiological optics and its matching to the neural apparatus.

In addition to the JCSMR Department of Physiology, with its definitive work on binocular vision, retina and optics, the Research School of Biological Sciences (RSBS) houses Adrian Horridge's insect vision group in the Department of Neurobiology and Richard Mark's vision group in the Department of Behavioural Biology. Across campus, in the Research School of Physical Sciences (RSPhys.S), the Department of Applied Mathematics contains a fibre optics research group headed by Allen Snyder and a group, also headed by him, which worked on theoretical aspects of insect vision. During the last decade the extensive collaborations between Simon Laughlin of RSBS, other colleagues and Allen Snyder dominated the study of insect visual optics by combining Snyder's 'brash' theory with precision experimental data (Waterman, 1981) to produce fundamental insights into insect eye design (Wehner, 1981).

In due course, a dialogue developed between Applied Mathematics and Physiology on the application to vertebrate systems of the insights gained from insects. Increasingly sophisticated techniques and the corresponding rapid growth of understanding of the vertebrate eye opened up the possibility of practical applications. Two paths were followed: one involved the setting up of a physiological optics group with the formal collaboration of a skilled optical physicst, Peter Sands of the Commonwealth Scientific and Industrial Research Organisation (CSIRO), close to ANU campus and with Melanie Campbell acting as a bridge between the various institutions and disciplines The other was a separate New Intitiative in the form of an ANU funded cross-schools collaboration between the Departments of Physiology, Applied Mathematics and Engineering Physics, in which Bill Jagger and Terry Bossomaeier provided interdisciplinary skills for the study of the optical/retinal interface in terms of communication theory.

The varied cross-disciplinary interactions have been both stimulating and fertile. It is regretted by many that attempts, strongly backed by Peter Bishop, to consolidate this community of vision and optical workers into an ANU Centre for the Visual Sciences have not yet come to fruition.

That the task of developing an analytical, computer-based, eye model is now well advanced and that its use has begun in laying the foundations of a science of quantitative ecological optics (Campbell & Hughes, 1981; Campbell, Hughes & Sands, 1982; Campbell, 1985*a*; Campbell, Sands & Hughes, 1985; Sands, 1985) owes a great deal to the efforts of Peter Bishop who made possible the formal basis of the collaborations which underlie the ANU/CSIRO physiological optics group and the New Initiative. Such debts cannot be repaid; all we can do is follow his lead and try to give similar support to the next generation of young scientists.

In addition to these acknolwedgements of my personal debt, I also take this opportunity on behalf of all those working in the field to recognize the important contribution made by the precision work of Bishop and his colleagues to the revival of interest in physiological optics. It is worth remembering that, prior to 1958 (Hermann, 1958) there had been no schematic eyes developed for mammals other than man.

References

Block, M. T. (1969) A note on the refraction and image formation of the rat's eye. *Vision Res.*, 9, 705–11

Glickstein, M. & Millodot, M. (1970) Retinoscopy and eye size *Science*, 168, 605–6

Helmholtz, H. von (1856) *Handbuck der Physiologischen Optik*. ed. A. Gullstrand, J. Kries, W. Nagel) 3rd edn (1909). Reprint: New York: Dover (1962), of translation by Southall J.P.C. for Am. Opt. Soc. (1924)

Hermann, G. (1958) Beiträge zur Physiologie des Rattenauges. *Z. Tierpsychol.*, 15, 462–518

Hughes, A. (1972) A schematic eye for the rabbit. *Vision Res.*, 12, 123–38

Hughes, A. (1976) A supplement to the cat schematic eye. *Vision Res.*, 16, 149–54

Hughes, A. (1977*a*) The refractive state of the rat eye. *Vision Res.*, 17, 927–39

Hughes, A. (1977*b*) The topography of vision in mammals of contrasting life style: comparative optics and retinal organisation. In *Handbook of Sensory Physiology*, vol 7/5, ed. F. Cresciteli, Berlin: Springer

Hughes, A. (1979*a*) The artefact of retinoscopy in the rat and rabbit eye has its origin at the retina/vitreous interface rather than in longitudinal chromatic aberration. *Vision Res.*, 19, 1293–4

Hughes, A. (1979*b*) A schematic eye for the rat. *Vision Res.*, 19, 56–588

Hughes, A. & Vaney, D. I. (1979) The refractive state of the rabbit eye: variation with eccentricity and correction for oblique astigmatism. *Vision Res.*, 18, 1351–5

Vakkur, G. J. & Bishop, P. O. (1963) The schematic eye in the cat. *Vision Res.*, 3, 357–81

Vakkur, G. J., Bishop, P. O. & Kozak, W. (1963) Visual optics in the cat, including posterior nodal distance and retinal landmarks. *Vision Res.*, 3, 289–314

Waterman, T. H. (1981) Polarization sensitivity. In *Handbook of Sensory Physiology Comparative Physiology and Evolution of Vision in Invertebrates B, Invertebrate Visual Centres and Behaviour, I*, ed. H. Autrum. Berlin: Springer

Wehner, R. (1981) Spatial vision in arthropods. In *Handbook of Sensory Physiology. Comparative Physiology and Evolution of Vision in Invertebrates. C. Invertebrate Visual Centres and Behaviour, II*, ed. H. Autrum. Berlin: Springer

P. O. BISHOP

Bibliography

1. Bishop, P. O. (1939) The nature of consciousness. *Sydney Univ. Med. J.*, **32**, 61–74
2. Bishop, P. O. (1940) 'What is Truth?' – said Pontius Pilate. *Sydney Univ. Med. J.*, **33**, 12–16
3. Bishop, P. O. & Harris, E. J. (1947) Electroencephalograph amplifier. *Wireless Engineer*, **24**, 375
4. Harris, E. J. & Bishop, P. O. (1948) Low-frequency noise from thermionic valves working under amplifying conditions. *Nature*, **161**, 971
5. Bishop, P. O. (1949) A note on interstage coupling for D.C. amplifiers. *Electron. Eng.*, **21**, 61
6. Harris, E. J. & Bishop, P. O. (1949) The design and limitations of D.C. amplifiers. Part I. *Electron. Eng.*, **21**, 332–5
7. Harris, E. J. & Bishop, P. O. (1949) The design and limitations of D.C. amplifiers. Part II. *Electron. Eng.* **21**, 355–9
8. Bishop, P. O. (1949) A high impedance input stage for a valve amplifier. *Electron. Eng.*, **21**, 469–70
9. Bishop, P. O. & Harris, E. J. (1950) A D.C. amplifier for biological application. *Rev. Sci. Instr.*, **21**, 366–77
10. Bishop, P. O. & Collin, R. (1951) Steel microelectrodes. *J. Physiol.*, **112**, 8–10P
11. Bishop, P. O., Brown, G. L. & Kearney, A. (1951) A modified Czermak head-holder. *J. Physiol.*, **114**, 19–20P
12. Bishop, P. O. (1953) Synaptic transmission. An analysis of the electrical activity of the lateral geniculate nucleus in the cat after optic nerve stimulation. *Proc. R. Soc. B*, **141**, 362–92
13. Bishop, P. O., Jeremy, D. & McLeod, J. G. (1953) Phenomenon of repetitive firing in lateral geniculate of cat. *J. Neurophysiol.*, **16**, 437–47
14. Bishop, P. I., Jeremy, D. & McLeod, J. G. (1953) Repetitive post-synaptic discharge in a sensory nucleus. *Nature*, **171**, 844–5
15. Bishop, P. O., Jeremy, D. & Lance, J. W. (1953) Properties of pyramidal tract. *J. Neurophysiol.*, **16**, 537–50
16. Bishop, P. O., Jeremy, D. & Lance, J. W. (1953) The optic nerve. Properties of a central tract. *J. Physiol.*, **121**, 415–32

17. Bishop, P. O. & Davis, R. (1953) Bilateral interaction in the lateral geniculate body. *Science*, 118, 241–3

18. Bishop, P. O. & McLeod, J. G. (1954) Nature of potentials associated with synaptic transmission in lateral geniculate of cat. *J. Neurophysiol.*, 17, 387–414

19. Bishop, P. O. (1955) Synaptic and neuromuscular transmission. *Bull. Post-grad. Comm. Med. Univ. Sydney*, 10, 263–78

20. Bishop, P. O. (1955) The future of physiology and its relation to medicine. *Sydney Univ. Med. J.*, 45, 14–19

21. Bishop, P. O. & Levick, W. R. (1956) Saltatory conduction in single isolated and non-isolated myelinated nerve fibres. *J. Cell Comp. Physiol.*, 48, 1–34

22. Bishop, P. O. & Evans, W. A. (1956) The refractory period of the sensory synapses of the lateral geniculate nucleus. *J. Physiol.*, 134, 538–57

23. Bishop, P. O., Field, G., Hennessy, B. L. & Smith, J. R. (1958) Action of d-lysergic acid diethylamide on lateral geniculate synapses. *J. Neurophysiol.*, 21, 529–49

24. Bishop, P. O., Burke, W. & Davis, R. (1958) Synapse discharge by single fibre in mammalian visual system. *Nature*, 182, 728–30

25. Bishop, P. O., Burke, W., Davis, R. & Hayhow, W. R. (1958) Binocular interaction in the lateral geniculate nucleus – a general review. *Trans. Ophthal. Soc. Aust.*, 18, 15–35

26. Bishop, P. O., Burke, W. & Davis, R. (1959) Activation of single lateral geniculate cells by stimulation of either optic nerve. *Science*, 130, 506–7

27. Bishop, P. O., Burke, W. & Hayhow, W. R. (1959) Repetitive stimulation of optic nerve and lateral geniculate synapses. *Exp. Neurol.*, 1, 534–55

28. Bishop, P., Burke, W. & Hayhow, W. R. (1959) Lysergic acid diethylamide block of lateral geniculate synapses and relief by repetitive stimulation. *Exp. Neurol.*, 1, 556–68

29. Bishop, P. O. (1959) Neurophysiological problems. Au niveau des voies et des centres. In *Mechanisms of Colour Discrimination – Proceedings of International Symposium on Fundamental Mechanisms of Chromatic Discrimination in Animals and Man*. London: Pergamon Press

30. Bishop, P. O. & Davis, R. (1960) The recovery of responsiveness of the sensory synapses in the lateral geniculate nucleus. *J. Physiol.*, 150, 214–38

31. Bishop, P. O. & Davis, R. (1960) Synaptic potentials, after-potentials and slow rhythms of lateral geniculate neurones. *J. Physiol.*, 154, 514–46

32. Bishop, P. O., Burke, W., Davis, R. & Hayhow, W. R. (1960) Drugs as tools in visual physiology with particular reference to (a) the effects of prolonged disuse, and (b) the origin of the electroencephalogram. *Trans. Ophthal. Soc. Aust.*, 20, 50–65

33. Levick, W. R., Bishop, P. O., Williams, W. O. & Lampard, D. G. (1961) Probability distribution analyser programmed for neurophysiological research. *Nature*, 192, 629–30

34. Bishop, P. O., Burke, W. & Davis, R. (1962) The identification of single units in central visual pathways. *J. Physiol.*, 162, 409–31

35. Bishop, P. O., Burke, W. & Davis, R. (1962) Single-unit recording from antidromically activated optic radiation neurones. *J. Physiol.*, 162, 432–50

36. Bishop, P. O., Burke, W. & Davis, R. (1962) The interpretation of the extracellular response of single lateral geniculate cells. *J. Physiol.*, 162, 451–72

37. Bishop, P. O., Kozak, W., Levick, W. R. & Vakkur, G. J. (1962) The determination of the projection of the visual field onto the lateral geniculate nucleus in the cat. *J. Physiol.*, 163, 503–39

38. Bishop, P. O., Kozak, W. & Vakkur, G. J. (1962) Some quantitative aspects of the cat's eye: axis and plane of reference, visual field co-ordinates and optics. *J. Physiol.*, 163, 466–502

39. Vakkur, G. J., Bishop, P. O. & Kozak, W. (1963) Visual optics in the cat, including posterior nodal distance and retinal landmarks. *Vision Res.*, 3, 289–314

40. Vakkur, G. J. & Bishop, P. O. (1963) The schematic eye in the cat. *Vision Res.*, 3, 357–81

41. Bishop, P. O. (1964) Properties of afferent synapses and sensory neurons in the lateral geniculate nucleus. *Int. Rev. Neurobiol.*, 6, 191–255

42. Bishop, P. O., Levick, W. R. & Williams, W. O. (1964) Statistical analysis of the dark discharge of lateral geniculate neurones. *J. Physiol.*, 170, 598–612

43. Bishop, P. O. (1965) The neurophysiological basis of form vision. *Trans. Asia-Pacif. Acad. Ophthal.*, 2, 198–210

44. Kozak, W., Rodieck, R. W. & Bishop, P. O. (1965) Response of single units in lateral geniculate nucleus of cat to moving visual patterns. *J. Neurophysiol.*, 28, 19–47

45. Bishop, P. O. (1965) The nature of the representation of the visual fields in the lateral geniculate nucleus. *Proc. Aust. Assoc. Neurologists*, 3, 15–25

46. Taub, A. & Bishop, P. O. (1965) The spinocervical tract: dorsal column linkage, conduction velocity, primary afferent spectrum. *Exp. Neurol.*, 13, 1–21

47. Bishop, P. O. & Rodieck, R. W. (1965) Discharge patterns of cat retinal ganglion cells. In *Information Processing in Sight Sensory Systems*, ed. P. W. Nye, pp. 116–27. Pasadena, California: California Institute of Technology

48. Ogawa, T., Bishop, P. O. & Levick, W. R. (1966) Temporal characteristics of responses to photic stimulation by single ganglion cells in the unopened eye of the cat. *J. Neurophysiol.*, 29, 1–30

49. Bishop, P. O. (1967) Central nervous system: afferent mechanisms and perception. *Ann. Rev. Physiol.*, 29, 427–84

50. Rodieck, R. W., Pettigrew, J. D., Bishop, P. O. & Nikara, T. (1967) Residual eye movement in receptive field studies of paralyzed cats. *Vision Res.*, 7, 107–10

51. Pettigrew, J. D., Nikara, T. & Bishop, P. O. (1968) Neural mechanisms concerned in the development of amblyopia ex anopsia. *Proc. Aust. Ass. Neurol.*, 5, 221–4

52. Nikara, T., Bishop, P. O. & Pettigrew, J. D. (1968) Analysis of retinal correspondence by studying receptive fields of binocular single units in-cat striate cortex. *Exp. Brain Res.*, 6, 353–72

53. Pettigrew, J. D., Nikara, T. & Bishop, P. O. (1968) Responses to moving slits by single units in cat striate cortex. *Exp. Brain Res.*, **6**, 373–90

54. Pettigrew, J. D., Nikara, T. & Bishop, P. O. (1968) Binocular interaction on single units in cat striate cortex: simultaneous stimulation by single moving slit with receptive fields in correspondence. *Exp. Brain Res.*, **6**, 391–410

55. Sanderson, K. J., Darian-Smith, I. & Bishop, P. O. (1969) Binoclar corresponding receptive fields of single units in the cat dorsal lateral geniculate nucleus. *Vision Res.*, **9**, 1297–303

56. Henry, G. H., Bishop, P. O. & Coombs, J. S. (1969) Inhibitory and sub-liminal excitatory receptive fields of simple units in cat striate cortex. *Vision Res.*, **9**, 1289–96

57. Kinston, W. J., Vadas, M. A. & Bishop, P. O. (1969) Multiple projection of the visual field to the medial portion of the dorsal lateral geniculate nucleus and the adjacent nuclei of the thalamus of the cat. *J. Comp. Neurol.*, **136**, 295–316

58. Bishop, P. O. (1970) Beginning of form vision and binocular depth discrimination in cortex. In *The Neurosciences: Second Study Program*, ed F. O. Schmitt, pp. 471–85. New York: Rockefeller University Press

59. Joshua, D. E. & Bishop, P. O. (1970) Binocular single vision and depth discrimination. Receptive field disparities for central and peripheral vision and binocular interaction on peripheral single units in cat striate cortex. *Exp. Brain Res.*, **10**, 389–416

60. Bishop, P. O. (1970) Seeing with two eyes. *Aust. J. Sci.*, **32**, 383–91

61. Bishop, P. O. & Henry, G. H. (1971) Spatial vision. *Ann. Rev. Physiol.*, **22**, 119–60

62. Bishop, P. O., Henry, G. H. & Smith, C. J. (1971) Binocular interaction fields of single units in the cat striate cortex. *J. Physiol. (Lond.)*, **216**, 39–68

63. Henry, G. H. & Bishop, P. O. (1971) Simple cells of the striate cortex. In *Contributions to Sensory Physiology*, vol. 5, ed. W. D. Neff, pp. 1–46. New York: Academic Press

64. Bishop, P. O., Coombs, J. S. & Henry, G. H. (1971) Responses to visual contours: spatio-temporal aspects of excitation in the receptive fields of simple striate neurones. *J. Physiol. (Lond.)*, **219**, 625–57

65. Bishop, P. O., Coombs, J. S. & Henry, G. H. (1971) Interaction effects of visual contours on the discharge frequency of simple striate neurones. *J. Physiol. (Lond.)*, **219**, 659–87

66. Bishop, P. O., Dreher, B. & Henry, G. H. (1971) Stimulus specificities of the discharge centre in the receptive field of simple striate neurones in the cat. *J. Physiol. (Lond.)*, **218**, 53–55P

67. Sanderson, K. J., Bishop, P. O. & Darian-Smith, I. (1971) The properties of the binocular receptive fields of lateral geniculate neurones. *Exp. Brain Res.*, **13**, 178–207

68. Henry, G. H., Bishop, P. O. & Coombs, J. S. (1971) The beginning of form recognition at the level of the simple striate neuron. From: Proceedings of the Workshops of the 9th International Conference on Medical and Biological Engineering, Melbourne, 1971; Workshop No. 7. In *Information Processing in the Visual Pathway*, pp. 5–9. Canadian Medical & Biological Engineering Society

69. Bishop, P. O. & Henry, G. H. (1972) Striate neurons: receptive field concepts. *Invest. Ophthal.*, **11**, 346–54

70. Henry, G. H. & Bishop, P. O. (1972) Striate neurons: receptive field organization. *Invest. Ophthal.*, **11**, 357–68

71. Bishop, P. O., Dreher, B. & Henry, G. H. (1972) Simple striate cells: comparison of responses to stationary and moving stimuli. *J. Physiol. (Lond.)*, **227**, 15–17P

72. Bishop, P. O. (1973) Neurophysiology of binocular single vision and stereopsis. In *Handbook of Sensory Physiology*, vol. vii/3A, ed. R. Jung, pp. 255–305. Berlin: Springer-Verlag

73. Bishop, P. O., Coombs, J. S. & Henry, G. H. (1973) Receptive fields of simple cells in the cat striate cortex. *J. Physiol. (Lond.)* **231**, 31–60

74. Henry, G. H., Bishop, P. O., Tupper, R. M. & Dreher, B. (1973) Orientation specificity and responses variability of cells in the striate cortex. *Vision Res.*, **13**, 1771–9

75. Bishop, P. O. (1974) Stereopsis and fusion. *Trans. Ophthal. Soc. N.Z.*, **26**, 17–27

76. Bishop, P. O., Goodwin, A. W. & Henry, G. H. (1974) Direction selective sub-regions in striate simple cell repetive field. *J. Physiol. (Lond.)* **238**, 25–27P

77. Bishop, P. O. (1974) Grafton Elliot Smith's contribution to visual neurology and the influence of Thomas Henry Huxley. In *Grafton Elliot Smith: The Man and his Work*, ed. A. P. Elkin & M. W. G. Macintosh, pp. 50–7. Sydney: Sydney University Press

78. Henry, G. H., Bishop, P. O. & Dreher, B. (1974) Orientation, axis and direction as stimulus parameters for striate cells. *Vision Res.*, **14**, 767–77

79. Henry, G. H., Dreher, B. & Bishop, P. O. (1974) Orientation specificity of cells in the cat striate cortex. *J. Neurophysiol.*, **37**, 1394–409

80. Bishop, P. O. (1975) Binocular vision. In *Adler's Physiology of the Eye: Clinical Application*, 6th edn, ed. R. A. Moses, pp. 558–614. St Louis: C. V. Mosby

81. Bishop, P. O. (1975) Visual disability and vision research. *Aust. J. Optom.*, June, 202–9

82. Goodwin, A. W., Henry, G. H. & Bishop, P. O. (1975) Direction selectivity of simple striate cells: properties and mechanism. *J. Neurophysiol.*, **38**, 1500–23

83. Goodwin, A. W., Henry, G. H. & Bishop, P. O. (1975) Inhibitory mechanism for direction selectivity in simple cells in striate cortex. *Proc. Aust. Physiol. Pharmacol. Soc.*, **6**, 205–6

84. Bishop, P. O., Kato, H. & Nelson, J. I. (1976) Position and orientation disparities as depth cues for striate neurons. *J. Physiol. (Lond.)*, **263**, 168–69P

85. Nelson, J. I., Kato, H. & Bishop, P. O. (1977) The discrimination of orientation and position disparities by binocularly-activated neurons in cat striate cortex. *J. Neurophysiol.*, **40**, 260–84

86. Bishop, P. O. (1977) Receptive field concepts. In *Spatial Contrast*, ed. L. H. van der Tweel & H. Spekreijse, pp. 19–23. Amsterdam: Royal Netherlands Academy of Arts and Science

87. Henry, G. H., Goodwin, A. W. & Bishop, P. O. (1978)

Spatial summation of responses in receptive fields of single cells in cat striate cortex. *Exp. Brain Res.*, **32**, 245–66

88. Bishop, P. O. (1978) Orientation and position disparities in stereopsis. In *Frontiers in Visual Sciences*, ed. S. J. Cool & E. L. Smith, III, pp. 336–50. New York: Springer-Verlag

89. Orban, G. A., Kato, H. & Bishop, P. O. (1978) Properties of striate hypercomplex cells in the cat. *Arch. Int. Physiol. Biochem.*, **86**, 1157–8

90. Kato, H., Bishop, P. O. & Orban, G. A. (1978) Hypercomplex and simple/complex cell classification in cat striate cortex. *J. Neurophysiol.*, **41**, 1071–95

91. Kulikowski, J. J., Bishop, P. O. & Kato, H. (1979) Sustained and transient responses by cat striate cells to stationary flashing light and dark bars. *Brain Res.*, **170**, 362–7

92. Orban, G. A., Kato, H. & Bishop, P. O. (1979) End-zone region in receptive fields of hypercomplex and other striate neurons in the cat. *J. Neurophysiol.*, **42**, 818–32

93. Orban, G. A., Kato, H. & Bishop, P. O. (1979) Dimensions and properties of end-zone inhibitory areas in receptive fields of hypercomplex cells in cat striate cortex. *J. Neurophysiol.*, **42**, 833–49

94. Bishop, P. O. (1979) Stereopsis and the random element in the organization of the striate cortex. *Proc. R. Soc. Lond. B*, **204**, 415–34

95. Bishop, P. O., Kato, H. & Orban, G. A. (1980) Direction selective cells in the complex family in cat striate cortex. *J. Neurophysiol.*, **43**, 1266–83

96. Bishop, P. O. (1981) Binocular vision. In *Adler's Physiology of the Eye: Clinical Application*, 7th edn, ed. R. A. Moses, pp. 575–649. St Louis: C. V. Mosby

97. Kato, H., Bishop, P. O. & Orban, G. A. (1981) Binocular interaction on monocularly-discharged lateral geniculate and striate neurons in the cat. *J. Neurophysiol.*, **46**, 932–51

98. Bishop, P. O. (1981) Neural mechanisms for binocular depth discrimination. In *Advances in Physiological Sciences*, vol. 16, *Sensory Functions*, ed. E. Grastyan & P. Molnar, pp. 441–9. Oxford: Pergamon Press

99. Kulikowski, J. J. & Bishop, P. O. (1981) Fourier analysis and spatial representation in the visual cortex. *Experientia*, 37, 160–3

100. Kulikowski, J. J., Bishop, P. O. & Kato, H. (1981) Spatial arrangements of responses by cells in the cat visual cortex to light and dark bars and edges. *Exp. Brain Res.*, **44**, 371–85

101. Kulikowski, J. J. & Bishop, P. O. (1981) Linear analysis of the responses of simple cells in the cat visual cortex. *Exp. Brain Res.*, **44**, 386–400

102. Kulikowski, J. J. & Bishop, P. O. (1982) Silent periodic cells in the cat striate cortex. *Vision Res.*, **22**, 191–200

103. Kulikowski, J. J., Marcelja, S. & Bishop, P. O. (1982) Theory of spatial position and spatial frequency relations in the receptive fields of simple cells in the visual cortex. *Biol. Cybern.*, **43**, 187–98

104. Yamane, S., Maske, R. & Bishop, P. O. (1982) [New observation on direction selectivities of simple cells in cat striate cortex.] *Bull. Electrotech. Lab.*, **46**, 336–42. (In Japanese)

105. Bishop, P. O. (1983) Physiology in Sydney in the 1950's. *Proc. Aust. Physiol. Pharmac. Soc.*, **14**, 48–57

106. Bishop, P. O. (1983) Vision with two eyes. *Nihon Sierigakukai Zasshi (J. Physiol. Soc. Japan)*, **45**, 1–18

107. Maske, R., Yamane, S. & Bishop, P. O. (1983) Simple striate cells: binocular receptive field properties for local stereopsis. *Neurosci. Lett. NELED*, Suppl. 11, 517

108. Bishop, P. O. (1984) Processing of visual information within the retinostriate system. In *Handbook of Physiology*, Section I: *The Nervous System*, vol. 3, *Sensory Processes*, part 1, ed. I. Darian-Smith, pp. 341–424. Bethesda, MD: American Physiological Society

109. Bishop, P. O. (1984) The striate cortex: feature detector or Fourier analyzer. *Proc. Aust. Physiol. Pharmac. Soc.*, **15**(1), 1–20

110. Bishop, P. O. & Everitt, A. V. (1984) The Medical Sciences: Physiology. In *Centenary Book of the University of Sydney Faculty of Medicine*, ed. J. A. Young *et al.*, pp. 238–65. Sydney: University of Sydney Press

111. Bishop, P. O. (1984) Terra Australis Incognita: the emergence of physiology in Australia. In *Frontiers of Physiological Research*, ed. D. G. Garlick & P. I. Korner. Canberra: Australian Academy of Science

112. Maske, R., Yamane, S. & Bishop, P. O. (1984). Binocular simple cells for local stereopsis: comparison of receptive field organizations for the two eyes. *Vision Res.*, **24**, 1921–9

113. Peterhans, E., Bishop, P. O. & Camarda, R. M. (1985). Direction selectivity of simple cells in cat striate cortex to moving light bars. I. Relation to stationary flashing bar and moving edge responses. *Exp. Brain Res.*, **57**, 512–22

114. Yamane, S., Maske, R. & Bishop, P. O. (1985). Direction selectivity of simple cells in cat striate cortex to moving light bars, II. Relation to moving dark bar responses. *Exp. Brain Res.*, **57**, 523–36.

115. Maske, R., Yamane, S. & Bishop, P. O. (1985). Simple and B-cells in cat striate cortex. Complementarity of responses to moving light and dark bars. *J. Neurophysiol.*, **53**, 670–85

116. Camarda, R. M., Peterhans, E. & Bishop, P. O. (1985). Spatial organization of subregions in receptive fields of simple cells in cat striate cortex as revealed by stationary flashing bars and moving edges. *Exp. Brain Res.*, **60**, 136–50

117. Camarda, R. M., Peterhans, E. & Bishop, P. O. (1985). Simple cells in cat striate cortex: responses to stationary flashing and to moving light bars. *Exp. Brain Res.*, **60**, 151–8

118. Yamane, S., Maske, R. & Bishop, P. O. (1985). Properties of end-zone inhibition of hypercomplex cells in cat striate cortex. *Exp. Brain Res.*, **60**, 200–3

119. Bishop, P. O. (1986). Binocular vision. In: *Adler's Physiology of the Eye: Clinical Application*, 8th edn, ed. R. A. Moses. St. Louis: C. V. Mosby. (In press)

120. Bishop, P. O. & Pettigrew, J. D. (1986). Neural mechanisms of binocular vision. *Vision Res.*, **26**. (In press)

121. Maske, R., Yamane, S. & Bishop, P. O. (1986). Stereoscopic mechanisms: binocular responses of cat striate cells to moving light and dark bars. *Proc. R. Soc. Lond. B.* (In press)

INDEX